MOS (Metal Oxide Semiconductor) Physics and Technology

E. H. NICOLLIAN
J. R. BREWS

Bell Laboratories
Murray Hill, New Jersey

A WILEY-INTERSCIENCE PUBLICATION

JOHN WILEY & SONS
New York · Chichester · Brisbane · Toronto · Singapore

Library of Congress Cataloging in Publication Data:

Nicollian, E. H.
  MOS (metal oxide semiconductor) physics and
technology.

  "A Wiley-Interscience publication."
  Includes bibliographical references and index.
  1. Metal oxide semiconductors. I. Brews, J. R.
II. Title.
TK7871.99.M44N52   621.3815'2   81-7607
ISBN 0-471-08500-6   AACR2

Printed in the United States of America

10 9 8

MOS (Metal Oxide Semiconductor) Physics and Technology

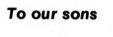

To our sons

Preface

Charges in the metal oxide semiconductor (MOS) system have assumed new importance with the emergence of integrated circuits using MOS field effect transistors MOSFETs as the active elements. These circuits are sensitive to the presence of minute traces of charge, charge densities of the order of 10^{10} charges/cm^2, or a mere thousand charges in a device of dimension 1 by 10 μm. To detect such low charge densities, very sensitive electrical measurements have been developed. Rather elaborate theoretical models have evolved, connecting the behavior of the charges to the measurements. The resulting knowledge of the charges in the MOS system has been used to improve fabrication techniques and to make better-performing, longer-lasting integrated circuits.

To simplify the study of the MOS system as prepared under realistic manufacturing conditions, the MOS capacitor is used extensively as a simple test structure. This book describes measurements using this structure and the theory behind them, measurement principles and instrumentation, and the elements of the material aspects of modern MOS technology. The purposes of this book are to (1) give an understanding of the charges in the MOS system from the viewpoint of integrated circuit technology, (2) give a critical review of the subject, (3) enable the reader to select a measurement technique suitable for his or her needs, with an understanding of its limitations, (4) assemble the appropriate instrumentation, interpret the results, and (5) set the foundation for studies of other semiconductor-insulator systems.

The material takes the reader up to the present state of the art. Not all the literature in this field is cited, but a representative portion is listed up to 1979, including the work of the pioneers.

To the authors' knowledge, this book is unique in its depth and choice of topics. The book extends the treatments by Grove[1] and Richman,[2] which are introductory, updates the works by Many et al.[3] and Frankl,[4] which have little relevance to modern MOS technology, enlarges on the necessarily brief treatment of the encyclopedic work of Sze[5] and goes into new topics untouched by other books in this field, such as full discussions of the admittance of the MOS capacitor at general frequencies, of drift, of oxidation, of hot electron effects, of ionizing radiation effects, and of instrumentation.

Among the discussions of special topics, we wish to mention specifically: (1) the derivation of the MOS admittance with bulk traps which is original, and tractable without computer calculation; (2) the treatment of the conductance method, an area in which the authors have contributed actively; (3) the treatment of interfacial charge nonuniformities, again an area in which the authors have first-hand experience; and (4) a very complete assessment of our knowledge of interface traps including strong experimental evidence for the continuum model of interface traps broadened by interfacial charge nonuniformities. Also noteworthy are the new, extensive discussions of charge profiling by photoinjection methods, of doping profiles, and of flatband voltage determination. Overall, the reader will find that virtually every chapter contains new material filling gaps in the literature and reassessments of prevalent ideas.

Because of its depth of treatment, this book should be useful to research workers in the MOS field. Because of its careful development, it should be accessible to graduate students. Because each chapter is written to stand alone, the book can be adapted easily to accent simple or advanced topics for a variety of courses. Independence of the chapters also assists the research worker who does not have time to read an entire book. Where a topic is subsidiary to the main theme of a chapter, a cross-reference is given to another chapter or to an appendix for a more detailed treatment.

The presentation requires a modest knowledge of semiconductor device physics but no MOS background. Depending on the topic, the necessary mathematical background varies from slight to considerable. In all cases an effort has been made to explain the physical principles, so that the mathematics extends intuition, rather than serving as incantation. Equivalent circuits have been used frequently to summarize the mathematical arguments, because it is easier to grasp the behavior of a circuit than of a complex algebraic expression.

Chapter 1 presents a qualitative introduction to the electrical properties of the MOS system and a brief history of the development of the MOS field. In Chapters 2–4 the basic small-signal theory of the MOS capacitor is developed, including bulk traps but excluding interface traps. In Chapters 5–11 the small-signal, steady-state methods for measuring the electrical properties of the MOS system and the understanding of these properties on the basis of these measurements are described. The transient methods are put into perspective in Chapter 8. Chapter 12 covers the required instrumentation and how it is used. Chapters 13–15 discuss the oxidation of silicon and how the understanding gained of the MOS system has led to the control of its electrical properties to produce stable, high performance integrated circuits. Chapter 16 discusses the morphology and the chemical composition of the silicon-silica interface and some of the speculative models for the electrically active centers in the MOS system.

We briefly consider (1) semiconductors other than silicon and insulators other than SiO_2, (2) charge transport through SiO_2 including dielectric breakdown, (3) quantum effects at low temperatures, (4) the vacuum-silicon and the metal-silicon interface, and (5) the operation of the MOSFET, charge-coupled device (CCD), bipolar transitor, and $p-n$ junction diode. Detailed discussions of these topics are excluded, although they are certainly of technological and scientific importance. Their exclusion allows an in-depth exposition of the topics selected while keeping this book within reasonable bounds.

This book crystalizes an amorphous mass of literature and provides direction for further growth. We have stated our view of the basic limitations on MOS measurements and of the experimental evidence supporting various models often used. Different measurement techniques have been evaluated and their niches established. Not everyone will share our views. However, we hope everyone will find the book useful and that disagreements will provide more stimulation than irritation.

In this book we use the nomenclature and symbols for the charges in the Si–SiO_2 system recommended by the joint ECS-IEEE committee on charge nomenclature chaired by Deal.[6] Table 1 lists this nomenclature, symbolism, and units.

The oxide fixed charge number density is $|N_f| = |Q_f|/|q|(cm^{-2})$, the oxide trapped charge number density is $|N_{ot}| = |Q_{ot}|/|q|(cm^{-2})$, the mobile ionic charge number density is $|N_m| = |Q_m|/|q|(cm^{-2})$, and the interface trap charge number density is $N_{it} = \pm Q_{it}/q(cm^{-2})$, only if interface traps are all donors or all acceptors. In addition to the charges in Table 1, we define interface charge density as $Q_f + Q_{it}$ and oxide charge density as $Q_o = Q_{ot} + Q_f + Q_{it}$.

The authors acknowledge Dr. R. M. Ryder, who was instrumental in obtaining Bell laboratories approval for the writing of this book; Dr. S. M. Sze, whose understanding and encouragement made the writing of this book possible; and Drs. M. P. Lepselter and G. E. Smith for providing the environment in which we have been welcomed and assisted. Finally, we thank our reviewers for many helpful suggestions, specifically, Drs. S. M. Sze, A. M. Voshchenkov, C. C. Chen, H. Seidel, J. R. Ligenza, B. E. Deal,

Table 1

Nomenclature	Symbol	Units
Oxide fixed charge density	Q_f	C/cm^2
Oxide trapped charge density	Q_{ot}	C/cm^2
Mobile ionic charge density	Q_m	C/cm^2
Interface trap charge density	Q_{it}	C/cm^2
Interface trap level density	D_{it}	$cm^{-2}\,eV^{-1}$

W. A. Pliskin, M. J. Rand, S. P. Murarka, G. E. Smith, A. B. Fowler, W. Fichtner, G. L. Miller, D. E. Aspnes, W. E. Dahlke, G. P. Schwartz, J. A. Cooper, Jr., G. W. Taylor, N. M. Johnson, D. R. Young, J. Maserjian, A. K. Sinha, L. C. Kimerling, E. Arnold, B. Schwartz, S. A. Schwarz, G. F. Derbenwick, and K. K. Thornber. We also thank Mrs. S. Ward and Mrs. P. C. Vencill, who did the typing during the early stages of the book; Mr. P. W. Imbimbo at the Murray Hill Word Processing Center, who typed much of the first draft himself and coordinated the remaining typing; Ms. K. C. Edwards and her staff at the Murray Hill Word Processing Center, who did the typing during the final editing; and Ms. A. Johnson of the Technical Documentation Department, for proofreading the galleys. We thank G. Holmfelt and the Murray Hill Drafting Department for drafting the figures. K. Drake of the Murray Hill Art Department did the dust jacket artwork.

All figures from books and journals are reprinted with permission.

E. H. NICOLLIAN
J. R. BREWS

Murray Hill, New Jersey
November 1981

1   A.S. Grove, *Physics and Technology of Semiconductor Devices*, Wiley, New York, 1967.
2   P. Richman, *MOS Field-Effect Transistors and Integrated Circuits*, Wiley, New York, 1973.
3   A. Many, Y. Goldstein, and N. B. Grover, *Semiconductor Surfaces*, North Holland, Amsterdam, 1965.
4   D. R. Frankl, *Electrical Properties of Semiconductor Surfaces*, Pergamon, New York, 1967.
5   S. M. Sze, *Physics of Semiconductor Devices*, Wiley, New York, 1969.
6   B. E. Deal, *J. Electrochem. Soc.*, **127**, 979 (1980); *IEEE Transact. Electron Devices*, **ED–27**, 606 (1980).

Contents

MOS (Metal Oxide Semiconductor) Physics and Technology

1

Introduction

1.1 INTRODUCTION

Control of the electrical properties of the MOS system has been one of the major factors that has led to stable and high performance silicon integrated circuits. This book tells how the electrical properties of the MOS system are measured, the results of these measurements and how these electrical properties are controlled. This book should enable the reader to monitor the electrical characteristics of the MOS system during integrated circuit fabrication, to develop new experiments, and to further analysis, particularly in the area of very large scale integrated (VLSI) circuits.

The MOS capacitor is used in both monitoring integrated circuit fabrication and studying the electrical properties of the MOS system. The MOS capacitor has the advantages of simplicity of fabrication and of analysis. Fabrication of the MOS capacitor uses the same processing used in fabricating the integrated circuit. Therefore, the MOS capacitor provides direct measurement and monitoring of the MOS system as it is actually fabricated and used in the integrated circuit.

Historically, the motivation behind the use of the MOS capacitor has been fabrication of stable and high performance devices and integrated circuits. For this reason, a brief history of the development of the integrated circuit starting from the discovery of rectification at the end of the nineteenth century is given in Section 1.4.

1

Three regions of the MOS system are important in integrated circuit technology: the bulk oxide, the Si–SiO$_2$ interface,* and the silicon itself. Charges in all three regions play a role in integrated circuits. In the oxide and at the Si–SiO$_2$ interface, these charges are undesirable because they adversely affect device performance and stability. In the silicon, charges such as dopant impurity ions are necessary for proper device operation.

Two major properties of the oxide are of paramount importance: the level of mobile ionized sodium contamination and the dielectric breakdown strength of the oxide. Also, there are two major electrical properties of the Si–SiO$_2$ interface: oxide fixed charge and interface trap level density. With the exception of oxide dielectric breakdown strength, these quantities are discussed in Sections 1.2 and 1.3 and throughout this book.

Mobile sodium ions, oxide fixed charge, and interface traps adversely affect silicon device performance and stability if present at high densities. Oxide dielectric breakdown strength must be high, or device operation will be adversely affected. The optimal values of these quantities achieved by modern integrated circuit technology are (1) a mobile ionized sodium density in the low 10^{10} cm^{-2} range, (2) a dielectric breakdown strength from 6×10^6 to 10^7 V/cm, (3) an oxide fixed charge density in the low 10^{10} cm^{-2} range, and (4) an interface trap level density near midgap in the low 10^{10} cm^{-2} eV^{-1} range. These values are adequate for most high performance integrated circuits. Under laboratory conditions, however, even lower ionized sodium and interface trap level densities can be achieved. The lowest ionized sodium densities reported are in the 10^9 cm^{-2} range, and the lowest interface trap level densities near midgap are in the low 10^9 cm^{-2} eV^{-1} range. Measurement and control of the densities of mobile sodium ions, oxide fixed charge, and interface traps are among the major topics discussed in this book.

Two silicon properties are discussed at length: the distribution of dopant ions in the silicon and minority carrier lifetime. Both properties influence devices used in integrated circuits. For example, the dopant ion distribution affects threshold voltage (the gate bias at which inversion occurs) and gate capacitance. Minority carrier lifetime affects dynamic memory cells and CCDs, determining how long a "one" or a "zero" can be maintained without "refreshing".

However, these two properties of the silicon also are important in interpreting MOS capacitor measurements. It is not possible to make quantitative measurements of interfacial properties without knowing the dopant ion distribution. For example, incorrect interface trap level densities may be inferred from the MOS capacitance if an incorrect profile is assumed. As another example, minority carrier response must be under-

---

*The two other interfaces of importance in integrated circuit technology are the p-n junction and the metal-silicon interface. A discussion of these two interfaces is beyond the scope of this book, and the reader is referred to Refs. 1–5 for further discussion.

stood to interpret the time dependence of MOS capacitor measurements. Otherwise, minority carrier response may be misinterpreted as an interface trap response.

1.2 SILICON SURFACE

Surfaces of solids and interfaces between materials play an important role in a very wide variety of phenomena extending from everyday experience (anyone who has painted a house knows the importance of surfaces) to catalysis of chemical reactions, corrosion, semiconductor devices, and to complex life processes. Because of the scientific importance and the economic impact of these phenomena, considerable effort has been devoted to their study. We concentrate here on a very narrow portion of this vast subject.

The properties of the semiconductor surface are sufficiently different from the properties of the semiconductor bulk that the study of surface properties can be regarded as one of the branches of semiconductor physics and chemistry. The branch of semiconductor surface science is new. There are several reasons for this. First, a study of semiconductor surface properties depends on an understanding of the bulk properties of the semiconductor. Therefore, the study and understanding of bulk properties had to precede the study of surface properties. Second, unlike the bulk, which can be regarded as an infinite uniform periodic structure amenable to theoretical treatment, the surface is an abrupt termination of the periodic lattice. This termination results in deformation of the crystal surface and a loss in periodicity. As a result, theoretical analysis becomes much more complicated, and it still is in its infancy. Third, atoms at the semiconductor surface are not completely surrounded by other atoms of the crystal like those in the bulk. As a result, all the valence electrons of the surface atoms are not satisfied. The unsaturated bonds of the surface atoms make the surface highly reactive chemically; thus the surface of a semiconductor usually is covered with one or more layers of foreign atoms. Therefore, a well-characterized experimental surface requires extremely careful preparation. Fourth, a surface constitutes only the first few monolayers of atoms on a sample so that measurement signals originate from very few atoms in surface experiments compared to bulk experiments. Attempts to increase surface : volume ratio, such as by powdering the sample, alter the surface being measured. Finally, because surface dimensions are on an atomic scale, only a few angstroms thick, there is a severe resolution problem.

In summary, theory and experiment both face severe difficulties in surface measurements, that are not encountered in bulk studies.

The difficulty of controlled measurements on semiconductor surfaces does not mean that surface influence on semiconductor device behavior is stood to interpret the time dependence of MOS capacitor measurements. Otherwise, minority carrier response may be misinterpreted as an interface trap response.

1.2 SILICON SURFACE

Surfaces of solids and interfaces between materials play an important role in a very wide variety of phenomena extending from everyday experience (anyone who has painted a house knows the importance of surfaces) to catalysis of chemical reactions, corrosion, semiconductor devices, and to complex life processes. Because of the scientific importance and the economic impact of these phenomena, considerable effort has been devoted to their study. We concentrate here on a very narrow portion of this vast subject.

The properties of the semiconductor surface are sufficiently different from the properties of the semiconductor bulk that the study of surface properties can be regarded as one of the branches of semiconductor physics and chemistry. The branch of semiconductor surface science is new. There are several reasons for this. First, a study of semiconductor surface properties depends on an understanding of the bulk properties of the semiconductor. Therefore, the study and understanding of bulk properties had to precede the study of surface properties. Second, unlike the bulk, which can be regarded as an infinite uniform periodic structure amenable to theoretical treatment, the surface is an abrupt termination of the periodic lattice. This termination results in deformation of the crystal surface and a loss in periodicity. As a result, theoretical analysis becomes much more complicated, and it still is in its infancy. Third, atoms at the semiconductor surface are not completely surrounded by other atoms of the crystal like those in the bulk. As a result, all the valence electrons of the surface atoms are not satisfied. The unsaturated bonds of the surface atoms make the surface highly reactive chemically; thus the surface of a semiconductor usually is covered with one or more layers of foreign atoms. Therefore, a well-characterized experimental surface requires extremely careful preparation. Fourth, a surface constitutes only the first few monolayers of atoms on a sample so that measurement signals originate from very few atoms in surface experiments compared to bulk experiments. Attempts to increase surface : volume ratio, such as by powdering the sample, alter the surface being measured. Finally, because surface dimensions are on an atomic scale, only a few angstroms thick, there is a severe resolution problem.

In summary, theory and experiment both face severe difficulties in surface measurements, that are not encountered in bulk studies.

The difficulty of controlled measurements on semiconductor surfaces does not mean that surface influence on semiconductor device behavior is minor. Surface properties can be the dominant influence on device performance and stability. This is undesirable because devices with surface-dominated characteristics usually have poorer performance and stability than devices with bulk-dominated characteristics. Surface properties are affected markedly by small amounts of contamination picked up from the environment during handling, processing, and use. For example, one ten-thousandth (0.0001) of a monolayer of contamination on a semiconductor device surface can dominate the electrical properties of the device. Because of this extreme sensitivity to contaminants, semiconductor device characteristics can change with time in an uncontrolled manner.

The surface sensitivity of semiconductor devices originates in two characteristics of semiconductors: (1) their low mobile carrier densities and (2) the extreme variations in mobile carrier density introduced by small variations in potential. Because carrier densities are low, even trace quantities of charge on the semiconductor surface have a major influence on carrier densities, an influence extending many microns into the semiconductor. These same factors make semiconductor devices sensitive to surface electrodes, making field effect devices possible. Therefore, this sensitivity must be exploited while avoiding unwanted surface effects.

The first attempt to solve this problem for bipolar transistors and junction diodes was to encapsulate them in a hermetically sealed metal can. However, encapsulation was not only costly and bulky, but device performance was not optimal. The most practical solution was to *passivate* the semiconductor surface.

Passivation is accomplished by forming a thin insulating layer on the semiconductor surface. Passivation minimizes the electrical activity of the device surface. The passivating layer also protects the surface from environmental contamination; thus passivation results in a marked improvement of device performance, stability, and uniformity of characteristics from device to device. By far the most successfully passivated semiconductor is silicon with a thermally grown silicon dioxide layer. The passivation of silicon devices by thermally grown silicon dioxide is so successful that it has no serious competition today.

Thermal oxidation results in more stable and better performing devices, giving workers in surface science a stable and reproducible interface to study and characterize. This circumstance, as well as the technological importance of the $Si–SiO_2$ interface, has led to intensive investigation of the MOS system.

Besides its passivating properties, SiO_2 has three more properties of importance in integrated circuit technology: (1) the oxide acts as a barrier against the diffusion of impurities into the silicon underneath; thus, by cutting holes into the oxide film, impurities can be diffused into the silicon through these holes and nowhere else; (2) the oxide provides insulation between the metallization pattern interconnecting devices in the circuit and the silicon substrate; and (3) the oxide insulates the gate from the silicon in field effect devices.

1.2.1 Examples of Interface Effects

A few examples illustrate the effect of uncontrolled interface charges. A silicon wafer with a thin insulating layer on top of it is shown in Fig. 1.1. A metallic electrode, called a *gate*, in the form of a circular dot has been deposited on the oxide. If the carriers in the silicon are holes, a positive charge on the gate repels the holes under it. When these holes leave, negative ions fixed in the silicon lattice are left behind. The number of these ions equals the number of charges on the gate, to preserve charge neutrality. If there are 10^{11} charges/cm^2 on the gate and there are 10^{15} ions/cm^3 in the silicon, holes are depleted to a depth of 1 μm for neutrality to prevail. If more than 10^{11} charges are present, depletion layer width does not increase because the voltage drop between the surface and the bulk silicon has become a large fraction of the bandgap. Instead, *negative* mobile carriers, electrons, appear near the interface in the region of greatest voltage. The silicon surface has *inverted* (i.e., has mobile carriers of opposite sign to the zero field situation). Because electrons are mobile, they can carry current in a thin layer near the surface called an *inversion layer* or a *channel*.

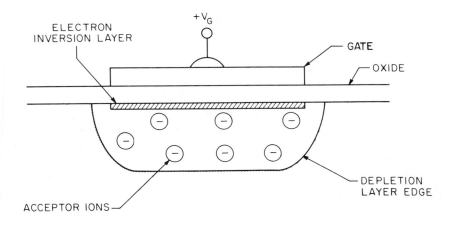

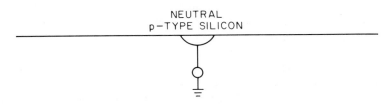

Fig. 1.1 Cross section of an MOS capacitor biased to inversion showing the inversion layer, depletion layer, and neutral bulk.

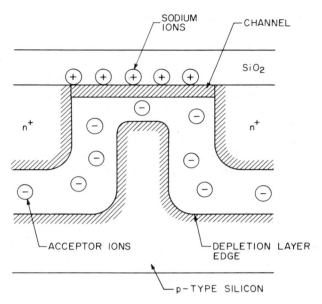

Fig. 1.2 Cross section of an oxidized silicon wafer that has two closely spaced n^+p junctions connected by a conducting surface channel created by sodium ions at the oxide-silicon interface. The channel is an inversion layer of electrons.

Now imagine two adjacent devices in an integrated circuit. For example, they may be two n^+-p junctions, as shown in Fig. 1.2. Instead of charges in a gate, suppose that the oxide between the two junctions has been contaminated by positive sodium ions from the ambient. If more than 10^{11} ions/cm^2 are in the oxide near the Si–SiO$_2$ interface, a channel will be formed between the two junctions, forming an unexpected short between the two. In a typical integrated circuit, these junctions may be spaced only $2\ \mu$m or less apart. They may be only $2\ \mu$m wide. Therefore, a short can be caused by only 4000 sodium ions, which is a number chemically undetectable.

Ions also can move in the oxide, causing the electrical behavior of circuits to vary with time. They also can congregate, causing one part of a circuit to behave differently from another, even though both parts were designed alike. When this happens, some devices (MOSFETs) are inadvertently turned on by the ions, even though no control voltages have been applied.

Ions in the oxide also can modify the breakdown behavior of p-n junctions by focusing electrical fields in small regions. In Fig. 1.3, a p^+-n junction is shown. The edges of the junction reach the silicon surface. Figure 1.3a shows the depletion region around the p^+ region continuing up to the surface. However, in Fig. 1.3b, the presence of positive ions in the

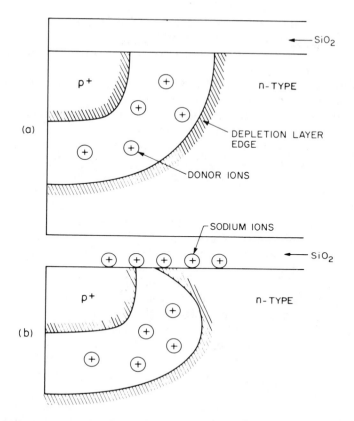

Fig. 1.3 Cross section of a p^+n junction showing that positive ions in the oxide can increase the lateral field in the depletion layer of a p^+n junction: (a) no ions in the oxide; (b) positive ions in the oxide pinching in the depletion layer near the Si–SiO₂ interface.

oxide at the Si–SiO₂ interface has attracted electrons in the silicon bringing them closer to the p^+ region and narrowing the depletion layer. Thus any voltage drop between the n- and p^+ regions leads to a higher field in the surface region than in the bulk. Before the voltage can be increased to the avalanche breakdown value for the "clean" junction shown in Fig. 1.3a, the field will become large enough to cause avalanche breakdown in the silicon at the corner region, due to ions in the oxide.

Besides ionic contamination, there are two other types of charge center at the interface. *Oxide fixed charge* refers to localized charge centers that cannot change their charge state by exchange of mobile carriers with the silicon. *Interface trap charge* refers to charge localized on centers that can change their charge state by exchange of mobile carriers with the silicon. In particular, the charge state of an interface trap site changes with gate bias if the interface trap level is moved past the Fermi level, causing its occupancy to change. Both oxide fixed charge centers and interface trap centers are defects related to the chemical structure of the interface.

Interface traps and oxide fixed charge are sensitive to the processing used to fabricate integrated circuits. In particular, interface trap level and oxide fixed charge densities are affected by oxidation conditions, annealing, and contaminants. Just like ionic contaminants, interface trap level and oxide fixed charge densities must be controlled and minimized. Control consists of careful design of the process and careful monitoring during the processing once established. Both design and monitoring require the measurement of oxide fixed charge and interface trap level densities. Determination of these quantities is indirect, as it is based on analysis and interpretation of measured electrical device characteristics in terms of physical models.

Measurement of interface trap charge in levels near midgap is possible by a variety of methods discussed in this book. Measurement of oxide fixed charge density alone is not possible because interface trap levels near or within the silicon-allowed energy bands cannot be measured independently. Consequently, measurement of oxide fixed charge density usually lumps together both oxide fixed charge density and interface trap charge density from these band edge levels.

Device behavior is governed not only by charges in the oxide or at the interface; the charges in the silicon itself, caused by traps there, are extremely important. There are traps distributed throughout the silicon crystal related to impurities. These traps have energy levels in the silicon bandgap and can exchange charge with the conduction or valence bands.

The two simplest types of trap are caused by dopant ions, called *donors* and *acceptors*. A donor has an energy level very near the conduction band edge, whereas an acceptor has an energy level very near the valence band edge. Because donors and acceptors are traps with very shallow energy levels, they will be ionized at room temperature, contributing mobile charged carriers to the silicon. Therefore, donors and acceptors are deliberately introduced into the silicon to control its conductivity. Device characteristics are critically dependent on this type of trap.

Another type of trap, which we call a *bulk* trap, is related to heavy-metal impurities inadvertently introduced into the silicon with the dopant donor or acceptor impurities. Bulk traps have one or more energy levels in the silicon bandgap, usually near midgap. Therefore, bulk traps can participate efficiently in generation and recombination of mobile carriers. Generation, which is the alternate emission of an electron and a hole by a bulk trap, or recombination, which is the alternate capture of an electron and a hole by a bulk trap, are the mechanisms by which the silicon reaches thermal equilibrium after a disturbance has been applied, such as an electric field. Mobile carriers, particularly in inversion layers, are of paramount importance in all active devices used in integrated circuits. Factors that govern minority carrier response to time varying voltages and their intimate relation to generation and recombination through interface and bulk traps thus are important in device operation and must be measured. All these matters are considered in this book.

1.3 STUDY OF THE MOS SYSTEM

Most semiconductor devices demonstrate surface sensitivity, as we have seen. Therefore, in principle, any device could be used in measurements to detect surface charges.

There are several different ways of investigating semiconductor surfaces and oxide-semiconductor interfaces. In the earliest studies the classical field effect experiment was the mainstay. With the advent of planar technology, where the oxidized silicon surface plays a major role, classical field effect experiments were displaced by studies using the MOSFET and the MOS capacitor. Of these, the MOS capacitor is the most powerful for investigating nearly all the electrical properties of the MOS system.

The superiority of the MOS capacitor rests on its simple structure, simple fabrication, and simplicity of analysis. The simplicity of analysis results because thermal equilibrium conditions obtain and because a one-dimensional treatment is accurate. Using the MOS capacitor, nearly all of the properties of interest in the SiO_2, at the Si–SiO_2 interface, and in the silicon can be measured.

The quantity measured in an MOS capacitor is its admittance as a function of gate bias and frequency. Exactly how information is extracted from the admittance is one of the main subjects of this book. In the next three subsections we briefly describe the problems associated with the classical field effect experiment, with MOSFET measurements, and with junction diode measurements of the electrical properties of the MOS system. We explain why the MOS capacitor is preferred.

1.3.1 The Classical Field Effect Experiment

The purpose of a field effect measurement is to measure field effect mobility and interface trap properties. In the classical field effect experiment, the change in conductance along a semiconductor surface is studied as a function of electric field applied normal to the surface. An electric field normal to the surface is produced by applying a bias to a gate parallel to but electrically insulated from the semiconductor. The conductivity of the semiconductor, modulated by the voltage applied to the gate, is measured between two ohmic contacts on either side of the gate or at the ends of the semiconductor sample. The experimentally observed conductance change is invariably smaller than expected on theoretical grounds. This deviation can be attributed to two factors: (1) some of the mobile excess charge attracted to the silicon surface by the applied field may become immobilized in interface traps so that the conductivity change is smaller than it would be otherwise and (2) mobility of mobile carriers near the silicon surface usually is smaller than in the bulk.

Interface traps affect surface mobility in an unknown manner. To extract interface trap and other interfacial properties from surface conductance changes in a field effect experiment, it is necessary to separate the influence of interface traps from the influence of other scattering mechanisms on surface mobility. Because these effects are not quantitatively understood, conclusions extracted from measurements of conductivity changes in a field effect experiment are uncertain at best. To separate the various influences on surface mobility, one approach has been to apply a normal magnetic field in the surface conductance field effect experiment.[7,8] Such experiments are involved and difficult to interpret. Thus the early field effect experiments were useful in showing the presence of interface traps and in distinguishing the accumulation, depletion, and inversion regions of the field induced surface charge. However, these measurements did not prove to be an effective way of extracting detailed interfacial properties.

1.3.2 MOSFET and Junction Diode Measurements

(a) *MOSFET*

In the MOSFET, conductivity changes of an inversion layer can be measured to determine channel mobility. As with the classical field effect experiment, some inference about interfacial properties can be made from such transport measurements. However, transport properties can be measured over a rather limited range of bandgap energy because the silicon surface must be inverted before a measurement can be made. The silicon surface is inverted over a limited range of potential, and once inverted, surface potential cannot be changed appreciably. This limitation of the MOSFET is not shared by the classical field effect experiment.

The inference of interfacial properties from transport measurements is uncertain because the transport equations contain the mobility. The mobility varies with applied bias and temperature, as well as with scattering at the interface. The mobility is not directly measured but is inferred from the I-V characteristics of the MOSFET. These characteristics can be calculated accurately only when the mobility is a known function of field, leading to a circular problem. In practice, such complications are ignored, and the departures of MOSFET characteristics from a fictitious device with uniform doping and an *a priori* form for the mobility are taken to reflect interfacial properties. Analysis is complex, requiring solution of a transport problem in a two-dimensional potential.

Information on the properties of the MOS system similar to that obtained from MOS capacitor measurements can be obtained using the

MOSFET by measuring gate to substrate capacitance with source and drain connected to the substrate. However, there is no advantage in using the MOSFET to obtain information equally accessible with an MOS capacitor, which is easier to fabricate. In particular, for C-V measurements, MOSFET test devices of large size must be used, just as with the MOS capacitor. The MOSFETs used in integrated circuits are so small that the measurement of C-V curves is inaccurate and dominated by parasitic capacitances. Even for large MOSFETs, measured capacitance has to be corrected for the parasitic capacitance caused by overlap of the gate over the source and drain junctions. These parasitic capacitances are not accurately measurable, so MOSFET capacitance is subject to a measurement error not found with the MOS capacitor.

The major advantages of the MOSFET in measuring interfacial properties are (1) measurement of channel mobility and how it is influenced by interfacial and oxide properties is possible, (2) majority and minority carrier capacitances can be measured independently at the same frequency, and (3) either the MOSFET or the gated diode, which is a MOSFET-like structure that has a single junction instead of two, or a MOSFET with source and drain connected together, is useful for measuring generation currents and surface recombination velocity. Thus, despite its limitations, measurements of interfacial properties with the MOSFET are important for (1) confirmation of results obtained from MOS capacitor measurements and (2) measurement of properties that cannot be measured with the MOS capacitor.

(b) Junction Diode

Another structure for investigating surface effects is the p-n junction, which also is sensitive to interfacial properties. However, the electric field configuration at the surface of a planar p-n junction is two dimensional, thus complicating analysis and limiting the usefulness of this structure in surface studies. Also, like the MOSFET, the p-n junction does not operate in thermal equilibrium, thus further complicating analysis.

1.3.3 Advantages of the MOS Capacitor

Using the MOS capacitor for measuring properties of the MOS system, the following properties can be obtained:

1 Surface band bending and depletion layer width in the silicon as a function of gate bias.
2 Voltage and field at avalanche breakdown in the silicon.
3 Doping profile in the silicon.
4 Interface trap level density as a function of energy in the bandgap.

5 Interface trap capture probability for both electrons and holes as a function of energy in the bandgap.

6 Lifetime in the bulk silicon.

7 Surface recombination velocity.

8 Oxide thickness.

9 Oxide breakdown field.

10 Charge configurations in the oxide such as oxide fixed charge and the charge at the interface between SiO_2 and another insulator deposited on top of it.

11 Nonuniformities in the oxide charge distribution and nonuniformities of surface potential caused by the discrete nature of charge in the oxide.

12 Work function differences between silicon and gate.

13 Ionic drift and polarization effects in SiO_2.

14 Diffusion of water into SiO_2.

15 Band-to-band tunneling in the silicon and tunneling into SiO_2.

16 Quantum effects in the inversion layer at low temperatures (surface quantization).

17 Results of thermally activated chemical reactions and electrochemical reactions in SiO_2.

18 Conductivity type of the silicon.

19 Dielectric constant of silicon.

20 Dielectric constant of SiO_2.

21 Properties of electron and hole traps in SiO_2.

The MOS capacitor also has been used in determining ways of controlling oxide fixed charge and interface trap level densities, effects of light on interface traps and surface charges, and in studying the effects of ionizing radiation on SiO_2. Internal photoemission measurements using the MOS capacitor have yielded information about the barrier height between the conduction bands of silicon and SiO_2 and about trapping centers distributed in the SiO_2 layer. We discuss most of these measurements in this book.

The goal of studying the MOS system is to improve the performance and the stability of devices such as the MOSFET used in integrated circuits. As just discussed, the MOS capacitor is simpler to use for these studies than the actual devices used in integrated circuits. The MOS capacitor is useful in such studies because any change in processing that improves the electrical properties of the MOS capacitor makes the same improvement on the actual device. Thus the simplicity and the versatility of the MOS capacitor can be exploited with confidence for optimizing integrated circuit processing.

Other insulators and semiconductors are being studied using the MOS capacitor measurement methods developed in the course of studies on the MOS system. Caution should be used in applying these methods to other semiconductors and insulators because significant modifications of interpretation are necessary in many cases.

1.4 HISTORICAL BACKGROUND

1.4.1 Early Work

(a) Rectification

The first reports of phenomena directly related to semiconductor surfaces date back to the end of the nineteenth century, when rectification was discovered in a semiconductor-to-metal contact. In 1874 Brown[9] observed that the I-V characteristics of a metal-galena contact were asymmetric with respect to voltage polarity. Similar effects were observed about the same time by Schuster[10] with copper-copper oxide contacts. The role of the interface in these phenomena was not appreciated until much later. By the 1920s copper oxide and selenium rectifiers were manufactured commercially. There was much art and little understanding of the underlying physical principles at that time. Understanding had to await the development of quantum mechanics and its application to metals and semiconductors by Sommerfeld,[11] Bloch,[12] and Wilson[13] in the late 1920s and early 1930s.

In the early 1930s it became apparent that rectification was related to the properties of the interface between the metal and the semiconductor. It was realized that rectification depended on a potential barrier created when the metal and the semiconductor made intimate contact. In early theories the height of this potential barrier was assumed to be equal to the difference in work function between the metal and the semiconductor, that is, to the energy required to move an electron from one material into the other. The theories of Schottky,[14] Mott,[15] and Davidov[16] proposed independently in 1939 were based on these considerations and accounted for rectification. They constituted a major advance in our understanding of interface phenomena. However, many investigators found the barrier height often did not depend on work function differences.* This inconsistency in the theory of rectification was clarified by Bardeen[17] in 1947. Bardeen proposed that the potential barrier between the metal and the semiconductor could be due to an electric field originating on charges in interface traps or electronic energy levels in the bandgap of the semiconductor belonging to wave functions localized at this interface, rather than as a result of work function differences. These interface traps explained the results obtained on covalent semiconductors such as ger-

---

*It turns out that in ionic semiconductors, barrier height is equal to work function difference, whereas it is nearly independent of work function difference in covalent semiconductors.

manium and silicon. The possibility of interface traps in the bandgap of a semiconductor had been pointed out theoretically by Tamm[18] in 1932 and Shockley[8] in 1939, but the important role played by interface traps in surface phenomena was not recognized until it was pointed out by Bardeen[17] in 1947. Considerable study of surfaces then ensued.

A milestone was reached when Brattain and Bardeen[19] in 1953 found that the barrier height at the free surface of a covalent semiconductor could be varied controllably, and more or less reversibly, by exposing the semiconductor surface to a sequence of controlled atmospheres (ozone, dry O_2, and water vapor). With this cycle, surface potential could be varied over a range of about 0.5 eV between a p-type extreme in ozone and an n-type extreme in water vapor. This technique, known as the *Bardeen–Brattain cycle*, was used widely in early studies of the free surface barrier.

(b) Surfaces

Liandrat[20] had proposed in 1935 that the conductivity of a thin semiconductor layer could be modulated by an external electric field. This effect, called the *field effect*, came to be one of the most powerful tools in surface studies. The existence of quantum states on a free semiconductor surface was demonstrated in a classical paper by Shockley and Pearson[21] in 1948 by a field effect experiment. It was in the course of surface studies that transistor action was discovered by Bardeen and Brattain[22] in 1947. Following this discovery, the main effort shifted from the surface to the bulk. Major advances were made in the preparation of high purity single-crystal semiconductors, particularly germanium and silicon, and the fundamental properties of these semiconductors was intensively studied.

Interest in the free surface was revived again in the early 1950s. This renewed interest was motivated primarily by problems encountered in bipolar transistor and junction diode performance and reliability. It was recognized that although bipolar transistor action was a bulk phenomenon, bulk properties alone could not account for the observed anomalous behaviour of p-n junction diodes and bipolar transistors and that many of these effects were associated with the surface. At the same time, the availability of germanium and silicon crystals of high purity and high lattice perfection and the better understanding of their bulk properties made these types of crystal ideal semiconductors for surface studies. It is mainly through work on germanium and silicon that surface research has become a quantitative discipline.

1.4.2 The Modern Era

Unlike today, where interest centers on the oxide passivated surface, work in the 1950s centered around two other types of semiconductor surface, *clean* and *real*. Clean surfaces are those that are as free as possible of contamination by foreign impurities. Clean semiconductor surfaces are highly reactive and thus are difficult to produce, maintain, and measure. However, they are the surfaces closest to the ideal crystal surface and thus should exhibit the fundamental properties of a crystal surface. Clean surfaces can be prepared by high vacuum techniques such as cleavage, ion bombardment, or heating at elevated temperatures. Once prepared, a surface can be maintained clean in an ultrahigh vacuum (10^{-10} to 10^{-9} Torr), thus preventing recontamination by foreign atoms. Work on clean surfaces is the least advanced because of the great difficulties involved and is not of major interest in integrated circuit technology.

Real surfaces were the so-called practical surfaces prior to the advent of oxide passivation. Real surfaces were prepared by the various device fabrication techniques involving wet chemistry such as mechanical polishing followed by etching to remove the polishing damage. Such a surface is covered by chemisorbed atoms, generally oxygen, and by other molecules or atoms from the surrounding ambient. Real surfaces have been studied extensively because they are easy to prepare and handle and were the surfaces encountered in practical device applications during the 1950s. However, real surfaces have surface films of unknown composition and structure that are not very stable or reproducible in most cases.

Real surfaces no longer are of great practical importance in semiconductor electronics, as they have been replaced during the 1960s by the stable and reproducible solid-solid interface between silicon and a grown oxide film of known composition and structure. We concentrate our attention on this third type of surface. However, during the period when real surfaces were being intensively studied, theoretical modeling of surface band bending and its consequences were formulated by Brown[23] in 1953, Kingston and Neustadter[24] in 1955, and Garrett and Brattain[25] in 1955. This theoretical groundwork was applied in the next decade to the silicon-silicon dioxide interface.

Initial work on surface passivation was undertaken by a group headed by Atalla, with junction diodes and bipolar transistors in mind. The first good-quality silicon dioxide films of high dielectric breakdown strength and low loss were oxidized in high pressure steam by Ligenza and Spitzer[26] in 1960–1961. In a classic paper, Atalla et al.[27] in 1959 laid the foundations for oxide passivation of junction diodes and bipolar transistors. Frosh and Derrick[28] in 1957 found that silicon dioxide acts as an effective barrier against many commonly used dopant impurities, thus preventing them from reaching the underlying silicon. All this work formed the basis for the planar process described next, which finally led to modern monolithic integrated circuits.

1.4.3 The Planar Process

The planar process was described first by Hoerni[29] in 1960. This process evolved from generations of earlier ones. This evolution as well as the essentials of the planar process are illustrated in Fig. 1.4, in which the major techniques of fabricating a *p-n* junction are shown.

Figure 1.4*a* shows the grown junction method first described by Teal et al.[30] in 1951 soon after the initial discovery of the transistor effect in 1948. In this method a semiconductor crystal is grown out of a melt, doped *p*-type in this example. This is the Czochralski method of growing a single crystal of semiconductor. Growth is started by contacting the melt contained in a crucible with a single crystal seed, thereby establishing a temperature gradient between the seed and the melt. As the seed slowly is drawn away from the melt, a single crystal is grown that has the same crystallographic orientation as the seed. The Czochralski method of crystal growth is widely used for growing single crystals of germanium and silicon. This method is advantageous for both silicon and germanium, which expand on solidification. Because the crystal is pulled out of the melt, it is not confined during solidification as it would be if it were solidified in a crucible. Solidifying single crystals in a crucible is suitable for materials that contract on solidification, such as copper.

To grow a *p-n* junction in the crystal as it is being pulled, the dopant

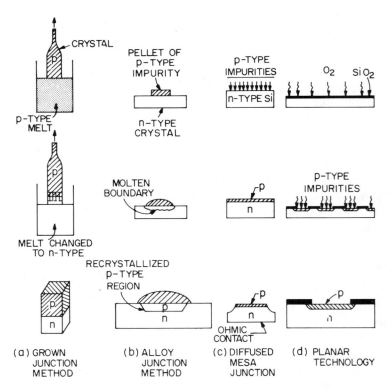

Fig. 1.4 Methods of fabricating *pn* junctions. After Grove.[2]

concentration in the melt suddenly is changed, for instance, by dropping a pellet containing donor-type impurities into the melt. The rest of the crystal then will grow n-type. When growth is completed, the crystal is sectioned into bars containing the p-n junction, as shown in Fig. 1.4a. This grown junction method was important in the early days following discovery of the bipolar transistor, mainly because diodes made this way were used to verify the theory of the I-V characteristics of p-n junctions. The grown junction method, however, was not as suitable for mass production as the alloy junction method reported in 1950 by Hall and Dunlap.[31]

The alloy junction method is illustrated in Fig. 1.4b, where a pellet containing acceptor-type impurities (e.g., aluminum) is heated until the pellet fuses or alloys with the silicon or the semiconductor. When the crystal is cooled, a recrystallized region, saturated with acceptors forms under the pellet, the p-n junction is formed at the recrystallized region and the unalloyed aluminum on top forms a metallic contact to the junction. The alloy junction method still is used in the mass production of diodes and transistors, mainly those made of germanium. The depth to which the recrystallized region penetrates into the semiconductor always has been difficult to control. This limits the performance of bipolar transistors, particularly at high frequencies where a narrow base width (distance between the emitter and collector junctions) requires very precise control.

The search for a method with better control of the location of a p-n junction in the semiconductor led to the development of diffused p-n junctions in 1956 by Tannenbaum and Thomas.[32] The first diffused junctions were the mesa type illustrated in Fig. 1.4c. Here the junction is formed by exposing the surface of the n-type wafer to acceptor impurities, commonly contained in a gas that flows past the wafer. However, unlike the alloy junction, no phase transformation takes place. Instead, the impurities penetrate into the semiconductor by solid-state diffusion, which can be very precisely controlled. The diode is completed by etching away semiconductor material to form a mesa containing the junction. Many junction diodes are made in this way on each wafer, although only a single one is illustrated in Fig. 1.4c.

Monolithic integrated circuits were made using mesa technology, but they were very expensive. Costly interconnections were made by thermocompression bonding of small wires from element to element. Reliability, performance, and uniformity of characteristics from device to device in the circuit were poor.

The planar process shown in Fig. 1.4d superseded the mesa technology for integrated circuits so successfully that all monolithic circuits are made by the planar process today. Frosh and Derrick's[28] discovery that a thin layer of SiO_2 can effectively mask against the diffusion of most important acceptor and donor impurities made possible precise control of both the depth of a diffused junction and its location on the silicon wafer. The location of devices on the silicon wafer could be delineated by covering the silicon with

SiO_2 and then exposing the silicon to the diffusing impurities only in selected areas, defined by photolithography, where the SiO_2 layer had been removed. The planar technology illustrated in Fig. 1.4d combines the advantages of junction formation by solid-state diffusion with the masking properties of SiO_2 for the precise definition of device geometry.

A later development is ion implantation, which is a well-controlled predeposition of dopant impurities. Much lower doping densities can be dependably predeposited by ion implantation than by the conventional diffusion process. Because dopant ions have a restricted penetration depth into SiO_2 depending on their energy, SiO_2 still is useful as a mask. After ion implantation, the predeposition step, the wafer is annealed to electrically activate and drive the dopant impurities deeper into the silicon. This drive-in step is analogous to the drive-in step in the conventional diffusion process.

Interconnections between devices are made by evaporating a metal over the SiO_2, making contact with the diffused or implanted regions through holes cut in the SiO_2. The metal is then cut into a particular pattern by photolithography to make the circuit. The insulating properties of the silicon dioxide film keep the metal leads isolated from the silicon substrate.

1.4.4 The MOSFET

In large scale integrated circuits the MOSFET is important because of its low fabrication cost, small size, and low power consumption. The bipolar transistor, the MOSFET, the p-n junction, and the Schottky barrier diode are the most widely used active elements in modern integrated circuits.

Although the idea of a field effect transistor is an old one, the modern MOSFET was made possible and practical with the development of the oxide passivation technology of silicon. The field effect, that is, a conductance change in a semiconductor induced by the application of a transverse electric field, was a subject of intensive studies by many people long before the discovery of transistor action. As described earlier, the discovery of the transistor effect was one outcome of studies on the field effect. As early as the 1920s and 1930s, proposals on amplifying devices, based on the field effect were made by Lilienfeld[33] and Heil.[34] However, there was little understanding of the physical phenomena involved, and practical means were not available for implementing these concepts at that time. An unequivocal demonstration of the field effect was made in the classic paper by Shockley and Pearson[21] in 1948. They showed that an external field caused appreciable modulation of conductance in the surface region of a semiconductor.

Figure 1.5 shows a basic field effect circuit. The semiconductor sample shown shaded is usually shaped in the form of a thin rectangular strip with most of the surface area confined to the two larger faces. A metal electrode is placed parallel to one of these faces and insulated from it by a spacer.

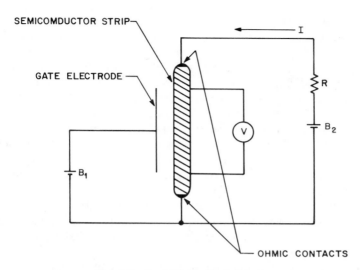

Fig. 1.5   Basic field effect circuit.

Thin sheets ($\approx 20\,\mu$m thick) of mylar or mica were used as spacers, as were single crystals of barium or strontium titanate. Barium or strontium titanate were used because their large dielectric constants helped to achieve a large electric field at the semiconductor surface. By applying a voltage across the capacitor formed by the metal electrode and the semiconductor with battery B_1, electric fields as high as 2×10^6 V/cm could be produced at the semiconductor surface. Dielectric breakdown of the spacers limited the field that could be applied.

The field perpendicular to the semiconductor surface produces a charge layer at the surface. Suppose that the semiconductor is p-type. When the metal electrode is made negative with respect to the semiconductor, this charge layer will consist of holes attracted to the surface. As the bias on the metal electrode is made less negative or more positive, holes will be depleted from the surface so that a depletion layer forms containing ionized acceptors. Finally, as bias is made even more positive, the charge layer will consist of a surface layer of electrons, on top of the depletion layer of ionized acceptors.

In general, the semiconductor strip consists of two conductors in parallel. One conductor is the surface charge layer, and the other is the remaining semiconductor bulk. The conductivity of the entire strip is measured by applying a voltage to ohmic contacts at each end with battery B_2 shown in Fig. 1.5. The voltage drop across the strip is read by the voltmeter labeled V. In this way, changes of conductance of the strip can be measured as a function of the bias applied to the metal electrode. In our example of a p-type semiconductor with ohmic contacts, conductance will be high when there is an accumulation of holes at the semiconductor surface and will reach a minimum when the surface is depleted of mobile carriers.[35] Using this method, interface traps were measured first by Shockley and Pearson,[21] as mentioned before, and majority carrier surface conductivity modulation also was measured this way. Thermal oxidation was shown to reduce interface trap level density using this method by Atalla et al.[27] Thus the field effect became a tool for studying the surface properties of semiconductors, mainly silicon and germanium.

The field effect subsequently was applied to various similar amplifying device configurations. These devices used majority carrier modulation; that is, the transverse electric field modulated the majority carrier (holes in our p-type example) density at the silicon surface resulting in conductance changes between the ohmic contacts at each end of the strip.

Useful devices based on the principle of majority carrier modulation are achievable only under severe geometric limitations. That is, a large portion of the conducting region must be modulated by the imposed field. However, the maximum field imposed is limited by avalanche breakdown of the semiconductor in some devices [junction field effect transistors (JFETs)] (5–7×10^5 V/cm in silicon) or insulator breakdown in others (6–10×10^6 V/cm in silica). Given this maximum field, Gauss's law determines the maximum charge density per unit area that can be controlled using the field effect. This maximum charge per unit area divided by majority carrier density per unit volume determines the depth to which majority carrier conductivity modulation can extend, provided inversion does not occur. For example, for 10^{16} carriers/cm^3 (typical for silicon), the maximum depth of depletion of carriers is about 3μm. If it is to be modulated, the entire conducting region cannot extend much beyond this depth. This depth can be increased by resorting to material where the carrier density is low. However, the use of high resistivity material leads to high impedance in amplifier circuits, thus limiting operation to low frequencies. To use low resistivity materials and improve performance, devices of practical interest based on the principle of majority carrier modulation are built around thin-film technology. The first thin-film field effect transistors were described by Weimer[36] in 1961 more than a decade after the Shockley-Pearson paper appeared.

Field effect devices that are not subject to the geometric constraints mentioned above are based on the principle of *inversion layer* conductance modulation. In such devices, majority carrier conductivity modulation is made irrelevant by using p-n junction contacts, not ohmic contacts, so no majority carrier current can flow. Instead, the conducting region is an inversion layer, a thin (30–300 Å) layer of minority carriers (electrons in p-type silicon) at the semiconductor surface. The inversion layer is created by the normal field, so 100% modulation is possible. The inversion layer is thin by nature, so no condition on sample thickness or restriction to thin film technology is imposed. However, the conductance of the inversion layer itself is dictated by the carrier density per unit area in the inversion layer, aside from mobility, and this carrier density is limited by Gauss's law and the breakdown field of the insulator; therefore, technology requirements for inversion layer conductance devices focus on the properties of this insulator.

Devices based on majority carrier conductivity modulation rather than minority carrier conductivity modulation appeared first, because the first field effect experiments were done with ohmic contacts to the semiconductor. Ohmic contacts permit majority carrier flow but are unable to supply large minority carrier currents. Therefore, even when an inversion layer is present, the fastest way the inversion layer can be supplied with minority carriers is from the bulk. Such supply is by diffusion in germanium and by thermal generation in silicon. These processes are slow, so that the inversion layer could not contribute appreciably to the first field effect experiments. The key requirement for an inversion layer conductance modulation device is a rectifying contact capable of rapidly supplying the required minority carriers, namely, a p-n junction.[35]

Systematic study of inversion layer conductance was prompted by reliability problems in bipolar transistors. Conductance between the emitter-collector junction of an n-p-n transistor, with base floating, was much larger than expected on the basis of reverse bias junction impedance and leakage across the junction. An inversion layer was postulated to exist between the two p-n junctions bridging the emitter and the collector. The inversion layer was thought to arise in unpassivated devices from ions adsorbed on the surface of the base region that repelled holes in the base, thus creating a depletion layer, and attracted electrons to the surface to create the inversion layer. Because the emitter and the collector were n-type, they were electrically connected by this ion induced inversion layer. The quantitative behavior of such an inversion layer was both predicted by this model and experimentally verified by Brown[23] in 1953. Subsequently, a significant proposal was made by Ross[37] in 1955 that such an inversion layer could be induced electrostatically by an electrode deliberately placed in the vicinity of the base region of a bipolar transistor. He further proposed that if the space between the control electrode or the gate and the surface of the base were filled with a ferroelectric, the inversion layer could be controllably set up and permanently maintained until the control gate was suitably excited to erase the inversion layer. Then an impedance meter connected between emitter (source) and collector (drain) would show whether an inversion layer existed. This is an electrostatic memory cell with nondestructive access.

A memory element in which a ferroelectric is used has not proven to be commercially practical. The idea of putting a simple insulator between the gate electrode and underlying semiconductor has turned out to be practical. This idea was implemented only after significant technological developments allowed the production of an oxide of high dielectric breakdown strength, low loss, low interface trap level and oxide fixed charge densities, and negligible drift.

In 1958 Wallmark[38] proposed structures grossly similar to those of Lilienfeld[33] and Heil[34] but incorporating the native oxide of germanium. Wallmark's structures operated by modifying the resistance of the semiconductor body rather than by creating a surface channel region as in Ross's device. Atalla[39] proposed that thermally grown SiO_2 on silicon single crystal surfaces should be used as the gate insulator. This suggestion emerged from studies on oxide passivation of silicon bipolar transistors. The first oxides used by Kahng and Atalla to fabricate the first MOSFET structures reported in 1960[39,40] were grown by Ligenza and Spitzer[26] in high pressure steam.

In 1961 Kahng[41] showed that the performance of the MOSFET was consistent with simple first-order theory. However, reproducibility of characteristics was poor, due to large interface trap level and oxide fixed charge densities as well as sodium contamination of the oxide. Practical MOSFET fabrication was accomplished only after the interfacial properties of the oxidized silicon surface and the nature of drift and ionic contamination had been understood and controlled.

Interface traps affect the gain of the device because every charge placed on the gate is not compensated by an additional free carrier in the inversion layer; instead, some gate charge is compensated by charged interface traps. Thus interface traps reduce conductivity modulation of the inversion layer. High interface trap level densities ($> 10^{11} \, cm^{-2} \, eV^{-1}$) also can cause increased noise as the fluctuating occupancy of interface traps modulate the inversion layer conductance. Also, charged interface traps and fixed charge influence the voltage needed to form the inversion layer. Finally, sodium contamination causes drift in device characteristics as well as changes with time of both interface trap level and oxide fixed charge densities.

The first published analysis of MOSFET characteristics was by Ihantola and Moll.[42] Similar theoretical work was reported by Hofstein and Heiman,[43] who published the first detailed comparison of theory and experiment.

In 1963, Grove, Deal,[44] and Snow developed the first commercially usable process for the construction of MOSFETs with stable characteristics. In the process developed by these workers, sodium levels were minimized by scrupulous care and cleanliness in every critical process step. Another advance in this period first reported by Kerr and Young[45] was the concept of gettering the sodium with phosphorus, which was an alternative solution to the sodium drift problem.

The next key to commercialization was the development in 1963 of both p-channel and n-channel devices with a wide variety of characteristics by Wanlass and Sah,[46] who had been challenged by the interesting but unstable field effect devices reported by Weimer[36] using cadmium sulfide. Wanlass and Sah[46] reported complementary MOS circuits constructed on separate chips of silicon for the n-channel and p-channel devices rather than on the same chip of silicon as done now.

1.4.5 The Integrated Circuit

The concept of an integrated circuit moved from discrete devices wired together on a single header to the monolithic integrated circuit known today. Several patents were filed in 1959 on different aspects of integrated circuits.[46-49] Accounts of the history of integrated circuit development have been given by Hogan[50] and Kilby.[51] A comprehensive review of the evolution of silicon technology during 1952–1977 was given by Deal and Early,[52] and Kahng[53] gave an historical review of the development of the MOSFET. Today, a polycrystalline silicon gate,[54,55] rather than a metal gate, and ion implantation[56-60] are widely used in integrated circuit manufacture because of the many advantages these technologies offer.

The intensive worldwide study of the MOS system over the past 20 years, given impetus by these developments, makes it one of the best understood solid-solid systems today. The understanding gained of this system, which can be studied by the refined and sensitive measurement techniques developed in the course of discrete device and integrated circuit work, is applicable not only to integrated circuit technology, but to other areas of technological importance involving insulators and insulator-semiconductor interfaces. This review brings us to the start of our story.

REFERENCES

1 S. M. Sze, *Physics of Semiconductor Devices*, Wiley, New York, 1969.

2 A. S. Grove, *Physics and Technology of Semiconductor Devices*, Wiley, New York, 1967.

3 P. Richman, *MOSFET'S and IC'S*, Wiley, New York, 1973; R. S. C. Cobbold, *Theory and Applications of FET'S*, Wiley, New York, 1970.

4 M. P. Lepselter and J. M. Andrews, in *Ohmic Contacts to Semiconductors*, B. Schwartz, Ed., Electrochemical Society, Princeton, N. J., 1969, pp. 159–186.

5 E. H. Nicollian and A. K. Sinha, in *Thin Films—Interdiffusion and Reactions*, J. M. Poate, K. T. Nu, and J. W. Mayer, Eds., The Electrochemical Society, Princeton, N. J.; Wiley, New York, 1978, Chapter 13.

6 N. B. Hannay, Ed., *Semiconductors*, Reinhold, New York, 1960, Chapter 8.

7 R. de L. Kronig and W. G. Penney, *Proc. Roy. Soc. (Lond.)*, **A130**, 499 (1931); J. Dekker, *Solid-State Physics*, Prentice-Hall, New York, 1957, Chapter 10.

8 W. Shockley, *Phys. Rev.*, **56**, 317 (1939).

9 F. Brown, *Pogg. Ann.*, **153**, 556 (1874); *Wied. Ann.*, **1**, 95 (1877); ibid., **4**, 476 (1878); ibid., **19**, 340 (1883).

10 A. Schuster, *Phil. Mag.*, **48** (4), 25 (1874).

11 A. Sommerfeld, *Z. Physik*, **47**, 1 (1928); A Sommerfeld and H. Bethe, in *Hanbuch der Physik*, Vol. 2412, H. Geiger and K. Scheel, Eds., Springer, Berlin, 1933, p. 333.

12 F. Bloch, *Z. Physik*, **52**, 555 (1928).

13 A.H. Wilson, *Proc. Soc. (Lond.)*, **A133**, 458 (1931); ibid., **A134**, 277 (1931).

14 W. Schottky, *Z. Physik* **223**, 367 (1939); ibid., **118**, 539 (1942).

15 N. F. Mott, *Proc. Roy. Soc. (Lond.)*, **A171**, 27 (1939).

16 B. Davidov, *J. Phys.* (*USSR*), **1**, 167 (1939).

17 J. Bardeen, *Phys. Rev.*, **71**, 717 (1947).

18 I. E. Tamm, *Z. Phys.*, **76**, 849 (1932); *Phys. Z.—Sowjet Union*, **1**, 733 (1932).

19 W. H. Brattain and J. Bardeen, *Bell Syst. Tech. J.*, **32**, 1 (1953).

20 G. Liandrat, conferences at the College of France, 1935.

21 W. Shockley and G. L. Pearson, *Phys. Rev.*, **74**, 232 (1948).

22 J. Bardeen and W. H. Brattain, *Phys. Rev.*, **74**, 230 (1948); ibid., **75**, 203 (1949).

23 W. L. Brown, *Phys. Rev.*, **91**, 518 (1953).

24 R. H. Kingston and S. F. Neustadter, *J. Appl. Phys.*, **26**, 718 (1955).

25 C. G. B. Garrett and W. H. Brattain, *Phys. Rev.*, **99**, 376 (1955).

26 J. R. Ligenza and W. G. Spitzer, *J. Phys. Chem. Solids*, **14**, 131 (1960); J. R. Ligenza, *J. Phys. Chem.*, **65**, 2011 (1961).

27 M. M. Atalla, E. Tannenbaum, and E. J. Scheibner, *Bell Syst. Tech. J.*, **38**, 749 (1959).

28 C. J. Frosh and L. Derrick, *J. Electrochem. Soc.*, **104**, 547 (1957).

29 J. A. Hoerni, IRE Electron Devices Meeting, Washington, D.C., 1960; U.S. Patents 3,025,589 (1962) and 3,064,167 (1962).

30 G. K. Teal, M. Sparks, and E. Buehler, *Phys. Rev.*, **81**, 637 (1951).

31 R. N. Hall and W. C. Dunlap, *Phys. Rev.*, **80**, 467 (1950).

32 M. Tannenbaum and D. E. Thomas, *Bell Syst. Tech. J.*, **35**, 23 (1956).

33 J.E. Lilienfeld, U.S. Patents 1,745,175 (filed in 1926, issued in 1930), 1,877,140 (filed in 1928, issued in 1932), and 1,900,018 (filed in 1928, issued in 1933).

34 O. Heil, British Patent 439457 (filed in 1935, issued in 1935).

35 A. Many, Y. Golstein, and N. B. Grover, *Semiconductor Surfaces*, North Holland, Amsterdam (1965), Chapter 6.

36 P. K. Weimer, IRE-AIEE Device Research Conference, Stanford University, Stanford, Calif., 1961.

37 I. M. Ross, U.S. Patent 2,791,760 (filed in 1955, issued in 1957).

38 J. T. Wallmark, U.S. Patent 2,900,531 (issued August 18, 1959).

39 M. M. Atalla, U.S. Patent 3,206,670 (filed in 1960, issued in 1965).

40 D. Kahng and M. M. Atalla, IRE-IEEE Solid-State Device Research Conference, Carnegie Institute of Technology, Pittsburgh, Pa., 1960; D. Kahng, U.S. Patent 3,102,230 (filed in 1960, issued in 1963).

41 D. Kahng, Bell Telephone Laboratories Technical Memorandum (January 16, 1961).

42 H. K. J. Ihantola and J. L. Moll, *Solid-State Electron.* **7**, 423 (1964).

43 S. R. Hofstein and F. P. Heiman, *Proc. IEEE*, **51**, 1190 (1963).

44 B. E. Deal, U.S. Patent 3,426,422 (filed in 1965, issued in 1969).

45 D. R. Kerr and D. R. Young, U.S. Patent 3,303,059 (filed in 1964, issued in 1967).

46 F. M. Wanlass and C. T. Sah, *ISSC Digest*, pp. 32–33 (February 1963); U.S. Patent 3,356,858 (filed in 1963, issued in 1967).

47 J. S. Kilby, U.S. Patent 3,138,743 (1964).

48 K. Lehovec, U.S. Patent 3,029,366 (1962).

49 R. N. Noyce, U.S. Patent, 2,981, 877 (1961).

50 C. L. Hogan, *Interface Age*, **2** (3), 24 (1977).

51 J. S. Kilby, *IEEE Transact. Electron Devices*, **ED-23**, 648 (1976).

52 B. E. Deal and J. M. Early, *J. Electrochem. Soc.*, **126**, 20c (1979).

53 D. Kahng, *IEEE Transact. Electron Devices*, **ED-23**, 655 (1976).

54 J. C. Serace, R. E. Kerwin, D. L. Klein, and R. Edwards, *J. Solid-State Electron.*, **11**, 653 (1968).

55 R. E. Kerwin, D. L. Klein, and J. C. Sarace, U.S. Patent 3,476,236 (filed in 1967, issued in 1969).

56 H. G. Dill, U.S. Patent 3,544,399 (filed in 1966, issued in 1970).

57 F. Faggin and T. Klein, *Solid-State Electron.*, **13**, 1125 (1970).

58 D. H. Lee and J. W. Mayer, *Proc. IEEE*, **62**, 1241 (1974).

59 J. Stephen, *Proc. Internat. Microelectron. Conf.*, Anaheim, Calif., February 11–13, 1975.

60 J. L. Stone and J. C. Plunkett, *Solid-State Technol.*, **19**, (6) 35 (1976).

2

Field Effect

2.1 INTRODUCTION

In this chapter we derive expressions for mobile charge densities at the silicon surface as functions of potential. Using these expressions, we solve the Poisson equation for potential and charge density at the silicon surface, both with and without applied bias. Although the mathematical derivations in this chapter can be used to derive the I-V characteristics of the MOSFET and the characteristics of other field effect devices, such derivations are beyond the scope of this book.

For the moment, we simplify discussion by considering only steady, time-independent gate voltage. Throughout this book time-independent gate voltage is termed gate *bias*. In later chapters transient effects are considered such as mobile carrier response to sudden changes in gate voltage. In Chapter 3 mobile carrier response to small-signal sinusoidal time variations of gate voltage is treated and used to derive C-V characteristics of the MOS capacitor.

We begin with neutral bulk silicon with no fields present. Density of states, bandgap, dopant ions, and so on are introduced. Then a semi-infinite slab of silicon with a single planar surface is treated. A time-independent external voltage is considered with no current flow. This voltage is applied uniformly everywhere on the silicon surface, causing carrier densities and potential to vary in the direction normal to the surface. To find the

26

potential and carrier densities within the silicon, the one-dimensional Poisson equation is solved and discussed.

2.2 SILICON PHYSICS

We treat only the semiconductor physics necessary to calculate silicon surface charge and to solve the Poisson equation. An exhaustive treatment of semiconductor physics is beyond the scope of this book. The reader is referred to textbooks on this subject.[1-12]

To solve the Poisson equation, first an expression for the charge density is needed. This charge density consists of charged ions fixed in the silicon lattice and electrons that are mobile. We begin by discussing the quantum energy levels available to the electrons and their occupancy according to the Fermi-Dirac distribution. Included is a discussion of the energy levels introduced by dopant ions and the occupancy of these levels. This last point determines the charge density of ions fixed in the silicon lattice.

2.2.1 Energy Bands

The most important result of applying quantum mechanics to electrons in a semiconductor is that electrons are allowed only in certain energy levels grouped in bands separated by an *energy gap*, also termed the *forbidden gap* or the *bandgap*. Electrons in the silicon can be regarded as traveling waves that undergo constructive and destructive interference as they interact with the ion cores of the silicon lattice. This interaction leads to the creation of an energy gap separating bands of allowed propagation.* The bandgap increases with the strength of the interaction between electrons and ion cores of the silicon lattice. In silicon, the two allowed bands that participate in the electrical conduction process are the *valence band* (mostly filled with electrons) and the higher lying (mostly empty) *conduction band*. The free carrier concentrations in the two bands depend on the density of allowed energy levels and the probability that given energy levels are occupied by electrons.

Electrons in the outermost shell of each silicon atom are called *valence* electrons. Each silicon atom has four valence electrons that it shares with four neighbors in the crystal by forming *covalent* bonds. Each covalent bond contains a pair of valence electrons of opposite spin. The valence electrons in covalent bonds are localized on silicon atoms in the lattice. These electrons occupy energy levels in the valence band that will be full at a temperature of $0°$ K. No electric conduction can take place at this temperature because there are no unoccupied energy levels in the valence

---

*The silicon is analogous to an electrical filter with stop bands and pass bands.

band for the electrons to move into. At temperatures above absolute zero, some valence electrons are excited into the conduction band by thermal agitation. The energy required to excite one electron is equal to the forbidden gap energy.

An electron excited into the conduction band is a negative charge free to migrate throughout the crystal under the influence of thermal agitation. The valence charge deficiency produced by excitation of an electron to the conduction band is a positive charge called a *hole* and is free to migrate throughout the crystal under the influence of thermal agitation. If an electric field is applied, the electron will drift in the direction opposite to the applied field, and the hole will drift in the direction of the applied field. The resulting current is the sum of the electron and hole drift components. Thus excitation of an electron from the valence to the conduction band constitutes the generation of a hole-electron pair, which preserves electrical neutrality within the crystal and contributes to the electrical conductivity of the crystal.

In Fig. 2.1 E_c represents the energy of an electron at rest, the minimum energy of an electron in the conduction band (apart from a neglected zero point energy). An electron at a higher energy than E_c would have to be moving and would have a kinetic energy in addition to its potential energy. Similar arguments hold for holes in the valence band at or below the energy E_v. Between these two energies is the forbidden gap of width $E_g = 1.12$ eV at 300°K. Above E_c the number of allowed energy levels for electrons increases quickly. This density of allowed energy levels in the conduction band is labeled $M_c(E)$, and $M_v(E)$ is the density of allowed energy levels in the valence band.

The number of energy levels within the conduction and valence bands is extremely large. In a silicon crystal there are 5×10^{22} silicon atoms/cm³. Thus in every cubic centimeter of the crystal there are 5×10^{22} energy levels in each band, and 10^{23} electrons can be accommodated in such a band as two electrons can occupy each level, one of each spin.

A mathematical means must be found for dealing with so many different energy levels or quantum states. We define $\Gamma(E)$ as the total number of quantum states up to an energy E per unit volume of the crystal. Then we count the number $\Delta\Gamma$ of quantum states per unit volume of the crystal within a small energy increment ΔE and divide by ΔE. In the limit of small ΔE this ratio becomes the *density* of quantum states per unit volume per unit energy $M(E)$. The definition is

$$M(E) \equiv \frac{d\Gamma}{dE}. \tag{2.1}$$

Usually, electrons will be found only near the minimum energy E_c of the conduction band and holes near the maximum (minimum for holes) energy E_v of the valence band. *Near the band edges* the bands can be considered

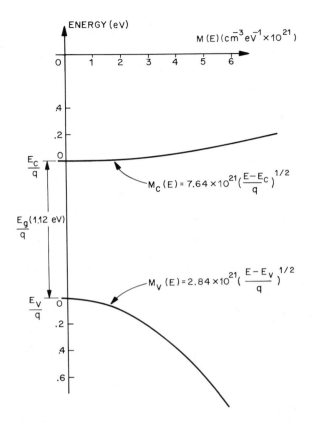

Fig. 2.1 Density of states diagram for nondegenerate silicon at 290°K. Values of M_c and M_v are calculated from (2.2) and (2.3). The faster change of the density of states with energy in the valence band is a consequence of the smaller coefficient for M_v. Energy values E_c, E_v, E_g, and so on are algebraic symbols from (2.2) and (2.3), with units in joules. The voltage equivalents are indicated as numbers on the axes in this figure.

to be *parabolic*, which means that energy is related quadratically to momentum[13].

Under these conditions the density of states can be written[13]

$$M_c(E) = 6(4\pi h^{-3})(2m_n)^{3/2}(E - E_c)^{1/2} \tag{2.2}$$

and

$$M_v(E) = (4\pi h^{-3})(2m_p)^{3/2}(E_v - E)^{1/2} \tag{2.3}$$

where h is Planck's constant and m_n and m_p are the density of states effective masses for electrons and holes, respectively, and m_n and m_p are different from the free electron mass and reflect the fact that an electron in a solid moves in the periodic force field of the lattice atoms.*

2.2.2 Equilibrium Free Carrier Densities

We calculate the *thermal equilibrium* free carrier densities in the field free semiconductor bulk in Sections 2.2.3 and 2.2.5. The state of thermal equilibrium is defined as that state of an *isolated* system in which there is no tendency of any of its *macroscopic* properties to change with time. That is, measurable macroscopic quantities characterizing the system such as temperature, potential, charge, and volume must not change with time. Practically speaking, time over which changes cannot occur must be long relative to the time scale of interest. Thermodynamics shows that this equilibrium state corresponds to a minimum in the free energy of the system, and statistical mechanics relates the free energy of the system to a maximum randomness in the distribution of the particles in the system, subject to the restrictions imposed by the macroscopic parameters of the system. In thermal equilibrium the probability of occupancy of an energy level is known from statistical mechanics, a great simplification over nonequilibrium conditions where transport equations must be solved. Thermal equilibrium prevails in the MOS capacitor, even when a gate bias is applied. Thus, using the thermal equilibrium probability of occupation, the charge at the silicon surface can be calculated. First we treat the field free case.

2.2.3 Intrinsic Case

In a pure semiconductor a number of hole-electron pairs are present at a given temperature. Because the semiconductor is not charged, the concentration of electrons n is equal to the concentration of holes p. These concentrations are intrinsic to the semiconductor and equal the intrinsic concentration n_i; that is, $n = p = n_i$.

Figure 2.1 shows the allowable energies of electrons and holes in silicon. How are electrons and holes distributed among the allowable energy levels in their respective bands? The most probable energy distribution of an ensemble of free carriers at thermal equilibrium, subject to the Pauli exclusion principle,† is the Fermi-Dirac distribution. That is, the probability f_n that an energy level E is occupied at thermal equilibrium by an

---

*A more comprehensive treatment would have considered ellipsoidal energy surfaces and related m_n and m_p to different effective masses along different crystal axes.[11] The factor of six in (2.2) refers to the six equivalent minima for electrons in silicon. The parameters $m_n = 0.327m_o$ and $m_p = 0.559m_o$ for silicon, where m_o = free electron mass.

†The Pauli exclusion principle states that no two electrons can have the same quantum state and that all electrons are indistinguishable.

electron is given by the Fermi-Dirac distribution function[14]

$$f_n = \left(1 + \exp\frac{E - E_F}{kT}\right)^{-1} \qquad (2.4)$$

where k is Boltzmann's constant $(8.617 \times 10^{-5}\,\text{eV}/^\circ\text{K})$, T is the absolute temperature of the system, and E_F is the Fermi energy or Fermi level. The energy E_F is the energy at which a state, if one is present, has a 50% chance of being occupied by an electron [i.e., when $E = E_F$, $f_n = \frac{1}{2}$ from (2.4)].

The probability f_p that a level E is vacant or that it is occupied by a positive hole is $f_p = 1 - f_n$ or

$$f_p = \left(1 + \exp\frac{E_F - E}{kT}\right)^{-1}. \qquad (2.5)$$

Both f_n and f_p are shown in Fig. 2.2 as functions of $E - E_F$. Because energy for holes increases in the downward direction and for electrons in the upward direction (see Fig. 2.2), the distribution function for holes is identical with that for electrons, except that energy is measured with respect to E_F in opposite directions. Figure 2.2 shows that the major change in occupation probability occurs in the neighborhood of E_F within a narrow energy range $|E - E_F|$ of a few kT. The dependence of the shape of the Fermi-Dirac function on temperature is illustrated in Fig. 2.3. Figure 2.4 shows how the energy bands of an intrinsic semiconductor are filled according to the Fermi-Dirac function.

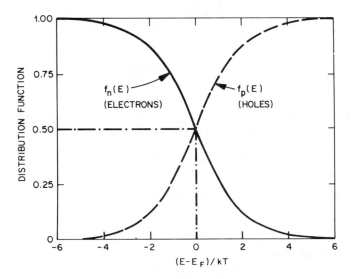

Fig. 2.2 Fermi-Dirac distribution functions $f_n(E)$ and $f_p(E)$ for electrons and holes, respectively. After Many et al.[14]

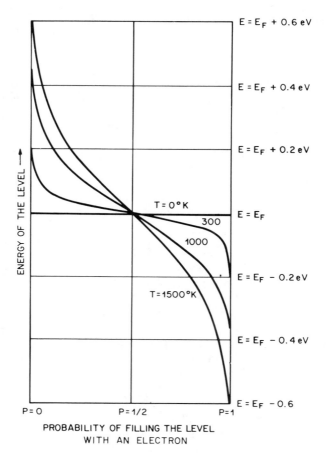

Fig. 2.3 Dependence of shape of the Fermi-Dirac function on temperature. After Shive.[15]

When the Fermi level E_F is far from either of the band edges, the condition $E - E_F \gg kT$ holds, so that (2.4) becomes

$$f_n = \exp\left[-\frac{(E - E_F)}{kT}\right].\tag{2.6}$$

Similarly for holes, $E_F - E \gg kT$, and (2.5) becomes

$$f_p = \exp\left[-\frac{(E_F - E)}{kT}\right].\tag{2.7}$$

Under these conditions the semiconductor is said to be nondegenerate, and (2.6) and (2.7) are the Maxwell-Boltzmann distributions of classical statistics. Classical statistics adequately describe the system because the conditions $E - E_F \gg kT$ and $E_F - E \gg kT$ mean that electron and hole den-

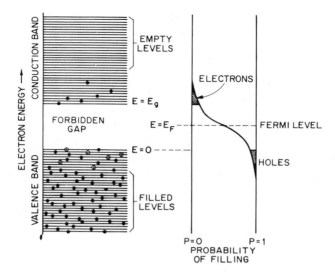

Fig. 2.4 Determination of electron and hole occupancy in conduction and valence bands in an intrinsic semiconductor by the Fermi-Dirac distribution function. After Shive.[15]

sities in the conduction and valence bands are so low that any restriction on the statistics is no longer significant.* Equations (2.6) and (2.7) are used extensively.

The total number of electrons in an allowed energy band of a semiconductor is given by

$$n_T = V \int_{E_{\min}}^{\infty} M(E)f_n(E - E_F)dE \qquad (2.8)$$

where $E_{\min}$ is the lowest energy in the allowed band, V is the volume of the crystal, and $M(E)dE$ is the number of energy states per unit volume with energies between E and $E + dE$. Thus the total density of electrons in the conduction band with no applied electric field is

$$n = \int_{E_c}^{\infty} M_c(E)f_n(E - E_F)dE \qquad (2.9)$$

where $M_c(E)$ is the density of states in the conduction band. Similarly, the

---

*That is, for low densities it does not matter whether the particles are Fermions (for which two are allowed per energy level, one of each spin) or Bosons (for which any number are allowed in each energy level).

total density of holes in the valence band is

$$p = \int_{-\infty}^{E_v} M_v(E) f_p(E_F - E) dE. \tag{2.10}$$

Equations (2.9) and (2.10) are to be integrated for the nondegenerate case.* Therefore, the Maxwell-Boltzmann approximation of f_n and f_p given in (2.6) and (2.7) is used. In addition, M_c and M_v are approximated by the parabolic approximations (2.2) and (2.3) because the Fermi function limits integration to a range of energies near the band edges. Substitution of (2.2), (2.3), (2.6), and (2.7) into (2.9) and (2.10) yields

$$n = 6(4\pi h^{-3})(2m_n)^{3/2} \int_{E_c}^{\infty} (E - E_c)^{1/2} \exp\left[-\frac{(E - E_F)}{kT}\right] dE \tag{2.11}$$

and

$$p = (4\pi h^{-3})(2m_p)^{3/2} \int_{-\infty}^{E_v} (E_v - E)^{1/2} \exp\left[-\frac{(E_F - E)}{kT}\right] dE. \tag{2.12}$$

Integrating (2.11) and (2.12), we obtain

$$n = M_c \exp\left[-\frac{(E_c - E_F)}{kT}\right] \tag{2.13}$$

and

$$p = M_v \exp\left[-\frac{(E_F - E_v)}{kT}\right] \tag{2.14}$$

where $M_c = 12(2\pi m_n kTh^{-2})^{3/2}$ and $M_v = 2(2\pi m_p kTh^{-2})^{3/2}$. Thus, for calculating equilibrium carrier densities, the conduction and valence bands can be regarded as two single levels of energies E_c and E_v of effective densities of states M_c and M_v, respectively.

The position of the Fermi level for intrinsic silicon can be obtained from the condition of charge neutrality $n = p$. Equating (2.13) to (2.14) and solving for E_F, we obtain

$$E_F = \frac{1}{2}(E_c + E_v) + \frac{kT}{2} \ln \frac{M_v}{M_c} = E_i. \tag{2.15}$$

Taking the valence band edge as the reference level, $\frac{1}{2}(E_c + E_v) = E_g/2$, where E_g is the silicon bandgap energy and E_i is the intrinsic Fermi level, so that (2.15) becomes $E_i = E_g/2 + (kT/2)\ln M_v/M_c$; thus E_i is very close to midgap. If $m_n \neq m_p$, (2.13), (2.14), and (2.15) show that E_i would be displaced from midgap by a term that is usually very small (a few kT/q in

*Evaluation for the general case has been discussed by Joyce and Dixon.[16]

silicon) because the difference between m_p and m_n enters logarithmically. For most purposes, the intrinsic Fermi level in silicon can be regarded as being in the middle of the bandgap.

2.2.4 The *pn* Product

An important relationship is obtained by forming the product of the hole and electron concentrations using (2.13) and (2.14)

$$pn = M_c M_v \exp\left[-\frac{(E_c - E_v)}{kT}\right] = M_c M_v \exp\left[-\frac{E_g}{kT}\right]. \qquad (2.16)$$

Also, because $p = n = n_i$ in intrinsic material, $pn = n_i^2$. If we use this relation in (2.16), the intrinsic concentration is

$$n_i = (M_v M_c)^{1/2} \exp\left[-\frac{E_g}{2kT}\right]. \qquad (2.17)$$

Using (2.2) and (2.3), $(M_v M_c)^{1/2} = 3.30 \times 10^{15} T^{3/2}$ for silicon. With this value, and taking $E_g = 1.12 \, \text{eV}$, (2.17) predicts $n_i = 3 \times 10^9 \, \text{cm}^{-3}$ at 290°K. This value is about a third of the experimental value $n_i = 10^{10} \, \text{cm}^{-3}$ at 290°K.[17] In general, (2.17) agrees with experimental values of n_i, such as the values given by Sze,[18] only as to order of magnitude. Part of this discrepancy is due to temperature dependence of the density of states effective masses in (2.2) and (2.3).[17] In addition, it is not clear how accurately E_g is known at high temperatures, and n_i is not known accurately near or below room temperature. Hence the adequacy of (2.17) is unclear. In this book we adopt the experimental number $n_i = 10^{10} \, \text{cm}^{-3}$ at 290°K.*

According to (2.16), the product of hole and electron concentrations, usually called the "*pn* product" ($pn = n_i^2$), is independent of E_F. Therefore, the *pn* product is independent of whether the silicon is *n*-type or *p*-type and independent of individual values of n and p. In fact, (2.17) always holds everywhere in the silicon crystal as long as the crystal is in thermal equilibrium. The *pn* product is useful in calculating minority carrier densities anywhere in the silicon as explained in Section 2.3.2(c).

2.2.5 Extrinsic Case

Now consider the extrinsic case, when elemental impurities called *dopants* are added to the silicon. Dopants are added to silicon to modify the free carrier densities of intrinsic silicon, controlling conductivity, and allowing fabrication of *p-n* junctions and various unipolar and bipolar structures. In particular, extrinsic silicon is used in MOS capacitors for measuring the properties of the MOS system.

---

*The temperature 290°K frequently is chosen for illustration because this temperature is near room temperature, and the thermal voltage $(kT/q) = 0.025 \, \text{V}$ is a simple number to remember.

Effective dopants are elemental impurities that enter the silicon lattice substitutionally; that is, they take the place of a silicon atom in the silicon lattice. Because silicon has a valence of 4, substitutional addition of elements from group V of the periodic table such as phosphorus, antimony, and arsenic, each of which have a valence of 5, results in an extra electron weakly bonded to the impurity atom. This extra electron occupies an energy level in the bandgap very close to the conduction band edge (0.05 eV).[18] The other four electrons are strongly bonded to neighboring silicon atoms. At normal temperatures this extra electron is thermally excited into the conduction band, leaving behind a positively charged impurity atom called a *donor*. The substitutional addition of elements from group III of the periodic table, such as boron and aluminum, which have a valence of 3, results in the acceptance of an additional electron from the valence band to complete the four valence electrons shared with the surrounding silicon atoms. Similarly, the addition of a group III element creates a shallow energy level in the bandgap very near the valence band edge (0.05 eV).[18] At ordinary temperatures an electron is thermally excited from the valence band to the energy level of the group III impurity, creating a hole in the valence band, and a negatively charged impurity atom called an *acceptor*.

As long as the system is nondegenerate and there is no applied electric field, the hole and electron concentrations will be given by (2.13) and (2.14). Now, however, E_F will not be equal to the intrinsic level. To accommodate the extra electrons from donors, E_F will move from the intrinsic level closer to the conduction band edge. Similarly, to accommodate extra holes, E_F will move closer to the valence band edge. The position of the Fermi level still is determined by the condition of charge neutrality, just as in the intrinsic case. However, the charged dopant ions must be included in the neutrality condition. Therefore, we must determine the number of charged dopant ions.

We begin by rewriting the expressions for the electron and hole concentrations. The Fermi energy E_F is measured with respect to the conduction band edge in (2.13) and the valence band edge in (2.14). It is convenient to write expressions for electron and hole concentrations with E_F measured with respect to the intrinsic Fermi level E_i. Thus we write (2.13) and (2.14) with $E_F = E_i$ for the intrinsic case

$$n_i = M_c \exp\left[-\frac{(E_c - E_i)}{kT}\right] = M_v \exp\left[-\frac{(E_i - E_v)}{kT}\right]. \tag{2.18}$$

Then, from (2.13), (2.14), and (2.18), we have

$$n = n_i \exp\frac{E_F - E_i}{kT} \tag{2.19}$$

and

$$p = n_i \exp\frac{E_i - E_F}{kT}. \tag{2.20}$$

To obtain the charge from ionized donors and acceptors from groups V and III elements, we know that donors introduce a shallow energy level in the bandgap at energy E_D just below E_c and that acceptors introduce a level at energy E_A just above E_v; $E_c - E_D$ and $E_A - E_v$ are typically the order of 0.05 eV.[18] The density of electrons n_D bound to donors and the density of holes p_A bound to acceptors is given by

$$n_D = N_D \left(1 + g \exp \frac{E_D - E_F}{kT} \right)^{-1} \qquad (2.21)$$

and

$$p_A = N_A \left(1 + g \exp \frac{E_F - E_A}{kT} \right)^{-1} \qquad (2.22)$$

where N_D is the donor density and N_A is the acceptor density per unit volume. One might expect, for a single energy level at energy E_D or E_A within the bandgap, that the results (2.21) and (2.22) from (2.9) or (2.10) could be obtained simply by replacing the density of states $M(E)$ with a delta function $N_A \, \delta(E - E_A)$ and integrating over the bandgap. However, this approach would lead to (2.21) and (2.22) with $g = 1$. To introduce the degeneracy factor g, the effect of multiple occupancy of a level by electrons or by holes must be included in the derivation of the occupancy functions f_n and f_p of (2.4) and (2.5).

The value of g is derived by comparison of a chemical picture of the substitutional donor or acceptor with a band picture. First consider a donor. The ionized state of the donor is the state most similar to the perfect silicon lattice. In the chemical picture the ionized donor atom is bonded in the silicon lattice just like a silicon atom, with all the valence bonds filled. The important difference is that the atomic core of the donor has one more positive charge. Going to the band picture, we approximate the silicon plus its ionized donor as a perfect crystal with a localized attractive potential well at one lattice site. Solving this quantum mechanical problem, this attractive potential is found to create an energy level in the silicon bandgap just below the conduction band edge. Within the band model, which applies to independent electrons, this level could be doubly occupied by two electrons of opposite spin. But reference to the chemical picture, which includes Coulombic repulsion between electrons, shows that double occupancy cannot occur. Occupancy by a second electron would reduce the binding energy of both occupant electrons, both by Coulombic repulsion and by screening of the atomic core. This reduced binding energy would lead to the escape of one of the two electrons. Therefore, we impose on the band model the constraint that the energy level below the conduction band edge can be occupied only singly. This constraint leads to $g = \frac{1}{2}$ in (2.21).[19] With $g = \frac{1}{2}$, the Boltzmann limit of (2.21) provides $n_D \approx (N_D/2) \exp[-(E_D - E_F)/kT]$. That is, only half as many electrons remain bound to donors as would be bound with $g = 1$. This is expected because restriction to single occupancy ($g = \frac{1}{2}$) should lead to half as many bound electrons as double occupancy ($g = 1$). Therefore, the fraction of filled levels (neutral donors) f_n is given by

$$f_n = \left(1 + \frac{1}{2} \exp \frac{E_D - E_F}{kT}\right)^{-1}. \tag{2.23}$$

Next, consider an acceptor. The ionized state of the acceptor (negatively charged) is the state most like the perfect silicon lattice. Again, all the valence bonds are filled. Again, in the band picture we imagine a perfect crystal, this time with a localized repulsive potential at the lattice site. From quantum mechanics, an energy level is found near the valence band edge.* In our band picture, this level could be doubly occupied. However, we appeal to the chemical model. Single occupancy of this level by an electron corresponds to an acceptor with all valence bonds satisfied. Hence, any additional electron at this site could not be bound within the valence bonds. To become bound, a second electron would have to go into another energy level. In such a level, a second electron would see an acceptor core screened by the first electron. The binding energy of a second electron would be low, and a second electron easily could leave the acceptor site. Again, the chemical picture shows that double occupancy of the level within the band picture is not allowed. Then, for the acceptor, the fraction of levels filled by electrons also is given by (2.23). Therefore, the fraction of *empty* levels (neutral acceptors), levels filled by holes f_p, is

$$f_p = (1 - f_n) = \left(1 + 2 \exp \frac{E_F - E_A}{kT}\right)^{-1}. \tag{2.24}$$

As a result, $g = \frac{1}{2}$ for donors and $g = 2$ for acceptors.

In practice, the value of g has little influence on MOS capacitor measurements. For deep traps, inaccuracy in the value of g is masked by inaccuracy in the energy level of the trap. For shallow donors or acceptors, all levels are ionized independent of the value of g, except for low temperatures ($\leq 50°K$) or large dopant ion densities. Low temperatures are not commonly used in MOS measurements, with the possible exception of emission time spectroscopy measurements (see Section 8.5.2).

After expressions for the various charge densities have been established, the position of the Fermi level can be found from the condition of charge neutrality. The density of negative charge is $-q(n + N_A - p_A)$, whereas the density of positive charge is $q(p + N_D - n_D)$. Charge neutrality requires that $q(p + N_D - n_D) - q(n + N_A - p_A) = 0$. Thus we have

$$n + n_D + N_A = p + p_A + N_D. \tag{2.25}$$

Equation (2.25) can be simplified because the dopant ion levels are about $2kT$ from the band edges and the Fermi level usually is many kT from the band

---

*The valence band edge in silicon is more complex than the conduction band edge because light and heavy mass bands are present. This complication is ignored here.

edges. Thus the conditions $E_D - E_F \gg kT$ and $E_F - E_A \gg kT$ will hold. These conditions mean that $n_D = p_A = 0$ as seen from (2.21) and (2.22), which is another way of saying that donors and acceptors are completely ionized. Equation (2.25) simplifies to

$$n + N_A = p + N_D. \tag{2.26}$$

Combining the condition of charge neutrality (2.26) with the equilibrium relationship $pn = n_i^2$ gives a quadratic equation in n and another in p. Solving these quadratic equations, we obtain the electron concentration in n-type silicon

$$n = \tfrac{1}{2}\{N_D - N_A + [(N_D - N_A)^2 + 4n_i^2]^{1/2}\} \tag{2.27}$$

and the hole concentration in p-type silicon

$$p = \tfrac{1}{2}\{N_A - N_D + [(N_A - N_D)^2 + 4n_i^2]^{1/2}\}. \tag{2.28}$$

When $N_D - N_A \gg n_i$, the usual case in an extrinsic semiconductor, (2.27) and (2.28) simplify to $n = N_D - N_A$ and $p = N_A - N_D$.

For the case where $N_D \gg N_A$ and $N_D > n_i$, $n = N_D$. Electron density will be a constant, independent of temperature, as long as the donors remain ionized* and we have an extrinsic n-type semiconductor.† From (2.19) and (2.27), the position of the Fermi level in n-type material (material in which $N_D > N_A$) is

$$E_F = E_i + kT \ln\left\{y\left[1 + \left(1 + \frac{1}{y^2}\right)^{1/2}\right]\right\} \tag{2.29}$$

where $y = \tfrac{1}{2}(N_D - N_A)/n_i$. For practical doping levels and temperatures, $y \gg 1$ so that a Taylor series expansion in $1/y^2$ is possible. Then

$$E_F = E_i + kT \ln\left[\frac{N_D - N_A}{n_i} + \frac{n_i}{N_D - N_A}\right]. \tag{2.30}$$

Usually, the first two terms of this expansion are adequate. Similarly, in p-type material where $N_A > N_D$, (2.20) and (2.28) produce

$$E_F = E_i - kT \ln\left[\frac{N_A - N_D}{n_i} - \frac{n_i}{N_A - N_D}\right]. \tag{2.31}$$

Thus the Fermi level lies above the intrinsic level in n-type and below it in p-type, as previously mentioned. Equations (2.30) and (2.31) show the dependence of E_F on temperature, and the temperature dependence of n_i is given in (2.17). The temperature dependence of E_F is illustrated in Fig. 2.5, which shows $E_F - E_i$ as a function of temperature with N_D and N_A as parameters.

---

*Donors and acceptors remain ionized at all temperatures down to about 45–50°K because they are so shallow.
†At elevated temperatures n_i becomes so large that $n_i > N_D$, and the semiconductor becomes intrinsic despite the dopant ions.

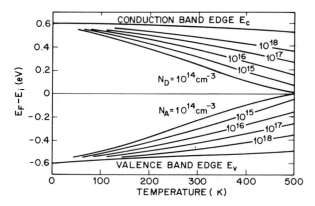

Fig. 2.5 Fermi level measured with respect to midgap energy as a function of absolute temperature with impurity concentration as parameter for silicon. These curves were calculated using (2.17), (2.30), and (2.31). The variation of the silicon bandgap with temperature, also included, is calculated from the relation[18]: $E_g(T) = 1.16 + 7.02 \times 10^4 \, T^2(T + 1108)^{-1}$. After Grove.[20]

2.3 MOS STRUCTURE

We now consider the carrier densities when an external electric field is applied to the silicon surface in an MOS capacitor. The MOS capacitor consists of a parallel plate capacitor with one electrode a metallic plate, called the *gate* and the other electrode, the *silicon*. The two electrodes are separated by a thin insulating layer of SiO_2. Such a structure is shown in Fig. 2.6a.

Figure 2.6b shows the forbidden energy gaps in SiO_2 and silicon for the MOS capacitor of Fig. 2.6a. Also shown is the position of the Fermi level in the gate and in the silicon. The forbidden gap in SiO_2 is very large (~ 8.8 eV), whereas that of silicon is much smaller (1.12 eV). Also shown is the Fermi level for an aluminum gate.*

Figure 2.6b shows that a large energy barrier exists between the metal and the oxide, with an equally large barrier between the silicon and the oxide. For example, an energy of 3.2 eV would be needed to get an electron from the Fermi level of the metal (in this case aluminum) E_{FM} to the lowest unoccupied states in the oxide,† and 4.3 eV would be needed to get an electron from the silicon valence band to the lowest unoccupied states in the oxide. These energy barriers have been determined by internal photo-emission measurements discussed in Section 10.3.

---

*Aluminum is a common gate metal because it is easy to evaporate and it adheres strongly to the oxide.

†Loosely speaking, these states are referred to as the SiO_2 *conduction band*. However, SiO_2 is an amorphous insulator where the band picture may not apply.

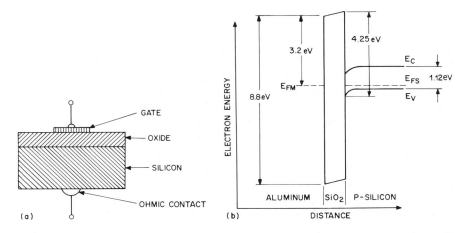

Fig. 2.6 (a) Cross section of an MOS capacitor. The gate is usually evaporated onto the oxide surface in the shape of a circular dot. (b) Energy-band diagram of the MOS capacitor showing the energy barrier between metal and SiO_2 and between silicon and SiO_2. The metal is aluminum and the silicon is p-type. Position dependence of the forbidden energy gap in the SiO_2 and the silicon is explained later.

The importance of these energy barriers is that they prevent the free flow of carriers from the metal to the silicon or vice versa. Thus the application of a bias across the MOS capacitor does not result in current flow (apart from transients related to charging the capacitor). Rather, an electric field is established in the oxide by surface charge layers that form in the metal and in the silicon.

The surface charge layer in the gate can be ignored in our analysis because it has no effect on MOS capacitor measurements. Because the surface charge layer in the metal responds instantly to ac voltage variations in the frequency range of interest (0–10 MHz), no loss contribution is made by the metal to the MOS admittance. The surface charge layer is very thin in the metal (~ 1 Å thick) as a result of the large carrier density there. Therefore, the capacitive contribution of this surface charge layer is undetectable in series with the oxide and silicon surface charge layer capacitances; thus no analysis is made of the surface charge layer in the gate.

We now discuss the silicon surface charge layer for an MOS capacitor using p-type silicon. We begin by placing a negative bias on the gate. The negative charges on the gate attract holes to the silicon surface to form an *accumulation* layer. The thickness of such a layer is comparable to a Debye length, typically 100–1000 Å thick, depending on bias and doping density. Now we increase gate bias. As negative charges are removed from the gate, holes leave the accumulation layer. At a gate bias called the *flatband* voltage, the silicon will be neutral everywhere.

As gate bias is made positive with respect to flatbands, holes are repelled from the silicon surface. Because the *pn* product must remain constant, electron density at the silicon surface builds up. However, for our *p*-type example, electron density is negligible for positive biases that are not too large. The positive gate charges are balanced not by electrons, but by negative acceptor ions in the silicon surface *depletion layer,* so called because holes have been depleted from this surface region. As gate bias increases, the depletion layer widens to provide more acceptor ions to balance the gate charge. When depletion prevails, the silicon surface charge layer consists of a region of ionized dopant ions extending about 0.1–10 μm deep, depending on bias and doping density. For distances greater than the depletion layer depth, the silicon still is neutral.

Finally, we increase the positive gate bias until electrons appear at the silicon surface in great numbers. The appearance of these electrons is a consequence of thermal equilibrium. As holes are repelled from the silicon surface by the applied gate bias, electron density must increase to keep the *pn* product constant. These electrons then form a thin *inversion layer* located very near the Si–SiO$_2$ interface, in a region 30–300 Å thick, depending on bias and doping density. Below the inversion layer, the depletion layer still exists, and below the depletion layer, the neutral silicon. Once inversion occurs, any further increase in positive gate charge is balanced almost entirely by the addition of electrons to the inversion layer. Consequently, the depletion layer no longer increases much in width.

Our objective is to provide mathematical support for this picture of the silicon surface charge layer and its variation with bias. To do this, we need to solve Poisson's equation for the MOS capacitor under an applied bias. The previous results for the charge densities, particularly (2.9) and (2.10), can be used because the MOS capacitor is in thermal equilibrium under applied bias. Because of its importance, thermal equilibrium is now discussed in detail.

2.3.1   Silicon Surface Charge—Mobile Carriers

(a)   *Equilibrium in the MOS Capacitor*

The Fermi level plays an important role in formulating equilibrium conditions when two different systems of different allowable energies are brought into contact. The combined system will be in thermal equilibrium only when E_F is the same in both parts, because all quantum levels at a given energy must have equal occupation probabilities at thermal equilibrium. If E_F in each system—relative to a common datum—is not initially equal before contact, then on contact there will be a flow of electrons from the system with the higher initial E_F to the system with the lower initial E_F. This electron flow will continue until equality of the Fermi energies of the two systems is achieved.

When no further net transfer of electrons occurs from one material to the other, the Fermi levels are equal. However, there also must be an energy balance; that is, no energy is gained by the transfer of electrons. Otherwise, electrons would transfer to gain energy. Evidently, there is a relation between balance of Fermi levels and balance of energy of transfer from one material to the other.

Let the work done in removing an electron from one material in isolation be μ_1 and from the other, μ_2. The quantities μ_1 and μ_2 are the *chemical potentials* of electrons in the two materials. Then, one might suppose that equilibrium on contact would occur when $\mu_1 = \mu_2$. However, for charged particles like electrons, transfer is accompanied by charging of the materials. As a result, the two materials acquire potentials ψ_1 and ψ_2. The work done on transfer of an electron of charge $(-q)$ now will be zero provided that

$$\mu_1 - q\psi_1 = \mu_2 - q\psi_2. \qquad (2.32)$$

In fact, statistical mechanics shows that this condition for equilibrium is identical to the condition of equality of Fermi levels because

$$E_F = \mu - q\psi. \qquad (2.33)$$

In chemical language, the Fermi level is the *electrochemical potential* of the electrons.

The application of a voltage to a system results in an "electrochemical potential" difference within the system. This difference between the Fermi levels within the system is equal to the applied voltage. Therefore, we adopt the terminology that *voltage* or *bias* is the difference between Fermi levels whereas *potential* is the electrostatic part of E_F.

Equation (2.33) is valid for the case of electrostatic potentials ψ, which vary very slowly with position.* This case is the one of interest in this book.

With no bias applied, the system, consisting of the MOS capacitor with the gate short-circuited to the silicon, will be in thermal equilibrium. No dc current flows through the system, there will be no net work done on a test charge moved from a given point in the system through the system and back again to the same point, and there are no changes in the macroscopic properties of the system with time.

When a bias is applied between the gate and the silicon, the isolated system, consisting of the MOS capacitor connected to a bias source such as a battery, still will be in thermal equilibrium except for a brief initial transient phase during which the MOS capacitor charges up to the battery voltage. After the initial transient, no current flows through the system

---

*For electrostatic potentials that vary so rapidly with position that quantum mechanics is needed to calculate the effect of ψ on the energy of the electrons, (2.33) must be modified. For a discussion of this case, see Brews and Hwang.[21]

because (1) current flow is blocked by the large energy barriers between the oxide and each electrode (metal and silicon) and (2) free carrier concentration in the oxide is infinitesimal; that is, the oxide is a nearly perfect insulator.*

There will be no net work done on a test charge moved from a given point in the system through the system and back again to the same point. For example, no work is done in moving a test charge in the electric field within the silicon depletion layer, because whatever work is done against the electric field is balanced by energy gained by moving from a region of high carrier density to one of low carrier density. That is, any tendency to move due to drift is balanced by an opposing tendency to move by diffusion. A similar balance exists with respect to movement of charge from the gate to the silicon, as discussed in Section 10.4. Because no energy advantage can be gained by charge movement, there are no changes in the macroscopic properties of the system with time. Even with leakage currents, free carriers in the silicon will be approximately in thermal equilibrium, provided that these leakage currents are too small to perturb the free carrier distributions.†

Because there is no dc current flow, no transport equations need to be solved. Because there is no time dependence, the Poisson equation alone governs the potential. All we need to know is the dependence on position of mobile carriers in the silicon surface region. For example, Fig. 2.7 shows thermal equilibrium hole and electron concentrations as a function of distance in the silicon. This free carrier surface charge results from an electric field perpendicular to the silicon surface plane pointing into the silicon and corresponds to a gate voltage of about 1 V and an oxide thickness of 1000 Å. This field repels holes from the silicon surface and attracts electrons. Therefore, hole density decreases from its value equal to N_A in the neutral bulk to a lower value at the silicon surface, and electron density increases from its value of n_i^2/N_A in the neutral bulk to a larger value at the silicon surface. Free carrier charge pileup at the Si–SiO$_2$ interface is possible because no dc current can flow through the oxide. As a result of these hole and electron gradients, ionized acceptors within the

---

*The SiO$_2$ has a finite conductivity, which means that some current always can flow. However, with an oxide resistivity of 10^{18}–10^{20} Ω-cm, the typical oxide resistance will be on the order of 10^{16}–10^{18} Ω, assuming an oxide thickness of 1000 Å and a metal electrode area of 10^{-3} cm^2. With an applied bias of 50 V, a current between 5×10^{-15} and 5×10^{-17} A will flow. Such currents are so small that the system can be considered to be at thermal equilibrium with negligible error, as long as the applied bias is below the breakdown strength of the oxide. In addition, the dielectric relaxation time of the oxide, which is the characteristic time required for the oxide to come to thermal equilibrium after the applied bias is changed, is very long—typically many days. This time is longer than that taken for bias changes in most applications and experiments. The resulting current, which flows as the oxide tends toward equilibrium, is negligible.

†For oxide layers thinner than 50 Å, thermal equilibrium can be disturbed by tunneling through the oxide [see Section 5.10].

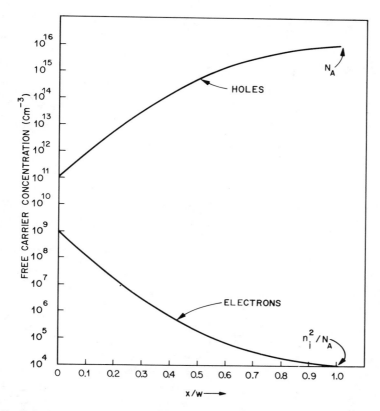

Fig. 2.7 Calculated free carrier concentrations as functions of normalized depth for p-type silicon that has an acceptor density of 10^{16} cm^{-3} and a surface charge density of 2×10^{11} cm^{-2}. This charge is produced by an electric field pointing into the silicon that repels holes and attracts electrons. Distance x is measured from the Si–SiO$_2$ interface into the silicon bulk, and w is depletion layer width. The calculation for these curves is given in Section 2.3.4(d).

region of these free carrier gradients will be uncompensated. Therefore, ionized impurities also contribute to surface charge.

(b) Comparison of the Field Induced Depletion Layer to a Metallurgical p-n Junction

The field induced depletion layer in the MOS capacitor is similar to half of a step junction. The major difference is that the field induced "junction" of the MOS capacitor is in thermal equilibrium whereas the metallurgical p-n junction is in steady state under an applied bias. Therefore, $pn = n_i^2$ everywhere in the field induced junction whereas in a p-n junction under bias, $pn \neq n_i^2$ anywhere in the junction depletion layer.

A polarity of applied bias that produces accumulation in the field induced junction forward biases the p-n junction. After forward bias is applied to a field induced junction, current will flow until an accumulation layer builds up and thermal equilibrium is achieved. Accumulation of majority carriers at the silicon surface is one consequence of the condition of thermal equilibrium. In a p-n junction, flow of forward current does not allow an accumulation layer to build up.

A polarity of applied bias which produces depletion and inversion in the field induced junction reverse biases the p-n junction. Reverse bias generation current flows in steady state in the p-n junction but flows only until thermal equilibrium is achieved in the field induced junction. Buildup of an inversion layer is a consequence of this current flow. Generation current flow, the dominant source of carriers for an inversion layer in silicon at room temperature, occurs by alternate emission of electrons and holes by traps that have energy levels near midgap. Defects such as impurities incorporated in the crystal can act as such traps (see Section 4.3).

2.3.2   Solution of the Poisson Equation

At this point the thermal equilibrium results (2.9) and (2.10) for the carrier densities has been justified. Now, the Poisson equation is to be formulated. First, a number of simplifications are discussed. The most interesting of these is the band-bending approximation, which enables us to adapt (2.9) and (2.10) to provide the carrier densities in the presence of a spatially varying potential, provided that thermal equilibrium exists. Once this adaptation is made, resulting in (2.37) and (2.38), and if the other simplifications now to be discussed are adopted, the Poisson equation can be solved for both uniformly and nonuniformly doped silicon.

(a) *Simplifying Conditions*

We solve the Poisson equation with certain simplifying conditions. These conditions are:

1   The Poisson equation will be solved in one dimension in the direction perpendicular to the plane of the Si–SiO$_2$ interface. It is reasonable to treat only one dimension because the field under the gate is uniform and perpendicular to the silicon surface. The fringing field at the gate edge is negligible, affecting an area that extends only an oxide thickness from the periphery of the gate. In practice, the gate diameter is orders of magnitude larger than oxide thickness, typically tens or hundreds of micrometers in extent compared to a few tenths of a micrometer. Therefore, edge effects would become important if the extent of the gate were comparable to oxide thickness, but this case is seldom

encountered. Other two-dimensional effects can be caused by charged interface traps and oxide fixed charges, which are treated in Section 6.3.

2 We assume that the impurity concentration in the silicon is uniform right up to the surface. We know from Section 14.4.2 that this assumption does not apply because thermal oxidation causes redistribution of the impurity concentration at the silicon surface.* However, the case of a uniform impurity distribution is a good introduction.

3 The Poisson equation is solved for the nondegenerate case. In the degenerate case, equilibrium free carrier concentration in the silicon is described by degenerate or Fermi-Dirac statistics. Devices such as the MOSFET and CCD normally are not operated in bias ranges that render the silicon surface degenerate. Moreover, little useful information about interfacial or oxide properties are gained from MOS capacitor characteristics in the degenerate bias region. We use the nondegenerate or Boltzmann approximation of the Fermi-Dirac statistics.

4 The Poisson equation will be solved using an approximate charge density. The charge of the dopant ions is accounted for approximately by smearing out the ion charge into a uniform background density. This smearing ignores the discrete nature of these ions as well as statistical variations in their spatial distribution. The charge of electrons and holes is treated in a self-consistent field approximation. That is, each electron or hole is treated as though it moved in an *average* field. This average field is computed as the field due to the *average* mobile charge density plus the field due to the smeared out dopant ion charge. Thus each electron or hole can be treated as an *independent* particle in an (approximate) average field. In this approach, the average carrier distribution is a function only of the average potential, ignoring all dependence on more detailed properties of the real potential. Conversely, the average field also is assumed to depend only on the *average* carrier and dopant ion densities, ignoring dependence on more detailed properties of the charge density. For very large electron or hole concentrations and also at low temperatures, the treatment of electrons and holes as independent particles may fail.[22] However, in the range of free carrier concentrations and temperatures of interest in this book, if a slight breakdown of the independent particle treatment occurs, it does so in extremely small volumes (atomic dimensions) and over very short time periods (fractional picoseconds). As such effects are not detected by the measurements discussed, breakdown of the independent particle treatment can be ignored without sacrificing accuracy.

5 Surface quantization is neglected. Surface quantization has no bearing on device or MOS capacitor characteristics under conditions of normal

*An implant also will give a nonuniform impurity distribution.

use, except through mobility in the MOSFET at high gate biases. However, surface quantization is an important part of our basic understanding of the behavior of silicon surfaces and becomes important when the silicon surface is degenerate or at very low temperatures (comparable to liquid helium temperature) or at very high gate biases.

Surface quantization, first pointed out by Schrieffer,[23] occurs as follows. Suppose that a high electric field ($> 10^6$ V/cm) is applied to an MOS capacitor, attracting a high concentration of free carriers to the silicon surface. Because these carriers cannot flow through the oxide, they are trapped in a narrow potential well at the surface of the silicon. Because the potential well is very narrow, carrier motion is restricted in a direction perpendicular to the interfacial plane. In this direction only standing wave patterns are possible, thus resulting in electric subbands. The energy separation between energy levels in these subbands is smaller than kT at room temperature and normal fields below 10^6 V/cm; thus surface quantization is not important under normal device operating conditions. Although free carriers are quantized for motion perpendicular to the plane of the interface, they are free to move and are not quantized in directions parallel to the interfacial plane.

Fang and Howard[24] were the first to demonstrate surface quantization experimentally. These experiments were done on an n-channel (electron) MOSFET with a (100)-oriented silicon surface. Quantization was confirmed using the MOS capacitor by Kaplit and Zemel[25] and studied using admittance measurements by Voshchenkov and Zemel.[26] Theoretical calculations that agreed quite well with experiment were made by Stern and Howard[27] for the electric subbands in electron inversion layers. Surface quantization in hole inversion layers was first investigated by von Klitzing et al.[28] A good review of surface quantization was given by Landwehr.[29]

(b) Band-Bending Approximation

Figure 2.8a illustrates the *band bending*, or the barrier height ψ_s at the surface of p-type silicon, and Fig. 2.8b illustrates band bending at the surface of n-type silicon.* The band-bending approximation assumes that the density of states in the conduction and valence bands is not changed by an electric field. In the band-bending approximation the only effect of an electric field is to shift all the energy levels in the conduction and valence bands by a constant amount determined by the potential at each given point in the silicon.

The band-bending approximation is valid over the temperature and electric field ranges of interest in many device applications and MOS

---

*The variable ψ_s is not a directly measurable quantity. It is calculated from the measured capacitance of an MOS capacitor.

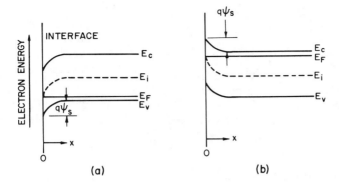

Fig. 2.8 (a) Band-bending diagram for p-type silicon. Band bending $\psi(x)$ is shown as a function of distance x measured from the Si–SiO₂ interface. Here the surface value of band bending ψ_s is positive, leading to a decrease in electron energy, causing the bands to bend downward. (b) Band-bending diagram for n-type silicon. Here $\psi_s < 0$, causing electron energy to increase and bending the bands upward.

capacitor measurements. As previously discussed in this chapter, one case where the band-bending approximation fails is surface quantization, where the density of states is changed by the electric field. Another case is very heavy doping, where the impurities can produce band tails. Neither of these cases is treated.

Using the band-bending approximation, we calculate the hole and electron density distribution at the silicon surface under an applied bias. Because band bending is a function of distance, as seen in Fig. 2.8, free carrier concentrations also are functions of distance. The distance x is measured in a direction perpendicular to the interfacial plane into the silicon bulk. To take this potential into account and calculate the electron concentration in the conduction band as a function of x in n-type silicon, (2.9) must be modified to

$$n(x) = \int_{E_c(x)}^{\infty} M_c(x) f_n(E - E_F) dE. \tag{2.34}$$

With bias applied, the band bending $\psi(x)$ is positive when the bands bend down, lowering the electron energy. Therefore, the density of states is shifted to lower energies and

$$M_c(x) = M_c[E + q\psi(x)]. \tag{2.35}$$

The sign of $q\psi(x)$ in (2.35) arises as follows. A positive band bending lowers the electron energy, so that at x the energy level $E(x)$ corresponding to the level E in the bulk $(x \to \infty)$ is $E(x) = E - q\psi(x)$.* Using (2.35), we obtain

$$M_c[E(x)] = M_c(E)$$

so that the density of states at the same energy relative to the band edge is independent of x, as it should be.

Approximating the Fermi-Dirac distribution by the Boltzmann approximation (2.6), substituting (2.35) into (2.34), and changing variables from E to $E + q\psi(x)$, we obtain

$$n(x) = \exp\frac{q\psi(x)}{kT} \int\limits_{E_c}^{\infty} M_c(E) \exp\left[-\frac{(E - E_F)}{kT}\right] dE. \qquad (2.36)$$

The integral in (2.36) is the same as (2.11) for the field free silicon bulk. For n-type, this integral simply is equal to N_D; thus (2.36) becomes

$$n(x) = N_D \exp\frac{q\psi(x)}{kT}. \qquad (2.37)$$

Similarly, for p-type, modifying (2.10) and using (2.7) and (2.35), the hole density in the valence band in the presence of an electric field is

$$p(x) = N_A \exp\left[-\frac{q\psi(x)}{kT}\right]. \qquad (2.38)$$

(c) Definition of Potentials and Sign Conventions

Figures 2.9 and 2.10 show the sign conventions of the various potentials to be defined. Figure 2.9a shows n-type silicon in accumulation, and Fig. 2.9b shows n-type silicon in depletion-inversion. Figures 2.10a, b show the same regions for p-type silicon. In both Figs. 2.9 and 2.10, electron energy increases in the upward direction so that an arrow pointing up denotes a negative potential and an arrow pointing down denotes a positive potential.

First, we define the potential $\phi(x)$ by the equation

$$q\phi(x) \equiv E_F - E_i(x) \qquad (2.39)$$

where E_F is the extrinsic Fermi level. The intrinsic energy level $E_i(x)$ is parallel to both bands everywhere in the silicon and coincides in the bulk with the intrinsic Fermi level whose exact position in the bandgap is given by (2.15). Deep in the bulk silicon $(x \to \infty)$, $\phi(x)$ is called the *bulk potential*

---

*This relation applies only in the case of potentials $\psi(x)$, which vary slowly with position x. This is the case of interest. If potential varies rapidly compared to an electron wavelength, quantum mechanics must be used to recompute the electron energies E. In addition, for rapidly varying potentials, the interaction of electrons with one another may be affected, thus violating our assumption of independent particles. That is, we may need to compute energy levels of the entire system of electrons quantum mechanically.

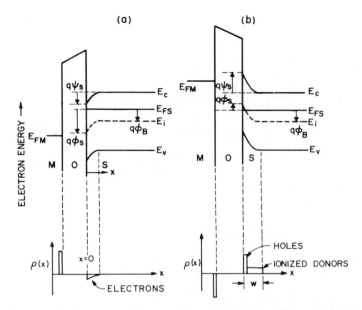

Fig. 2.9 (a) Energy diagram of the MOS system for n-type silicon in accumulation. The various potentials defined in the text are shown along with their sign conventions as indicated by the arrows. An arrow pointing down denotes positive potential; an arrow pointing up denotes negative potential. The charge distribution diagram below the energy diagram shows that surface charge in accumulation is made up entirely of electrons. (b) Same as (a), except depletion-inversion is shown. Charge distribution diagram below the energy diagram shows that surface charge in depletion-inversion is made up of holes and ionized donors.

ϕ_B, and at the silicon surface ($x = 0$), $\phi(x)$ is called the *surface potential* ϕ_s.

Ordinarily, energy levels are located within the silicon bandgap by stating their distance in electron volts from the intrinsic level. Therefore, $\phi(x)$ is very useful because it denotes the position of those energy levels coincident with the Fermi level at position x. A special case is that of interface trap levels. In this case interface trap levels located opposite the Fermi level are at an energy ϕ_s above the intrinsic level at the silicon surface. As discussed in Chapter 5, these interface traps are important to the small-signal ac response of the MOS capacitor.

The *band bending* $\psi(x)$ is defined as

$$\psi(x) \equiv \phi(x) - \phi_B \qquad (2.40)$$

where $\psi(x)$ represents the potential at any point x in the depletion layer with respect to its value in the bulk. In particular, the *barrier height* $\psi_s = \phi_s - \phi_B$ is the total potential difference between the silicon surface and the bulk; that is, ψ_s is the total band bending. It is convenient to define

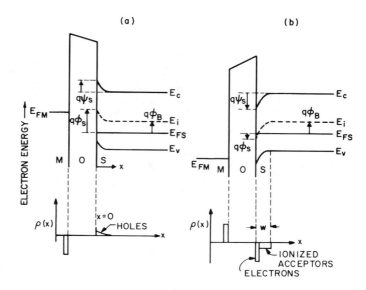

Fig. 2.10 (a) Energy diagram of the MOS system for p-type silicon in accumulation. The various potentials defined in the text are shown along with their sign conventions as indicated by the arrows. An arrow pointing down denotes positive potential; an arrow pointing up denotes negative potential. Charge distribution diagram below the energy diagram shows that surface charge in accumulation is made up entirely of holes. (b) Same as (a), except depletion-inversion is shown. Charge distribution diagram below the energy diagram shows that surface charge in depletion-inversion is made up of electrons and ionized acceptors.

dimensionless potentials $u(x)$ and $v(x)$ by the equations

$$u(x) = \frac{q\phi(x)}{kT} \quad and \quad v(x) = \frac{q\psi(x)}{kT}. \tag{2.41}$$

Using the definitions of potentials in (2.39), (2.40), and (2.41), (2.37) and (2.38) can be written

$$n(x) = n_i \exp[u(x)] = N_D \exp[v(x)] \tag{2.42}$$

where

$$\frac{N_D}{n_i} = \exp(u_B) \tag{2.43}$$

and

$$p(x) = n_i \exp[-u(x)] = N_A \exp[-v(x)] \tag{2.44}$$

where

$$\frac{N_A}{n_i} = \exp(-u_B). \tag{2.45}$$

At the silicon surface, $u(0) = u_s$ and $v(0) = v_s$. The electron and hole densities at the silicon surface n_s and p_s, respectively, are from (2.42) and (2.44)

$$n_s = n_i \exp(u_s) = N_D \exp(v_s) \quad and \quad p_s = n_i \exp(-u_s) = N_A \exp(-v_s).$$
(2.46)

Equations (2.42) and (2.44) describe the electron density distribution in n-type silicon and the hole distribution in p-type silicon, respectively. To calculate the minority carrier distribution in each type, the relation $pn = n_i^2$ must be used. For example, the hole density at x in n-type, using (2.42) and the pn product relationship, would be $p(x) = (n_i^2/N_D) \exp[-v(x)]$.

The sign convention for the potentials for n-type in Fig. 2.9 are: (1) $\phi_B > 0$; (2) $\psi_s > 0$ and $\phi_s > 0$ in accumulation; (3) when $\psi_s = 0$ and $\phi_s = \phi_B$, the bands will be flat; (4) $\psi_s = -\phi_B < 0$ is the intrinsic point at the surface ($n_s = p_s$); and (5) $\psi_s \lesssim -2\phi_B < 0$ in inversion. Figures 2.10a, b show p-type silicon in accumulation and inversion respectively. The sign conventions for p-type in Fig. 2.10 are: (1) $\phi_B < 0$; (2) $\phi_s < 0$ and $\psi_s < 0$ in accumulation; (3) when $\phi_s = \phi_B$ and $\psi_s = 0$, the bands will be flat; (4) $\psi_s = -\phi_B > 0$ is the intrinsic point ($p_s = n_s$); and (5) $\psi_s \gtrsim -2\phi_B > 0$ in inversion. The sign convention for bias applied between the gate and the silicon will be the polarity on the gate with respect to the silicon.

Consider Fig. 2.9a, where $V_G > 0$ (V_G is the applied gate bias). Positive bias on the gate attracts electrons to the silicon surface. The bands will bend down so that $u(x) > 0$ and $v(x) > 0$. Equation (2.42) shows that when $u(x) > 0$, $n(x) > p(x)$ at all values of x, and when $v(x) > 0$, $n(x) > N_D$ at all values of x. Thus surface charge will consist of an accumulation layer of electrons with a density at the surface greater than in the bulk.

When $V_G = 0$, $u_s = u(x) = q\phi_B/kT = u_B$, and $v(x) = v_s = 0$, the bands will be flat throughout the silicon. Hole and electron densities now will be uniform throughout the silicon and will be the same at the surface as they are in the bulk, that is, $n_s = N_D$ and $p_s = n_i^2/N_D$. This condition is called the *flatband condition.*

Making $V_G < 0$ or negative with respect to the silicon repels electrons from the surface and attracts holes to the surface so that the bands now bend up as shown in Fig. 2.9b. In this case $u(x) < 0$ and $v(x) < 0$. Equation (2.42) shows that $n(x) < N_D$ at all values of x, and $p(x) = n_i^2/n(x)$. As the magnitude of $-v$ is increased, $n(x)$ will decrease as given by (2.42), and $p(x)$ will increase as $n_i^2/n(x)$ at all values of x until $n(x) = p(x) = n_i$, which occurs at the surface when $u_s = 0$ and $v_s = -q\phi_B/kT = -u_B$. The surface is then intrinsic.

As the magnitude of $-v$ is further increased, $p_s > n_s$ and an inversion layer of holes begins to form. When $v_s = -2u_B$, minority carrier density (holes) at the surface equals majority carrier density (electrons) in the bulk, that is, $p_s = N_D$. When $|v_s| < 2u_B$, surface hole density exceeds electron density in the bulk and the surface is inverted.

Figures 2.10a, b show p-type silicon, and the discussion is similar to n-type except that the signs of the potentials are different, and (2.44) is used instead of (2.42). Charge distributions in the gate and in the silicon are also shown in Figs. 2.9 and 2.10.

(d) The Poisson Equation

With a physical description of the field induced depletion layer, definitions of potentials, and expressions for free carrier densities, we are ready to discuss the surface charge region quantitatively. We treat the silicon as a semi-infinite homogeneous crystal in thermal equilibrium. The silicon surface is represented by the plane at $x = 0$ and the bulk by positive values of x. This problem is one dimensional, so that potentials are functions of x only.

The Poisson equation will be solved under the simplifying conditions listed in Section (a) above and where the band-bending approximation is valid.* Surface potential as a function of x is given by the Poisson equation in one dimension

$$\frac{d^2\phi(x)}{dx^2} = -\frac{\rho(x)}{\epsilon_s} \tag{2.47}$$

where $\rho(x)$ is the charge density (coul/cm^3) composed of immobile ionized donors and acceptors and mobile holes and electrons and $\epsilon_s = 1.04 \times 10^{-12}$ F/cm is the dielectric permittivity of silicon. Now

$$\rho(x) = q[p(x) - n(x) + N_D - N_A] \tag{2.48}$$

and (2.47) becomes

$$\frac{d^2\phi(x)}{dx^2} = \frac{-q[p(x) - n(x) + N_D - N_A]}{\epsilon_s}. \tag{2.49}$$

The condition of charge neutrality must exist in the bulk, that is, far from the surface where $\phi(\infty) = \phi_B$ and $\rho(x) = 0$. Therefore, (2.48) becomes, deep in the bulk $(x \to \infty)$

$$N_D - N_A = n(\infty) - p(\infty). \tag{2.50}$$

When (2.42) and (2.44) are used, (2.50) becomes

$$N_D - N_A = n_i[\exp(u_B) - \exp(-u_B)]. \tag{2.51}$$

As $[\exp(u_B) - \exp(-u_B)] = 2 \sinh(u_B)$, (2.51) becomes

$$N_D - N_A = 2n_i \sinh u_B. \tag{2.52}$$

In general, for any value of x, we write from (2.42) and (2.44) that

$$n(x) - p(x) = 2n_i \sinh u(x). \tag{2.53}$$

---

*Some of the literature concerning the integration of the Poisson equation is given in Refs. 21 and 30–35. This work has been reviewed by McDonald.[35]

Substituting (2.52) and (2.53) into (2.49), Poisson's equation becomes, expressing potential in the dimensionless form

$$\frac{d^2 u(x)}{dx^2} = \lambda_i^{-2}[\sinh u(x) - \sinh u_B] \tag{2.54}$$

where λ_i, called the *intrinsic Debye length*, is defined as[36]

$$\lambda_i = \left(\frac{\epsilon_s kT}{2q^2 n_i}\right)^{1/2} \text{(cm)}. \tag{2.55}$$

The value of λ_i, at $T = 290°K$, is 2.84×10^{-3} cm using $n_i = 10^{10}$ cm^{-3}. The complicated nature of (2.54) is due to the mutual interdependence of carrier density on potential through the Maxwell-Boltzmann relations (2.42) and (2.44) and the dependence of potential on free carrier density through Poisson's equation. The appropriate boundary conditions are that at the surface ($x = 0$), $u = u_s$ and in the bulk ($x \to \infty$), $u = u_B$. This last condition ensures charge neutrality in the bulk.

To integrate (2.54), we use an integrating factor given by

$$\frac{d}{dx}\left(\frac{du}{dx}\right)^2 = 2\frac{du}{dx}\left(\frac{d^2 u}{dx^2}\right). \tag{2.56}$$

Multiplying (2.54) by $2(du/dx)$, we obtain

$$2\frac{du}{dx}\left(\frac{d^2 u}{dx^2}\right) = 2\lambda_i^{-2}[\sinh u(x) - \sinh u_B]\frac{du}{dx} = \frac{d}{dx}\left(\frac{du}{dx}\right)^2. \tag{2.57}$$

Integrating from the surface to the bulk, we have

$$\int\limits_{du_s/dx}^{0} d\left(\frac{du}{dx}\right)^2 = 2\lambda_i^{-2}\int\limits_{u_s}^{u_B} [\sinh u(x) - \sinh u_B]du \tag{2.58}$$

where $du_s/dx = (q/kT)F_s$, and F_s is the field at the semiconductor surface. Integrating each term in (2.58) yields

$$\int\limits_{E_s}^{0} d\left(\frac{du}{dx}\right)^2 = -\left(\frac{qF_s}{kT}\right)^2, \tag{2.59}$$

$$\int\limits_{u_s}^{u_B} \sinh u(x)du = \cosh u_s - \cosh u_B, \tag{2.60}$$

and

$$\sinh u_B \int\limits_{u_s}^{u_B} du = (u_B - u_s)\sinh u_B. \tag{2.61}$$

The result of the integration is

$$F_s = \text{Sgn}(u_B - u_s)(2)^{1/2} \frac{kT}{q\lambda_i} [(u_B - u_s)\sinh u_B - (\cosh u_B - \cosh u_s)]^{1/2}. \quad (2.62)$$

For $u_s < u_B$, $\text{Sgn}(u_B - u_s)$ is positive, and for $u_s > u_B$, $\text{Sgn}(u_B - u_s)$ is negative. We also get the field in the depletion layer by integrating (2.58) from a point x in the depletion layer to the bulk in a manner similar to that already done. The result is

$$F(x) = \text{Sgn}(u_B - u_s)(2)^{1/2} \frac{kT}{q\lambda_i} \{[u_B - u(x)]\sinh u_B - [\cosh u_B - \cosh u(x)]\}^{1/2}.$$

$$(2.63)$$

To simplify (2.62), we write

$$F_s = \text{Sgn}(u_B - u_s) \frac{kT}{q\lambda_i} F(u_s, u_B) \quad (2.64)$$

where $F(u_s, u_B)$ is a *dimensionless* electric field given by

$$F(u_s, u_B) = (2)^{1/2}[(u_B - u_s) \sinh u_B - (\cosh u_B - \cosh u_s)]^{1/2}. \quad (2.65)$$

A plot of $F(u_s, u_B)$ is given for n-type in Fig. 2.11 for values of u_s between -24 and 24 and of u_B between 0 and 24. Also shown is the sign of F that, if positive, gives an electric field pointing from the bulk to the surface and positive charge. The identity $F(u_s, u_B) = -F(-u_s, -u_B)$ allows Fig. 2.11 to be used for negative values of u_B. This figure is handy for calculating silicon surface field without evaluating (2.62). To use Fig. 2.11, impurity and intrinsic carrier concentrations must be known. Then, u_B can be

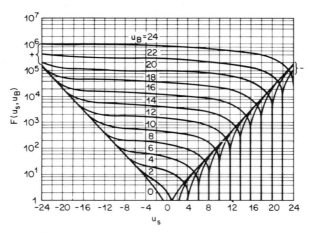

Fig. 2.11 Graph of $F(u_s, u_B)$ as a function of u_s calculated from (2.65) over the range $u_s = \pm 24$ with u_B as parameter over the range $0 < u_B < 24$. After Young.[30]

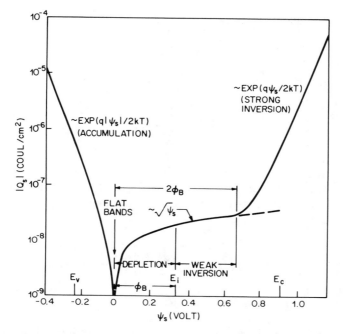

Fig. 2.12 Plot of (2.66) showing variations of silicon surface charge density $|Q_s|$ as a function of barrier height ψ_s for p-type silicon having $N_A = 4 \times 10^{15} \, cm^{-2}$, $T = 300°K$, and $\phi_B = 0.335 \, V$ calculated from (2.45). After Sze.[18]

calculated using (2.43) or (2.45). The independent variable is u_s. Figure 2.11 shows how electric field at the silicon surface should vary with u_s in an MOS capacitor as gate bias is varied. To check agreement, values of surface field and u_s can be extracted from a measured C-V curve once doping concentration is known.

To get total charge per unit area, we use Gauss's law

$$Q_s = \epsilon_s F_s = \text{Sgn}(u_B - u_s) C_o \left(\frac{kT}{q} \right) F(u_s, u_B) \tag{2.66}$$

where $C_o = \epsilon_s / \lambda_i$ is an effective semiconductor capacitance per unit area. Figure 2.12 is a plot of (2.66).

2.3.3 Shape of the Potential Barrier

To calculate the shape of the surface potential and the depletion layer width at the silicon surface, we write

$$F(x) = \frac{kT}{q} \left(\frac{du}{dx} \right). \tag{2.67}$$

When (2.63) is used, (2.67) becomes

$$\frac{du}{dx} = \frac{1}{\lambda_i} F(u, u_B).$$ (2.68)

Solving (2.67) for x yields

$$\frac{x}{\lambda_i} = \int_{u_s}^{u} \frac{du}{F(u, u_B)}.$$ (2.69)

Integration of (2.69) results in the desired dependence of surface potential on distance in the form of a relation between the dimensionless quantities x/λ_i and u. Figure 2.13 shows a plot of (2.69). Integration of (2.69) must be done numerically, except in the intrinsic case where $u_B = 0$. For this intrinsic case, (2.69) becomes

$$\frac{x}{\lambda_i} = \int_{u_s}^{u} \frac{du}{[\cosh(u) - 1]^{1/2}} = \ln\left[\frac{\tanh(u_s/4)}{\tanh(u/4)}\right].$$ (2.70)

2.3.4 The Depletion Approximation

In Section 2.3.2(d) the Poisson equation was solved by taking free carrier concentrations into account. Thus (2.65), (2.66), (2.67), and (2.69) describe the charge density, field, and potential distribution at the silicon surface for all values of surface potential in the nondegenerate case. However, to a good approximation in depletion, $p = n = 0$ everywhere in the depletion layer. The depletion approximation approximates the free carrier distribution in the transition region between the depletion region and the neutral bulk by a step function. That is, $p = n = 0$ in the depletion layer and

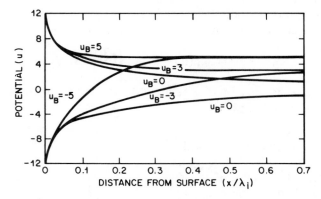

Fig. 2.13 Plot of (2.69). Normalized surface potential versus distance in multiples of the intrinsic Debye length λ_i for $u_s = \pm 12$ and various bulk potentials u_B for silicon. After Wallmark and Johnson.[37]

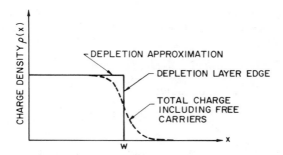

Fig. 2.14 Silicon surface charge density in the depletion approximation. Depletion layer width is *w*. Total charge density including free carriers is shown as a dotted curve. The dotted line drops below the solid line for *x* < *w* because the sign of majority carriers is always opposite the sign of ionized impurities, so that total charge is the difference between the two.

charge neutrality prevails beyond the depletion layer edge. This step function approximation greatly simplifies the solution of the Poisson equation in depletion and is widely used.

Figure 2.14 shows total charge density as a function of distance from the silicon surface. Also shown is silicon surface charge density, in the depletion approximation. In the vicinity of the depletion layer edge, the true charge density of majority carriers only approximates an abrupt step, the transition from depletion to neutrality occurring over a distance comparable to an extrinsic Debye length.* An extrinsic Debye length is the characteristic screening length of a small charge in an extrinsic semiconductor. Because the extrinsic Debye length varies as $N_B^{-1/2}$ [see (2.87) or (2.88)], where N_B is the doping concentration, the mobile charge concentration more closely approximates the charge distribution for larger values of N_B, and the depletion approximation becomes more accurate. Thus in Fig. 2.14, as doping concentration increases, the transition from the depletion region to the neutral bulk region becomes more abrupt.

(a) *General Case—Nonuniform Impurity Distribution*

Consider the general case where impurity concentration varies with distance from the silicon surface. This situation can arise, for example, in a diffused or ion implanted surface, or as a result of impurity redistribution during thermal oxidation, described in Section 14.4.2. Whether a closed-form solution can be found for the general, nonuniform case will depend on the specific functional dependence of impurity concentration on distance.

---

*This point can be established for uniform doping using (2.69) for $\psi(x)$ and (2.37) and (2.38) for n and p. The Debye length is defined in (2.87) and (2.88).

For the general case where donor and acceptor concentrations are not uniform, the depletion layer charge density in the depletion approximation is

$$\rho(x) = q[N_D(x) - N_A(x)].$$

Poisson's equation, (2.47), becomes

$$\frac{d^2\phi(x)}{dx^2} = -\frac{q[N_D(x) - N_A(x)]}{\epsilon_s}. \tag{2.71}$$

By solving (2.71), we obtain band bending as a function of x. We transform variables. From (2.40), $\psi(x) = \phi(x) - \phi_B$ so that (2.71) becomes

$$\frac{d^2\psi(x)}{dx^2} = -\frac{q[N_D(x) - N_A(x)]}{\epsilon_s}. \tag{2.72}$$

Integrating (2.72) from the bulk toward the surface, we obtain

$$\int_0^{F(x)} d\left(\frac{d\psi(x)}{dx}\right) = \frac{q}{\epsilon_s}\int_w^x [N_A(x') - N_D(x')]dx' \tag{2.73}$$

where w is the depletion layer width. Integrating the left side of (2.73) to obtain the electric field yields

$$F(x) = \frac{q}{\epsilon_s}\int_w^x [N_A(x') - N_D(x')]dx'. \tag{2.74}$$

To evaluate (2.74), the specific functional dependence on x of the donor and acceptor concentrations in the silicon must be known. Continuing with the general case, the potential in the depletion layer is obtained by integrating (2.74) again

$$\int_0^{\psi(x)} F(x')dx' \doteq \psi(x) = \frac{q}{\epsilon_s}\int_w^x\int_w^{x''} [N_A(x') - N_D(x')]dx'\,dx''. \tag{2.75}$$

By interchanging the order of integration (doing the x'' integral before the x' integral) (2.75), can be simplified further

$$\psi(x) = \frac{q}{\epsilon_s}\int_w^x dx'(x - x')[N_A(x') - N_D(x')]. \tag{2.76}$$

(b)   Specific Case—Uniform Impurity Distribution

For a uniform impurity distribution, $q(N_A - N_D) = $ constant, and (2.74) can be written

$$F(x) = \frac{q}{\epsilon_s}(N_A - N_D)\int_w^x dx'. \tag{2.77}$$

Integrating (2.77) yields

$$F(x) = \frac{q}{\epsilon_s}(N_A - N_D)(x - w).$$ (2.78)

The field at the silicon surface ($x = 0$) will be given by

$$F_s = -\frac{q}{\epsilon_s}(N_A - N_D)w.$$ (2.79)

Equations (2.78) and (2.79) show that the field in the silicon depletion layer decreases linearly from E_s at the surface to zero at $x = w$, the edge of the depletion layer. Substituting (2.78) into (2.76) and integrating from the bulk to the surface to obtain the potential in the depletion layer yields

$$\int_0^{\psi(x)} F(x')dx' = \frac{q}{\epsilon_s}(N_A - N_D)\int_w^x (x' - w)dx'.$$ (2.80)

The result of integrating (2.80) is

$$\psi(x) = \frac{1}{2}\frac{q}{\epsilon_s}(N_A - N_D)(w - x)^2.$$ (2.81)

At the surface, $x = 0$, and (2.81) becomes

$$\psi_s = \frac{1}{2}\frac{q}{\epsilon_s}(N_A - N_D)w^2.$$ (2.82)

For n-type, $N_D > N_A$ so that $\psi_s < 0$, whereas for p-type, $N_A > N_D$ so that $\psi_s > 0$. From (2.81) and (2.82), we write

$$\psi(x) = \psi_s\left(1 - \frac{x}{w}\right)^2.$$ (2.83)

Equations (2.81) or (2.83) show that band bending in the depletion layer has a parabolic shape as it varies from ψ_s at the surface to zero in the bulk. This parabolic shape is illustrated in Fig. 2.15.

(c) *Maximum Depletion Layer Width*

One of the important applications of the depletion approximation is in estimating maximum depletion layer width in strong inversion. Depletion layer width increases with applied bias until the silicon surface becomes strongly inverted. Depletion layer width then increases only slowly with further increases in bias because the inversion layer shields the silicon from further penetration of the applied field. That is, nearly all the field lines now terminate on surface inversion layer charges rather than on ionized impurities.

Maximum depletion layer width can be calculated from (2.82) if the band bending corresponding to this width is known. An approximate criterion for the onset of weak inversion is that the minority carrier concentration

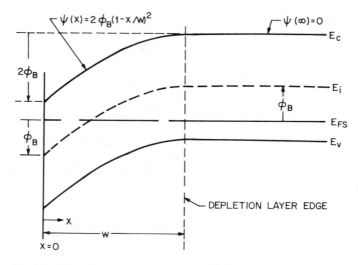

Fig. 2.15 Band-bending diagram for p-type silicon showing that $\psi(x)$ is parabolic where x is measured from the interface and $\psi(\infty) = 0$. The band bending corresponds to strong inversion, namely, $\psi_s = 2\phi_B$. Because the silicon surface is inverted, $\psi(x)$ deviates from the parabolic law within the inversion layer. This deviation is small, occurring over the last few kT/q of potential, which is much less than $2\phi_B$ in a width small compared to w. This small region is not shown.

near the surface equals the majority carrier concentration at the depletion layer edge, that is, when $n_s = N_A$ for p-type silicon. Under this condition, for p-type silicon, the Fermi level at the silicon surface will be above the intrinsic Fermi level by the same amount as it is below the intrinsic Fermi level in the silicon bulk as shown in Fig. 2.15. Therefore, the total band bending at the onset of strong inversion will be $\psi_s(\text{inv}) = 2\phi_B$. To derive this result mathematically for a p-type sample, we start with the surface hole concentration $p_s = n_i \exp(-u_s)$ from (2.46). Substituting p_s into the relation $p_s n_s = n_i^2$ and solving for n_s gives the electron concentration at the silicon surface $n_s = n_i \exp(u_s)$. When $u_s = u_B$, $n_s = N_A$ from (2.46). The relation between band bending and surface potential is $\psi_s = \phi_B + \phi_s(\text{inv})$ using the sign convention in Fig. 2.10b for the depletion-inversion region so that $\psi_s(\text{inv}) = 2\phi_B$ when $\phi_s = \phi_B$.

Although strong inversion can be considered to occur for $\psi_s(\text{inv}) = 2\phi_B$, this value of band bending is only approximate. Another criterion for the onset of strong inversion was proposed by Lindner,[38] who suggested that band bending at the onset of strong inversion should make the minority carrier contribution to the surface field match the dopant ion contribution. This match occurs, using (2.62) for p-type material with $|u_B| \gg 1$ (extrinsic silicon) and $u_s > |u_B|$ (inverted surface), when the band bending is v_L, where

$$\exp(v_L)\exp(-2|u_B|) = v_L - 1. \qquad (2.84)$$

Here v_L is the Lindner band bending and $\psi_s(\mathrm{inv}) = (kT/q)v_L$. The Lindner band bending is closer to the value at which minority carriers pin the surface band bending in inversion.

An adequate approximation to (2.84) is

$$v_L = 2.10u_B + 2.08. \tag{2.85}$$

For $v_s > v_L$, minority carriers dominate the capacitance. Thus any further increase of gate charge draws more electrons to the surface (for our p-type example) rather than exposing more negative acceptor ions. Consequently, depletion layer width saturates as a function of gate bias once $v_s > v_L$. Equation (2.85) shows that v_L differs from $2u_B$ by only a few kT/q, and hence the difference is small.

This maximum depletion layer width is the sum of the inversion and depletion layer thicknesses for $v_s = v_L$. If we neglect inversion layer thickness (which is only a small fraction of a Debye length) and simply use the depletion approximation (2.82) for the depletion layer width, we find

$$w_{\max} = \left(\frac{2\epsilon_s \psi_L}{q(N_A - N_D)} \right)^{1/2} \tag{2.86}$$

where ψ_L is given by (2.84), $\psi_L = (kT/q)v_L$. Equation (2.86) is more obviously related to a length if we introduce the extrinsic Debye lengths λ_n and λ_p [39]

$$\lambda_n \equiv \left(\frac{\epsilon_s kT}{q^2 N_D} \right)^{1/2} \tag{2.87}$$

$$\lambda_p \equiv \left(\frac{\epsilon_s kT}{q^2 N_A} \right)^{1/2}. \tag{2.88}$$

For n-type silicon, $N_D \gg N_A$ and $\psi_s < 0$, so (2.86) becomes

$$w_{\max} = (2)^{1/2} \lambda_n (-v_L)^{1/2} \tag{2.89}$$

and for p-type

$$w_{\max} = (2)^{1/2} \lambda_p (v_L)^{1/2}. \tag{2.90}$$

Maximum depletion layer width from (2.90) is plotted in Fig. 2.16a as a function of doping concentration. The resulting depletion layer widths are within a few percent of the most accurate expressions available.[40] Figure 2.16b shows the normalized Lindner band bending v_L as a function of impurity concentration obtained from (2.84).

(d) Calculation of Surface Charge in the Depletion Approximation

Surface charge in the depletion layer is

$$Q_s = \epsilon_s F_s. \tag{2.91}$$

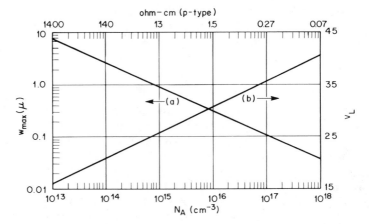

Fig. 2.16 (a) Plot of maximum depletion layer width as a function of impurity concentration for silicon. Maximum depletion layer width occurs in strong inversion where shielding by the inversion layer prevents further increase of depletion layer width with gate bias. Calculated for $T = 290°K$ from (2.86). (b) Plot of normalized Lindner band bending v_L at onset of strong inversion as a function of impurity concentration. Calculated iteratively for $T = 290°K$ from (2.84) in the form $v_L = 2u_B + \ln(v_L - 1)$.

Substituting (2.79) into (2.91), we obtain

$$Q_s = -q(N_D - N_A)w. \qquad (2.92)$$

To determine Q_s in terms of ψ_s, solve (2.82) for w and substitute into (2.92) to obtain

$$Q_s = -[2\epsilon_s q(N_A - N_D)\psi_s]^{1/2}. \qquad (2.93)$$

With the development of the equations for the depletion approximation, we can apply them to calculate the free carrier concentrations as functions of distance shown in Fig. 2.7.

Hole density given by (2.38) for p-type silicon is $p(x) = N_A \exp[-q\psi(x)/kT]$, and electron density using the pn product is $n(x) = (n_i^2/N_A) \exp[q\psi(x)/kT]$. Next we obtain $\psi(x)$. With a surface charge density of 2×10^{11} cm^{-2} and an acceptor concentration of 10^{16} cm^{-3}, the silicon surface is in depletion, and surface charge consists almost entirely of ionized acceptors. Therefore, we can use the depletion approximation for $\psi(x)$ that is from (2.81): $\psi(x) = \frac{1}{2}q(N_A/\epsilon_s)w^2(1 - x/w)^2$. Thus the free carrier concentrations as explicit functions of x are: $p(x) = N_A \exp\{-[w/(2\lambda_p)]^2(1 - x/w)^2\}$ and $n(x) = (n_i^2/N_A) \exp\{-[w/(2\lambda_n)]^2(1 - x/w)^2\}$, using (2.87) and (2.88), where w is 2×10^{-5} cm from (2.92). Then, from (2.82), $\psi_s = 0.31$ V and from (2.45), $\phi_B = 0.35$ V.

(e) *Calculation of the Field and Potential in Silicon and Silicon Dioxide*

Consider the field in the SiO_2 and the silicon. Assuming no charge at the interface, displacement across the Si–SiO$_2$ interface is continuous so that

$$\epsilon_{ox}F_{ox} = \epsilon_s F_s \tag{2.94}$$

where ϵ_{ox} is the dielectric permittivity of SiO_2 ($\epsilon_{ox} = 0.34\ \text{pF/cm}$) and F_{ox} is the electric field in the SiO_2. Because we have assumed that there are no charges in the SiO_2, the field in the SiO_2 is constant and equal to $F_{ox} = V_{ox}/x_o$, where V_{ox} is the voltage across the SiO_2. The field configuration in the SiO_2 and in the silicon [given by (2.78)] is illustrated in Fig. 2.17a. The

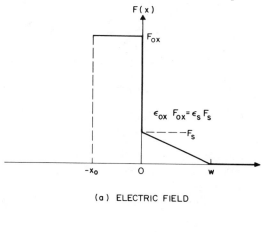

(a) ELECTRIC FIELD

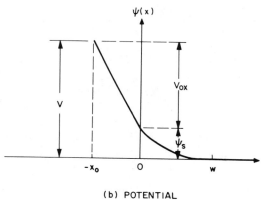

(b) POTENTIAL

Fig. 2.17 (a) Electric field configuration in the SiO_2 and silicon surface calculated from (2.78) as a function of x. (b) Potential distribution in the SiO_2 and the silicon surface calculated from (2.83) corresponding to the electric field configuration in (a).

potential configuration in the SiO_2 and the silicon [given by (2.83)] is illustrated in Fig. 2.17*b*.

(f) Errors in the Depletion Approximation

Equations (2.74) and (2.76) are not exact. In fact, more accurate expressions for a semi-infinite sample in depletion are

$$F(x) = \frac{-q}{\epsilon_s} \int_x^\infty dx'\{N_A(x') - N_\infty \exp[-v(x')]\} \tag{2.95}$$

and

$$\psi(x) = \frac{q}{\epsilon_s} \int_x^\infty dx'(x' - x)\{N_A(x') - N_\infty \exp[-v(x')]\} \tag{2.96}$$

where we assume $N_A(x) \gg N_D(x)$ for all x, and for sufficient distances from the interface, doping becomes uniform at a level N_∞. That is, as x approaches infinity, $N_A(x) \to N_\infty$ and $v(x) \to 0$ so that charge neutrality prevails at large distances. The approximate formulas (2.74) and (2.76) result if we approximate the integrals in (2.95) and (2.96) by

$$N_A(x) - N_\infty \exp[-v(x)] = N_A(x); \qquad x < w$$
$$= 0; \qquad x > w. \tag{2.97}$$

That is, (2.97) *assumes charge neutrality for* $x > w$. Now, in fact, charge neutrality cannot exist for $x > w$ unless $N_A(x) = N_A(w)$ for all $x > w$. There is a built-in potential gradient inevitably present when the doping profile is not uniform. Therefore, the first error of the depletion approximation is:

1 The depletion approximation ignores any built-in potential or any field at the depletion layer edge due to lack of charge neutrality in the region $x > w$.

The second error in (2.97) is that an *abrupt* transition from full depletion to full charge neutrality is assumed. In fact, this transition is not a transition to full charge neutrality and also cannot occur more rapidly than within a local Debye length, $[kT\epsilon_s/q^2 N_A(w)]^{1/2}$ (for *p*-type). Because the width of this transition region varies with w when $N_A(w)$ varies with w, the potential drop across this transition region also will vary. Hence a second error in the depletion approximation is as given in 2 (below).

2 The depletion approximation ignores variations in the potential drop across the transition region at the depletion layer edge as the position of the depletion layer edge is varied.

Finally, in extreme cases, there may be a potential barrier to majority carrier flow due to built-in junction effects. Such regions of low majority carrier density result in (1) long majority carrier response times, limiting the frequency at which a purely capacitive response is possible, and (2) very extended transition regions or nearly discontinuous jumps in w with increasing ψ_s when low majority carrier density regions are near the depletion layer edge. Thus, in extreme cases, a third error of the depletion approximation is as given in 3 (below).

3 The depletion approximation ignores the possibility of very low majority carrier density regions due to built-in potential effects.

In practice, how serious are these errors? For gradual profile variations, such as thermal redistribution effects, the errors are small, and the depletion approximation can be used with confidence. For implanted structures, particularly those in which the depletion layer edge is within the implant, the errors can be serious. Such cases are not discussed.

2.4 SUMMARY OF IMPORTANT EQUATIONS

To make it easier to use the results of this chapter, the equations used most often are concisely listed in this section.

1 Fermi function for electrons:

$$f_n = \left(1 + \exp\frac{E - E_F}{kT}\right)^{-1}.$$

2 Fermi function for holes:

$$f_p = \left(1 + \exp\frac{E_F - E}{kT}\right)^{-1}.$$

3 Density of states in the conduction band:

$$M_c = 12(2\pi m_n h^{-2}kT)^{3/2}.$$

4 Density of states in the valence band:

$$M_v = 2(2\pi m_p h^{-2}kT)^{3/2}.$$

5 Intrinsic carrier density:

$$n_i = (M_v M_c)\exp\left(-\frac{E_g}{kT}\right) = 3.02 \times 10^{15}T^{3/2}\exp\left(-0.55\frac{q}{kT}\right)$$

$$E_F - E_i = kT \ln\frac{N_D}{n_i} = q\phi_B \quad \text{for } n\text{-type}$$

$$E_i - E_F = kT \ln\frac{N_A}{n_i} = q\phi_B \quad \text{for } p\text{-type.}$$

6 Surface potential:

$$q\phi(x) = E_F - E_i(x).$$

7 Band bending:

$$\psi(x) = \phi(x) - \phi_B.$$

8 Normalized surface potential:

$$u(x) = \frac{q\phi(x)}{kT}.$$

9 Normalized band bending:

$$v(x) = \frac{q\psi(x)}{kT}.$$

10 Electron density in the surface charge layer:

$$n(x) = n_i \exp[u(x)] = N_D \exp[v(x)].$$

11 Hole density in the surface charge layer:

$$p(x) = n_i \exp[-u(x)] = N_A \exp[-v(x)].$$

12 Intrinsic Debye length:

$$\lambda_i = \left(\frac{\epsilon_s kT}{2q^2 n_i}\right)^{1/2}.$$

13 Extrinsic Debye length:

$$\lambda_n = \left(\frac{\epsilon_s kT}{q^2 N_D}\right)^{1/2} \quad \text{for } n\text{-type}$$

$$\lambda_p = \left(\frac{\epsilon_s kT}{q^2 N_A}\right)^{1/2} \quad \text{for } p\text{-type}.$$

14 Dimensionless electric field:

$$F(u_s, u_B) = (2)^{1/2}[(u_B - u_s) \sinh u_B - (\cosh u_B - \cosh u_s)]^{1/2}.$$

15 Surface electric field:

$$F_s = \text{Sgn}(u_B - u_s) \frac{kT}{q\lambda_i} F(u_s, u_B).$$

16 Surface charge density:

$$Q_s = \text{Sgn}(u_B - u_s) \frac{\epsilon_s}{\lambda_i} \frac{kT}{q} F(u_s, u_B).$$

17 Electric field in the surface charge layer in the depletion approximation:

$$F(x) = \frac{q}{\epsilon_s}(N_A - N_D)(x - w).$$

18 Band bending in the surface charge layer in the depletion approximation:

$$\psi(x) = \frac{1}{2}\frac{q}{\epsilon_s}(N_A - N_D)(w - x)^2.$$

19 Maximum depletion layer width in the depletion approximation:

$$w_{max} = \left(\frac{2\epsilon_s \psi_L}{q(N_A - N_D)}\right)^{1/2} \quad \text{where} \quad \psi_L = 2.10\left(\frac{q}{kT}\right)\phi_B + 2.08.$$

20 Surface charge in the depletion approximation:

$$Q_s = -q(N_D - N_A)w = -[2\epsilon_s q(N_D - N_A)\psi_s]^{1/2}.$$

REFERENCES

1 F. Seitz, *Modern Theory of Solids*, McGraw-Hill, New York, 1940.

2 W. Shockley, *Electrons and Holes in Semiconductors*, Van Nostrand, Princeton, N.J., 1950.

3 J. M. Ziman, *Principles of the Theory of Solids*, Cambridge University Press, New York, 1964.

4 R. E. Peierls, *Quantum Theory of Solids*, Oxford University Press, London, 1955.

5 C. Kittel, *Introduction to Solid-State Physics*, 2nd ed., Wiley, New York, 1956.

6 A. J. Dekker, *Solid State Physics*, Prentice Hall, New York, 1957.

7 F. Seitz and D. Turnbull, Eds., *Solid State Physics*, Vols. 1–15 plus supplements, Academic, New York, 1950–1963.

8 R. Tolman, *The Principles of Statistical Mechanics*, Oxford University Press, London, 1938.

9 J. E. Mayer and M. G. Mayer, *Statistical Mechanics*, Wiley, New York, 1940.

10 W. A. Harrison, *Electronic Structure and Properties of Solids*, Freeman, San Francisco, 1980.

11 R. A. Smith, *Semiconductors*, 2nd ed., Cambridge University Press, New York, 1978.

12 A. F. Ioffe, *Physics of Semiconductors*, Academic, New York, 1960.

13 A. Many, Y. Goldstein, and N. B. Grover, *Semiconductor Surfaces*, Wiley, New York, 1965, Chapter 2, pp. 34–37.

14 An excellent description of the Fermi-Dirac statistics and the derivation of the Fermi-Dirac distribution function can be found in Max Born, *Atomic Physics*, 5th ed., Hafner, New York, 1935, Chapter VIII, p. 261.

15 J. N. Shive, *The Properties, Physics, and Design of Semiconductor Devices*, Van Nostrand, Princeton, N.J., 1959.

16 W. B. Joyce and R. W. Dixon, *Appl. Phys. Lett.*, **31**, 354 (1977).

17 H. D. Barber, *Solid-State Electron.*, **10**, 1039 (1967); T. Wasserrab, *Z. Naturforsch.*, **31**, 505 (1976).

18 S. M. Sze, *Physics of Semiconductor Devices*, Wiley, New York, 1969.

19 A. H. Wilson, *Theory of Metals*, 2nd ed., Cambridge University Press, New York, 1958.

20 A. S. Grove, *Physics and Technology of Semiconductor Devices*, Wiley, New York, 1967.

21 J. R. Brews and C. J. Hwang, *J. Chem. Phys.*, **54**, 3263 (1971).

22 D. Redfield and M. Afromowitz, *Phil. Mag.*, **19**, 831 (1969).

23 J. R. Schrieffer, in *Semiconductor Surface Physics*, R. G. Kingston, Ed., University of Pennsylvania Press, Philadelphia, 1957, p. 55.

24 F. F. Fang and W. E. Howard, *Phys. Rev. Lett.*, **16**, 797 (1966).

25 M. Kaplit and J. N. Zemel, *Phys. Rev. Lett.*, **21**, 212 (1968).

26 A. M. Voshchenkov and J. N. Zemel, *Phys. Rev.*, **B-9**, 4410 (1974).

27 F. Stern and W. E. Howard, *Phys. Rev.*, **163**, 816 (1967).

28 K. von Klitzing, G. Landwehr, and G. Dorda, *Solid-State Commun.*, **14**, 387 (1974).

29 G. Landwehr, "Quantum Transport in Silicon Inversion Layers," in *Festkörperprobleme* (Advances in Solid-State Physics), Vol. XV, H. J. Queisser, Ed., Pergamon/Vieweg, Braunschweig, West Germany, 1975, p. 49.

30 C. E. Young, *J. Appl. Phys.*, **32**, 329 (1961).

31 C. G. B. Garrett and W. H. Brattain, *Phys. Rev.*, **99**, 376 (1955).

32 R. G. Kingston and S. F. Neustadter, *J. Appl. Phys.*, **26**, 718 (1955).

33 G. C. Dousmanis and R. C. Duncan, Jr., *J. Appl. Phys.*, **29**, 1927 (1958).

34 D. R. Frankl, *J. Appl. Phys.*, **31**, 1752 (1960).

35 J. R. Mcdonald, *Solid-State Electron.*, **5**, 11 (1962).

36 W. Shockley, *Bell Syst. Tech. J.*, **28**, 435 (1949).

37 J. T. Wallmark and H. Johnson, Eds., *Field-Effect Transistors—Physics, Technology, and Applications*, Prentice-Hall, New York, 1966, Chapter 2.

38 R. Lindner, *Bell Syst. Tech. J.*, **41**, 803 (1962).

39 A. Many, Y. Goldstein, and N. B. Grover, *Semiconductor Surfaces*, North-Holland, Amsterdam, 1965, p. 138.

40 J. R. Brews, *Solid-State Electron.*, **17**, 447 (1974).

3

Metal Oxide Silicon Capacitor at Low Frequencies

3.1 INTRODUCTION

In this chapter we derive the low frequency C-V curve of the MOS capacitor using the dc characteristics derived in Chapter 2 as the starting point. This chapter, along with Chapter 4, lays the groundwork for the later chapters that discuss measurements using the MOS capacitor.

This chapter introduces the *steady state* characteristics of the MOS capacitor. In physics, steady state usually means that all macroscopic parameters of a system are time independent. Steady state differs from equilibrium because dissipation can occur, such as generation of heat in a resistor. In electrical engineering and in this book, a wider definition of steady state is used. Here steady state means the macroscopic parameters of a system are either time independent or varying sinusoidally in time with time independent amplitudes.

An MOS capacitor consists of an oxide layer usually thermally grown, an electrode called a *gate* (usually in the shape of a round disk) made by vacuum deposition of a metal or by the deposition of polysilicon and an

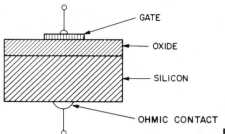

Fig. 3.1   Cross section of an MOS capacitor.

ohmic contact to the silicon substrate.* A cross section of the MOS capacitor is shown in Fig. 3.1.

The MOS capacitor first was proposed as a voltage-dependent capacitor by a number of workers.[1-4] Soon afterward it was used to study the properties of the Si–SiO$_2$ interface by Terman,[5] who calculated energy levels and relaxation times of interface traps from an analysis of experimental C-V characteristics measured over a wide range of frequencies. Lehovec et al.[6] performed a similar study. These workers pioneered the investigation of the Si–SiO$_2$ interface using the MOS capacitor. However, Grove et al.[7] firmly established the MOS capacitor as the major tool in the study of the MOS system. Much of the present understanding of the MOS system is based on MOS capacitor measurements. This work has led to improved device (bipolar transistor, MOSFET, junction diode, and CCD) performance and stability. In addition, this work has application to MOS analogue circuits such as analogue-to-digital converters, attenuators, codecs, active filters, and switched capacitor filters.

3.2   ONE-DIMENSIONAL CHARACTERISTICS

The equilibrium C-V curves of the MOS capacitor are now derived. Two capacitances can be defined for the MOS capacitor. The *static* capacitance is defined as $C_{stat} \equiv Q_T/V_G$, where Q_T is the total charge density on the capacitor and V_G is the bias applied to it. The *differential* capacitance is defined as $C \equiv dQ_T/dV_G$. Because charge on an MOS capacitor can vary *nonlinearly* with voltage, these two capacitances will be different.† Of the two capacitances, the differential capacitance is the most important in

---

*An ohmic contact between a metal and silicon must have a low resistance (fraction of an ohm), have a resistance independent of polarity, and obey Ohm's law at all temperatures and at all reasonable values of electric field. An ohmic contact can be made if the substrate has a degenerate layer, or if it does not, by first abrading the substrate surface with emery cloth before evaporating a metal to complete the contact.

†For a capacitor with a linear charge-voltage characteristic, such as a tuning capacitor in a radio receiver, the static and differential capacitances are identical.

MOS capacitor measurements because small-signal measurements determine the rate of change of charge with voltage. In this chapter we derive expressions for the small-signal differential capacitance of the MOS capacitor. The term capacitance is used throughout this book to mean differential capacitance.

To measure capacitance as a function of gate bias in steady state, a small alternating component of voltage or ac voltage is superimposed on the gate bias as shown in Fig. 3.2. The amplitude of this ac voltage always must be within the *small-signal* range to agree with the theory developed in this chapter. The small-signal range is the range of applied signal amplitude producing a linear response of ac current to ac voltage. Interface traps and interfacial charge nonuniformities influence the small-signal range because they alter how rapidly the admittance of the MOS capacitor varies with gate bias. Under such variable conditions, it is practical to determine the small-signal range by experiment as explained in Section 12.3.2.

3.2.1 Small-Signal Equivalent Circuit

We next derive the small-signal equivalent circuit of the MOS capacitor. The equivalent circuit consists of the frequency independent resistances and capacitances that respond to the ac voltage and is a snapshot of these values at a particular value of gate bias. At a different value of gate bias, those elements that are bias dependent change their values. To complete the picture, the gate bias dependence of these elements must be known

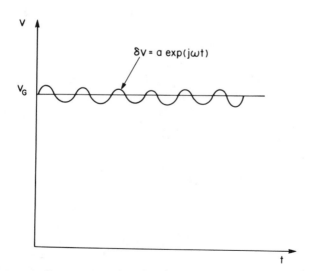

Fig. 3.2 Small ac voltage of amplitude a superposed on the gate bias applied to the terminals of the MOS capacitor to measure its capacitance or admittance as a function of gate bias.

independently. In this chapter we develop the equivalent circuit representation of the MOS capacitor with no interface traps, oxide charge, or work function difference (ideal case) and derive mathematical expressions for the circuit elements as a function of gate bias, temperature, and ionized dopant impurity concentration.

Using Gauss's law, we relate the charge per unit area in the silicon $Q_s(t)$ to the field in the oxide, $[V_G(t) - \psi_s(t)]/x_o$, by [see (2.94)]

$$C_{ox}[V_G(t) - \psi_s(t)] = -Q_s(t) \tag{3.1}$$

where $C_{ox} \equiv \epsilon_{ox}/x_o$ is the oxide capacitance per unit area; x_o is oxide thickness; $V_G(t)$ and $\psi_s(t)$ are the time-dependent gate voltage and silicon band bending, respectively; and $Q_s(t)$ is the time-dependent silicon surface charge density. Equation (3.1) applies at any time t, even when gate voltage is time varying. Superposing a small ac voltage $\delta V_G(t)$ on the gate bias, V_G we obtain

$$V_G(t) = V_G + \delta V_G(t). \tag{3.2}$$

Because gate voltage is time varying, band bending also will have an ac variation; $\delta \psi_s(t)$; thus

$$\psi_s(t) = \psi_s + \delta \psi_s(t) \tag{3.3}$$

where ψ_s is the time-independent band bending established by V_G. In addition, there will be a small-signal variation in Q_s, $\delta Q_s(t)$. If δQ_s is small enough to be in the small-signal regime, it can be estimated using only the first term of a Taylor series expansion of $Q_s(\psi_s)$

$$Q_s(t) = Q_s[\psi_s + \delta \psi_s(t)]$$
$$= Q_s(\psi_s) + \left(\frac{dQ_s}{d\psi_s}\right) \delta \psi_s(t) \tag{3.4}$$

where $Q_s(\psi_s)$ is the silicon surface charge density established by V_G. That is, from (3.4)

$$\delta Q_s(t) = \frac{dQ_s}{d\psi_s} \delta \psi_s(t)$$
$$= -C_s(\psi_s) \delta \psi_s(t) \tag{3.5}$$

where the low-frequency differential capacitance per unit area of the silicon $C_s(\psi_s)$ is

$$C_s(\psi_s) \equiv -\frac{dQ_s}{d\psi_s}. \tag{3.6}$$

Equation (3.5) is valid only at low frequencies because it assumes that $Q_s(t)$ is given by the equilibrium expression $Q_s(\psi_s)$ by substituting $\psi_s(t) = \psi_s + \delta \psi_s(t)$. At high frequencies this assumption fails because minority carrier density cannot change rapidly enough to maintain equilibrium. The high frequency case is discussed in Section 4.4.

Equations (3.2), (3.3), and (3.4) are substituted into (3.1). Then, the time independent terms are eliminated by subtracting (3.1) from its equilibrium version. The equilibrium version of (3.1), that is, with no ac excitation, is

$$C_{ox}(V_G - \psi_s) = -Q_s. \tag{3.7}$$

Making these substitutions and subtracting (3.7) from (3.1), we obtain

$$C_{ox}(\delta V_G - \delta\psi_s) = C_s(\psi_s)\delta\psi_s. \tag{3.8}$$

Rearranging terms, we find

$$\delta\psi_s = C_{ox}[C_{ox} + C_s(\psi_s)]^{-1}\delta V_G. \tag{3.9}$$

Finally, the total capacitance per unit area C, which relates δQ_s to the small-signal ac gate voltage, becomes

$$
\begin{aligned}
C &\equiv -\frac{\delta Q_s}{\delta V_G} \\
&= -\frac{\delta Q_s}{\delta\psi_s}\frac{\delta\psi_s}{\delta V_G} \\
&= C_s(\psi_s)C_{ox}[C_{ox} + C_s(\psi_s)]^{-1}
\end{aligned} \tag{3.10}
$$

or

$$\frac{1}{C} = \frac{1}{C_s(\psi_s)} + \frac{1}{C_{ox}}. \tag{3.11}$$

Equation (3.11) expresses the total capacitance per unit area of the MOS capacitor as the series combination of the silicon capacitance per unit area and the oxide capacitance per unit area. From (3.11), the equivalent circuit of the MOS capacitor at low frequencies is that given in Fig. 3.3. In Fig. 3.3 $C_s(\psi_s)$ is shown as a variable capacitor because it is bias dependent.

3.2.2 Calculation of Low Frequency Capacitance as a Function of Gate Bias

We calculate differential capacitance per unit area as a function of surface potential for the case where minority carriers follow the ac gate voltage at all values of gate bias. This case applies at low frequencies in silicon. We treat the case of uniform dopant impurity concentration in the silicon using the one-dimensional solution of the Poisson equation derived in Section 2.3.2(d). The differential capacitance C_s is the derivative of Q_s with respect to ψ_s. According to Gauss's law, Q_s is proportional to the field at the silicon surface. Rewriting (2.65) for the dimensionless electric field, we obtain

$$F(u_s, u_B) = 2^{1/2}[(u_B - u_s)\sinh u_B - \cosh u_B + \cosh u_s]^{1/2}. \tag{3.12}$$

As $F(u_s, u_B)$ is the dimensionless electric field in the silicon, it can be used

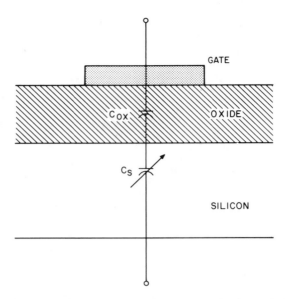

Fig. 3.3 Cross section of an MOS capacitor showing a simple equivalent circuit. The capacitor labeled C_s is shown variable to denote its bias dependence.

to calculate the silicon surface charge density Q_s obtained from (2.66)

$$Q_s = \frac{\epsilon_s}{\lambda_i}\left(\frac{kT}{q}\right) \text{Sgn}(u_B - u_s) F(u_s, u_B) \tag{3.13}$$

where $\text{Sgn}(u_B - u_s)$ has the sign of $(u_B - u_s)$. That is, $\text{Sgn}(u_B - u_s)$ is positive when $u_B - u_s$ is positive, and $\text{Sgn}(u_B - u_s)$ is negative when $u_B - u_s$ is negative.

(a) Discussion of Various Terms Contributing to the Field

Equation (3.13) appears to be rather complicated. Before proceeding with the calculation of C_s from (3.13), we digress for a moment to clarify the meaning of the terms in (3.12). From (2.43) and (2.45), $\exp(u_B) = n_B/n_i$ and $\exp(-u_B) = p_B/n_i$, where n_B and p_B are the electron and hole densities in the silicon bulk respectively so that

$$\sinh u_B = \frac{n_B - p_B}{2n_i} \tag{3.14}$$

and

$$\cosh u_B = \frac{n_B + p_B}{2n_i}. \tag{3.15}$$

From (2.46), $\exp(u_s) = n_s/n_i$ and $\exp(-u_s) = p_s/n_i$ so that

$$\cosh u_s = \frac{n_s + p_s}{2n_i}. \tag{3.16}$$

Substituting (3.14), (3.15), and (3.16) into (3.12), we obtain

$$F(u_s, u_B) = \left(\frac{n_B}{n_i}\right)^{1/2}\left[(u_B - u_s - 1) - \frac{p_B}{n_B}(u_B - u_s + 1) + \frac{n_s}{n_B} + \frac{p_s}{n_B}\right]^{1/2}. \tag{3.17}$$

For n-type, $n_B = N_D$ and $p_B/N_D = 0$ so that (3.17) becomes

$$F(u_s, u_B) = \left(\frac{N_D}{n_i}\right)^{1/2}\left[(u_B - u_s - 1) + \frac{n_s}{N_D} + \frac{p_s}{N_D}\right]^{1/2}. \tag{3.18}$$

Substituting (3.18) into (3.13) yields

$$Q_s = \frac{\epsilon_s}{\lambda_i}\left(\frac{kT}{q}\right)\text{Sgn}(u_B - u_s)\left(\frac{N_D}{n_i}\right)^{1/2}\left[(u_B - u_s - 1) + \frac{n_s}{N_D} + \frac{p_s}{N_D}\right]^{1/2}. \tag{3.19}$$

The first term in the bracket of (3.19), $(u_B - u_s - 1)$, is derived from the term $[(u_B - u_s)\sinh u_B - \cosh u_B]$ in (3.12). This term is dominant when n_s/N_D and p_s/N_D are small, which occurs in depletion. Neglecting these mobile carrier terms, (3.19) becomes

$$Q_s = Q_D = (2)^{1/2}qN_D\lambda_n(u_B - u_s - 1)^{1/2} \tag{3.20}$$

where λ_n is the extrinsic Debye length for n-type silicon defined by (2.87), namely

$$\lambda_n = \frac{kT}{q^2}(2N_Dn_i)^{-1/2}\frac{\epsilon_s}{\lambda_i}$$

$$= \left(\frac{kT}{q}\frac{\epsilon_s}{qN_D}\right)^{1/2}. \tag{3.21}$$

The depletion layer charge per unit area Q_D can be written in terms of depletion layer width w defined by

$$w = (2)^{1/2}\lambda_n(u_B - u_s - 1)^{1/2}$$

$$= (2)^{1/2}\lambda_n(-v_s + 1)^{1/2}. \tag{3.22}$$

Ordinarily $|v_s| \gg 1$, so that (3.22) agrees with (2.82) from the depletion approximation. Using (3.22) in (3.20), Q_D becomes

$$Q_D = qN_Dw. \tag{3.23}$$

The second and third terms in the brackets of (3.19), $n_s/N_D + p_s/N_D$, are derived from the $\cosh u_s$ term in (3.12). The second term, n_s/N_D, becomes important in accumulation. In strong accumulation ($u_s > u_B$), the second term is dominant and gives the accumulation layer charge density.

The third term, p_s/N_D, becomes important in inversion. In strong inversion ($u_s > 2u_B$), the third term is dominant and gives the inversion layer charge density. In inversion, Q_s is the sum of the inversion layer charge density qN_I and the depletion layer charge density Q_D

$$Q_s = qN_I + Q_D. \tag{3.24}$$

If (3.24) is combined with (3.20) for Q_D, with (3.19) for Q_s, and with (3.21) for λ_n, then qN_I becomes

$$qN_I = (2)^{1/2}qN_D\lambda_n\left\{\left[u_B - u_s - 1 + \frac{p_s}{N_D}\right]^{1/2} - (u_B - u_s - 1)^{1/2}\right\}. \qquad (3.25)$$

Equation (3.25) is approximate because the depletion layer is not quite as wide in inversion as (3.22) predicts because of the rapid change in potential within the inversion layer, which leaves a potential smaller than u_s across the depletion layer. Nonetheless, (3.25) is accurate to within 5% in weak inversion and is extremely accurate in strong inversion.[8]

For p-type, (3.12) becomes

$$F(u_s, u_B) = \left(\frac{N_A}{n_i}\right)^{1/2}\left[-(u_B - u_s + 1) + \frac{n_s}{N_A} + \frac{p_s}{N_A}\right]^{1/2} \qquad (3.26)$$

and the meaning of the terms are similar except that p_s/N_A gives rise to the accumulation layer charge density and n_s/N_A, to the inversion layer charge density.

For p-type silicon, the expression for depletion layer width, (3.22), becomes

$$w = (2)^{1/2}\lambda_p(u_s - u_B - 1)^{1/2}$$
$$= (2)^{1/2}\lambda_p(v_s - 1)^{1/2} \qquad (3.27)$$

where the extrinsic Debye length λ_p is

$$\lambda_p = \left(\frac{kT}{q}\frac{\epsilon_s}{qN_A}\right)^{1/2}. \qquad (3.28)$$

By analogy, the depletion layer charge per unit area is now

$$Q_D = -qN_Aw \qquad (3.29)$$

and also

$$-qN_I = -qN_A(2)^{1/2}\lambda_p\left\{\left[(u_s - u_B - 1) + \frac{n_s}{N_A}\right]^{1/2} - (u_s - u_B - 1)^{1/2}\right\}. \qquad (3.30)$$

(b) Calculation of the Low Frequency Capacitance of Silicon

Surface differential capacitance per unit area is

$$C_s = -\left(\frac{\delta Q_s}{\delta u_s}\right)\left(\frac{\delta u_s}{\delta \phi_s}\right) = -\left(\frac{q}{kT}\right)\left(\frac{\delta Q_s}{\delta u_s}\right). \qquad (3.31)$$

Rigorously, δQ_s is the ac variation in Q_s. We propose to use (3.13) for Q_s, which assumes that the carriers are in equilibrium with u_s. That is, (3.13) requires that both majority and minority carriers follow changes in ac band bending. This requirement restricts the validity of the derivation to low frequencies because minority carriers can follow only at low frequencies, as described in Section 4.3.

From (3.13) and (3.31), we obtain

$$C_s = -\operatorname{Sgn}(u_B - u_s)\left(\frac{\epsilon_s}{\lambda_i}\right)\frac{\delta}{\delta u_s}\,[F(u_s, u_B)]. \tag{3.32}$$

At low frequencies, ac changes are so slow that equilibrium is maintained. Therefore, the ac variation of (3.12) can be found by differentiation. Differentiating (3.12), we obtain

$$\frac{\delta}{\delta u_s}\,F(u_s, u_B) = (2)^{-1/2}\frac{-\sinh u_B + \sinh u_s}{[(u_B - u_s)\sinh u_B - \cosh u_B + \cosh u_s]^{1/2}}. \tag{3.33}$$

or

$$\frac{\delta}{\delta u_s}\,F(u_s, u_B) = \frac{\sinh u_s - \sinh u_B}{F(u_s, u_B)}. \tag{3.34}$$

Substituting (3.34) into (3.32) yields

$$C_s = -\operatorname{Sgn}(u_B - u_s)\left(\frac{\epsilon_s}{\lambda_i}\right)\frac{\sinh u_s - \sinh u_B}{F(u_s, u_B)}. \tag{3.35}$$

Equation (3.35) is the expression for the silicon capacitance to be used in (3.11).

The four equations that describe the ideal MOS capacitor are (3.11), (3.13), (3.35), and

$$V_G = -\frac{Q_s}{C_{ox}} + \frac{kT}{q}\,(u_B - u_s). \tag{3.36}$$

Equation (3.36) expresses how the applied gate bias divides between the SiO_2 and the silicon. The variables in (3.36) are the time-independent, quiescent components of the variables in (3.1).

To calculate a C-V curve from (3.11), (3.13), (3.35), and (3.36), it is simpler to choose v_s or ψ_s rather than V_G as the *independent variable*, although in measurements V_G is the independent variable. Equations (3.35) and (3.36) are functions of surface potential. It is convenient to express them as functions of band bending using the relation $v_s = u_s - u_B$ from (2.40) and (2.41). To calculate a C-V curve to compare to experiment, (3.35) is calculated as a function of band bending. Then, this result is used to calculate total capacitance as a function of band bending from (3.11). Finally, band bending is related to gate bias using (3.36) to obtain the C-V curve. Surface charge is calculated from (3.13) by substituting (3.18) into (3.13) for n-type and (3.26) into (3.13) for p-type.

Figure 3.4 shows a plot of ψ_s as a function of V_G with oxide thickness as parameter for both n-type and p-type. The curves in Fig. 3.4 were calculated from (3.36). For p-type, the curves saturate at higher values of positive bias once an inversion layer of electrons forms. Saturation occurs because this inversion layer charge electrostatically screens the depletion layer charge from further penetration of the applied electric field. Because

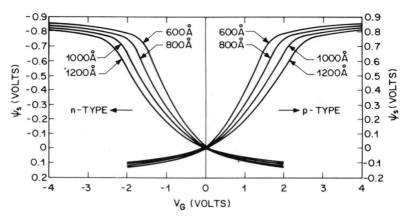

Fig. 3.4 Band bending as a function of gate bias with oxide thickness as parameter for both n-type and p-type silicon calculated from (3.36) and (3.13) for $N_A = N_D = 10^{16}$ cm^{-3}, with $T = 300°$K. Adapted from Goetzberger.[9] Copyright (1966), American Telephone and Telegraph Company. Reprinted by permission.

this screening depends on the inversion layer carrier density, which increases exponentially with band bending, band bending does not increase very much with increasing gate bias. When bias becomes negative, holes accumulate at the silicon surface and the surface charge is made up entirely of holes. This accumulation layer charge screens the silicon surface. Again, the accumulation layer charge density increases exponentially with band bending; thus band bending increases very slowly with increasing bias. The fastest change in ψ_s occurs between these two extremes when the silicon surface charge is made up almost entirely of ionized acceptors. To satisfy charge neutrality, band bending must increase to repel holes from the silicon surface and increase the ionized acceptor charge. However, the depletion layer width varies as the square root of band bending [see (3.27)], so band bending must increase with gate bias. The n-type is similar, except the inversion layer consists of holes rather than electrons, the ionized impurities are donors, the accumulation layer charge consists of electrons, and the valence band effective density of states is smaller than that of the conduction band. Thus the signs of ψ_s and V_G are opposite to what they are for p-type.

Figure 3.4 also shows that the change in ψ_s with V_G is faster the thinner the oxide. The reason for this dependence on oxide thickness is that a given electric field at the silicon surface, which determines the magnitude of the surface charge, occurs at a lower bias, the thinner the oxide film.

Figure 3.5 shows a calculated C-V curve for n-type with oxide thickness as parameter. Figure 3.5 shows that the thinner the oxide, the more rapidly capacitance changes with gate bias. The reason is that ψ_s changes more rapidly with gate bias is seen in Fig. 3.4. In Fig. 3.5 each calculated

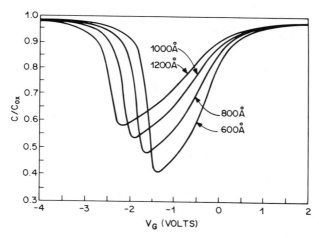

Fig. 3.5 Normalized capacitance as a function of bias with oxide thickness as parameter calculated using (3.11), (3.36), and (3.37) for $N_D = 10^{16}\,cm^{-3}$. Adapted from Goetzberger.[9] Copyright (1966), American Telephone and Telegraph Company. Reprinted by permission.

value of C has been divided by $C_{ox}A_g$, the total oxide layer capacitance (A_g = gate area). The resulting quantity is called the *normalized capacitance* of the MOS capacitor and is used frequently. Normalized capacitance is useful in comparing C-V curves because it is independent of gate area.

Figure 3.6 illustrates the operation of the MOS capacitor mathematically described by (3.11), (3.35), and (3.36). The C-V curve in Fig. 3.6a is divided into four different regions with the energy-band diagram for each region shown in Figs. 3.6b–e. In the energy-band diagrams, electrons fall down in energy and holes rise up as in Figs. 2.9 and 2.10. Here p-type silicon is used as the illustrative example, although the arguments are similar for n-type.

3.2.3 Description of the Low Frequency C-V Characteristics

The MOS capacitor is in thermal equilibrium at all values of gate bias below the oxide breakdown field. That is, no dc current flows when gate bias is applied, for reasons described in Section 2.3. Thus the Fermi level in the silicon always is flat all the way to the Si–SiO$_2$ interface as shown in Figs. 3.6b–e. At low frequencies the MOS capacitor also is in thermal equilibrium under small-signal ac excitation, provided that minority carriers can respond to variations in the ac field to prevent energy loss.*

---

*In practice, there will be some hole and electron traps at the Si–SiO$_2$ interface and in the bulk silicon. The system including these traps still will be in thermal equilibrium if all these traps immediately respond to the ac voltage.

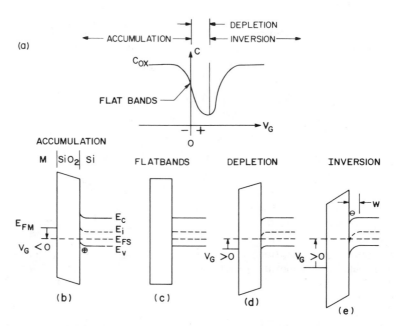

Fig. 3.6 To illustrate the operation of an MOS capacitor, capacitance as a function of bias, and the corresponding energy-band diagrams are shown in accumulation, depletion, inversion, and at flatbands. Energy values E_c and E_v are the conduction and valence band edges, respectively; E_i is the intrinsic Fermi level; E_{FS} is the Fermi level in the silicon; E_{FM} is the Fermi level in the metal; and w is depletion layer width.

Starting with a negative gate bias, the first region of the C-V curve (Fig. 3.6b) is called *accumulation*. Negative gate bias attracts holes to the silicon surface and the silicon bands bend up as shown in Fig. 3.6b. At very large negative gate bias, hole density at the silicon surface will greatly exceed hole density in the bulk. The large hole charge density at the silicon surface will contribute a large differential capacitance; that is, C_s will be large and $C_s \gg C_{ox}$ so that $C = C_{ox}$ from (3.11). As gate bias is made less negative, surface hole density will decrease, making C_s smaller. As a result, C becomes less than C_{ox}. When gate bias decreases to zero, we are at the *flatband* point on the C-V curve. Figure 3.6c shows the energy-band diagram at the flatband point.

As gate bias is made positive, holes are repelled from the silicon surface, resulting in the formation of a depletion layer of ionized acceptors. This bias range is called the *depletion region*, and the bands bend down as shown in Fig. 3.6d. As gate bias is made increasingly more positive, the depletion layer widens, making C_s smaller. Therefore, C becomes smaller [see (3.11)]. As gate bias is made more positive, surface hole density decreases whereas surface electron density increases, keeping the pn product a constant at the surface ($p_s n_s = n_i^2$). When the Fermi level crosses midgap (or, to be more exact, the intrinsic Fermi level E_i), surface hole and electron densities are both equal to n_i ($p_s = n_s = n_i$). This point is called the *onset of inversion*. As gate bias is made even more positive, surface electron density exceeds surface hole density and an *inversion layer* of electrons is formed. Figure 3.6*e* shows band bending in inversion. When $\psi_s \geq 2\phi_B$ and inversion layer electron density exceeds bulk acceptor density (i.e., $n_s \geq N_A$), the differential capacitance of the inversion layer becomes comparable to and then exceeds C_{ox} (i.e., $C_s \gg C_{ox}$) and C approaches C_{ox} asymptotically.

It is useful to divide the inversion regime into two parts because different phenomena characteristic of each are observed. The two parts are called *weak inversion* and *strong inversion*. Weak inversion starts at the onset of inversion when $n_s = p_s = n_i$ or $\psi_s = \phi_B$ and ends when minority carrier density equals ionized dopant impurity density or $\psi_s = 2\phi_B$. Strong inversion starts at $\psi_s = 2\phi_B$ and extends for $\psi_s > 2\phi_B$.

The general expression for C_s in (3.35) takes on simpler forms in accumulation, at flatbands, in depletion, and in inversion. To derive these simpler forms for n-type, substitute (3.18) for $F(u_s, u_B)$, (3.14) for $\sinh u_B$, and $\sinh u_s = (n_s - p_s)/2n_i$ from (2.46) into (3.35), to obtain

$$C_s = \frac{-\operatorname{Sgn}(u_B - u_s)C_{FBS}}{\sqrt{2}} \frac{(n_s/N_D - p_s/N_D - 1)}{[(u_B - u_s - 1) + n_s/N_D + p_s/N_D]^{1/2}}. \tag{3.37}$$

The parameter C_{FBS} introduced in (3.37) is called the silicon *flatband capacitance*. Comparison of (3.37) and (3.35) shows C_{FBS} to be given by

$$C_{FBS} = \frac{\epsilon_s}{\lambda_n} \tag{3.38}$$

where C_{FBS} is the value of C_s at flatbands, that is, when $v_s = 0$ or $u_s = u_B$, to be shown shortly.

By concentrating on accumulation, depletion, and inversion, we not only replace (3.37) with simpler expressions, but illustrate the underlying concepts. Moreover, the simplified relations are important because a great deal of the information about the MOS system is obtained from measurements confined to one or the other of these regimes. Equation (3.37) must be used without simplification in the transition regions between accumulation and depletion and in weak inversion.

(a) *Flatbands* ($v_s = 0$)

At flatbands, $p_s/N_D = 0$; thus (3.37) becomes

$$C_s = \frac{-\operatorname{Sgn}(u_B - u_s)C_{FBS}}{\sqrt{2}} \frac{(n_s/N_D - 1)}{[(u_B - u_s - 1) + n_s/N_D]^{1/2}}. \tag{3.39}$$

At flatbands, $v_s = 0$. From (2.46), $n_s/N_D = \exp(v_s)$ so that $n_s = N_D$ when $v_s = 0$. These values lead to the indeterminate result $C_s = 0/0$ when substituted into (3.39). Therefore, to calculate flatband capacitance, the exponential in (2.46) must be expanded in a power series about $v_s = 0$. Before making the power series expansion, (3.39) must be written as an explicit function of v_s. If we use (2.46) and the relation $u_s = u_B + v_s$, (3.39) becomes

$$C_s = \frac{\text{Sgn}(v_s)C_{FBS}}{\sqrt{2}} \frac{[\exp(v_s) - 1]}{[(-v_s - 1) + \exp(v_s)]^{1/2}}. \tag{3.40}$$

Expanding the exponential in (3.40) around $v_s = 0$ and retaining the first three terms, we have

$$\exp(v_s) = 1 + v_s + \frac{v_s^2}{2} + \cdots . \tag{3.41}$$

Substituting (3.41) into (3.40), we obtain

$$C_s = C_{FBS}\left(\frac{v_s}{2} + 1\right). \tag{3.42}$$

If we set $v_s = 0$, (3.42) becomes

$$C_s = C_{FBS} \tag{3.43}$$

$$C_{FBS} = \frac{\epsilon_s}{\lambda_p} \quad \text{for } p\text{-type}$$

and

$$C_{FBS} = \frac{\epsilon_s}{\lambda_n} \quad \text{for } n\text{-type}. \tag{3.44}$$

(b) Strong Accumulation

In strong accumulation, $n_s/N_D \gg p_s/N_D$, $n_s/N_D \gg 1$, and $n_s/N_D \gg |u_B - u_s - 1|$, so that (3.37) becomes

$$C_s = \frac{C_{FBS}}{\sqrt{2}}\left(\frac{n_s}{N_D}\right)^{1/2} = \frac{C_{FBS}}{\sqrt{2}} \exp\frac{v_s}{2} = C_A; \qquad v_s > 0. \tag{3.45}$$

Similarly, for p-type

$$C_A = \frac{C_{FBS}}{\sqrt{2}} \exp\frac{-v_s}{2}; \qquad v_s < 0. \tag{3.46}$$

(c) Depletion $(|v_s| \leq |u_B|)$

In depletion, $n_s/N_D = p_s/N_D = 0$, and if we use $u_s = u_B + v_s$, (3.37) becomes

$$C_s = \frac{C_{FBS}}{\sqrt{2}}(-v_s - 1)^{-1/2} = \frac{\epsilon_s}{w} = C_D; \qquad v_s < 0. \tag{3.47}$$

Equation (3.47) also applies for p-type in depletion, $v_s > 0$.

(d) Strong Inversion ($|v_s| \geq 2|u_B|$)

Finally, in strong inversion, $p_s/N_D \gg n_s/N_D$ and $p_s/N_D \gg (u_B - u_s - 1)$, so that (3.37) becomes

$$C_s = \frac{C_{FBS}}{\sqrt{2}} \left(\frac{p_s}{N_D}\right)^{1/2} = \frac{C_{FBS}}{\sqrt{2}} \frac{n_i}{N_D} \exp \frac{-v_s}{2} = C_I; \qquad v_s < 0. \qquad (3.48)$$

Similarly, for p-type

$$C_I = \frac{C_{FBS}}{\sqrt{2}} \frac{n_i}{N_A} \exp \frac{v_s}{2}; \qquad v_s > 0. \qquad (3.49)$$

(e) Summary

Figure 3.7 shows a plot of (3.37), where $C_o = \epsilon_s/\lambda_i$, with the various regions indicated. In strong inversion, (3.48) and (3.49) are good approximations, and C_s increases exponentially with v_s. In depletion, (3.47) is

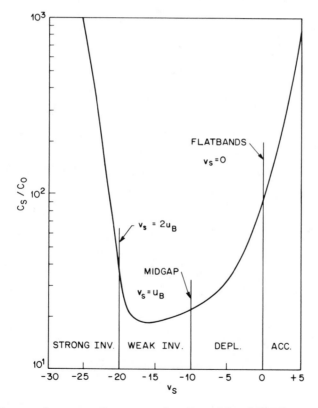

Fig. 3.7 Silicon surface capacitance as a function of band bending calculated from (3.37) for a donor concentration of 2.8×10^{14} cm^{-3} ($C_0 = \epsilon_s/\lambda_i$; and $T = 300°$K).

a good approximation, and C_s varies as the square root of v_s. At flatbands, C_s is given by (3.43) with C_{FBS} from (3.43) or (3.44). Finally, in accumulation, (3.45) and (3.46) are good approximations, and C_s increases exponentially with $-v_s$.

Flatband capacitance is useful for comparing measured and calculated C-V curves. It is a universally accepted reference point on the C-V curve, although there are many problems in experimentally determining it accurately as described in Section 10.5.

Figure 3.8 shows the normalized flatband capacitance C_{FB}/C_{ox} as a function of oxide thickness with acceptor density as parameter. For p-type, $C_{FBS} = \epsilon_s/\lambda_p$, and from (3.11), $C_{FB}/C_{ox} = C_{FBS}/(C_{FBS} + C_{ox})$. This expression shows that C_{FB}/C_{ox} will increase with increasing oxide thickness as seen in Fig. 3.8.

3.2.4 Capacitance Minimum

The silicon surface capacitance as a function of v_s in Fig. 3.7 goes through a minimum. In this section we calculate the normalized surface potential $u_{s\,min}$ at which capacitance is a minimum and the minimum capacitance as a function of $u_{s\,min}$. $u_{s\,min}$ is found from the condition; $dC_s(u_{s\,min})/du_s = 0$. Differentiating (3.35) at $u_s = u_{s\,min}$ and using (3.34), we obtain the following relation[10] between $u_{s\,min}$ and u_B

$$(\cosh u_{s\,min})^{1/2} = \frac{\sinh u_{s\,min} - \sinh u_B}{F(u_{s\,min}, u_B)}. \tag{3.50}$$

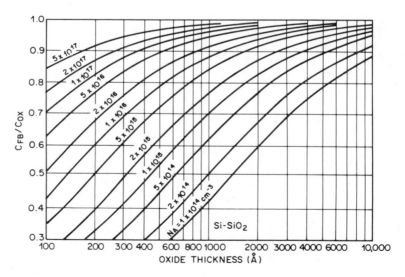

Fig. 3.8 Normalized flatband capacitance as a function of oxide thickness with acceptor concentration as parameter calculated using (3.11) and (3.43) ($T = 300°K$). After Goetzberger.[9] Copyright (1966), American Telephone and Telegraph Company. Reprinted by permission.

Equation (3.50) shows that $u_{s\,min}$ is a transcendental function of u_B or doping concentration.

To calculate the minimum capacitance,[10] we substitute (3.50) into (3.35) evaluated at $u_s = u_{s\,min}$ and obtain

$$C_s(u_{s\,min}) = \frac{\epsilon_s}{\lambda_i}(\cosh u_{s\,min})^{1/2}. \qquad (3.51)$$

Equation (3.51) depends implicitly on doping concentration through the dependence of $u_{s\,min}$ on doping concentration in (3.50).

Figure 3.9 is a plot of C_s as a function of v_s calculated from (3.35) using $v_s = u_s - u_B$ with acceptor concentration as parameter at 300°K. Figure 3.9 shows that $v_{s\,min}$ increases with increasing doping concentration. The variable $u_{s\,min}$ is related to $v_{s\,min}$ by the relation $u_{s\,min} = u_B + v_{s\,min}$. Also from Fig. 3.9: (1) the minimum capacitance increases with increasing doping concentration; and (2) a larger spread of v_s is needed to make the silicon

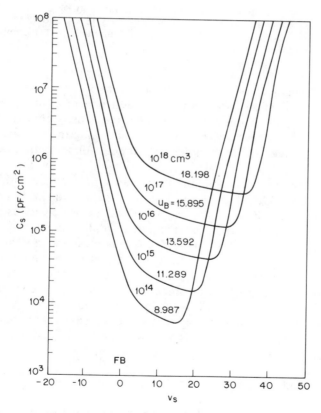

Fig. 3.9 Silicon surface capacitance as a function of total band bending with acceptor concentration as parameter calculated using (3.37) ($T = 300$°K). The value of u_B corresponding to each value of N_A also is shown (*FB* denotes flatbands). After Sah.[10]

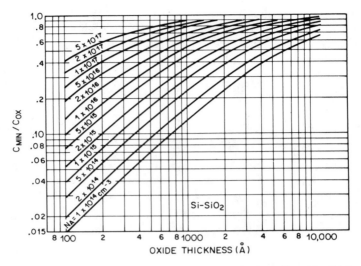

Fig. 3.10 Normalized minimum capacitance as a function of oxide thickness with acceptor concentration as parameter calculated using (3.11) and (3.51) ($T = 300°$K). After Goetzberger.[9] Copyright (1966), American Telephone and Telegraph Company. Reprinted by permission.

surface go from accumulation to inversion as doping concentration increases. This last effect is simply a consequence of Gauss's law, which requires a greater v_s change for a greater charge density change. Therefore, the measured C-V curve will be stretched out along the voltage axis more and more, at a given oxide thickness, as doping concentration is increased as seen from (3.11).

Figure 3.10 shows total minimum capacitance normalized with respect to oxide capacitance C_{min}/C_{ox}, the ratio of the smallest to the largest capacitance, as a function of oxide thickness with acceptor density as parameter. Then $C_{s\,min}$ is calculated from (3.51) and then total capacitance is calculated using (3.11). Figure 3.10 shows that C_{min}/C_{ox} increases with oxide thickness for a given acceptor concentration, which is just a consequence of (3.11). Figure 3.11 shows the bias V_{min} at which capacitance goes through a minimum as a function of oxide thickness with acceptor concentration as parameter. Increasing oxide thickness increases the gate bias at which total capacitance passes through a minimum. The reason for this gate bias dependence on oxide thickness is that a given band bending, in this case $v_{s\,min}$, occurs at a higher gate bias as seen in Fig. 3.4.

3.2.5 Intrinsic Case

For intrinsic silicon, $u_B = 0$ and the Fermi level is, for all practical purposes, at midgap. When $u_B = 0$, (3.50) will be satisfied only for $u_{s\,min} =$

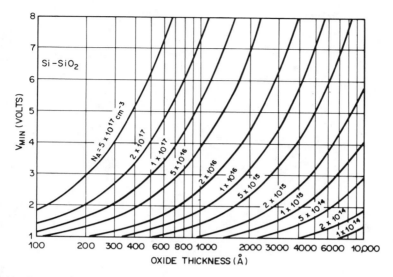

Fig. 3.11 Bias at which capacitance goes through a minimum as a function of oxide thickness with acceptor concentration as parameter. After Goetzberger.[9] Copyright (1966), American Telephone and Telegraph Company. Reprinted by permission.

$u_B = 0$ or at flatbands. Thus the minimum capacitance in intrinsic silicon occurs at the flatband point; that is, the silicon surface is intrinsic at the flatband point. Thus the three points on the capacitance curve, the flatband point, the intrinsic surface point, and the capacitance minimum all coincide.

Minimum capacitance for the intrinsic case is $C_{s\,min} = \epsilon_s/\lambda_i$ from (3.51). This result is analogous to (3.43) for the extrinsic case, except that the Debye length in the formula, instead of being the extrinsic Debye length λ_n, is the intrinsic Debye length λ_i.

3.2.6 Degenerate Range of Bias

All the relationships derived are for the nondegenerate case. It is possible to apply sufficiently large gate bias to bend the bands enough to make the silicon surface degenerate either in strong accumulation or strong inversion as shown in the inserts in Fig. 3.12 for p-type silicon. Degeneracy occurs in accumulation when the Fermi level is near the valence band edge and in inversion when the Fermi level is near the conduction band edge at the surface.

The bias range over which the silicon surface is degenerate will not be a region of interest for either device application or measurements of interfacial and oxide properties. However, it is important to know the limits over which the approximation of nondegeneracy is valid. We ignore

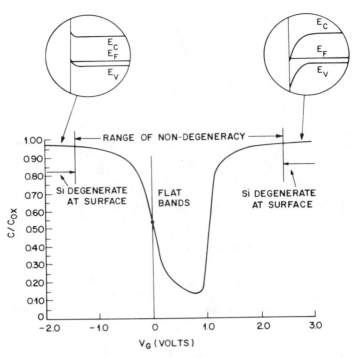

Fig. 3.12 Normalized capacitance as a function of bias, calculated for an acceptor concentration of 2×10^{16} cm^{-3} and an oxide thickness of 100 Å, showing the gate bias range over which the C-V curve is nondegenerate ($T = 300°$K). The inserts show band bending at the onset of degeneracy in strong accumulation and strong inversion.

surface quantization effects in this treatment, as the criterion for the onset of degeneracy allows for a sufficient definition of the useful extent of the C-V curve.

The portion of the C-V curve in the degenerate regime is very small compared to the portion in the nondegenerate regime. To compare, we calculate C_s at the onset of degeneracy in a p-type sample in strong inversion. Electron density at the onset of degeneracy at 300°K is 6×10^{19} cm^{-3}.[11] When $n_s = 6 \times 10^{19}$ cm^{-3} in strong inversion, C_s from (3.37) will be

$$C_s = \frac{\epsilon_s}{2\lambda_i}\left(\frac{n_s}{n_i}\right)^{1/2} = 8.5 \times 10^{-6} \text{ F/cm}^2 \tag{3.52}$$

using $n_i = 10^{10}$ cm^{-3} and $\lambda_i = 4.8 \times 10^{-3}$ cm at 290°K.

From (3.11), we write

$$\frac{C}{C_{ox}} = \frac{C_s}{C_s + C_{ox}}. \tag{3.53}$$

For a $1000 \,\text{Å}$ thick oxide, C_{ox} will be $C_{ox} = 3.4 \times 10^{-13}/1 \times 10^{-5} = 3.4 \times 10^{-8} \,\text{F/cm}^2$. Substituting the values of C_s and C_{ox} just calculated into (3.53), we obtain $C/C_{ox} = 0.993$. For an oxide thickness of $100 \,\text{Å}$, $C_{ox} = 3.4 \times 10^{-7} \,\text{F/cm}^2$ and $C/C_{ox} = 0.934$. Thus the thinner the oxide, the smaller the nondegenerate bias range. Hole density at the onset of degeneracy at the silicon surface in accumulation will not be exactly the same as electron density at degeneracy. Because the electron density of states is 2.7 times the hole density of states, the degeneracy point has to be recalculated for holes.

Figure 3.12 shows an ideal C-V curve calculated from (3.35) and (3.11) and also shows the bias points at which the silicon surface becomes degenerate calculated using (3.45) and (3.49). Although oxide thicknesses of $1000 \,\text{Å}$ or more are often used, an oxide thickness of $100 \,\text{Å}$ was chosen in calculating the C-V curve in Fig. 3.12 as a worst case. Figure 3.12 shows that the silicon is nondegenerate at the surface over practically the entire gate bias range. For oxide thickness ($\le 1000 \,\text{Å}$), the gate bias range over which the silicon is degenerate at the surface decreases somewhat at the extreme ends of the C-V curve.

3.2.7 Impurity Ionization

As temperature is reduced below room temperature, the Fermi level for any initial doping concentration moves closer to the majority carrier band edge as seen in Fig. 2.5. At temperatures in the 40–50°K range, for normal doping concentrations used in most integrated circuit and device applications, the Fermi level will pass through the dopant impurity level, causing the dopant impurity atoms to go from the completely ionized state to the completely neutral state. The exact temperature at which dopant impurities begin de-ionizing will increase with increasing initial doping concentration and with the dopant impurity energy level. We do not consider temperature ranges where partial ionization of donor or acceptor impurities becomes important. The reason for this neglect is that for device applications, the properties of the MOS system can be adequately measured at temperatures well above the de-ionization level. This state of affairs is fortunate because the greater complexity of analysis, in the low de-ionization temperature range, would make it considerably more difficult to interpret measurements, and the characteristics of any devices in an integrated circuit would be radically altered.

We have calculated C_s and Q_s for the case where donor and acceptor impurities are fully ionized. A more general treatment of the surface charge problem is given by Seiwatz and Green,[12] who take into account partial ionization of dopant impurities both in the neutral bulk and in the surface charge region in solving the Poisson equation. This treatment is valid at low temperatures and moderate dopant levels. However, their treatment is not valid for the degenerately doped case because dopant impurities no longer introduce a discrete level in the bandgap. Rather, at heavy doping densities, the dopant impurity level is broadened to form a narrow band of energy levels, thus invalidating the treatment due to Seiwatz and Green as pointed out by Voshchenkov.[13]

Gray and Brown[14] have measured the effect of dopant impurity de-ionization on the C-V characteristics of an MOS capacitor that has a p-type silicon substrate, and Sah[10] has calculated the effect of dopant impurity de-ionization on the C-V characteristics of a silicon MOS capacitor.

3.3 SILICON BAND BENDING

It is useful to know silicon band bending as a function of gate bias to (1) determine the energy scale for interface trap level density and capture probability, (2) calculate interface trap capture probability, (3) locate the bias value corresponding to flatbands, (4) delineate accumulation, depletion, weak inversion, and strong inversion, and (5) verify the expression for silicon surface capacitance.

There are two ways of directly measuring silicon band bending as a function of gate bias: the Q-V method and the low frequency C-V method. An indirect method, the high frequency capacitance method, uses comparison of a theoretical high frequency C-ψ_s curve with a measured high frequency C-V curve. This method is discussed later, in Section 8.2, as it requires more background.

3.3.1 *Q-V Method*

In the Q-V method silicon surface band bending is obtained as a function of gate bias directly from a measurement of the charge on the MOS capacitor as a function of gate bias.[15] In common with the low frequency C-V method discussed next, the Q-V method requires knowledge of neither the doping profile, nor the interface trap level density, nor the oxide charge density. The Q-V method is less convenient than the low frequency C-V method, but it has the potential advantage of greater accuracy.

The Q-V method is used as follows. A bias-independent capacitor C_i is connected in series with the MOS capacitor [see Section 12.5.2(c)]. There is a parasitic capacitance across the MOS capacitor, C_w, consisting of the capacitance of the sample holder and the connecting wires. A bias V_a is applied across C_i in series with the parallel combination of C_w and the MOS capacitor (see Fig. 12.8). After steady state is reached, V_a and the voltage across C_i, V_i, is measured as described in Section 12.5.2(c). From the known values of V_a and V_i, V_G is

$$V_G = V_a - V_i. \qquad (3.54)$$

The band bending ψ_s can be found in terms of measured voltages and capacitances. To begin from a definite state of the system, all the capacitors are short circuited so that $V_a = V_i = 0$. Then there is an unknown charge on each capacitor due to work function differences. We define ΔQ_i, ΔQ_w, and ΔQ_G as the charges on the three capacitors relative to these initial values. Then charge conservation at the node common to all three capacitors requires that

$$\Delta Q_i = \Delta Q_w + \Delta Q_G. \tag{3.55}$$

The definitions of the capacitances C_i and C_w are

$$\Delta Q_i = C_i V_i \tag{3.56}$$

$$\Delta Q_w = C_w (V_a - V_i). \tag{3.57}$$

Substituting (3.56) and (3.57) into (3.55), we obtain the basic relation of the Q-V method

$$\Delta Q_G = C_i V_i - C_w (V_a - V_i). \tag{3.58}$$

Using (3.58), the charge on the MOS capacitor relative to its value in the short circuited condition, ΔQ_G, can be plotted versus gate bias, (3.54). By numerical differentiation of (3.58) with respect to gate bias, the low frequency capacitance can be found.

The plot of ΔQ_G versus V_G is the only experimental information obtained from the Q-V method. However, theory relates ΔQ_G to band bending at the silicon surface, ψ_s, allowing a ψ_s versus V_G plot to be made. Integrating Poisson's equation in the oxide we find that

$$Q_G = C_{ox}(\psi_G - \psi_s) - \frac{1}{x_o} \int_0^{x_o} dx\, x\, \rho_{ot}(x) \tag{3.59}$$

where $\rho_{ot}(x)$ is the volume density of oxide trapped charge a distance x into the oxide from the Si–SiO$_2$ interface and ψ_G is the gate potential, related to the gate bias by

$$V_G = \psi_G + W_{ms} \tag{3.60}$$

where W_{ms} is the gate-silicon work function difference described in Section 10.4. Using (3.60) in (3.59) and setting $V_G = 0$, we obtain the initial value of Q_G, $Q_G(0)$

$$Q_G(0) = C_{ox}[- W_{ms} - \psi_s(0)] - \frac{1}{x_o} \int_0^{x_o} dx\, x\, \rho_{ot}(x) \tag{3.61}$$

where $\psi_s(0)$ is the band-bending when $V_G = 0$. Subtracting (3.61) from (3.59), ΔQ_G is given by

$$\Delta Q_G = C_{ox}[V_G - \psi_s + \psi_s(0)]. \tag{3.62}$$

Using (3.62) in (3.58) yields the band bending as a function of gate bias as

$$\psi_s = \psi_s(0) + V_G\left(1 + \frac{C_w}{C_{ox}}\right) - V_i\left(\frac{C_i}{C_{ox}}\right). \tag{3.63}$$

Thus ψ_s versus V_G is determined to within the additive constant $\psi_s(0)$ by varying V_a and using (3.54) for V_G and (3.63) for ψ_s.

If $\psi_s(0)$ could be measured, the Q-V method would be the most accurate way to determine ψ_s versus V_G. Regardless of the error in $\psi_s(0)$, the Q-V method is the most accurate means to obtain the *shape* of the ψ_s versus V_G curve [see Section 10.5.6].

3.3.2 *C-V* Method

Silicon surface band bending also can be obtained from a thermal equilibrium C-V curve[16] for arbitrary interface trap level density and doping profile. Differentiating (3.62) with respect to V_G yields

$$\frac{d\psi_s}{dV_G} = 1 - \frac{C_{LF}}{C_{ox}} \tag{3.64}$$

where $C_{LF} = dQ_G/dV_G$. Equations (3.58) and (3.64) are valid only when the system is in thermal equilibrium, that is, when both interface traps and minority carriers follow ac gate voltage and changes in gate bias. Therefore, C_{LF} either must be calculated from a Q-V measurement using (12.10) or must be measured by the low-frequency C-V method described in Sections 8.2.5 and 12.5.2. Integrating (3.64) yields

$$\psi_s(V_G) - \psi_s(V_{FB}) = \int_{V_{FB}}^{V_G} \left[1 - \frac{C_{LF}}{C_{ox}}\right] dV \tag{3.65}$$

where the *flatband voltage*, V_{FB}, is defined by $\psi_s(V_{FB}) = 0$.

Integrating (3.65) from V_{FB} to gate biases in accumulation and then from V_{FB} to gate biases in depletion and inversion gives the experimental ψ_s versus V_G curve, provided that V_{FB} is independently determined.* The integration of (3.65) must be done numerically, usually an accurate procedure.

The C-V method is convenient, but the Q-V method can be more accurate under conditions listed in Section 12.5.2(c). The comparative advantages and disadvantages of these techniques lie in the instrumentation and how measurements are made rather than in the theory developed here. The advantages and the disadvantages are discussed in Section 12.5.2(c).

---

*Equation (3.65) may be used starting from any gate bias, as discussed in Section 8.2.5. The determination of V_{FB} is discussed in Section 10.5.

3.3.3 Determination of Silicon Bandgap Energy

In Section 8.2.5 it is shown how interface trap level density D_{it} can be determined as a function of gate bias using the high-low frequency capacitance method. In Section 5.5 it is shown how D_{it} and the interface trap time constants τ_p or τ_n can be found as functions of gate bias using the conductance method. To complete the interface trap analysis, the position of the Fermi level with respect to the majority carrier band edge at the silicon surface must be determined as a function of gate bias. Figure 3.13 shows how a given band bending ψ_s is related to a given position in the silicon bandgap for n-type. From the doping density, the Fermi level is a distance $\phi_B = (kT/q) \ln N_D/n_i$ from (2.40) above the intrinsic level E_i in the bulk silicon. The intrinsic level is everywhere parallel to the conduction and valence bands. Thus the intrinsic level bends toward the Fermi level a distance ψ_s at the silicon surface. Consequently, at the interface, the Fermi level is $\psi_s - \phi_B$ above the intrinsic level. Finally, the intrinsic level is approximately half the bandgap $E_g/2q$ below the majority carrier band edge (conduction band in the illustrative example). Therefore, the energy position opposite the Fermi level at the interface is located at $E_F = E_T$ where

$$\frac{E_c - E_T}{q} = \frac{E_g}{2q} + \psi_s - \phi_B. \tag{3.66}$$

The bandgap is scanned by varying ψ_s, which is accomplished by changing gate bias. The final problem is to relate $(E_c - E_T)/q$ to gate bias. This relation is determined from ψ_s versus V_G obtained as described in Sections 3.3.1 or 3.3.2. For p-type, $(E_c - E_T)/q$ in (3.66) becomes $(E_T - E_v)/q$.

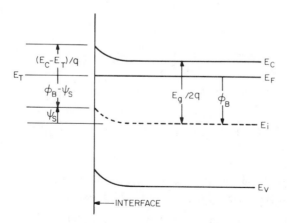

Fig. 3.13 Band-bending diagram of n-type silicon showing how the position of the Fermi level at the silicon surface is related to band bending. Arrows pointing up denote negative potential; arrows pointing down denote positive potential.

3.4 SUMMARY OF USEFUL EQUATIONS

This section contains a summary of the most useful equations derived in this chapter.

1 Total capacitance:

$$C = \frac{C_s C_{ox}}{C_s + C_{ox}}.$$

2 Dimensionless electric field:

$$F(v_s, u_B) = \left(\frac{N_D}{n_i}\right)^{1/2}\left[-(v_s + 1) + \exp(v_s) + \left(\frac{n_i}{N_D}\right)^2 \exp(-v_s)\right]^{1/2} \quad \text{for } n\text{-type.}$$

and

$$F(v_s, u_B) = \left(\frac{N_A}{n_i}\right)^{1/2}\left[(v_s - 1) + \exp(-v_s) + \left(\frac{n_i}{N_A}\right)^2 \exp(v_s)\right]^{1/2} \quad \text{for } p\text{-type.}$$

3 Silicon surface charge:

$$Q_s = \text{Sgn}(-v_s)\frac{\epsilon_s}{\lambda_i}\left(\frac{kT}{q}\right)F(v_s, u_B).$$

4 Dependence of applied bias on band bending:

$$V_G = -\frac{Q_s(\psi_s)}{C_{ox}} + \psi_s.$$

5 Extrinsic Debye length:

$$\lambda_n = \left(\frac{\epsilon_s kT}{q^2 N_D}\right)^{1/2} \quad \text{for } n\text{-type,}$$

and

$$\lambda_p = \left(\frac{\epsilon_s kT}{q^2 N_A}\right)^{1/2} \quad \text{for } p\text{-type.}$$

6 Low frequency silicon surface capacitance:

$$C_s = \frac{C_{FBS}}{\sqrt{2}}\frac{\exp(v_s) - (n_i/N_D)^2 \exp(-v_s) - 1}{[-(v_s + 1) + \exp(v_s) + (n_i/N_D)^2 \exp(-v_s)]^{1/2}} \quad \text{for } n\text{-type,}$$

and

$$C_s = \frac{C_{FBS}}{\sqrt{2}}\frac{1 - \exp(-v_s) + (n_i/N_A)^2 \exp(v_s)}{[(v_s - 1) + \exp(-v_s) + (n_i/N_A)^2 \exp(v_s)]^{1/2}} \quad \text{for } p\text{-type.}$$

7 Silicon surface capacitance in strong accumulation:

$$C_A = \frac{C_{FBS}}{\sqrt{2}}\exp\frac{v_s}{2}; \qquad v_s > 0 \quad \text{for } n\text{-type,}$$

and

$$C_A = \frac{C_{FBS}}{\sqrt{2}} \exp\frac{-v_s}{2}; \qquad v_s < 0 \quad \text{for } p\text{-type.}$$

8   Silicon surface capacitance at flatbands:

$$C_{FBS} = \frac{\epsilon_s}{\lambda_n}; \qquad v_s = 0 \quad \text{for } n\text{-type,}$$

and

$$C_{FBS} = \frac{\epsilon_s}{\lambda_p}; \qquad v_s = 0 \quad \text{for } p\text{-type.}$$

9   Silicon surface capacitance in depletion:

$$C_D = \left(\frac{\epsilon_s}{(2)^{1/2}\lambda_n}\right)[-(v_s + 1)]^{-1/2}; \qquad v_s < 0 \quad \text{for } n\text{-type,}$$

$$= \frac{\epsilon_s}{w}$$

and

$$C_D = \left(\frac{\epsilon_s}{(2)^{1/2}\lambda_p}\right)[(v_s - 1)]^{-1/2}; \qquad v_s > 0 \quad \text{for } p\text{-type}$$

$$= \frac{\epsilon_s}{w}.$$

10   Silicon surface capacitance in strong inversion (low frequency):

$$C_I = \frac{C_{FBS}}{\sqrt{2}} \frac{n_i}{N_D} \exp\frac{-v_s}{2}; \qquad v_s < 0 \quad \text{for } n\text{-type,}$$

$$C_I = \frac{C_{FBS}}{\sqrt{2}} \frac{n_i}{N_A} \exp\frac{v_s}{2}; \qquad v_s > 0 \quad \text{for } p\text{-type.}$$

REFERENCES

1   J. L. Moll, *Institute of Radio Engineers Wescon Convention Record*, Part 3 (1959), p. 32.
2   W. G. Pfann and C. G. B. Garrett, *Proc. Inst. Radio Eng.*, **47**, 2011 (1959).
3   D. R. Frankl, *Solid-State Electron.*, **2**, 71 (1961).
4   R. Lindner, *Bell Syst. Tech. J.*, **41**, 803 (1962).
5   L. M. Terman, *Solid-State Electron.*, **5**, 285 (1962).
6   K. Lehovec, A. Slobodskoy, and J. L. Sprague, *Phys. Stat. Solids*, **3**, 447 (1963).
7   A. S. Grove, B. E. Deal, E. H. Snow, and C. T. Sah, *Solid-State Electron.*, **8**, 145 (1965).
8   J. R. Brews, *Solid-State Electron.*, **21**, 345 (1978).
9   A. Goetzberger, *Bell Syst. Tech. J.*, **45**, 1097 (1966).

10   C. T. Sah, Solid State Electronics Laboratory Technical Report No. 1, Electrical Engineering
     Laboratory, University of Illinois (1964).
11   W. Shockley, *Electrons and Holes in Semiconductors*, D. Van Nostrand, Princeton, N.J.,
     1950, Chapter 10.
12   R. Seiwatz and M. Green, *J. Appl. Phys.*, **29**, 1034 (1058).
13   A. M. Voshchenkov, private communication.
14   P. V. Gray and D. M. Brown, *Appl. Phys. Lett.*, **13**, 247 (1968).
15   K. Ziegler and E. Klausmann, *Appl. Phys. Lett.*, **26**, 400 (1975).
16   C. N. Berglund, *IEEE Transact. Electron Devices*, **ED-13**, 701 (1966).

4

Metal Oxide Silicon Capacitor at Intermediate and High Frequencies

4.1 INTRODUCTION

In Chapter 2 the depletion layer charge density, the electron and hole densities, and the band bending in the silicon substrate of an MOS capacitor were determined as functions of gate bias. Then in Chapter 3 a small ac voltage was superposed on the gate bias and the C-V characteristics of the MOS capacitor were determined under the condition that both electron and hole distributions in the silicon responded instantaneously to the ac gate voltage.

In this chapter we consider in more detail the response of both majority and minority carriers to the ac gate voltage. This response determines the high frequency C-V characteristics of the MOS capacitor in inversion, as described in Section 4.4. Thus Chapter 3 and this chapter are necessary for understanding many of the measurement methods using high and low frequency C-V curves described in later chapters.

Much of this chapter is about bulk traps. In the absence of an external source of minority carriers, and near or below room temperature, it is bulk traps that determine minority carrier response. In addition, in inversion, bulk traps dominate the ac loss in the MOS capacitor. Thus, bulk traps determine the behavior of the conductance in strong inversion.

Our treatment of bulk traps is very similar to the treatment of interface traps in Chapter 5. Thus, using the same formalism we are able to discuss the conductance in strong inversion, where bulk traps dominate, and in depletion, where interface traps usually dominate. The formalism also covers weak inversion, where both types of traps contribute, but this regime is complex and is not discussed in detail.

In Section 4.2 we show that majority carriers respond instantaneously to the ac gate voltage for frequencies of interest in this book (0–1 MHz). This fast response is important to both capacitance and equivalent parallel conductance versus gate bias characteristics.

Minority carrier response is much slower and certainly not instantaneous over the frequency range of interest. Minority carrier response is important mainly in inversion. At room temperature, minority carrier response is governed by bulk traps having energy levels in the silicon bandgap near midgap and spatially distributed throughout the silicon depletion layer. In Section 4.3 we discuss how bulk traps determine minority carrier response. Above room temperature, diffusion determines minority carrier response, as discussed in Section 4.3.

In Sections 4.3 and 4.4, free carriers are assumed to be in equilibrium with the gate bias. This assumption is valid for slow gate bias changes. However, when minority carrier lifetime is very long ($> 10 \, \mu\text{sec}$), when gate bias changes are very rapid, or at low temperatures, minority carriers do not remain in equilibrium under gate bias changes. Therefore, we have included a discussion of minority carrier response to gate bias changes in Section 4.5.

The equivalent circuit for the MOS capacitor is used throughout this chapter because it is easier to relate measured admittance to an equivalent circuit than to a complex algebraic formula. All analyses are restricted to *small-signal steady-state* response.* This response is the simplest to evaluate. Except for Section 4.5, those special cases where large signals are used or transient measurements are made are treated in later chapters.

Analysis is for uniform doping concentration only. Although nonuniform doping profiles result from ion implantation or from impurity redistribution during oxidation as discussed in Section 14.4.2, the uniform case is of such major importance that it warrants detailed discussion. The effect of nonuniform impurity distributions on the *C-V* curve is discussed briefly in Section 4.4.4.

Some relations are derived for *n*-type and others for *p*-type. Arguments for each type are the same, so discussion of both types is not necessary.

---

*Small-signal means analysis may be linearized in all ac amplitudes. For a definition of steady state, see the introduction to Chapter 3.

4.2 MAJORITY CARRIER RESPONSE TIME

In accumulation and depletion, capacitance arises from *majority* carrier flow in and out of the silicon depletion layer in response to an ac gate voltage. Majority carrier charge will follow ac gate voltage as long as the period of the ac voltage is much longer than the *dielectric relaxation time* of silicon,[1] τ_D, ($\tau_D = \Omega \epsilon_s$ where Ω = silicon resistivity). That is, majority carriers will respond when $1/\omega \gg \tau_D$, where ω is the angular frequency of the ac voltage.

To derive majority carrier response time from a simple physical argument, imagine a localized fluctuation in potential of one thermal unit kT/q. This fluctuation results in nonequilibrium; thus majority carriers must rearrange themselves to restore equilibrium. The disturbance extends over a volume of about a local Debye length in radius λ where

$$\lambda = \left(\frac{kT\epsilon_s}{q^2 n} \right)^{1/2} \tag{4.1}$$

and n = local majority carrier density per unit volume. The field driving the carriers is about $(kT/q)/\lambda$. The time for majority carriers to flow a distance λ from the undisturbed environment to the center of disturbance is the response time τ_{maj}:

$$\begin{aligned} \tau_{maj} &= \frac{\lambda^2}{\mu(kT/q)} \\ &= \frac{\epsilon_s}{q\mu n} \\ &= \frac{\epsilon_s}{\sigma} \end{aligned} \tag{4.2}$$

where μ = carrier mobility (cm^2/V-sec) and $\sigma = q\mu n$ is the conductivity of a region with n carriers per unit volume.

To estimate τ_{maj} for silicon, we use $\epsilon_s = 1.04 \times 10^{-12}$ F/cm. A typical surface mobility for electrons is 600 cm^2/V-sec. Consequently, in a depleted MOS capacitor (n-type) biased near midgap, we find, near the interface, that $n \approx 10^{10}$ cm^{-3} and τ_{maj} (interface) $\approx 10^{-6}$ sec. Near the depletion layer edge where $n \approx$ dopant density $= N_D = 10^{16}$ cm^{-3}, for example, τ_{maj} (depletion layer edge) $\approx 10^{-12}$ sec. For this case, where $n = N_D$, τ_{maj} equals the dielectric relaxation time τ_D.[1]

The frequencies of interest in this book extend to the megahertz range, so from midgap to inversion those majority carriers near the interface may not follow the ac gate voltage. However, this failure will not affect the high frequency capacitance. This capacitance is dominated by majority carriers near the depletion layer edge that do follow. In conductance measurements, for band bending from midgap to weak inversion the measurement frequencies of interest are low, less than 1 kHz, so even majority carriers at the interface follow the ac gate voltage. Thus the approximation of *instantaneous* majority carrier response is satisfactory for all the measurements considered in this book.*

To establish the estimate given by (4.2) for τ_{maj} more rigorously, the equations governing response of the silicon to a localized disturbance must be solved. These equations are

1 Poisson's equation for band bending v is

$$\nabla^2 v = -\left(1 - \frac{n}{N_D}\right)\lambda_n^{-2}, \qquad (4.3)$$

where λ_n is the extrinsic Debye length given by (2.87). The Poisson equation (4.3) is written for three dimensions because the charge fluctuation is localized, leading to a potential that decays with distance in every direction. The charge density in (4.3) is the dopant ion density less the majority carrier density, as can be shown using (2.87).

2 The continuity equation relating current density per unit area I to the rate of change of charge density is

$$\nabla \cdot I = q\frac{\partial n}{\partial t}. \qquad (4.4)$$

3 The current density equation or "Ohm's law" for a semiconductor, relating current to the gradient in *quasi-Fermi* level, u_{Fn} is

$$I = \beta^{-1}q\mu n\nabla u_{Fn} \qquad (4.5)$$

where $\beta = (kT/q)^{-1}$. The quasi-Fermi level used in (4.5) introduces an approximate treatment of transport in the silicon. When transport occurs the approximation is made that the mobile carrier density still is given by the equilibrium Fermi distribution (2.4) but with the Fermi level u_F replaced by the quasi-Fermi level u_{Fn}. The quasi-Fermi level varies with time and position, adjusting the carrier density at each time and position to accomodate the current that is flowing, including both drift and diffusion.†

---

*Although τ_D is quite short for silicon, it is very long for SiO$_2$. The resistivity of good quality thermally grown SiO$_2$ is of the order of $10^{18}\,\Omega$ cm or more and $\epsilon_{ox} = 3.4 \times 10^{-13}$ F/cm so that $\tau_D = 3.4 \times 10^5$ sec or about 4 days. Thus, it makes no sense to draw a Fermi level in the SiO$_2$: the SiO$_2$ is not in equilibrium with the changes in potential that occur during most experiments or during usual device operation. Thus a very tiny current must flow most of the time in the system as the few mobile carriers present in the SiO$_2$ rearrange themselves. However, this current is so small that the MOS system can be considered to be in thermal equilibrium.

†In principle, the current density can be determined using the Boltzmann transport equation, taking into account all the scattering mechanisms that limit transport. However, for the case of interest here, the case of small departures from equilibrium in response to a small-signal external voltage variation at a frequency low compared to the scattering rate, the Boltzmann approach simplifies to (4.5) with all the scattering effects contained in the mobility.

Equations (4.3)–(4.5) are to be solved for a *small* disturbance from equilibrium. Consequently, the equilibrium versions of (4.3)–(4.5) can be written and subtracted from the general expressions (4.3)–(4.5). The results can be *linearized* in the departures from equilibrium. As an example, let a zero subscript denote equilibrium. (The zero subscript denotes equilibrium values of all variables throughout this chapter.) Then, *in equilibrium*, (4.3) becomes

$$\nabla^2 v_o = -\left(1 - \frac{n_o}{N_D}\right)\lambda_n^{-2}.$$

(4.6)

Subtracting (4.6) from (4.3) and letting $\delta v \equiv v - v_o$, and $\delta n \equiv n - n_o$, we find

$$\nabla^2 \delta v = \frac{\delta n}{N_D \lambda_n^2}.$$

(4.7)

Similarly, (4.4) and (4.5) become

$$\nabla \cdot \boldsymbol{\delta I} = q \frac{\partial \delta n}{\partial t}$$

(4.8)

and

$$\boldsymbol{\delta I} = \beta^{-1} q \mu n_o \nabla \, \delta u_{Fn}.$$

(4.9)

As the term involving $\delta n \nabla \delta u_{Fn}$ is second order in the disturbance, it was dropped during linearization of (4.9). Finally,

$$\delta n \equiv n - n_o = n_o[\exp(v - v_o - u_{Fn} + u_{Fno}) - 1]$$
$$\approx n_o(\delta v - \delta u_{Fn}).$$

(4.10)

Combining (4.8) and (4.9) and using (4.10) for δn, one finds

$$\nabla^2 \delta u_{Fn} + \nabla v_o \cdot \nabla \delta u_{Fn} = -\left(\frac{\beta}{\mu}\right) \frac{\partial(\delta v - \delta u_{Fn})}{\partial t}.$$

(4.11)

Subtracting (4.11) from (4.7)

$$\nabla^2(\delta v - \delta u_{Fn}) + \nabla v_o \cdot \nabla(\delta v - \delta u_{Fn}) - \left(\frac{n_o}{N_D \lambda_n^2}\right)(\delta v - \delta u_{Fn})$$
$$= \left(\frac{\beta}{\mu}\right) \frac{\partial(\delta v - \delta u_{Fn})}{\partial t} + \nabla v_o \cdot \nabla \delta v.$$

(4.12)

Simultaneous solution of (4.7) and (4.12) is unnecessarily complex for our purposes. Where the equilibrium field due to the gate bias $(-\nabla v_o)$ is small, for instance, near the depletion layer edge, (4.12) becomes a standard diffusion equation

$$\nabla^2(\delta v - \delta u_{Fn}) - \left(\frac{n_o}{N_D \lambda_n^2}\right)(\delta v - \delta u_{Fn}) = \left(\frac{\beta}{\mu}\right)\frac{\partial(\delta v - \delta u_{Fn})}{\partial t}.$$

(4.13)

For a sinusoidal time variation $\exp(j\omega t)$, (4.13) is not time dependent if

$$\left(\frac{\beta}{\mu}\right)\omega \ll \frac{n_o}{N_D \lambda_n^2}.$$

(4.14)

Equation (4.14) is the condition that the right side of (4.13) is negligible compared to the second term on the left of (4.13). Rewriting (4.14), we have

$$\omega \ll \left(\frac{1}{\tau_D}\right)\left(\frac{n_o}{N_D}\right). \tag{4.15}$$

Near the depletion layer edge $n_o = N_D$, so that τ_D determines majority carrier response time.

Where (∇v_o) is not negligible, it tends to speed up the majority carrier response. Physically, this effect is apparent for an MOS capacitor biased in depletion. Any change in gate bias causes a nearly instantaneous motion of the depletion layer edge (where majority carrier response is rapid), in turn altering band bending and assisting the flow of majority carriers to the disturbed region. Consequently, the estimate given by (4.14) is conservative for majority carriers at the interface, further supporting the conclusion that the approximation of instantaneous majority carrier response is adequate.

4.3 MINORITY CARRIER RESPONSE TIME

In inversion, *minority* carrier response governs the steady-state C-V characteristics of the MOS capacitor.* This section describes the physical processes governing minority carrier response and develops the equivalent circuit of the MOS capacitor in inversion. The elements of this equivalent circuit are calculated as functions of band bending and gate bias, permitting the calculation of C-V and G-V curves.

The following subsections modify the equivalent circuit of Fig. 3.3 to include the dependence of minority carrier response on frequency. This frequency dependence occurs when minority carriers lag behind small-signal ac gate voltage changes. This lag produces an *energy loss* supplied by the ac voltage source. In terms of a circuit, minority carrier phase lag can be related to an R-C time constant called the *minority carrier response time* with the energy loss occurring in the *resistance*. The central problem is to calculate the resistance that regulates the supply of minority carriers to and from the silicon surface from the thermal mechanisms determining minority carrier response time.

4.3.1 Minority Carriers Generated by Light

Before solving this problem, we digress for a moment to consider generation of minority carriers by light at wavelengths shorter than about

---

*In accumulation and depletion, minority carrier response to the ac gate voltage is unimportant because minority carrier density is negligible compared to majority carrier density at the silicon surface.

1 μm. If the gate electrode is opaque* and has no scratches or pinholes, minority carriers generated within a diffusion length of the gate edge will form an inversion layer under the gate if gate bias is of the right magnitude and polarity. The response time of the light generated inversion layer will depend on light intensity. Because exposure to light dominates the thermal processes that otherwise govern minority carrier response time, we restrict the discussion to the usual case where the MOS capacitor is in a lightproof box, as described in Section 12.8.1.

4.3.2 Thermal Minority Carrier Response to the ac Gate Voltage

Minority carriers follow an applied ac gate voltage as long as the period of the applied ac voltage is much longer than the minority carrier response time, τ_R. Response time for minority carriers in silicon at room temperature is typically 0.01–1 sec in strong inversion. Because this response time is very long, inversion layer capacitance will be frequency dependent, except at very low frequencies. This frequency dependence is illustrated in Fig. 4.1 which shows capacitance measured as a function of gate bias at 27°C with frequency of the ac gate voltage as parameter. Interface trap level

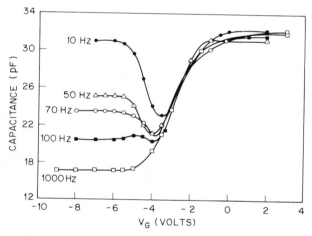

Fig. 4.1 Capacitance as a function of bias measured at 300°K with frequency as parameter. Sample is n-type oriented in the (100) direction. Gate diameter is 370 μm, donor density is 1.2×10^{16} cm^{-3}, and oxide layer capacitance is 2.84×10^{-8} F/cm^2. After Goetzberger and Nicollian[3]. Copyright (1967), American Telephone and Telegraph Company. Reprinted by permission.

---

*An example is an aluminum gate 3000 Å or more thick. Because the diameter of the gate is so large (tens of micrometers) compared to oxide thickness (less than a micrometer), the area under the gate exposed to scattered light is negligible. Sometimes, particularly when metal is evaporated through a shadow mask, a thinning of the metal near the gate edge occurs, reducing the diameter of opaque metal. A mild etch can remove the thin edge.

density was low in this sample, so the portions of the C-V curve in accumulation and depletion were identical at all frequencies within experimental error. Significant differences in the illustrated C-V curves arise in inversion. Figure 4.1 shows that inversion layer charge responds immediately to frequencies below 10 Hz. As long as $1/\omega \gg \tau_R$ and $1/\omega \gg \tau_D$, the entire experimental C-V curve will be described accurately by (3.10), with C_s given by (3.35) for uniformly doped silicon.

Because the condition $1/\omega \gg \tau_R$ applies at very low frequencies in silicon, as seen in Fig. 4.1, the C-V curve described by (3.10) and (3.35) is called the *low frequency* C-V curve. Figure 4.1 shows that at frequencies near 1 kHz, capacitance in strong inversion *saturates at a minimum value* and does not go back to C_{ox} as it does at low frequencies. This curve is called the *high frequency* C-V curve. It is described accurately by (3.10) and (3.35) in accumulation and depletion, but only with drastically increasing error in inversion.

To understand the high frequency C-V curve, consider first a slow increase in gate bias. In strong inversion, the silicon surface charge density is made up of a minority carrier component and an ionized impurity charge component, as described in Section 2.3. The minority carrier component increases exponentially with band bending [see (3.30)]. The ionized impurity component increases as the square root of band bending [see (3.20)]. As band bending is increased beyond the onset of inversion, the rapid exponential increase of minority carrier density dominates the slower increase of depletion layer charge. Therefore, charge neutrality is satisfied almost entirely by the increase of minority carrier density, and ionized impurity charge hardly increases at all. That is, for *slow* gate bias variations, depletion layer width hardly changes in strong inversion.

Consider next a rapidly varying ac gate voltage. In the inversion layer established by the gate bias, minority carrier density per unit area does not change with ac gate voltage because minority carrier generation does not follow the ac voltage. Therefore, minority carriers contribute no capacitance. However, charge neutrality must be satisfied for the ac gate voltage. As majority carriers can respond immediately to the ac voltage, ac charge neutrality will be satisfied by majority carrier flow in and out of the boundary between the depletion layer and the bulk silicon, that is, by an ac movement of the depletion layer edge. The capacitance measured at these high frequencies is the depletion layer capacitance in series with C_{ox}.

As noted, the depletion layer capacitance does not change with slow variations in gate bias because depletion layer width is approximately constant, as it is screened from the gate by the inversion layer charge. Therefore, the measured high frequency capacitance in inversion is constant, almost independent of gate bias. Also, this capacitance is the *lowest* of the entire C-V curve, because the depletion layer width is at its maximum value.

The two extreme frequency limits in Fig. 4.1 are the low frequency C-V

curve and the high frequency C-V curve. For the sample in Fig. 4.1, a low frequency C-V curve is measured below 10 Hz and a high frequency C-V curve above 1 kHz. Therefore, minority carrier response time in device grade silicon at room temperature is rather long. We now discuss the origin of this response.

(a) *Mechanisms Determining Minority Carrier Response*

There are three mechanisms for getting minority carriers to and from the inversion layer;[2-6] (1) generation of minority carriers at the back contact, followed by diffusion through the quasi-neutral region of the silicon, and then drift through the depletion layer;[6] (2) generation and recombination of minority carriers in the depletion layer;[4] and (3) supply from an inversion layer beyond the gate electrode.[3,4]*

The dominant mechanism depends on temperature. At room temperature, generation and recombination determine minority carrier response time in silicon (mechanism 2). As temperature is raised, diffusion becomes dominant (mechanism 1). In the particular silicon sample for which measurements are presented, diffusion was dominant for supplying minority carriers above 140°C[5] while generation-recombination in the depletion layer was dominant below 140°C.[5] Hofstein and Warfield[4] first identified depletion layer generation-recombination as controlling minority carrier response time at lower temperatures, with diffusion dominating only at higher temperatures.

The exact temperature at which transition is made from generation-recombination dominance to diffusion dominance depends on the density of bulk traps that have energy levels near midgap and on doping concentration. These two quantities determine the transition temperature at a given gate bias. Generation-recombination current increases with increasing midgap trap level density (which tends to increase with doping density because of increased contamination). Minority carrier diffusion current increases with decreasing doping concentration because minority carrier density increases.

An exception occurs when an inversion layer exists beyond the gate electrode (mechanism 3). If a strong external inversion layer exists, it will be the dominant source of minority carriers at all temperatures of interest. An external inversion layer may arise in several ways.*

1 It may be caused by charge in the oxide layer. Because oxide charge is invariably positive, this mechanism is of importance mainly when the silicon substrate is p-type. Positive oxide charge can be minimized by

---

*Contact to the silicon is ohmic for majority carriers but is blocking for minority carriers, so no minority carriers are injected from the external circuit. However, in structures such as the gated diode, a nearby p-n junction can act as a source of minority carriers. This case is analogous to supply from an external inversion layer.

the processing described in Sections 15.3 and 15.4, so that inversion beyond the gate does not occur.

2   An external inversion layer also may form as a result of the buildup of a potential along the oxide surface. This effect is important in a moist ambient for both n-type and p-type substrates and is discussed in Section 4.3.12(b).

3   In an integrated circuit the silicon surface beneath an interconnecting lead from a gate electrode can become inverted by the potential applied to that lead. This inversion usually is avoided either by making the oxide under the lead thicker than under the gate or by the use of a heavily doped region.

We start our discussion with the diffusion mechanism, follow with the depletion layer generation-recombination mechanism, and conclude with the response time governed by an inversion layer beyond the gate electrode.

(b) Diffusion from the Bulk

Figure 4.2 illustrates minority carrier charging and discharging of the inversion layer in an MOS capacitor for diffusion and depletion layer

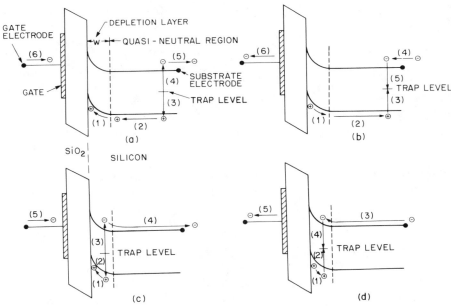

Fig. 4.2 Parts (a) and (b) are energy band diagrams that show how ac minority carrier current flows through the MOS capacitor by diffusion. Parts (c) and (d) are energy-band diagrams that show how ac minority carrier current flows through the MOS capacitor by generation and recombination in the silicon depletion layer. Width of the silicon depletion layer is w.

generation-recombination dominant current flow. An n-type silicon substrate is taken as an illustration. One contact of the MOS capacitor is the gate and the other is the ohmic contact to the silicon substrate. Because the substrate contact is ohmic for electrons, it is shown connected to the conduction band.

Figure 4.2a shows the negative half of the ac cycle, which requires an increase of inversion layer charge. During the negative half of the cycle, minority carriers (holes in the illustrative example) in the depletion layer are attracted toward the silicon surface by the field in the depletion layer produced by the ac voltage. These holes drift across the depletion layer as shown by arrow 1 in Fig. 4.2a, thus increasing inversion layer charge. The deficit of holes produced in the depletion layer causes a concentration gradient of holes to build up in the quasi-neutral region of the silicon, the region between the back contact to the silicon and the depletion layer edge. As shown by arrow 2 in Fig. 4.2a, this concentration gradient causes holes to diffuse across the quasi-neutral region toward the depletion layer.

When the neutral region of the silicon is in thermal equilibrium, the rates of hole capture and emission by traps are equal.* Similarly, the rates of electron capture and emission are equal. However, as holes drift across the depletion layer, leaving the quasi-neutral region for the silicon surface, a deficit of holes in the quasi-neutral region is created, and the system is no longer in thermal equilibrium. In the quasi-neutral region, capture rates of holes and electrons are diminished, as there are fewer holes available for capture, leaving hole and electron emission dominant. The traps alternately emit a hole, arrow 3 in Fig. 4.2a, and an electron, arrow 4 in Fig. 4.2a. This process in which an electron makes a transition from the valence to the conduction band by way of a trap level is called *generation*. As shown by arrow 5 in Fig. 4.2a, the electron emitted into the conduction band flows out of the ohmic substrate contact. Arrival of the hole at the silicon surface attracts an electron to the gate, as shown by arrow 6 in Fig. 4.2a. This final arrow completes the flow of displacement current through the MOS capacitor during the charging half of the ac cycle.

Figure 4.2b shows the positive half of the ac cycle, which requires a decrease of inversion layer charge. As shown by arrow 1 in Fig. 4.2b, holes drift back across the depletion layer under the repulsive action of the field produced by the ac gate voltage. The arrival of holes at the depletion layer edge produces a concentration gradient in the quasi-neutral region, causing holes to diffuse across the quasi-neutral region toward the substrate ohmic contact. This flow is shown by arrow 2 in Fig. 4.2b. The excess of holes in the quasi-neutral region causes capture to dominate. Excess hole capture by a trap is shown by arrow 3 in Fig. 4.2b. An electron coming into the conduction band from the ohmic contact is shown by arrow 4 in Fig. 4.2b.

---

*These traps may be near the ohmic back contact where their density is large compared to the density in the depletion layer. Therefore, diffusion rather than generation limits the current.

Capture of this electron by a trap is shown by arrow 5 in Fig. 4.2b. Alternate capture of a hole and an electron by a trap in which an electron makes a transition from the conduction to the valence band is called *recombination*. Departure of a hole from the silicon surface repels an electron from the gate, the process shown by arrow 6 in Fig. 4.2b. This final arrow completes the flow of displacement current through the MOS capacitor during the discharging half of the ac cycle.

By these processes, holes are provided to (or removed from) the inversion layer, even though the ohmic contact in this n-type example can transmit only electrons to or from the external circuit.

(c) Generation–Recombination in the Depletion Layer

Figures 4.2c,d illustrate the case where generation and recombination in the depletion layer are dominant for charging and discharging the inversion layer. Figure 4.2c shows the negative half of the ac cycle, which requires an increase of inversion layer charge. During the negative half of the cycle, holes in the depletion layer are driven to the silicon surface by the field produced by the ac voltage. As shown by arrow 1 in Fig. 4.2c, the arrival of holes leads to an increase in inversion layer charge, and a hole deficit in the depletion layer. Therefore, generation dominates in the depletion layer. Traps alternately emit a hole, shown by arrow 2 in Fig. 4.2c, and an electron, shown by arrow 3 in Fig. 4.2c. The generated hole reduces the hole deficit in the depletion layer. The electron emitted into the conduction band flows out of the ohmic substrate contact. This process is shown by arrow 4 in Fig. 4.2c. As shown by arrow 5 in Fig. 4.2c, an electron flows into the gate, attracted by a hole arriving at the silicon surface. Arrow 5 completes the flow of displacement current during the charging half of the ac cycle. The trap returns to its original charge state (on the average) so that steady-state occupancy of the traps remains unchanged.

Figure 4.2d shows the positive half of the ac cycle, which requires a decrease of inversion layer charge. As shown by arrow 1 in Fig. 4.2d, holes drift back across the depletion layer under the repulsive action of the field produced by the ac voltage. Now there is an excess of holes in the depletion layer. Excess hole capture by a trap is shown by arrow 2 in Fig. 4.2d. An electron entering the conduction band through the ohmic contact is shown by arrow 3 in Fig. 4.2d. Capture of this electron by a trap is shown by arrow 4 in Fig. 4.2d. Finally, an electron leaving the gate is shown by arrow 5 in Fig. 4.2d, to complete displacement current flow through the MOS capacitor during the discharge half of the ac cycle.

When diffusion is dominant, generation and recombination occur near the back contact to the silicon, and minority carriers are transported by diffusion across the quasi-neutral region. When generation-recombination is dominant, generation and recombination occur in the depletion layer and majority carriers are transported across the quasi-neutral region.

There are two types of trap in the silicon distinguished by their spatial location. The first type is *interface traps* located at or near the Si–SiO$_2$ interface. In device grade oxides, interface traps are associated with defects at the Si–SiO$_2$ interface that contribute energy levels distributed over the silicon bandgap. The second type is *bulk traps*, which usually are distributed uniformly throughout the silicon substrate. Bulk traps are associated with impurity atoms that contribute one or more energy levels in the silicon bandgap. Bulk traps with energy levels near midgap, rather than interface traps, are responsible for generation and recombination in strong inversion.

Two conditions are required for efficient generation and recombination through a trap. Both are fulfilled by bulk traps only.

1 The trap energy level must be within a few kT/q of the Fermi level.* If this condition is fulfilled, a small change of band bending will cause a large change in the equilibrium occupancy of the traps. As a result, a small-signal band-bending variation will result in capture and emission of significant numbers of free carriers as the traps attempt to regain equilibrium with the ac voltage. If this condition is not satisfied, the traps are either nearly full or nearly empty, and a small shift in Fermi level hardly changes occupancy; that is, the ac voltage causes an amount of capture and emission too small to detect.

2 Emission and capture rates for both holes and electrons should be the same. Equality of these rates leads to efficient generation and recombination, as seen by following these processes step by step. For example, recombination at a given trap may consist of hole capture, followed by electron capture. If the rate of hole emission is high compared to the rate of electron capture, then there is a good chance that a captured hole will be reemitted before an electron is captured, thus preventing recombination. Therefore, for efficient recombination, the electron and hole capture rates should be equal. This statement can be rephrased as a requirement for equality of electron and hole emission rates because capture rates and emission rates are very nearly equal for small-signal excitations (i.e., small deviations from equilibrium).

Condition 2 is satisfied approximately for trap levels near midgap because electron emission rates vary exponentially with energy measured from the trap level to the conduction band edge.† Similarly, hole emission rates vary

---

*Although the system is not in thermal equilibrium because of the small-signal excitation applied to the gate, the departure is so small that trap occupancy can be considered to be the equilibrium value for this discussion.
†Such an exponential dependence is a consequence of thermal equilibrium between electrons in the traps and those in the conduction band, assuming Boltzmann statistics.

exponentially with energy measured from the valence band edge to the trap level. Consequently, if the proportionality factors for holes and electrons are comparable, equality of electron and hole emission rates requires a trap level near midgap.

In reality, the proportionality factors are not the same because capture cross sections for holes and electrons differ. However, because of the exponential dependence of emission rates upon energy, even an order of magnitude difference in cross sections will place the most efficient recombination center only (kT/q) ln 10 V from midgap. The preceding arguments for recombination can be repeated exactly for generation.

In strong inversion, either condition 1 or condition 2 is satisfied by interface traps, but not both. Therefore, interface traps are inefficient generation and recombination centers in strong inversion. To illustrate, an arbitrary distribution of interface traps over the bandgap, with density D_{it}, was chosen in Fig. 4.3a. As shown in Fig. 4.3a, in strong inversion, the Fermi level at the silicon surface is near the valence band edge. As a result, for interface traps satisfying condition 1, hole emission becomes very probable, violating condition 2. On the other hand, interface trap levels near midgap in Fig. 4.3a, which satisfy condition 2, are very far from the Fermi level, violating condition 1. Thus interface traps will not contribute significantly to generation and recombination in strong inversion, even though their level density might be high.

Figure 4.3b shows that bulk traps in the depletion layer satisfy both requirements for efficient generation and recombination, provided that the bulk trap level is near midgap. In strong inversion, there will be a point within the depletion layer where the Fermi level crosses bulk trap levels near midgap, called the *crossover point*. Consequently, such bulk trap levels will satisfy condition 1. Because the active bulk trap levels are located near midgap, they also satisfy condition 2. Thus minority carrier response time in the silicon MOS capacitor near room temperature will be determined by generation and recombination through bulk traps.

Silicon always has some bulk trap levels located near midgap. These bulk traps originate from residual impurities, such as nickel, iron, manganese, zinc, copper, and gold that, in device-grade silicon, usually have densities orders of magnitude smaller than the density of the dominant ionized donor or acceptor impurities. Thus residual impurities, even if ionized, do not contribute significantly to the doping density. On the other hand, ionized donor and acceptor impurities do not contribute to generation and recombination because they have energy levels very near the band edges and do not satisfy both conditions 1 and 2.

In strong inversion, depletion layer width is fixed independent of gate bias. Therefore, the crossover point does not move relative to the Si–SiO$_2$ interface as gate bias is varied. Hole and electron densities at crossover also do not change with bias. Thus generation and recombination current

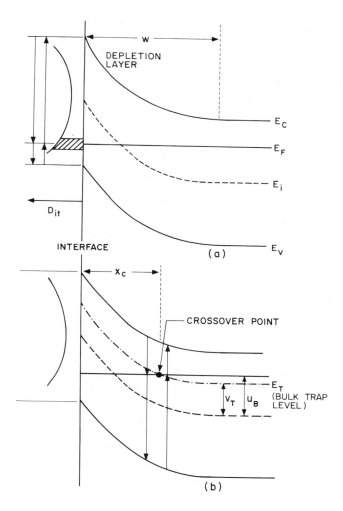

Fig. 4.3 (a) Energy-band diagram of an MOS capacitor with an n-type substrate in strong inversion showing generation and recombination through interface traps. Energies E_c and E_v are the conduction and valence edges, respectively; E_i is the intrinsic Fermi level; D_{it} is the interface trap level density; and E_F is the Fermi level. Arrows pointing up show transition of an electron from the valence to the conduction band through an interface trap within an energy interval a few kT/q wide centered around the Fermi level at the silicon surface. Arrows pointing down show transition of an electron from the conduction to the valence band through an interface trap located near the Fermi level at the silicon surface. Similar arguments apply for p-type silicon. (b) Energy-band diagram of an MOS capacitor with an n-type substrate in strong inversion showing generation and recombination through silicon bulk traps in the depletion layer. Arrows pointing up show transition of an electron from the valence to the conduction band through a bulk trap level near crossover and represent the generation process. Arrows pointing down show transition of an electron from the conduction to the valence band through a bulk trap level near crossover and represent recombination.

will be independent of gate bias, as long as strong inversion is maintained.*

For interface trap level densities greater than $\sim 10^{10}\,\mathrm{cm}^{-2}\,\mathrm{eV}^{-1}$, there will be a range of gate bias where generation and recombination through interface traps, rather than through bulk traps, will be dominant. In this bias range, the Fermi level at the surface will be between midgap and several kT/q past midgap into weak inversion so that interface traps will satisfy conditions 1 and 2. Therefore, between a bias in strong inversion and a bias near midgap, there will be a transition range where neither interface traps nor bulk traps are dominant. This complex bias range is not treated, and is avoided in measurements.

(d) Capture and Emission Rates of Electrons and Holes

Next we derive the equivalent circuit of the MOS capacitor in inversion taking generation and recombination into account. To begin, we derive expressions for capture and emission rates of electrons and holes based on the generation-recombination theory due to Hall[7] and Shockley and Read[8] applied to a single-level bulk trap.

There are three conditions to be met before trapping can be described accurately by the Shockley-Read-Hall (SRH) theory: (1) a bulk trap must change occupancy by only one unit of charge; that is, in changing occupancy the charge state of a donor-type trap can go from neutral to one unit of positive charge, and vice versa or, for an acceptor-type trap, from neutral to one unit of negative charge and vice versa; (2) each bulk trap has a single energy level that must be independent of occupancy; therefore, no lattice relaxation results from a change in bulk trap occupancy; and (3) capture of an electron or hole must be instantaneous.

Because of the above conditions, the treatment in terms of a capture probability in the SRH theory is approximate. Dynamics (such as lattice readjustments, phonon emission associated with each capture event, and energy loss mechanisms, in general, associated with individual capture events) do not control the capture process and cannot be taken into account. Nevertheless, predictions of the SRH theory agree well with experiment.

Figure 4.4 illustrates the four processes of capture and emission of electrons and holes. Process *a* in Fig. 4.4 is the capture of an electron, a transition from the conduction band to an unoccupied bulk trap level. Process *b* in Fig. 4.4 is the emission of an electron, a transition from an occupied bulk trap level to the conduction band. Process *c* in Fig. 4.4 is the capture of a hole by the bulk trap. What actually occurs is a transition of an electron from the occupied bulk trap level to the valence band. Process *d* in Fig. 4.4 is emission of a hole by the bulk trap. What actually occurs is

---

* For uniform doping and a bulk trap density independent of depth, generation-recombination current is independent of gate bias even outside of strong inversion, provided that crossover occurs.

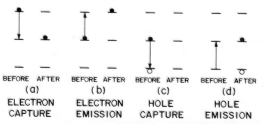

Fig. 4.4 Illustration of capture and emission of electrons and holes through intermediate centers. Arrows designate direction of electron transition. After Grove.[10]

the transition of an electron from the valence band to the unoccupied bulk trap level. In the typical bulk trap found in silicon, and adequately described by the SRH theory, only one electron can occupy a given bulk trap. Thus once an electron is captured by that bulk trap, it cannot capture another electron.

4.3.3 Analysis of Generation-Recombination Controlled Response Time

To measure the admittance of the MOS capacitor in strong inversion, a gate bias is applied to hold the silicon surface in strong inversion. Then a small sinusoidal ac gate voltage is superposed. The fraction of bulk trap levels occupied by electrons now is time dependent and is given by the bulk trap level occupancy function $f(x, t)$. This occupancy function enters all the rate expressions for the various processes illustrated in Fig. 4.4, as discussed next.

(a) *Electron Capture and Emission Rates*

The rate of electron capture by bulk trap levels (process a in Fig. 4.4) will be proportional to the density of conduction band electrons in the depletion layer $n(x, t)$ and the concentration of bulk trap levels not occupied by electrons $n_T[1 - f(x, t)]$, where n_T is the total concentration of bulk trap levels per unit volume. Electron density in the depletion layer is a function of x because potential is a function of x in this region as explained in Section 2.3. The rate of electron capture is

$$r_a(x, t) = c_n n(x, t) n_T [1 - f(x, t)] \quad (\text{cm}^{-3}\text{-sec}^{-1}) \qquad (4.16)$$

where c_n, the electron capture probability in units of cm^3/sec, is the constant of proportionality and is the average over all occupied states in the conduction band.

As process b in Fig. 4.4 is the transition of an electron from an occupied bulk trap level to the conduction band, the rate of electron emission is proportional to the ooncentration of bulk trap levels occupied by electrons

$n_T f(x, t)$.

$$r_b(x, t) = e_n n_T f(x, t) \quad (\text{cm}^{-3}\text{-sec}^{-1}) \tag{4.17}$$

where e_n, the emission probability in units of sec^{-1}, is the constant of proportionality. Electron emission probability depends upon the density of unoccupied states in the conduction band and the position of the bulk trap level in the bandgap.

(b) Hole Capture and Emission Rates

As process c in Fig. 4.4 is the transition of an electron from a bulk trap level to the valence band, the hole capture rate is proportional to the concentration of bulk trap levels occupied by electrons $n_T f(x, t)$ and the concentration of empty states in the valence band or hole concentration $p(x, t)$; that is,

$$r_c(x, t) = c_p p(x, t) n_T f(x, t) \quad (\text{cm}^{-3}\text{-sec}^{-1}) \tag{4.18}$$

where c_p, the hole capture probability in units of cm^3/sec, is the constant of proportionality.

As process d in Fig. 4.4 is the transition of an electron from the valence band to the bulk trap level, hole emission rate will be proportional to the concentration of empty bulk trap levels $n_T[1 - f(x, t)]$

$$r_d(x, t) = e_p n_T[1 - f(x, t)] \quad (\text{cm}^{-3}\text{-sec}^{-1}) \tag{4.19}$$

where e_p, the hole emission probability in units of sec^{-1}, is the constant of proportionality. Hole emission probability depends on the concentration of occupied states in the valence band and the position of the bulk trap level in the bandgap.

In the next subsections we calculate first the admittance due to electron capture and emission and then the admittance due to hole capture and emission.

(c) Electron Capture and Emission Admittance

Electron current density flowing at point x in the depletion layer at time t is

$$i_n(x, t) = q[r_a(x, t) - r_b(x, t)] \quad (\text{A/cm}^3) \tag{4.20}$$

or from (4.16) and (4.17)

$$i_n(x, t) = q n_T c_n[1 - f(x, t)]n(x, t) - q e_n n_T f(x, t). \tag{4.21}$$

As shown in part a of Appendix I, making the small-signal approximation the admittance is

$$Y_n(x, t) = j\omega \frac{q^2 n_T f_o(1 - f_o)}{kT[1 + j\omega f_o/c_n n_o(x)]} \tag{4.22}$$

where $j = \sqrt{-1}$, ω is the angular frequency of the ac voltage, sec^{-1}, k is Boltzmann's constant in V-coul/°K, T is the absolute temperature, in degrees Kelvin, $n_o(x)$ is the electron density per unit volume at x established by the gate bias, f_o is the equilibrium bulk trap level occupancy Fermi function established by the gate bias, $f_o = [1 + \exp(v - v_T + u_B)]^{-1}$, and v_T is the bulk trap level potential measured from midgap in units of kT/q. The Fermi function is used rather than the Boltzmann approximation because the Fermi level at crossover is very near the bulk trap level.

Equation (4.22) is the admittance of a series R-C network with capacitance

$$C_T = \frac{q^2}{kT} n_T f_o (1 - f_o) \quad \text{(F/cm}^3) \tag{4.23}$$

and conductance

$$G_n(x) = \frac{q^2}{kT} n_T (1 - f_o) c_n n_o(x) \quad \text{(mhos/cm}^3). \tag{4.24}$$

(d) Hole Capture and Emission Admittance

The calculation parallels that for the electron capture and emission admittance. Hole current density flowing at point x in the depletion layer at time t is

$$i_p(x, t) = q[r_c(x, t) - r_d(x, t)] \quad \text{(A/cm}^3) \tag{4.25}$$

or from (4.18) and (4.19), we obtain

$$i_p(x, t) = q n_T c_p p(x) f(x, t) - q n_T e_p [1 - f(x, t)]. \tag{4.26}$$

Making the small-signal approximation, the admittance, as shown in Section (b) in Appendix I, is

$$Y_p(x, t) = j\omega \frac{q^2 n_T f_o (1 - f_o)}{kT[j\omega(1 - f_o)/c_p p_o(x) - 1]} \tag{4.27}$$

where the parameters are similar to those in (4.22) except that $p_o(x)$ is the hole density per unit volume at x established by the gate bias.

Equation (4.27) is the admittance of a series R-C network with the same capacitance as given by (4.23) and conductance

$$G_p(x) = \frac{q^2}{kT} n_T f_o c_p p_o(x) \quad \text{(mhos/cm}^3). \tag{4.28}$$

4.3.4 The Equivalent Circuit in Strong Inversion: Phenomenological Derivation

We derive the equivalent circuit of the MOS capacitor in strong inversion taking generation and recombination through bulk traps into account. We connect the equivalent circuit of the bulk trap to the equivalent circuit of the MOS capacitor originally derived without generation and recombination centers.

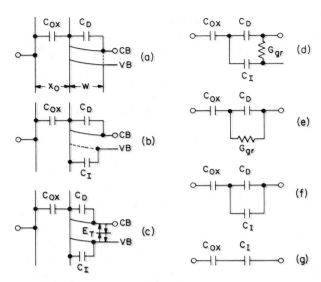

Fig. 4.5 Equivalent circuits of the MOS capacitor showing how various elements of the circuit communicate with the conduction and valence bands; Symbols CB and VB denote conduction and valence band edges, respectively. External terminals are designated by circles. The substrate is n-type.

Figure 4.5a depicts the MOS capacitor biased only to depletion. This circuit is identical to Fig. 3.3, except the conduction and valence bands are shown to clarify development of the circuit. Continuing with an n-type silicon substrate as example, one contact to the MOS capacitor is to the gate and the other, the ohmic back contact, is to the conduction band of the silicon. The valence band is not connected in the circuit because there is no way for it to communicate with either contact. The depletion layer capacitance C_D given by (3.47) is connected to the conduction band because C_D is charged and discharged by the flow of majority carriers in and out of the depletion layer edge region by way of the ohmic back contact. Suppose that we now bias the MOS capacitor into strong inversion. The inversion layer capacitance C_I given by (3.48) must be in contact with the interface (where it is capacitively coupled to the gate through the oxide capacitance C_{ox}) and also in contact with the valence band. The equivalent circuit becomes that of Fig. 4.5b. As the valence band does not communicate with the back contact, C_I makes no contribution to the total capacitance. The capacitance across the contacts is just C_D in series with C_{ox}.

Let us introduce generation and recombination centers into the silicon. Figure 4.5c shows generation and recombination through bulk traps with a single bulk trap energy level located near midgap (a common situation in practice). Next, we replace the physical process of generation and recombination shown in Fig. 4.5c with an equivalent circuit representation.

When recombination or generation can occur, an energy loss can arise from the ac change in occupancy of the inversion layer. For example, consider an n-type substrate biased in inversion and subjected to a small-signal ac gate voltage. On the negative half of the ac cycle, holes are demanded by the inversion layer. These holes are provided by generation through bulk traps in the depletion layer. The energy to create electron-hole pairs initially comes from the silicon lattice, but this energy is returned to the lattice when the generated holes thermalize in the inversion layer and the generated electrons thermalize in the bulk. At low frequencies all the holes demanded by the inversion layer can be generated just when they are needed, and no net loss occurs. At higher frequencies the demand will exceed the supply. Then the inversion layer holes have an average energy *below* that of the holes in the bulk silicon because the ac voltage has lowered the inversion layer hole energy. Holes supplied to the inversion layer fall to this lower level, losing energy to the lattice. Hence, the ac signal source must provide this extra energy loss.

Similarly, on the positive half cycle, holes are ejected from the inversion layer. The silicon lattice provides enough energy for the holes to return to the crossover point, and this energy is returned to the lattice on recombination through bulk traps. However, if the ac frequency is too high, recombination cannot keep pace with ejection from the inversion layer. Then the energy of holes in the inversion layer rises above that of holes in the bulk, increased by the ac gate voltage. After some delay, when recombination does occur, this extra energy is lost to the lattice. Again energy must be provided by the ac signal source.

Because of this energy loss, generation and recombination can be represented by a conductance G_{gr} connected between the conduction and valence bands, as shown mathematically in Section 4.3.5. It is shown in Section 4.3.5 that capacitive effects associated with capture and emission can be neglected for small bulk trap level densities. The circuit now becomes that of Fig. 4.5d. This equivalent circuit can be simplified further in strong inversion where $\omega C_I \gg G_{gr}$ (see Fig. 3.7). Then C_I can be treated as an ac short circuit in series with G_{gr} to give the circuit of Fig. 4.5e. This circuit describes the strong inversion case in Fig. 4.1 at frequencies intermediate between high and low.

Minority carrier response time, or the characteristic time needed to supply minority carriers, is $\tau_R = G_{gr}^{-1}C_D$. However, this response time is not measured directly but is modified by the presence of C_{ox}. In measuring the admittance of an MOS capacitor, a capacitance and an equivalent parallel conductance will be measured in strong inversion. Minority carrier response time is obtained by extracting G_{gr} and C_D from the measured admittance using the equivalent circuit in Fig. 4.5e as shown in Section 4.3.11.

At sufficiently low frequencies $\omega C_I \ll G_{gr}$, generation or recombination is sufficiently rapid to keep the inversion layer in equilibrium, so the energy loss drops to zero and the equivalent circuit in Fig. 4.5d becomes that in Fig. 4.5f. In strong inversion, $C_I \gg C_D$ so that the equivalent circuit in Fig. 4.5f becomes that in Fig. 4.5g. In strong inversion $C_I \gg C_{ox}$, so the measured low frequency capacitance will be equal to C_{ox}.

At very high frequencies $\omega\tau_n$ and $\omega\tau_p \gg 1$, generation and recombination cannot follow the ac signal at all, and G_{gr} becomes frequency dependent, dropping off as ω^{-1}, as is seen later [in (4.105)]. The transition from intermediate to high-frequency behavior in G_{gr} occurs at a frequency ω_c introduced in (4.114). For frequencies above ω_c, generation-recombination loss becomes less important than capture and emission loss. This energy loss results from delay in the change of occupancy of the bulk trap levels as the crossover point (see Fig. 4.3) moves closer and further away from the Si–SiO$_2$ interface in response to the ac voltage. For low densities of bulk traps, ω_c is well above the frequency range where inversion layer response can be seen, as explained in Section 4.3.7b. Hence, for bulk traps this loss usually can be ignored.*

Bulk series resistance between the edge of the depletion layer and the back contact has been neglected in the equivalent circuits in Fig. 4.5. Bulk series resistance in strong inversion usually is negligible compared to $G_{gr}^{-1} = R_{gr}$ and so does not affect minority carrier response time. However, as discussed in Section 5.7, bulk series resistance, unless minimized, causes a frequency dependent capacitance in accumulation, and to a much lesser extent in depletion.

The C-V curves in Fig. 4.1 can be calculated from the equivalent circuit of Fig. 4.5d using (3.36) for the relation between gate bias and band bending, the expression for C_D from (3.47) and for C_I from (3.48). At moderately high frequencies and in strong inversion, where the circuit of Fig. 4.5e applies, the saturation value of capacitance can be calculated using (3.47), assuming that $\omega\tau_R \gg 1$, with about 10% error. A more accurate value of high frequency capacitance in strong inversion is derived in Section 4.4.

The preceding arguments provide an intuitive basis for the circuits in Figs. 4.5d,e. However, these circuits are only approximate, and a mathematical formula for G_{gr} remains unexpressed. A more complete circuit is derived now, with mathematical formulas for the circuit elements, particularly for G_{gr}.

4.3.5   Derivation of the Complete Equivalent Circuit

To derive the circuit in Fig. 4.5d from the admittance calculated in Sections 4.3.3(c) and (d), consider an element located in the depletion layer

---

*The origin of this loss is explained at length in Chapter 5. For interface traps, this loss dominates recombination-generation loss, because interface trap loss usually is measured in depletion, where generation-recombination loss is small.

a distance x from the Si–SiO$_2$ interface of width Δx and of unit area. From Section 4.3.3(c), the equivalent circuit for electron capture and emission in element Δx is the bulk trap capacitance C_T in series with the conductance $G_n(x)$ connected to the conduction band. Similarly, from Section 4.3.2(d), the equivalent circuit for hole capture and emission in element Δx is C_T in series with $G_p(x)$ connected to the valence band. Figure 4.6 is the combined equivalent circuit for generation and recombination in element Δx at position x.

The bulk trap capacitance C_T is driven by variations in band bending at point x. That is, C_T is capacitively coupled to the interface through the portion of the depletion layer between the interface and the bulk trap by the capacitance denoted by $C_x = \epsilon_s/x$ in Fig. 4.6. The dashed-line representation of the conduction band between the bulk trap and the interface in Fig. 4.6 is simply a reflection of the very low majority carrier density in the depletion region, which allows the neglect of any capacitive coupling of these carriers to the gate, and also the neglect of any conductive coupling between the bulk trap and the interface through the conduction band. The bulk trap communicates conductively through the conduction band with the neutral bulk and the ohmic back contact. The bulk trap also communicates conductively with the inversion layer through the valence band. Communication with the bulk by the diffusion mechanism is ignored in Fig. 4.6, which thus does not apply at high temperatures.

The capacitive divider made up of C_x and C_{w-x} complicates derivation of the admittance for bulk traps. Inclusion of this divider means that Pois-

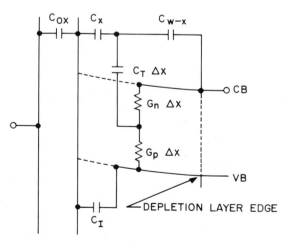

Fig. 4.6 Equivalent circuit for generation and recombination in an element of width Δx in the depletion layer located a distance x from the silicon surface. The element is of unit area, and CB and VB are the conduction and valence band edges, respectively. The silicon is n-type. Circles denote the terminals.

son's equation for band bending must be solved to obtain the capacitive coupling between bulk traps and the interface.

Fortunately, only bulk traps near x_c, whose levels cross the Fermi level ("crossover") contribute to recombination-generation, as shown in the mathematical derivation of the next section. Therefore, virtually all bulk trap levels far from crossover can be neglected, and for those near crossover, a single capacitive divider made up of C_{x_c} and C_{w-x_c} will serve. This approximation results in the equivalent circuit for the MOS capacitor with bulk traps at all distances shown in Fig. 4.7.*

The capacitive divider made up of C_{x_c} and C_{w-x_c} was not included in the analysis of Baccarani et al.[2] Although this neglect is not important to their results, this fact is not known in advance of a complete derivation. The general transmission line equivalent circuit of Sah[9] does include this divider. However, this general circuit is useful mainly for numerical analysis. A more suitable derivation now will be presented for the bulk trap dominated case, following the approach introduced by Garrett[6] for one interface trap level fed by diffusion from the bulk. A different approach was adopted by Mar et al.[11] The two methods begin with same assumptions but introduce different approximations. In our treatment the .approximations used are explained carefully.

The mathematical formulation that follows is useful, not only as a justification of the circuit of Fig. 4.7, but as an illustration of an approximate treatment of the MOS capacitor equations basic enough to be convincing, and yet simple enough to avoid numerical analysis. This

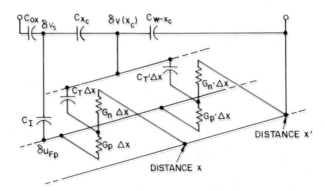

Fig. 4.7 Approximate equivalent circuit for the MOS capacitor with many bulk traps. The branches in parallel represent bulk traps at different distances from the interface. These branches all are coupled to the interface through the capacitive divider at the crossover point x_c. The small-signal voltages δv_s, $\delta v(x_c)$ and δu_{Fp} are indicated at three points. External terminals are denoted by open circles.

*This convenient simplification may not apply to *large-signal* measurements.

treatment also is applied to interface trap admittance in Section 5.3 and to the high frequency inversion layer capacitance in Section 4.4.

4.3.6   Derivation of the Admittance

The admittance of the MOS capacitor is the ratio of the ac gate current to the ac gate voltage. To find the admittance, we must find the ac current or, equivalently, the ac charge variation within the silicon. This charge variation has three components: (1) a majority carrier component, as expressed by an ac motion of the depletion layer edge, (2) a minority carrier component, as expressed by an ac variation in inversion layer charge, and (3) an ac variation in the charge state of the bulk traps responsible for generation and recombination.

All three ac charge components are coupled. That is, the ac variation in field across the oxide is balanced by the sum of all three ac charge variations in the silicon. For example, if the ac frequency is high, generation and recombination cannot follow. The ac inversion layer charge will not follow. Hence all the ac field in the oxide is balanced by the ac majority carrier response at the depletion layer edge. If the ac frequency is low, generation and recombination can follow, and so can the ac inversion layer charge. Then the ac field variation in the oxide is balanced in part by these ac charge components, and the ac motion of the depletion layer edge is reduced. In fact, in strong inversion at low frequencies, virtually no ac motion of the depletion layer edge occurs.

Not all the bulk traps are important. As explained in Section 4.3.2(c), only trap levels near the Fermi level can contribute. The important bulk traps are those located in the depletion layer where the trap levels cross the Fermi level. Their ac response depends on the ac band bending at this location, the crossover point. Therefore, we must find this ac band bending by solving Poisson's equation.

A set of coupled equations results, relating the bulk trap response to the ac band bending at the crossover point, the ac inversion layer charge to the ac bulk trap response, and the ac band bending to the combined ac response of the inversion layer, of the bulk traps and of the depletion layer edge.

(a)   Bulk Trap and Inversion Layer Dynamics:

The heuristic arguments leading to the circuit in Fig. 4.7 can be based on the equations governing generation and recombination. The approximations needed to reach this equivalent circuit then are clear. Quantitative formulas for the admittance of an MOS capacitor with bulk traps at all distances from the interface can be derived.

The analysis is based on SHR centers. To be definite, we treat donor centers of density n_T per unit volume. These centers participate in the two reactions

$$n_T^+ + e \rightleftharpoons n_T^o \qquad (4.29)$$

$$n_T^o + h \rightleftharpoons n_T^+ \qquad (4.30)$$

where n_T^+ and n_T^o represent volume densities of positive and neutral bulk traps, respectively, with e = electron and h = hole. The rate expressions for these processes have been discussed in connection with (4.16)–(4.19). Combining these results, the rate of change of positive bulk trap density is

$$\frac{dn_T^+}{dt} = -c_n n_T^+ n + n_T^o e_n + c_p n_T^o p - n_T^+ e_p \qquad (4.31)$$

where the symbols are the same as those in (4.16)–(4.19), with $n_T^o = n_T f$ and $n_T^+ = n_T (1 - f)$. The first pair of terms on the right of (4.31) represent the process in (4.29), and the second pair represent the process in (4.30).

In addition to the bulk trap dynamics given by (4.31), we need the inversion layer dynamics. We assume that all minority carriers generated by bulk traps are collected by the inversion layer. Conversely, all minority carriers leaving the inversion layer are trapped. Then, denoting distance from the Si–SiO$_2$ interface by the coordinate x

$$\frac{dN_I}{dt} = \int_0^w dx \, [-c_p p n_T^o + e_p n_T^+] \qquad (4.32)$$

where N_I = area density of inversion layer carriers (assumed holes in our example) and w = depletion layer width. Equation (4.32) also assumes that no holes enter the inversion layer by diffusion from the bulk. Therefore, (4.32) does not apply at elevated temperatures, nor for very thin samples.

Besides (4.31) and (4.32), we need Poisson's equation relating potential to charge. It is assumed that majority carrier response is so rapid in times compared to a period of the ac gate voltage that majority carriers are in equilibrium with the bulk. That is, the majority carrier quasi-Fermi level is the same as the bulk Fermi level, independent of position in the depletion layer as justified earlier. Similarly, for minority carriers in the inversion layer, it is assumed that the collection of holes generated by bulk traps is instantaneous, as is the transport of holes from the inversion layer to the traps. Consequently, the hole quasi-Fermi level also is independent of depth between the inversion layer and crossover. This approximation makes sense because the field in the depletion layer ensures a rapid sweeping of holes between the bulk traps and the inversion layer.

However, the hole quasi-Fermi level is not the bulk value. Rather, the hole quasi-Fermi level adjusts itself relative to the band bending to ensure the correct density of carriers in the inversion layer. This *steady-state* inversion layer carrier density is determined by the collection of holes generated by bulk traps, during one half-cycle of the ac gate voltage, and collection by bulk traps of holes lost from the inversion layer on the other half-cycle. An additional factor is screening of the bulk traps from the ac gate voltage by the intervening inversion layer. That is, the amplitude of the ac excitation at the bulk trap site is a function of the occupancy of the inversion layer, determined by the hole quasi-Fermi level.

(b) Small-Signal Equations

The equations needed are the equations governing bulk trap occupancy, (4.31), inversion layer occupancy, (4.32), and the Poisson equation relating band bending at any time and position to charge density. Because of the approximation of spatial independence of quasi-Fermi levels, we need not solve the continuity equations relating current to time rate of change of charge.

Rather than write the full set of equations in their most general form, we describe the small-signal versions of these equations obtained by setting, for example

$$v = v_o + \delta v \tag{4.33}$$

$$u_{Fp} = u_B - \delta u_{Fp} \tag{4.34}$$

where v_o is the equilibrium value of band bending and δv is the ac component. Using (4.33) and (4.34), we obtain [see (4.10)]

$$\delta p = p_o(\delta u_{Fp} - \delta v) \tag{4.35}$$

$$\delta n = n_o \delta v. \tag{4.36}$$

Equation (4.36) agrees with (4.10) because $\delta u_{Fn} \cong 0$ at frequencies of interest, that is, frequencies satisfying (4.15). Also, we set

$$f = f_o + \delta f \tag{4.37}$$

so that

$$\delta n_T^o = n_T \delta f \tag{4.38}$$

$$\delta n_T^+ = -n_T \delta f. \tag{4.39}$$

Equations (4.38) and (4.39) add to zero because their sum is the change in the total number of bulk traps, which is zero.

(c) Bulk Trap Occupancy

Assuming small-signal variations have a sinusoidal time dependence proportional to $\exp(j\omega t)$, the small-signal version of (4.31) is

$$j\omega \delta f = -c_n(1 - f_o)n_o\,\delta v + c_n n_o\,\delta f + e_n\,\delta f$$
$$+ c_p p_o\,\delta f + c_p p_o f_o(\delta u_{Fp} - \delta v) + e_p\,\delta f. \tag{4.40}$$

To simplify notation and to make the relation to an equivalent circuit more transparent, we introduce the bulk trap capacitance described by Lehovec and Slobodskoy[12] C_T from (4.23)

$$C_T = \beta q n_T (1 - f_o) f_o \tag{4.41}$$

and the capture conductances of Shockley and Read[8] from (4.24) and (4.28)

$$G_n \equiv R_n^{-1} = \beta q n_T c_n n_o (1 - f_o). \tag{4.42}$$

$$G_p \equiv R_p^{-1} = \beta q n_T c_p p_o f_o \tag{4.43}$$

Note that in equilibrium, the time derivatives of (4.31) and (4.32) are zero. If bulk traps at each depth are independently at equilibrium with the inversion layer, the integrand in (4.32) vanishes. Therefore

$$c_p p_o + e_p = (C_T R_p)^{-1}. \tag{4.44}$$

Also, combining (4.44) with (4.31) in equilibrium yields

$$c_n n_o + e_n = \frac{c_n n_o}{f_o} = (C_T R_n)^{-1}. \tag{4.45}$$

Therefore, (4.40) can be rewritten as

$$j\omega \beta q n_T \delta f = j\omega C_T [(G_n + G_p)\delta v - G_p \delta u_{Fp}] D^{-1} \tag{4.46}$$

where

$$D \equiv j\omega C_T + G_n + G_p. \tag{4.47}$$

(d) Inversion Layer Occupancy

The small-signal version of (4.32) is

$$j\omega \beta q \delta N_I = \delta \int_0^w dx [- c_p p_o n_{To}^o + e_p n_{To}^+]$$

$$= \int_0^w dx [- G_p (\delta u_{Fp} - \delta v) - \beta q n_T \delta f (C_T R_p)^{-1}] \tag{4.48}$$

where no term in the depletion layer width variation δw enters because the crossover point is far from the depletion layer edge $[n_T^+(w) \approx 0]$ and hole density there is negligible $[p_o(w) \approx 0]$.

(e) Small-Signal Poisson Equations

Equations (4.46) and (4.48) are two of the necessary equations. Two other equations are needed. One relates inversion layer carrier density variation to interface potential, namely

$$j\omega \beta q \delta N_I = j\omega C_I (\delta u_{Fp} - \delta v_s). \tag{4.49}$$

The other is the small-signal Poisson equation

$$\nabla^2 \delta v = (N_D \lambda_n^2)^{-1} (n_o \delta v + n_{To} \delta f). \tag{4.50}$$

Equation (4.50) already has been simplified by the omission of minority carriers, which have been replaced by a charge sheet (i.e., a very thin inversion layer) through the use of (4.49). Two further simplifications of (4.50) are made now. First, majority carrier density in depletion is small, less than $n_i \approx 10^{10}\,\text{cm}^{-3}$ between the interface and the point where the bulk Fermi level crosses midgap. The major influence of these carriers is near the depletion layer edge, where their effect can be modeled as a depletion layer edge variation $w = w_o + \delta w$. Therefore, the first approximation to (4.50) is to eliminate n_o and treat it as an influence on the field at the depletion layer edge [see (4.76)].

The second simplification of (4.50) is less justified. In material of device quality, bulk trap density is low. In fact, it is possible for this density to be so low that the average spacing between bulk traps is comparable to or larger than a depletion layer width. For example, the bulk trap density for which the spacing is a depletion layer width is $n_T = 2 \times 10^{12}\,\text{cm}^{-3}$ for $N_D = 10^{15}\,\text{cm}^{-3}$. Consequently, the localized nature of bulk traps is apparent at these densities, and a three-dimensional solution of (4.50) with point charges representing bulk traps would be desirable. However, this is not done here. As justified shortly, we replace the discrete bulk traps by a smeared out, uniform sheet of bulk traps of density $n_T(x)$. Then (4.50) can be replaced by a one-dimensional Poisson equation, involving only the coordinate x. This equation is

$$\frac{d^2 \delta v}{dx^2} = (N_D \lambda_n^2)^{-1} n_{To} \delta f \tag{4.51}$$

subject to the boundary conditions

$$\delta v(x = 0) = \delta v_s \tag{4.52}$$

$$\delta v(x = w) = 0. \tag{4.53}$$

For bulk trap levels near midgap at moderate densities, use of (4.51) is justified because the bulk traps are much less numerous than the dopant ions. As a result, the band bending is controlled by the electrodes and the dopant ions, which can be treated in a one dimensional approximation.

Equation (4.51) can be integrated to find δv. The result is

$$\delta v = \delta v_s \left(1 - \frac{x}{w}\right) - (N_D \lambda_n^2)^{-1} \left[x \int_x^w dx'\, n_{To}(x') \delta f \right.$$

$$\left. + \int_0^x dx'\, x' n_{To}(x') \delta f - x \int_0^w dx' \left(\frac{x'}{w}\right) n_{To}(x') \delta f \right]. \tag{4.54}$$

To check that (4.54) is the correct solution, note that it satisfies conditions (4.52) and (4.53) and that differentiating twice yields (4.51). Application of a theorem of ordinary differential equations[13] implies that only one solution of (4.51) satisfying (4.52) and (4.53) exists, so (4.54) is this solution.

The integrands in (4.54) all involve δf, which, from (4.46), involves C_T and hence the product $f_o(1 - f_o)$. This product is sharply peaked near crossover. Consequently, we may approximate the integrands in (4.54) by their values at crossover, except for the factor δf, which varies rapidly in the vicinity of crossover. In this way, (4.54) becomes

$$\delta v \approx \delta v_s \left(1 - \frac{x}{w}\right) - (N_D \lambda_n^2)^{-1} n_{To}(x_c)$$

$$\times \left[(x - x_c) \int_x^w dx' \delta f + x_c \left(1 - \frac{x}{w}\right) \int_0^w dx' \, \delta f\right]. \tag{4.55}$$

From (4.55), δv is a reasonably slowly varying function of x. For later use, we are interested in its value at crossover. Set $x = x_c$ in (4.55) and introduce the (one-dimensional) capacitances of the regions between crossover and the interface, C_{x_c}, and between crossover and the depletion layer edge C_{w-x_c}

$$C_{x_c} = \frac{\epsilon_s}{x_c} \tag{4.56}$$

$$C_{w-x_c} = \frac{\epsilon_s}{w - x_c}. \tag{4.57}$$

Then (4.55) at $x = x_c$ becomes

$$\left(\frac{\delta v}{\delta v_s}\right)_{x_c} = \left(\frac{C_D}{C_{w-x_c}}\right)\left[1 - \beta q n_T C_{x_c}^{-1}\left(\frac{\delta H}{\delta v_s}\right)\right] \tag{4.58}$$

where

$$\delta H \equiv \int_0^{} dx' \delta f. \tag{4.59}$$

(f) The "Crossover" Capacitances

We evaluate δH using (4.46)

$$\beta q n_T \delta H = \int_0^w dx \, C_T[(G_n + G_p)\delta v - G_p \delta u_{Fp}]D^{-1}$$

$$\approx (C_{TXn} + C_{TXp}) \delta v(x_c) - C_{TXp} \delta u_{Fp} \tag{4.60}$$

where δv has been evaluated at crossover, where C_T is peaked, and where we have introduced the "crossover" capacitances C_{TXn} and C_{TXp} defined by

$$C_{TXn} \equiv \int_0^w dx \, C_T G_n D^{-1} \tag{4.61a}$$

$$C_{TXp} \equiv \int_0^w dx \; C_T G_p D^{-1} \tag{4.61b}$$

and D is defined in (4.47).

The meaning of these capacitances is clarified by an approximate evaluation. Because $f_o \equiv [1 + \exp(v_T - v - u_B)]^{-1}$, we find

$$f_o(1 - f_o) = \frac{df}{dv}. \tag{4.62}$$

Also, for depleted n-type material containing donor traps, $(dv/dx) > 0$, $f(x = 0) = 0$, and $f(w) = 1$; therefore (4.61a) becomes

$$C_{TXn} = \int_0^1 df \; \frac{\beta q n_T \, G_n}{dv/dx} \, D^{-1}$$

$$\approx C_{bt} \int_0^1 df G_n D^{-1}. \tag{4.63}$$

The factor C_{bt} outside the integral is the capacitance of the number of bulk trap levels per unit area within kT/q of crossover, as shown in Fig. 4.8.

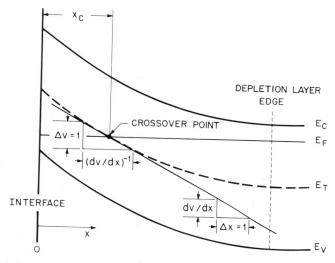

Fig. 4.8 Band bending versus distance from the interface into the silicon. This diagram shows bulk traps to be active in a region of width $(dv/dx)^{-1}$ near the crossover point (large dot). This active width corresponds to the region where bulk trap levels (dashed line) are within kT/q of the Fermi level E_F as inferred from the two similar triangles drawn on the line tangent to the crossover point. This line has slope $(-dv/dx)$, evaluated at crossover, as shown by the lower of the two similar triangles.

That is, C_{bt} is defined by

$$C_{bt} \equiv \beta q n_T(x_c)\left(\frac{dv}{dx}\right)_{x_c}^{-1}. \tag{4.64}$$

The capacitance C_{bt} can be estimated using Gauss's law to find dv/dx. Band bending at crossover is evaluated using Fig. 4.3. The bulk trap level is at a potential v_T above the intrinsic level in the bulk. The Fermi level is at a potential u_B above the intrinsic level. Therefore, band bending at $x = x_c$, is $v = -(u_B - v_T)$. Hence, the distance from crossover to the depletion layer edge is $\sqrt{2}\,\lambda_n(u_B - v_T)^{1/2} = w(1 - v_T/u_B)^{1/2}/\sqrt{2}$. The dopant ion charge in this region is then $qN_D w(1 - v_T/u_B)^{1/2}/\sqrt{2}$, and from Gauss's law

$$\left(\frac{dv}{dx}\right)_{x_c} = \beta\epsilon_s^{-1} qN_D w\left(1 - \frac{v_T}{u_B}\right)^{1/2}\!\!/\sqrt{2}$$

$$= \frac{\beta q N_D\left(1 - \dfrac{v_T}{u_B}\right)^{1/2}}{\sqrt{2}C_D}. \tag{4.65}$$

Using this result in (4.64), C_{bt} is given by

$$C_{bt} = \sqrt{2}\, C_D\left(\frac{n_T}{N_D}\right)\left(1 - \frac{v_T}{u_B}\right)^{-1/2}. \tag{4.66}$$

The integral (4.63), using (4.47) for D, is

$$\int_0^1 df\, G_n[j\omega C_T + G_n + G_p]^{-1} = \tau_n^{-1} H_{C_n} \tag{4.67}$$

where H_{C_n} is evaluated in Appendix II, and

$$\tau_n^{-1} = c_n n_o(x_c) \tag{4.68a}$$

is the capture time for electrons at crossover. Similarly,

$$\tau_p^{-1} = c_p p_o(x_c). \tag{4.68b}$$

Substituting (4.68) and (4.67) in (4.63) we find

$$C_{TXn} = C_{bt}\tau_n^{-1} H_{C_n} \tag{4.69a}$$

and

$$C_{TXp} = C_{bt}\tau_p^{-1} H_{C_p} \tag{4.69b}$$

where H_{C_p} is given by (4.67) with subscripts n and p interchanged, and is evaluated in (26) of Appendix II.

Using the results of Appendix II for angular frequency ω satisfying

$$\omega\tau_p \quad and \quad \omega\tau_n \ll 1 \tag{4.70}$$

we find

$$C_{TXn} = C_{bt} \frac{\tau_p}{\tau_p - \tau_n} \left[1 - \frac{\tau_n}{\tau_p - \tau_n} \ln \left(\frac{\tau_p}{\tau_n} \right) \right]. \qquad (4.71)$$

For C_{TXp}, τ_p and τ_n are interchanged.

We evaluate (4.71) in the special case of an efficient recombination center, for which hole and electron emission rates are equal

$$\tau_p = \tau_n = \tau = [(c_p c_n)^{1/2} n_i]^{-1}. \qquad (4.72)$$

In the limit $\tau_p \rightarrow \tau_n$, (4.71) requires careful evaluation. Let $\tau_p/\tau_n = 1 + \epsilon$. Then, the bracket in (4.71) becomes (for small ϵ) $1 - \epsilon^{-1} \ln(1 + \epsilon) = \epsilon/2$ and as $\epsilon \rightarrow 0$, (4.71) becomes

$$C_{TXn} = C_{TXp} = \tfrac{1}{2} C_{bt}. \qquad (4.73)$$

In general, for arbitrary values of τ_n and τ_p, $(C_{TXp} + C_{TXn}) = C_{bt}$.

(g) The Admittance

The admittance of the silicon portion of the MOS capacitor, Y_s is defined as

$$Y_s \equiv - j\omega\beta \frac{\delta Q_T}{\delta v_s} \qquad (4.74)$$

where δQ_T is the small-signal variation of total silicon surface charge per unit area, taken as

$$\delta Q_T = \delta Q_D + q\delta N_I - q \int_0^w dx' n_{To}(x') \, \delta f. \qquad (4.75)$$

In (4.75), the term δQ_D is the depletion layer charge variation, which we model as a depletion layer edge effect

$$\beta \, \delta Q_D = \epsilon_s \left(\frac{d\delta v}{dx} \right)_{x=w}. \qquad (4.76)$$

If (4.55) is used for $\delta v(x)$ and (4.60) for the variation in trapped charge, (4.76) becomes

$$\beta \, \delta Q_D = - C_D \, \delta v_s + \left(\frac{C_D}{C_{x_c}} \right) \int_0^w dx \, \beta q n_{To} \, \delta f$$

$$\approx - C_D \, \delta v_s + \left(\frac{C_D}{C_{x_c}} \right) [(C_{TXn} + C_{TXp})\delta v(x_c) - C_{TXp}\delta u_{Fp}]. \qquad (4.77)$$

If (4.77), (4.59), and (4.60) are used for the integrals over δf and (4.49) for

δN_I, the admittance (4.74) becomes

$$Y_s = j\omega C_D + j\omega C_I \left(1 - \frac{\delta u_{Fp}}{\delta v_s}\right) + \left(\frac{C_D}{C_{w-x_c}}\right)$$

$$\times \left[j\omega (C_{TXn} + C_{TXp}) \left(\frac{\delta v}{\delta v_s}\right)_{x_c} - j\omega C_{TXp} \left(\frac{\delta u_{Fp}}{\delta v_s}\right)_{x_c} \right] \qquad (4.78)$$

where the crossover capacitances C_{TXp} and C_{TXn} are defined in (4.61).

The admittance Y_s is known from (4.78), provided that $(\delta u_{Fp}/\delta v_s)$ and $(\delta v/\delta v_s)_{x_c}$ are known. Combining (4.48) and (4.49) and eliminating δf using (4.46), we obtain one relation between these quantities

$$\frac{\delta u_{Fp}}{\delta v_s} \left[j\omega C_I + \int_0^w dx\, G_p(j\omega C_T + G_n)D^{-1} \left(\frac{\delta u_{Fp}(x)}{\delta u_{Fp}}\right) \right]$$

$$= j\omega C_I + \int_0^w dx\, j\omega C_T G_p D^{-1} \left(\frac{\delta v}{\delta v_s}\right)$$

or

$$\frac{\delta u_{Fp}}{\delta v_s} \approx \frac{j\omega C_I + \left(\frac{\delta v}{\delta v_s}\right)_{x_c} j\omega C_{TXp}}{j\omega C_I + j\omega C_{TXp} + G_{gr}}. \qquad (4.79)$$

where the *generation-recombination* conductance per unit area is given by

$$G_{gr} \equiv \int_0^w dx\, G_p G_n D^{-1}$$

$$\approx C_{bt} \tau_p^{-1}(x_c) \tau_n^{-1}(x_c) H_G \qquad (4.80)$$

and the integral H_G defined by

$$H_G = \int_0^1 df [j\omega f(1-f) + f\tau_p^{-1} + (1-f)\tau_n^{-1}]^{-1} \qquad (4.81)$$

is evaluated in Appendix II.

A second relation between $(\delta v/\delta v_s)_{x_c}$ and $(\delta u_F/\delta v_s)$ is obtained from (4.58) and (4.60)

$$\left(\frac{\delta v}{\delta v_s}\right)_{x_c} = \frac{\left(1 + \frac{C_{TXp}}{C_{x_c}} \frac{\delta u_{Fp}}{\delta v_s}\right) \frac{C_D}{C_{w-x_c}}}{1 + \left(\frac{C_D}{C_{w-x_c}}\right)\left(\frac{C_{TXp} + C_{TXn}}{C_{x_c}}\right)}. \qquad (4.82)$$

Combining (4.79) and (4.82), both $(\delta v/\delta v_s)_{x_c}$ and $(\delta u_{Fp}/\delta v_s)$ are determined. Admittance of the equivalent circuit in Fig. 4.7 is given by (4.78), as shown in Appendix III.

(h) Simplification for Low Bulk Trap Densities

Rather than use (4.79) and (4.82) to evaluate Y_s in the general case, we limit consideration to bulk traps such that

$$(C_{TXp} + C_{TXn}) \ll C_{x_c}. \tag{4.83a}$$

The inequality (4.83a) implies either low bulk trap densities or bulk trap levels with a crossover point near the interface.

More exactly, using (4.66) and (4.73), the inequality (4.83a) becomes

$$\sqrt{2}\, C_D \left(\frac{n_T}{N_D}\right) \ll \frac{\epsilon_s}{x_c}, \tag{4.83b}$$

assuming a midgap trap with $v_T = 0$ for illustration. For a bulk trap level near midgap [see discussion prior to (4.65)]

$$x_c = \left(1 - \frac{1}{\sqrt{2}}\right) w. \tag{4.84}$$

Therefore, (4.83) becomes

$$(\sqrt{2} - 1) n_T(x_c) \ll N_D \tag{4.85}$$

which is satisfied in device-grade material for bulk traps with levels near midgap. When (4.83) is used, (4.82) becomes

$$\left(\frac{\delta v}{\delta v_s}\right) = \frac{C_D}{C_{w-x_c}}. \tag{4.86}$$

Equation (4.86) means that the ac band bending variation at crossover is not affected by the ac variation in bulk trap charge. That is, the dominant ac response of the silicon is a mixture of ac majority carrier variation near the depletion layer edge and ac minority carrier variation in the inversion layer. The first component of this mixture is dominant at high frequencies, the second at low frequencies.

When (4.86) and (4.79) are used in (4.78), the admittance becomes

$$Y_s = j\omega C_D + j\omega C_{TXn} \left(\frac{C_D}{C_{w-x_c}}\right)^2 + \frac{j\omega (C_I + C_{TXp}(C_D/C_{w-x_c})^2] G_{gr}}{j\omega (C_I + C_{TXp}) + G_{gr}}$$

$$+ \left(\frac{x_c}{w}\right)^2 \frac{j\omega C_I\, j\omega C_{TXp}}{j\omega (C_I + C_{TXp}) + G_{gr}}. \tag{4.87}$$

In (4.87) C_{TXn} and C_{TXp} are purely capacitive at frequencies satisfying $\omega\tau_n$ and $\omega\tau_p \ll 1$. The multiplier $(C_D/C_{w-x_c})^2$ expresses the reduction of the ac signal at the bulk trap site due to its location a distance x_c from the silicon surface. For a midgap trap, $(x_c/w)^2 = (1 - C_D/C_{w-x_c})^2 \approx (1 - 1/\sqrt{2})^2 = 0.086$, so the last term in (4.87) often is negligible.

(i) Strong Inversion

In strong inversion C_I is large and satisfies the condition

$$C_I \gg C_{bt}. \tag{4.88}$$

Then (4.87) becomes

$$Y_s = j\omega C_D + j\omega C_{TXn}\left(\frac{C_D}{C_{w-x_c}}\right)^2 + \frac{j\omega C_I G_{gr}}{j\omega C_I + G_{gr}}. \tag{4.89}$$

The last term in (4.87) is small compared to the terms retained in (4.89).

At frequencies satisfying $\omega\tau_n$ and $\omega\tau_p \ll 1$, C_{TXn} and C_{TXp} are given by (4.71) or (4.73). In this case, using (4.66) and the condition $n_T(x_c) \ll N_D$, we find

$$C_D \gg C_{TXp}, C_{TXn}. \tag{4.90}$$

Using (4.90), (4.89), becomes

$$Y_s \approx j\omega C_D + \frac{j\omega C_I G_{gr}}{j\omega C_I + G_{gr}}. \tag{4.91}$$

The admittance given by (4.91) corresponds to the simplified equivalent circuit for strong inversion in Fig. 4.5d. This circuit is valid even in the limit of very low frequencies.

For frequencies satisfying the condition $\omega C_I \gg G_{gr}$, the circuit of Fig. 4.5d can be simplified further. In strong inversion, this condition can be achieved even at quite low frequencies. For this case of intermediate ω, (4.91) becomes

$$Y_s \approx j\omega C_D + G_{gr} \tag{4.92}$$

in agreement with the circuit of Fig. 4.5e.

When (4.92) is used, the measured admittance of the entire MOS capacitor in strong inversion, including C_{ox}, becomes

$$Y_m = j\omega C_m + G_m \tag{4.93}$$

where

$$C_m = C_{ox}\left[1 - \frac{\omega^2 C_{ox}(C_{ox} + C_D)}{G_{gr}^2 + \omega^2(C_{ox} + C_D)^2}\right] \tag{4.94}$$

and

$$G_m = G_{gr}\frac{(\omega C_{ox})^2}{G_{gr}^2 + \omega^2(C_{ox} + C_D)^2}. \tag{4.95}$$

(j) The Transition Frequency

Using (4.94) we can define a useful *transition frequency* that marks the change from low frequency behavior ($C_m \approx C_{ox}$) to high frequency behavior [$C_m = C_{ox}C_D(C_{ox} + C_D)^{-1}$]. This transition frequency, ω_m, is the angular frequency for which C_m is midway between C_{ox} and C_{HF}, where C_{HF} is the value of the high frequency capacitance in strong inversion.* The midway

---

*This transition frequency also is the frequency for which G_m/ω is a maximum, where G_m is given by (4.95).

value of the measured capacitance C_{mid} is

$$C_{mid} = C_{ox}\left(1 - \frac{1}{2}\frac{C_{ox}}{C_{ox} + C_D}\right). \tag{4.96}$$

Comparing (4.96) with (4.94), we find the transition frequency ω_m to be

$$\omega_m = \frac{G_{gr}}{C_{ox} + C_D}.$$

$$= \frac{G_{gr}}{C_{ox}}\left(1 - \frac{C_{HF}}{C_{ox}}\right). \tag{4.97}$$

It is easy to measure ω_m and thereby determine G_{gr} using the known values of C_{ox} and the maximum to minimum capacitance ratio, C_{ox}/C_{HF}.

Equation (4.97) provides an estimate of ω_m provided G_{gr} is known. G_{gr} is evaluated in the next section and used to show that ω_m is a low angular frequency, such that $\omega_m \tau_p$ and $\omega_m \tau_n \ll 1$. These inequalities allow great simplification of (4.80) for G_{gr}. They also justify the neglect of the imaginary components of C_{TXp} and C_{TXn}, an important approximation needed to derive the equivalent circuits of Fig. 4.5d and 4.5e.

4.3.7 Further Discussion of the Generation–Recombination Conductance, G_{gr}

The influence of bulk traps on the admittance in strong inversion stems from G_{gr} in (4.91) and the equivalent circuit of Fig. 4.5d. This quantity was introduced in (4.80) and approximately evaluated as

$$G_{gr} = C_{bt}\tau_p^{-1}(x_c)\tau_n^{-1}(x_c)H_G. \tag{4.98}$$

As just discussed, for frequencies satisfying $\omega C_I \gg G_{gr}$, the equivalent circuit in Fig. 4.5d becomes that in Fig. 4.5e. For strong inversion, where C_I is large, this condition on ω can be satisfied for any reasonable frequency.

First, we consider intermediate frequencies satisfying $G_{gr}/C_I \ll \omega \ll 1/\tau_n$ and $1/\tau_p$. Simplification of (4.98) for this case shows that G_{gr} is independent of frequency and varies with temperature as n_i. Second, we consider high frequencies satisfying $G_{gr}/C_I \ll 1/\tau_n$, $1/\tau_p \ll \omega$. Simplication of (4.98) in this case shows that G_{gr} no longer determines the loss, which becomes dominated by the dissipative component of C_{TXn} in (4.89).

(a) Intermediate Frequencies: $G_{gr}/C_I \ll \omega \ll 1/\tau_n$ and $1/\tau_p$

For frequencies that satisfy (4.70), G_{gr} can be evaluated in the same way that (4.71) for crossover capacitance was obtained from (4.63). Equation (4.98) becomes

$$G_{gr} \approx C_{bt}(\tau_p - \tau_n)^{-1}\ln\left(\frac{\tau_p}{\tau_n}\right). \tag{4.99}$$

For the important case of an efficient recombination center, for which

$\tau_n = \tau_p = \tau$, (4.99) can be simplified just as for (4.73) to provide

$$G_{gr} = C_{bt}/\tau. \tag{4.100}$$

When (4.100) and (4.66) are used, we can estimate ω_m from (4.97) for a bulk trap level at midgap

$$\omega_m = \tau^{-1} \frac{C_{bt}}{(C_{ox} + C_D)}$$
$$= \tau^{-1}\sqrt{2}\left(\frac{n_T}{N_D}\right)\left(\frac{C_{HF}}{C_{ox}}\right). \tag{4.101}$$

Because $n_T \ll N_D$, (4.101) predicts $\omega_m \tau \ll 1$, thereby justifying the neglect of the imaginary components of C_{TXn} and C_{TXp} in deriving the equivalent circuits of Fig. 4.5.

We rewrite (4.99) to display its temperature dependence, due primarily to the carrier densities at crossover. From the discussion of (4.65), the band bending at crossover is $(v_T - u_B)$. Therefore, the carrier densities at crossover are

$$n_o(x_c) = n_i \exp(v + u_B) = n_i \exp(v_T) \tag{4.102a}$$

and

$$p_o(x_c) = n_i \exp[-(v + u_B)] = n_i \exp(-v_T) \tag{4.102b}$$

where $v_T > 0$ for a bulk trap level in the upper half of the gap. Using (4.68) and (4.102), in (4.99) we find*

$$G_{gr} = C_{bt}\left(\frac{n_i}{n_T}\right)\frac{\ln(c_p/c_n) - 2v_T}{\tau_{Tn}\exp(-v_T) - \tau_{Tp}\exp(v_T)} \tag{4.103}$$

where the bulk hole and electron *lifetimes*, τ_{Tp} and τ_{Tn}, have been introduced.[8]

$$\tau_{Tp} = [c_p n_{To}(x_c)]^{-1} \tag{4.104a}$$

and

$$\tau_{Tn} = [c_n n_{To}(x_c)]^{-1}. \tag{4.104b}$$

Unlike τ_p and τ_n, τ_{Tp} and τ_{Tn} are at most weakly temperature dependent. These lifetimes, typically the order of 10^{-6} sec, represent the time for a

---

*In this low frequency case, evaluation of G_{gr} can be done more carefully, including the variation of carrier densities with band bending, but retaining the removal of the field at crossover from the integration. If this is done, one obtains the result of Sah, Noyce, and Shockley[14] for the low-bias case. Equation (4.103) agrees with Ref. 14, provided band bending at the interface v_s satisfies the condition

$$\exp\left(\frac{v_T + v_s - u_B}{2}\right) \ll 1$$

that is, provided that crossover is not very near the interface. Hence (4.103) is not accurate if the crossover region in which bulk traps are active reaches the interface, causing a reduction in the size of the active region.

carrier density fluctuation to decay to its equilibrium concentration by recombination through bulk traps of density $n_{To}(x_c)$. Minority carrier response time [see (4.116)] is much longer than lifetime but is related to it.

Equation (4.103) shows that as v_T becomes larger, G_{gr} decreases exponentially with v_T because generation and recombination become less efficient for bulk trap levels further from midgap. In practice, if bulk trap levels are present primarily at one energy v_T, this exponential dependence on v_T allows extraction of v_T from the measured temperature dependence of G_{gr}, which will depart from that of n_i, the other major source of temperature dependence in G_{gr}.

(b) High Frequencies: $G_{gr}/C_I \ll 1/\tau_n$ and $1/\tau_p \ll \omega$.

At high frequencies, $\omega\tau_p$ and $\omega\tau_n \gg 1$, using (21) in Appendix II, G_{gr} from (4.98) is given by

$$G_{gr} \approx \frac{\pi C_{bt}\tau_p^{-1}\tau_n^{-1}}{2\omega} \tag{4.105}$$

where a reactive term involving $\ln[2(\tau_p^{-1} + \tau_n^{-1})^{-1}\omega]$ has been dropped, using (4.90). Substituting (4.68) and (4.102) in (4.105) yields

$$G_{gr} = \pi C_{bt}c_p c_n n_i^2/(2\omega). \tag{4.106}$$

From (4.106), at high frequencies the temperature dependence of G_{gr} is that of n_i^2, unlike that of (4.103) for intermediate frequencies, which is that of n_i for small v_T.

According to (4.105) and (4.106), for frequencies such that $\omega\tau_n$ and $\omega\tau_p \gg 1$, G_{gr} falls off with frequency as ω^{-1}. However, we now show that $j\omega C_{TXn}$ has a real component that *increases* with frequency in this frequency range. Thus C_{TXn} dominates the loss at high frequencies. In fact, $j\omega C_{TXn} + G_{gr}$ is a monotonic function increasing with ω.

To determine the frequency at which the dissipative component of C_{TXn} becomes important, we use (27) of Appendix II and (4.69) to find

$$C_{TXn} = C_{bt}\left[\frac{1}{2j\omega\tau_n}\ln\frac{\tau_n}{\tau_p} + \frac{1}{2}\left(1 + \frac{1}{j\omega\tau_n} - \frac{1}{j\omega\tau_p}\right)H_G/\tau_n\right]. \tag{4.107}$$

From (4.80)

$$G_{gr} = C_{bt}\tau_p^{-1}\tau_n^{-1}H_G. \tag{4.108}$$

From (15) of Appendix II, for the case $\tau_p \gg \tau_n$,

$$H_G \approx \frac{\tau_n}{(1 + j\omega\tau_n)}\left[\ln\left(\frac{\tau_p}{\tau_n}\right) + 2\ln(1 + j\omega\tau_n)\right]. \tag{4.109}$$

Using (4.109) in (4.107) we find

$$j\omega C_{TXn} \simeq \left(\frac{C_{bt}}{\tau_n}\right)\ln(1 + j\omega\tau_n)$$

$$= \left(\frac{C_{bt}}{\tau_n}\right)\left\{\frac{1}{2}\ln[1 + (\omega\tau_n)^2] + j\tan^{-1}(\omega\tau_n)\right\}. \tag{4.110}$$

The expression (4.110) is the same as the Lehovec admittance (5.46) for interface traps, and expresses the loss due to delay between capture and emission of electrons by bulk trap levels.

Substituting (4.109) into (4.108), we find

$$G_{gr} = \left(\frac{C_{bt}}{\tau_n}\right)\left(\frac{\tau_n}{\tau_p}\right)\frac{\ln\left(\frac{\tau_p}{\tau_n}\right) + 2\ln(1 + j\omega\tau_n)}{1 + j\omega\tau_n}. \tag{4.111}$$

Thus for frequencies satisfying $\omega\tau_n \ll 1$, $j\omega C_{TXn}$ is negligible and the loss is given by the low $\omega\tau_n$ limit of (4.111). That is, for $\tau_p \gg \tau_n$ and $\omega\tau_n \ll 1$, (4.111) becomes

$$G_{gr} = \left(\frac{C_{bt}}{\tau_p}\right)\ln\left(\frac{\tau_p}{\tau_n}\right) \tag{4.112}$$

in agreement with (4.99). Also, (4.89) becomes (4.91). However, at high frequencies satisfying $\omega\tau_n \gg 1$, G_{gr} from (4.111) falls off as ω^{-1}, becoming negligible compared to $j\omega C_{TXn}$ of (4.110), which increases monotonically with $\omega\tau_n$.

The angular frequency ω_c for which the onset of important loss occurs due to ac delay between capture and emission, is estimated from the expression

$$j\omega C_{TXn} + G_{gr} = \left(\frac{C_{bt}}{\tau_n}\right)\left[\left(\frac{\tau_n}{\tau_p}\right)\ln\left(\frac{\tau_p}{\tau_n}\right) + \ln\left(1 + j\omega\tau_n\right)\right] \tag{4.113}$$

obtained from (4.110) and (4.111). The sum (4.113) represents the bulk trap contribution to the admittance (4.89) for $\omega \gg G_{gr}/C_I$ which, in strong inversion, is satisfied even for low frequencies.* The second, or Lehovec term in (4.113) becomes equal to the generation-recombination term when $\omega = \omega_c$ where, for $\tau_p \gg \tau_n$

$$\omega_c = \frac{1}{\tau_n}\left[\left(\frac{\tau_p}{\tau_n}\right)^2\left(\frac{\tau_n}{\tau_p}\right) - 1\right]^{1/2} \tag{4.114}$$

The condition (4.114) leads to $1.0 > \omega_c \tau_n > .3$ for $\tau_p/\tau_n < 100$. For n-type silicon, usually $\tau_p \ll \tau_n$.† Therefore, we consider this case.

*The capacitive divider C_D/C_{w-x_c} can be taken as unity in (4.89) for the purpose of estimating ω_c.

†For $\tau_p \ll \tau_n$, H_G is given by (4.109) with τ_p and τ_n interchanged. In this case, (4.110) becomes

$$j\omega C_{TXn} = \left(\frac{C_{bt}}{\tau_n}\right)\left[j\omega\tau_p \ln\left(\frac{\tau_n}{\tau_p}\right) - (1 - j\omega\tau_p)\ln(1 + j\omega\tau_p)\right](1 + j\omega\tau_p)^{-1},$$

and (4.111) becomes

$$G_{gr} = \left(\frac{C_{bt}}{\tau_n}\right)\left[\ln\left(\frac{\tau_n}{\tau_p}\right) + 2\ln(1 + j\omega\tau_p)\right](1 + j\omega\tau_p)^{-1}.$$

Now C_{TXn} is negligible for $\omega\tau_p \ll 1$ and (4.91) applies using (4.112) with τ_p and τ_n interchanged. To estimate ω_c, we use the sum $j\omega C_{TXn} + G_{gr}$ and find

$$\omega_c = \frac{\tau_n}{\tau_p^2}.$$

This case usually would apply to a bulk trap level below midgap, so that $p_o(x_c) \gg n_o(x_c)$.

The carrier densities at crossover are evaluated using (4.102), and (4.114) is replaced by

$$\omega_c = \left(\frac{c_n^2}{c_p}\right) n_i \exp(-3v_T).$$ (4.115)

For τ_p to satisfy $\tau_p \ll \tau_n$, we expect the bulk trap level to be in the lower half of the gap ($v_T < 0$). For example, if the bulk trap level lies $3\,kT/q$ below midgap, (4.115) predicts that $f_c = \omega_c/(2\pi) = 130\,\text{kHz}$ for $c_p = c_n = 10^{-8}\,\text{cm}^3/\text{sec}$.

Available bulk trap measurements concentrate upon frequencies near ω_m of (4.97). Combining (4.66), (4.112), and (4.115), when $\tau_p \ll \tau_n$ we find

$$\frac{\omega_m}{\omega_c} = \sqrt{2}\left(\frac{\tau_p}{\tau_n}\right)^2\left(\frac{n_T}{N_D}\right)\left(\frac{C_{HF}}{C_{ox}}\right)\frac{\ln(\tau_n/\tau_p)}{(1 + v_T/u_B)^{1/2}}.$$

Accordingly, $\omega_m \ll \omega_c$, and the available measurements do not extend to frequencies high enough to cause an increase in bulk trap conductance due to the Lehovec contribution. However, Chan et al.[15] have observed a *drop* in conductance beginning at frequencies as low as 30 Hz. This drop is unexpected,* and the frequency of 30 Hz is much closer to ω_m of (4.97) than to ω_c. A drop in conductance at low frequencies is not observed in samples with thermally grown oxides too thick for tunneling.[2,3] The results of Chan et al.[15] may be a result of their use of MNOS structures. Tunneling and nitride conduction are expected in such samples, and can affect the measured conductance.

4.3.8 Response Time in Terms of Basic Silicon Parameters

Minority carrier response time from Section 4.3.7 is $\tau_R = C_D/G_{gr}$. Combining (4.103) for G_{gr} with (4.66), we obtain

$$\tau_R = \frac{C_D}{G_{gr}} = \frac{1}{\sqrt{2}}\left(\frac{N_D}{n_i}\right)\left(1 - \frac{v_T}{u_B}\right)^{1/2}\left[\frac{\tau_{Tn}\exp(-v_T) - \tau_{Tp}\exp(v_T)}{\ln(c_p/c_n) - 2v_T}\right]$$ (4.116a)

which, for $\tau_n = \tau_p = \tau$, becomes

$$\tau_R = \frac{1}{\sqrt{2}}\left(\frac{N_D}{n_i}\right)\tau_T\left(1 - \frac{v_T}{u_B}\right)^{1/2}$$ (4.116b)

with $\tau_T = (\tau_{Tn}\tau_{Tp})^{1/2}$. With $N_D = 10^{16}\,\text{cm}^{-3}$, $\tau_T = 10^{-6}\,\text{sec}$ and $v_T = 0$, (4.116) results in $\tau_R = 0.7\,\text{sec}$, which is rather long. The amplifying factor (N_D/n_i) makes possible measurement of rather short lifetimes, τ_T, by observation of the much longer minority carrier response time τ_R.

---

*Chan et al.[15] interpreted their results using the theory of Mar et al.[11] However, this theory neglects the Lehovec admittance contained in C_{TXn} and is based entirely upon G_{gr} of (4.106). Such neglect is not justified.

In deriving (4.116) and in the equivalent circuit, spatial redistribution of inversion layer charge in response to the ac voltage has been neglected. This effect is taken into account in deriving an expression for the high frequency C-V curve in Section 4.4.

4.3.9 Experimental Test of the Calculation

The generation-recombination conductance G_{gr} given by (4.103) is exponentially dependent on temperature because of its dependence on n_i. The other temperature-dependent factors such as lifetimes and C_{bt} in (4.103) vary so slowly with temperature that the exponential temperature dependence of n_i dominates.

One test of the theory is to measure the temperature dependence of the MOS capacitance. Figure 4.9 shows the capacitance of an MOS capacitor measured at 100 kHz as a function of bias with temperature as parameter. Figure 4.9 shows that there is little effect on the C-V characteristics in accumulation and depletion over the temperature range shown. The major effect of temperature occurs in inversion.

At room temperature (27°C) minority carriers do not follow the 100 kHz signal; thus a high frequency curve is measured. However, as temperature is increased, minority carriers begin to follow because generation and recombination rates increase with temperature [G_{gr} increases so that $\tau_R = C_D/G_{gr}$ decreases; C_D varies more slowly with temperature than does G_{gr}, see (3.47)] and the transition is made from a high frequency curve to a low frequency curve.

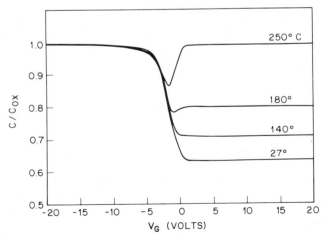

Fig. 4.9 Normalized capacitance as a function of bias with temperature as parameter measured at 100 kHz. Acceptor concentration was 1.45×10^{16} cm^{-3} and oxide thickness 2000 Å. After Grove et al.[5] Copyright (1965), Pergamon Press, Ltd. Reprinted with permission.

To define a transition frequency,[5] we regard any C-V curve exhibiting a minimum capacitance, such as the curve in Fig. 4.9 measured at 180°C, as a low frequency curve even though capacitance in strong inversion does not increase to C_{ox}. This definition is merely a matter of convenience and is arbitrary.* Those C-V curves without a distinct minimum, such as the curve measured at 140°C in Fig. 4.9, are considered to be of the high frequency type.

With this definition, a distinct transition from high-frequency to low frequency characteristics can be observed at about 160°C. By systematically varying both temperature and measurement frequency, the temperature dependence of the transition frequency can be determined. The results of such experiments,[5] for both p-type and n-type silicon, including a gold doped sample, are shown in Fig. 4.10. Figure 4.10 shows that the transition frequency has an activation energy of about half the bandgap energy in all three cases. This activation energy is that of n_i. Thus the observed temperature dependence of the transition frequency agrees with

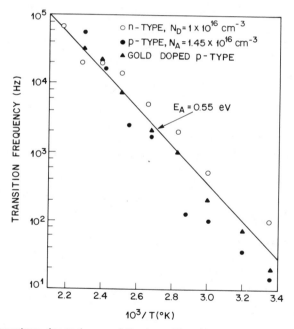

Fig. 4.10 Temperature dependence of the transition frequency between high and low frequency C-V curves. Oxide thickness was 2000 Å. The gold doped p-type sample was doped to a gold concentration of about 10^{15} cm^{-3}. After Grove et al.[5] Copyright (1965), Pergamon Press, Ltd. Reprinted with permission.

---

*The transition frequency in Ref. 5 could be taken more usefully as the frequency ω_m of (4.97) for which a theoretical expression is known.

(4.97) using (4.103) for G_{gr}. Therefore, the dominant mechanism for controlling minority carrier response must be generation and recombination through bulk traps in the measured samples over the temperature range 21–182°C.

(a) Loss Due to Interface Traps in Inversion

We argued in Section 4.3.1(c) that interface traps would not contribute significantly to generation-recombination current at *room* temperature because the Fermi level at the silicon surface was too far away from midgap in strong inversion. However, as temperature is increased, the Fermi level moves closer to midgap. Nonetheless, Fig. 4.10 shows that interface traps still do not play a significant role. If interface traps were dominant, τ_R would increase with increasing temperature rather than decrease as observed.

To be quantitative, examine the interface trap generation and recombination resistance that, in strong inversion for an n-type substrate, is given by (5.65)

$$G_{grs} = qD_{it}N_Dc_n \exp\left(\frac{q\psi_s}{kT}\right) \ln\left(\frac{\tau_n}{\tau_p}\right) \quad \text{(mhos/cm}^2) \qquad (4.117)$$

where D_{it} is total interface trap level density per unit area per electron volt. The temperature dependence of (4.117) is dominated by the exponential term. Increasing temperature increases G_{grs} at constant gate bias because $\psi_s < 0$. However, also for this reason, the exponential is so small that (4.117) is negligible compared to the bulk trap contribution. Therefore, bulk traps account entirely for the behavior of Fig. 4.10.

(b) Loss Due to Bulk Traps in Depletion

Interface traps are not important in strong inversion, where bulk traps dominate the conductance. In depletion, however, interface traps dominate the conductance, as discussed in Chapter 5. How important are bulk traps in depletion?

In depletion and weak inversion, C_I becomes small and (4.88) does not apply at any frequency. The simplified equivalent circuit of Fig. 4.5d does not apply, and analysis must be based upon (4.87).

For bulk trap loss to occur, crossover also must occur. For a bulk trap level below midgap and n-type material, crossover will not occur in depletion. Hence we consider only bulk trap levels above midgap. Then the condition $\tau_p \gg \tau_n$ usually applies, and C_{TXn} is given by (4.110), and G_{gr} by (4.111). C_{TXp} can be found using (4.69b) and (26) of Appendix II

$$C_{TXp} = C_{bt}\left(\frac{\tau_n}{\tau_p}\right)(j\omega\tau_n)^{-1}(1 + j\omega\tau_n)^{-1}\left[j\omega\tau_n \ln\left(\frac{\tau_p}{\tau_n}\right) + (j\omega\tau_n - 1)\ln(1 + j\omega\tau_n)\right].$$
$$(4.118)$$

The term in C_{TXn} from (4.110) represents the Lehovec admittance of the bulk traps. Comparison of C_{TXn} from (4.110) with the other terms in (4.87) shows that C_{TXn} dominates the bulk trap loss at all frequencies.

For a bias that places crossover near the silicon surface, bulk traps and interface traps could respond at the same frequencies. Assuming that $(n_T/N_D) \approx 0.1$, we find that C_{bt}/q from (4.66) is $5 \times 10^9 \, \text{cm}^{-2} \, \text{eV}^{-1}$ at $N_D = 10^{14} \, \text{cm}^{-3}$ for $v_T = \ln 10$. Hence, particularly for low densities of interface trap levels, the bulk trap contribution to the conductance should be checked experimentally before it is ignored in interface trap measurements.

In weak inversion interface traps can make a significant contribution to generation and recombination current as explained in Section 4.3.1(c). Because of the complexity when both interface traps and bulk traps are important, no attempt is made here to extract information about interfacial, bulk, or oxide properties from the weak inversion region of the MOS capacitor characteristics.

4.3.10 Analysis of Diffusion-Controlled Response

Next, the equivalent circuit of the MOS capacitor in strong inversion is derived when diffusion controls minority carrier response. It is shown by comparing this analysis to experiment that τ_R is controlled by generation-recombination at low temperatures and by diffusion as temperature increases.

The equivalent circuit for diffusion controlled minority carrier response is shown in Fig. 4.11a. In this equivalent circuit, G_d is the diffusion conductance for hole generation and diffusion through the quasi-neutral region of the silicon. In series with G_d is G_p, the hole conductance of the depletion layer. The ac current flows through these conductances between the conduction band and the inversion layer capacitance. To compare the temperature dependence of diffusion controlled to generation-recom-

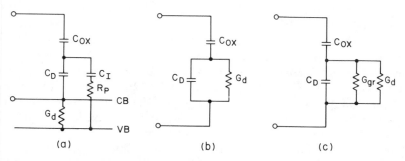

(a)   (b)   (c)

Fig. 4.11 (a) Equivalent circuit of the MOS capacitor in inversion for diffusion controlled minority carrier response. Symbols CB and VB are conduction and valence band edges, respectively; open circles denote the external terminals; G_d is the diffusion conductance for hole diffusion through the quasi-neutral region of the silicon; and R_p is the resistance of the depletion layer to hole flow. (b) Simplified version of (a). (c) Equivalent circuit of the MOS capacitor in strong inversion including both diffusion and generation-recombination.

bination controlled minority carrier response time, it is necessary to calculate G_d. The relation for G_d, calculated by Garrett,[6] by Lehovec and Slobodskoy,[12] and in Section (a) of Appendix IV, is

$$G_d = \frac{q\mu_p n_i^2}{L_p N_D} \quad (\text{mhos/cm}^2) \tag{4.119}$$

where μ_p is bulk hole mobility (in cm^2/V-sec) and $L_p = (D_p \tau_p)^{1/2}$ is the hole diffusion length (in centimeters).*

It is shown in Section (b) of Appendix IV that $G_p \gg G_d$ so that G_p can be treated as a short-circuit and the equivalent circuit in Fig. 4.11a becomes that of Fig. 4.11b. Minority carrier response time when diffusion controls is $\tau_R = C_D/G_d$ as seen from Fig. 4.11b.

The temperature dependence of G_d is determined by n_i^2 from (4.119) so that the activation energy will be E_g, compared to $E_g/2$ for G_{gr}.

Minority carrier response time is C_D/G_d. Combining the expression for C_D used in calculating (4.116) with (4.119) yields

$$\tau_R = \left(\frac{q\epsilon_s N_D^3}{\phi_B}\right)^{1/2} \frac{L_p}{2\mu_p n_i^2}. \tag{4.120}$$

4.3.11   Overall Temperature Dependence of Minority Carrier Response Time

Combining the equivalent circuit of Fig. 4.5e in which G_{gr} represents the recombination-generation resistance in the depletion layer with the equivalent circuit in Fig. 4.11b for diffusion in the quasi-neutral region yields the overall equivalent circuit of Fig. 4.11c which relates G_d and G_{gr} to the measured admittance.

To obtain the temperature ranges over which τ_R is controlled by G_{gr} and by G_d, the admittance of the MOS capacitor is measured as a function of gate bias with temperature as parameter. Capacitance C_m and equivalent parallel conductance, G_m, measured at 6 kHz as a function of bias with temperature as parameter are shown in Figs. 4.12a,b, respectively. Equivalent parallel conductance as a function of temperature peaks at 120°C in Fig. 4.12b. The dashed lines in Fig. 4.12b are on the high and the solid lines on the low temperature side of the peak. Both C_m and G_m saturate in inversion at negative bias. The small bumps appearing in the curves in depletion are caused by interface traps.

From the equivalent circuit in Fig. 4.11c, G_I, which consists of G_{gr} and G_d in parallel, and C_D can be extracted from the measured admittance Y_m. Then the overall value, $\tau_R = C_D/G_I$ can be calculated. The value of τ_R^{-1} is the sum of the reciprocals of (4.116) and (4.120). To make this calculation,

---

* A more complete analysis which applies when the diffusion length L_p is not small compared to the sample thickness L multiplies (4.119) by coth (L/L_p).[2,6]

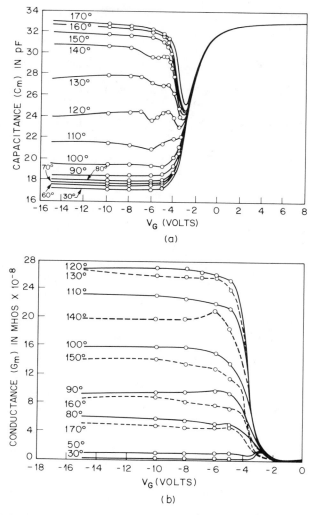

Fig. 4.12 (a) Capacitance as a function of gate bias with temperature as parameter measured at 6 kHz. Sample is the same as in Fig. 4.1. (b) Equivalent paralleled conductance as a function of gate bias with temperature as parameter measured at 6 kHz on the same sample as part (a). After Goetzberger and Nicollian.[3] Copyright (1967), American Telephone and Telegraph Company. Reprinted with permission.

the measured admittance, $Y_m = G_m + j\omega C_m$, first is converted to an impedance, the reactance of C_{ox} subtracted, and the resulting impedance is converted back to an admittance. The real part of the admittance calculated in this way is

$$G_I = \frac{\omega^2 C_{ox}\tau_o(1 + \omega^2\tau_m^2)}{\omega^2\tau_o^2 + [\omega^2\tau_m(\tau_o - \tau_m) - 1]^2} \tag{4.121}$$

where $\tau_m = C_m/G_m$ and $\tau_o = C_{ox}/G_m$. From the imaginary part of the admittance, C_D is given by

$$C_D = \frac{[\omega^2\tau_m(\tau_o - \tau_m) - 1]C_{ox}(1 + \omega^2\tau_m^2)}{\omega^2\tau_o^2 + [\omega^2\tau_m(\tau_o - \tau_m) - 1]^2}. \tag{4.122}$$

The conductance G_l in Fig. 4.13 is calculated using (4.121).* The values of G_m and C_m measured at -15 V and C_{ox} all are obtained from Fig. 4.12. Figure 4.13 is an Arrhenius plot of G_l versus $1/T$ obtained from (4.121).

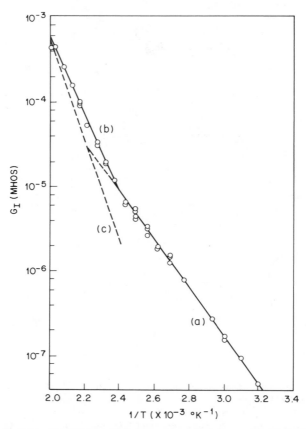

Fig. 4.13 Equivalent parallel conductance calculated from admittance measured at 6 kHz and a gate bias of -15 V as a function of reciprocal degrees Kelvin. The experimental points indicated by circles were obtained from Fig. 4.12b. Multiple circles at a given temperature represent several runs. The solid lines are the best fit to the experimental points. Curve c is obtained by subtracting the values of G_l in curve b from the extrapolation of curve a at each temperature. After Goetzberger and Nicollian.[3] Copyright (1967), American Telephone and Telegraph Company. Reprinted with permission.

*Alternatively, ω_m from (4.97) can be determined, where G_{gr} becomes $G_l = (G_{gr} + G_d)$ when both generation-recombination and diffusion play a role.

For temperatures below 140°C, curve a in Fig. 4.13 shows that the activation energy is 0.56 eV. This value is close to $E_g/2$ or the activation energy of n_i, which is in agreement with (4.102) for generation-recombination in the depletion layer. Above 140°C a new process dominates, as shown by the break in the $1/T$ curve. If the high-temperature points in Fig. 4.13 are connected by a straight line (curve b), an activation energy of 0.908 eV is obtained. This energy is lower than the expected activation energy for n_i^2 from (4.119), which is $E_g = 1.12$ eV. The discrepancy can be removed by correcting the high-temperature points (curve b in Fig. 4.13) by subtracting the contribution of depletion layer generation and recombination. This correction is done by extrapolating curve a into the high-temperature region and subtracting the extrapolated values of G_I from the values of G_I obtained from curve b. The result is curve c, which has an activation energy of 1.17 eV, which is very close to that expected for diffusion controlled minority carrier response time, namely, E_g.

By taking the inversion conductance G_I, from Fig. 4.13 at room temperature and calculating C_D from the least value of C_{HF} in Fig. 4.12a as $C_D^{-1} = C_{ox}^{-1}(C_{ox}/C_{HF} - 1)$, minority carrier response time is calculated to be $\tau_R = C_D/G_I = 1.6$ msec. This τ_R is much shorter than calculated previously from (4.116) because lifetime in the epitaxial silicon layer was much less than the typical value of $1\,\mu$sec assumed earlier. This value of τ_R corresponds to a transition frequency from (4.97) of 51 Hz. Figure 4.1, measured on the same sample as Figs. 4.12 and 4.13, demonstrates that the midway capacitance of 25 pF occurs near 50 Hz. That is, the estimate based upon (4.97) and the capacitance is consistent with the estimate based upon (4.121) and (4.122).

The transition temperature at which minority carrier response time changes from control by depletion layer generation-recombination to control by diffusion depends on such parameters as ionized impurity concentration and lifetime. Therefore, the transition temperature differs from sample to sample. For example, the transition temperature for the samples in Fig. 4.10 is above 180°C, whereas that for the sample in Fig. 4.13 is 140°C. When (4.119), (4.120), and the relation $L_p = (D_p\tau_p)^{1/2}$ are used with the measured data, high temperature lifetime, and diffusion length can be calculated.

The transition temperature, T_T can be estimated theoretically from the condition that the response times due to both mechanisms be equal. This condition is

$$G_{gr} = G_d . \tag{4.123}$$

When (4.100) is used for G_{gr}, (4.66) for C_{bt}, (4.72) for τ, and (4.119) for G_d, (4.123) becomes

$$n_i(T_T) = \frac{\sqrt{2}\,C_D L_p}{q\mu_p \tau_T} . \tag{4.124}$$

The temperature dependence of n_i is taken from (2.17) to find

$$\left(\frac{kT_T}{q}\right) = \frac{\dfrac{E_g}{2q}}{\ln\left[\dfrac{q(M_v M_c \beta \mu_p \tau_T)^{1/2}}{C_D}\right]} \ . \tag{4.125}$$

Using $(M_v M_c)^{1/2} = 3.3 \times 10^{15} T^{3/2}\,\text{cm}^{-3}$, $\mu_p = 500\,\text{cm}^2\,\text{V}^{-1}\text{sec}^{-1}$, $N_D = 10^{15}\,\text{cm}^{-3}$ and $\tau_T = 10^{-6}\,\text{sec}$, we find $kT_T/q = 0.032$ or $T_T = 371°\text{K}$, ignoring the temperature dependence of mobility and lifetime.

Because the temperature dependence of the mobility is uncertain, and that of n_i is not given accurately by (2.17), estimates of T_T based upon (4.125) will not be accurate. However, (4.125) does display the qualitative dependence of T_T upon the various sample parameters.

The dominant mechanism determining minority carrier response time can be different for semiconductors other than silicon. For example, diffusion is dominant at room temperature for narrow-gap semiconductors like germanium, and depletion layer generation-recombination is dominant at temperatures below room temperature. Because n_i is so large in germanium, it is possible to obtain a low frequency C-V curve at room temperature even at 100 kHz. In contrast, measurement of a room temperature C-V curve for silicon at 100 kHz would almost certainly result in a high frequency curve.

For semiconductors with a wider bandgap than silicon, generation and recombination will dominate up to much higher temperatures. In fact, for semiconductors that have very large bandgap energies, n_i becomes so small that all sources of minority carriers, except from an inversion layer beyond the gate or generation by light, may be so slow that an inversion layer will take a long time to form after the application of an appropriate gate bias.

4.3.12 Response Time Governed by an Inversion Layer Beyond the Gate

Generation-recombination, diffusion, and light are not the only mechanisms that determine minority carrier response in an MOS capacitor. If the silicon surface beyond the gate is inverted, minority carrier response time will be determined by this inversion layer at all reasonable temperatures.* It has been experimentally shown[4,16] how an inversion layer beyond the gate affects minority carrier resonse time and the C-V characteristics in inversion for an MOS capacitor made on p-type silicon.

The silicon surface can become inverted by charge within the oxide or by a potential buildup caused by current flow in the presence of moisture along the oxide surface. Typically, charges within the oxide are sodium ions or oxide fixed charge characteristic of thermal oxidation. The exact

*See footnote in Section 4.3.2(a), p. 107.

mechanism for moisture induced current flow along the oxide surface is unknown. However, the prevailing model is that this current is carried by residual contaminants that become ionized and mobile in the presence of an adsorbed water layer.*

Usually, the area of inverted surface beyond the gate is large compared to the area of the gate itself. Therefore, the inversion layer beyond the gate is a large source of minority carriers that can move laterally into and out of the MOS capacitor very rapidly. The inversion layer beyond the gate was produced originally by generation and recombination. Once in existence, the inversion layer becomes a source and sink for minority carriers. Dominance of minority carrier response time by an inversion layer beyond the gate can occur on both n-type and p-type substrates. Inversion caused by charges within the oxide, is most common on p-type substrates because oxide charge is usually positive. Inversion beyond the gate is important for very lightly doped p-type substrates and when processing is not done carefully enough either to avoid significant sodium contamination or to minimize oxide fixed charge.

After gate bias has been applied, (positive for a p-type substrate and negative for an n-type substrate), the flow of moisture induced current between the gate and the silicon can invert the silicon beyond the gate by the buildup of a potential across the oxide surface. Now, instead of dropping abruptly to zero at the edge of the gate, the potential spreads out across the oxide surface.

This moisture effect usually is undesirable in MOS capacitor measurements. It can be made negligible by placing the MOS capacitor in a box where humidity is kept low by directing a stream of dry nitrogen into it, as discussed in Section 12.8.1 The effect of oxide charge and moisture are undesirable in integrated circuits as well. For example, potential buildup along the oxide between adjacent leads at different potentials can invert the silicon surface, resulting in undesirable current flow paths. The moisture problem can be kept under control either by placing the integrated circuit in a hermetic encapsulation, or by coating the integrated circuit with silicone rubber. Silicone rubber is permeable to water but prevents the formation of a continuous adsorbed water film on the oxide surface between electrodes through which a moisture induced current could flow.

Another situation where neither generation and recombination nor diffusion is dominant occurs when an island of conductivity type opposite to that of the substrate is implanted under the gate. If the dimensions of this implanted island are comparable to the size of the gate, carriers in the substrate can couple easily to the inversion layer produced by gate bias in the island. The effect on minority carrier response time is similar to that

* Because water has a high dielectric constant (80), the electrostatic forces binding compounds such as salts together are weakened sufficiently that the compound becomes ionized into a separate cation and anion.

produced by an inversion layer beyond the gate caused either by charges in the oxide or current flow along the oxide surface.

In the next section we first discuss the effect on minority carrier response time, and thus on the C-V curve, of an inversion layer beyond the gate produced by charge within the oxide. Then we discuss the problem of current flow along the oxide surface in a moist ambient.

(a)   *Charges in the Oxide*

The C-V characteristics of a p-type MOS capacitor are affected when the surface beyond the gate is inverted by oxide charge. Oxide charge beyond the gate does not affect the capacitance in accumulation and depletion. However, when oxide charge beyond the gate inverts the surface, instead of saturating in strong inversion, the capacitance rises to C_{ox}, as for a low frequency C-V curve, even at high frequencies ($f > 1$ kHz).

Such a C-V curve, measured at 100 kHz, is illustrated in Fig. 4.14. In this sample, undoubtedly a small component of the oxide charge was sodium ions at the Si–SiO$_2$ interface. With the amounts of interface charge and ionized sodium density in this sample, the silicon surface was strongly inverted everywhere.

Electrons (minority carriers) in the inversion layer cannot follow the 100 kHz signal. However, the inversion layer is a conductor. Consequently, an ac current flows along the inversion layer beyond the gate, then down across the depletion layer into the quasi-neutral silicon, and out the back contact. This ac current flow, for zero gate bias, is depicted schematically in Fig. 4.15.

As a result of lateral spreading of the ac current along the inversion layer beyond the gate, ac current flows over an area several times larger than the

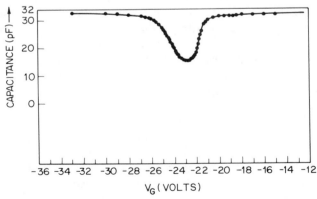

Fig. 4.14   Capacitance measured at $T = 300°$K and $f = 100$ kHz as a function of gate bias. Sample had an acceptor concentration of 2×10^{15} cm^{-3}, oxide thickness was 1300 Å, the aluminum gate diameter was 380 μm, and oxide charge density was 10^{12} cm^{-2}. After Nicollian and Goetzberger.[16] Copyright (1965), IEEE.

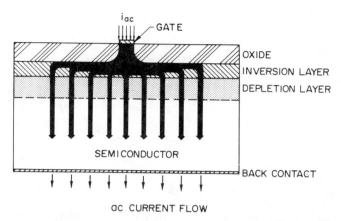

Fig. 4.15 Cross section of an MOS capacitor schematically showing ac current flow at zero bias from the gate, through the oxide, out along the inversion layer, beyond the gate, across the depletion layer, into the quasi-neutral silicon, and out the back contact. The radial decrease in current along the inversion layer is depicted. There is negligible ac current flow beyond a critical or cut-off radius, depending on inversion layer sheet resistance, depletion layer capacitance, and frequency. After Nicollian and Goetzberger.[16] Copyright (1965), IEEE.

area of the gate. Therefore, the additional capacitance associated with the area beyond the gate is very large. Consequently, the measured capacitance rises to C_{ox}. The frequency response of the MOS capacitor, when the surface under the gate is inverted, is now governed by the resistance of the inversion layer and the capacitance of the depletion layer beyond the gate.

To see how ac current couples to the inversion layer beyond the gate, note that ac current must first flow through the inversion layer under the gate before it can spread out into the inversion layer beyond. Gate bias determines the resistance of the inversion layer underneath the gate, controlling the ac current that flows into the inversion layer beyond. This control allows displacement current flowing through the MOS capacitor to couple gradually into the inversion layer beyond the gate and results in the gate bias dependence of the C-V curve in inversion seen in Fig. 4.14. The greatest influence exerted by gate bias is from the capacitance minimum to saturation at C_{ox}. Saturation occurs as the inversion layer under the gate becomes stronger.

The fixed inversion layer beyond the gate, established by the oxide charge, is a distributed low-pass filter that determines the frequency dependence of the C-V characteristics when the surface under the gate is inverted. That is, this distributed network determines minority carrier response time. Figure 4.16 shows the equivalent circuit of the combined capacitor and extended inversion layer, including the distributed low-pass filter representation of the inversion layer beyond the gate, the bias-variable series resistance of the inversion layer under the gate R_f, and the depletion layer capacitance under the gate C_D.

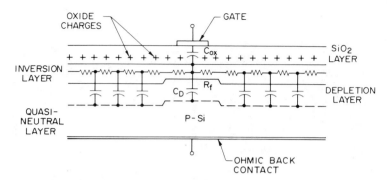

Fig. 4.16 Cross section of an MOS capacitor having a p-type substrate that is inverted by positive charge in the oxide. An equivalent circuit is superposed showing the distributed network low-pass filter line whose cut-off frequency determines the frequency dependence of the C-V characteristic in the gate bias region where the silicon surface under the gate is inverted. The value R_f is the inversion layer resistance under the gate coupling the ac current to the inversion layer beyond the gate.

The frequency dependence of the C-V characteristic, determined by the low-pass filter characteristics of a very strong inversion layer beyond the gate, is shown in Fig. 4.17. Figure 4.17 shows that the C-V characteristics are independent of frequency, within experimental error, in accumulation and depletion. However, in inversion, the C-V curves become strongly frequency dependent. The cut-off frequency of the distributed line is about 5 MHz, considerably higher than the value of 51 Hz in Fig. 4.1, where response time is governed by generation and recombination. The cut-off frequency of the low-pass filter is determined by the inversion layer charge and silicon doping densities. Thus curves similar to Fig. 4.17 would be

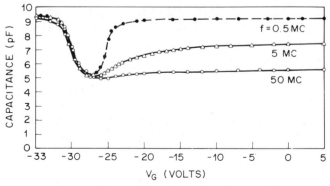

Fig. 4.17 Measured capacitance as a function of gate bias with frequency as parameter ($T = 300°K$). The sample had an acceptor density of 2.3×10^{15} cm^{-3}, an oxide thickness of 2000 Å, and a gate diameter of 250 μm. Positive oxide charge has inverted the silicon surface beyond the gate. After Nicollian and Goetzberger.[16] Copyright (1965), IEEE.

obtained at a fixed frequency with inversion layer charge density beyond the gate as parameter.

(b) Surface Leakage Current

To study the effect of potential buildup beyond the gate quantitatively, it is necessary to fix the potential across the oxide at two points. For this purpose, a ring-dot metallization pattern evaporated over the oxide will suffice. Figure 4.18 shows the cross section of a ring-dot MOS structure. The silicon substrate is p-type, and the entire surface of the silicon is assumed to be inverted initially by positive charges within the oxide. On application of a negative bias V_R to the ring with respect to the dot, the potential distributes itself along the oxide surface as shown schematically in Fig. 4.18. Figure 4.18 assumes the surface conductance of the oxide is uniform and that the inversion layer will be pinched off for voltages below a value V_p called the *pinch-off voltage*. That is, V_p is sufficiently large to overcome the field, and the silicon surface is no longer inverted. The distance r is measured from the center of the dot, and r_p is the pinch-off radius.

To calculate $V(r)$[16] we use the oxide surface resistance per unit radius $R_\square/2\pi r$, where $R_\square$ is the sheet resistance of the oxide surface in ohms per square. Taking radial geometry into account, total oxide surface resistance

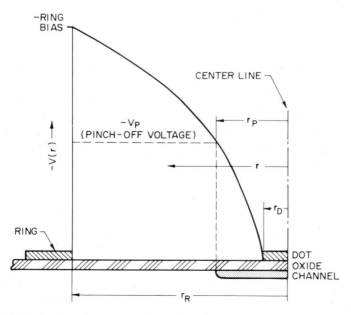

Fig. 4.18 Equilibrium potential distribution with bias on the ring between ring and dot plotted from (4.129). After Nicollian and Goetzberger.[16] Copyright (1965), IEEE.

is

$$R_T = \frac{R_\square}{2\pi} \int\limits_{r_D}^{r_R} \frac{dr}{r} = \frac{R_\square}{2\pi} \ln\left(\frac{r_R}{r_D}\right) \tag{4.126}$$

and oxide surface resistance at r is

$$R(r) = \frac{R_\square}{2\pi} \int\limits_{r_D}^{r} \frac{dr}{r} = \frac{R_\square}{2\pi} \ln\left(\frac{r}{r_D}\right). \tag{4.127}$$

Current flowing is $I = V_R/R_T$ so that

$$V(r) = IR(r) = V_R\left[\frac{R(r)}{R_T}\right]. \tag{4.128}$$

Substituting (4.126) and (4.127) into (4.128) yields

$$V(r) = V_R \frac{\ln(r/r_D)}{\ln(r_R/r_D)}. \tag{4.129}$$

The potential distribution shown in Fig. 4.18 is based on (4.129). Evaluating (4.129) at $r = r_p$, and $V(r_p) = V_p$, we obtain the pinch-off radius

$$r_p = r_D\left(\frac{r_R}{r_D}\right)^{V_p/V_R}. \tag{4.130}$$

Measured dot capacitance C_{dot} is given by

$$\frac{1}{C_{dot}} = \frac{1}{C_p} + \frac{1}{C_{ox}} \tag{4.131}$$

where C_p is the depletion layer capacitance in inversion limited by r_p. With a strong inversion layer, a high frequency, and a limited radius, the inversion layer coupling resistance R_f (see Fig. 4.16) and the inversion layer resistance can be neglected. For negligible inversion layer resistance, the distributed capacitance shown in Fig. 4.16 becomes a lumped capacitance in parallel with the depletion layer capacitance under the dot. Total lumped channel and dot depletion layer capacitance is

$$C_p = \frac{\epsilon_s \pi r_p^2}{w}. \tag{4.132}$$

Combining (4.130), (4.131), and (4.132), yields

$$\frac{1}{C_{dot}} = \frac{w}{\epsilon_s \pi r_D^2}\left(\frac{r_D}{r_R}\right)^{2V_p/V_R} + \frac{1}{C_{ox}}. \tag{4.133}$$

To verify this model and analysis, C_{dot} at zero bias was measured as a function of V_R. Figure 4.19 shows a plot of $1/C_{dot}$ as a function of $(r_D/r_R)^{2V_p/V_R}$ with $V_p = -5$ V obtained from the capacitance minimum of the dot capacitance. The linear dependence shown in Fig. 4.19 verifies the model. The slope

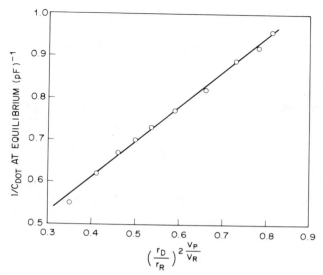

Fig. 4.19 Plot of (4.133) showing linearity of measured equilibrium data. This plot verifies the buildup of a potential distribution beyond the electrode to which bias has been applied. The pinch-off voltage is 5 V. After Nicollian and Goetzberger.[16] Copyright (1965), IEEE.

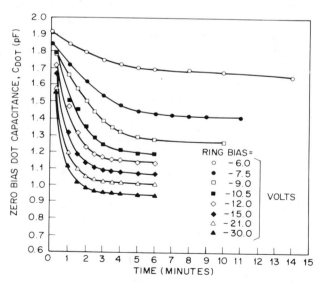

Fig. 4.20 Zero bias dot capacitance as a function of time with ring bias as parameter. These data were taken in room ambient at 300°K. After Nicollian and Goetzberger.[16] Copyright (1965), IEEE.

in Fig. 4.19 also agrees with the constant $(w/\epsilon_s \pi r_D^2)$ in (4.133). The slope in Fig. 4.19 is the steady-state value.

Dependence of zero bias dot capacitance on the time elapsed after application of several values of ring bias is shown in Fig. 4.20. It takes zero bias dot capacitance about 5 min, depending on ring bias, to reach steady state. Moisture induced conduction along the oxide surface depends critically on ambient and surface conditions. Thus Fig. 4.20 represents only one particular case measured in room ambient ($\sim 50\%$ relative humidity) at 27°C.

Minority carrier response time can be extracted from admittance measurements of an MOS capacitor that has an inverted surface beyond the gate. This extraction requires a quantitative analysis of the problem which has been done by Nicollian and Goetzberger[16] for a circular gate geometry.

4.4 HIGH FREQUENCY CAPACITANCE

A quantitative expression for low frequency capacitance as a function of surface potential has been derived in Section 3.2 for uniform doping. This derivation required that free carriers follow both ac voltage and gate bias changes. Silicon surface capacitance as a function of surface potential is derived from (3.35) for n-type and p-type silicon. Equations (3.11), (3.13), (3.35), and (3.36) can be used to calculate low frequency C-V curves for n-type and p-type substrates. Figure 4.21 shows a low frequency C-V curve calculated this way. The various regions, accumulation, depletion, weak inversion, and strong inversion and the surface band bending associated with each are shown. At high frequencies, where $\omega^{-1} \ll \tau_R$, minority

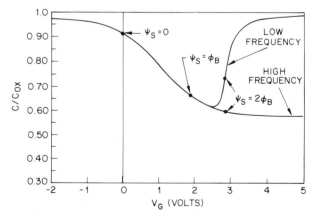

Fig. 4.21 Normalized high and low frequency capacitance as a function of gate bias calculated for an acceptor concentration of 2×10^{16} cm^{-3}, oxide thickness 1000 Å, and $T = 300°$K.

carriers do not follow the ac gate voltage but do follow gate bias changes. The low and high frequency $C\text{-}V$ curves are practically identical in accumulation, depletion, and most of weak inversion because in these regions, minority carrier concentration in the depletion layer is negligibly small compared to majority carrier concentration. Therefore, it does not matter whether minority carriers respond to the ac voltage.* The major difference between low and high frequency $C\text{-}V$ curves occurs in weak to strong inversion where minority carrier concentration in the depletion layer becomes comparable to and exceeds majority carrier concentration. Minority carrier effects cannot be neglected in these regions.

4.4.1 Calculation of the High Frequency Capacitance

This section discusses the calculation of high frequency capacitance in inversion and its smooth connection with the $C\text{-}V$ curve in weak inversion, depletion, and accumulation. Two effects must be considered in obtaining the high frequency inversion capacitance. First, the total number of minority carriers is fixed by the quiescent gate bias and does not change in response to ac gate voltage. Second, minority carriers can move spatially at the silicon surface in response to the high frequency gate voltage.[17-21] That is, the inversion layer becomes wider and narrower during each cycle of the ac gate voltage. This spatial rearrangement of inversion layer carriers amounts to a change in volume density of the inversion layer charge with no change in the total density per unit area. This rearrangement also may be viewed as a *polarization* of the inversion layer.

Spatial rearrangement of the inversion layer charge accounts for no more than a 7% contribution to the capacitance.[17,19] That is, neglect of this effect would introduce an error of no more than 7%, depending on bias, in the calculated capacitance. Although this error is small, the shape of the $C\text{-}V$ curve and the magnitude of the capacitance is influenced, especially between weak and strong inversion.

The high frequency capacitance can be calculated with relations already derived in this chapter, if spatial redistribution is not taken into account. However, to take the spatial redistribution into account, the capacitance will be rederived starting from the solution of the Poisson equation. This rederivation must be done because the depletion layer charge density is altered by this effect.

Although an accurate solution involves a numerical integration, which is neither difficult nor time consuming, a simpler closed-form approximation also will be described that introduces some error. This error amounts to 1.5% or less in inversion and is negligibly small in other regions, as is seen later.

---

*The very small difference between the high and low frequency $C\text{-}V$ curves in these regions barely can be detected with an admittance bridge, one of the most sensitive and accurate instruments for measuring capacitance.

We calculate the high frequency small-signal capacitance for an MOS capacitor with a p-type substrate using the analysis due to Brews.[19] A small-signal solution of the Poisson equation is made. The procedure in Section 2.3 is followed except the spatial redistribution of the inversion layer electrons, in response to the ac gate voltage, is taken into account whereas the total number of inversion layer electrons remains at a constant value set by the gate bias.

At high frequencies, for which net inversion layer charge per unit area is fixed, depletion layer width varies with ac voltage. Consequently, inversion layer width also varies. The inversion layer narrows as the depletion layer widens, in response to the increased field at the interface, and vice versa. This narrowing and widening of the inversion layer also means that the electrons as a whole are moving their center of charge closer and further away from the interface during each cycle, a spatial rearrangement of the inversion layer.

When the inversion layer narrows, electrons must fill states in the conduction band at the silicon surface to a higher level above the band edge. Otherwise, the narrowed inversion layer could not accommodate the same net electron density per unit area. Similarly, when the inversion layer widens, on the other half cycle, electrons must fill states in the conduction band at the silicon surface to a level lower than the equilibrium level. Such a variation in the level to which the inversion layer is filled is described by allowing the Fermi level to possess an alternating component. This *quasi-Fermi level* moves up and down in each cycle to adjust the level of occupancy in the conduction band so the net electron density per unit area always is fixed, independent of time. The ac electron quasi-Fermi level E_{Fn} is considered spatially uniform within the inversion layer where the electron density is significant.

The assumption that E_{Fn} is spatially uniform introduces negligible error. In an exact calculation, in which the spatial uniformity of E_{Fn} was not assumed, Baccarani and Severi[17] found that E_{Fn} was spatially uniform from the silicon surface practically to the point where $n(x, t)$ had dropped to n_i. Beyond that point, E_{Fn} rapidly converged to E_{Fp}, as in Fig. 4.22. In the region of convergence and beyond, electron density is so small that the spatial variation of E_{Fn} has little effect on the surface capacitance.

The hole quasi-Fermi level E_{Fp} is considered spatially uniform throughout the silicon substrate. As holes are the majority carriers, they are in thermal equilibrium so that the hole quasi-Fermi level is identical to the bulk Fermi level.

Because E_{Fn} is an ac quantity, it will have components both in phase and out of phase with the ac gate voltage. Over the same distance into the silicon that E_{Fn} is constant, E_{Fn} is in phase with the ac gate voltage as shown in Fig. 4.22. Therefore, the phase of E_{Fn} can be neglected without significant error.

Electron density is calculated as in Section 2.3.2(b), except that E_F in

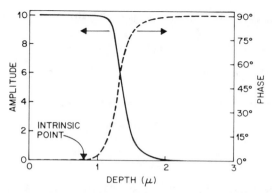

Fig. 4.22 Variation of amplitude and phase of the ac quasi Fermi level with depth. This variation is ($E_{Fn} - E_F$), where E_F is the equilibrium, bulk value of the Fermi level. The point where the equilibrium minority carrier density has dropped to the intrinsic value n_i is indicated. For this calculation, $N_A = 10^{14}$ cm^{-3}, $v_s = 24$, $T = 300°$K, and the frequency is 1 MHz. After Baccarani and Severi.[17] Copyright (1974), IEEE.

(2.36) is replaced by E_{Fn}. This substitution yields

$$n(x, t) = \exp\left[\frac{q\psi(x, t) + E_{Fn}(t)}{kT}\right] \int_{E_c}^{\infty} M_c(E) \exp\left(\frac{-E}{kT}\right) dE. \quad (4.134)$$

The integral in (4.134) is equal to $n_i \exp(- E_i/kT)$ so that (4.134) becomes

$$n(x, t) = n_i \exp[v(x, t) + u_{Fn}(t)] \quad (4.135)$$

where $v(x, t) = (kT/q) \, \psi(x, t)$ and

$$u_{Fn} \equiv \frac{E_{Fn} - E_i}{kT}. \quad (4.136)$$

For p-type, charge density per unit volume is

$$\rho = q[p - N_A - n]. \quad (4.137)$$

From (2.40), (2.41), (2.44), (2.45), and (4.135), we have

$$p - N_A - n = n_i\{\exp(- u_B) [\exp(- v) - 1] - \exp(v + u_{Fn})\}. \quad (4.138)$$

When (2.45), (2.87), and (4.135) are used, the Poisson equation is

$$\frac{d^2v}{dx^2} = \lambda_p^{-2}[1 - \exp(- v) + \exp(v + u_{Fn} + u_B)]. \quad (4.139)$$

Equation (4.139) now takes into account the redistribution of electrons at the silicon surface in response to the ac voltage. Because u_{Fn} is a constant independent of x, (4.139) can be integrated easily. Integrating (4.139) from $v = 0$ to $v = v_s$, following the same procedure as in section 2.3.2(d), the dimensionless electric field $(dv/dx)_{x=0}$ at the silicon surface is (for $v_s > 0$)

$$\left|\frac{dv}{dx}\right|_{x=0} = 2^{1/2}\lambda_p^{-1}\{v_s + \exp(-v_s) - 1 + \exp(u_B + u_{Fn})[\exp(v_s) - 1]\}^{1/2}$$

$$= F(v_s, u_{Fn}, u_B)\lambda_p^{-1}. \tag{4.140}$$

We introduce next the small-signal quantities defined by

$$u_{Fn} = u_B + \delta u_{Fn} \tag{4.141}$$

and

$$v_s = v_{so} + \delta v_s \tag{4.142}$$

where v_{so} is the total band bending set by the gate bias and δv_s is the ac band bending at $x = 0$. The silicon surface capacitance is

$$C_s = \epsilon_s \frac{\delta}{\delta v_s}\left(\frac{dv}{dx}\right)_{x=0}. \tag{4.143}$$

Taking the derivative of the dimensionless electric field [given by (4.140)] with respect to v_s, the silicon surface capacitance [from (4.143)] becomes (for $v_s > 0$)

$$C_s = C_{FBS}\left\{1 - \exp(-v_s)\right.$$

$$\left. + \left(\frac{n_i}{N_A}\right)^2\left[(\exp(v_s) - 1)\left(1 - \frac{\delta u_{Fn}}{\delta v_s}\right) + 1\right]\right\} F^{-1}(v_s, u_B, u_B) \tag{4.144}$$

where $C_{FBS} = \epsilon_s\lambda_p^{-1}$ from (3.43). In obtaining (4.144), the small-signal approximation has been made that $u_{Fn} \approx u_B$ from (4.141), and the relation $\delta/\delta v_s [\exp(u_{Fn})] = (\delta u_{Fn}/\delta v_s) \exp(u_{Fn})$ has been used.

The next step is to evaluate $\delta u_{Fn}/\delta v_s$. This evaluation is done by applying the restriction that total inversion layer charge or inversion layer charge density per unit *area* is fixed by gate bias and does not change in response to the ac voltage. Remember that it is *volume* density, and not surface density, that changes. In applying this constraint on minority carrier charge, we need consider only the excess minority carrier charge, located near the silicon surface, over the bulk minority carrier charge, located deeper in the neutral silicon, because only this excess charge is affected by the ac gate voltage. This excess minority carrier charge Q_n is given by

$$Q_n = q \int_0^\infty [n(x) - n_B]dx \tag{4.145}$$

where n_B is minority carrier or electron density in the neutral bulk. Because the neutral bulk can be considered to be in thermal equilibrium, the condition for thermal equilibrium can be applied to obtain $n_B = n_i^2/N_A$ where from (2.45), $n_i^2/N_A = n_i \exp(u_B)$. When this result and (4.135) are used, (4.145) becomes

$$Q_n = qn_i \int_0^\infty \{\exp[v(x) + u_{Fn}] - \exp(u_B)\}dx. \tag{4.146}$$

To find the excess electron charge set by gate bias alone (i.e., with no ac excitation), we use the same relation for n_B as in (4.146), and $n(x) = n_i^2/p(x)$ where $p(x)$ is given by (2.44). Thus (4.146) becomes

$$Q_n = qn_i \int_0^\infty \{\exp[v_o(x) + u_B] - \exp(u_B)\}dx \tag{4.147}$$

where $v_o(x)$ is the band bending at x established by the gate bias. The constraint on Q_n is obtained by setting (4.147) equal to (4.146). Thus we obtain

$$\int_0^\infty \{\exp[v(x) + u_{Fn}] - \exp(u_B)\} dx = \int_0^\infty \{\exp[v_o(x) + u_B] - \exp(u_B)\} dx. \tag{4.148}$$

Introducing the small-signal quantities from (4.141) and (4.142) into (4.148) and making a small-signal analysis described in Appendix V, one finds

$$\frac{\delta u_{Fn}}{\delta v_s} = \frac{1}{1 + \Delta} \tag{4.149}$$

where Δ is a fraction taking into account the constancy of inversion layer charge and its spatial redistribution. The particular form of (4.149) has been chosen deliberately to avoid round-off error in the capacitance when $\delta u_{Fn}/\delta v_s \approx 1$, which will occur in strong inversion. The expression for Δ derived in Appendix V is

$$\Delta \approx \frac{F(v_{so}, u_B)}{\exp(v_{so}) - 1} \left\{ \int_0^{v_{so}} dv_s \left[\frac{\exp(v_s) - \exp(-v_s) - 2v_s}{F^3(v_s, u_B)} \right] - 1 \right\}. \tag{4.150}$$

Now we can rewrite (4.144) in terms of Δ, to obtain

$$C_s = 2C_{FBS}\left\{ 1 - \exp(-v_{so}) + \left(\frac{n_i}{N_A}\right)^2 \left[(\exp(v_{so}) - 1)\frac{\Delta}{1 + \Delta} + 1 \right] \right\} F^{-1}(v_{so}, u_B),$$

where $F(v_{so}, u_B) \equiv F(v_s, u_B, u_B)$. \hfill (4.151)

Equation (4.151) is an accurate expression for the high frequency capacitance. To see how Δ affects the high frequency capacitance at different values of v_{so}, (4.150) must be evaluated. Figure 4.23 shows a plot of $\delta u_{Fn}/\delta v_{so}$ as a function of v_{so} with three different acceptor concentrations as parameter. Figure 4.23 illustrates the range of v_{so} over which the

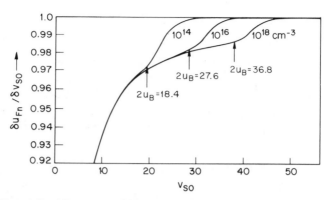

Fig. 4.23 Plot of (4.149) calculated using (4.150) for three different acceptor concentrations as a function of band bending (v_{so}) established by the gate bias. Values of $2u_B$ corresponding to each value of acceptor concentration are shown. These curves show that the redistribution of inversion layer charge in response to the high frequency ac gate voltage is important when $v_{so} > 2u_B$. The analysis is no longer valid for $v_{so} < u_B$.

redistribution effect is important and was obtained by a numerical integration of (4.150). The arrows in Fig. 4.23 denote when $v_{so} = 2u_B$ for each value of N_A. In the range of v_s, and $v_{so} \leq 2u_B$, where the curves for the three acceptor concentrations merge into a common curve and $\delta u_{Fn}/\delta v_{so}$ becomes independent of N_A, the redistribution effect is negligible because minority carrier concentration has become small compared to acceptor concentration. The value of $\delta u_{Fn}/\delta v_{so}$, in this region, only reflects the constraint on the supply of minority carriers. Because the inversion layer charge density is negligibly small in this region, minority carrier terms in (4.151) can be neglected so that (4.151) becomes (for $v_{so} > 0$)

$$C_s = C_{FBS}[1 - \exp(-v_{so})] \, F^{-1}(v_{so}, u_B); \qquad v_{so} \leq 2u_B \qquad (4.152)$$

where from (2.65) for p-type

$$F(v_{so}, u_B) = 2^{1/2}[v_{so} - 1 + \exp(-v_{so})]^{1/2}. \qquad (4.153)$$

For $v_{so} \geq 2u_B$, minority carrier concentration is comparable to or higher than N_A and redistribution becomes important. In strong inversion, where redistribution is important, Fig. 4.23 shows that $\delta u_{Fn}/\delta u_{so}$ is close to unity so that $\Delta \ll 1$, and (4.151) becomes (for $v_{so} > 0$)

$$C_s = C_{FBS}\left\{1 - \exp(-v_{so})\right.$$

$$\left. + \left(\frac{n_i}{N_A}\right)^2 [(\exp(v_{so}) - 1)\Delta + 1]\right\} F^{-1}(v_{so}, u_B).$$

$$(4.154)$$

4.4.2 Closed Form Approximation of the High Frequency Capacitance

Exact values of high frequency capacitance can be obtained by integrating (4.150) and substituting the result into (4.151). In this section closed form expressions for high frequency capacitance are developed, using the same approach as Lindner[22] except that minority carrier redistribution is taken into account.[20] In this approximation capacitance values do not deviate more than 1.5% from the true values in inversion, depending on gate bias. The procedure uses (4.152) in accumulation, depletion, and most of weak inversion, and a gate bias-independent capacitance in strong inversion. The gate bias-dependent capacitance from (4.152) is made to join the bias independent capacitance at a certain value of band bending v_m called the *match point*. This match point minimizes the maximum error between true and approximate capacitance in inversion.

Figure 4.24 shows a plot of C_s/C_{FBS} as a function of v_{so} that illustrates this procedure. From $v_{so} = 10$ to $v_{so} = v_m$, the curve is calculated from (4.152). The horizontal solid line joining the curve from (4.152) at $v_{so} = v_m$ is the constant capacitance value.

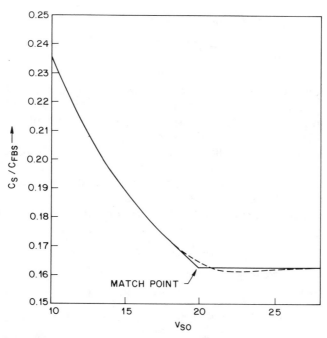

Fig. 4.24 Plot of C_s/C_{FBS} as a function of v_{so}. Dotted curve was calculated using (4.150) and (4.151); solid curve is the approximate capacitance calculated from (4.155), (4.156), and (4.161). An acceptor concentration of 10^{14} cm^{-3} was used in these calculated curves. The match point corresponds to $v_m = v_L - 0.75$.

The closed form formula for high frequency capacitance is

$$C_s(v_{so}) = C_L(v_{so}); \qquad v_{so} < v_m \qquad (4.155)$$

and

$$C_s(v_{so}) = C_L(v_m); \qquad v_{so} > v_m \qquad (4.156)$$

where C_L from (4.152) is, for p-type

$$C_L = 2^{-1/2} \, \text{Sgn}(v_{so}) C_{FBS} [1 - \exp(- v_{so})][(v_{so} - 1) + \exp(- v_{so})]^{-1/2} \qquad (4.157)$$

where Sgn is defined following (2.62). Similarly, for n-type

$$C_L = 2^{-1/2} \text{Sgn}(v_{so}) C_{FBS} [\exp(v_{so}) - 1][- (v_{so} + 1) + \exp(v_{so})]^{-1/2}. \qquad (4.158)$$

The next problem is to determine the match point v_m. Lindner[22] used a match point v_L in calculating high frequency capacitance that we use as a first estimate of the optimal match point v_m. To define v_L, Lindner used the condition that at $v_{so} = v_L$, ionized impurities and minority carriers contribute equally to the square of the total surface charge density. Mathematically expressed, this condition is

$$\exp(v_L - 2u_B) = v_L - 1. \qquad (4.159)$$

Equation (4.159) can be solved by iteration or an adequate approximation is

$$v_L = 2.10 \, u_B + 2.08. \qquad (4.160)$$

The match point v_L depends on doping concentration. Figure 4.25 shows a plot of v_L calculated from (4.160) as a function of ionized acceptor concentration. A plot of $2u_B = 2 \ln(N_A/n_i)$ also is shown in Fig. 4.25 for comparison. To find the error involved in using (4.155) and (4.156), we compare the true value of silicon capacitance in inversion calculated from (4.150) and (4.151) and shown as a dotted line in Fig. 4.24, with the approximate capacitance calculated from (4.156) and shown in Fig. 4.24 as a solid line after selecting the match point to minimize the maximum error in inversion. The optimal match point for any doping concentration is found to be

$$v_m = v_L - 0.75 \qquad (4.161)$$

or $v_m = 2.10 \, u_B + 1.33$ using (4.160).

The error between accurate and approximate capacitance is defined as

$$\text{Error} \equiv \frac{C_s(\text{accurate}) - C_s(\text{approximate})}{C_s(\text{accurate})}. \qquad (4.162)$$

Figure 4.26 shows a plot of (4.162) (in percent) as a function of v_{so} for three different values of acceptor concentration. The shape of these error curves can be explained by reference to Fig. 4.24. The accurate capacitance curve

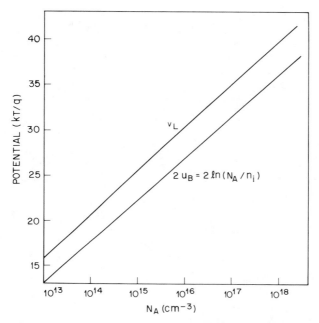

Fig. 4.25 Potential as a function of acceptor concentration. The curve marked v_L was calculated using (4.159). A curve of $2u_B$ is shown for comparison.

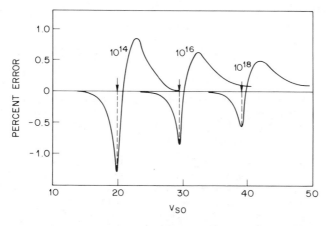

Fig. 4.26 Percent error as a function of v_{so}. The error is calculated from (4.162) for three different values of acceptor concentration using the optimum match point from (4.161). The arrows indicate the point of maximum error for each acceptor concentration which occurs at $v_{so} = v_m$.

crosses the approximate capacitance curve resulting in a change in sign of the error as seen in Fig. 4.26. Figure 4.26 shows that the error does not exceed 1.5% anywhere in the inversion range and the maximum error, which occurs at $v_{so} = v_m$, indicated by arrows in Fig. 4.26, becomes smaller as doping concentration increases.

Figures 4.24 and 4.26 show that the difference in shape between true and approximate C-V curves in inversion is responsible for most of the error. In accumulation, depletion, and most of weak inversion, where (4.152) applies, the error due to neglecting minority carriers is negligibly small ($< 10^{-5}$ C_{FBS}).[23] Thus the approximate closed-form equations for high frequency capacitance are adequate for many applications.

4.4.3 Comparison of Theory to Experiment

The accurate differential capacitance curve calculated from (4.150) and (4.151) is found to fit an experimentally measured C-V curve.[18] Figure 4.27 shows the fit for an n-type sample.* An n-type substrate was chosen to

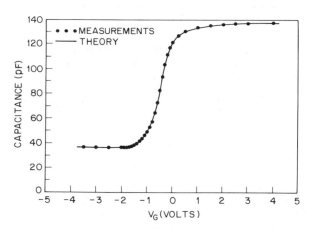

Fig. 4.27 High frequency capacitance as a function of gate bias. The dots are capacitance values measured at 100 kHz on an n-type sample oriented in the (100) direction, with a donor concentration of 4.55×10^{15} cm^{-3}, an oxide capacitance of 6.5×10^{-8} F/cm^2, and a gate diameter of 521 μm. The penumbra of thin aluminum at the gate perimeter resulting from evaporation through a shadow mask was removed by a brief etch to permit a more accurate determination of gate area. Wafers were oxidized in an $O_2 \rightarrow$ steam $\rightarrow N_2$ sequence at 950°C to a thickness of about 500 Å and annealed in forming gas at 400°C for 1 hr following aluminum metalization. The sample was processed to minimize impurity redistribution (see Section 15.3.2) and to give a low interface trap level density (see Section 15.4.3). The solid curve has been calculated using (4.151). After Berman and Kerr.[18] Copyright (1974), Pergamon Press, Ltd. Reprinted with permission.

*Equations (4.150) and (4.151) were derived for p-type silicon. The relationships for n-type are obtained simply by reversing signs in (4.150) and (4.151) as the curve for n-type is simply the mirror image of the curve for p-type.

avoid an inversion layer in the region beyond the gate caused by positive oxide charge as described in Section 4.3.11(a).

The experimental curve was shifted along the voltage axis to make the flatband points of the experimental and calculated curves coincide. Such a shift was necessary because some positive oxide fixed charge is unavoidable with present technology.

To relate the calculated C-V curve to a measured C-V curve, the ionized dopant impurity concentration was taken as an adjustable parameter, chosen to fit theory to experiment in strong inversion. The fitted value was compared with the value obtained from the slope of the C-V curve in depletion. (See Section 9.4.1). The fitted value of N_D was 4.55×10^{15} cm^{-3}, only 1.5% higher than the value of 4.60×10^{15} cm^{-3} calculated from the slope of the C-V curve in depletion. Thus the two values are consistent, and the overall fit is excellent, thus supporting the theoretical expressions.

In relating calculated C-V curves to measured C-V curves, there are certain independently measured parameters such as ϵ_s, ϵ_{ox}, N_D, x_o, and gate area that must be used in calculating a C-V curve. The accuracy and precision with which these parameters are known determines, in the absence of other effects such as interface traps, how closely the calculated C-V curve will fit the measured C-V curve. Of these parameters, N_D is known least accurately, typically within 10–20%. The acceptor concentration in p-type also is known to the same accuracy. This inaccuracy results mainly from impurity redistribution during thermal oxidation as described in Section 14.4.2. This error becomes smaller as doping concentration becomes smaller, because depletion layer width in strong inversion is wider, making the region of nonuniform doping a smaller fraction of total depletion layer width. For doping concentrations in the 10^{15}–10^{16} cm^{-3} range, commonly used in device applications, the uniform doping assumption results in a negligibly small error in depletion and inversion. However, for heavy doping, for implants, or where oxides are repeatedly etched off and regrown, a large error can result.

Errors can be considered in two ways: (1) by determining the error involved in the calculation and (2) by determining the error involved in the measurement of a C-V curve and in the independently measured parameters used in calculating the C-V curve. In this section emphasis has been on errors involved in the calculation, but it must be remembered that calculation errors of 5% probably are negligible in many applications.

4.4.4 Effect of Nonuniform Doping Profiles

In practice, thermal redistribution of dopant impurities during oxidation and deliberate alterations of doping by ion implantation or other means, result in doping profiles that vary with distance into the silicon from the interface. Such nonuniform profiles affect the shape of the C-V curve, causing departures from the ideal curve for uniform doping. We consider only depletion to illustrate the effect on the C-V curve of a nonuniform doping profile in the silicon. Fortino and Nadan[24] have made computer calculations of the entire C-V curve for nonuniform doping profiles but these numerical calculations are not given here. Their calculated C-V curves are shown in Fig. 4.28. Figure 4.28a shows the effect on the C-V

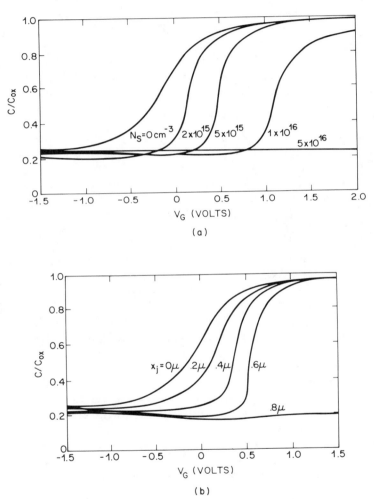

Fig. 4.28 (a) Normalized C-V curves calculated for implanted MOS structures with various implant densities and a fixed implant depth of 0.5 μm; p-type implants on an n-type substrate are illustrated. Substrate donor concentration 10^{15} cm^{-3}, $x_o = 1000$ Å, frequency 1 MHz. After Fortino and Nadan.[24] (b) Normalized C-V curves calculated for implanted MOS structures with various implant depths and a fixed p-type implant density of 5×10^{15} cm^{-3} and an n-type substrate density of 10^{15} cm^{-3}, $x_0 = 1000$ Å, and frequency 1 MHz. After Fortino and Nadan.[24] Copyright (1977), IEEE.

curves of a compensating surface implant density at fixed implant depth, and Fig. 4.28b shows the effect of implant depth at fixed compensating surface implant density.

In Fig. 4.28a, the uniformly doped curve is on the upper left labeled $N_s = 0$. The remaining curves correspond to implanted structures with steadily increasing doping density, holding the implant depth to about $0.5 \mu m$. Because the samples are n-type, accumulation occurs at positive gate biases.

As the implanted impurity density increases, the C-V curves translate to more positive voltages, indicating a p-type implant whose acceptor ions tend to maintain the device in inversion at gate biases below 0V. At the maximum implant density of $5 \times 10^{16} \, \mathrm{cm^{-3}}$, the device cannot be pulled out of inversion at the maximum gate bias employed.

Also as implant density increases, the capacitance tends to go from inversion to accumulation more quickly as gate bias is increased. Such a sudden contraction in depletion layer depth is expected when the depletion layer edge moves abruptly from the n-type substrate through the built-in surface p-n junction and into the p-type surface region.

For sufficiently high implant densities, such as $10^{16} \, \mathrm{cm^{-3}}$, there is apparently a sufficient built-in potential barrier against electron injection that the silicon surface is partly inaccessible to electrons for the range of biases explored. That is, electrons are held away from the interface by the implanted ions, in effect increasing oxide thickness by the thickness of this inaccessible region. As gate bias is increased further, the region slowly reduces in size, but the capacitance very slowly approaches C_{ox}.

Figure 4.28b is similar to Fig. 4.28a, but here the implant depth has been increased, rather than the implant density. The discussion of these capacitance curves is analogous to Fig. 4.28a.

A mathematical treatment of the C-V characteristics of an MOS capacitor with a nonuniform doping profile is given by Bartelink.[25]

(a) Depletion Approximation for Nonuniform Doping

In depletion the capacitance for a nonuniformly doped structure often is taken to be

$$C_D = \frac{\epsilon_s}{w}. \tag{4.163}$$

Assuming p-type material, we estimate the depletion layer charge density to be

$$Q_s = -q \int_0^w dx \, N_A(x). \tag{4.164}$$

We check that (4.163) and (4.164) are compatible with the definition of the differential capacitance

$$C_D \equiv -\frac{dQ_s}{d\psi_s}. \tag{4.165}$$

Using (4.164) in (4.165), we find

$$C_D = qN_A(w)\frac{dw}{d\psi_s}. \tag{4.166}$$

Equation (4.166) is consistent with (4.163) if the band bending, ψ_s is given by

$$\psi_s = \frac{q}{\epsilon_s}\int_0^w dx\, xN_A(x). \tag{4.167}$$

Equation (4.167) is the expression for band bending in the depletion approximation. Therefore, (4.166) is a *consistent* expression for depletion layer capacitance within the depletion approximation.

Several errors are involved in the use of the depletion approximation for nonuniform doping. These errors are discussed in Section 2.3.4(f).

4.5 MINORITY CARRIER RESPONSE TO GATE BIAS CHANGES

Until now we have considered minority carrier response only to the ac gate voltage, assuming that minority carriers follow changes in gate bias. In this section, we consider the case where minority carriers follow neither ac gate voltage nor gate bias sweep.

Usually, a C-V curve is measured by automatically sweeping gate bias. If sweep rate is too rapid for minority carriers to follow, the system no longer will be in thermal equilibrium with respect to gate bias and the resulting C-V curve will differ from the thermal equilibrium curve. At room temperature, the minority carrier generation rate will be much smaller than the recombination rate. Thus in many cases, particularly for silicon that has a long minority carrier lifetime, the sweep rates normally used are too rapid for generation to follow but are sufficiently slow for recombination to follow. At room temperature the system usually is not in equilibrium when gate bias is swept in the direction of increasing inversion, but it is in equilibrium when gate bias is swept in the direction of decreasing inversion.

When gate bias is swept in the direction of increasing inversion too rapidly for minority carriers to follow, a net generation current of minority carriers will flow into the silicon surface. If the bias sweep is stopped in inversion, generation current will continue to flow until the system reaches thermal equilibrium.

In the return sweep toward inversion, if the sweep rate is so high that a minority carrier concentration in excess of the thermal equilibrium concentration exists, a small forward bias will develop across the field induced junction. Minority carriers then are injected into the silicon, increasing the recombination current in the middle of the depletion layer.[26] Injection becomes important at low temperatures.

4.5.1 Deep Depletion

Figure 4.29 illustrates both the nonequilibrium high frequency C-V curve measured when sweeping gate bias into inversion too fast for minority carriers to follow and also the equilibrium C-V curve measured when sweeping out of inversion.

First, consider the n-type sample. From (4.116) and τ_{T_p} from the caption of Fig. 4.29, minority carrier response time in this sample is about 3.4 sec. Response is too slow for minority carriers to follow the gate bias sweep of 0.1 V/sec. Consequently, as gate bias sweeps into inversion, no inversion layer forms. Therefore, charge neutrality must be satisfied by ionized donors alone, the depletion layer width becomes wider than in thermal

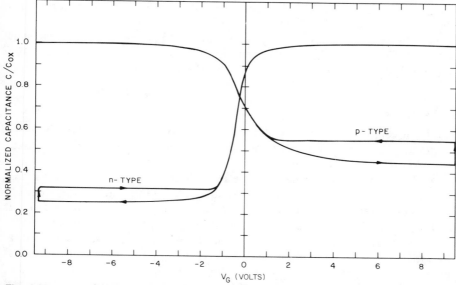

Fig. 4.29 Normalized capacitance measured at 1 MHz and 300°K as a function of gate bias showing the case where minority carriers do not respond either to ac gate voltage or gate bias sweep. The n-type sample had a donor concentration of 1.5×10^{15} cm^{-3} and a hole lifetime of $\tau_{T_p} \approx 0.15$ msec. The p-type sample had an acceptor concentration of 1.5×10^{16} cm^{-3} and an electron lifetime of about 7 μsec. Both samples had an oxide thickness of 1000 Å. Interface trap level density was in the 10^9 cm^{-2} eV^{-1} range; thus interface traps had a negligible effect. After Sinha.[27] Reprinted by permission of the publisher, The Electrochemical Society, Inc.

equilibrium, and the capacitance decreases below its thermal equilibrium saturation value. This nonequilibrium condition is called *deep depletion* because the silicon surface is depleted to a greater depth than it would be in thermal equilibrium. Because minority carrier density at the surface is smaller than in thermal equilibrium, generation exceeds recombination as the system tries to restore thermal equilibrium. As gate bias is made increasingly negative, depletion layer width continues to widen and capacitance continues to decrease. This decrease continues, either until avalanche breakdown occurs in the silicon, or until generation in the widened depletion layer has increased sufficiently to result in a steady-state balance. That is, steady-state will result if generation in the widened depletion region has increased sufficiently to balance the charge added to the gate by the gate bias sweep, and if the sweep provides a constant rate of addition of charge to the gate. Lifetime measurements based on observation of this steady-state balance are discussed in Section 9.5.

If gate bias sweep is stopped before avalanche breakdown occurs in the silicon, as in Fig. 4.29, net generation eventually will restore the inversion layer to thermal equilibrium.[27] This restoration causes the increase in capacitance at about -9 V in Fig. 4.29, as the depletion layer narrows to its equilibrium width. When gate bias is returned to less negative values, the equilibrium, saturated, high frequency capacitance is measured and rejoins the remainder of the C-V curve, in depletion and accumulation.

For the p-type sample in Fig. 4.29, minority carrier response time was 3.4 sec from (4.107), again too slow to follow a sweep rate of 0.1 V/sec. The sample deep depletes in inversion, as did the n-type sample.

4.5.2   The Stagnant Inversion Layer–Behavior at Low Temperatures

The inversion layer at low temperature is of interest in extending our understanding of inverted silicon surfaces and in interpreting low temperature C-V measurements.

At liquid nitrogen temperature it is difficult to obtain a smooth, high frequency C-V curve in inversion. At such a low temperature, the *stagnant inversion layer*[28] is encountered. The inversion layer is stagnant because it follows gate bias very slowly, if at all, when swept in or out of inversion. Not only is formation of the inversion layer by generation very slow but its disappearance by recombination also is very slow compared to practical gate bias sweep rates. Figure 4.30 shows the inversion, nonequilibrium, C-V characteristics of a p-type MOS capacitor measured at 79°K and 1 MHz for both directions of gate bias sweep. Seven features of this plot are designated by the numbers shown in Fig. 4.30 and explained next.

1   The depletion part of the characteristic shows no unusual behavior. The sample had such a low interface trap level density that interface trap effects need not be considered.

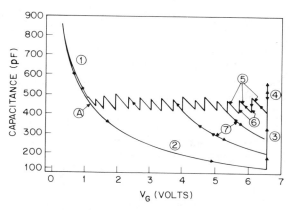

Fig. 4.30 Inversion region of *C-V* curve of an MOS capacitor measured at a frequency of 1 MHz and a temperature of 79°K illustrating the stagnant inversion layer. The sample was thermally oxidized to a thickness of 600 Å, acceptor concentration was 1.9×10^{15} cm^{-3}, and gate area was 6×10^{-2} cm^2. Numbers in the circles are explained in the text. After Goetzberger.[28] Copyright (1967), IEEE.

2   At 79°K n_i is a factor of 10^{30} lower than at room temperature so that minority carrier response time resulting from generation-recombination is extremely long [see (4.116)]. As a result, the silicon goes into deep depletion as gate bias is swept beyond point *A* in Fig. 4.30.

3   When gate bias is kept constant in the deep depletion condition, capacitance increases very slowly as the tiny generation current supplies electrons to the silicon surface. Because thermal generation of electrons at 79°K is so small (probably negligible), very long times are required to restore thermal equilibrium. Equilibrium was found to be achieved in several hours most likely by generation caused by stray light.

4   The capacitor is very light sensitive. If room light is admitted, capacitance increases rapidly to a value greater than the thermal equilibrium value in the dark by increased steady-state carrier concentration generated by the light. When light is removed, capacitance drifts back toward its thermal equilibrium value. Time constants for this process are about a factor of 10 shorter than those required for buildup of the inversion layer in the dark (process 3).

5   This part of the curve is obtained by reducing gate bias in approximately equal steps. After each bias step, the silicon is allowed to relax. Capacitance increases with decreasing bias and the slope of each step labeled 5 is the same. This slope is also the same as that of the depletion curve at the same capacitance in depletion around point *A*. Thus capacitance is determined by depletion effects, and the inversion layer remains unchanged during gate bias reduction. That is, very little

inversion layer charge recombines as gate bias is decreased. At room temperature inversion layer charge is injected at a very low forward bias so that equilibrium is rapidly established. At 79°K a much larger forward bias is required before a comparable current can flow. To get an idea of the magnitudes involved, the injected carrier recombination rate in the middle of the depletion layer is $U = n_i/\tau \exp(q\psi_F/2kT)$, where ψ_F is the forward bias developed across the field induced junction.[28] This relation shows that as temperature is lowered, ψ_F must increase to maintain the same value of U. This increase is due primarily to the decrease of n_i with temperature [see (2.17) for the temperature dependence of n_i]. At 79°K forward bias is estimated to be about 0.25 V before injection of the inversion layer becomes noticeable. For forward bias less than 0.25 V, depletion layer width changes with gate bias whereas inversion layer charge remains unchanged. When forward bias exceeds 0.25 V, injection increases rapidly and the C-V curve becomes horizontal provided gate bias is changed continuously rather than in steps. For a continuous decrease in gate bias, a steady state displacement current, $I = C_{ox}(dV_{ox}/dt)$, flows through the capacitor so that the steady-state capacitance depends on the rate of change of gate bias.

6   At every voltage step in Fig. 4.30, gate bias was kept constant for about 2 min allowing the silicon to approach equilibrium.

7   At any time while decreasing gate bias in steps, gate bias can be increased again. Increasing gate bias resulted in the curves labeled 7, which can be traced reversibly. These curves support the concept of the stagnant inversion layer. Increasing gate bias causes the depletion layer width to increase, resulting in a decrease in measured capacitance. Inversion layer charge density does not change; hence its only influence is to shift the C-V curve along the voltage axis by a constant amount. The curves labeled 7 in Fig. 4.30 are deep depletion C-V curves translated parallel to the original deep depletion curve, labeled 2, by a constant voltage.

To obtain an equilibrium high frequency C-V curve in inversion, it would be necessary to obtain equilibrium at a gate bias in strong inversion and then reduce gate bias slowly. As such a slow, continuous reduction in gate bias is impractical, gate bias can be reduced in steps allowing equilibrium to be established between each step. The lowest points of the resulting sawtooth curve would approximate the true equilibrium C-V curve.

The stagnant inversion layer at low temperatures has been observed on a variety of samples (both n-type and p-type) with different silicon resistivities.

REFERENCES

1   W. Shockley, *Electrons and Holes in Semiconductors*, Van Nostrand, Princeton, N.J., 1950, p. 58.

2   G. Baccarani, C. A. Baffoni, M. Rudan, and G. Spadini, *Solid-State Electron.*, **18**, 1115 (1975).

3   A. Goetzberger and E. H. Nicollian, *Bell Syst. Tech. J.*, **46**, 513 (1967).

4   S. R. Hofstein and G. Warfield, *Solid-State Electron.*, **8**, 321 (1964).

5   A. S. Grove, B. E. Deal, E. H. Snow, and C. T. Sah, *Solid-State Electron.*, **8**, 145 (1965).

6   C. G. B. Garrett, *Phys. Rev.*, **107**, 478 (1957).

7   R. N. Hall, *Phys. Rev.*, **87**, 387 (1952).

8   W. Shockley and W. T. Read, *Phys. Rev.*, **87**, 835 (1952).

9   C. T. Sah, *Solid-State Electron.*, **13**, 1547 (1970).

10 A. S. Grove, *Physics and Technology of Semiconductor Devices*, Wiley, New York, 1967, Chapter 5.

11 H. A. Mar, J. G. Simmons, and G. W. Taylor, *Solid-State Electron.*, **20**, 241 (1977).

12 K. Lehovec and A. Slobodskoy, *Solid-State Electron.*, **7**, 59 (1964).

13 I. Stakgold, *Boundary Value Problems of Mathematical Physics*, Vol. 1, MacMillan, New York, 1967, p. 85.

14 C. T. Sah, R. N. Noyce, and W. Shockley, *Proc. IRE*, **45** 1228 (1957).

15 V. W. Chan, J. G. Simmons, and H. A. Mar, *Solid-State Electron.*, **20**, 249 (1977).

16 E. H. Nicollian and A. Goetzberger, *IEEE Transact. Electron Devices*, **ED-12**, 108 (1965).

17 G. Baccarani and M. Severi, *IEEE Transact. Electron Devices*, **ED-21**, 122 (1974).

18 A. Berman and D. R. Kerr, *Solid-State Electron.*, **17**, 735 (1974).

19 J. R. Brews, *J. Appl. Phys.*, **45**, 1276 (1974).

20 J. R. Brews, *Solid-State Electron.*, **20**, 607 (1977).

21 M. J. McNutt and C. T. Sah, *Solid-State Electron.*, **17**, 377 (1974).

22 R. Lindner, *Bell Syst. Tech. J.*, **41**, 803 (1962).

23 J. R. Brews, *Solid-State Electron.*, **17**, 447 (1974).

24 A. G. Fortino and J. S. Nadan, *IEEE Transact. Electron Devices*, **ED-24**, 1137 (1977).

25 D. J. Bartelink, "Capacitance-Voltage Technique in Nonuniformly Doped Semiconductors," in *Integrated Circuit Process Models*, J. D. Meindl, Ed., to be published.

26 A. S. Grove, *Physics and Technology of Semiconductor Devices*, Wiley, New York, 1967, Chapter 6, p. 182.

27 A. K. Sinha, *J. Electrochem. Soc.*, **123**, 65, (1976).

28 A. Goetzberger, *IEEE Transact. Electron Devices*, **ED-14**, 787 (1967).

5

Extraction of Interface Trap
Properties from the Conductance

5.1 INTRODUCTION

The purpose of this chapter is to describe the relation between the measured admittance of the MOS capacitor and interface trap properties. Interface trap admittance is not measured directly, but it can be extracted from measured admittance using an equivalent circuit. After various theoretical assumptions are made, a general equivalent circuit is derived for the MOS capacitor, on the basis of capture and emission of mobile carriers by interface trap levels distributed throughout the silicon bandgap.

Using this circuit, it is shown how the parameters governing interface trap capture and emission are extracted from the measured conductance and capacitance.

The behavior observed in depletion, near midgap, and in inversion all are discussed. Quite different behavior is found in each regime, and each is well described using the same basic model. This success supports the assumptions of the model, as pointed out later in this chapter and in greater detail in Chapter 7. Thus the conductance method described here is fundamental to our understanding of interface traps.

The conductance method is simplest in depletion because minority carrier effects are not important. That is, in depletion interface trap occupancy changes by capture and emission of majority carriers (holes in p-type and electrons in n-type). Minority carrier capture is unimportant because the minority carrier density is very low in depletion for biases not too close to midgap. Similarly, minority carrier emission is negligible because the small-signal measurement disturbs equilibrium only slightly, and so the emission rate is comparable to the negligible capture rate.

Weak inversion also is discussed. In weak inversion, not too close to midgap, the major effect of minority carriers is to couple most of the interface trap levels directly to the inversion layer. This coupling introduces some dependence on minority carriers, complicating the extraction of interface trap parameters compared to depletion. In addition, very low measurement frequencies must be used, which is an experimental drawback.

Within about $3kT/q$ on either side of midgap (dependent on interface trap level density and capture cross sections), recombination and generation of mobile carriers through interface traps influences the admittance. Recombination is very sensitive to band bending and interface trap parameters. Uncertainty in any one parameter causes large uncertainty in the rest. Consequently, midgap is not used for extracting interface trap parameters and is discussed only briefly.

In Sections 5.2–5.4 the admittance of the MOS capacitor, including interface traps, is derived as a function of frequency for depletion, midgap, weak inversion, and strong inversion. On the basis of this theoretical foundation, Section 5.5 discusses the conductance method.[1,2] The conductance method extracts interface trap level density, capture probability, and time constant dispersion from the real component of the admittance. It is the most accurate and sensitive of the small-signal steady-state methods. For these reasons, our present knowledge of the electrical properties of interface traps, described in Chapter 7, is based almost entirely on conductance measurements. On the other hand, the small-signal steady-state capacitance methods of Chapter 8[3-5] are more suitable for rapid evaluation of interface trap level density. Because of their convenience these methods are useful in studies of the chemical nature of interface traps, their annealing properties, and so on.

The resolution of the small-signal steady-state* conductance and capacitance methods is discussed in Section 5.6, series resistance and bulk trap energy loss is discussed in Section 5.7, and the errors and limitations of the conductance method are discussed in Sections 5.8 and 5.9. Finally, the extraction of interface trap properties using an MOS capacitor that has an oxide sufficiently thin for direct tunneling to occur is discussed in Section 5.10.

Chapters 5–8 deal mainly with the small-signal steady-state methods that are well established for measuring interface trap properties. However, there are a number of other methods for measuring the electrical properties of interface traps. These methods are (1) transient methods—emission time spectroscopy,[6] thermal dielectric relaxation current,[7] isothermal dielectric relaxation current,[8] and charge transfer loss in a charge coupled device,[9, 10] (2) low temperature C-V measurements,[11] (3) charge pumping,[12] and (4) optical methods.[13, 14] Transient methods are discussed in Section 8.5. These methods are not as well developed as the small-signal steady-state methods and still are not widely used. The Gray-Brown method, discussed in Section 8.4, and the charge pumping method[12] are understood too poorly to be useful for interface trap studies at present.

Optical methods give information on photon capture probability supplementary to the electron and hole capture probabilities yielded by the conductance method. However, optical methods are in an early stage of development and are not discussed.

5.1.1 Qualitative Discussion of Small-Signal Energy Loss Due to Interface Traps

Interface traps are defects located at the Si–SiO$_2$ interface, each of which has one or more energy levels within the silicon bandgap. These defects can exchange charge with the silicon. Specifically, they can interact with the silicon conduction band by capturing or emitting electrons and with the valence band by capturing or emitting holes. Capture or emission occurs when interface traps change *occupancy*.

Changes in occupancy can be produced by changes in gate bias as illustrated in Fig. 5.1 for a *p*-type substrate. In Fig. 5.1a, no gate bias is applied and band bending is caused by charged interface traps and other contributions to flatband voltage discussed in Section 10.5. Interface trap levels below the Fermi level are full, and those above are empty. In Fig. 5.1b a negative gate bias has been applied, moving the valence band edge at the silicon surface toward the Fermi level. Interface trap levels empty by capturing majority carriers (holes in Fig. 5.1) until the equilibrium condition in Fig. 5.1b is reached. In Fig. 5.1c, a positive gate bias has been applied, moving

---

Small-signal means a linearized analysis is valid. Experimental determination of the small-signal range of ac gate voltage amplitudes is discussed in Section 12.4.1(b). *Steady-state* is defined in Section 3.1.

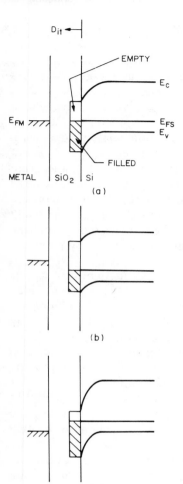

Fig. 5.1 Band-bending diagram showing how interface traps change occupancy with gate bias. An arbitrary uniform distribution of interface trap levels is shown for illustration. The sample is p-type. (a) No gate bias; (b) negative gate bias; (c) positive gate bias.

the conduction band edge at the silicon surface toward the Fermi level. Interface trap levels fill by emitting majority carriers until the equilibrium condition in Fig. 5.1c is reached. If negative gate bias is sufficient to invert the surface, equilibrium involves minority carriers as well as majority carriers.

In the conductance method, interface trap levels are detected through the loss resulting from changes in their occupancy produced by small variations of gate voltage. A small ac voltage applied to the gate of an MOS capacitor alternately moves the band edges toward or away from the Fermi level. Majority carriers are captured or emitted, changing occupancy of interface trap levels in a small energy interval a few kT/q wide centered about the Fermi level. This capture and emission of majority carriers causes an energy loss observed at all frequencies except the very lowest (to which interface traps immediately respond) and the very highest (to which no interface trap response occurs).

To explain this energy loss, consider n-type silicon as an example. The electrons are viewed as a statistical ensemble, and the energy loss arises from changes in the average energy of the ensemble.

In the positive half cycle of the applied ac gate voltage, the silicon conduction band moves toward the Fermi level at the silicon surface. The average energy of electrons at the silicon surface immediately* increases. At frequencies where an energy loss is observed, interface traps do not respond immediately but lag behind the ac gate voltage. Thus there are empty interface trap levels below the Fermi level in the silicon. An energy loss occurs when electrons at a higher average energy in the silicon are captured by interface trap levels at a lower average energy, eventually making the average energy of trapped electrons the same as that of the free electrons. The energy lost during electron capture is taken up by phonons, heating the lattice.

On the negative half of the ac cycle, the silicon conduction band moves away from the Fermi level at the silicon surface. Electrons in filled interface trap levels above the Fermi level in the silicon will be at a higher average energy than electrons in the silicon. As electrons are emitted by interface traps into the silicon, the electrons lose energy again until the energy of trapped electrons becomes equal to that of the free electrons. The energy needed to emit electrons from interface trap levels to the silicon conduction band is provided by phonons. Emitted electrons enter the high energy tail of the Fermi distribution, raising the average energy of electrons in the silicon momentarily. The ensemble of electrons in the silicon then returns to its original average energy through phonons, heating the lattice.

Thus there will be an energy loss on both halves of the ac cycle that must be supplied by the signal source. This energy loss is measured as an equivalent parallel conductance G_P.

In addition to an energy loss associated with capture and emission, interface traps also can hold an electron for some time after capture. That is, interface traps store charge. Therefore, there will be a capacitance C_{it} proportional to interface trap level density.

At any given frequency of ac gate voltage the loss depends both on the speed of response of interface traps, determined by their capture probability, and on the interface trap level density near the Fermi level at the silicon surface. Knowing the loss at all frequencies and biases from experiment, we can find capture probability and interface trap level density for energy levels within the silicon bandgap.

---

*In this context "immediately" means within several dielectric relaxation times, which is an extremely short time compared to the period of the ac gate voltage (5–7 orders of magnitude shorter).

There are two steps in obtaining interface trap level density and capture probability as functions of silicon bandgap energy in the conductance method: (1) measuring admittance as a function of gate bias and frequency and extracting interface trap level density and time constants as functions of gate bias from the equivalent parallel conductance and (2) measuring a low frequency C-V curve or a Q-V curve as a function of gate bias as described in Section 3.3, to extract silicon surface band bending as a function of gate bias. Combining these two steps gives the interface trap properties as a function of energy within the bandgap. We concentrate on step 1 of this procedure in this chapter.

5.2 ADMITTANCE OF SINGLE-LEVEL INTERFACE TRAPS

A quantitative expression for interface trap admittance is derived by modeling interface traps using the Shockley-Read-Hall (SRH) model for bulk traps described in Section 4.3.2(d). In Section 7.3 the SRH model is shown to fit the experimentally determined properties of interface traps.

Because interface traps follow the SRH model, they are completely characterized electrically by their level density and their capture probabilities for both electrons and holes as functions of silicon bandgap energy. These interface trap properties are determined by oxidation conditions and subsequent annealing described in Section 15.4.3.

Experimentally, interface trap sites are found to be widely separated from one another on the interfacial plane. Interface trap levels are found to be closely spaced in energy and distributed throughout the entire silicon bandgap. Electron and hole capture probabilities are found to be independent of energy within experimental error. These observations apply over most of the silicon bandgap in device-grade oxides. Only the regions near both band edges, where small-signal steady-state methods become unreliable, are uncertain.

We now derive the admittance contributed by interface traps and relate this admittance to the measured admittance. In this way, interface trap level density and capture probability are related to the measured admittance, and the model can be put to experimental test.

We start with interface traps all with the same energy level; that is, all interface traps are identical. Paralleling the development in Chapter 4, the equivalent circuit can be constructed intuitively and the admittance derived. In Section 5.2.1 this circuit is simplified for depletion, to bring out the basic behavior of the admittance.

As in Section 4.3, we assume that departures from thermal equilibrium are small under steady-state small-signal ac excitation. Such an assumption is reasonable because interface trap occupancy is perturbed only slightly from its equilibrium value by small ac gate voltages. In thermal equilibrium, the probability $f_o(E_T)$ that an interface trap level is occupied is determined by the Fermi function

$$f_o(E_T) = \frac{1}{1 + g \exp[(E_T - E_F)/kT]} \tag{5.1}$$

where E_T is the interface trap energy level and E_F is the Fermi energy. The degeneracy factor g is the same as in Section 2.2.5.

Because our model of an interface trap assumes an SRH center located at the Si–SiO$_2$ interface, the equivalent circuit for an MOS capacitor with interface traps at a single energy within the bandgap (Fig. 5.2) resembles the circuit shown in Fig. 4.6 for a single-level bulk trap. The differences are:

1 As the interface trap is located at the interface, the capacitive divider in Fig. 4.6 is replaced by the depletion layer capacitance because C_x is an infinite susceptance, in other words, a short circuit.

2 Interface trap level density is expressed per unit *area*, and not per unit *volume*. Therefore, $C_T \Delta x$ and the like are replaced by C_T, G_n, and G_p per unit area.

Therefore, for single-level interface traps, the equivalent circuit is that in Fig. 5.2 where [see (4.41)–(4.43)]

$$C_T = \beta q N_T f_o (1 - f_o) \tag{5.2}$$

$$G_n = \beta q N_T c_n n_s (1 - f_o) \tag{5.3}$$

$$G_p = \beta q N_T c_p p_s f_o \tag{5.4}$$

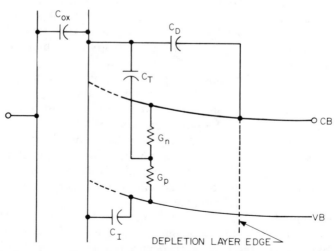

Fig. 5.2 Equivalent circuit of the MOS capacitor with single-energy level interface traps. The silicon substrate is n-type; hence capture and emission of electrons is the dominant process. Circles denote the external terminals.

and N_T is the density of interface traps per unit area, n_s and p_s are the electron and hole densities per unit volume at the interface, and f_o is given by (5.1).

Circuit analysis shows that the admittance of the equivalent circuit in Fig. 5.2 is the series combination of $j\omega C_{ox}$ and the semiconductor admittance Y_s

$$Y_s = j\omega C_D + G_n[j\omega(C_I + C_T)G_p + j\omega C_I j\omega C_T]\{j\omega(C_I + C_T)G_p$$
$$+ j\omega C_I G_n + G_n G_p + j\omega C_I j\omega C_T\}^{-1}. \tag{5.5}$$

The admittance given by (5.5) also can be derived in a fundamental way from the basic small-signal equations. Such a derivation is made later for the more realistic case of interface trap levels distributed in energy through the silicon bandgap. Now (5.5) is simplified for gate biases corresponding to depletion.

5.2.1 Depletion

In depletion the equivalent circuit in Fig. 5.2 is simpler because only majority carriers are important. We develop the equivalent circuit in depletion and describe the behavior of the admittance with single-level interface traps. In the next section we modify the equivalent circuit for the more realistic case of a distribution of interface trap levels and then treat the case when both types of carrier are important. These generalizations are used to discuss the admittance near midgap and in both weak and strong inversion in Sections 5.3.5–5.3.7.

In depletion, $C_I = \beta q N_I$, $N_I \ll N_T$, and $C_I \ll C_T$ for biases near the peak of $f_o(1 - f_o)$. Using these inequalities in (5.5), we find that Y_s becomes

$$Y_s = j\omega C_D + j\omega C_T G_n[G_n + j\omega C_T]^{-1}. \tag{5.6}$$

Equation (5.6) shows that Y_s consists of the depletion layer capacitance shunted by the series combination of C_T and G_n. That is, the equivalent circuit in Fig. 5.2 simplifies to the equivalent circuit in Fig. 5.3a. For a single-level interface trap in depletion, the equivalent parallel conductance G_P and capacitance C_P corresponding to (5.6) are

$$\frac{G_P}{\omega} = C_T\omega\tau[1 + (\omega\tau)^2]^{-1} \quad \text{(F/cm}^2\text{)} \tag{5.7a}$$

and

$$C_P = C_T[1 + (\omega\tau)^2]^{-1} + C_D \quad \text{(F/cm}^2\text{)} \tag{5.7b}$$

where

$$\tau \equiv \frac{C_T}{G_n} = f_o[c_n n_s]^{-1}.$$

Figure 5.3b is a plot of the dimensionless quantities $G_P/(\omega C_T)$ and $(C_P - C_D)/C_T$ as functions of $\omega\tau$ calculated from (5.7a) and (5.7b). Figure

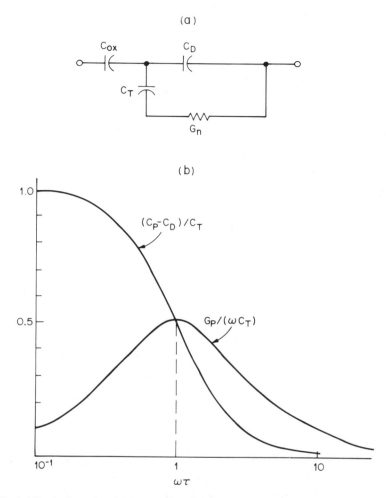

Fig. 5.3 (a) Equivalent circuit of the MOS capacitor in depletion for a single-level interface trap. Capacitance and equivalent parallel conductance are given by (5.7b) and (5.7a). Silicon is n-type. (b) Plot of $(C_P - C_D)/C_T$ from (5.7b) and $(G_P/(\omega C_T))$ from (5.7b) as functions of $\omega\tau$.

5.3b illustrates the behavior of (5.7a) and (5.7b) represented by the equivalent circuit in Fig. 5.3a.

Consider first the behavior of the capacitance in Fig. 5.3b. At low values of $\omega\tau$ ($\omega\tau \to 0$), interface traps immediately change occupancy in response to the ac gate voltage. Then, from (5.7b), $(C_P - C_D)/C_T \approx 1$, that is, $C_P = C_D + C_T$. As $\omega\tau$ increases, interface traps no longer change occupancy immediately in response to the ac gate voltage, but lag behind. Then $(C_P - C_D)/C_T$ decreases with increasing $\omega\tau$ until, at very large values of $\omega\tau$

$(\omega\tau \to \infty)$, interface trap occupancy changes very little in response to the ac gate voltage, and $(C_P - C_D)/C_T \approx 0$, that is, $C_P = C_D$.

Consider next the behavior of the equivalent parallel conductance in Fig. 5.3b. At low $\omega\tau$, interface traps change occupancy in phase with the ac gate voltage, maintaining equilibrium. Therefore, there is no energy loss, and (5.7a) predicts $G_P/(\omega C_T) \approx 0$. As $\omega\tau$ increases, $G_P/(\omega C_T)$ increases because interface traps lag behind the ac gate voltage, producing an energy loss. Then $G_P/(\omega C_T)$ increases with $\omega\tau$ until it reaches a peak value of $G_P/(\omega C_T) = \frac{1}{2}$ when $\omega\tau = 1$. As $\omega\tau$ increases further, $G_P/(\omega C_T)$ decreases from its peak value because some interface traps do not respond at all, thus decreasing the energy loss. At very large values of $\omega\tau$, interface traps hardly change occupancy in response to the ac gate voltage, there is no longer any energy loss, and $G_P/(\omega C_T) = 0$ again.

5.3 ADMITTANCE OF A DISTRIBUTION OF SINGLE-LEVEL INTERFACE TRAPS

Single-energy level interface traps are not observed in device-grade interfaces. What is observed are many interface trap levels so closely spaced in energy over the silicon bandgap that they cannot be distinguished in the measurements as separate levels. Interface trap levels far from the band edges do not form a band, however, because interface trap sites are spaced too far apart on the interfacial plane for the wave functions of electrons localized on neighboring interface traps to overlap. Transitions do not occur directly between interface trap levels on different sites, even for levels close in energy. Rather, transitions occur only between the interface trap levels and the silicon bands.

Section 5.3.1 develops the equivalent circuit in depletion for interface traps with energy levels distributed throughout the bandgap. Again, an intuitive construction of the equivalent circuit is presented, combining branch circuits from the earlier single-level discussion.

Having provided the intuitive background, in Section 5.3.2 we provide a mathematical derivation of the admittance for depletion and weak inversion, on the basis of the small-signal equations. This derivation leads to an equivalent circuit that appears different from the one intuitively constructed. However, the intuitive circuit and the derived circuit are related by a Y-Δ or T-Π transformation and possess the same admittance.

The admittance from the general circuit is simplified for various special cases in Sections 5.3.3–5.3.7.

5.3.1 Equivalent Circuit in Depletion

For a distribution of interface trap levels, and in depletion, the equivalent circuit in Fig. 5.3a for a single-level interface trap becomes the

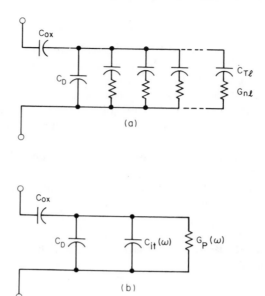

(a)

(b)

Fig. 5.4 (a) Equivalent circuit of the MOS capacitor in depletion for a distribution of single-level interface traps. The silicon is n-type. (b) Lumped parallel equivalent of circuit in (a).

equivalent circuit in Fig. 5.4a. This circuit is a parallel combination of G_n-C_T branches like that in Fig. 5.3a, with one branch for each level. Each interface trap level is labeled by the index l. Interface traps with the same l-value have energy levels at the same energy in the silicon bandgap. In Fig. 5.4a, C_{Tl} is the capacitance of the lth level, and G_{nl}^{-1} is the corresponding electron capture resistance.

In Fig. 5.4a, each interface trap level is denoted by a series R-C circuit. It is convenient to express the circuit for each interface trap level as the parallel equivalent of the series R-C circuit, to aid comparison with the mathematical derivation of a quantitative expression for the admittance and to aid in comparison with the measured admittance. Expressing each R-C circuit as its parallel equivalent and adding all the equivalent parallel capacitances and the equivalent parallel conductances from Fig. 5.4a, we obtain the lumped equivalent parallel circuit in Fig. 5.4b, where $C_{it}(\omega)$ is the equivalent parallel interface trap capacitance and $G_P(\omega)$ is the equivalent parallel conductance. Expressions for $G_P(\omega)$ and $C_{it}(\omega)$ are derived later [see (5.49a–d)].

In analogy with the discussion connecting Fig. 5.3a with Fig. 5.4a, we obtain Fig. 5.5 from Fig. 5.2. Figure 5.5 is the equivalent circuit that applies for a distribution of interface trap levels, including transitions to both bands. This circuit is a parallel array of T-sections, each of which resem-

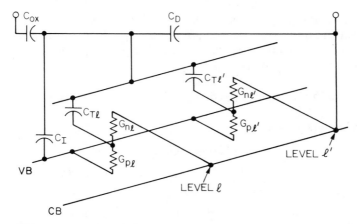

Fig. 5.5 Equivalent circuit of the MOS capacitor with a distribution of interface trap levels over the silicon bandgap. The silicon substrate is n-type. Symbols VB and CB denote valence and conduction band, respectively.

bles the T-section in Fig. 5.2. In depletion, Fig. 5.5 becomes Fig. 5.4a, for the same reasons given for simplification of the circuit of Fig. 5.2 to that in Fig. 5.3a [i.e., because of the inequalities leading to (5.6)]. In Section 5.3.4 the admittance of the circuit in Fig. 5.5 in depletion is shown to become that in Fig. 5.4b.

5.3.2 Mathematical Basis for the General Equivalent Circuit

Next, an expression is derived for the admittance, including transitions to both bands, providing the mathematical basis for the equivalent circuit of Fig. 5.5. This derivation supports the intuitive considerations leading to the circuit of Fig. 5.5 and also provides quantitative expressions for the various circuit elements. In addition, the assumptions needed to justify the circuit of Fig. 5.5 are made clear. The derivation parallels that of Section 4.3 for bulk traps.

We suppose there are N_{Tl} interface trap levels per unit area with energy level E_{Tl}. The basic small-signal equation governing interface trap occupancy is [see (4.46)]

$$j\omega\beta q N_{Tl}\,\delta f_l = j\omega C_{Tl}[(G_{nl} + G_{pl})\delta v_s - G_{pl}\,\delta u_{Fp}]D_l^{-1} \tag{5.8}$$

where

$$D_l = j\omega C_{Tl} + G_{nl} + G_{pl} \tag{5.9}$$

and the l subscript simply has been added to the quantities in (5.2)–(5.4) to denote a variety of levels. Here, δf_l is the small-signal occupancy variation, δv_s the small-signal band-bending variation, and δu_{Fp} the small-signal variation in quasi-Fermi level of holes in the inversion layer. Unlike bulk traps, where the Poisson equation had to be solved to determine band bending at the bulk trap site, for interface traps this band bending is v_s.

The variation in inversion layer carrier density per unit area δN_I corresponding to δv_s is [see (4.48)]

$$j\omega\beta q\,\delta N_I = -(\delta u_{Fp} - \delta v_s)\sum_l G_{pl} - \sum_l \beta q N_{Tl}\,\delta f_l(C_{Tl}R_{pl})^{-1}. \quad (5.10)$$

We also have [see (4.49)]

$$j\omega\beta q\,\delta N_I = j\omega C_I(\delta u_{Fp} - \delta v_s). \quad (5.11)$$

Using (5.11) in (5.10) and substituting for δf_l from (5.8), we solve for $\delta u_{Fp} - \delta v_s$ [see (4.79)]

$$\delta u_{Fp} - \delta v_s = -\delta v_s\, G_{gr}[j\omega(C_I + C_{Tp}) + G_{gr}]^{-1} \quad (5.12)$$

where the generation-recombination conductance G_{gr} is

$$G_{gr} \equiv \sum_l G_{grl} \equiv \sum_l G_{nl} G_{pl} D_l^{-1} \quad (5.13)$$

and

$$C_{Tp} \equiv \sum_l C_{Tpl} \equiv \sum_l C_{Tl}\, G_{nl} D_l^{-1} \quad (5.14a)$$

where D_l is given by (5.9). For later use, we define C_{Tn} as

$$C_{Tn} \equiv \sum_l C_{Tnl} \equiv \sum_l C_{Tl} G_{pl} D_l^{-1}. \quad (5.14b)$$

Substituting (5.12) into (5.8) yields

$$j\omega\beta q N_{Tl}\,\delta f_l = \delta v_s\{j\omega C_{Tl} G_{nl} D_l^{-1}$$
$$+ j\omega C_{Tl} G_{pl} D_l^{-1} G_{gr}[j\omega(C_I + C_{Tp}) + G_{gr}]^{-1}\}. \quad (5.15)$$

Examination of (5.15) shows that every energy level l has an occupancy variation, δf_l, with a frequency dependence that differs from other energy levels with different index l. In other words, each interface trap level finds its own occupancy individually, independent of the other levels.

Suppose, instead, that the interface traps jointly determined their occupancy, as an ensemble. Then, as in equilibrium, their occupancy would be governed by (5.1), but the Fermi level in (5.1) would be replaced by a time-varying interface trap *quasi*-Fermi level. The ac variation of this interface trap quasi-Fermi level would have a single frequency dependence, the same for all interface trap levels. The frequency dependence of the occupancy of different interface trap energy levels would not differ, as it does in (5.15). Therefore, in (5.1), an interface trap quasi-Fermi level approximation for interface trap occupancy is too crude.

With (5.12) and (5.15), the small-signal variation in silicon surface charge can be written [see (4.75)]

$$\delta Q_T = \delta Q_D + q\,\delta N_I - q \sum_l N_{Tl}\,\delta f_l \qquad (5.16)$$

where now [see (4.77)]

$$\beta\delta Q_D = -C_D\,\delta v_s. \qquad (5.17)$$

The admittance of the silicon portion of the MOS capacitor Y_s is found from (5.16) using (5.17) for δQ_D, (5.10) for δN_I, and (5.15) for δf_l. The result is

$$
\begin{aligned}
Y_s &\equiv -j\omega\beta\,\frac{\delta Q_T}{\delta v_s} \\
&= j\omega C_D + j\omega C_{Tn} + j\omega(C_I + C_{Tp})G_{gr}[j\omega(C_I + C_{Tp}) + G_{gr}]^{-1}
\end{aligned}
\qquad (5.18)
$$

where G_{gr}, C_{Tp}, and C_{Tn} are defined in (5.13) and (5.14) [see (4.87)].

At low frequencies, such as those near the loss peak in weak inversion, G_{gr} is a conductance and C_{Tn} and C_{Tp} are capacitances, as the notation indicates. Hence, the equivalent circuit leading to Y_s is the lumped circuit shown in Fig. 5.6 for low frequencies.

Although the circuit in Fig. 5.6 leads to the admittance of (5.18), and although the derivation of (5.18) is straightforward, the relation between the circuits in Fig. 5.6 and Fig. 5.5 is not clear. To relate these circuits, we must consider (5.18) for arbitrary frequencies.

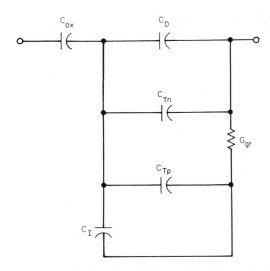

Fig. 5.6 Equivalent circuit of a distribution of interface trap levels closely spaced in energy over the silicon bandgap. At low frequencies C_{Tn} and C_{Tp} are pure capacitances, and G_{gr} is a pure conductance. At higher frequencies all three components have both capacitive and conductive parts.

For the more general case of arbitrary frequency, $j\omega C_{Tn}$, $j\omega C_{Tp}$, and G_{gr} must be considered as admittances with both conductive and susceptive components given by (5.13), (5.14a), and (5.14b). From these equations, each admittance is a sum over all interface trap levels of a parallel array of branch admittances (denoted by G_{grl}, C_{Tpl}, and C_{Tnl}), one branch contributed by each interface trap level. Such an interpretation yields the equivalent circuit in Fig. 5.7. The admittance of this circuit is given by (5.18) and can be arrived at directly. By a T-Π (or Y-Δ) transformation, the circuit in Fig. 5.5 is obtained. This circuit is made up from the connection of individual T-sections from Fig. 5.2, one for each energy level. Rather than using a T-Π transformation, the admittance of the circuit in Fig. 5.5 can be shown directly to be given by (5.18) by following the procedure for the bulk trap circuit described in Appendix III.

The admittance [see (5.18)] can be simplified greatly for particular cases. Corresponding simplifications result for the equivalent circuit in Fig. 5.5. Understanding also is simplified. Therefore, we now evaluate G_{gr}, C_{Tn}, and C_{Tp} in (5.18) and specialize the results for low frequencies (Section 5.3.3), for depletion (Section 5.3.4), for biases near midgap (Section 5.3.5), and for weak and strong inversion (Sections 5.3.6 and 5.3.7). The most important of these cases is depletion because this bias range is used for extracting interface trap parameters from the admittance. Weak inversion is of interest because the admittance behavior becomes qualitatively different. The ability of (5.18) to predict the full range of behavior of the admittance throughout the entire bias range provides strong support for the basic assumptions behind our derivation.

In interpreting (5.18), it must be remembered that G_{gr}, C_{Tn}, and C_{Tp} are conductances and capacitances only at low frequencies. Therefore, (5.18)

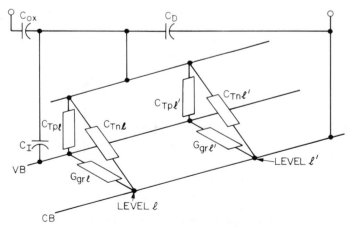

Fig. 5.7 Parallel equivalent circuit of the MOS capacitor following a T-Π (or Y-Δ) transformation of the T-sections in Fig. 5.5. Silicon is n-type. Symbols VB and CB denote valence and conduction band, respectively.

requires further examination when the frequency-dependent expression for these quantities is needed. Explicit frequency-dependent expressions for G_{gr}, C_{Tn}, and C_{Tp} are derived now by evaluating the summations involved in their definitions.

(a) Converting Sums Over Interface Trap Levels to Integrals

Because individual interface trap levels cannot be distinguished experimentally, the summation over all interface trap levels can be replaced by an integral. Let $D_{it}(\zeta_s)$ be the *probability* per unit area that an interface trap level is present with energy (in electron volts) between ζ_s and $\zeta_s + d\zeta_s$; $D_{it}(\zeta_s)$ commonly is called the *density of interface trap levels per unit area per electron volt.*

The argument of D_{it}, ζ_s, denotes an energy (in electron volts) in the bandgap, measured from the intrinsic level at the silicon surface. When a band bending ψ_s is present, the bands bend down at the surface, and thus D_{it} also moves down. However, the usual reference for energies is the intrinsic level in the *bulk* silicon.

Figure 5.8 illustrates the notation. The Fermi level, a solid horizontal line, is a distance ϕ_B above the intrinsic level in the bulk (dashed-dotted line). An arbitrary level of energy in the bulk, shown a distance ζ (in electron volts) above the intrinsic level, is related by a dashed line to the corresponding energy level at the silicon surface. This energy level at the surface is located a distance ζ_s above the intrinsic level at the surface.

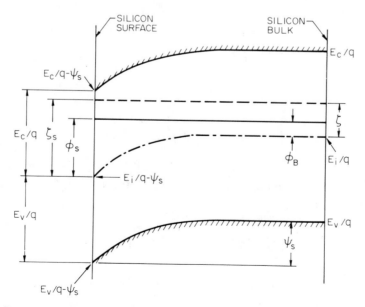

Fig. 5.8 Band bending versus depth from silicon surface illustrating the notation ζ_s and ϕ_s for energies measured from the intrinsic level at the silicon surface.

How are ζ and ζ_s related? Because the bands are bent an amount ψ_s, the intrinsic level at the silicon surface is ψ_s below its level in the bulk. Therefore, $\zeta_s = \zeta + \psi_s$. Figure 5.8 also shows that as ζ_s varies from the valence band edge to the conduction band edge, it varies from E_v/q to E_c/q. For the particular value $\zeta_s = \phi_s$, the corresponding energy level at the silicon surface coincides with the Fermi level.

With the use of $D_{it}(\zeta + \psi_s)$, the interface trap charge density per unit area $Q_{it}(\psi_s)$ can be defined at any band bending ψ_s. This definition depends on the donor or acceptor nature of the interface traps.

Donor interface traps are positive when empty, and neutral when full. For donor interface traps (superscript d)

$$Q_{it}^d(\psi_s) = q \int_{E_v/q - \psi_s}^{E_c/q - \psi_s} d\zeta [1 - f_o(\zeta - \phi_B)] D_{it}^d(\zeta + \psi_s) \qquad (5.19)$$

where ϕ_B is the Fermi level measured from the intrinsic level in the bulk silicon, $f_o(\zeta - \phi_B)$ is the Fermi function, and $(E_c/q - \psi_s)$ and $(E_v/q - \psi_s)$ locate the band edges at the silicon surface relative to the intrinsic level in the bulk silicon.

Acceptor interface traps are neutral when empty and negative when full. For acceptor interface traps (superscript a)

$$Q_{it}^a(\psi_s) = -q \int_{E_v/q - \psi_s}^{E_c/q - \psi_s} d\zeta \, f_o(\zeta - \phi_B) D_{it}^a(\zeta + \psi_s). \qquad (5.20)$$

The net interface trap charge per unit area $Q_{it}(\psi_s)$ is given by

$$
\begin{aligned}
Q_{it}(\psi_s) &= Q_{it}^d(\psi_s) - Q_{it}^a(\psi_s) \\
&= q \int_{E_v/q - \psi_s}^{E_c/q - \psi_s} d\zeta \, D_{it}^d(\zeta + \psi_s) \\
&\quad - q \int_{E_v/q - \psi_s}^{E_c/q - \psi_s} d\zeta \, f_o(\zeta - \phi_B) \{D_{it}^a(\zeta + \psi_s) + D_{it}^d(\zeta + \psi_s)\}. \qquad (5.21)
\end{aligned}
$$

If the variables of integration are changed from ζ to $\zeta_s = \zeta + \psi_s$, (5.21) becomes

$$
\begin{aligned}
Q_{it}(\psi_s) &= q \int_{E_v/q}^{E_c/q} d\zeta_s \, D_{it}^d(\zeta_s) \\
&\quad - q \int_{E_v/q}^{E_c/q} d\zeta_s \, f_o(\zeta_s - \phi_B - \psi_s) \{D_{it}^a(\zeta_s) + D_{it}^d(\zeta_s)\}. \qquad (5.22)
\end{aligned}
$$

Using (5.22), the change in Q_{it}, dQ_{it}, under a very slow incremental change in band bending $d\psi_s$ is

$$
\frac{dQ_{it}}{d\psi_s} = q \int_{E_v/q}^{E_c/q} d\zeta_s \left(\frac{kT}{q}\right) f_o(\zeta_s - \phi_B - \psi_s)[1 - f_o(\zeta_s - \phi_F - \psi_s)]
$$
$$
\times [D_{it}^a(\zeta_s) + D_{it}^d(\zeta_s)] \tag{5.23}
$$

where the following identity has been used

$$
\frac{df_o(\zeta_s - \phi_B - \psi_s)}{d\psi_s} = \frac{q}{kT} f_o(1 - f_o). \tag{5.24}
$$

Equation (5.24) is obtained using (5.1) and expressing energy in electron volts. Equation (5.23) shows the change in interface trap charge is determined by the *sum* of probabilities for donor and acceptor interface traps. We define the total probability density of interface trap energy levels $D_{it}(\zeta_s)$ by

$$
D_{it}(\zeta_s) = D_{it}^a(\zeta_s) + D_{it}^d(\zeta_s). \tag{5.25}
$$

In general, where several species of interface trap are present, $D_{it}(\zeta_s)$ is the sum of probability densities for each species, supposing these probabilities are not correlated.

Because (5.24) is sharply peaked near zero argument, that is, at $\zeta_s = \phi_B + \psi_s \equiv \phi_s$, (5.23) is dominated by $D_{it}(\phi_s)$. That is, it is primarily the interface trap levels opposite the Fermi level at the silicon surface that respond to a variation in band bending $d\psi_s$.

Experimentally, it has been shown [see Section 7.3.6(c)] that interface traps in the upper half of the silicon bandgap are donor type in device grade oxides. There are no reliable measurements to determine whether interface traps in the lower half of the silicon bandgap are donor or acceptor type.

Using D_{it}, integrations replace summations over l. For example

$$
\int_{E_v/q}^{E_c/q} d\zeta_s D_{it}(\zeta_s) = \sum_l N_{Tl}. \tag{5.26}
$$

(b) Expression for G_{gr}, C_{Tn}, and C_{Tp}

First consider G_{gr}, defined in (5.13). The definition of G_{gr} from (5.13) becomes [using (5.3), (5.4), (5.9), and the integration variable of (5.26)]

$$
G_{gr} = \beta q \int_{E_v/q}^{E_c/q} d\zeta_s D_{it}(\zeta_s) \tau_n^{-1} \tau_p^{-1} f_o(1 - f_o)
$$
$$
\times [j\omega f_o(1 - f_o) + f_o \tau_p^{-1} + (1 - f_o)\tau_n^{-1}]^{-1} \tag{5.27}
$$

where, as in Section 4.3.6(f) [see (4.68)]

$$
\tau_n^{-1} = c_n n_s \tag{5.28}
$$

$$\tau_p^{-1} = c_p p_s. \tag{5.29}$$

Experiment shows the probability density $D_{it}(\zeta_s)$ to be a slowly varying function of energy, with the possible exception of energies near the band edges. The same is true for c_n and c_p. Hence these quantities can be evaluated at the position in the bandgap opposite the Fermi level, where $f_o(1 - f_o)$ is a sharply peaked function of energy. Therefore, approximating the slowly varying portions of the integrand by their values at $\zeta_s = \phi_s$ (the Fermi level at the silicon surface is located ϕ_s eV from the intrinsic level at the silicon surface), and using (5.24), we find that (5.27) becomes

$$G_{gr} = qD_{it}(\phi_s)\tau_n^{-1}(\phi_s)\tau_p^{-1}(\phi_s)H_G \tag{5.30}$$

where from (4.81)

$$H_G \equiv -\int_0^1 df[j\omega f(1-f) + f\tau_p^{-1} + (1-f)\tau_n^{-1}]^{-1}. \tag{5.31}$$

Now consider C_{Tp}, defined in (5.14a). Following the same procedure outlined above, we obtain

$$C_{Tp} = qD_{it}(\phi_s)\tau_p^{-1}(\phi_s)H_{Cp} \tag{5.32}$$

where H_{Cp} is defined as

$$H_{Cp} \equiv \int_0^1 df\, f[j\omega f(1-f) + f\tau_p^{-1} + (1-f)\tau_n^{-1}]^{-1}. \tag{5.33}$$

The value C_{Tn} is given by (5.14b), which becomes

$$C_{Tn} = qD_{it}(\phi_s)\tau_n^{-1}(\phi_s)H_{Cn} \tag{5.34}$$

where H_{Cn} is given by

$$H_{Cn} \equiv \int_0^1 df(1-f)[j\omega f(1-f) + f\tau_p^{-1} + (1-f)\tau_n^{-1}]^{-1}. \tag{5.35}$$

The integrals (5.31), (5.33), and (5.35) are evaluated in Appendix II.

Using these evaluations for G_{gr}, C_{Tp}, and C_{Tn} and the expressions for the integrals H_G, H_{Cp}, and H_{Cn}, from Appendix II, we discuss various limiting forms of the admittance Y_s in (5.18).

5.3.3 Low Frequency Regime: $\omega\tau_n$ and $\omega\tau_p \ll 1$

At low frequencies we find using (17), (31), and (32) from Appendix II and (5.30) that

$$G_{gr} = qD_{it}(\phi_s)(\tau_p - \tau_n)^{-1} \ln\frac{\tau_p}{\tau_n} \tag{5.36}$$

$$C_{Tp} = qD_{it}(\phi_s)\tau_n(\tau_n - \tau_p)^{-1}\left[1 + \tau_p(\tau_n - \tau_p)^{-1}\ln\frac{\tau_p}{\tau_n}\right] \quad (5.37)$$

$$C_{Tn} = qD_{it}(\phi_s)\tau_p(\tau_p - \tau_n)^{-1}\left[1 + \tau_n(\tau_p - \tau_n)^{-1}\ln\frac{\tau_n}{\tau_p}\right] \quad (5.38)$$

and Y_s is given by (5.18). In this case we rewrite Y_s as

$$Y_s = j\omega(C_D + C_{Tn} + C_{Tp} + C_I) - j\omega(C_{Tp} + C_I)\frac{j\omega(C_{Tp} + C_I)}{j\omega(C_{Tp} + C_I) + G_{gr}}. \quad (5.39)$$

The sum $C_{Tn} + C_{Tp}$ is the *interface trap capacitance* C_{it}

$$C_{it} \equiv \frac{dQ_{it}}{d\psi_s} \approx qD_{it}(\phi_s) \quad (5.40)$$

as follows by adding (5.37) to (5.38). As $\omega \to 0$, the last term in (5.39) becomes negligible, so Y_s becomes the low frequency capacitance of the MOS capacitor with interface traps, namely, at low ω

$$Y_s = j\omega[C_D + C_I + C_{it}]. \quad (5.41)$$

The admittance (5.41) corresponds to the low frequency equivalent circuit in Fig. 5.9. In Fig. 5.9 C_D and C_I are combined as $C_s \equiv C_D + C_I$. For (5.41) to be valid

$$\left[\frac{j\omega(C_{Tp} + C_I)}{j\omega(C_{Tp} + C_I) + G_{gr}}\right] \ll 1$$

or $G_{gr} \gg \omega(C_{Tp} + C_I)$. Therefore, in strong inversion, where C_I is large, ω must be very low. In weak inversion, where $p_s \gg n_s$, from (5.28) and (5.29) we find $\tau_n/\tau_p \gg 1$. Thus the condition for the validity of (5.41) is

$$\omega\tau_p \ll \ln\frac{\tau_n}{\tau_p}. \quad (5.42)$$

In depletion, where $n_s \gg p_s$ and $\tau_p/\tau_n \gg 1$, the condition for the validity of (5.41) is

$$\omega\tau_n \ll \frac{\tau_p}{\tau_n}\ln\frac{\tau_p}{\tau_n}. \quad (5.43)$$

This bound is much less strict than (5.42), so much lower frequencies are needed to maintain the lossless condition in weak inversion than in depletion.

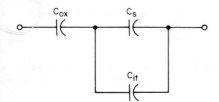

Fig. 5.9 Low frequency equivalent circuit of the MOS capacitor with a distribution of interface trap levels over the bandgap. The silicon capacitance per unit area is $C_s = C_D + C_I$. The interface trap capacitance per unit area C_{it} is given by (5.40).

5.3.4 Depletion: $\tau_n^{-1} \gg \tau_p^{-1}$

In depletion, for n-type, we find using (15) from Appendix II and (5.30) that

$$G_{gr} = -C_{it}\tau_p^{-1}(1 + j\omega\tau_n)^{-1}\ln\left[\frac{(\tau_n/\tau_p)}{(1 + j\omega\tau_n)^2}\right].\qquad (5.44)$$

Also, using (29) and (30) from Appendix II (5.32) and (5.34) become

$$C_{Tp} = -C_{it}\frac{1}{j\omega\tau_p}\left\{j\omega\tau_n\ln\frac{\tau_n}{\tau_p} + \tfrac{1}{2}(1 - j\omega\tau_n)\right.$$

$$\left. \times\{\ln[1 + (\omega\tau_n)^2] + 2j\tan^{-1}(\omega\tau_n)\}\right\}(1 + j\omega\tau_n)^{-1}\qquad (5.45)$$

and

$$C_{Tn} = C_{it}\frac{1}{2j\omega\tau_n}\{\ln[1 + (\omega\tau_n)^2] + 2j\tan(\omega\tau_n)\}.\qquad (5.46)$$

Comparing (5.46) with (5.45) and (5.44), we see that

$$\omega C_{Tn} \gg \omega C_{Tp} \quad and \quad G_{gr}.\qquad (5.47)$$

Therefore, in depletion, Y_s from (5.18) becomes

$$\begin{aligned}
Y_s &= j\omega C_D + j\omega C_{Tn}\\
&= j\omega C_D + C_{it}(2\tau_n)^{-1}\{\ln[1 + (\omega\tau_n)^2] + 2j\tan^{-1}(\omega\tau_n)\}\\
&\equiv j\omega C_P + G_P
\end{aligned}\qquad (5.48)$$

where the silicon equivalent parallel capacitance C_P and conductance G_P are defined by (5.48) to be

$$C_P \equiv C_D + C_{it}(\omega\tau_n)^{-1}\tan^{-1}(\omega\tau_n)\qquad (5.49a)$$

$$\frac{G_P}{\omega} \equiv C_{it}(2\omega\tau_n)^{-1}\ln[1 + (\omega\tau_n)^2].\qquad (5.49b)$$

Equations (5.49a) and (5.49b) were derived first by Lehovec.[15]

The admittance (5.48) agrees with that of the simple equivalent circuit in Fig. 5.4b. The equivalent parallel capacitance of (5.49a) can be expressed as

$$C_P = C_D + C_{it}(\omega)\qquad (5.49c)$$

which defines the equivalent parallel interface trap capacitance $C_{it}(\omega)$ as

$$C_{it}(\omega) \equiv C_{it}(\omega\tau_n)^{-1}\tan^{-1}(\omega\tau_n)\qquad (5.49d)$$

where $C_{it}(\omega)$ expresses the capacitive response of interface traps at arbitrary frequencies. In particular, as $\omega\tau_n \to 0$, $C_{it}(\omega) \to C_{it}$, the low frequency interface trap capacitance.

For a distribution of interface trap levels over the silicon bandgap, transitions occur between the majority carrier band and interface trap levels in an energy interval a few kT wide about the Fermi level. Each interface trap level in this energy interval contributes a different energy loss depending on its distance in energy from the Fermi level. As a result, each interface trap level in this energy interval has a different time constant. The equivalent circuit is shown in Fig. 5.4. The capacitance and equivalent parallel conductance of the distributed R-C network of the interface trap branch of the equivalent circuit in Fig. 5.4 is given by (5.49a) and (5.49b).

The frequency dependence of the components in the two circuits in Figs. 5.3a and 5.4b is compared in Fig. 5.10. The solid curve in Fig. 5.10a is calculated from (5.49a) and the dotted curve is calculated from (5.7b) for comparison. The solid curve in Fig. 5.10b is calculated from (5.49b) and the dotted curve is calculated from (5.7a), again for comparison. The behavior of the capacitance and equivalent parallel conductance as functions of $\omega\tau_n$ for the distribution of single-level interface traps over the silicon bandgap is qualitatively similar to that for the single-level interface trap. However, a comparison of the solid and dotted curves in Figs. 5.10a,b shows that the capacitance and equivalent parallel conductance for the distributed interface trap levels have been broadened along the $\omega\tau_n$ axis. Moreover, the conductance peak has been shifted from $\omega\tau_n = 1$ to $\omega\tau_n = 1.98$,* and the peak magnitude has decreased from $G_P/(\omega C_T) = 0.5$ to $G_P/(\omega C_{it}) = 0.4$ for the distributed interface trap levels as shown in Fig. 5.10b.

From the discussion of Figs. 5.3, 5.4, and 5.10, we can see how interface trap properties can be extracted from the measured admittance. We start by extracting C_P and G_P from the measured admittance. Frequency and gate bias also are measured quantities. The value $\omega\tau_n$ can be varied either by changing gate bias with frequency as parameter [τ_n is a function of band bending from (5.28)] or by changing frequency with gate bias as parameter. Which course is chosen will depend on the method used, as described later.

The two quantities sought from the measured quantities are τ_n and D_{it}. *Two independent* equations are needed to determine these two unknowns. The first independent equation is obtained from the condition $\omega\tau_n = 1.98$ corresponding to the peak value of G_P/ω, which yields $\tau_n = 1.98/\omega$, the first unknown in terms of the measured frequency. The second equation, which follows from the values of the conductance or capacitance when $\omega\tau_n = 1.98$, is either $G_P/\omega = 0.40qD_{it}$ from (5.49b) evaluated at the peak frequency or $C_P = C_D + 0.56qD_{it}$ from (5.49a), which gives the second unknown in terms

---

*The value $\omega\tau_n = 1.98$ is found by setting the derivative $\partial(G_P/\omega)/\partial(\omega\tau_n)$ to zero. An iterative solution to this equation is, for the $(i+1)$-trial, $x^{i+1} = \exp\{2/[1 + (1/x^i)]\} - 1$, with $\omega\tau_n = \sqrt{x}$. At this peak value of $\omega\tau_n$, $G_P/\omega = 0.402C_{it}$ and $C_P = C_D + 0.557C_{it}$.

of measured conductance or capacitance.* The value D_{it} is obtained from the equivalent parallel conductance with no need to know the silicon doping density. If D_{it} is to be obtained from the capacitance, C_D must be obtained by experiment from a high frequency C-V curve or by calculation using the measured silicon doping profile.

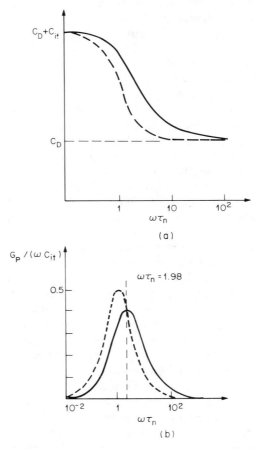

Fig. 5.10 (a) Representative plot of C_P from (5.49a) as a function of log $\omega\tau_n$. Solid line; distribution of energy levels; dashed line; single energy level. (b) Representative plot of (5.49b) showing G_P/ω as a function of log $\omega\tau_n$. Solid and dashed lines as in (a). Vertical dashed line locates the maximum for the distributed interface trap level case, which occurs at $\omega\tau_m = 1.98$.

*Capacitance and equivalent parallel conductance are not independent quantities. They are related to each other by the Kronig-Kramers relations. Rather than going through a mathematical proof of this, note that the capacitance and the conductance of any circuit made up of passive elements, such as the circuit in Fig. 5.3, satisfy the Kronig-Kramers relations.[16]

Having found D_{it} and τ_n at the gate bias corresponding to the peak in G_P/ω, the band bending corresponding to this gate bias can be found as explained in Section 3.3. Given this band bending, D_{it} can be located in the silicon bandgap as explained in Section 3.3.3, and τ_n leads to the capture probability by way of (5.28). In calculating τ_n and in locating D_{it} in the bandgap, the silicon doping profile must be known.

5.3.5 Region Near Midgap: $\tau_n = \tau_p = \tau$

Again using (12) and (28) from Appendix II, we obtain

$$G_{gr} = -2C_{it}\frac{1}{\tau}\frac{1}{j\omega\tau(1+4/j\omega\tau)^{1/2}}\ln\left\{\frac{(1+4/j\omega\tau)^{1/2}-1}{(1+4/j\omega\tau)^{1/2}+1}\right\} \tag{5.50}$$

$$C_{Tp} = \frac{1}{2}G_{gr}\tau \tag{5.51a}$$

$$C_{Tn} = \frac{1}{2}G_{gr}\tau \tag{5.51b}$$

and

$$Y_s = j\omega C_D + \frac{1}{2}G_{gr}j\omega\tau\left\{1 + \frac{1+2C_I/(G_{gr}\tau)}{1+(j\omega\tau/2)[1+2C_I/(G_{gr}\tau)]}\right\}. \tag{5.52}$$

For $\omega\tau \ll 1$, $G_{gr} \approx C_{it}/\tau$, and (5.52) becomes

$$Y_s = j\omega(C_D + C_I + C_{it}). \tag{5.53}$$

Equation (5.53) agrees with (5.41) and with the low frequency equivalent circuit in Fig. 5.9.

For $\omega\tau \gg 1$, $G_{gr} \approx 2(C_{it}/\tau)(j\omega\tau)^{-1}\ln(1+j\omega\tau)$, and (5.52) becomes

$$Y_s \approx j\omega C_D + \frac{1}{2}G_{gr}j\omega\tau = j\omega C_D + \frac{C_{it}}{\tau}\left[\ln(\omega\tau) + j\frac{\pi}{2}\right]. \tag{5.54}$$

The interface trap term agrees with (5.49) for depletion for $\omega\tau_n \gg 1$, if τ_n is replaced by τ.

Under what circumstances is C_I negligible compared to C_{it}? Now $C_I = \beta q N_I$ and near midgap, $N_I \approx 4 \times 10^7\,\mathrm{cm}^{-2}$ for $N_D = 10^{17}\,\mathrm{cm}^{-3}$, and $N_I \approx 2 \times 10^9\,\mathrm{cm}^{-2}$ for $N_D = 10^{14}\,\mathrm{cm}^{-3}$. Hence the inequality $G_{gr}\tau \approx C_{it} \gg 2C_I$ requires that $D_{it} \gg 2\beta N_I \approx 3 \times 10^9$ to $2 \times 10^{11}\,\mathrm{cm}^{-2}\,\mathrm{eV}^{-1}$. This inequality is satisfied marginally for device-grade interfaces, for which $D_{it} \approx 10^9$ to $10^{10}\,\mathrm{cm}^{-2}\,\mathrm{eV}^{-1}$.

5.3.6 Weak Inversion: $\tau_p^{-1} \gg \tau_n^{-1}$

From Appendix II

$$G_{gr} = C_{it}\tau_n^{-1}(1+j\omega\tau_p)^{-1}\left\{\ln\frac{\tau_n}{\tau_p} + \ln[1+(\omega\tau_p)^2] + 2j\tan^{-1}(\omega\tau_p)\right\} \tag{5.55a}$$

$$C_{Tp} = C_{it} \frac{1}{2j\omega\tau_p} \{\ln[1 + (\omega\tau_p)^2] + 2j \tan^{-1}(\omega\tau_p)\} \tag{5.55b}$$

and

$$C_{Tn} = C_{it} \left\{ \left(\frac{\tau_p}{\tau_n}\right) (1 + j\omega\tau_p)^{-1} \ln \frac{\tau_n}{\tau_p} - (2j\omega\tau_n)^{-1}(1 - j\omega\tau_p)(1 + j\omega\tau_p)^{-1} \right.$$

$$\left. \times \{\ln[1 + (\omega\tau_p)^2] + 2j \tan^{-1}(\omega\tau_p)\} \right\}. \tag{5.55c}$$

In this case, as N_I increases exponentially with band bending in weak inversion, the inequality $C_{it} \gg 2C_I$ becomes reversed for band bendings $2–4kT/q$ further toward inversion from midgap. Also, for such biases, $\tau_n \gg \tau_p$ so that $\omega C_{Tp} \gg G_{gr}$. Therefore, (5.18) becomes, for frequencies such that $\omega(C_I + C_{Tp}) \gg G_{gr}$

$$Y_s = j\omega C_D + j\omega C_{Tn} + G_{gr}$$

$$= j\omega C_D + C_{it}\tau_n^{-1} \left\{ \ln \frac{\tau_n}{\tau_p} + \frac{1}{2} \{\ln[1 + (\omega\tau_p)^2] + 2j \tan^{-1} \omega\tau_p\} \right\}$$

$$= j\omega C_D + C_{it}\tau_n^{-1} \ln \frac{\tau_n}{\tau_p} + j\omega \frac{\tau_p}{\tau_n} C_{it}(\omega) + \frac{\tau_p}{\tau_n} G_P(\omega). \tag{5.56}$$

The equivalent circuit corresponding to (5.56) is shown in Fig. 5.11a. In (5.56) and Fig. 5.11a, the equivalent parallel interface trap capacitance $C_{it}(\omega)$ is given by

$$C_{it}(\omega) = C_{it}(\omega\tau_p)^{-1} \tan^{-1}(\omega\tau_p) \tag{5.57a}$$

and the equivalent parallel interface trap conductance by

$$\frac{G_P}{\omega} = \frac{C_{it}}{2\omega\tau_p} \ln[1 + (\omega\tau_p)^2]. \tag{5.57b}$$

Equations (5.57a) and (5.57b) resemble the Lehovec terms $C_{it}(\omega)$ and $G_P(\omega)$ from (5.49) with τ_n replaced by τ_p.

For lower frequencies, $\omega\tau_p \ll 1$, and from (5.55)

$$G_{gr} \approx C_{it}\tau_n^{-1} \ln \frac{\tau_n}{\tau_p} \tag{5.58}$$

$$C_{Tp} \approx C_{it}. \tag{5.59}$$

Using these results, (5.18) becomes

$$Y_s = j\omega C_D + \frac{j\omega[C_I + C_{it}]C_{it}\tau_n^{-1} \ln(\tau_n/\tau_p)}{j\omega[C_I + C_{it}] + C_{it}\tau_n^{-1} \ln(\tau_n/\tau_p)}. \tag{5.60}$$

The equivalent circuit corresponding to (5.60) is shown in Fig. 5.11b, which can be compared with the circuit in Fig. 5.3a and the admittance given by (5.6). Such a comparison shows the two circuits to be the same, except the

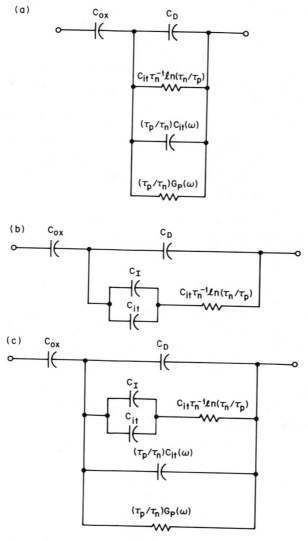

Fig. 5.11 Equivalent circuit in weak inversion for an MOS capacitor with a distribution of interface trap levels throughout the silicon bandgap. Silicon is n-type. (a) Circuit for $\omega\tau_p \gg 1$; $C_{it}(\omega)$ and G_P are given by (5.57) and are contributed by interface trap levels located in the bandgap between the conduction band edge and the interface trap level for which $R_n = R_p$. (b) Circuit for $\omega\tau_p \ll 1$. This admittance is due to interface trap levels between the valence band and the energy level for which $R_n = R_p$. Here, C_{it} is given by (5.40). (c) Approximate combined circuit for all frequencies. This circuit interpolates between those of (a) and (b) and may not be very accurate for $\omega\tau_p \approx 1$.

conductance G_P from (5.60) is given by (5.7a) with C_T replaced by $(C_I + C_{it})$ and τ replaced by the characteristic time τ_c, where

$$\tau_c = \tau_n \frac{(C_I + C_{it})}{C_{it} \ln(\tau_n/\tau_p)}. \tag{5.61}$$

The time τ_c increases as $\exp(v_s)$ due to τ_n and is proportional to $\exp(2v_s)$ once C_I becomes important during approach to strong inversion. Also, $\tau_c \gg \tau_p$ because $\tau_n \gg \tau_p$.

The capacitance C_P from (5.60) is given by (5.7b) with the same replacements.

From this comparison, the frequency dependence of the capacitance and equivalent parallel conductance in weak inversion is that of Fig. 5.3b. This type of frequency dependence is called *single time constant* behavior, to distinguish it from a broadened frequency dependence, such as that in (5.56), (5.57a), and (5.57b).

Combining (5.56) and (5.60), the weak inversion regime possesses a small magnitude continuum-like admittance, (5.56), at higher frequencies near τ_p^{-1} and a dominant single time constant admittance, (5.60), at lower frequencies near $\tau_c^{-1} \ll \tau_p^{-1}$.

The first term in C_{it} of (5.56) is the high frequency limit of (5.60). Thus (5.56) and (5.60) can be combined in the approximate equivalent circuit for weak inversion shown in Fig. 5.11c. In the low frequency branch of this circuit the parallel combination of C_I and C_{it} is in series with the conductance $C_{it}\tau_n^{-1} \ln \tau_n/\tau_p$. The high frequency branch consists of $(\tau_p/\tau_n)C_{it}(\omega)$ and $(\tau_p/\tau_n)G_P$. The values $C_{it}(\omega)$ and G_P are given by (5.57). The coefficient $\tau_p/\tau_n \ll 1$.

The equivalent circuit in Fig. 5.11c can be understood through the following argument. We start again by considering a single interface trap level. The majority carrier (electron) capture resistance from (5.3) is

$$R_n \equiv G_n^{-1} = \left(\frac{kT}{q^2}\right) [c_n n_s N_T (1 - f_o)]^{-1}. \tag{5.62}$$

The minority carrier (hole) capture resistance is from (5.4)

$$R_p \equiv G_p^{-1} = \left(\frac{kT}{q^2}\right) (c_p p_s N_T f_o)^{-1}. \tag{5.63}$$

Substituting $p_s = n_i \exp(-u_s)$ and $n_s = n_i \exp(u_s)$ from (2.46), $f_o = \{1 + g \exp[(E_T - E_F)/kT]\}^{-1}$ from (5.1), (5.62), and (5.63) are plotted in Fig. 5.12 as a function of $E_T - E_F$ for several values of band bending. These curves can be understood from the behavior of f_o. For $E_T < E_F$, f_o is near unity; therefore, R_p is almost a constant for these interface trap levels and is at its lowest value, $R_p = kT/q^2[c_p p_s N_T]^{-1}$. For $E_T > E_F$, $f_o \approx \exp[-(E_T - E_F)/kT]$, and R_p increases exponentially; R_n, which depends on $(1 - f_o)$, has the reverse behavior.

In weak inversion, where $p_s \gg n_s$, R_p is smaller than R_n for interface trap

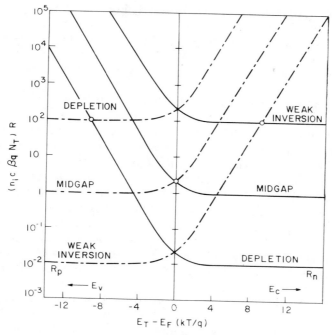

Fig. 5.12 The values R_n and R_p were taken from (5.62) and (5.63) as a function of position of the interface trap level E_T relative to the Fermi level E_F. Three regimes are shown for n-type silicon: depletion, midgap, and inversion. The resistance R_n is represented by the dash-dot curves and R_p by the solid curves; R_n and R_p are normalized to $c_n n_i \beta q N_T$ and $c_p n_i \beta q N_T$, respectively. For $c_n = c_p$, the interface trap level for which recombination and generation are most pronounced (for which $R_n = R_p$) is indicated by the circled intersection of the R_n and R_p curves for the same bias.

levels located between the valence band edge and the interface trap level for which $R_n = R_p$. To a first approximation, R_p for all these interface trap levels can be taken to be zero. For the remaining interface trap levels, $R_n < R_p$. These interface trap levels are located between the conduction band edge and the interface trap level for which $R_n = R_p$. To a first approximation, take R_p as infinite for these interface trap levels.

When the equivalent circuit in Fig. 5.5 is used, the approximation of R_p just described divides the interface trap levels into two groups. For interface trap levels in the lower part of the bandgap, G_{pl} becomes a short circuit, connecting C_{Tl} for these interface trap levels directly to C_l. For interface trap levels in the upper part of the bandgap, G_{pl} becomes an open circuit, disconnecting C_l from these interface trap levels.

The admittance of this approximate circuit can be evaluated just as for (5.18), except the summations over all interface trap levels must be split into two summations over the appropriate range of interface trap levels. As a result, when evaluating these sums, as in (5.31) or (5.33), the integrals are divided into two ranges of occupancy.* For the range, $f = 0$ to $f = (1 + \tau_n/\tau_p)^{-1}$, the G_{pl} are open circuits (see Fig. 5.11a). From $f = (1 + \tau_n/\tau_p)^{-1}$ to $f = 1$ the G_{pl} are short circuits (see Fig. 5.11b). The results have been reported by Cooper and Schwartz.[17] The admittance derived in this intuitive way agrees with the mathematically derived equivalent circuit in Fig. 5.11c.

Generation and recombination through bulk traps has been omitted in this discussion and in Fig. 5.11 for clarity. Bulk trap generation and recombination may contribute to the admittance at frequencies near τ_c^{-1} in (5.61). If bulk traps are important, an additional conductance, G_{gr} in Fig. 4.5e, must be added in parallel with the other branches in Fig. 5.11. In this case, sensitivity of measurement to interface trap properties in weak inversion will be limited by bulk trap loss as described in Section 5.7.2.†

The preceding analysis requires the additional assumption that the lateral resistance of the inversion layer between interface trap sites is small compared to minority carrier capture resistance. However, as interface trap site density decreases and the mean separation between interface trap sites increases, this condition may not hold. Therefore, interface trap properties extracted ignoring this effect will be in error.

Extracting interface trap properties from admittance measurements in depletion as described in Section 5.4 is not subject to these problems and limitations. Therefore, it is preferable to measure both an n-type and a p-type sample over the depletion range rather than extend the energy range by extracting interface trap properties in weak inversion when bulk traps and lateral inversion layer resistance have to be taken into account.

5.3.7 Strong Inversion

In strong inversion (5.60) still applies, but now $C_I \gg C_{it}$. Therefore, (5.60) simplifies, becoming

$$Y_s \approx j\omega C_D + C_{it}\tau_n^{-1} \ln \frac{\tau_n}{\tau_p}. \qquad (5.64)$$

Using (5.28) in (5.64), the equivalent parallel conductance due to generation and recombination through interface trap levels is

$$G_{grs} = C_{it}c_n N_D \ln \frac{\tau_n}{\tau_p} \exp v_s. \qquad (5.65)$$

The corresponding capacitance in (5.64) is C_D.

Because the band bending v_s in (5.65) is large and negative in strong

---

*The occupancy of the interface trap level for which $R_n = R_p$ is found by equating (5.62) to (5.63) and solving for f_o. The result is $f_o = (1 + \tau_n/\tau_p)^{-1}$.

†Because the circuit of Fig. 4.5e has been established only for strong inversion, the treatment of bulk traps using this equivalent circuit is only illustrative in weak inversion.

inversion, G_{grs} generally is negligible compared to recombination and generation through bulk trap levels at room temperature. Therefore, G_{grs} is difficult to measure, and no information can be gained about interface traps in strong inversion.

5.3.8 Summary

The various equivalent circuits can be understood in terms of the effects of R_n and R_p, shown in Fig. 5.12, on the general equivalent circuit in Fig. 5.5. In depletion, not too near midgap, R_n is low for interface trap levels near the Fermi level (see Fig. 5.12). To a first approximation, generation and recombination can be ignored and R_p treated as infinite. Hence, G_{pl} in Fig. 5.5 can be eliminated, leading to the simple circuit of Fig. 5.4a.

This approximation is viable in depletion because majority carrier density at the silicon surface is several orders of magnitude greater than minority carrier density at the silicon surface. Therefore, the time constant for minority carrier transitions is much greater than for majority carrier transitions. The time constant for majority carrier transitions is comparable to the period of the gate voltage frequency, and ac current flows into and out of interface trap levels as a result of majority carrier transitions between interface trap levels near the Fermi level and the majority carrier band edge. Minority carriers cannot follow the ac gate voltage at all. Because each interface trap level communicates individually with the conduction band, time constant dispersion is large.

Near midgap, the condition $R_n = R_p$, shown by circles in Fig. 5.12, moves closer to the Fermi level. Interface traps for which $R_n = R_p$ are efficient generation-recombination centers. When these interface trap levels are near the Fermi level, the ac gate voltage causes large generation-recombination currents. Consequently, near midgap, generation and recombination dominate the admittance of (5.52). No simple equivalent circuit is available.

For weak inversion, as seen in the last section, the interface trap levels divide naturally into two groups, distinguished by different approximations for R_p. For one group, $R_p < R_n$. These levels are located in the lower half of the bandgap, to the left of the circled intersection in Fig. 5.12. At frequencies $\omega < \tau_p^{-1}$, this group of interface trap levels is dominant. For the other group, $R_p > R_n$. These levels are located in the upper half of the bandgap, above the circled intersection in Fig. 5.12. At frequencies where $\omega > \tau_p^{-1}$, this group makes a small amplitude contribution to the admittance.

First consider the group of interface trap levels with $R_p < R_n$. There is a range of frequencies over which the time constant for majority carrier transitions is comparable to the period of the ac gate voltage frequency. Because the majority carrier time constant is much longer than the minority carrier time constant for this group of interface trap levels, there is negligible loss associated with minority carrier transitions. The ac current flows into and out of interface trap levels, controlled by the majority carrier capture resistance. The main effect of minority carriers on the equivalent circuit for this group of levels is to tie all interface trap capacitances and capture resistances together. The corresponding admittance is given by (5.60). These interface trap levels respond as a unit, with a single time constant, τ_c given by (5.61). This single time constant behavior is dominant at low frequencies in weak inversion and is quite different from the dispersion seen in depletion.

For the other group of interface trap levels located near the conduction band edge, $R_p > R_n$. These interface trap levels communicate mainly with the conduction band and respond at higher frequencies $\omega \approx \tau_p^{-1}$. The corresponding admittance is given by (5.56). The amplitude of this response is small, proportional to τ_p/τ_n. Because these interface trap levels cannot communicate easily with the inversion layer, they exhibit time constant dispersion similar to that seen in depletion, as shown by comparing (5.57) with (5.49).

(a) *Properties of Observed C-V and G-V Curves*

Figure 5.13 shows capacitance and equivalent parallel conductance as functions of gate bias measured at 5 and 100 kHz. Capacitance varies from an upper to a lower value over the illustrated bias range as predicted by

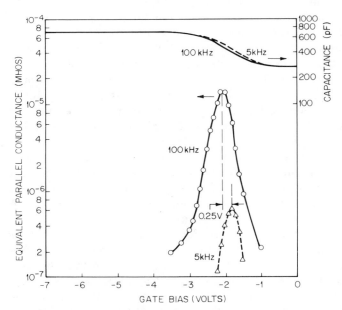

Fig. 5.13 Capacitance and equivalent parallel conductance measured at 5 and 100 kHz on a p-type substrate having $N_A = 2.1 \times 10^{16}$ cm^{-3} and an interface trap level density in the 10^{11} cm^{-2} eV^{-1} range. After Nicollian and Goetzberger.[2] Copyright (1967), American Telephone and Telegraph Company. Reprinted with permission.

(5.49a), whereas equivalent parallel conductance goes through a peak and approaches zero on either side as predicted by (5.49b). In addition, Fig. 5.13 shows that (1) the peak magnitude of the equivalent parallel conductance increases with increasing frequency because of C_{ox} in series with the interface trap R-C network, (2) the conductance peak moves toward a gate bias closer to the flatband point (which in Fig. 5.11 is at -2.6 V) as frequency is increased, as expected from (5.49b), the dependence of τ_n on band bending (5.28), and the condition at the peak $\omega\tau_n = $ constant, and (3) the capacitance is frequency dependent from 5–100 kHz as expected because $C_{it}(\omega)$ is in parallel with C_D from (5.49a).

5.4 INTERFACE TRAP TIME CONSTANT DISPERSION

Figure 5.14 shows that although measured G-V curves follow the trends predicted by (5.7a) and (5.49b), these equations do not quantitatively fit the observed curves. The reason is that the observed G_P/ω versus log $\omega\tau$ curve is broader than predicted by (5.49b). Curve a in Fig. 5.14 represents an experimentally obtained G_P/ω versus log $\omega\tau$ curve for a steam-grown oxide

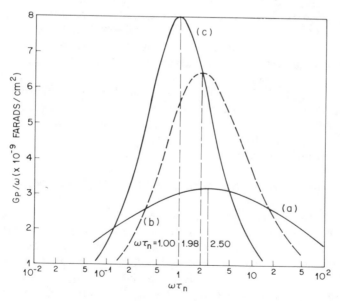

Fig. 5.14 (a) Plot of G_P/ω versus log $\omega\tau_n$ corresponding to a measured curve illustrating the observed interface trap time constant dispersion. (b) Plot of (5.49b) for an interface trap level continuum of constant density and capture probability. (c) Plot of (5.7a) for a single-level interface trap. All curves correspond to the same interface trap level density of 10^{11} cm^{-2} eV^{-1}. After Nicollian and Goetzberger.[2] Copyright (1967), American Telephone and Telegraph Company. Reprinted with permission.

that has an interface trap level density of $10^{11}\,\mathrm{cm}^{-2}\,\mathrm{eV}^{-1}$. Curve b is calculated from (5.49b) for an interface trap level density of $10^{11}\,\mathrm{cm}^{-2}\,\mathrm{eV}^{-1}$ and represents the time constant dispersion caused by a distribution of interface trap levels closely spaced in energy over the silicon bandgap. Curve c is calculated from (5.7a) for $N_T = 10^{11}\,\mathrm{cm}^{-2}$ and represents interface traps with a single energy level and no time constant dispersion. Figure 5.14 shows that time constant dispersion broadens the G_P/ω versus $\log \omega\tau$ curve, reduces the peak height for the same value of interface trap level density, and increases the value of $\omega\tau$ at which the curve peaks.

Figure 5.14 shows that a significant error can be made in extracting interface trap level density and time constant from a measured G_P/ω versus $\log \omega\tau$ curve if interface trap time constant dispersion is not taken into account.

The observed time constant dispersion could be treated by fitting the variance of a Gaussian distribution of time constants. However, the behavior of the variance resulting from this fit, as a function of bias, interface charge density, doping density, and so forth, can be explained only with a model for the dispersion.

Two models that have been proposed to explain the observed time constant dispersion are (1) a distribution of band bending over the interfacial plane caused by a random distribution of discrete charges in the oxide [because τ_n or τ_p depends exponentially on band bending as shown by (5.28) and (5.29), small variations of v_s will produce large variations of τ_n and τ_p] and (2) a distribution of capture probabilities related to the chemical environment of the interface trap sites. A detailed discussion of the theoretical and experimental aspects of these two models is given in Sections 6.5 and 7.3.5. The reasons for using the first model (distribution of band bending) for deriving the analytical expression needed to extract interface trap properties are also given in these sections. Section 5.4.1 extends the equivalent circuit once more to include band-bending fluctuations originating from localized interface charges.

5.4.1 Equivalent Circuit With Band-Bending Fluctuations

To obtain the equivalent circuit of the MOS capacitor with small-scale interfacial charge nonuniformities, the entire gate area of the MOS capacitor is divided into patches. In each patch the depletion layer width w is nearly uniform. Because the depletion layer edge is a distance w from the Si–SiO$_2$ interface, it responds only to spatial variations of interface charge that extend over distances comparable to w. Therefore, the division into patches will use patches of about a size w or larger in diameter. In Section 6.3 an analysis of small-amplitude interface charge variations suggests that the minimum patch area α should be $\alpha = 4\pi w^2$ [see discussion following (6.70)].

In Fig. 5.15a the equivalent circuit of the MOS capacitor is shown as a parallel array of elementary capacitors. Each branch corresponds to one of the patches just discussed. The depletion layer width of the silicon capacitance in a particular branch is given by the usual one-dimensional Gauss's law relation, with an interface charge density given by the average value for the particular patch chosen. Thus the branches in the circuit in Fig. 5.15a take into account *long* range (larger than w) variations in interface charge.

Within each branch of the circuit in Fig. 5.15a is an average interface trap admittance $\langle Y_{it} \rangle$, given later in (5.68). This admittance is averaged over

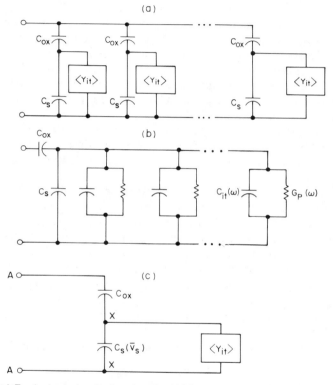

Fig. 5.15 (a) Equivalent circuit showing the MOS capacitor broken up into patches, one branch circuit for each patch. The dc band bending of each branch differs as a result of spatial variations in interface charge over distances larger than a depletion layer width: C_{ox} is the oxide layer capacitance per unit area; C_s is the silicon capacitance per unit area. (b) Equivalent circuit for one branch of the circuit in (a), showing time constant dispersion caused by short range spatial variations in interface charge density: $C_{it}(v_s)$ and $\tau_n(v_s)$ are interface trap capacitance and time constant respectively, averaged over bandgap energy at a representative band bending. Bulk generation-recombination and series resistance are zero. (c) Simplified version of (a) corresponding to the measured admittance: $C_s(\bar{v}_s)$ is the capacitance per unit area corresponding to the band bending for the average interface charge density, and $\langle Y_{it} \rangle$ is the interface trap admittance per unit area corresponding to the average band bending; $\langle Y_{it} \rangle$ is frequency dependent.

short range spatial variations in interface charge. These short range spatial variations are those of wavelength less than a patch dimension. Such short wavelength charge variations are present in a Poisson distribution of localized charges, for example. In general, these variations arise from the fine details of the interface charge, which correspond to multipole contributions to the band bending.

Multipole contributions decay rapidly with distance and so do not affect the depletion layer edge. Hence, they do not affect depletion layer capacitance nor the division into patches in Fig. 5.15a. However, as they are located at the Si–SiO$_2$ interface, interface traps certainly are sensitive to these multipole contributions to the band bending. The angular brackets about the interface trap admittance $\langle Y_{it} \rangle$ in Fig. 5.15a indicate that this admittance has been calculated in each patch taking into account the short range spatial variations of band bending within that patch.

An equivalent circuit for the interface trap admittance of one patch from the ensemble of patches is represented in Fig. 5.15b. Each R-C circuit in Fig. 5.15b represents interface trap levels averaged over silicon bandgap energy, as in Fig. 5.4b. Although the silicon surface is represented as an ac equipotential plane in Fig. 5.15b, the dc band bending varies from one branch to the next with the probability given later in (5.69). The neglect of spatial variations in the ac component of band bending is justified in Section 6.3.

Finally, as discussed in Section 6.3 for a Poisson distribution of point charges across the Si–SiO$_2$ interface, the large area fluctuations treated by the equivalent circuit of Fig. 5.15a are so weak and so improbable that they may be ignored. Thus we can treat all the patches as though they were all the same as the most probable patch. This most probable patch corresponds to a patch with the mean charge density and has a depletion layer width corresponding to the most probable band bending $\bar{v}_s$ in (6.31). Therefore, the approximate equivalent circuit used for analyzing conductance measurements becomes that in Fig. 5.15c for the most probable branch in Fig. 5.15a.

We introduce the interface trap admittance Y_{it} by the relation

$$Y_s \equiv j\omega C_D + Y_{it} \tag{5.66}$$

where Y_s is defined in (5.48). In the presence of band-bending fluctuations, it is shown in Section 6.4.2 that (5.66) becomes

$$Y_s = j\omega C_D(\bar{v}_s) + \langle Y_{it} \rangle \tag{5.67}$$

where, from (6.102), $\langle Y_{it} \rangle$ is given by

$$\langle Y_{it} \rangle = q \int_{-\infty}^{\infty} dv_s \, P(v_s) \overline{Y}_{it}(v_s) \tag{5.68}$$

where $\overline{Y}_{it}(v_s)$ is the interface trap admittance integrated over the silicon bandgap and averaged over the interfacial distribution of interface trap levels,* and $\langle Y_{it} \rangle$ is the interface trap admittance after further averaging over band-bending variations.†

Figure 5.15c represents the equivalent circuit in terms of (5.67) and (5.68). Equations (5.67) and (5.68) incorporate the average over band bending of the R-C circuits in Fig. 5.13b, as drawn for the most probable patch.

A random spatial distribution of discrete interface charges results in a *spatial* distribution of band bending over the Si–SiO$_2$ interfacial plane. Therefore, the integrand of (5.68) is a function of position on the interfacial plane through its dependence on band bending, and $\langle Y_{it} \rangle$ in (5.68) results from averaging over the entire interfacial plane. In (5.69) $P(v_s)$, the probability that the band bending is v_s, is from (6.101)

$$P(v_s) = (2\pi\sigma_s^2)^{-1/2} \exp\left[-\frac{(v_s - \langle v_s \rangle)^2}{2\sigma_s^2}\right] \qquad (5.69)$$

where σ_s^2 is the variance of band bending in units of kT/q given by (6.70), and $\langle v_s \rangle$ is the mean value of band bending, (6.56).

With the use of (5.49b), valid for depletion, the real part of (5.68) for p-type is

$$\frac{\langle G_P \rangle}{\omega} = \frac{q}{2} \int_{-\infty}^{\infty} \frac{D_{it}}{\omega\tau_p} \ln(1 + \omega^2\tau_p^2)P(v_s)dv_s \qquad (5.70)$$

and the imaginary part of (5.69) using (5.49a) is

$$\langle C_P \rangle = C_D(\bar{v}_s) + q \int_{-\infty}^{\infty} \frac{D_{it}}{\omega\tau_p} \tan^{-1}(\omega\tau_p)P(v_s)dv_s \qquad (5.71)$$

where $C_D(\bar{v}_s)$ is the depletion layer capacitance corresponding to the band bending in a capacitor with the mean interface charge density given by (6.31).

The integration of (5.68), made more explicit in (5.70) and (5.71), is simplified if interface trap level density and majority carrier capture probability do not vary over a potential range equal to the standard deviation of band bending. Typically, the standard deviation is $2.5kT/q$ and interface trap level density and capture probability vary slowly enough to be considered constant over this potential range[1,2] as shown in Section 7.3 for device-grade oxides. Therefore, interface trap level density can be taken out of the integral in (5.70) and (5.71). To integrate (5.70) and (5.71),

---

*The bar superscript denoting an average over the interfacial distribution of interface traps is suppressed elsewhere in this chapter.

†We use the symbol $\langle \rangle$ to denote average over band-bending variations in the remainder of this chapter and in Chapter 6 but drop this symbol in the other chapters for simplicity.

substitute (5.29) and (5.69) into these equations to obtain the integrands as explicit functions of band bending. Thus (5.70), for p-type becomes

$$\frac{\langle G_P \rangle}{\omega} = \frac{1}{2} q D_{it} (2\pi\sigma_s^2)^{-1/2} \left(\frac{c_p N_A}{\omega}\right) \int_{-\infty}^{\infty} \exp\left[-\frac{(v_s - \langle v_s \rangle)^2}{2\sigma_s^2}\right] \exp(-v_s)$$

$$\times \ln[1 + \omega^2 (c_p N_A)^{-2} \exp(2v_s)] dv_s \tag{5.72}$$

and (5.71) becomes $\langle C_P \rangle = C_D(\bar{v}_s) + \langle C_{it} \rangle$ with

$$\langle C_{it} \rangle = q D_{it} (2\pi\sigma_s^2)^{-1/2} \left(\frac{c_p N_A}{\omega}\right) \int_{-\infty}^{\infty} \exp\left[-\frac{(v_s - \langle v_s \rangle)^2}{2\sigma_s^2}\right] \exp(-v_s)$$

$$\times \tan^{-1}[\omega(c_p N_A)^{-1} \exp(v_s)] dv_s \tag{5.73}$$

where $\langle C_{it} \rangle$ is the average over band bending in (5.49d).

The measured characteristic time constant now corresponds approximately to mean band bending and is given by

$$\tau_n = \frac{1}{c_n N_D} \exp(-\langle v_s \rangle) \quad \text{for electrons} \tag{5.74a}$$

and

$$\tau_p = \frac{1}{c_p N_A} \exp(\langle v_s \rangle) \quad \text{for holes} \tag{5.74b}$$

where c_n and c_p are the electron and hole capture probabilities, respectively.

From (5.74b), we write

$$\xi = \omega\tau_p \cong \omega(c_p N_A)^{-1} \exp(\langle v_s \rangle). \tag{5.75}$$

Eliminating $(c_p N_A)$ from (5.72) using (5.75) yields

$$\frac{\langle G_P \rangle}{\omega} = \frac{q D_{it} (2\pi\sigma_s^2)^{-1/2}}{2\xi} \int_{-\infty}^{\infty} \exp\left(-\frac{\eta^2}{2\sigma_s^2}\right) \exp(-\eta) \ln(1 + \xi^2 \exp 2\eta) d\eta \tag{5.76}$$

where $\eta = v_s - \langle v_s \rangle$.

Figure 5.16 shows a plot of the dimensionless quantity $\langle C_{it} \rangle / q D_{it}$ from (5.73) as a function of ξ from (5.75) with σ_s as parameter. Figure 5.16 illustrates the behavior of the capacitive component of the interface trap branch of the equivalent circuit. However, we make no further use of this component in this chapter.

5.5 CONDUCTANCE METHOD[2]

To extract accurate values of D_{it}, τ_n, and τ_p from the equivalent parallel conductance, the equivalent circuit has been extended in Section 5.4.1 to

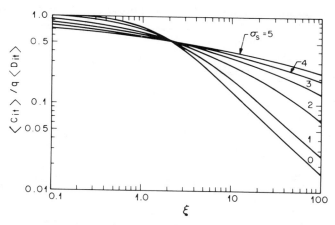

Fig. 5.16 Plot of $C_{it}/(qD_{it})$ as a function of ξ_p with σ_s as parameter calculated from (5.73). After Goetzberger et al.[20] Copyright (1976), The Chemical Rubber Co., CRC Press. Reprinted with permission.

incorporate the experimentally observed time constant dispersion. With this final extension of the equivalent circuit, a comparison of the theoretical and measured admittances can be made. Section 5.5.1 describes the extraction from conductance measurements of the variance of band-bending fluctuations, interface trap level density, and the characteristic interface trap time constant for majority carrier capture. A systematic procedure, described in Sections 5.5.1–5.5.4, is necessary to avoid needless trial and error in parameter extraction.

5.5.1 Interface Trap Parameter Extraction in Depletion[18, 19]

To measure interface trap level density and capture probability as functions of gate bias and interface trap time constant dispersion, admittance must be measured as a function of both gate bias and frequency. The introduction of frequency as a variable increases the amount and complexity of data taking and of analysis, the price paid for obtaining a more detailed account of interfacial properties. The frequency range over which interface trap properties can be extracted with high sensitivity and accuracy is 50 Hz–1 MHz for reasons discussed in Section 12.6.1.

In this section we show how to extract $\langle C_{it} \rangle$ and τ_p as functions of gate bias from the admittance measured in depletion as a function of bias and frequency. The first step is to extract from the capacitance and equivalent parallel conductance measured across terminals A-A in Fig. 5.15c, the conductive component of $\langle Y_{it} \rangle$ across terminals x-x in Fig. 5.15c. The resulting $\langle G_P \rangle / \omega$ is related to D_{it} and τ_p by (5.76). The admittance of the MOS capacitor measured across terminals A-A in Fig. 5.15c is $G_m + j\omega C_m$, where G_m is measured equivalent parallel conductance and C_m is the measured capacitance. Converting this admittance to an impedance, sub-tracting the reactance of the oxide capacitance, and converting back to an admittance yields for the real part

$$\frac{\langle G_P \rangle}{\omega} = \frac{\omega C_{ox}^2 G_m}{G_m^2 + \omega^2 (C_{ox} - C_m)^2} \tag{5.77}$$

where C_{ox} is measured in strong accumulation.

It is important to use (5.77) to obtain $\langle G_P / \omega \rangle$ rather than using G_m / ω. The value C_{ox} makes the time constant measured across terminals A-A in Fig. 5.15c significantly shorter than that of the interface trap branch of the equivalent circuit alone, even though $C_D(\bar{v}_s)$ tends to lengthen it, that is, $C_m / G_m < \langle C_{it} \rangle / \langle G_P \rangle$. Therefore, the corrected $\langle G_P \rangle / \omega$ curve obtained from (5.77) will peak at a lower value of ξ and have a higher peak magnitude than the measured G_m / ω curve. That is, a comparison of G_m / ω with $\langle G_p \rangle / \omega$ shows that $\langle G_P \rangle / \omega$ will peak at a lower frequency when admittance is measured as a function of frequency with gate bias as parameter or at a gate bias closer to flatbands when admittance is measured as a function of gate bias with frequency as parameter.

From (5.76), only $\langle G_P \rangle / \omega$ is a function of ξ when frequency is the variable, whereas both $\langle G_P \rangle$ and $\langle G_P \rangle / \omega$ are functions of ξ when gate bias is the variable. Therefore, both $\langle G_P \rangle$ and $\langle G_P \rangle / \omega$ will peak as a function of gate bias, but only $\langle G_P \rangle / \omega$ will peak as a function of frequency.

The next step is to determine the standard deviation of band bending σ_s. The standard deviation of band bending is a measure of the width of curve a in Fig. 5.14. The width of the $\langle G_P \rangle / \omega$ versus log frequency curve along the frequency axis depends only on σ_s. However, the $\langle G_P \rangle / \omega$ versus gate bias curve will be spread over a bias range determined by σ_s and D_{it}. The interface trap contribution to the width is caused by the change in occupancy with gate bias.

Figure 5.17 shows a typical $\langle G_P \rangle / \omega$ versus log frequency curve from which σ_s is to be determined. The procedure is to measure the amplitude change of this curve either between the points f_p and f_p / n or between f_p and nf_p where f_p is the frequency corresponding to the peak value of $\langle G_P \rangle / \omega$. The ratio between $(\langle G_P \rangle / \omega)_{f_p / n}$ and $(\langle G_P \rangle / \omega)_{f_p}$ from (5.76) is

$$\frac{(\langle G_P \rangle / \omega)_{f_p / n}}{(\langle G_P \rangle / \omega)_{f_p}} = n \frac{\displaystyle\int_{-\infty}^{\infty} \exp\left(-\frac{\eta^2}{2\sigma_s^2}\right) \exp(-\eta) \ln[1 + \xi_p^2 \exp(2\eta)/n^2]\, d\eta}{\displaystyle\int_{-\infty}^{\infty} \exp\left(-\frac{\eta^2}{2\sigma_s^2}\right) \exp(-\eta) \ln(1 + \xi_p^2 \exp 2\eta)\, d\eta} \tag{5.78}$$

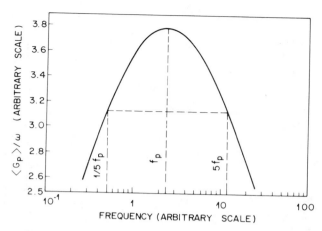

Fig. 5.17 A typical $\langle G_p \rangle / \omega$ versus log frequency curve obtained from measurement by correcting for C_{ox} using (5.77), where f_p is the frequency corresponding to the peak value of $\langle G_p \rangle / \omega$. The points at $f_p/5$ and $5f_p$ define the width of the curve on the low and high frequency side of f_p, respectively. After Nicollian et al.[18] Copyright (1969), Pergamon Press, Ltd. Reprinted with permission.

The ratio between $(\langle G_P \rangle / \omega)_{nf_p}$ and $(\langle G_P \rangle / \omega)_{f_p}$ from (5.76) is

$$\frac{(\langle G_P \rangle / \omega)_{nf_p}}{(\langle G_P \rangle / \omega)_{f_p}} = \frac{1}{n} \frac{\displaystyle\int_{-\infty}^{\infty} \exp\left(-\frac{\eta^2}{2\sigma_s^2}\right) \exp(-\eta) \ln(1 + n^2 \xi_p^2 \exp 2\eta) d\eta}{\displaystyle\int_{-\infty}^{\infty} \exp\left(-\frac{\eta^2}{2\sigma_s^2}\right) \exp(-\eta) \ln(1 + \xi_p^2 \exp 2\eta) d\eta} \qquad (5.79)$$

where $\xi_p = \omega_p (c_p N_A)^{-1} \exp(\langle v_s \rangle)$.

Equations (5.78) and (5.79) are the single-valued functions of σ_s plotted in Fig. 5.18. Choosing $n = 5$ is reasonable because the values of $\langle G_p \rangle / \omega$ at these frequencies generally will not be too low or too close to the peak value so that accurate values of these ratios can be found. Using the experimental ratios, the corresponding σ_s is found from Fig. 5.18.

The asymmetry of the $\langle G_P \rangle / \omega$ versus log frequency curves about f_p results in high and low frequency ratios that differ for values of σ_s below 1.8, as shown in Fig. 5.18. This asymmetry is due to the influence of the Fermi function weight factor on the loss. For values of σ_s greater than 1.8, the Gaussian weight factor given by (5.69) dominates, making the $\langle G_P \rangle / \omega$ versus log frequency curves symmetrical about f_p, indicated by the merging of the $f_p/5$ and $5f_p$ curves in Fig. 5.18.

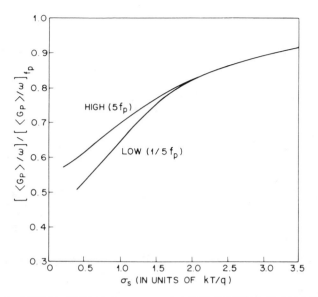

Fig. 5.18 Plot of $[\langle G_P \rangle / \omega] / [\langle G_P \rangle / \omega]_{f_p}$ versus standard deviation of band bending. This curve is used to get σ_s from the width of a $\langle G_P \rangle / \omega$ versus log f curve such as in Fig. 5.17 measured at a given gate bias. Determination of the frequency f_p corresponding to the peak of the measured curve and either the low or high frequency side will suffice in using this curve. The range of σ_s given will cover most cases met in practice. For values of σ_s greater than 3.5, the peak of the $\langle G_p \rangle / \omega$ versus log f curve becomes so broad that a large error will be made in determining f_p. After Nicollian et al.[18] Copyright (1969), Pergamon Press, Ltd. Reprinted with permission.

To calculate D_{it}, we evaluate (5.76) at $\xi = \xi_p$ and obtain the peak value $(\langle G_P \rangle / \omega)_{f_p}$

$$\left(\frac{\langle G_P \rangle}{\omega} \right)_{f_p} = \frac{q D_{it} (2 \pi \sigma_s^2)^{-1/2}}{2 \xi_p} \int_{-\infty}^{\infty} \exp\left(-\frac{\eta^2}{2\sigma_s^2} \right) \exp(-\eta) \ln(1 + \xi_p^2 \exp 2\eta) d\eta.$$

(5.80)

Solving (5.80) for D_{it}, we obtain

$$D_{it} = \left(\frac{\langle G_P \rangle}{\omega} \right)_{f_p} [f_D(\sigma_s) q]^{-1}$$

(5.81)

where

$$f_D(\sigma_s) = \frac{(2 \pi \sigma_s^2)^{-1/2}}{2 \xi_p} \int_{-\infty}^{\infty} \exp\left(-\frac{\eta^2}{2\sigma_s^2} \right) \exp(-\eta) \ln(1 + \xi_p^2 \exp 2\eta) d\eta.$$

(5.82)

Equation (5.82) is plotted in Fig. 5.19, which is used with (5.81) to determine D_{it} once σ_s is determined from Fig. 5.18.

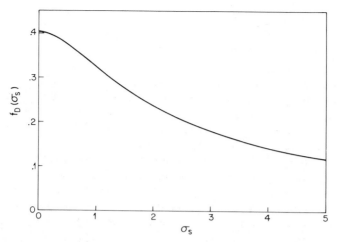

Fig. 5.19 Plot of the universal function $f_D(\sigma_s)$ defined in (5.82) as a function of σ_s; f_D is used in (5.81) to calculate interface trap level density.

To determine τ_p, we use the condition

$$\frac{d}{d\xi}\left(\frac{\langle G_P\rangle}{\omega}\right)_{f_p} = 0. \tag{5.83}$$

Applying this condition to (5.76) yields[20]

$$\int_{-\infty}^{\infty} \exp\left(-\frac{\eta^2}{2\sigma_s^2}\right) \exp(-\eta)\left\{\frac{2\xi_p^2\exp 2\eta}{1+\xi_p^2\exp 2\eta} - \ln(1+\xi_p^2\exp 2\eta)\right\} d\eta = 0. \tag{5.84}$$

Solving (5.84) numerically yields ξ_p as a function of σ_s. Figure 5.20 is a plot of ξ_p versus σ_s obtained this way. Then, from the relation $\xi_p = \omega_p\tau_p$, we

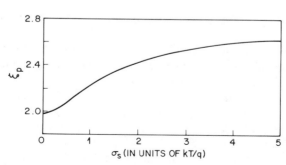

Fig. 5.20 Plot of $\xi_p = \omega_p(c_pN_A)^{-1}\exp\langle v_s\rangle$ as a function of σ_s calculated by solving (5.84); ξ_p is used to calculate interface trap response time from (5.85). After Goetzberger et al.[20] Copyright (1976), The Chemical Rubber Co., CRC Press. Reprinted with permission.

write

$$\tau_p = \frac{\xi_p}{\omega_p}. \tag{5.85}$$

The value of $\omega\tau$ corresponding to the peak of the G_P/ω curve $(\omega\tau)_p$ depends only on the broadening. For a distribution of interface trap levels but no band-bending fluctuations, the condition at the peak is $(\omega\tau)_p = 1.98$ (see discussion of Fig. 5.10) independent of interface trap level density, capture probability, and silicon doping density. Equation (5.84) shows that $(\omega\tau)_p$ depends only on the width of the G_P/ω versus log $\omega\tau$ curve, which is a measure of time constant dispersion, because σ_s is the only parameter entering this equation. The increase in $(\omega\tau)_p$ with increasing time constant dispersion is caused by the asymmetric influence of the weight factor $\exp(-\eta)$ that favors $\eta < 0$, forcing an increase in $(\omega\tau)_p$ to keep the factor in large brackets in (5.84) near zero (i.e., to keep $\xi_p \exp \eta \approx 1.98$). Once σ_s is found from Fig. 5.18, we find ξ_p from Fig. 5.20 and calculate τ_p from (5.85).

In summary, we need only two points on a $\langle G_P \rangle/\omega$ versus log frequency curve to determine D_{it} and τ_p, namely, $[\langle G_P \rangle/\omega]_{f_p}$ and $[\langle G_P \rangle/\omega]_{5f_p}$ or $[\langle G_P \rangle/\omega]_{f_p/5}$. Because it not possible in advance to know these two frequencies, it is not possible to measure the admittance at only these two points; a complete admittance versus frequency curve must be measured at fixed bias. Then a curve such as that in Fig. 5.17 is obtained from (5.76), the ratio between $[\langle G_P \rangle/\omega]_{f_p/5}/[\langle G_P \rangle/\omega]_{f_p}$ or between $[\langle G_P \rangle/\omega]_{5f_p}/[\langle G_P \rangle/\omega]_{f_p}$ is calculated, and σ_s is found from Fig. 5.18. Using this value of σ_s, $f_D(\sigma_s)$ is found from Fig. 5.19 and D_{it} is calculated from (5.81). Finally, using the same value of σ_s, ξ_p is found from Fig. 5.20 and τ_p calculated from (5.85). Gate bias then is changed, and the process is repeated until complete curves of D_{it} and τ_p are obtained as a function of gate bias, within the limitations on the range of gate bias discussed in Section 5.8. To cover a wider range of bandgap energy, both n-type and p-type samples must be measured.

5.5.2 Interface Trap Parameter Extraction in Weak Inversion

The preceding section described the conductance method as applied in depletion. Typical G_p/ω curves in this region are single peaked and broadened by interface trap time constant dispersion. In weak inversion a more complex behavior is observed, as shown in Fig. 5.21.

Figure 5.21 is a comparison of theoretical (lines) and experimental (points) G_P/ω versus log ω curves. For this n-type sample, $u_s = 0$ is midgap, $u_s < 0$ is weak inversion, and $u_s > 0$ is depletion, where $u_s = (q/kT)\phi_s$ is as defined in Fig. 5.8. In depletion, the theory described in Section 5.4.1 is adequate. That is, a peak broadened by band-bending fluctuations superimposed on the Lehovec model fits experiment. Well into weak inversion, single time constant behavior is observed, as predicted and

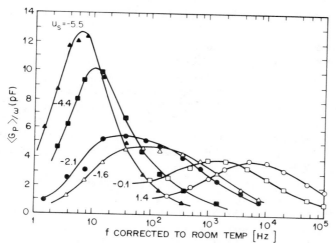

Fig. 5.21 Plot of $\langle G_P \rangle / \omega$ versus ln f in the range of biases from weak inversion ($u_s < 0$) through midgap ($u_s = 0$) into depletion ($u_s > 0$). The silicon substrate was oriented in the (100) direction and had a donor density of 3×10^{15} cm^{-3}. The experimental data (points) were taken at 60°C and corrected for temperature to 20°C. The theoretical results (lines) are a computer fit of $\langle G_P \rangle / \omega$ using (5.18) extended to include an integration over band-bending fluctuations, as was done for depletion in (5.76). After Cooper and Schwartz.[17] Copyright (1974), Pergamon Press, Ltd. Reprinted with permission.

observed by Nicollian and Goetzberger,[2] following the behavior of (5.56) and (5.58) extended to include band-bending variations. That is, a narrow peak is observed, which is described by Fig. 5.3b. A continuum of interface trap levels and band-bending fluctuations do not broaden this peak but do affect its time constant, as described shortly. Between these extremes, a combination of the two types of behavior is seen.

To accomplish the extension of (5.56) and (5.58) to include band-bending fluctuations, interface trap levels at a given energy in the bandgap are denoted with index l, as before. However, the number of interface trap levels with given l now depends on band-bending fluctuations as well as the probable density of interface trap levels, $D_{it}(\zeta_s)$, of (5.25). Consequently, summations over l are reinterpreted as sums over energy in the bandgap [as in (5.26)] *and* sums over band bending, weighted by the probability of a given band bending $P(\psi_s)$.

This reinterpretation of the l-summations does not change the formal appearance of the equations for admittance, but now G_{gr}, C_{Tp}, and C_{Tn} will involve double integrations. In particular, (5.55a) and (5.55b) for G_{gr} and C_{Tp}' must be integrated over band-bending fluctuations. The admittance given by (5.60) retains single time constant behavior, but the characteristic time τ_c of (5.61) becomes

$$\tau_c = \int d\psi_s \, P(\psi_s)(C_I + C_{it}) \left\{ \int d\psi_s \, P(\psi_s) \tau_n^{-1} C_{it} \ln \frac{\tau_n}{\tau_p} \right\}^{-1} \qquad (5.86)$$

where $P(\psi_s)$ is the probability of the band bending being between ψ_s and $\psi_s + d\psi_s$.*

The excellent fit of theory to experiment suggests that the basic ideas behind the equivalent circuit developed in this chapter are valid. However, there are a number of problems associated with extracting interface trap parameters from measurements in weak inversion that limit its practicality. We have already mentioned that bulk trap loss becomes important in this bias region because bulk trap levels near midgap cross the Fermi level, making efficient generation recombination centers; therefore, this contribution to the conductance must be known before the interface trap contribution can be found.

Additional difficulty in analysis is presented, compared to depletion, because both majority and minority carriers must be considered (see Sections 5.3.5 and 5.3.6). This consideration entails computer calculation to obtain a fit to measured curves, as discussed by Cooper and Schwartz.[17]

The distribution of band bending over the interface adds additional complexity in weak inversion because some regions of the interface will be in depletion and others in weak inversion. Because majority carrier density is much larger in the depleted regions than in the weakly inverted regions, the majority carrier capture resistances of interface traps in depleted regions is much smaller than the capture resistances of interface traps in weakly inverted regions. Therefore, interface traps in the depleted areas make a disproportionately large contribution to the conductance. Interface trap admittance in the depleted regions shunts interface trap admittance in the weakly inverted regions. Therefore, the interface trap level density extracted becomes sensitive to the tails of the Gaussian. This tail region is, in turn, sensitive to the details of the statistical model, which are uncertain. It is only the regions of the Gaussian away from the tail, in the vicinity of the maximum, that are model insensitive. Consequently, interpretation of measurements in weak inversion may depend on features of the interface trap time constant dispersion not otherwise known and not independently measured.

In summary, measurement in weak inversion offers the potential advantages of extending measurement of interface trap parameters over a larger range of energy and of obtaining minority as well as majority carrier capture cross sections. However, interpretation and measurement are more difficult than in depletion, and further work is needed to establish the uniqueness and accuracy of the fitted parameters in this regime. As matters now stand, measurement in depletion is subject to less complication and uncertainty.

---

*Equation (5.86) was given by Cooper and Schwartz[17] for the case $C_I \ll C_{it}$.

5.5.3 Determination of Capture Probability and Cross Section

Capture cross section is related to capture probability by the relation $\sigma_n = c_n/\bar{v}$ for electrons and $\sigma_p = c_p/\bar{v}$ for holes, where $\bar{v}$ is the thermal velocity of the carriers, typically 10^7 cm/sec in silicon at room temperature.

Capture probability is calculated from interface trap response time using (5.74). This calculation requires knowledge of both $\langle v_s \rangle$ and doping density. Once $\langle \psi_s \rangle$ versus V_G is determined, capture probability can be obtained as a function of $\langle \psi_s \rangle$ from interface trap response time versus V_G. Because of the exponential dependence of capture probability on $\langle \psi_s \rangle$, any error introduced by V_{FB} (flatband voltage) in the values of $\langle \psi_s \rangle$ will be magnified.* Therefore, measured capture probability is considerably less accurate than interface trap level density. The use of V_{FB} in calculating $\langle \psi_s \rangle$ versus V_G from (3.63) or (3.65) introduces a *fictitious* dependence of capture probability on temperature, interface trap level density, doping profile, and oxidation conditions.

There is an additional uncertainty in the determination of $\langle \psi_s \rangle$, namely uncertainty in the doping profile. The doping profile is needed all the way to the Si–SiO$_2$ interface. Unfortunately, this profile cannot be measured, for example because of the interface proximity limitation described in Section 9.4.3 (b). Therefore, doping profile near the Si–SiO$_2$ interface must be theoretically estimated, introducing some error.

For all these reasons, capture probabilities and cross sections determined by the methods described here should be regarded as only order of magnitude estimates.

5.6 RESOLUTION OF SMALL-SIGNAL STEADY-STATE METHODS

When variations of band bending over the interfacial plane are present, interface trap levels quite far away from the Fermi level will contribute significantly to the measured admittance. Figure 5.18 shows that the width of the $\langle G_P \rangle/\omega$ versus log frequency curve is dominated by the Gaussian distribution of band bending $P(v_s)$ for $\sigma_s > 1.8$. Because values of $\sigma_s > 1.8$ are common in practice, the *resolution* of admittance measurements will be the half width of the Gaussian band bending distribution, which is about $\sqrt{2}\sigma_s$. Therefore, determination of σ_s in the conductance method really determines the resolution with which interface trap level density is probed by any of the small-signal steady-state methods. For example, $\sigma_s = 2.5$ is commonly found. Resolution in this case is $3.5kT/q$, or 88 mV. If interface trap level density or capture probability change rapidly with bandgap energy over the interval equal to $\sqrt{2}\sigma_s$, the values of interface trap level density and

---

*See Section 10.5 for a method of determining V_{FB} and Section 3.3 for how to determine $\langle \psi_s \rangle$ versus V_G.

capture probability extracted will be in error (see Section 5.9). This condition can be checked by extracting both of these parameters from the measured admittance to see whether they are nearly constant over $\sqrt{2}\sigma_s$. A more detailed discussion is given in Section 8.3.2.

5.7 SERIES RESISTANCE AND BULK TRAP LOSS

There are three important sources of small-signal energy loss in the MOS capacitor: (1) changes in interface trap level occupancy, (2) changes in the occupancy of bulk trap levels, and (3) series resistance. Current leakage through the oxide or along its surface also produces an energy loss, but leakage should be eliminated in a well-controlled experiment. The measurement of equivalent parallel conductance can be helpful in detecting these effects.

5.7.1 Series Resistance

Series resistance can cause a serious error in the extraction of interfacial properties and doping profiles from admittance measurements. It also can limit the sensitivity of the small-signal steady-state methods. To avoid this error and the sensitivity limitation, series resistance can be minimized by (1) sample fabrication, (2) making measurements at low frequencies so that the effect of series resistance is negligible, or (3) measuring series resistance and applying a correction to the measured admittance before the desired information is extracted. In this section we show how to minimize and correct the measured admittance for series resistance.

Series resistance can arise from five different sources: (1) the contact made by the probe wire to the gate; (2) the back contact to the silicon; (3) a dirt film or particulate matter between the back contact and the pedestal; (4) the resistance of the quasi-neutral bulk silicon between the back contact to the silicon and the depletion layer edge at the silicon surface underneath the gate; and (5) an extremely nonuniform doping distribution in the silicon underneath the gate. Extreme variations in doping profile result in highly resistive regions where majority carrier density is very low as a result of a built-in potential. This effect is not important for the usually small impurity redistribution that occurs during thermal oxidation.

Ways of minimizing series resistance from the first three sources are described in Sections 12.6.1, 12.8.2, and 12.9.1. In nearly all cases series resistance arising from these sources can be made negligible. It also is possible to make bulk series resistance negligible by using a thin epitaxial layer of low resistivity grown on a degenerate substrate. However, it may not always be possible or desirable to make a special substrate like this. Also, this solution leads to series resistance caused by a nonuniform doping profile, and this resistance is difficult to estimate. For these two

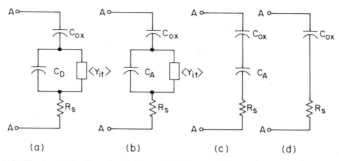

Fig. 5.22 (a) Equivalent circuit of the MOS capacitor in depletion including series resistance R_s; (b) Equivalent circuit of the MOS capacitor in strong accumulation; (c) simplified version of (b); (d) simplified version of (c) used in extracting values of C_{ox} and R_s from the admittance measured in strong accumulation.

reasons, it is not possible to make series resistance negligible in every case. The most general approach is to measure series resistance and then apply it as a correction to the measured admittance.

To measure series resistance, consider the equivalent circuit of the MOS capacitor in depletion, shown in Fig. 5.22a. Figure 5.22a is taken from Fig. 5.15c but includes series resistance R_s, and R_s can cause error in both measured capacitance and equivalent parallel conductance. This error depends on gate bias and frequency.

At a given frequency, the greatest effect on the measured admittance occurs in *strong* accumulation. Figure 5.22b shows the equivalent circuit of the MOS capacitor in strong accumulation. For this bias, the accumulation layer capacitance C_A is in parallel with $\langle Y_{it} \rangle$. The equivalent circuit can be further simplified because C_A is large, shunting $\langle Y_{it} \rangle$; that is, $\omega C_A \gg \langle Y_{it} \rangle$. The equivalent circuit becomes that in Fig. 5.22c. Because $C_A \gg C_{ox}$, Fig. 5.22c can be simplified further to Fig. 5.22d.

To determine R_s, the MOS capacitor is biased into strong accumulation. The admittance Y_{ma} across terminals A-A in Fig. 5.22d in strong accumulation in terms of the capacitance C_{ma} and equivalent parallel conductance G_{ma} is

$$Y_{ma} = G_{ma} + j\omega C_{ma}. \tag{5.87}$$

Series resistance is the real part of the impedance $Z_{ma} = 1/Y_{ma}$ or

$$R_s = \frac{G_{ma}}{G_{ma}^2 + \omega^2 C_{ma}^2}. \tag{5.88}$$

Capacitance C_{ox} is obtained by substituting R_s from (5.88) into the relation $C_{ma} = C_{ox}/(1 + \omega^2 R_s^2 C_{ox}^2)$ derived from the equivalent circuit in Fig. 5.22d and solving for C_{ox}, which yields

$$C_{ox} = C_{ma} \left[1 + \left(\frac{G_{ma}}{\omega C_{ma}} \right)^2 \right]. \tag{5.89}$$

Corrected capacitance C_c and corrected equivalent parallel conductance G_c at the frequency of interest are

$$C_c = \frac{(G_m^2 + \omega^2 C_m^2)C_m}{a^2 + \omega^2 C_m^2} \tag{5.90}$$

and

$$G_c = \frac{(G_m^2 + \omega^2 C_m^2)a}{a^2 + \omega^2 C_m^2}. \tag{5.91}$$

respectively, where $a = G_m - (G_m^2 + \omega^2 C_m^2)R_s$ and C_m and G_m are the capacitance and the equivalent parallel conductance measured across the terminals of the MOS capacitor.

Figure 5.23 shows the effect of series resistance on admittance for a sample with an unusually high series resistance and a low interface trap level density measured at 1 MHz. The absence of a peak in the G-V curve means that series resistance produced the dominant loss, completely masking the interface trap loss. No interface trap properties can be extracted from the ac response on this sample.

Although R_s is independent of gate bias, the measured equivalent parallel conductance varies with gate bias. This variation of G_m arises from C_c [from (5.90)] in the simple series R-C circuit relation, $G_m = \omega^2 C_c^2 R_s / (1 + \omega^2 R_s^2 C_c^2)$, where C_c is a function of gate bias. Using $C_{ma} = 99 \text{ pF}$ and $G_{ma} = 330 \text{ } \mu\text{mho}$ from Fig. 5.23, $C_{ox} = 126.9 \text{ pF}$ from (5.89), and $R_s = 665.6 \text{ } \Omega$ from (5.88). The corrected capacitance C_c obtained from (5.90) is

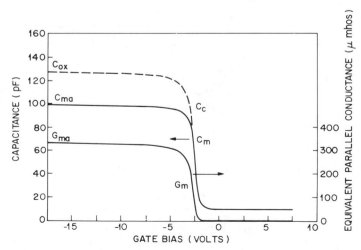

Fig. 5.23 Capacitance and equivalent parallel conductance measured at 1 MHz as a function of gate bias at a sweep rate of 10 mV/sec. The capacitance corrected for series resistance is shown as the dashed curve. Gate area is $1.12 \times 10^{-2} \text{ cm}^2$, oxide thickness is 300 Å, and acceptor density is $2 \times 10^{15} \text{ cm}^{-3}$.

shown as the dotted curve in Fig. 5.23. The greatest error in the capacitance occurs in accumulation and a portion of the depletion region, but as inversion is approached and in inversion itself, series resistance produces very little error in the capacitance because C_c has become so small that now $(\omega R_s C_c)^2 \ll 1$.

A series resistance of 665.6 Ω is so high that the sample in Fig. 5.23 has limited usefulness. However, even higher values of R_s can arise if care is not exercised.

Series resistance is critical if low interface trap level densities are measured. Figure 5.24 shows capacitance and equivalent parallel conductance measured at 1 MHz (solid lines) as functions of gate bias. In this sample, series resistance does not completely mask interface trap loss. Dashed curve a in Fig. 5.24 is the equivalent parallel conductance corrected for series resistance using (5.91). The value of series resistance (18.8 Ω in this sample) is too small to affect the capacitance, showing that the equivalent parallel conductance is much more sensitive to series resistance than the capacitance. Dashed curve b is calculated from the real part of a series R-C circuit admittance, $G_m = \omega^2 C_c^2 R_s / (1 + \omega^2 R_s^2 C_c^2)$, to show the series resistance component of the equivalent parallel conductance.

It is emphasized that both capacitance and conductance measurements can be affected by series resistance, particularly at higher frequencies and that care should be exercised both in the preparation of samples and in the

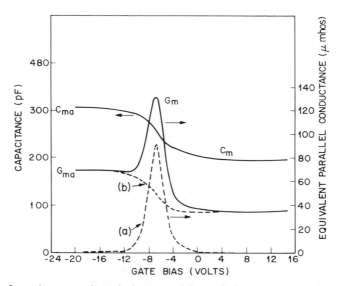

Fig. 5.24 Capacitance and equivalent parallel conductance measured at 1 MHz as a function of gate bias. The dashed conductance curve has been corrected for series resistance.

techniques used in measurement. Even then, for reliable results, series resistance should be measured and results corrected. Correction for series resistance is particularly important in conductance measurements. Often failure to obtain a conductance peak after correcting for C_{ox} using (5.77) is due to neglecting series resistance.

5.7.2 Bulk Trap Loss[2]

Bulk trap loss becomes detectable in weak inversion, usually when interface trap level density is low ($<10^{10}$ cm^{-2} eV^{-1}). When the silicon surface becomes inverted, the Fermi level will cross bulk trap levels near midgap at a point in the silicon depletion layer shown in Fig. 4.3b. Generation and recombination through these bulk trap levels as well as through interface trap levels can contribute significantly to the measured admittance. Bulk trap loss limits sensitivity when interface trap properties are extracted in weak inversion. However, this limitation is usually not important in depletion [see Section 4.3.9(b)]. Bulk trap loss can be used to study bulk trap properties, as described in Chapter 4.

To interpret admittance measurements and distinguish between interface trap loss and bulk trap loss, we discuss the differences in conductance arising from these two sources. The two are easily distinguished when one or the other is dominant. Although both capacitance and equivalent parallel conductance are affected, the shape of the *G-V* curve is strikingly different whereas the shape of the *C-V* curve is similar in each case. Therefore, these two cases are distinguished by the gate bias dependence of the equivalent parallel conductance.

When generation and recombination through interface trap levels dominates the loss, the equivalent parallel conductance goes through a peak as a function of gate bias in weak inversion because interface trap time constant varies inversely with majority carrier density at the silicon surface as shown by (5.74). This case is illustrated in Fig. 5.25a, where equivalent parallel conductance goes through a peak in weak inversion and drops to a very low value in strong inversion.

When generation and recombination through bulk trap levels dominate the loss, equivalent parallel conductance does not go through a peak as a function of gate bias in weak inversion because the loss due to this process is independent of gate bias. This bias independence results from the bias independence of hole and electron densities at the crossover point (see Fig. 4.3b), where the Fermi level crosses bulk trap levels located near midgap. This case is illustrated in Fig. 5.25b. The slight indication of a peak in the conductance curve means that generation and recombination through interface trap levels make a slight contribution to the loss. However, the primary effect is that of a gate bias independent loss. The variation of measured conductance with gate bias is mainly due to the variation of capacitance with gate bias.

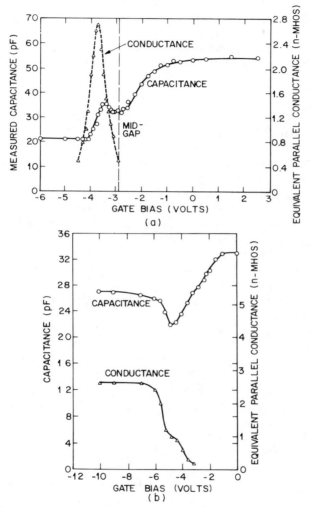

Fig. 5.25 (a) Capacitance and equivalent parallel conductance measured at 50 Hz versus gate bias. Silicon orientation was (111), gate diameter was 3.8×10^{-2} cm², donor density was 1.2×10^{16} cm⁻³, oxide thickness was 670 Å, and interface trap level density in the 10^{11} cm⁻² eV⁻¹ range. After Nicollian and Goetzberger.[2] (b) Capacitance and equivalent parallel conductance measured at 50 Hz versus gate bias. Gate diameter was 3.8×10^{-3} cm², donor density was 5×10^{15} cm⁻³, silicon orientation was (100), and oxide thickness was 1200 Å. The oxide was grown in steam and then annealed in H₂ to produce an interface trap level density in the 10^{10} cm⁻² eV⁻¹ range. After Nicollian and Goetzberger.[2] Copyright (1967), American Telephone and Telegraph Company. Reprinted with permission.

227

Figure 5.25a is measured on a sample that has an interface trap level density in the $10^{11}\,\mathrm{cm}^{-2}\,\mathrm{eV}^{-1}$ range, whereas Fig. 5.25b is measured on a sample that has an interface trap level density in the $10^{10}\,\mathrm{cm}^{-2}\,\mathrm{eV}^{-1}$ range. The difference in shape between the capacitance curves is attributable to interface traps. Figure 5.25 demonstrates that samples can be made in which one or the other process dominates the loss and that the dominant process can be determined by inspection of a measured G-V curve.

5.8 ERRORS IN EXTRACTING INTERFACE TRAP PARAMETERS FROM THE CONDUCTANCE

Both capacitance and equivalent parallel conductance contain the same interface trap information as seen from (5.49a, b). However, there are two major advantages in extracting interface trap properties from equivalent parallel conductance rather than capacitance. For a given error in the measured admittance, these advantages are: (1) greater sensitivity to the interface trap parameters and (2) greater accuracy.

The conductance method is more sensitive than the capacitance methods discussed in Chapter 8 because $\langle C_{it} \rangle$ is measured directly, not as a difference between C_P and C_D. We make an error analysis of the conductance method similar to the one done for the high-low frequency capacitance method in Section 8.3.5.

To get the relative error in $\langle G_P \rangle$, we rewrite (5.77) as

$$\langle G_P \rangle = \frac{\omega^2 C_{ox}^2 G_m}{G_m^2 + \omega^2 (C_{ox} - C_m)^2}. \tag{5.92}$$

For device-grade oxides where D_{it} is in the $10^{11}\,\mathrm{cm}^{-2}\,\mathrm{eV}^{-1}$ range or less, the loss angle is very small so that over most of the depletion region, $G_m^2 \ll \omega^2 (C_{ox} - C_m)^2$. Under this condition, (5.92) simplifies to

$$\langle G_P \rangle = \frac{C_{ox}^2 G_m}{(C_{ox} - C_m)^2}. \tag{5.93}$$

Differentiating (5.93) and dividing the result by (5.93) yields the relative error in $\langle G_P \rangle$

$$\frac{d\langle G_P \rangle}{\langle G_P \rangle} = -2 \left(\frac{C_n}{1 - C_n} \right) \frac{dC_{ox}}{C_{ox}} + \frac{2C_n}{1 - C_n} \left(\frac{dC_m}{C_m} \right) + \frac{dG_m}{G_m} \tag{5.94}$$

where $C_n = C_m / C_{ox}$.

Rather than $\langle G_P \rangle$, we need $\langle G_P \rangle / \omega$ to obtain σ_s from Fig. 5.18 and then D_{it} and time constant from Figs. 5.19 and 5.20. The relative error in $\langle G_P \rangle / \omega$ is

$$\frac{d(\langle G_P \rangle / \omega)}{\langle G_P \rangle / \omega} = \frac{d\langle G_P \rangle}{\langle G_P \rangle} + \frac{d\omega}{\omega}. \tag{5.95}$$

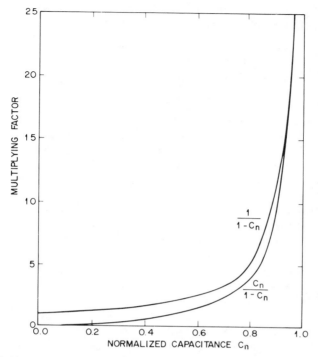

Fig. 5.26 Multiplying factors as functions of normalized capacitance. After Zaininger.[21]

The first term in (5.95) is more important than the second term, which will be negligibly small if a frequency counter is used. Equation (5.94) shows that the relative error in $\langle G_P \rangle$ increases rapidly as C_n approaches unity as shown in Fig. 5.26. This error limits, in principle, how near the majority carrier band edge D_{it} can be accurately extracted. As with the high-frequency capacitance measurement of C_D, this limit seldom is reached. Rather, the limitation is set by the breakdown of the equivalent circuit that is the basis for the theoretical expression relating G_P/ω to D_{it}, derived in Section 5.4. This relation, which must be used to extract D_{it} from the values of $\langle G_P \rangle/\omega$ extracted from the measured admittance, is valid only in depletion, whereas in many cases the relative error limit given by (5.94) occurs in accumulation. The equivalent circuit for depletion is valid to within a few kT/q of flatbands and to within a few kT/q of midgap (see Section 6.4). Therefore, the conductance method yields accurate values of D_{it} and capture probability in this range of energies. In this energy range the relative error with which interface trap properties can be obtained is limited by the relative error with which C_m, G_m, C_{ox}, and ω in (5.94) and (5.95) can be measured. Generally, the relative error with which these quantities can be measured with an admittance bridge and frequency counter will be a fraction of a percent.

5.9 ERRORS DUE TO RAPID VARIATION WITH ENERGY OF D_{it} AND CROSS SECTIONS

5.9.1 Interface Trap Level Density

The simple extraction of D_{it} from equivalent parallel conductance using (5.70) depends on D_{it} being constant over a range of band bending of several σ_s, where σ_s is commonly $2-3kT/q$.* Figure 5.27 shows the theoretical results of this approximation on the conductance in weak inversion. Both curves use the rapidly varying D_{it} in Fig. 8.6, but curve a employs the approximation $D_{it} = D_{it}(\phi_B + \langle \psi_s \rangle)$ whereas curve b employs an exact evaluation using a numerical integration, leaving D_{it} inside the integral. The variance in kT/q is $\sigma_s = 1.81$ and is taken to be bias independent. Both peak position and amplitude are affected by the rapid variation of D_{it} in the weak inversion range of Fig. 8.6.

In depletion, the interface trap level density in Fig. 8.6 does not vary rapidly with energy, and no difficulty with the approximation $D_{it} = D_{it}(\phi_B + \langle \psi_s \rangle)$ arises.

5.9.2 Interface Trap Capture Cross Section

Capture cross sections also can vary rapidly with energy, especially near the conduction band edge. Deuling et al.,[22] assuming a slowly varying D_{it} and an exponential variation of capture cross section given by

$$\sigma(\phi) = \sigma_o \exp(-\gamma\phi) \qquad (5.96)$$

found the results shown in Fig. 5.28. The variance of band-bending fluctuations σ_s was taken to be zero. A similar test was made by Cooper and Schwartz[17] with $\sigma_s = 1.8$ and $\gamma = 0.5$. The distribution is narrower in both cases, with the peak position hardly affected.

The primary complication introduced by (5.96) is that the estimate of σ_s from the width of the conductance peak will be grossly in error if a rapid variation in capture cross section, such as (5.96), is ignored.

As already stated, capture cross section behavior such as (5.96) usually is found near the band edge, where the variance of band bending σ_s^2 itself varies rapidly with bias. This rapid variation of σ_s^2 is due to the strong bias dependence of the mobile carrier density, which causes a strongly bias dependent screening of interface charges. Whenever σ_s^2 varies rapidly with bias, the spatial variation of the ac band bending becomes important, a factor that is neglected in deriving the equivalent circuit that relates measured admittance to interface trap admittance. Thus the conductance method must be developed further before band edge measurements can be interpreted.

---

*A more careful evaluation can be obtained from (6.70).

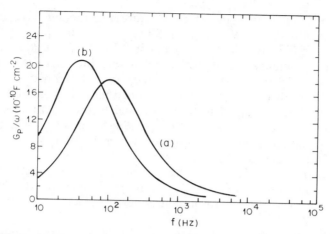

Fig. 5.27 Effect of a rapid variation in D_{it} with energy on G_P/ω in weak inversion. The interface trap level density is the same as that in Fig. 8.6. Curve a assumes $D_{it} = D_{it}(\phi_B + \langle \psi_\sigma \rangle)$ and takes D_{it} outside all integrals. In curve b the integrals are computed numerically for $T = 60°C$, $u_s = 3.89$, and $\sigma_s = 1.8$. After Cooper and Schwartz.[17] Copyright (1974), Pergamon Press, Ltd. Reprinted with permission.

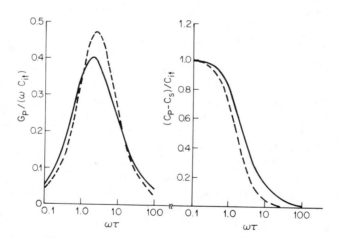

Fig. 5.28 Effect of an exponential variation of interface trap capture cross section with energy on G_p/ω assuming $\sigma_s = 0$. Solid lines, energy-independent interface trap capture cross section; dashed lines, $\gamma = 0.5$ in (5.96). After Deuling et al.[22] Copyright (1972), Pergamon Press, Ltd. Reprinted with permission.

5.10 THIN OXIDES

Until now, only oxides too thick for direct tunneling have been considered. The possibility of using an MOS tunnel structure* to measure interface trap properties first was proposed by Gray.[23] Dahlke and Sze[24] reported the influence of interface traps on the characteristics they measured on MOS tunnel structures. Freeman and Dahlke[25] showed in a theoretical study that in the extreme cases of thin oxides and thick oxides, interface traps were coupled to the metal and to the silicon, respectively.

For extremely thin oxides ($x_o \leq 15$ Å), *direct* tunneling will occur between gate metal and interface trap levels when the Fermi level in the gate is opposite the silicon bandgap. The resulting large tunnel current strongly couples interface trap levels to the Fermi level in the gate. Now, the interface trap contribution to the admittance becomes frequency independent, and very little information about interface trap properties can be extracted.

At slightly greater oxide thicknesses (15–30 Å), direct tunnel current between the gate and interface trap levels becomes greatly reduced, thus decoupling interface trap levels from the gate Fermi level. Interface traps now predominantly exchange charge with the silicon and the interface trap contribution to the admittance becomes frequency dependent. The small-signal steady-state methods based on charge exchange between interface trap levels and the silicon can be used to extract interface trap properties in this thickness range.

Kar and Dahlke[26] used the conductance method to measure interface trap properties on MOS tunnel structures that had an oxide thickness in the range greater than 15 Å. Card and Roderick[27] used the conductance method in the 15–30 Å oxide thickness range and found that accurate values of interface trap properties could be obtained only over a limited silicon bandgap energy range. The reason was that minority carriers did not maintain equilibrium under changes in gate bias. As a result, the silicon surface went into deep depletion as the majority carrier Fermi level was made to approach the minority carrier band edge by the gate bias. Therefore, most of the gate bias appeared across the silicon rather than the oxide, preventing the Fermi level in the gate from approaching a position opposite the minority carrier band edge. For oxide thicknesses in the range 40–60 Å there was no longer a problem. Ma and Barker[28] used the conductance method to extract interface trap properties in the 40–60 Å range. Hunter et al.[29] and Eaton and Sah[30] used the low frequency capacitance method to measure interface trap level density on oxides 80 Å thick.

No advantage was found in MOS tunnel structures compared to thicker oxide MOS structures. The basic problem is that a large tunnel current

---

*An MOS tunnel structure is an MOS capacitor with such a thin oxide that *direct* tunneling can occur between gate and silicon.

flows between the gate and the silicon when gate bias moves the Fermi level in the gate toward a position opposite either the majority or minority carrier band edge. This tunnel current results in a large, frequency independent ac conductance near the band edges that must be subtracted from the total conductance to obtain the interface trap contribution. This limitation occurs in the same bias ranges as the limitation on thick oxide MOS capacitors, namely, the limitation imposed by a rapidly increasing accumulation layer or inversion layer capacitance near the band edges. Therefore, interface trap properties cannot be extracted significantly closer to the band edges with the MOS tunnel structure than with the thick oxide MOS structure. Accurate results are limited to about the same bandgap energy range.

There are three additional disadvantages in using the MOS tunnel structure: (1) there is always some direct tunnel current between interface trap levels and the Fermi level in the gate at every bandgap energy; this background current must be subtracted, thus reducing sensitivity and accuracy compared to the thick oxide MOS structure, which needs no such correction; (2) very thin oxides require greater care to fabricate, continue to grow in room ambient, and are not as reproducible as thick oxides; and (3) metal atoms evaporated to form the gate may penetrate nearly to the silicon* in a thin oxide and act as the dominant interface traps.[24] This effect obscures interface traps resulting from thermal oxidation or heating treatments. Interface traps measured on MOS tunnel structures may bear little resemblance to interface traps found in real thick-oxide device structures. Exceptions will occur when the sample is kept near room temperature during deposition or if a polysilicon or a mercury gate electrode is used. If measurements relating interface trap properties to device characteristics are desired, an oxide thickness comparable to the oxide thickness in the device should be used in the MOS test structure.

REFERENCES

1   E. H. Nicollian and A. Goetzberger, *Appl. Phys. Lett.*, **7**, 216 (1965).
2   E. H. Nicollian and A. Goetzberger, *Bell Syst. Tech. J.*, **46**, 1055 (1967).
3   I. M. Terman, *Solid-State Electron.*, **5**, 285 (1962).
4   C. N. Berglund, *IEEE Transact. Electron Devices*, **ED-13**, 701 (1966).
5   R. Castagné and A. Vapaille, *Surface Sci.*, **28**, 557 (1971).
6   M. Schulz and N. M. Johnson, *Appl. Phys. Lett.*, **31**, 622 (1977).
7   J. G. Simmons and L. S. Wei, *Solid-State Electron.*, **17**, 117 (1974).
8   J. G. Simmons and G. W. Taylor, *Solid-State Electron.*, **17**, 125 (1974).
9   D. L. Heald, *Solid-State Electron.*, **20**, 657 (1977).

*Depth of penetration of the metal atoms into the oxide depends on the temperature of the oxide during metal deposition.

10   R. J. Kriegler, T. F. Devenyi, K. D. Chik, and J. Shappir, *J. Appl. Phys.*, **50**, 398 (1979).

11   P. V. Gray and D. M. Brown, *Appl. Phys. Lett.*, **8**, 31 (1966).

12   J. S. Brugler and P. G. A. Jespers, *IEEE Transact. Electron Devices*, **ED-16**, 297 (1969); J. Golder and E. Baldinger, *Helv. Phys. Acta*, **44**, 387 (1971); M. Declercq and P. Jespers, *Revue HF (Belg.)*, **9**, 244 (1974).

13   E. Kamieniecki, *Solid-State Electron.*, **16**, 1487 (1973); R. F. Pierret, *Solid-State Electron.*, **19**, 577 (1976); R. F. Pierret and B. B. Roesner, *Solid-State Electron.*, **19**, 593 (1976).

14   E. Kamieniecki and R. Nitecki, in *The Physics of SiO₂ and Its Interfaces*, S. T. Pantelides, Ed., Pergamon, New York, 1978, pp. 417–420; W. E. Dahlke and D. W. Greve, *Solid-State Electron.*, **22**, 893 (1979).

15   K. Lehovec, *Appl. Phys. Lett.*, **8**, 48 (1966).

16   A. R. Von Hipple, *Dielectrics and Waves*, Wiley, New York, 1954, Appendix I.

17   J. A. Cooper, Jr. and R. J. Schwartz, *Solid-State Electron.*, **17**, 641 (1974).

18   E. H. Nicollian, A. Goetzberger, and A. D. Lopez, *Solid-State Electron.*, **12**, 937 (1969).

19   J. J. Simonne, *Solid-State Electron.*, **16**, 121 (1973).

20   A. Goetzberger, E. Klausmann, and M. J. Schulz, "Interface States on Semiconductor/Insulator Surfaces," CRC Critical Reviews in Solid-State Science (1976).

21   K. H. Zaininger, "Experimental Study of Metal-Oxide-Semiconductor Capacitors," Ph.D. thesis, Department of Electrical Engineering, Solid-State Electronics Laboratory, Princeton University, Princeton, N.J. (1964).

22   H. Deuling, E. Klausmann, and A. Goetzberger, *Solid-State Electron.*, **15**, 559 (1972).

23   P. V. Gray, *Phys. Rev.*, **A140**, 179 (1965).

24   W. E. Dahlke and S. M. Sze, *Solid-State Electron.*, **10**, 865 (1967).

25   L. B. Freeman and W. E. Dahlke, *Solid-State Electron.*, **13**, 1483 (1970).

26   S. Kar and W. E. Dahlke, *Solid-State Electron.*, **15**, 221 (1972).

27   H. C. Card and E. H. Roderick, *Solid-State Electron.*, **15**, 993 (1972).

28   T. P. Ma and R. C. Barker, *Solid-State Electron.*, **17**, 913 (1974).

29   W. R. Hunter, D. H. Eaton, and C. T. Sah, *Appl. Phys. Lett.*, **17**, 211 (1970).

30   D. H. Eaton and C. T. Sah, *Physica Status Solidi*, **A12**, 95 (1972).

6

Interfacial Nonuniformities

6.1 INTRODUCTION

In a simplified, idealized model of the oxide-silicon interface, oxide fixed charge and interface trap charges are considered to be smeared-out, uniform charge sheets.* In fact, both oxide fixed charge and charged interface traps are localized, not sheets. In addition,, mobile alkali metal ions may be present. The general idea is illustrated in Fig. 6.1.

How important is the localization of oxide fixed charge and interface trap charge? Does this localization affect electrical measurements? Answers depend on both the measurement considered and the bias range. For example, the depletion layer edge of a device is remote from the interface and thus is sensitive not to very detailed variations in interface properties, but only to rather long range variations, variations of a size comparable to the depletion layer width itself; C-V curves are distorted by such *long range* nonuniformities, and measurement of this distortion leads directly to a probability distribution of flatband voltage shifts. As another example,

---

*In this chapter "oxide fixed charge" refers to localized charges unable to change charge state by exchange of mobile carriers with the silicon. Interface trap charge refers to localized charge on centers that can change charge state by exchange of mobile carriers with the silicon. These two charges are treated as separate entities.

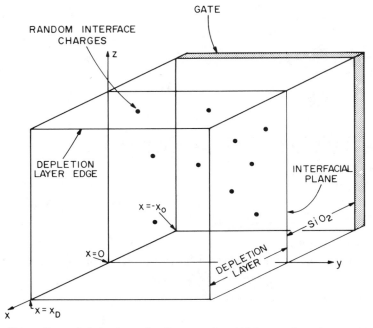

Fig. 6.1 Three-dimensional schematic diagram of an MOS capacitor showing a random distribution of oxide fixed charges across the Si–SiO$_2$ interface. After Brews.[1]

capture by interface traps is very sensitive to the mobile carrier densities near the interface trap. These carrier densities are affected by the details of charges nearby. Localization is important here. Experimentally, the width of conductance peaks leads directly to information on the variance of band bending caused by *short range*, localized nonuniformities. An intermediate case is the behavior of the inversion layer itself. Minority carriers are close to the interface, and in weak inversion they respond to the localized nature of interface trap and oxide fixed charge. Experimentally, these effects can be observed in transport measurements in MOSFETs not discussed here. However, in strong inversion, minority carrier density becomes so large that localized charges are screened out, and most minority carriers remain oblivious to these interface imperfections.*

Interfacial charge nonuniformity caused by the localized nature of interface traps and oxide fixed charge is only one aspect of charge nonunifor-

---

*A review of experimental and theoretical findings on the effects of nonuniformities on transport in MOSFETs can be found in Ref. 2. The number of minority carriers affected by nonuniformities varies slowly with carrier density as carrier density increases in strong inversion. Hence the ratio of the carriers affected to the total carrier density becomes small as the total carrier density increases.[3]

mity in real MOS devices.* We consider briefly two other types of charge nonuniformity: gross nonuniformities (e.g., edge effects and decoration of large defects that collect or nucleate clusters of ions) and chemical in-homogeneity (stretched, bent, or broken bonds, and oxide compositional variations).

From a practical standpoint, charge nonuniformities are undesirable. Their detection is the first step in a program to eliminate them. For example, for large MOSFETs in digital applications at room temperature, charge nonuniformity effects can be eliminated as a practical matter. For small MOSFETs, analogue application, or cryogenic operation, the prac-tical elimination of charge nonuniformities is not always possible; experience here still is at an early stage.

For example, oxide charging occurs near the drain of small MOSFETs, introducing a gross nonuniformity that influences threshold and the sym-metry of the device under interchange of source and drain. In CCDs microscopic charge nonuniformities affect the end of the charge transfer process. In cryogenic operation the turnoff of a MOSFET is controlled by percolation of current through a channel made tortuous by microscopic nonuniformities. Our emphasis is on how interfacial charge nonunifor-mities affect the interpretation of small-signal steady-state MOS capacitor measurements and how they can be detected.

Our discussion begins in Section 6.2 with large-scale nonuniformities, treating the uniform device as an assembly of *independent* smaller, uniform devices. The consequences for MOS capacitance measurements are derived.

Section 6.3 is a discussion of microscopic interfacial charge nonunifor-mities. This problem requires a three-dimensional treatment. The model based on an assembly of independent smaller devices is no longer useful, and interactions between the small devices become important. Expressions for mean band bending and the variance of band bending σ_s^2 are derived in Section 6.3. The dependence of σ_s^2 on gate bias and interface charge density is explored.

In Section 6.4 an equivalent circuit for the nonuniform MOS capacitor with a Poisson distribution of localized interface charges is derived. This circuit relates σ_s^2 to the measured admittance of the MOS capacitor, making possible the interpretation of admittance measurements in terms of σ_s^2 and interface trap parameters. These results are applied in Chapters 5 and 8, where methods for extracting interface trap properties are described.

Statistical matters are explored at various points. In any treatment of nonuniformities, their *distribution* across the Si–SiO$_2$ interfacial plane plays a role. For the microscopic case, a Poisson (random) distribution

---

*Dopant ions also are nonuniformly distributed in MOS devices. No theoretical treatment of fluctuations in dopant ion density is available, but experiments do not show this effect to be important.

seems appropriate.* For the macroscopic case, the spatial distribution varies with the source of nonuniformity. Examples with differing distributions are edge effects, oxide charging effects due to electron injection near the drain of a MOSFET, oxide charging due to electron injection in an MOS capacitor with water-related traps near the Si–SiO$_2$ interface (see Section 11.5.1), and charged defects decorated by alkali metal ions diffused to the Si–SiO$_2$ interface.

Finally, in Section 6.5 we consider the possible influence of chemical inhomogeneities at the Si–SiO$_2$ interface on interface trap properties.

6.2 GROSS NONUNIFORMITIES

The first published treatment of interfacial charge nonuniformities in the MOS system using an assembly of independent small, uniform devices was made by Nicollian and Goetzberger.[5] These authors modeled an MOS capacitor with a nonuniform oxide fixed charge density as a parallel array of smaller, uniform capacitors. Each elementary capacitor had its own oxide fixed charge density and, therefore, its own flatband voltage. Moreover, each elementary capacitor was treated according to the customary one-dimensional MOS capacitor formulas, equivalent to assuming *no interaction* between elementary capacitors.†

Another application of the model was made by Castagné and Vapaille.[6] These authors analyzed C-V curves of grossly nonuniform MOS capacitors. Because the depletion layer width is not very sensitive to fine details of the interface charge, the parallel-array model is suitable for the depletion layer capacitance. These authors showed empirically that such a model fitted the measured capacitance curves of samples intentionally contaminated with lithium. Nonuniformity resulted when lithium ions were drifted to the Si–SiO$_2$ interface and held near this interface at elevated temperatures for an hour or so.

Some typical results obtained by Castagné and Vapaille are shown in Fig. 6.2. Here, curves a are low frequency (solid-line) and high frequency (dashed-line) C-V curves for an uncontaminated oxide grown on an n-type substrate. Curves b resulted after lithium implantation through an aluminum gate. Curves c, d, and e apply after various bias-temperature treatments. The symptoms of gross interfacial charge nonuniformity are:

---

*The only statistics background needed is a knowledge of the Gaussian approximation to the Poisson distribution, see Ref. 4.

†It should be noted that although the results of Ref. 5 for D_{it}, σ_n, σ_p, and so forth are correct, the discussion of nonuniformities in Ref. 5 is in error because the parallel array model does not apply to *microscopic* nonuniformities. That is, the variance of Ref. 5 [see (6.44) of this chapter] applies to low amplitude gross nonuniformities. Equation (6.70) of this chapter applies to microscopic nonuniformities and would have been used in Ref. 5 if this result had been available when that work was done.

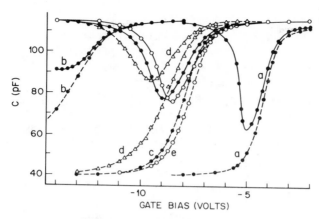

Fig. 6.2 Plot of C-V curves measured before and after lithium contamination and drift. Curves a; prior to lithium implant; curves b; after an implant; curves c, after − 15 V at 90°C for 1 hr; curves d, after + 16 V at 60°C for 1 hr; and curves e; at − 10.5 V at 70°C for 1 hr. Doping density $\approx 9 \times 10^{14}$ cm^{-3}, oxide thickness $\approx$ 1500 Å, (100) orientation. After Castagné and Vapaille.[6]

(1) a stretchout of the C-V curves along the voltage axis and (2) a rise in the minimum low frequency capacitance. Such effects conceivably could be caused by interface traps, but, as discussed in Section 6.2.5, tests to eliminate this possibility can be performed.

Castagné and Vapaille assumed their observations were due to a Poisson distribution of the lithium ions introduced into their samples. This statistical interpretation is not completely correct. However, before getting into statistical matters, let us consider the parallel array model in more detail.

6.2.1 Parallel Array Model for the Nonuniform MOS Capacitor

To understand the parallel array model, we follow the discussion of Brews and Lopez.[7] Figure 6.3a is a schematic diagram of an MOS capacitor with a spatially varying oxide fixed charge density, taken as a sinusoidal spatial variation for simplicity. Because the wavelength of the spatial variation is short relative to the depletion layer width, the depletion layer width responds only to the overall average oxide fixed charge density, and the depletion layer edge is flat. Figure 6.3b is a schematic diagram of an MOS capacitor with a slowly varying oxide fixed charge density with a wavelength comparable to the depletion layer width. Now the depletion layer width responds to the oxide fixed charge density, and the depletion layer edge is modulated.

Also shown in Fig. 6.3 are dotted lines, indicating a partition of the entire capacitor into an array of smaller, *elementary* capacitors. For each ele-

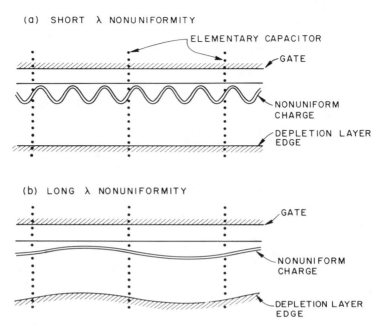

Fig. 6.3 Nonuniform MOS capacitor with a sinusoidal spatial variation in oxide fixed charge density (double line). (a) For short-wavelength variations of oxide fixed charge, the depletion layer edge is flat and a parallel array model is useless. (b) For long-wavelength interfacial charge nonuniformities, the depletion layer width is slowly modulated and a parallel array model is useful. After Brews and Lopez.[7] Copyright (1973), Pergamon Press Ltd., Reprinted with permission.

mentary capacitor to be independent of the rest, the size of these elementary capacitors must be larger than a depletion layer width to avoid strong interaction of adjacent capacitors through their fringing fields. In Fig. 6.3a the resulting elementary capacitors are so large that all are identical, and thus no advantage is gained by subdivision; in Fig. 6.3b each capacitor is different, with its own unique depletion layer width and oxide fixed charge density. Figure 6.3 shows that a parallel array model of independent elementary capacitors has meaning for the slowly varying case, but not for the rapidly varying case.

A realistic interface charge distribution can be Fourier analyzed as a sum of sinusoids of various wavelengths. For large defects, such as charge islands formed by charge decoration of large defects, a parallel array model is suitable because only long wavelength components enter its Fourier analysis. However, for localized charges, the parallel array model is not satisfactory for the shorter wavelength components of the charge distribution.

Because depletion layer width is gate bias dependent, the depletion layer width is sensitive to smaller interfacial charge nonuniformities when the device is biased near accumulation, where depletion layer width is small, than when it is biased in depletion, where depletion layer width is large. Therefore, the simplest case is that of interfacial charge nonuniformities large enough to permit use of the parallel array model at the largest depletion layer width. For this case, we can neglect the problem of incorporating the changing sensitivity of the depletion layer width to the charge distribution as depletion layer width is varied. However, we consider the general case when we discuss statistical matters.

6.2.2 Capacitance in the Parallel Array Model

Suppose that the capacitance per unit area of the nth elementary capacitor at gate bias V_G is $C(V_{Gn})$, where V_{Gn} is the gate bias corrected for the flatband voltage shift V_n in the nth capacitor. That is

$$V_{Gn} = V_G - V_n \tag{6.1}$$

where

$$V_n = -\frac{(\bar{x}/x_o)Q_{ot}(n)}{C_{ox}}. \tag{6.2}$$

In (6.2), $Q_{ot}(n)$ is the oxide trapped charge per unit area in the nth capacitor, with centroid $\bar{x}$ measured with respect to the gate. The band bending ψ_n in the nth capacitor is obtained by solving

$$C_{ox}V_G - C_{ox}\psi_n = C_{ox}V_n - Q_s(\psi_n) - Q_{it}(\psi_n) \tag{6.3}$$

where $Q_s(\psi_n)$ is the silicon surface charge per unit area for band bending ψ_n and $Q_{it}(\psi_n)$ is the interface trap charge per unit area. From (3.19), for uniform acceptor doping per unit volume N_A we obtain

$$Q_s(\psi_n) = -\text{Sgn}(\psi_n)\sqrt{2}qN_A\lambda_p \left[\exp(-\beta\psi_n) + \beta\psi_n - 1 \right.$$

$$\left. + \left(\frac{n_i}{N_A}\right)^2 (\exp\beta\psi_n - \beta\psi_n - 1) \right]^{1/2} \tag{6.4}$$

where $\text{Sgn}(\psi_n)$ is positive when ψ_n is positive and negative when ψ_n is negative. The measured capacitance per unit area of the whole device C_m for gate bias V_G is then

$$C_m(V_G) = \sum_n \alpha_n C(V_{Gn}) \tag{6.5}$$

where α_n = fraction of total area with oxide trapped charge density $Q_{ot}(n)$. Thus to generate a C-V curve from the parallel array model, one must know: (1) the gate bias V_{Gn} corrected for the various flatband voltage shifts V_n of (6.2); (2) the fractional area associated with each flatband voltage shift α_n; (3) the functional dependence of the silicon surface charge density on band bending $Q_s(\psi_n)$, [(e.g. for uniform doping; see (6.4)]; and (4) the high or low frequency capacitances as a function of $V_G - V_n$, namely, $C(V_{Gn})$.

For example, consider the low frequency capacitance for uniform doping. Then, (6.4) applies. Therefore, the low frequency silicon capacitance is [see (3.37)]

$$C_s(\psi_n) = \frac{-dQ_s}{d\psi_n}$$

$$= -\beta(qN_A\lambda_p)^2\frac{[1 - \exp(-\beta\psi_n) + (n_i/N_A)^2(\exp\beta\psi_n - 1)]}{Q_s(\psi_n)} \qquad (6.6)$$

and the low frequency capacitance per unit area of an elementary device is

$$C(V_{Gn}) = [C_{ox}^{-1} + C_s^{-1}(\psi_n)]^{-1} \qquad (6.7)$$

where ψ_n is related to V_n by (6.3).

The values of V_n and α_n must now be determined. This part of the problem has led to the most confusion.

In the simple case of an edge effect, V_n and α_n are determined by the theory of the phenomena under consideration. For example, for the case of sodium drift in the fringing field of a capacitor,[8] the radial distribution of sodium along the interface was assumed to be proportional to the fringing field of the capacitor. This fringing field could be calculated from electrostatics.

The most commonly used procedure[6,9] has been to *assume* a statistical distribution of V_n, namely, $P(V_n)$, assume the device large enough that $\alpha_n = P(V_n)$, and use (6.5) to compute the measured capacitance C_m

$$C_m = \int_{-\infty}^{\infty} dV_n\, P(V_n)C(V_G - V_n) \qquad (6.8)$$

where the function C is given by (6.7). Then C_m is compared to experiment. Usually the distribution function $P(V_n)$ contains a number of parameters that can be varied to fit C_m to the experimental curve.

Recently Chang[10] pointed out that (6.8) is simply a convolution of $P(V_n)$ and $C(V_G - V_n)$. Consequently, a Fourier transform technique can determine $P(V_n)$ directly from $C_m(V_G)$. Chang[10] and Chang and Johnson[11] have presented an approximate, simpler method for obtaining $P(V_n)$ from a combination of high and low frequency measurements of $C_m(V_G)$. These methods are not described here.

6.2.3 Parallel Array Model and the Poisson Distribution

Once $P(V_n)$ is found by one of the preceding methods, the question of its interpretation arises. That is, does $P(V_n)$ provide any information about the origin of the nonuniformity?

Most commonly, it has been assumed that the interfacial charge nonuniformity is the result of a Poisson distribution of point charges.[5,6,9] This distribution results inevitably from the simple presence of point charges if they are randomly dispersed. For C-V measurements that show marked distortion, we show that this interpretation is untenable. That is, a Poisson distribution of oxide fixed charge is so fine grained that it will not result in much modulation of the depletion layer width. For example, the capacitance of an MOS capacitor with a Poisson distribution of localized oxide fixed charges differs from that of an MOS capacitor with a uniformly smeared out sheet of oxide fixed charge only very close to accumulation or inversion. To support these assertions, we examine the case of a Poisson distribution of oxide fixed charge in more detail.

First, we determine the area α of a typical elementary capacitor. We call α the *characteristic area* of the array of charge patches. Let the nth elementary capacitor contain an oxide fixed charge density per unit area Q_{fn}. Also, the nth capacitor will have a depletion layer width w_n. For the one-dimensional analysis to apply to this nth capacitor, we must have

$$\alpha \geq \pi (2w_n)^2. \tag{6.9}$$

The radius $2w_n$ comes from the long-wavelength limit of a three-dimensional calculation, but it suffices here to note that the radius must be somewhat larger than w_n to avoid interaction effects between adjacent elementary capacitors. The reader may refer again to Fig. 6.3.

Inequality (6.9) will hold for all elementary capacitors if α is chosen large enough. However, a very large α unduly minimizes the fluctuations in Q_{fn} and w_n from one elementary capacitor to another. Hence we choose α so that (6.9) is valid for *most* of the elementary capacitors, but α is not chosen so large that even the exceptional cases with very large depletion layer widths are included. At the other extreme, α is not chosen so small that (6.9) fails for most of the elementary capacitors, because then our parallel array model collapses.

For a Poisson distribution, the probability P_n of n charges in an elementary capacitor is

$$P_n = \frac{\gamma^n}{n!} \exp(-\gamma) \tag{6.10}$$

where, to ensure that the mean oxide fixed charge density per unit area is $\bar{Q}_f$, we require that

$$\gamma = \frac{\bar{Q}_f}{q} \alpha. \tag{6.11}$$

Using (6.10), the variance of oxide fixed charge σ_q^2 (the mean square departure of Q_{fn} from $\bar{Q}_f$) is

$$\sigma_q^2 \equiv \sum_n P_n (Q_{fn} - \bar{Q}_f)^2 = \frac{q\bar{Q}_f}{\alpha}. \tag{6.12}$$

In the limit of many charges in an elementary capacitor, approximately two-thirds* of the elementary capacitors will possess Q_{fn} such that

$$\bar{Q}_f - \left(\frac{q\bar{Q}_f}{\alpha}\right)^{1/2} \leq Q_{fn} \leq \bar{Q}_f + \left(\frac{q\bar{Q}_f}{\alpha}\right)^{1/2}. \tag{6.13}$$

Therefore, our parallel array model will be adequate for about two-thirds of the elementary capacitors if we choose $\alpha = \pi(2w_+)^2$, where w_+ is the depletion layer width of a capacitor with $\bar{Q}_{f+} = \bar{Q}_f + (q\bar{Q}_f/\alpha)^{1/2}$.†

Thus a self-consistency problem arises. On one hand, the potential of our "worst *probable* case" capacitor $\beta\psi_+$ depends on α because from (6.3)

$$C_{ox}(V_G - \psi_+) = -\frac{\bar{x}}{x_o}\left[\bar{Q}_f + \left(\frac{q\bar{Q}_f}{\alpha}\right)^{1/2}\right] - Q_s(\psi_+) - Q_{it}(\psi_+) \tag{6.14}$$

where in our example $\bar{x} = x_o$ because the charge is at the interface. On the other hand, ψ_+ determines the depletion layer width, and hence $\alpha = \pi(2w_+)^2$. We choose the relation between depletion layer width and potential as

$$w = \frac{\epsilon_s}{C_{maj}(\psi)} \tag{6.15}$$

where, for uniform doping from (3.40)

$$C_{maj}(\psi) = C_{FBS}\frac{|1 - \exp(-\beta\psi)|}{\{2[\exp(-\beta\psi) + \beta\psi - 1]\}^{1/2}} \tag{6.16}$$

is the majority carrier silicon capacitance per unit area and $C_{FBS} = \epsilon_s/\lambda_p$ is the silicon flatband capacitance. A desirable property of (6.15) is that in depletion, w is the depletion layer width and at flatbands or in accumulation, w is the extrinsic Debye length taking into account screening of majority carriers. The joint solution of (6.14) and (6.15) for ψ_+ and α is plotted in Fig. 6.4 for $Q_{it} = 0$ assuming $\bar{x} = x_o$ (i.e., for charges at the interface).

The most important point in Fig. 6.4 is that the potential ψ_+ of our "worst probable case" differs by only a kT/q or so from that of the elementary capacitors with the average oxide fixed charge density corresponding to $\bar{\psi}$. This small difference prevails even for an oxide fixed charge density of $10^{12}\,\mathrm{cm}^{-2}$, an extreme level for this charge density. That is, the band-bending fluctuations due to interfacial charge nonuniformity are very small for the case of a simple Poisson distribution of point charges.

The second point of importance in Fig. 6.4 is the rather large size of α compared to the experimental values in Refs. 6 and 9. Even in accumulation ($\bar{\psi} < 0$), α exceeds $10^{-10}\,\mathrm{cm}^2$ and in depletion, $\alpha = 10^{-7}\,\mathrm{cm}^2$.

---

*This factor (two-thirds) approximates the integral of the Gaussian distribution from $+\sigma_q$ to $-\sigma_q$, which is more accurately 0.683 (see Ref. 4, Fig. 1, p. 165).

†We have assumed a p-type substrate and positive oxide fixed charge so that w_+ is the largest depletion layer width in the range of (6.11). For an n-type substrate and positive oxide fixed charge, we would choose w corresponding to $Q_f - (q\bar{Q}_f/\alpha)^{1/2}$.

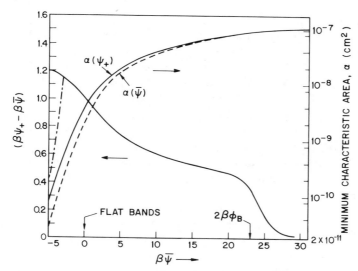

Fig. 6.4 The characteristic area α (right-hand logarithmic scale) and band-bending variation (left-hand scale in kT/q) as a function of the band bending $\beta\bar{\psi}$ corresponding to a mean oxide fixed charge density of 10^{12} cm^{-2}. Oxide thickness was 1000 Å, and $N_A = 10^{15}$ cm^{-3}. The solid-line curve for $\alpha(\psi_+)$ is computed so that approximately two-thirds of the elementary capacitors are large enough to validate the parallel array model. The dashed-line curve for $\alpha(\bar{\psi})$ is the value required to validate the model for all elementary capacitors with oxide fixed charge $\leq$ mean value. The dashed-dotted curve indicates the drop of band-bending fluctuations in accumulation when the limit, (6.17), on minimum depletion layer width is imposed. The flatband point and the point of strong inversion ($\bar{\psi} = 2\phi_B$) are shown by arrows.

These large values of α completely rule out the Poisson distribution of oxide fixed charge or of positive ions as the explanation of the experimental observations of Castagné and Vapaille,[6] and of later work by Baccarani et al.[9] These authors find $\alpha \approx 10^{-11}$ cm^2. Such a value of α is several orders of magnitude too small to agree with a Poisson distribution of single point charges.

Some other features of Fig. 6.4 are:

1 Despite the screening of majority carriers, the band-bending fluctuations are largest in accumulation and near flatbands. This effect is due to the small characteristic area in this region that leads to a large σ_q [see 6.12].*

2 Throughout depletion, the band-bending fluctuations are fairly constant, but the bias dependence of α still is important enough to affect a fit to experimental data.

---

*The screening of majority carriers has not been treated very accurately in our one-dimensional approximation in this bias range.

3 In strong inversion, silicon band bending tends to become clamped as a result of the exponential dependence of minority carrier density on band bending. The clamping of band bending also clamps the depletion layer width, so that α saturates and $(q\bar{Q}_f/\alpha)^{1/2}$ saturates. With $(q\bar{Q}_f/\alpha)^{1/2}$ clamped, and with the exponential variation of $Q_s(\psi)$ with ψ, the fluctuations of band bending $\beta(\psi_+ - \bar{\psi})$ drop rapidly for $\bar{\psi} > 2\phi_B$.

Finally, there are some reservations about Fig. 6.4. In accumulation and in inversion, the mobile carriers are very close to the oxide fixed charges. Consequently, the parallel array model is suspect in these regions because it does not treat the screening of the point charges properly at close approach. In accumulation, we expect the band-bending fluctuations to drop rather than saturate as in Fig. 6.4.

We can "patch up" the one-dimensional parallel array model in accumulation by introducing a physical effect contained in a three-dimensional treatment, but omitted from the one-dimensional model. This effect sets a limit on how small α can become. To illustrate, consider a positive point charge embedded in an unbounded neutral p-type semiconductor. The positive point charge repels mobile holes, thus creating a spherical depletion region surrounding the point charge. The size of this depletion sphere is fixed by charge neutrality. That is, the negative acceptor charge within the depletion sphere must balance the point charge. Mathematically, if the radius of the depletion sphere is r and the point charge is of magnitude Zq, where q is the magnitude of electronic charge, then

$$qN_A \left(\frac{4}{3}\right) \pi r^3 = qZ$$

or

$$r = \left[\frac{3Z}{(4\pi N_A)}\right]^{1/3}. \tag{6.17}$$

Interpreting r as the minimum depletion layer width possible in the parallel array model, we use (6.15) only when it predicts $w > r$. Whenever (6.15) predicts $w < r$, we take instead $w = r$.

As a result of this very crude modification of (6.15), the characteristic area α in (6.9) never can be smaller than $\alpha_{min} \approx 4\pi r^2$. For example, for $N_A = 10^{15}\,\text{cm}^{-3}$, $\alpha \geq 4.8 \times 10^{-10}\,\text{cm}^2$. Because α cannot decrease below α_{min}, $(q\bar{Q}_f/\alpha)$ cannot increase above $(q\bar{Q}_f/\alpha_{min})$. Hence, as the MOS capacitor is driven into accumulation beyond the point where α first equals α_{min}, the band bending ψ_+ from (6.14) will approach $\bar{\psi}$ more closely. That is, the exponential dependence of majority carrier density on band bending means only a small departure of ψ_+ from $\bar{\psi}$ will serve to compensate for $(q\bar{Q}_f/\alpha_{min})^{1/2}$ in (6.14). Hence the band-bending fluctuation $(\psi_+ - \bar{\psi})$ decreases as the bands are bent further into accumulation.

In Fig. 6.4 the result of introducing this minimum size for α is shown by a dash-dotted curve in accumulation for $\beta\bar{\psi} < -3$. Quantitatively, (6.17) is a crude limitation on how small α can become. However, the argument serves to show that a lower bound on α results in a decrease of band-bending fluctuations in accumulation. For an attractive interaction between the point charge and the mobile carriers, a quantum mechanical argument is needed to establish this lower bound on α.

6.2.4 Clustering Due to Defect Decoration or Nucleation

Because a simple Poisson distribution of point charges cannot lead to the small value of α observed by Castagné and Vapaille,[6] nor to that of Baccarani et al.,[9] we must search elsewhere for an explanation. The most probable explanation stems from experimental observations by DiStefano,[12] Williams and Woods,[13] and Bottoms et al.[14] These authors performed ion drift experiments similar to those of Castagné and Vapaille[6] but used sodium. This sodium either was present in their oxides as grown or, in some cases, was artificially introduced. (See Section 10.2.3 for a discussion of sodium in SiO_2.) The contaminated capacitors were scanned with a laser beam and the resulting photocurrent measured. Because charged ions in the oxide lower the barrier height for photoemission,[15] the measured photocurrent provides a map of the ion density at the $Si–SiO_2$ interface. (See Section 10.3 for a discussion of photoemission.) An example from the work of DiStefano[12] is shown in Fig. 6.5. Sodium contamination was deliberately introduced. Nonuniformity developed after an hour at elevated temperature ($150°C$) under positive gate bias. The gate bias drifted the ionized sodium to the $Si–SiO_2$ interface, where it congregated in clusters at the elevated temperature. This clustering continued to increase up to 20 hr, the longest time reported. Williams and Woods[13] performed similar measurements on lightly doped ($7 \times 10^{14}\,cm^{-3}$) samples that were not intentionally sodium contaminated. Similar results were obtained. It was found[12, 13] that these clusters sometimes occur near weak spots where oxide breakdown occurs at high gate bias.

These experiments provide a possible resolution of the paradox raised by the work of Castagné and Vapaille[6] and of Baccarani et al.[9] This paradox is that although a Poisson formula for the distribution of charge fits their experimental data, the variance ($q\bar{Q}_m/\alpha$) needed for this fit is far too large. That is, the characteristic area α of the fitted distribution is orders of magnitude smaller than expected from a simple Poisson distribution of the ionic charges.

A reasonable resolution of this paradox, a resolution compatible with the observations of DiStefano[12] and Williams and Woods,[13] is to suppose that *nucleation centers* of the charged clusters are Poisson distributed, not the charges themselves. Each nucleation center then accumulates, or is *decorated* by, a number of charges. A general model of this sort leads to what statisticians call a "compound Poisson distribution."[4] Because the nucleation centers are fewer in number than the number of charges present,

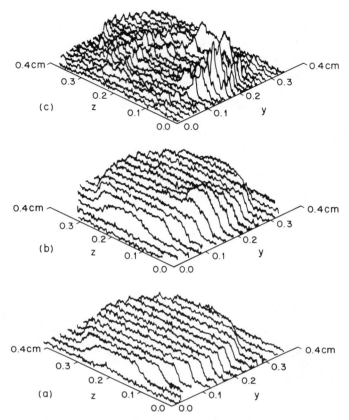

Fig. 6.5 Internal photoemission maps of the Si–SiO₂ interface scanned with a 20 μm diameter light spot from a laser: (a) initial map; (b) after 10 V bias at 150°C for 1 hr; (c) after 10 V bias at 150°C for 20 hr. The scale in (c) is (one-tenth) that of (a) and (b) and shows large nonuniformity effects. The sample was thermally oxidized to a thickness of 1000 Å, and $N_A = 10^{20}$ cm^{-3}. After DiStefano.[12]

the variance of their distribution is far greater than that of a Poisson distribution of the ionic charges themselves. This greater variance is reflected in the fictitiously small fitted value of α in Refs. 6 and 9.

As an example, a variety of nucleation centers could be envisioned. Each type of center could be labeled by the number of charges Z_j it can trap. Each type of *center* could be individually Poisson distributed with a distribution

$$P(k, \nu_j) = \exp(-\nu_j) \frac{\nu_j^k}{k!} \tag{6.18}$$

where ν_j is the mean number of nucleation centers in area α with Z_j charges and $P(k, \nu_j)$ is the probability of finding k such centers in area α.

We now ask for the probability of finding n *charges*, not centers, in area α irrespective of the type of center involved. The result can be found in Feller.[4] Here, we content ourselves with noting that the mean charge density $\bar{Q}_m$ is

$$\bar{Q}_m = q \sum_{j=1}^{N} Z_j \frac{\nu_j}{\alpha} \qquad (6.19)$$

(i.e., the sum of the means for each type of center alone) and the variance σ_q^2 is

$$\sigma_q^2 = q^2 \sum_{j=1}^{N} Z_j^2 \frac{\nu_j}{\alpha^2} \qquad (6.20)$$

(i.e., the sum of the variances for each type of center alone). The results (6.17) and (6.18) are particularly simple for the case of only one type of nucleation center capable of decoration by Z charges. Then, introducing the mean *density per unit area of nucleation sites* $\bar{\nu}$, $\bar{\nu} \equiv \nu/\alpha$, we have, in place of (6.19)

$$\bar{Q}_m = qZ\bar{\nu} \qquad (6.21)$$

and the variance, σ_q^2 [see (6.12)] is

$$\sigma_q^2 = q^2 Z^2 \frac{\bar{\nu}}{\alpha} = q\bar{Q}_m \frac{Z}{\alpha}. \qquad (6.22)$$

Comparison of (6.12) and (6.22) indicates that the parameter obtained in Refs. 6 and 9 by fitting experiment was *not* α, but (α/Z) and, therefore, an area smaller than α.

Let us analyze these experiments using the nucleation model to find what values for Z and $\bar{\nu}$ result. Flatband voltage V_{FB} and the variance $(\bar{x}/x_o)^2 \sigma_q^2$ are measured (see Section 10.5), where

$$V_{FB} = \frac{(\bar{x}/x_o)\bar{Q}_m}{C_{ox}}. \qquad (6.23)$$

For $\bar{Q}_m$ at the Si–SiO$_2$ interface, $\bar{x} = x_o$. In this case

$$V_{FB} = \frac{\bar{Q}_m}{C_{ox}}. \qquad (6.24)$$

Therefore, to infer an approximate value of $\bar{Q}_m$ experimentally, one can drift the mobile charge to the Si–SiO$_2$ interface at high temperature and afterward presume $\bar{x} = x_o$. This was the procedure of Baccarani et al.[9]

Besides V_{FB}, the other parameter obtained from experiment is the variance of the flatband voltage distribution. To determine this variance, we assume the parallel array model of elementary capacitors with various V_{FB}, assume some probability density (usually a Gaussian approximation to the Poisson distribution), and generate [from (6.8)] a family of C-V curves with the variance as a parameter. A comparison of these theoretical curves with experiment determines the "best-fit" value of the variance, $(\bar{x}/x_o)^2\sigma_q^2$. For example, Baccarani et al.[9] found the parameter

$$\left(\frac{\bar{x}}{x_o}\right)^2 \sigma_q^2 = q^2 Z^2 \left(\frac{\bar{x}}{x_o}\right)^2 \left(\frac{\bar{\nu}}{\alpha}\right) \tag{6.25}$$

as a function of mean flatband voltage V_{FB}

$$V_{FB} = \frac{(\bar{x}/x_o)\bar{Q}_m}{C_{ox}} = \frac{(\bar{x}/x_o)qZ\bar{\nu}}{C_{ox}}. \tag{6.26}$$

Combining (6.25) and (6.26), yields

$$\frac{\bar{x}}{x_o}\sigma_q = \left[\frac{C_{ox}qZ(\bar{x}/x_o)}{\alpha}\right]^{1/2} V_{FB}^{1/2}. \tag{6.27}$$

Hence a plot of $\ln[(\bar{x}/x_o)\sigma_q]$ versus $\ln V_{FB}$ should have a slope of $\frac{1}{2}$ and an intercept at $V_{FB} = 1\,\mathrm{V}$ of $\frac{1}{2}\ln[C_{ox}qZ(\bar{x}/x_o)/\alpha]$. Such a plot from Ref. 9 is shown in Fig. 6.6. Figure 6.6 shows that the square root dependence of (6.27) is obeyed. The intercept at $V_{FB} = 1\,\mathrm{V}$ [viz., $(\bar{x}/x_o)\sigma_q = 6 \times 10^{-9}\,\mathrm{coul/cm}^2$] provides an estimate of (α/Z). From (6.27):

$$\frac{x_o}{\bar{x}}\frac{\alpha}{Z} = \frac{qC_{ox}V_{FB}}{[(\bar{x}/x_o)\sigma_q]^2} = 1.5 \times 10^{-10}\,\mathrm{cm}^2. \tag{6.28}$$

After drifting all the charge to the Si–SiO$_2$ interface, it was estimated that $(\bar{x}/x_o) = 0.1$. Hence $(\alpha/Z) = 1.5 \times 10^{-11}\,\mathrm{cm}^2$. From Fig. 6.4 we estimate that

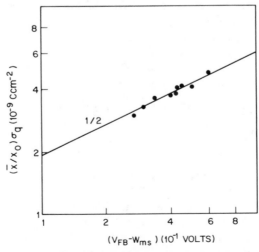

Fig. 6.6 Experimental test of the square root dependence of $(\bar{x}/x_o)\sigma_q$ on flatband voltage V_{FB}. Samples were (100)-oriented n-type silicon, $N_D = 4 \times 10^{14}\,\mathrm{cm}^{-3}$ and $x_o = 1000\,\text{Å}$; V_{FB} has been corrected for the work function difference W_{ms}. After Baccarani et al.[9]

$\alpha \approx 10^{-7}\,\text{cm}^2$ in depletion ($\beta\psi \cong 2\beta\phi_B$). Hence $Z \approx 6600$ charges per nucleation site. Also, after drift, it was found that $\bar{Q}_f/q = 2 \times 10^{12}\,\text{cm}^{-2}$. Hence from (6.21), $Z\bar{\nu} = \bar{Q}_f/q = 2 \times 10^{12}\,\text{cm}^{-2}$ and there are $\bar{\nu} \approx 3 \times 10^8\,\text{sites/cm}^2$.

The data of Castagné and Vapaille[6] (for curve e in Fig. 6.2) lead to $(\bar{x}/x_o)\sigma_q = 1.5 \times 10^{-8}\,\text{coul/cm}^2$, and (assuming the charges responsible for the initial 4 V of V_{FB} before implant do not participate in decoration) $C_{ox}V_{FB}/q \approx 4.5 \times 10^{11}\,\text{cm}^{-2}$. Also, we take $(\bar{x}/x_o) = 1$ in this case, because the lithium ions were drifted to the Si–SiO$_2$ interface under positive gate bias at elevated temperatures. Then $(\alpha/Z) = 5.1 \times 10^{-11}\,\text{cm}^2$, $Z \approx 2000$ charges per site, and $\bar{\nu} = 2.3 \times 10^8$ sites per cm^2.

It is likely that not all the ions participate in decoration. Restricting $\bar{Q}_m$ in (6.21) to only those charges participating in decoration rather than using all the charge contributing to V_{FB} leads to a larger estimate for Z [see (6.27)] and a smaller estimate for $\bar{\nu}$ [see (6.26)] for a given σ_q. Also, α may be larger than our estimate of $10^{-7}\,\text{cm}^2$, thus increasing Z and decreasing $\bar{\nu}$. If $\alpha = 10^{-7}\,\text{cm}^2$ is correct, then the above experiments indicate 10–20 decorated defect sites per characteristic area.

The preceding discussion shows that the C-V measurements of Refs. 6 and 9 can be interpreted in terms of defect decoration. For these grossly nonuniform MOS capacitors, the discussion of band-bending fluctuations is based not on a Poisson distribution of oxide charge, but on a Poisson distribution of decorated defects. Let us reexamine the parallel array model for the case of charge clusters. The treatment of band-bending fluctuations in this case takes on a different aspect depending on the size assumed for decorated defects.

If the decorated sites are small compared to a depletion layer width, then we can repeat the analysis of Fig. 6.4 with $(q\bar{Q}_f/\alpha)^{1/2}$ replaced by the much larger $(q\bar{Q}_f Z/\alpha)^{1/2}$. Naturally, because Z is a large number, a much larger band-bending fluctuation is found, thus making a nonlinear treatment mandatory. Also, because depletion layer width narrows as gate bias is reduced toward flatbands, at lower biases very few defects will be present in a characteristic area. Therefore, the Gaussian approximation to the Poisson distribution will fail. In addition, the detailed structure of the defect may influence the depletion layer width and a three-dimensional analysis may become necessary.

If the decorated sites are large relative to a depletion layer width w, they will occupy areas larger than $4\pi w^2$. Defects may merge or overlap. For example, the results in Fig. 6.5 indicate that charge islands on the order of 0.1 cm in diameter can occur under extended bias-temperature annealing. In a zero-order model we might replace the defect with a patch of uniform charge of the same area as the defect, and with the same net charge. In this model the characteristic area of an elementary capacitor becomes a typical defect area, unrelated to depletion layer width (but necessarily much larger) and independent of gate bias. Consequently, $\sigma_q = (q\bar{Q}_f Z/\alpha)^{1/2}$

also is independent of bias, as was assumed in the model used by Castagné and Vapaille.[6]

For those cases where σ_q is large, it is essential to use the probability distribution of flatband voltage shifts, rather than a Gaussian approximation to the distribution of band bending. As was emphasized by McNutt and Sah,[16] the distribution of band bending can be a multiple-peaked, complex distribution (because of the nonlinear relation between charge and band bending) even though the charge clusters are distributed simply according to a Poisson distribution. Castagné and Vapaille[6] computed their C-V curves correctly, and Baccarani et al.,[9] who used a distribution of band bending, treated samples where $(\bar{x}/x_o)\sigma_q$ was small enough that nonlinearities were not important.

All this discussion has presumed interfacial charge nonuniformities alone were present.* Interface traps also can cause superficially similar distortions of C-V curves. We now discuss the problem of distinguishing between these two effects.

6.2.5 Distinguishing Between Gross Nonuniformities and Interface Trap Effects

(a) MOS Capacitor Methods

As pointed out in connection with Fig. 6.3, interfacial charge nonuniformities of a size comparable to or greater than a depletion layer width cause distortions of C-V curves. In particular, the low frequency capacitance minimum is raised and both the high and low frequency C-V curves are stretched out along the gate bias axis. Interface traps can give rise to similar effects (see Chapters 5 and 8).

It is possible to distinguish between gross interfacial charge nonuniformity and interface trap effects because the two effects have a different frequency response. Gross interfacial charge nonuniformities modify majority carrier response to gate bias changes, a response so rapid that the effect of interfacial charge nonuniformities is instantaneous in the customary frequency range of a few megahertz or less. However, interface traps capture and emit majority carriers rather slowly, especially in depletion, so that their influence is virtually eliminated at 1 MHz, except in or near accumulation (see Chapter 5). Consequently, measurement of both high and low frequency C-V curves provides enough data to distinguish between the two effects.

There are several ways to use these data. One approach is to use high and low frequency C-V curves to compute an "apparent" interface trap level density, as done by Castagné and Vapaille.[6] Then, to determine if this

---

*Radial variations in dopant ion density across a silicon wafer also are a gross nonuniformity but occur over such long distances that they have no effect on an MOS capacitor, which normally is of a size $\lesssim 800\ \mu$m.

interface trap level density is real, we compute the gate bias stretch out these "apparent" interface traps would produce. For an interface trap level density that is real, both the experimental high and low frequency C-V curves are reproduced. For an interface trap level density that is only apparent, agreement will not be found.[10]

A method that requires less computation was proposed by Brews and Lopez,[7] who noted that the doping profile, when interface traps are present, can be obtained from high and low frequency C-V curves using (9.21) due to Brews.[17] However, if interfacial charge nonuniformities are present, the interface trap correction built into this formula automatically incorporates the "apparent" interface trap level density due to interfacial charge nonuniformity. As a result, an "apparent" doping profile is obtained, which is not in agreement with the true doping profile.

This test has been used to show that the distortion of a C-V curve due to negative bias-temperature stress is primarily an interface trap effect,[7] created when the oxide field is large (see Section 15.6). This test also will show whether decoration occurs, as in the sodium drift studies already mentioned, and can provide a check on the uniformity of charging due to an irradiating beam in studies using ultraviolet (UV) light or electron beam radiation.

(b) MOSFET Capacitance Method

A very simple means of distinguishing between interfacial charge nonuniformity and interface traps is to use a MOSFET instead of an MOS capacitor. The MOSFET has a very short minority carrier response time, because these carriers are provided by the source and drain diffusions and not by thermal generation in the depletion region. Consequently, a 1 MHz C-V curve for an ideal MOSFET measured between gate and substrate with source and drain connected to the substrate will look like a low frequency ideal MOS capacitor C-V curve. If interface traps are present, the MOSFET 1 MHz C-V curve will be virtually unaltered from the ideal curve in the vicinity of the capacitance minimum, because the interface traps cannot follow the ac signal, but the minority carriers can. If interfacial charge nonuniformities are present, minority and majority carrier response is affected, and the MOSFET 1 MHz C-V curve will be distorted. In particular, its minimum value will depart from the ideal. The simple observation of the MOSFET 1 MHz C-V curve minimum and comparison with the ideal minimum value, calculated from Section 3.2.4, provides a test for nonuniformity that involves almost no data manipulation.

Some caution is needed, however. If the gate of the MOSFET is made too large in area in an effort to increase its capacitance (and to make measurement easier), the response time of minority carriers is lengthened. Consequently, there is a limit on how large the MOSFET gate area can be made, imposed by the requirement that minority carrier response must be very much faster than interface trap response. In practice, this constraint is not too restrictive, allowing MOSFETs with source-to-drain spacings of 100 μm to be used. Of course, for a rectangular geometry, the *width* of the MOSFET can be increased without degrading minority carrier response time.

6.3 NONUNIFORMITIES DUE TO RANDOM POINT CHARGES

Device-grade interfaces contain localized charges, such as ions or charged interface traps, scattered across the interfacial plane. If individual charges are distributed randomly and no defect decoration occurs, we have seen in the last section that these charges hardly will affect the capacitance of the MOS capacitor in depletion. Fluctuations in interface charge density over distances comparable with a depletion layer width are so uncommon and so weak that the depletion layer width is sensitive only to the average charge density.

However, even in depletion there are some mobile carriers near the Si–SiO$_2$ interface, and these carriers are sensitive to microscopic interfacial charge nonuniformity. Although mobile carrier density near the Si–SiO$_2$ interface is too low to affect the admittance of the MOS capacitor directly, mobile carrier density does affect interface trap response to an applied ac gate voltage variation. Moreover, measurements of interface trap response in depletion contain information not only about interface trap parameters, but also about interface charge nonuniformity.

Charges very far from an interface trap do not affect mobile carrier density in its vicinity as much as charges close to the interface trap. Also, large, dense clusters of charges affect interface trap response more than small, dilute clusters. Thus the localization of interface charge and also its distribution affect interface trap response.

To deal with both the localization and the distribution of interface charge, it is convenient to Fourier analyze the interface charge distribution, representing this charge as a superposition of waves varying sinusoidally with distance across the interfacial plane.[1] Then we can consider the response to each such spatial sinusoid individually and allow the amplitude of this sinusoid to be determined later. This amplitude will incorporate the effects of both localization and distribution.

Consider the effect on an interface trap of a single spatial sinusoid of interface charge for various choices of wavelength. For wavelengths larger than a depletion layer width or so, we already have seen in the discussion of gross nonuniformities that the depletion layer width will be modulated. That is, the mobile carriers beyond the depletion layer edge follow these long wavelength spatial variations, resulting in a parallel array model of *one-dimensional* capacitors. In this array each capacitor reflects only the average of the charge sinusoid over its own area, over a dimension comparable to a depletion layer width. Each capacitor is independent of adjacent capacitors. Moreover, for a Poisson distribution of point charges, the amplitude of such a variation in depletion layer width was shown to be so small that it could be ignored. Thus long wavelength variations in interface charge, comparable to or larger than a depletion layer width, are *screened out* by the action of mobile carriers beyond the depletion layer edge. Such screening limits the *longest* wavelength of interface charge variation that can contribute to interface trap response. To be effective, an interface charge variation must have a wavelength no larger than a depletion layer width or so.*

There is also a *smallest* wavelength. For example it is not the carrier density exactly at the interface trap that affects interface trap response; rather, it is the carrier density averaged over some small volume containing the interface trap. The size of this volume is of the same order as the spread of the wave function of a carrier that occupies a bound state of the interface trap, say, a length λ. Moreover, interfacial charge variations of the same order as λ or smaller tend to be averaged out over a volume of dimension λ because a trapped carrier cannot respond to the resulting rapid variations in potential.

For shallow interface trap levels (near the silicon band edges), comparison with bulk traps suggests that $\lambda \approx 100$–200 Å. For midgap interface trap levels, estimates of λ are sensitive to the atomic model chosen for the interface trap. Because no atomic model has been established, even for the simpler case of bulk trap levels at midgap, estimates of λ based on such models are uncertain. Because the atomic nature of interface traps is not known sufficiently to allow calculation of λ from quantum mechanics, we treat λ as an empirical parameter to be obtained by experiment. Experiments described in Section 7.3.6 suggest $\lambda \approx 75$ Å.

Another reason for introducing a minimum length is that the concept of a *point* charge within a short distance λ of the interface trap becomes doubtful. Such a nearby charge may modify the nature of the interface trap itself. The treatment of interface trap and nearby oxide fixed charge must be combined at a chemical and quantum level, involving bonding and wave functions, and cannot be treated by way of a macroscopic electrostatic model of point charges plus interface traps. That is, an interface trap associated with oxide fixed charges in very close proximity can be viewed as a different species of interface trap. In this way the treatment of interface traps plus point charges at all distances from the interface traps is divided into two problems: (1) the electrostatic treatment of several species of interface traps plus point charges distant λ or more from the interface traps and (2) the quantum mechanical problem of determining what species of interface trap exist, as a result of interface traps plus various configura-

---

*The electrons in the gate also assist in screening out long wavelength interface charge variations. However, the depletion layer width usually is larger than the oxide thickness, so our estimate of the greatest wavelength is conservative.

tions of nearby charges within a distance λ of the interface trap. Problem 2 is far more difficult and should be extended to include the effects on an interface trap of chemical inhomogeneities of the environment, such as the proximity of oxygen atoms or partially oxidized silicon atoms. That is, problem 2 is quantum mechanical and incorporates influences on capture cross section and estimates of energies of interface trap levels. These considerations are explored briefly in Section 6.4, in connection with the tunneling model for interface traps.

In summary, interface trap response is sensitive to a range of the spatial variation of interface charge density, extending from a maximum distance of the order of a depletion layer width down to the minimum distance λ. The upper end (most slowly varying end) of this range is dictated by screening due to mobile carriers, which varies with gate bias. The lower end of the range runs into quantum mechanical considerations. Such considerations determine how short range charge variations smaller than λ alter the behavior of a given type of interface trap. However, at present, too little is known about any interface trap or its neighborhood to allow such a computation.

6.3.1 Band-Bending Fluctuations Near the Interface

Our objective is to examine how interface trap response may reveal interfacial charge nonuniformities. To do this, we would like to know how mobile carriers respond to these nonuniformities, because it is the mobile carrier density that modulates the interface trap capture rate.

The density of mobile carriers has an *exponential* dependence upon band bending. Therefore, we begin with the influence of interfacial charge nonuniformities on band bending, that is, how point charges on the interfacial plane affect band bending near the interfacial plane. The effect of the *average* interface charge density already is included in the one-dimensional MOS capacitor, by treatment of the average charge as a smeared-out sheet that simply alters the flatband voltage. However, because real charges are localized, the actual band bending is not the same everywhere on the Si–SiO$_2$ interface. On a given plane located in the silicon a distance x from the Si–SiO$_2$ interface, band bending will vary from one location to another, depending on the location chosen and the configuration of point charges nearby.

In principle, the actual band bending in the presence of localized interface charges could be calculated exactly, which is a very difficult problem. Instead, we imagine a hierarchy of simpler problems. The simplest of all is the one-dimensional treatment already made. In this case, only averages of band bending and charge density are treated. These averages are determined so they are all self-consistent. That is, the average band bending is consistent with the average oxide fixed charge density and the average mobile carrier density. The simplest next step is to treat self-consistently not only the averages, but also the variances or the fluctuations about these average values. We do this next step. In principle, this hierarchy could be continued to find higher and higher order moments self-consistently. These additional steps are not treated, for reasons given later.

We find the *mean* and the *variance* of band bending, given the spatial distribution of oxide fixed charge and interface traps. If a Gaussian approximation to the distribution is adequate, the distribution of band bending then is specified completely.

To describe the effects of interfacial charge nonuniformities on interface trap measurements, a three-dimensional treatment is necessary. After all, the interface charge distribution is a three-dimensional, random array of point charges (see Fig. 6.1). These charges are scattered across the interfacial plane, which we take as the y-z plane. The resulting potential varies not only in the y-z directions, but also in the x-direction. The y-z variation is expected because of the corresponding y-z variation of the interfacial charge. The x-variation results in part from the screening effect of the gate and the the depletion layer edge and in part from the rapid decrease in multipole contributions to the potential with distance from the originating charges.

For simplicity, we begin with a quasi-uniform model, which is a small fluctuation version of the gross interfacial charge nonuniformity model already discussed, but extended to include interface traps. The quasi-uniform model lays a basis for discussion and agrees with the three-dimensional model in the limit of long wavelength interfacial charge nonuniformities.

(a) Quasi-Uniform Case [1]

We consider an MOS capacitor with a patchwork interface. Each interface patch has its own flatband voltage that varies randomly from patch to patch over a narrow range of values. We assume patch dimensions much larger than a depletion layer width, so the whole capacitor may be viewed as a parallel array of smaller, elementary capacitors. Each elementary capacitor has a uniform flatband voltage.

Within the parallel array model the mean band bending and the variance of band bending can be related to the average densities of oxide fixed charge and interface trap charge and to the variance of these densities. To find the mean and the variance, Poisson's equation is used to relate band bending to interface charge density. Each elementary capacitor satisfies the customary one-dimensional Poisson equation. Integrating this equation once, we find from Gauss's law (6.3) for the nth elementary capacitor

$$C_{ox}V_G - C_{ox}\psi_{sn} = -Q_{fn} - Q_s(\psi_{sn}) - Q_{it}(\psi_{sn}). \qquad (6.29)$$

For any Q_{fn}, (6.29) determines a corresponding ψ_{sn}, with one complication. The interface traps also are distributed randomly. Therefore, $Q_{it}(\psi_{sn})$ also is a statistical quantity, not only because of its dependence on ψ_{sn}, but also because the number of interface traps varies from one elementary capacitor to another. Of course, a variation in the number of interface traps would cause variations in ψ_{sn} by way of (6.29) even if Q_{fn} were uniform.

To find the distribution of band bending, we need the distribution of oxide fixed charge and the distribution of interface traps of each energy level. We suppose that the oxide fixed charge is Poisson (randomly) distributed across the interfacial plane and that the interface traps of each energy also are so distributed, independent of both the interface traps at other energies and the oxide fixed charge.

Mean Band Bending   Let $Q_{it}(N_T, \psi_{sn})$ be the interface trap charge density when N_T interface traps per unit area are present [N_T may represent the array $\{N_{Tl}\}$ for the various interface trap levels l; we suppress this elaboration of symbols here.] If P_{Tn} is the probability that $N_T = N_{Tn}$ and if P_{fn} is the probability that $Q_f = Q_{fn}$, we can average (6.29) by multiplying it with P_{Tn} and P_{fn} and summing over all capacitors. Denoting this average by angular brackets, we find that

$$C_{ox}V_G - C_{ox}\langle\psi_s\rangle = -\langle Q_f\rangle - \langle Q_s\rangle - \langle Q_{it}\rangle. \tag{6.30}$$

Equation (6.30) expresses the mean band bending $\langle\psi_s\rangle$ in a complex way. The functional dependence of $\langle Q_{it}\rangle$ on $\langle\psi_s\rangle$ is not known. To solve (6.30), we introduce the band bending $\bar\psi_s$ corresponding to the average interface charge density and average interface trap level density. The band bending $\bar\psi_s$ is the *most probable* value of band bending because it corresponds to the most probable interface charge density

$$C_{ox}V_G - C_{ox}\bar\psi_s = -\langle Q_f\rangle - Q_s(\bar\psi_s) - \bar Q_{it}(\bar\psi_s) \tag{6.31}$$

where $\bar Q_{it}$ denotes the average of Q_{it} over all interface traps.*

The average band bending $\langle\psi_s\rangle$ and the band bending corresponding to the average charge density $\bar\psi_s$ are different because Q_s and Q_{it} are non-linear functions of band bending. Thus, for example, $Q_{it}(\langle\psi_s\rangle)$ is neither $\langle Q_{it}(\psi_s)\rangle$ nor $\bar Q_{it}(\langle\psi_s\rangle)$.

Unlike (6.30), (6.31) can be solved for $\bar\psi_s$ because Q_s and $\bar Q_{it}$ are functions known from experiment. Subtracting (6.30) for $\langle\psi_s\rangle$ from (6.31), we obtain

$$C_{ox}[\langle\psi_s\rangle - \bar\psi_s] = [\langle Q_s\rangle - Q_s(\bar\psi_s)] + [\langle Q_{it}\rangle - \bar Q_{it}(\bar\psi_s)]. \tag{6.32}$$

Now we make the approximation of low amplitude fluctuations, so that $(\langle\psi_s\rangle - \bar\psi_s)$ is small. Then Q_s and Q_{it} can be expanded in a Taylor series about $\bar\psi_s$. Terms up to second order are retained because our formulation should include all such terms to be consistent. Then (6.32) becomes

$$[C_{ox} + C_s(\bar\psi_s) + \bar C_{it}(\bar\psi_s)][\langle\psi_s\rangle - \bar\psi_s] = -\frac{d(C_s + \bar C_{it})}{d\psi_s}\frac{\langle(\psi_{sn} - \bar\psi_s)^2\rangle}{2} \tag{6.33}$$

---

*Because normal usage of Q_{it} and C_{it} already implies this average over interface trap sites, the bar superscript is omitted in other chapters.

where C_s and $\bar{C}_{it}$ are the silicon and interface trap capacitances per unit area. On the right of (6.33), $\bar{\psi}_s$ may be replaced by $\langle \psi_s \rangle$ because the error introduced is higher than second order. With this replacement, (6.33) becomes

$$\langle \psi_s \rangle = \bar{\psi}_s - \left(\frac{\sigma_s^2}{2}\right)\frac{d(C_s + \bar{C}_{it})}{d\bar{\psi}_s}[C_{ox} + C_s(\bar{\psi}_s) + \bar{C}_{it}(\bar{\psi}_s)]^{-1} \qquad (6.34)$$

where σ_s^2 is the variance of band bending defined by

$$\sigma_s^2 \equiv \beta^2 \langle (\psi_{sn} - \langle \psi_s \rangle)^2 \rangle. \qquad (6.35)$$

With (6.31) for $\bar{\psi}_s$, (6.34) determines $\langle \psi_s \rangle$ up to terms of order σ_s^2. The average band bending $\langle \psi_s \rangle$ is shifted from the band bending corresponding to the mean charge $\bar{\psi}_s$ by an amount dependent on σ_s^2.* This shift affects capture cross sections, which depend exponentially on $\langle \psi_s \rangle$ (see Section 5.5.3).

Variance of Band Bending   Next we find the variance of band bending σ_s^2. Equation (6.30) is used again, but now we use a Taylor series expansion of Q_s and Q_{it} about the mean band bending $\langle \psi_s \rangle$. We retain only linear terms, and show later that higher-order terms lead to corrections of higher order in σ_s^2. Linearizing Q_s and Q_{it}, we obtain

$$Q_s(\psi_{sn}) \cong Q_s(\langle \psi_s \rangle) - C_s(\langle \psi_s \rangle)[\psi_{sn} - \langle \psi_s \rangle] \qquad (6.36)$$

and

$$Q_{it}(N_T, \psi_{sn}) \cong Q_{it}(N_T, \langle \psi_s \rangle) - C_{it}(N_T, \langle \psi_s \rangle)[\psi_{sn} - \langle \psi_s \rangle]. \qquad (6.37)$$

Subtracting (6.29) from (6.30) and using the linearizations (6.36) and (6.37), we find the band-bending variation in the nth capacitor to be

$$[C_{ox} + C_s(\langle \psi_s \rangle) + C_{it}(N_T, \langle \psi_s \rangle)][\psi_{sn} - \langle \psi_s \rangle]$$
$$= Q_{fn} - \langle Q_f \rangle + Q_{it}(N_T, \langle \psi_s \rangle) - \langle Q_{it} \rangle. \qquad (6.38)$$

We can rewrite $C_{it}(N_T, \langle \psi_s \rangle)$ in (6.38) as

$$C_{it}(N_T, \langle \psi_s \rangle) = \bar{C}_{it}(\langle \psi_s \rangle) + [C_{it}(N_T, \langle \psi_s \rangle) - \bar{C}_{it}(\langle \psi_s \rangle)] \qquad (6.39)$$

where $\bar{C}_{it}$ is the average over interface traps at fixed band bending

$$\bar{C}_{it}(\langle \psi_s \rangle) \equiv \sum_n P_{Tn} C_{it}(N_{Tn}, \langle \psi_s \rangle).$$

The term in the square brackets in (6.39) is a fluctuation in C_{it} and hence small. In (6.38) this fluctuation multiplies the fluctuation in band bending and thus is a second-order correction that can be dropped. As a result,

---

*Experimentally it is $\langle \psi_s \rangle$ that is found by integration of the low frequency C-V curve from a quasi-uniform capacitor.[7] In principle, (6.34) must be used to obtain $\bar{\psi}_s$ before comparison with theoretical $\bar{\psi}_s$ versus V_G obtained from (6.31).

(6.38) becomes

$$\psi_{sn} - \langle \psi_s \rangle = \frac{\delta Q_f + \delta Q_{it}}{C_{ox} + C_s(\langle \psi_s \rangle) + \bar{C}_{it}(\langle \psi_s \rangle)} \tag{6.40}$$

where $\delta Q_f \equiv Q_f - \langle Q_f \rangle$, $\delta Q_{it} \equiv Q_{it}(N_T, \langle \psi_s \rangle) - \langle Q_{it} \rangle$ are the fluctuations in oxide fixed charge and interface trap charge densities for a fixed band bending. From (6.35), the *variance* of band bending is found by squaring (6.40) and averaging. We find, expressing potential in kT/q, that the variance in (6.35) becomes

$$\sigma_s^2 = \beta^2 [C_{ox} + C_s + \bar{C}_{it}]^{-2} [\langle \delta Q_f^2 \rangle + \langle \delta Q_{it}^2 \rangle]. \tag{6.41}$$

For a Poisson distribution [see (6.10) and (6.12)], we obtain

$$\langle \delta Q_f^2 \rangle = \frac{q \langle Q_f \rangle}{\alpha} \tag{6.42}$$

and

$$\langle \delta Q_{it}^2 \rangle = \frac{q \langle Q_{it} \rangle}{\alpha} \tag{6.43}$$

where to lowest order in the fluctuations, $\langle Q_{it} \rangle \approx \bar{Q}_{it}(\langle \psi_s \rangle)$ and is evaluated at the mean band bending $\langle \psi_s \rangle$ from (6.34). Substituting (6.42) and (6.43) into (6.41), we obtain the final expression for the variance in a quasi-uniform model

$$\sigma_s^2 = \sigma_q^2 [C_{ox} + C_s(\langle \psi_s \rangle) + \bar{C}_{it}(\langle \psi_s \rangle)]^{-2} \tag{6.44}$$

where the variance of interface charge density σ_q^2 is

$$\sigma_q^2 = \frac{q \langle Q_f \rangle}{\alpha} + \frac{q \bar{Q}_{it}(\langle \psi_s \rangle)}{\alpha}. \tag{6.45}$$

Equation (6.44), the expression for the variance of band bending, was derived first by Nicollian and Goetzberger[5] without interface traps.

In using (6.44) for σ_s^2, two precautions must be noted: (1) the analysis is based on *small* fluctuations, so (6.44) is inapplicable to the work of Castagné and Vapaille,[6] for example; and (2) (6.44) applies only for characteristic areas α of dimensions comparable to, or greater than, a depletion layer width, so (6.44) is inapplicable to the treatment of fluctuations in band bending caused by a Poisson distribution of localized interface charges. As this last problem is the one of interest in measurements of interface traps, we must extend the preceding treatment of mean and variance of band bending to include short range nonuniformities. This can be done in much the same way as the quasi-uniform case, but now the linearized charge-potential relation involves localized charges, rather than large-area sheets of charge.

(b) Three Dimensional Treatment[1,3]

The treatment of the quasi-uniform case now is extended to a three-dimensional treatment. First we find the mean band bending and then the variance of band bending.

Mean Band Bending   The mean band bending $\langle v_s \rangle$, is related to the band bending $\bar{v}_s$ corresponding to the mean oxide fixed charge density and the mean interface trap level density. The band bending $v(r, x) = \beta \psi(r, x)$ (in thermal units) satisfies the three-dimensional Poisson equation

$$\nabla^2 v(\mathbf{r}, x) = \lambda_n^{-2}[\exp(v) - 1] + \frac{\beta}{\epsilon_s}[Q_f(\mathbf{r}, x) + Q_{it}(\mathbf{r})]\delta(x) \qquad (6.46)$$

where $\mathbf{r} \equiv (y, z)$. Here, $Q_f(\mathbf{r})$ is the oxide fixed charge density approximated by an array of N point charges located at positions $\{\mathbf{R}_n\}$ on the interface

$$Q_f(\mathbf{r}) = q \sum_{n=1}^{N} \delta(\mathbf{r} - \mathbf{R}_n). \qquad (6.47)$$

If these charges are Poisson distributed and if a Gaussian approximation to the Poisson distribution is adequate, the average of (6.46) over the Si–SiO$_2$ interface becomes

$$\frac{d^2}{dx^2}\langle v(x) \rangle = \lambda_n^{-2}\left[\exp\left(\langle v \rangle + \frac{\sigma^2}{2}\right) - 1\right] + \frac{\beta}{\epsilon_s}[\langle Q_f \rangle + \langle Q_{it} \rangle]\,\delta(x) \qquad (6.48)$$

where we have used the result

$$\langle \exp(v) \rangle = [2\pi\sigma^2(x)]^{-1/2} \int_{-\infty}^{\infty} dv\, \exp\left[-\frac{(v - \langle v \rangle)^2}{2\sigma^2}\right] \exp(v)$$

$$= \exp(\langle v \rangle)\, \exp\frac{\sigma^2(x)}{2} \qquad (6.49)$$

where $\sigma^2(x)$ is the variance of band bending on the plane $x = $ constant.* To evaluate (6.48), we introduce the band bending $\bar{v}$ corresponding to the mean oxide fixed charge density $\langle Q_f \rangle$

$$\frac{d^2\bar{v}}{dx^2} = \lambda_n^{-2}[\exp(\bar{v}) - 1] + \frac{\beta}{\epsilon_s}[\langle Q_f \rangle + \bar{Q}_{it}(\bar{v})]\,\delta(x). \qquad (6.50)$$

Subtracting (6.50) from (6.48), an equation for the difference Δ defined by

$$\Delta(x) \equiv \langle v(x) \rangle - \bar{v}(x) \qquad (6.51)$$

---

*To establish (6.49), the dummy variable of integration is changed to $\eta = [v - \langle v \rangle - \sigma^2(x)]$. The integral then becomes $\exp[\langle v \rangle + \sigma^2(x)/2][2\pi\sigma^2]^{-1/2} \int_{-\infty}^{\infty} d\eta\, \exp\{-\eta^2/[2\sigma^2(x)]\} = \exp[\langle v \rangle + \sigma^2(x)/2]$.

is obtained, namely

$$\frac{d^2\Delta}{dx^2} = \lambda_n^{-2} \exp(\bar{v}) \left[\exp\left(\Delta + \frac{\sigma^2}{2}\right) - 1 \right] + \frac{\beta}{\epsilon_s} \left[\bar{C}_{it}(\bar{v}_s)\Delta_s + \bar{C}'_{it}(\bar{v}_s) \frac{\sigma_s^2}{2} \right] \delta(x)$$

(6.52)

where $[\langle Q_{it} \rangle - \bar{Q}_{it}]$ has been approximated by a Taylor series to second order [see (6.33)], where $\bar{C}'_{it} \equiv d\bar{C}_{it}(\bar{v}_s)/d\bar{v}_s$ and where Δ_s and σ_s^2 respectively denote Δ and σ^2 evaluated at the interface $x = 0$. Equation (6.52) can be integrated once to obtain

$$-\frac{d\Delta}{dx} = \lambda_n^{-2} \int_x^\infty dx \, \exp(\bar{v}) \left[\exp\left(\Delta + \frac{\sigma^2}{2}\right) - 1 \right]; \qquad (x > 0).$$

(6.53)

Because $d^2\Delta/dx^2$ vanishes in the oxide $x < 0$ assuming no oxide trapped charge, Δ must be linear in the oxide and vanish at the gate where $\langle v \rangle = \bar{v}$. Therefore

$$\Delta = \Delta_s \left(1 + \frac{x}{x_o} \right).$$

(6.54)

The discontinuity condition on $(d\Delta/dx)$ at $x = 0$, from the oxide $(x = 0^-)$ to the silicon $(x = 0^+)$, is

$$\epsilon_s \frac{d\Delta}{dx}\bigg|_{0^+} - \epsilon_{ox} \frac{d\Delta}{dx}\bigg|_{0^-} = \bar{C}_{it}(\bar{v}_s)\Delta_s + \bar{C}'_{it}(\bar{v}_s) \frac{\sigma_s^2}{2}.$$

(6.55)

Using (6.53) and (6.54) and treating Δ and σ^2 as small so $[\exp(\Delta + \sigma^2/2) - 1]$ $\approx \Delta + \sigma^2/2$, (6.55) becomes

$$\left[\epsilon_s \lambda_n^{-2} \int_0^\infty dx \, \exp(\bar{v}) \frac{\Delta(x)}{\Delta_s} + \frac{\epsilon_{ox}}{x_o} + \bar{C}_{it}(\bar{v}_s) \right] \Delta_s$$

$$= -\frac{1}{2} \bar{C}'_{it}(\bar{v}_s)\sigma_s^2 - \frac{1}{2} \epsilon_s \lambda_n^{-2} \int_0^\infty dx \, \exp(\bar{v})\sigma^2(x).$$

That is

$$\Delta_s \equiv \langle v_s \rangle - \bar{v}_s = -\frac{1}{2} \left[\bar{C}'_{it}(\bar{v}_s) \, \sigma_s^2 + \epsilon_s \lambda_n^{-2} \int_0^\infty dx \, \exp(\bar{v}) \, \sigma^2(x) \right] [C_{ox} + C_s + \bar{C}_{it}(\bar{v}_s)]^{-1}$$

(6.56)

which agrees with (6.34) in the limit of long wavelengths where $\sigma^2(x) \to \sigma_s^2[\delta\bar{v}(x)/\delta\bar{v}_s]^2$. To obtain (6.56), we have used the results

$$\frac{\Delta(x)}{\Delta_s} = \frac{\delta\bar{v}(x)}{\delta\bar{v}_s}$$

and

$$C_s(\bar{v}_s) = \epsilon_s \lambda_n^{-2} \int_0^\infty dx \, \exp[\bar{v}(x)] \frac{\delta\bar{v}(x)}{\delta\bar{v}_s}.$$

(6.57)

In the expression (5.75) for capture cross section, the factor $\exp(\langle v_s \rangle)$ occurs. However, integration of a low frequency C-V curve produces $\langle v_s \rangle$ from (6.34), which includes the influence of only long wavelength interfacial charge nonuniformities. Therefore, in principle, one must use (6.34) to find $\bar{v}_s$ and then use (6.56) to obtain the correct factor $\exp(\langle v_s \rangle)$ in a capture cross section determination. In practice, this correction has never been made, thus introducing an error in the available capture cross section determinations.

Variance of Band Bending Now we turn to the evaluation of the variance σ_s^2. Assume that the band bending on plane x due to one point charge at the Si–SiO$_2$ interface is $G(R, x)$, where

$$G(R, x) = \frac{q}{4\pi\bar{\epsilon}} \frac{1}{(R^2 + x^2)^{1/2}}; \quad (R < R_o)$$
$$= 0; \qquad\qquad\qquad (R > R_o). \qquad (6.58)$$

Equation (6.58) is derived by solving the electrostatic problem of a point charge at the interface between two dielectrics. One dielectric $(x < 0)$ is the silica with dielectric constant ϵ_{ox}, and the other $(x > 0)$ is the silicon with dielectric constant ϵ_s. The parameter $\bar{\epsilon}$ is

$$\bar{\epsilon} = \frac{\epsilon_{ox} + \epsilon_s}{2}. \qquad (6.59)$$

The other parameter in (6.58), R_o, is the radius at which screening becomes effective. That is, the gate, the bulk semiconductor, the inversion layer, and the interface traps all conspire to reduce the range of influence of the point charge. Equation (6.58) makes this screening zero for $R < R_o$ and completely effective for $R > R_o$. Physically, such an abrupt change is unrealistic. However, as shown in the following paragraphs, this approximation is adequate for estimating σ_s^2 near the Si–SiO$_2$ interface.

In (6.58), $(R^2 + x^2)^{1/2}$ is the separation between the point charge and the observation point at which band bending is to be evaluated. For simplicity, we choose the origin for R at the observation point. Then R becomes the radial coordinate of the point charge at the Si–SiO$_2$ interface as shown in Fig. 6.7.

Now consider computing the variance of band bending for a distribution of charge per unit area, $Q(\mathbf{R})$ on the interfacial plane. If (6.58) is used, the band bending is

$$v(x) = \left(\frac{q}{kT}\right) \frac{1}{4\pi\bar{\epsilon}} \int d^2R_1 \frac{Q(\mathbf{R}_1)}{(R_1^2 + x^2)^{1/2}}. \qquad (6.60)$$

The integral in (6.60) extends over a disk on the interfacial plane of radius R_o. The dummy variable of integration has been labeled R_1 for later convenience when the variance of band bending is calculated. The

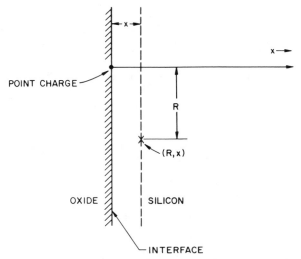

Fig. 6.7 Point charge at the Si–SiO₂ interface and observation point in the silicon distant $(R^2 + x^2)^{1/2}$ from the charge. The band bending at the observation point is to be calculated. After Brews[3].

departure of $v(x)$ from the mean value $\langle v \rangle$ is now

$$v(x) - \langle v \rangle = \left(\frac{q}{kT}\right) \frac{1}{4\pi\bar{\epsilon}} \int d^2R_1 \frac{Q(\mathbf{R}_1) - \langle Q \rangle}{(R_1^2 + x^2)^{1/2}}. \tag{6.61}$$

The variance of $v(x)$ is then

$$\sigma_s^2(x) \equiv \langle [v(x) - \langle v \rangle]^2 \rangle$$
$$= \left(\frac{q}{kT}\right)^2 \frac{1}{(4\pi\bar{\epsilon})^2} \int d^2R_1 \int d^2R_2 \frac{\langle [Q(\mathbf{R}_1) - \langle Q \rangle][Q(\mathbf{R}_2) - \langle Q \rangle]\rangle}{(R_1^2 + x^2)^{1/2}(R_2^2 + x^2)^{1/2}}. \tag{6.62}$$

For a Poisson distribution of interface charge

$$\langle [Q(\mathbf{R}_1) - \langle Q \rangle][Q(\mathbf{R}_2) - \langle Q \rangle]\rangle = q\langle Q \rangle \delta(\mathbf{R}_1 - \mathbf{R}_2). \tag{6.63}$$

Equation (6.63) is derived in Appendix VI. Substituting (6.63) into (6.62) yields

$$\sigma_s^2(x) = \left(\frac{q}{kT}\right)^2 \frac{q\langle Q \rangle}{8\pi\bar{\epsilon}^2} \int_0^{R_o} dR\, R \frac{1}{(R^2 + x^2)}$$
$$= \left(\frac{q}{kT}\right)^2 \frac{q\langle Q \rangle}{16\pi\bar{\epsilon}^2} \ln\left[1 + \frac{R_o^2}{x^2}\right]. \tag{6.64}$$

Equation (6.64) is the required variance of band bending on the plane $x = $ constant, provided that R_o can be evaluated.

The radius R_o is chosen so that (6.58) provides the bext approximation to screening that is possible within the limitations of the assumed abrupt cutoff at $R = R_o$.[3] Comparison is made with a more complex calculation of G from Appendix VII. Then R_o is chosen so that the average value of (6.58) over the Si–SiO$_2$ interface agrees with the average value of the G from Appendix VII. Using (6.58), we obtain

$$\langle G \rangle = \frac{q}{2\bar{\epsilon}} \int_0^{R_o} dR\, R\, \frac{1}{(R^2 + x^2)^{1/2}}$$

$$= \frac{q}{2\bar{\epsilon}} [(R_o^2 + x^2)^{1/2} - x]. \qquad (6.65)$$

From Appendix VII

$$\langle G \rangle = \int d^2R\, G(R, x)$$

$$= \int d^2R \int \frac{d^2k}{(2\pi)^2} \exp(i\mathbf{k} \cdot \mathbf{R}) G_k(x)$$

$$= G_{k=0}(x) \qquad (6.66)$$

where we have taken $x' = 0$, $R = |\mathbf{r} - \mathbf{r}'|$ in equation (2) in Appendix VII. Equations (15) and (19) in Appendix VII, in the limit $k \to 0$, provide

$$G_{k=0}(x) = \left(\frac{q}{\epsilon_s}\right) \frac{\epsilon_s x_o k(x_D - x)}{\epsilon_{ox}kx_D + \epsilon_s kx_o + (\bar{C}_{it} + C_I)kx_o x_D}$$

$$= \frac{q(1 - x/x_D)}{C_{ox} + C_D + \bar{C}_{it} + C_I} \qquad (6.67)$$

where the oxide capacitance per unit area is $C_{ox} = \epsilon_{ox}/x_o$, the depletion layer capacitance per unit area is $C_D = \epsilon_s/x_D$, $\bar{C}_{it}$ = interface trap capacitance per unit area, and C_I = inversion layer capacitance per unit area. For $x \ll R_o$, (6.65) provides

$$\langle G \rangle \cong \left(\frac{q}{\epsilon_s + \epsilon_{ox}}\right) R_o \left(1 - \frac{x}{R_o}\right). \qquad (6.68)$$

Comparing (6.68) and (6.67) for $x \ll R_o$, x_D, we obtain

$$R_o = \frac{\epsilon_s + \epsilon_{ox}}{C_{ox} + C_D + \bar{C}_{it} + C_I}. \qquad (6.69)$$

Consequently, the variance in (6.64) becomes[1]

$$\sigma_s^2(x) = \left(\frac{q}{kT}\right)^2 \frac{q\langle Q \rangle}{4\pi(\epsilon_s + \epsilon_{ox})^2} \ln\left[1 + \left(\frac{C_x}{C_{ox} + C_s + \bar{C}_{it}}\right)^2\right] \qquad (6.70)$$

with C_x given by

$$C_x \equiv \frac{\epsilon_s + \epsilon_{ox}}{x} \qquad (6.71)$$

and the semiconductor capacitance C_s by

$$C_s = C_D + C_I. \qquad (6.72)$$

For large x, (6.70) becomes (6.44) with $\alpha = 4\pi x^2$. Equation (6.70) also can be established directly, using G from Appendix VII.[1,3]

The value of x that should be used to characterize a measurement will depend on the distance from the Si–SiO$_2$ interface probed by the measurement. Thus, for interface trap loss measurements, x is approximately the distance a trapped carrier wave function extends into the silicon. For an inversion layer response measurement, such as measurement of the MOSFET source-to-drain current, x should be the average distance of carriers in the inversion layer from the Si–SiO$_2$ interface.[2,3] No meaningful x can be smaller than the cutoff parameter λ in Section 6.3.

Equation (6.70) shows that $\sigma_s^2(\lambda)$ has a dependence on interface traps. The major dependence on interface traps is through the interface charge density term $\langle Q \rangle$, which includes both oxide fixed and interface trap charge densities. Both oxide fixed and interface trap charge densities are treated on an equal footing here. We imagine that interface trap sites with a given energy level are distributed randomly across the interface. As a result, interface trap sites *and* levels are randomly distributed. Also, interface trap charge will be randomly distributed, as will oxide fixed charge. Then $\langle Q \rangle$ in (6.70) is given by

$$\langle Q \rangle = \bar{Q}_{it}(\langle v \rangle) + \langle Q_f \rangle. \tag{6.73}$$

A secondary dependence on interface traps in (6.70) is by way of $\bar{C}_{it}$ in the logarithm. This term reflects the screening effect of interface traps. When a fluctuation occurs near an interface trap, its energy is shifted by this band-bending variation, thereby changing the occupancy of the interface trap in a direction tending to screen out the fluctuation. Statistically, on the average over the Si–SiO$_2$ interface, (6.70) shows that screening due to this occupancy change will be governed by the average interface trap capitance $\bar{C}_{it}$.

In summary, we have solved for potential in such a way that for *small* departures from uniform interface charge density, both the mean and the variance of potential are self-consistent. It is quite possible to extend this work to higher order moments. However, higher order moments of the potential lead to more detailed dependence on the statistics of the interface charge distribution. In particular, the non-Gaussian nature of the statistics enters, as does the short-range screening of positive and negative point charges. Therefore, the dependence of these higher-order moments on λ is strong, and quantum mechanical aspects of screening enter.

Information about higher-order moments is measurable only near or in inversion and accumulation, where the spatial variations of the ac band bending also become important. Therefore, this problem becomes complex, and its treatment is unwarranted in view of the experimental difficulties involved in determining the resulting parameters.

6.3.2   Discussion of $\sigma_s^2(x)$

Using (6.70), we now discuss the behavior of σ_s^2 as a function of bias for various values of doping density, oxide thickness, and so forth. Following this general discussion, (6.70) is related to theoretical work on σ_s^2 at low temperatures and high fields. This comparison shows the generality of (6.70), although experimental confirmation of (6.70) to date is confined to conductance measurements in the range 90–300°K.

(a)   General Behavior

Figure 6.8 is a plot of the square root of the variance, (6.70), or the standard deviation, illustrating the dependence of the standard deviation of band bending on the mean band bending for several values of x. For C_s, the low frequency silicon surface capacitance from (3.35) is used

$$C_s = C_{FBS}\left\{[1 - \exp(-v_s)] + \left(\frac{n_i}{N_A}\right)^2 [\exp(v_s) - 1]\right\}$$

$$\times \left\{2\left[\exp(-v_s) + v_s - 1 + \left(\frac{n_i}{N_A}\right)^2 [\exp(v_s) - v_s - 1]\right]\right\}^{-1/2} \qquad (6.74)$$

where $C_{FBS} \equiv \epsilon_s/\lambda_p$. Strictly speaking, our derivation of (6.70) does not extend into accumulation, but we assume that this limitation could be removed, as already done for inversion. Also, (6.74) employs the band bending v_s required by the derivation of (6.70).

Figure 6.8 shows that the fluctuations of band bending peak just before the onset of strong inversion, $v_s = 2\beta\phi_B$. At this point the depletion layer width is near its greatest value, reducing screening by *majority* carriers to

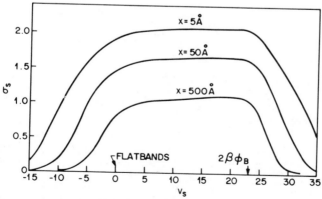

Fig. 6.8   Standard deviation of band bending σ_s as a function of band bending v_s with x as parameter. The dependence on x is not strong; a variation of two decades in x results in a kT/q change in band-bending fluctuations. Parameters are $C_{it} = 0$, $Q_f/q = 2 \times 10^{11}$ cm^{-2}, $x_o = 1000$ Å, $N_B = 10^{15}$ cm^{-3}, and $kT/q = 0.025$ V.

its lowest point. For larger values of band bending, corresponding to inversion, *minority* carrier screening reduces $\sigma_s^2(x)$. Similarly, if band bending is reduced, depletion layer width narrows, increasing screening by majority carriers until $\sigma_s(x)$ is near zero for band bending in accumulation.

Figure 6.8 shows that the dependence of $\sigma_s(x)$ on x is weak. This weak dependence has two consequences: (1) uncertainty in the value of x does not introduce much uncertainty in the theoretical prediction; and (2) as shown in Chapter 7, the theoretical result for $\sigma_s(x)$ fits experiment. This fit is meaningful because adjustment of the uncertain parameter x cannot improve the fit significantly. In what follows, we adopt $x = \lambda$, for interface traps.* A value of 5 Å seems a rather low value for λ, although conceivably a midgap interface trap level could have a wave function extending over only an interatomic spacing. A value of 500 Å is rather large, even for shallow interface trap levels. A value of 50 Å is probably reasonable. The work of Ziegler (discussed in Section 7.3.5) suggests 75 Å, which is about this size. The exact value of λ is not critical.

Silicon doping density naturally affects depletion layer width, and in Fig. 6.9 σ_s is plotted against v_s for several values of doping density. The reduction of doping density increases σ_s because depletion layer width increases. Reduction of doping density also delays the reduction of σ_s to a point further into accumulation because the Debye length is increased, thus lowering flatband capacitance. Also, inversion occurs earlier at lower doping density. Finally, lower doping density leads to a flatter σ_s versus v_s curve in depletion, because the oxide capacitance (screening by the gate) limits σ_s for large depletion layer widths.

In Fig. 6.10 the effect of oxide thickness is explored further. Here, for a gate bias corresponding to midgap, $(v_s = \beta\phi_B)$, the σ_s versus x_o curve is

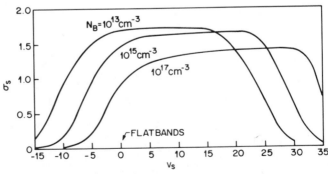

Fig. 6.9 Square root of the variance σ_s versus band bending v_s for several different doping densities; $x = 50$ Å and the other parameters are the same as those in Fig. 6.8.

---

*In inversion layer transport (not to be discussed) we would adopt $x =$ inversion layer depth.[2,3]

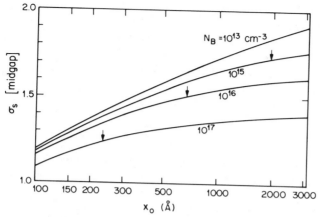

Fig. 6.10 Square root of variance σ_s for a band bending at midgap $v_s = u_B$ versus oxide thickness with doping density as parameter. Arrows indicate the oxide thickness for which $C_{ox} = C_s$. Below this thickness the gate dominates the screening. Parameters are the same as those in Figs. 6.8 and 6.9.

plotted with doping density as parameter. The arrows indicate the oxide thickness for which $C_{ox} = C_s$. For oxide thicknesses below this value, screening by the gate is dominant. For larger thicknesses, screening due to majority carriers at the depletion layer edge is dominant.

(b) Fluctuations in Strong Inversion

The behavior of σ_s^2 seen in Figs. 6.8–6.10 is observed most easily in depletion and weak inversion. The relation of σ_s^2 to measurement in these regimes is described in Section 6.4. However, some theoretical work has concentrated on the behavior of σ_s^2 in strong inversion, for temperatures near absolute zero. This work was motivated by basic questions about two-dimensional systems, and by the hope that transport measurements on inversion layers at low temperatures could illustrate new phenomena, such as Anderson localization, Wigner lattices, Mott hopping, and so forth. Unfortunately, the experimental results are very complex, and none of these matters has been explored successfully.

However, this theoretical work on σ_s^2 can be related to (6.70). In strong inversion the silicon capacitance C_s becomes very large and is dominated by the inversion layer capacitance C_I. For this case, $C_s \approx C_I \gg C_{ox}$, $\bar{C}_{it}$, and C_x. Hence the logarithm in (6.70) can be expanded, and σ_s^2 becomes

$$\sigma_s^2(x) = \left(\frac{q}{kT}\right)^2 \frac{q\langle Q\rangle}{4\pi(\epsilon_s + \epsilon_{ox})^2}\left(\frac{C_x}{C_I}\right)^2. \qquad (6.75)$$

If (6.71) is used for C_x, (6.75) becomes

$$\sigma_s^2(x) = \left(\frac{q}{kT}\right)^2 \frac{q\langle Q\rangle}{4\pi x^2}\left(\frac{1}{C_I}\right)^2. \qquad (6.76)$$

To compare (6.76) with other work, an expression for C_I is needed. In the extreme case of absolute zero and high fields, the quantum nature of the inversion layer becomes evident, and the minority carrier wave functions are plane waves for directions parallel to the interface. For the direction normal to the interface, where the carriers are trapped in the potential well that holds the carriers in the inversion layer, the wave functions are standing waves. The energy of a carrier E is now

$$E = E_o + \frac{p^2}{2m^*} \tag{6.77}$$

where E_o is the energy of the transverse standing wave and the remainder is the energy of the traveling wave, with momentum p and effective mass m^*.

The density of states per unit energy interval above E_o corresponding to (6.77), $N(E)$, is now a constant

$$\begin{aligned} N(E) &= \frac{2}{h^2} \int dp_x \int dp_y \, \delta\left(E - \frac{p^2}{2m^*}\right) \\ &= \frac{4\pi}{h^2} \int_0^\infty dp \, p\, \delta\left(E - \frac{p^2}{2m^*}\right) \\ &= \frac{4\pi m^*}{h^2}; \qquad (E \geq E_o) \end{aligned} \tag{6.78}$$

where h is Planck's constant. The value $N(E)$ is zero for energies below E_o, which is the band edge energy at the interface.* Thus, to within some zero-point constant, for p-type silicon, we have

$$E_o = E_c - q\psi_s \tag{6.79}$$

where E_c is the band edge energy in the bulk and ψ_s is the band bending.

If (6.78) and (6.79) are used, the inversion layer carrier density per unit area N_I can be derived. For a band bending that places the Fermi level above the band edge N_I becomes†

$$N_I = \int_{E_c - q\psi_s}^{E_F} \left(\frac{4\pi m^*}{h^2}\right) dE. \tag{6.80}$$

Hence the inversion layer capacitance per unit area is

$$C_I = \frac{d(qN_I)}{d\psi_s} = \frac{4\pi m^* q^2}{h^2}. \tag{6.81}$$

---

*Equation (6.78) assumes that only one standing wave is occupied. At lower band bendings or at temperatures above $T = 0$, many standing waves play a role.

†No integral over the depth of the inversion layer appears because this integral is unity, as it is the normalization integral of the standing wave portion of the wave function.

Substituting (6.81) in (6.76) and using $\sigma_s^2 \equiv \beta^2 \langle (\psi_s - \langle \psi_s \rangle)^2 \rangle$, we find

$$\langle (\psi_s - \langle \psi_s \rangle)^2 \rangle = \frac{1}{64\pi^3} \left(\frac{h^2}{m^* q} \right)^2 \frac{\langle Q \rangle / q}{x^2} \qquad (6.82)$$

in agreement with Brews,[3] Stern,[18] and Arnold.[19] Equation (6.82) shows that in the limit of inversion strong enough to make the inversion layer two dimensional, the band-bending fluctuations become temperature and bias independent. This was noted by Arnold.[20]

The residual value of the fluctuations given by (6.82) can be evaluated for electrons using $m^* = 0.19 m_o$ (m_o is the free electron mass) and $x \approx$ 75 Å. The result is

$$\langle (\psi_s - \langle \psi_s \rangle)^2 \rangle = \left\langle \frac{Q}{q} \right\rangle / (4.4 \times 10^{16} \, \text{cm}^{-2} \, \text{eV}^{-2}). \qquad (6.83)$$

Equation (6.83) predicts a standard deviation of band bending of about 5 mV at $\langle Q/q \rangle = 10^{12}$ cm^{-2}, a large value for the interface charge.

From this discussion, the behavior of σ_s^2 shown in Figs. 6.8–6.10, calculated ignoring degeneracy and quantum effects, should be very slightly altered in strong inversion and in accumulation. For example, for 10^{12} oxide fixed charges per square centimeter, instead of falling to zero at the extremes of bias, σ_s should fall to a value of approximately $5 \, \text{mV}/(kT/q) \approx$ 0.2 at room temperature.

6.4 RELATION OF σ_s^2 TO MEASUREMENT

The variance of band bending is observed most directly through its effect on interface traps. Because interface traps are located where band-bending fluctuations are largest (i.e., at the Si–SiO$_2$ interface) and because their response depends *exponentially* on band bending, interface traps are very sensitive to band-bending fluctuations. We derive the equivalent circuit relating this interface trap response to the measured MOS capacitor admittance.

In addition to their effect on interface traps, band-bending fluctuations also affect the capacitance directly, although to a lesser extent. The effect on capacitance is largest in accumulation and inversion, where band bending affects mobile carrier densities exponentially and where nearness of carriers to the Si–SiO$_2$ interface accents sensitivity to band-bending fluctuations. Unfortunately, analysis is complicated in these bias regimes for three reasons. First, when mobile carrier densities are large, screening of interface charge occurs over small distances. Therefore, screening is sensitive to the atomic details of the charges and of the carrier wave functions, bringing in quantum mechanical effects. Second, screening becomes a *nonlinear* phenomenon because the potential near a localized charge is large and rapidly varying with distance. In our analysis based on

Fourier decomposition of interface charge, a cutoff parameter λ has been introduced to avoid such problems. However, this cutoff approach is successful only for the mean and variance of potential, because higher moments of the distribution of potential depend sensitively on λ. These higher-order moments become important in accumulation and inversion, where the Gaussian approximation to the distribution fails. Third, statistical considerations are complicated because the number of charges within a screening radius is small. As a result, band-bending fluctuations are dominated by small clusters of charge. On such a small scale, the charges may not be randomly distributed. For example, such nearby charges may interact during oxide growth or annealing, and the configuration of a small group of ions may be governed by these interactions, rather than being a random arrangement.

For these reasons, the bias ranges most accessible to analysis are depletion and weak inversion. This is the bias range used in small-signal admittance measurements of interface traps. Therefore, we concentrate on deriving the admittance of the MOS capacitor in depletion and the relation of σ_s^2 to this admittance.

If we were to include band-bending fluctuations in the procedure used to find the admittance in Section 5.3, two complications would arise. First, equilibrium interface trap occupancy would vary from point to point across the interface, even in the case of interface traps with a single energy level, because the interface trap level $(v_T + v_s)$ (v_T = interface trap level energy in kT/q, v_s = band bending with *no* fluctuations) becomes $v_T + v_s(\mathbf{R})$, where $\mathbf{R}$ is the position of the interface trap and $v_s(\mathbf{R})$ varies with $\mathbf{R}$ because of the random interface charge. Therefore, the first problem is to describe the *equilibrium* interface trap charge density.

In the previous section the mean and variance of band bending were derived for the case of small fluctuations. This was an equilibrium (dc) analysis and depended on linearization of the dc charge-potential relation. In this way, the first problem (i.e., determination of the equilibrium interface trap occupancy) was solved approximately.

The second problem is one of determining ac band bending at the interface trap site. This ac band bending modulates the mobile carrier density at the interface trap site, driving electron or hole capture and determining the energy loss. In Chapter 5 this ac band-bending variation was δv_s. Now it becomes $\delta v_s(\mathbf{R})$ because the small-signal band bending also may vary from point to point across the Si–SiO$_2$ interface. For long wavelength interface charge nonuniformities, the variation with position of $\delta v_s(\mathbf{R})$ is found to be important. For short wavelength nonuniformities, of dimension less than a depletion layer width, variation of δv_s with position is found to be negligible in depletion or weak inversion.

Now we attack the ac problem of determining δv_s. First, the capacitance with oxide fixed charge but without interface traps is discussed. The equivalent circuit is derived, and the predicted capacitance is discussed.

Second, interface traps are introduced and the equivalent circuit modified accordingly. With this equivalent circuit, admittance measurements on the MOS capacitor can be related to interface trap properties and to the band-bending fluctuations. The results from analyzing experiments using this equivalent circuit are presented in Section 7.3.

6.4.1 Equivalent Circuit without Interface Traps

As an introduction to the equivalent circuit, consider an MOS capacitor with localized oxide fixed charge, but with no interface traps. We begin by reexamining Fig 6.3 to compare the effects of short and long wavelength interfacial charge nonuniformities on this MOS capacitor. Figure 6.3 shows that in depletion, only the longer wavelength components of the interface charge variations influence the depletion layer width. Therefore, the capacitance also will be affected only by these longer wavelength components of the interface charge variation because the capacitance in depletion depends on mobile carriers near the depletion layer edge and their response to an ac gate voltage.

Suppose that the interface charge were distributed quasi-uniformly, over patches of dimension much greater than a depletion layer width. Then the entire MOS capacitor could be partitioned, as shown in Fig. 6.3b, with a parallel array of independent, one-dimensional MOS capacitors, as in Section 6.3.1(a). The probability P_n of an oxide fixed charge density, Q_{fn} in one of these elementary capacitors leads to an average surface charge density in the silicon $\langle Q_s \rangle$ given by

$$\langle Q_s \rangle = \sum_{n=1}^{N} P_n Q_s(v_{sn}) \tag{6.84}$$

where v_{sn} is the band bending in thermal units in a capacitor with $Q_f = Q_{fn}$. Differentiating (6.84) with respect to gate bias, the measured capacitance C_m is given by

$$
\begin{aligned}
C_m &\equiv \frac{-d\langle Q_s \rangle}{dV_G} \\
&= -\sum_{n=1}^{N} P_n \frac{d\beta Q_s}{dv_{sn}} \frac{dv_{sn}}{d\beta V_G} \\
&= \sum_{n=1}^{N} P_n \frac{C_s(v_{sn})C_{ox}}{C_s(v_{sn}) + C_{ox}} \\
&= \sum_{n=1}^{N} P_n C_{m,n}
\end{aligned}
\tag{6.85}
$$

where $\beta \equiv q/(kT)$ and $C_{m,n}$ is the capacitance that would be measured on a uniform MOS capacitor with a uniform charge sheet of oxide fixed charge density Q_{fn}. In (6.85), the low frequency ac band bending variation dv_{sn} varies from one elementary capacitor to the next.

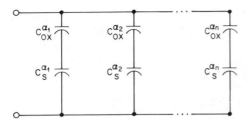

Fig. 6.11 Small-signal ac equivalent circuit for the nonuniform MOS capacitor without interface traps. The patch areas α_n are approximated in the text by equal area patches given by (6.90).

Equation (6.85) is the sum of a parallel array of capacitances $C_{m,n}$, each of which occurs P_n times. Therefore, (6.85) is represented by the equivalent circuit of Fig. 6.11. In Fig. 6.11 the entire quasi-uniform MOS capacitor is broken up into elementary capacitors made up of patches of sizes $\alpha_1, \dots, \alpha_n$. Each elementary capacitor consists of a series combination of an elementary oxide capacitance $C_{ox}^{\alpha_n}$ and an elementary, *one-dimensional* semiconductor capacitance $C_s^{\alpha_n}$. Each branch has its own oxide fixed charge density Q_{fn}, and this value occurs in a fraction P_n of the branches.

How is this picture modified for the case of *localized* oxide fixed charge, rather than large area patches of uniform oxide fixed charge? In general the silicon surface charge density in such an MOS capacitor is

$$\langle \beta Q_s \rangle = \frac{1}{A_g} \int d^2R \, \frac{\epsilon_s}{\lambda_n^2} \int_0^\infty dx \, [\exp v(\mathbf{R}, x) - 1] \tag{6.86}$$

where the integral over R and division by gate area A_g is an area average and the integral over x is the majority carrier density corresponding to the band bending $v(\mathbf{R}, x)$ integrated over the silicon depletion layer width. The area average in (6.86) can be replaced by an integral over all values of band bending on the plane $x = \text{constant}$, weighted by the probability $P_x(v)$ that $v(\mathbf{R}, x) = v$ on this plane. That is, (6.86) is equivalent to

$$\langle \beta Q_s \rangle = \frac{\epsilon_s}{\lambda_n^2} \int_0^\infty dx \int_{-\infty}^\infty dv \, P_x(v)[\exp(v) - 1]. \tag{6.87}$$

For convenience, let $P_x(v)$ be a Gaussian with mean $\langle v(x) \rangle$ and variance $\sigma^2(x)$

$$P_x(v) = (2\pi)^{-1/2}\sigma(x)^{-1} \exp\left\{ -\frac{[v - \langle v(x) \rangle]^2}{2\sigma^2(x)} \right\}. \tag{6.88}$$

Then, by substituting $\eta = v - \langle v \rangle - \sigma^2(x)$, and integrating over η in (6.88) we find

$$\langle \beta Q_s \rangle = \frac{\epsilon_s}{\lambda_n^2} \int_0^\infty dx \left\{ \exp[\langle v(x) \rangle] \exp \frac{\sigma^2(x)}{2} - 1 \right\}. \tag{6.89}$$

Equation (6.89) is a general formula for the average charge density $\langle Q_s \rangle$, which is valid for any variation in band bending $v(\mathbf{R}, x)$ if the Gaussian approximation (6.87) is valid. Using (6.89), we obtain the equivalent circuit for the semiconductor capacitance. For example, in Appendix VIII it is shown that (6.89) leads to (6.85) for the limiting case of long wavelength oxide fixed charge variations.

Now, we could apply (6.89) to find the equivalent circuit for the case of localized oxide fixed charges. To use (6.89), expressions for $\langle v(x) \rangle$ and $\sigma_s^2(x)$ are needed. These could be obtained using the Green's function approach in Appendix VII. Equation (6.89) then could be evaluated numerically and the capacitance found by numerical differentiation of (6.89) with respect to gate bias. However, this program has not been carried out in detail because the effects on the capacitance are slight, except near accumulation or inversion. In the latter bias regimes the simplifying assumptions that lead to (6.89) break down, as already discussed. Therefore, we content ourselves with the following qualitative discussion.

In depletion and weak inversion, the mobile carrier density in (6.89) is not important except near the depletion layer edge. Therefore, what is needed in (6.89) is $\langle v(x) \rangle$ and $\sigma^2(x)$ in the vicinity of the depletion layer edge. Here, only the longer wavelength oxide fixed charge variations contribute. For such long wavelengths, the quasi-uniform model in Section 6.3.1(a) is applicable. Therefore, (6.89) will lead to (6.85) with the area of a typical elementary capacitor given by [see the long wavelength limit of (6.70)]

$$\alpha \approx 4\pi w^2. \tag{6.90}$$

The corresponding equivalent circuit is that shown in Fig. 6.11, with equal area patches of size α given by (6.90).

6.4.2 Validity of the Equivalent Circuit

For a Poisson distribution of localized charges and for biases in depletion or weak inversion, the band-bending fluctuations are weak enough that the circuit in Fig. 6.11 is valid. However, what happens near flatbands or in accumulation or inversion?

In depletion or weak inversion, a Gaussian approximation to the distribution of band bending is adequate. However, as flatbands or accumulation is approached, the number of oxide fixed charges in an elementary area becomes small, and the Poisson distribution of charges must be used directly. If the elementary area is α, the probability of n charges P_n is

$$P_n = \left(\frac{\alpha \langle Q_f \rangle}{q} \right)^n \frac{\exp(-\alpha \langle Q_f \rangle / q)}{n!}. \tag{6.91}$$

A one-dimensional formulation for the capacitance often is used. That is, the band bending v_{sn} in an elementary capacitor with n charges is given by

$$\beta V_G - v_{sn} = \frac{\beta q n}{C_{ox}\alpha} - Q_s(v_{sn}).$$

(6.92)

Therefore, the capacitance becomes

$$C_m = \sum_n P_n \frac{C_{ox}C_{sn}}{(C_{ox} + C_{sn})}$$

(6.93)

where $C_{sn} = C_s(v_{sn})$ with v_{sn} from (6.92).

As in the earlier discussion of gross interfacial charge nonuniformities, a question of self-consistency arises. That is, the size of an elementary area depends on how many charges are in it. Let us take the one-dimensional formulation of this condition using (6.92). Thus if some patch size α is chosen, we know that the probability of n charges residing in this patch is P_n. For each n, (6.92) determines v_{sn} and hence the *self-consistent* area $\alpha_n = 4\pi w^2(v_{sn})$. If the original choice of area α is the same as α_n, the patch is self-consistent. If not, this patch does not occur.

Suppose that we choose some α and find the probability P_n for a self-consistent patch $\alpha_n = \alpha$. One way to do this is to set some value of $v_s = v_{sn}$. Then the corresponding n is known from (6.92) with α_n from (6.90). Consequently, for $\alpha = \alpha_n$, we find the probability P_n from (6.91). This probability is the fraction of patches of area α that are self-consistent. Then we can vary v_{sn} and find this probability as a function of patch size α. Figure 6.12 illustrates the result of such a calculation for two gate biases corresponding to band bending of mean charge $\bar{v}_s = 3$ and $\bar{v}_s = 5$, and for two choices of average oxide fixed charge density, $\bar{Q}_f/q = 10^{11}\,\text{cm}^{-2}$ (solid lines) and $\bar{Q}_f/q = 10^{12}\,\text{cm}^{-2}$ (dashed lines). The fraction of the capacitor area that is self-consistent peaks for one choice of α, but even this value of α results in only 1% or less of self-consistent elementary capacitors. Of those elementary capacitors that are not self-consistent, those with narrower depletion layer widths would contribute more to band bending fluctuations (because for them α should be smaller), whereas those with wider depletion layer width would contribute less and are inadequately treated in a one-dimensional approximation because their diameters are smaller than the separation of their plates.

Figure 6.12 shows that there is something wrong with this one-dimensional approach to self-consistency because a patch of size α is unlikely to contain exactly the required number of charges for self-consistency. What is missing in this formulation is the inability of the depletion layer width to respond abruptly to abrupt changes in interface charge. That is, the depletion layer width reacts only to an average of the interface charge distribution, an average over an area approximately given by (6.90). As a result, a small cluster of interface charge of diameter less than the average

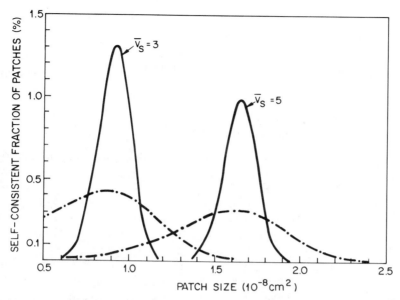

Fig. 6.12   The fraction of patches of area α that are self-consistent versus α. All patches in the parallel array model are forced to have the same characteristic α; $N_B = 10^{15}\,\text{cm}^{-3}$.

depletion layer width will affect the depletion layer width far less than (6.92) would suggest.

Figure 6.12 is based on an inadequate formulation of self-consistency. At present, no satisfactory assessment of the patchwork model exists for the case where depletion layer width modulation is significant. Fortunately, in depletion and weak inversion the small fluctuation case is sufficient for our purposes. For this case, the circuit in Fig. 6.11 is valid.

6.4.3   Equivalent Circuit with Interface Traps

The inclusion of interface traps in the circuit in Fig. 6.11 is straightforward. Long-wavelength variations in interface charge variation affect only the mean band bending in each elementary capacitor of Fig. 6.11. Therefore, only wavelengths shorter than a patch dimension need be included in analyzing interface trap response. Also, all elementary capacitors are independent. Therefore, only charge variations within a single elementary capacitor need be considered; there is no effect from charges in adjacent capacitors.

With attention restricted to only one elementary capacitor, we estimate interface trap response for charge nonuniformities within this capacitor. In the absence of interfacial charge nonuniformities, Lehovec[21] showed that the interface trap admittance was Y_{it} (as derived in Section 5.3.4)

$$Y_{it} = qD_{it}\frac{1}{\tau}\left\{\frac{1}{2}\ln[1 + (\omega\tau)^2] + j\tan^{-1}(\omega\tau)\right\} \tag{6.94}$$

where the interface trap capture time τ is defined by (for hole capture in p-type)

$$\begin{aligned}\tau &= (c_p p_s)^{-1}\\ &= [c_p N_A \exp(-v_s)]^{-1}.\end{aligned} \tag{6.95}$$

When fluctuations in band bending ψ_s are present with probability $P(\psi_s)$, we modify D_{it} in (6.94). In Section 5.3.2(a) $D_{it}(\zeta + \psi_s)$ was introduced as the probable number of interface trap levels per unit area with energy (in electron volts) between ζ and $\zeta + d\zeta$, where ζ is measured from the intrinsic level in the bulk silicon. The distribution $D_{it}(\zeta + \psi_s)$ is fixed relative to the band edges at the silicon surface. When the bands are bent, this distribution moves with the bands.

Now suppose that fluctuations in band bending occur. If band bending between ψ_s and $\psi_s + d\psi_s$ occurs with probability $P(\psi_s)$, the probability of the density $D_{it}(\zeta + \psi_s)$ occurring also is $P(\psi_s)$. As a result, the probable number of interface traps per unit area with energy levels between $(\zeta + \psi_s)$ and $(\zeta + d\zeta + \psi_s + d\psi_s)$ is $\tilde{D}_{it}$, where

$$\tilde{D}_{it}(\zeta + \psi_s)d\zeta\, d\psi_s \equiv D_{it}(\zeta + \psi_s)P(\psi_s)d\zeta\, d\psi_s. \tag{6.96}$$

Equation (6.96) assumes *no correlation* between band bending and interface trap energy level distribution.

To understand (6.96) better, recall that in Section 5.3.2(a), where no band-bending fluctuations were considered, introduction of the probability density D_{it} allowed sums over interface trap levels to be replaced by an integral. For example, at flatbands the interface trap charge per unit area Q_{it} is [assuming donor interface traps as an example; see (5.19)]

$$\begin{aligned}Q_{it} &= \sum_l qN_{Tl}[1 - f_o(\zeta_l - \phi_B)]\\ &= \int_{E_v/q}^{E_c/q} d\zeta\, qD_{it}^d(\zeta)[1 - f_o(\zeta - \phi_B)].\end{aligned} \tag{6.97}$$

The transition from a sum to an integral is made by treating ζ as a random variable, so that a sum over all interface trap levels is tantamount to an area average of the quantity summed, with the summation turning up the value $\zeta = \zeta_l$ exactly $D(\zeta_l)$ times.

If now at the site l the band bending is ψ_{sl}, the sum (6.97) becomes

$$Q_{it} = \sum_l qN_{Tl}[1 - f_o(\zeta_l - \psi_{sl} - \phi_B)] \tag{6.98}$$

where we use the fact that the energy level ζ_l moves to $\zeta_l - \psi_{sl}$ in the presence of the band bending ψ_{sl}.

Suppose that ψ_{sl} is *not* correlated with ζ_l. That is, in summing over l, any ψ_{sl} might occur with any ζ_l, and, in fact, ψ_{sl} occurs equally often with *every* ζ_l. Then we may make ψ_{sl} and ζ_l independent random variables. Introducing the probability $P(\psi_s)$ that ψ_s lies between ψ_s and $\psi_s + d\psi_s$, we find

$$Q_{it} = \int_{-\infty}^{\infty} d\psi_s\, P(\psi_s) \sum_l q N_{Tl}[1 - f_o(\zeta_l - \psi_s - \phi_B)]. \tag{6.99}$$

Then, using D_{it} for the remaining sum over l, just as in (6.87), (6.89) becomes

$$Q_{it} = \int_{-\infty}^{\infty} d\psi_s P(\psi_s) \int_{E_v/q}^{E_c/q} d\zeta\, D_{it}(\zeta)[1 - f_o(\zeta - \psi_s - \phi_B)]$$

$$= \int_{-\infty}^{\infty} d\psi_s\, P(\psi_s) \int_{E_v/q - \psi_s}^{E_c/q - \psi_s} d\zeta\, D_{it}(\zeta + \psi_s)[1 - f_o(\zeta - \phi_B)]. \tag{6.100}$$

Therefore, for band-bending fluctuations that are as likely at one interface trap site as any other, regardless of the interface trap energy level at that site, the density $D_{it}(\zeta + \psi_s)$ of the zero fluctuation case is replaced by the new density $\tilde{D}_{it}(\zeta + \psi_s)$ of (6.96).

The assumption that there is no correlation between interface trap levels and band bending is not strictly accurate because (1) the *average* band bending depends on average interface trap occupancy because interface traps contribute to the average interface charge density, (2) the average fluctuation is affected by fluctuations in interface trap charge, and (3) the average fluctuation will be damped by the average interface trap capacitance because fluctuations in band bending change interface trap occupancy in a direction tending to reduce the fluctuations. All these averaged effects have been included in our self-consistent treatment of small fluctuations. However, higher-order correlations have been ignored, and their importance is unknown if the fluctuations are not small.

If we try to use the modified density given by (6.96), we do not know the distribution of band bending $P(\psi_s)$. All we know is its mean and variance. The most common distribution characterized by only its mean and variance is the Gaussian. Moreover, when a *large* number of *independent* events (charges) contribute to v_s, a Gaussian distribution results from the law of large numbers.[4] Therefore, for the probability $P(v_s)$ that v_s lies within the interval v_s and $v_s + dv_s$, we adopt

$$P(v_s) = [2\pi\sigma_s^2(\lambda)]^{-1/2} \exp \frac{(v_s - \langle v_s \rangle)^2}{2\sigma_s^2(\lambda)}. \tag{6.101}$$

Then

$$\langle Y_{it} \rangle = q \int_{-\infty}^{\infty} dv_s \, P(v_s) \bar{Y}_{it}(v_s). \tag{6.102}$$

In (6.102) the angular brackets, $\langle \cdots \rangle$, denote the area average, whereas the bar superscript denotes an average over the interface trap level distribution for any given band bending v_s. Equation (6.102) cannot be integrated in closed form, but must be evaluated numerically. The results of such an integration have been presented in Section 5.4.

Unlike the low frequency admittance $Y_{it} = j\omega \, qD_{it}$, the admittance given by (6.94) at general frequencies is highly nonlinear because of the time τ that varies exponentially with band bending according to (6.95). Therefore, the use of a linearized, dc treatment of fluctuations for $P(v_s)$ requires justification.

The use of a dc analysis is justified as follows. Because we are dealing with a small-signal excitation, interface trap occupancy remains near its equilibrium value, and only those interface trap levels near the Fermi level respond to the ac gate voltage. This ac interface charge hardly affects the band-bending fluctuations, which are dominated by dc charge nonuniformities due to oxide fixed charge and to interface trap charge held in energy levels at all distances from the Fermi level.*

A linearized treatment is justified in the dc case provided that the higher-order moments of the band-bending fluctuations are small. Our linearized analysis shows that the screening by the majority carriers near the depletion layer edge and by the gate determines σ_s^2 in depletion and weak inversion. These sources of screening are only weak functions of band bending, so ac variation of this screening is weak for the small-signal regime. An exception occurs for gross charge nonuniformity, as already discussed. Other exceptions occur in or near accumulation and in inversion and for unusual interface trap level densities that vary rapidly within the bandgap.

In summary, the fluctuations are a dc phenomenon, adequately described by a linearized analysis in depletion and weak inversion. Any ac variation of the fluctuations is weak, again for biases in depletion or weak inversion. In particular, the nonlinear ac interface trap response does not affect the fluctuations; rather, the fluctuations spatially modulate the ac admittance through its dependence on dc band bending. Because ac phenomena have no part in determining the fluctuations, the Gaussian distribution for the dc fluctuations can be used to describe the effect of fluctuations on the nonlinear ac admittance given by (6.94).

Now, we need the relation between $\langle Y_{it} \rangle$ from (6.102) for one elementary

---

*In the unusual case of a large interface trap level density that varies very rapidly with energy near the Fermi level, this argument might not apply.

capacitor and the total admittance of this elementary capacitor. Then the admittance of the entire MOS capacitor will be the sum of all the elementary admittances because each elementary admittance is in parallel with and independent of the rest.

To derive the equivalent circuit, a small-signal variation of gate voltage $\delta V_{G\omega}$ at frequency ω is imposed. Then we calculate the resulting small-signal charge variation δQ_ω. The admittance $Y_m(\omega)$ is then given by

$$Y_m(\omega)\delta V_{G\omega} \equiv j\omega \frac{q}{kT}\delta Q_\omega. \tag{6.103}$$

Such a calculation was made in Ref. 22 and is not repeated here. The only complication met in this analysis is the question of whether nonuniformity in the ac band bending across the Si–SiO$_2$ interface must be taken into account. The dc band-bending variations must be included, as in our derivation of $\langle Y_{it}\rangle$ from (6.102), so it may seem that the ac variations also are important. However, for a small-signal analysis, valid up to terms in σ_s^2, the nonuniformity in ac band-bending variations ordinarily can be ignored. These terms are of the form of a derivative of σ_s^2 with respect to gate bias, multiplied by the small ac gate voltage. For short wavelength interfacial charge nonuniformities due to a Poisson distribution of point charges, σ_s^2 is a slowly varying function of gate bias so that terms in $d\sigma_s^2/dV_G$ are small

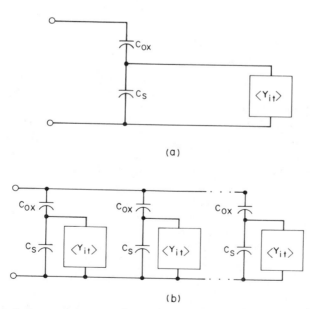

(a)

(b)

Fig. 6.13 Small-signal ac equivalent circuit for the nonuniform MOS capacitor with localized oxide fixed charge and interface traps: (a) circuit for an elementary capacitor; (b) array for entire MOS capacitor—this circuit is valid only in depletion and weak inversion.

and may be neglected in depletion and weak inversion. However, in accumulation and inversion, $d\sigma_s^2/dV_G$ is not small. Also, higher-order moments of the statistical distribution of band bending are significant. Therefore, we restrict ourselves to an equivalent circuit for depletion and weak inversion. Within this restriction the equivalent circuit for an elementary capacitor is shown in Fig. 6.13a, and the entire MOS capacitor is shown in Fig. 6.13b.

The final form for the admittance, using the circuit in Fig. 6.13b, is then

$$Y_m(\omega) = \sum_n P_n \frac{C_{ox}[C_s^n + \langle Y_{it}^n \rangle]}{C_{ox} + C_s^n + \langle Y_{it}^n \rangle} \qquad (6.104)$$

where P_n is the probability of interface charge density Q_i^n in a patch of area $\alpha = 4\pi w^2$, and $\langle Y_{it}^n \rangle$ is the average interface trap admittance for interface traps in elementary capacitors with $(Q_i^n/q)4\pi w^2$ randomly distributed point charges.

For a Poisson distribution over such large area patches, P_n will be so sharply peaked about the average charge density that all the contributing elementary admittances in (6.104) are identical, and each is the admittance of an elementary capacitor with the mean interface charge density. The sum in (6.104) then becomes the admittance of this elementary capacitor multiplied by the sum over P_n, which is unity. The equivalent circuit in Fig. 6.13b then reduces to that in Fig. 6.13a for an elementary device with the mean charge density. The admittance $\langle Y_{it} \rangle$ is given by (6.102), and $C_s = C_s(\bar{v}_s)$.

6.5 CHEMICAL INHOMOGENEITY

As discussed in Section 6.3, point charges within a distance λ of an interface trap may result in quantum mechanical modifications of interface trap behavior, which could include modifications of capture cross section. In addition, there may be variations of interface trap behavior due to varying chemical surroundings, such as the presence of nearby silicon or oxygen or complexes of silicon and oxygen in various stages of oxidation or of stretched or bent bonds. In short, because of chemical inhomogeneity, a variety of capture cross sections may prevail among the interface traps, even if the interface traps would appear to be identical if they were placed in identical surroundings.

A treatment of capture cross section variations was given by Preier[23] for the special case of interface traps distributed into the oxide (i.e., the "chemical inhomogeneity" considered was how much SiO_2 separated the interface trap from the interface). Preier's model was an offshoot of a treatment of interface traps due to Heiman and Warfield.[24] To summarize these studies, it was found that an interface trap a distance x in the oxide from the interface had a capture cross section that was reduced by

$\exp(-x/x_d)$ from its value at the interface, where x_d was ≈ 1 Å and depended on the barrier height between silica and silicon.

Using the Preier model to fit experimental data, the maximum depth x_d at which interface traps could be present is 5–10 Å, a distance of the same order as atomic spacings[23,25] (see Section 7.3.5). Consequently, the Preier model must be modified to include atomic scale variations in the silica-silicon interface region and then averaged over the statistical distribution of such variations. Such a program is impossible at present, given our limited knowledge of this interface. In our view, the Preier model could be interpreted as an empirical treatment of chemical inhomogeneity effects on capture cross section, which must await further experiments and analysis for vindication.

Because of its basis at an uncertain atomic level, the Preier model cannot be subjected to the same rigorous tests that can be applied to the band-bending fluctuation model. The parameters of the Preier model must remain empirical fitting parameters, whereas the parameters of the band-bending model are measurable and have well-defined bias and oxide fixed charge density dependencies that can be checked experimentally. In Section 7.3.6 the available experimental data are discussed in the light of the theory presented here, along with the theory of the conductance method presented in Section 5.5.

REFERENCES

1   J. R. Brews, J. Appl. Phys., **43**, 2306 (1972).

2   G. J. Declerck and R. J. Van Overstraeten, "Characterization of the MOSFET Operating in Weak Inversion," in *Advances in Electronics and Electron Physics*, Vol. 47, L. Marton, Ed., Academic, New York, 1978, pp. 197–265.

3   J. R. Brews, J. Appl. Phys., **46**, 2181 (1975).

4   W. Feller, *An Introduction to Probability Theory and Its Applications*, 2nd ed., Wiley, New York, 1957, p. 270.

5   E. H. Nicollian and A. Goetzberger, Bell Syst. Tech. J., **46**, 1055 (1967).

6   R. Castagné and A. Vapaille, *Surface Sci.* **28**, 157 (1971); *Electron. Lett.* **6**, 691 (1970); *C. R. Acad. Sci. (Paris)*, **B270**, 1347 (1970).

7   J. R. Brews and A. D. Lopez, Solid-State Electron., **16**, 1267 (1973).

8   A. Goetzberger, Solid-State Electron., **9**, 871 (1966).

9   G. Baccarani, M. Severi, and G. Soncini, Appl. Phys. Lett., **23**, 265 (1973).

10 C. C. Chang, "Study of Material Nonuniformities and Interface States in MIS Structures," Ph.D. dissertation, Department of Electrical Engineering, Princeton University, Princeton, N. J., 1976.

11 C. C. Chang and W. C. Johnson, IEEE Transact. Electron Devices, **ED-25**, 1368 (1978).

12 T. H. DiStefano, Appl. Phys. Lett., **19**, 280 (1971); J. Appl. Phys., **44**, 527 (1973).

13 R. Williams and M. H. Woods, J. Appl. Phys., **43**, 4142 (1977).

14 W. R. Bottoms, P. Roitman, and D. C. Guterman, CRC Crit. Rev. Solid State Phys. 297–311 (October 1975).

15   T. H. DiStefano and J. E. Lewis, *J. Vac. Sci. Technol.*, **11**, 1020 (1974).

16   M. J. McNutt and C. T. Sah, *J. Appl. Phys.*, **45**, 3916 (1974).

17   J. R. Brews, *J. Appl. Phys.*, **44**, 3228 (1973).

18   F. Stern, *J. Vac. Sci. Technol.*, **11**, 962 (1974).

19   E. Arnold, *Surface Sci.*, **58**, 60 (1976).

20   E. Arnold, private communication.

21   K. Lehovec, *Appl. Phys. Lett.*, **8**, 48 (1966).

22   J. R. Brews, *J. Appl. Phys.*, **43**, 3451 (1972).

23   M. Preier, *Appl. Phys. Lett.*, **10**, 361 (1967).

24   F. P. Heiman and G. Warfield, *IEEE Transact. Electron Devices*, **ED-12**, 167 (1965).

25   H. Deuling, E. Klaussmann, and A. Goetzberger, *Solid-State Electron.*, **15**, 559 (1972).

7

Experimental Evidence for Interface Trap Properties

- INTRODUCTION, 285
- PRESENT-DAY VIEW OF THE Si–SiO₂ INTERFACE, 285
- INFORMATION FROM CONDUCTANCE MEASUREMENTS, 286

7.1 INTRODUCTION

Extraction of interface trap properties using the conductance method was discussed in Chapters 5 and 6. This discussion presupposed several features of the interface without justification. The purpose of this chapter is to present experimental data supporting these assumptions. First, we present the accepted picture of the Si–SiO₂ interface of thermally grown device-grade oxides and point out its critical features.

7.2 PRESENT-DAY VIEW OF THE Si–SiO₂ INTERFACE

The prevalent view of the electrical properties of the Si–SiO₂ interface hinges on five features:

1 The Si–SiO₂ interface has charge centers called *oxide fixed charge*. These centers are predominantly positive, although a small number of compensating negative centers may be present. They are immobile under an applied electric field and do *not* exchange charge with the silicon when gate bias is varied. This charge ordinarily is dispersed randomly across the interface.

2 The interface has traps. These traps, called *interface traps*, change occupancy with gate bias changes and have energy levels distributed throughout the bandgap. These traps cannot communicate directly with one another and do not form an energy band.

3 Interface potential varies with position from point to point across the interface. These variations are the inevitable consequence of the random distribution of the very localized charged interface traps and of the oxide fixed charges.

4 Individual interface trap capture cross sections may be distributed over a range of values due to bent or stretched bonds or to other variations in chemical environment. Measured interface trap capture cross sections for both holes and electrons are average values independent of energy over most of the silicon bandgap.

5 Interface traps characteristic of thermal oxidation are *donor* type in the upper half of the bandgap.*

Identification of these five features of the Si–SiO₂ interface that affect electrical measurements has evolved gradually. The first feature, oxide fixed charge, is discussed in Sections 11.4.1 and 11.4.2, where it is noted that etch-off[1] experiments locate this charge in the oxide within 200 Å of the interface, and photoinjection experiments locate it within 34 Å of the interface.[2,3] The remaining features concern interface traps, and these features are discussed here.

Experiment does not reveal whether oxide fixed charges are interface trap levels near the silicon band edges, interface trap levels opposite levels in the silicon bands, or levels too far away from the Si–SiO₂ interface for charge exchange to occur. It is not known whether oxide fixed charge and interface trap charge have the same or a different chemical origin. Because we cannot measure interface trap charges in levels near the band edges or opposite the bands, in interpreting experiments these charges are lumped with oxide fixed charge.

7.3 INFORMATION FROM CONDUCTANCE MEASUREMENTS

7.3.1 Overall Behavior of Conductance Measurements in Depletion

The small-signal equivalent parallel conductance of the MOS capacitor G_P displays different behavior depending on the dominant loss mechanism in the sample under study.[4] The particular behavior of G_P that concerns us commonly is attributed to interface traps with the features already men-

*There are no experiments on the donor or acceptor nature of interface traps in the lower half of the bandgap.

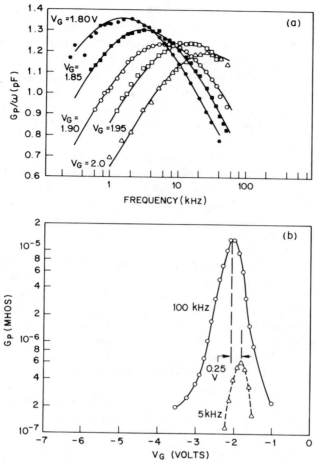

Fig. 7.1 Typical conductance peak behavior due to interface traps. (a) Plot of G_P/ω versus log ω for several biases. The sample had a p-type (111)-oriented substrate, $N_A = 2.24 \times 10^{15}$ cm^{-3}, $x_o = 1360$ Å, dry-grown oxide at 1200°C, annealed after metallization at 475°C. The samples were then bias-temperature stressed at 150°C for 10 min $V_G = -26$ V. Cross section is $\sigma_p \approx 5 \times 10^{-17}$ cm^2, $D_{it} \approx 10^{10}$ cm^{-2} eV^{-1}, $\sigma_s \approx 2.6(kT/q)$. Solid line is the theoretical fit using the full model with broadening. After Declerck et al.[5] Copyright (1973), Pergamon Press Ltd. Reprinted with permission. (b) Equivalent parallel conductance versus gate bias for a p-type sample with $N_A = 2.08 \times 10^{16}$ cm^{-3} and $D_{it} \approx 10^{11}$ cm^{-2} eV^{-1}. Steam-grown oxide. The solid line is the curve drawn to fit the data. After Nicollian and Goetzberger.[4] Copyright (1967), American Telephone and Telegraph Company. Reprinted with permission.

tioned, namely, a distribution of single-level interface traps with energies throughout the bandgap, isolated from one another, subject to band-bending variations, and with a range of individual capture cross sections. This model has evolved to explain the behavior of the measured equivalent parallel conductance shown in Fig. 7.1. The important features are:

1 The G_P/ω versus $\log \omega$ curves show a maximum, where ω is the angular frequency in radians per second, as shown in Fig. 7.1a.

2 The G_P versus V_G curves show a maximum, where V_G is gate bias, as shown in Fig. 7.1b.

3 The amplitude of the peak in the G_P/ω versus $\log \omega$ curves is a *weak* function of bias, as shown in Fig. 7.1a.

4 The frequency at which the peak in the G_P/ω versus $\log \omega$ curves occurs is a very strong function of bias, as shown in Fig. 7.1a.

The combination of observations 1–4 cannot be predicted on the basis of simple models, and to date only the interface trap model has been successful in explaining all four features quantitatively. To illustrate the difficulties in fitting all four observations, we consider two other models.

7.3.2 Bulk Trap Loss

The most obvious suggestion would be that bulk traps* could explain the loss. If the peak in the G_P/ω versus $\log \omega$ curves were due to single-level bulk traps, the magnitude of the peak would vary with gate bias *only* if bulk trap level density varied with depth. That is, as the device was biased further into depletion, the responding bulk traps would be deeper in the silicon, located always at a fixed distance l from the depletion layer edge as shown in Fig. 7.2. Therefore, the observed weak bias dependence of the amplitude of maximum loss would be explained.

However, the frequency corresponding to the maximum would not vary with gate bias. This frequency is dependent on the capture rates for carriers near the point where the bulk trap levels cross the Fermi level. This crossover point is always at the same distance from the depletion layer edge, so that the carrier densities at crossover are always the same regardless of gate bias. Therefore, the capture rates do not vary, and the loss peaks at the same frequency at all biases, contrary to Fig. 7.1a (see Section 4.3). In addition, the preceding argument shows that at a fixed frequency the loss will be the same at all values of gate bias. That is, no peak in the loss will occur as a function of gate bias, contrary to Fig. 7.1b.

7.3.3 Interface Dipole Loss

Another possibility is that dipoles exist near the Si–SiO$_2$ interface, such as polar molecules, or ions that can move between closely associated sites (e.g., bond flexing models) or electrons that can shift between closely spaced traps, in response to the ac gate voltage and so on. These particular

---

*The term "bulk traps" means traps spatially distributed throughout the silicon crystal having energy levels in the silicon bandgap.

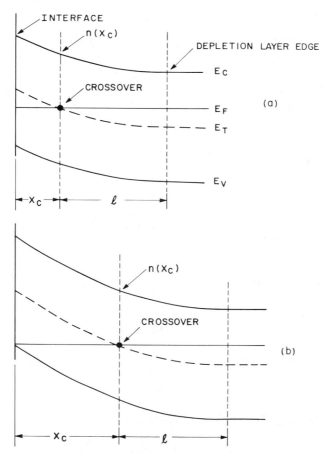

Fig. 7.2 Illustration showing active bulk traps at the crossover point remain a fixed distance l from the depletion layer edge, independent of gate bias. Consequently, majority carrier density at crossover $n(x_c)$ also is independent of gate bias for uniform doping. The sample shown is n-type silicon; (a) corresponds to a smaller negative gate bias than (b).

models are discussed because they have been considered by many people. In this subsection we show that none explain the observed loss.

The basic elements of all these models are contained in a simple two potential well model, shown in Fig. 7.3. Two potential wells are illustrated in Fig. 7.3, separated by a barrier. We may imagine an ion that moves from one well into the other, a polar molecule, or an electron that moves under the influence of the applied field. In any case, as gate bias is changed, the relative energy of the two wells varies. The energy difference between the wells is the field at the interface F_s times the well separation d. Equilibrium population of the two wells then changes with gate bias. Under ac excitation the ac transfer of population from one well to the

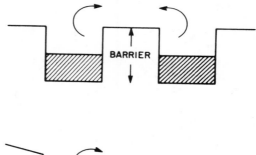

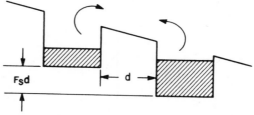

Fig. 7.3 A simple two-well dipole model that simulates either bond flexing, polar molecules, or ion or electron exchange between nearby sites.

other varies with gate bias as a result of the change in equilibrium occupancy.

It is shown in Appendix IX that this process yields an admittance

$$Y = q\beta N_T \frac{j\omega}{1 + j\omega\tau} \qquad (7.1)$$

where N_T is the number of dipoles and the time constant τ is

$$\tau^{-1} \equiv 2w_o \cosh\left[\frac{\beta(\psi_o - \psi_s)}{2}\right] \qquad (7.2)$$

where $(\psi_o - \psi_s)$ is the potential drop between the well on the oxide side of the dipole and the well on the silicon side and w_o is the probability of a transition from one well to the other at zero bias. The real part of Y is

$$\frac{G}{\omega} = \beta q N_T (\omega\tau)[1 + (\omega\tau)^2]^{-1} \qquad (7.3)$$

and the capacitive part is

$$C = \beta q N_T [1 + (\omega\tau)^2]^{-1}. \qquad (7.4)$$

The behavior of (7.3) and (7.4) is exactly the same single time-constant behavior as a single-level bulk trap. However, unlike the bulk trap, the time constant given by (7.2) now varies with gate bias, offering some hope of fitting experiment.

To estimate how much variation with gate bias is expected from (7.2), note that the voltage difference $(\psi_s - \psi_o)$ is given by $F_s d$. A reasonable dipole length is about 50 Å or less, as larger dipoles would lead to difficulties in communication between wells, and possibly to hysteresis effects in capacitance that are not observed. The interface field might vary from near zero at flatbands to a value $\epsilon_s F_s \approx q N_B w(2\phi_B)$ near inversion. [By $w(2\phi_B)$ we denote the depletion layer width at the onset of inversion.] This field, for a doping density of $N_B = 10^{16}\,\text{cm}^{-3}$, is about $4.6 \times 10^4\,\text{V/cm}$. Therefore, $(\psi_o - \psi_s) \le 23\,\text{mV}$ and τ varies by a factor of about $\cosh(0.5)/\cosh(0) = 1.13$ as bias is swept from flatbands to inversion. In contrast, experiment shows a variation of 6 orders of magnitude in half this gate bias range. The dipole model is incorrect, and must be discarded.

7.3.4 Interface Traps

The interface trap model can explain all four observed aspects of G_P/ω behavior listed in Section 7.3.1.

(a) Maximum in G_P/ω versus log ω

Consider interface trap levels at a particular energy in the bandgap. For an appropriate band bending, the Fermi energy will lie at these interface trap energy levels. For this band bending, a particular majority carrier density prevails at the silicon surface that determines the capture rate of the chosen interface trap levels. If the applied small-signal frequency corresponds to this capture rate, a peak loss due to these interface trap levels will occur. If the frequency is higher or lower, the loss is reduced because the interface trap levels either fail to respond completely (at higher frequencies) or they can respond in an equilibrium (lossless) manner (at lower frequencies). Consequently, a peak in the loss as a function of frequency occurs.

(b) Maximum in G_P/ω versus V_G

For a chosen ac frequency, let us vary gate bias from accumulation to midgap. In accumulation, majority carrier density is very large near the Si–SiO$_2$ interface, so that interface trap capture rates are very rapid compared to the ac frequency. Interface trap levels respond immediately to the ac voltage, and no loss occurs. In depletion, majority carrier density at the Si–SiO$_2$ interface is reduced.* Capture rates slow down, and interface trap levels cannot keep in phase with the ac voltage. A loss occurs. Still further in depletion, near midgap, majority carrier density becomes so low that interface trap levels hardly respond. The capture rate is so slow that almost no carriers are exchanged between interface trap levels and the silicon. Hence the loss is low.

---

*Because the silicon is in thermal equilibrium $p_s n_s = n_i^2$, where p_s is the hole density and n_s is the electron density at the silicon surface and n_i is the intrinsic carrier density, so that for p-type, $p_s \gg n_s$ and for n-type, $n_s \gg p_s$ in depletion. Thus minority carriers are always negligible in depletion.

In short, interface trap loss goes through a peak as a function of gate bias. Maximum loss occurs for the gate bias where majority carrier density makes the interface trap capture rate comparable to the ac frequency.

(c) Weak Bias Dependence of the Amplitude of Maximum Loss

Single-Level Interface Trap   A spatial distribution of single-level interface traps would not lead to the weak bias dependence of the amplitude of maximum loss seen in Fig. 7.1a. Instead, a rapidly diminished maximum would occur for gate biases that place the Fermi level more than several kT/q away from the interface trap energy level. This point is illustrated in Fig. 7.4. Here, theoretical G_P/ω versus log frequency curves for a single-level interface trap are shown for values of band bending that place the Fermi level opposite the interface trap level (labeled zero) and at $1-3kT/q$ away from the interface trap level. Figure 7.4 shows that at $3kT/q$ the loss peak has dropped about a factor of 6 below the value that occurred when the Fermi level coincided with the interface trap level. A range of capture cross sections would broaden these curves, but the amplitudes would still

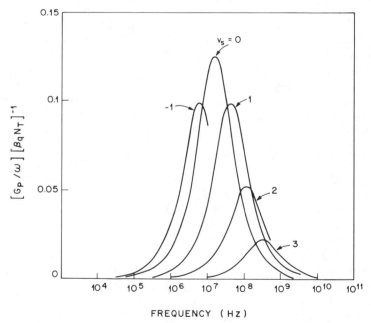

FREQUENCY (Hz)

Fig. 7.4   Plot of G_P/ω versus log ω normalized by division of $\beta q N_T$. This theoretical plot of single-level interface trap response shows, for the case of coincidence of Fermi level and interface trap level ($v_s = 0$), that the amplitude of maximum energy loss is large and drops off rapidly as the Fermi level is separated from the interface trap level. A cross section of 10^{-15} cm^2 was assumed, and the interface trap level was taken to coincide with the Fermi level at flatbands, with $N_D = 10^{16}$ cm^{-3} and $\bar{v} = 10^7$ cm/sec.

decrease with gate bias on either side of coincidence. Therefore, a variety of energy levels is needed to reproduce experiment.

Distribution of Interface Trap Energy Levels The necessary interface trap level distribution throughout the bandgap is typically a rather smooth function of energy as shown in Fig. 7.5. That is, the individual interface trap levels must be so close to one another in energy that they cannot be resolved, and only a smooth, structureless, gate bias variation of the amplitude of maximum loss is seen. Such a distribution often is called a *continuum* of interface trap levels, although individual interface traps are isolated spatially from one another, and no reference to an impurity band or the like is intended.

The postulate of a continuum of interface trap levels is not easy to accept. An attempt to explain such a continuum as due to a random distribution of oxide fixed charge was made by Goetzberger et al.[6]. Their concept is analogous to the Halperin-Lax[7] theory of band-tailing due to heavy doping in bulk semiconductors. That is, with rather low probability, large clusters of oxide fixed charge may occur in small areas. Where this happens, a localized state is pulled out of the conduction or valence band and placed within the bandgap. Naturally, the closer the interface trap level is to midgap, the more unlikely it becomes that the necessary cluster of oxide fixed charge will occur.* Hence the interface trap level density is large near the band edges and decays into the gap.

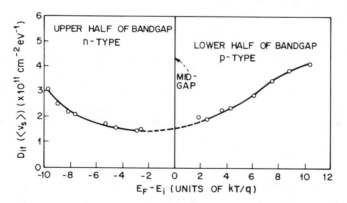

Fig. 7.5 A typical experimental interface trap density distribution. This curve was obtained using the conductance method with an energy resolution of about $\sigma_s = 2.6\,kT/q$ (see Section 5.6). Samples are the (111) samples in Fig. 7.7. After Nicollian and Goetzberger.[4] Copyright (1967), American Telephone and Telegraph Company. Reprinted with permission.

---

*For interface trap levels near the band edge, a low density cluster of few oxide fixed charges will suffice. Such clusters are common. For interface trap levels near midgap a high density cluster of many oxide fixed charges is needed. Such clusters are rare.

A deficiency of this model is that no theoretical estimate of the expected interface trap level density has been calculated. Experience in the case of band-tails in heavily doped bulk material suggests an enormous oxide fixed charge density would be needed to produce the observed interface trap level densities of 10^{10}–10^{12} cm^{-2} eV^{-1} at 0.5 V from the band edge. Hence this model is not very promising for explaining deep interface trap levels, although the presence of the interface may make tails of states more likely than in the bulk. This model may account for some interface trap levels near the band edges (see Section 16.5).

A more likely explanation is that the continuum of interface trap levels could result from chemical inhomogeneity. The energy level of an interface trap is very sensitive to the atoms in its vicinity and to their position. Again compared to the bulk, minor variations in the position of atoms neighboring an interface trap can cause variations in energy level of a large fraction of a volt.[8] Because the Si–SiO$_2$ interface is a region of great disorder, variations in the chemical environment of an interface trap such as the proximity of ions and stretched, bent, or broken bonds might account for spreading interface trap levels through the entire bandgap. Further discussion may be found in Section 16.5.

No detailed model of such effects has been worked out. Even for bulk traps, such calculations are at a rudimentary stage of development.[8,9] Therefore, we are forced to accept the continuum hypothesis without theoretical support.

(d) Strong Bias Dependence of the Frequency of Maximum Loss

Experimentally, Nicollian and Goetzberger[4] found that the frequency of maximum loss varied with gate bias as $\exp(\beta\psi_s)$. Their experimental results are shown in Figs. 7.6–7.8. In Fig. 7.6 the time constant at the maximum τ_m determined as $\omega_m\tau_m = 2.5$ is plotted against band bending ψ_s at two temperatures. The bias and the temperature dependence fit the theoretical exponential dependence $\exp(\beta\psi_s)$ indicated by the solid lines. Both p-type and n-type samples are compared for (111) orientation in Fig. 7.7 and for (100) orientation in Fig. 7.8.

The exponential dependence of τ_m on $\beta\psi_s$ (see Figs. 7.7 and 7.8) shows that the loss is proportional to the *majority carrier density* at the interface, as expected on the basis of an interface trap model.

7.3.5 Band-Bending Fluctuations and Capture Cross Section Variations

With the hypothesis of a continuous distribution of interface trap levels throughout the bandgap, the four features of Fig. 7.1 can be quantitatively explained, with one discrepancy: the width of the loss peak of a G_P/ω versus log ω curve cannot be explained.

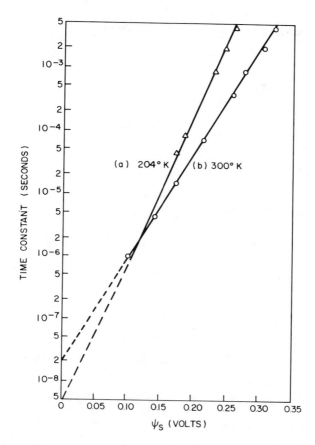

Fig. 7.6 Experimental measurement of interface trap time constant corresponding to maximum energy loss as a function of band bending ψ_s and temperature. The sample is p-type (111), and $N_A = 2.08 \times 10^{16}$ cm^{-3}. The oxide was grown in steam to a thickness $x_o \approx 600$ Å. After Nicollian and Goetzberger.[4] Copyright (1967), American Telephone and Telegraph Company. Reprinted with permission.

Lehovec[10] showed that the assumption of spatially *isolated* interface traps leads to a wider loss peak than does a continuum of communicating interface traps. The reason for a wider peak in the case of isolated interface traps is that each interface trap level independently responds to the ac excitation, rather than all interface trap levels contributing jointly to a single quasi-Fermi level. Consequently, each interface trap level has its own frequency response, independent of other levels, thus widening the range of frequencies over which loss occurs. Nonetheless, even with this added width, the observed width of the loss peak is far greater than predicted.

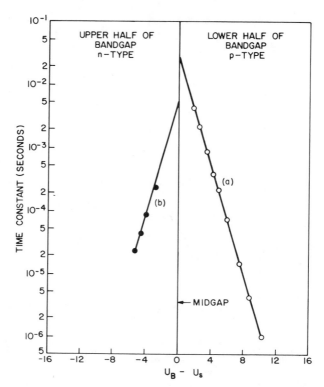

Fig. 7.7 Interface trap time constant versus band bending for both p-type and n-type (111) samples. The p-type sample is that of Fig. 7.6. The n-type sample had $N_D = 1.15 \times 10^{16}$ cm^{-3}. The oxide was grown in steam to a thickness of 680 Å. From the intercept of curve a on the τ-axis at $u_B = u_s$, $\sigma_p = 2.2 \times 10^{-16}$ cm^2 and from the intercept of curve b, $\sigma_n = 1.7 \times 10^{-15}$ cm^2. After Nicollian and Goetzberger.[4] Copyright (1967), American Telephone and Telegraph Company. Reprinted with permission.

At typical total interface trap densities of 10^{10}–10^{12} cm^{-2}, the spacing between interface traps is, on the average, 10^3–10^2 Å. Only at the high density end of this range would there be significant communication between interface traps because of overlap of trapped carrier wave functions. Consequently, the assumption of isolated interface traps is natural.

A striking confirmation of the model of isolated interface traps is found at low interface trap charge densities, where broadening of conductance peaks is minimized. In Fig. 7.9a a measured conductance curve is shown. The curve is folded to show its asymmetry. In Fig. 7.9b a similar folding of a conductance curve calculated from the Lehovec model is shown. Both curves show asymmetry, higher loss occurring on the high frequency side of the peak than on the low frequency side. Without the asymmetry of the Lehovec model, a theoretical fit to Fig. 7.9a would not be possible using

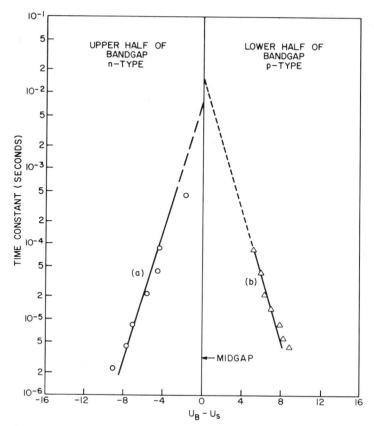

Fig. 7.8 Interface trap time constant versus band bending for both p-type and n-type (100) samples. For the p-type sample, $N_A = 1.19 \times 10^{16}$ cm^{-3}. The oxide was grown in steam to a thickness of 699 Å. For the n-type sample, $N_D = 1.44 \times 10^{16}$ cm^{-3}. The oxide was grown in steam to a thickness of 814 Å. From the intercept of curve a on the τ-axis at $u_B = u_s$, $\sigma_n = 1.2 \times 10^{-15}$ cm^2; from the intercept of curve b, $\sigma_p = 4.0 \times 10^{-16}$ cm^2. After Nicollian and Goetzberger.[4] Copyright (1967), American Telephone and Telegraph Company. Reprinted with permission.

the approach in Chapters 5 and 6. Thus the observed asymmetry supports the hypothesis of isolated interface traps with a distribution of energy levels.

When interface charge density is not very low, the observed G_P/ω versus log ω curve is considerably broader than predicted by the Lehovec model. This broadening is illustrated in Fig. 5.14, which shows a G_P/ω versus log ω curve calculated for a distribution of isolated interface traps (labeled b) and a measured G_P/ω versus log ω curve (labeled a). The G_P/ω versus log ω curve calculated for a single-level interface trap is shown as curve c in Fig. 5.14 for comparison. The single interface trap level, even if statistically broadened by the mechanisms now to be discussed, is unac-

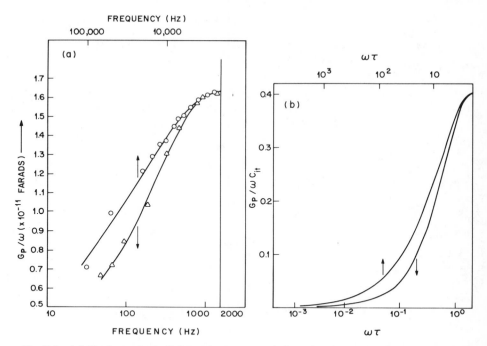

Fig. 7.9 (a) Equivalent parallel conductance peak for a low interface trap level density sample ($D_{it} \approx 10^{10}$ cm^{-2} eV^{-1}). The peak is folded to show its asymmetry. The solid line is a fit using the statistically broadened model of a continuum of isolated interface trap levels. The energy loss is higher on the high frequency side of the peak. After Nicollian and Goetzberger.[11] (b) A folded peak from the Lehovec model, normalized to C_{it}, without broadening. The same type of asymmetry appears as in (a), and it is this asymmetry that makes possible the solid line fit to the data shown in (a).

ceptable because it fails to provide the observed slow gate bias variation of the amplitude of maximum loss. Therefore, we consider statistical broadening of the Lehovec model.

Two sources of additional width in the experimental curves have been proposed: (1) band-bending fluctuations[4] and (2) capture cross section variations.[12] In either case the source of additional width of the loss peak is the ability of interface traps to contribute to the loss over an extended range of frequencies at any given gate bias.

If the Si–SiO$_2$ interface is not an equipotential, potential pockets will exist for which band bending is greater than the mean. Here majority carrier density will be higher. Consequently, response of these interface traps is faster. They will extend the loss to higher frequencies than that of the average interface trap. Similarly, regions of lower band bending will extend the loss to lower frequencies. In this way the loss peak is broadened in frequency.

The same effect can be obtained if a range of capture cross sections is allowed. Interface traps with larger cross sections extend the loss to higher frequencies and those with lower cross sections, to lower frequencies.

Can these models be distinguished? First, consider the capture cross section model. This model requires a quantum mechanical calculation to determine its parameters. Otherwise, the distribution of capture cross sections must be empirically determined by fitting experiment. Such a fit does little to confirm the model unless we know what parameter values are "reasonable." To illustrate these points, consider an example of such a model.

One very restricted form of capture cross section model is that due to Preier.[12] This model postulates traps not only at the interface, but distributed a short distance into the oxide. Because mobile carrier wave functions decay exponentially with distance into the oxide, the capture cross section of such an interface trap decays exponentially with its distance into the oxide. Consequently, the fastest capture occurs at the interface, and arbitrarily slow capture rates are obtained by assuming that interface traps are distributed to arbitrary depths into the oxide.

The Preier model does not attempt to calculate interface trap capture cross sections themselves, but the distribution of capture cross sections. This distribution determines the width of the measured conductance peak and depends on two parameters. First, a decay length x_d is needed to relate the capture cross section at the silicon surface to the depth of a trap in the oxide. Decay of the capture cross section was taken as $\exp(-x/x_d)$, where $x_d \approx 1\,\text{Å}$ on the basis of a decay of electron wave functions into the forbidden gap of the oxide. The second parameter is the length δ to which traps are distributed into the oxide from the silicon surface. The width of the measured conductance peak then is related to δ/x_d.

A fit of this model to experiment is shown in Fig. 7.10. This fit requires a rectangular interface trap distribution extending into the oxide to a maximum depth δ of 6–10 Å. Such a depth is suspiciously small. That is, the atomic nature of the Si–SiO$_2$ interface should be evident on such a scale, invalidating the treatment of decay of wave functions based on a simple one-dimensional barrier and invalidating the treatment of the oxide as a semi-infinite, bulk material with bandlike properties. Consequently, from a theoretical standpoint, this model must be regarded as too primitive a treatment of the capture cross section variation mechanism.

A test of the temperature dependence expected from the Preier model also was made by Deuling et al.[13] Figure 7.11 shows the depth of the interface trap distribution δ needed to fit experiment at various energies with temperature as parameter. This depth is a rather strong function of energy (bias) at a given temperature, tending to become shallower with approach to the majority carrier band edge. Also, at lower temperatures the depth increases. Neither of these effects is explained by the Preier model. Conceivably, the energy dependence of depth could be incorporated in a more elaborate Preier model, although a theoretical basis has not been

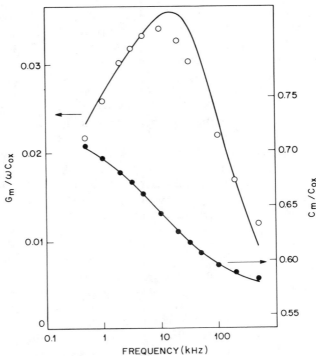

Fig. 7.10 A fit of the Preier tunneling model to experiment at a gate bias of − 3.75 V. The points are experiment and the solid lines theory. After Deuling et al.[13] Copyright (1972), Pergamon Press, Ltd. Reprinted with permission.

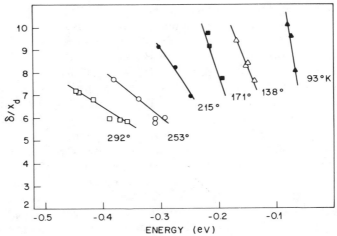

Fig. 7.11 Depth of the oxide trap distribution δ normalized to decay length x_d needed to fit conductance measurements at various energies with temperature as parameter. After Deuling et al.[13] Copyright (1972), Pergamon Press, Ltd. Reprinted with permission.

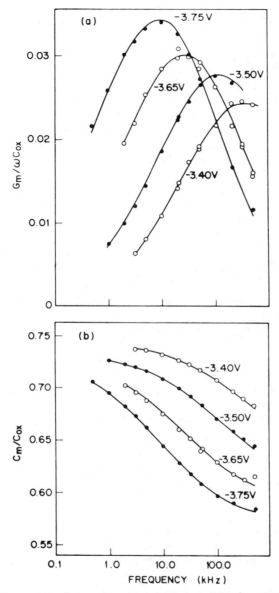

Fig. 7.12 Fit of the statistically broadened continuum model to experiment at several gate biases: (a) conductance versus frequency; (b) capacitance versus frequency. The sample is n-type, (111), dry oxidized at 1140°C to a thickness of 1000 Å, and postmetallization annealed 30 min at 350°C in dry H_2. After Deuling et al.[13] Copyright (1972), Pergamon Press, Ltd. Reprinted with permission.

advanced. The temperature dependence of this depth seems inexplicable because the interface trap positions should be fixed and independent of temperature.

Now consider the band-bending fluctuation model. Just as with the Preier model, the capture cross section is found empirically. Emphasis is on calculating the distribution of capture times. This distribution is related to the statistical variation in location of localized interface charges, variations that cause spatial fluctuations in band bending across the interfacial plane. Band-bending fluctuations due to the localized nature of interface charges *must* exist if the charges exist. Hence this mechanism is present, and its existence is not an additional assumption.

Unlike the capture cross section model, the parameters of this model can be estimated independently except for the parameter λ, which is related to the distance from an interface trap at which quantum mechanical effects must be considered in computing trap-charge interactions. However, the dependence on this parameter is logarithmic [see (7.5)], so that even an uncertainty in λ of orders of magnitude does not affect the results greatly. The width of the measured conductance peak is given by the fluctuations in band bending σ_s^2. Well-defined dependencies of σ_s^2 on gate bias, interface charge density, and temperature are predicted virtually independent of the value assumed for λ.

Unlike the unexplained variations of the width parameter δ/x_d in Fig. 7.11, the predicted variations of the width parameter σ_s^2 fit experiment, as explained in the next section. In anticipation, the reader might compare Figs. 7.11 and 7.18a. In both figures the ordinate is the parameter determining the broadening of the conductance peak and the dependence on energy and temperature is similar. However, the behavior in Fig. 7.11 is not explained within the Preier model, whereas that in Fig. 7.18a is explained within the band-bending fluctuation model.

Figure 7.12 shows the quality of fit to both conductance and capacitance obtainable with the statistically broadened continuum model. By fitting experiment this way, one obtains the density of interface trap levels per unit energy per unit area D_{it}, the capture cross section of the interface traps σ_p for holes or σ_n for electrons, and the variance of the statistical broadening σ_s^2. The behavior of all these parameters as a function of temperature, gate bias, and interface charge density then can be obtained. Of all these parameters, σ_s^2 is the only one for which a satisfactory model exists to describe its dependencies, namely, the band-bending fluctuation model. The comparison of this model with experiment is discussed next.

7.3.6 Observation of Band-Bending Fluctuations

Consider the measurement of the width of the loss peak in depletion and its variation with gate bias, oxide fixed charge density, interface trap charge density, and temperature. These dependencies are compared to the simple formula of Brews[14] [see (6.70)]

$$\sigma_s^2 = \frac{N_f^+ + N_{it}^+ + N_f^- + N_{it}^-}{N_o} \ln\left[1 + \left(\frac{C_\lambda}{C_s + C_{ox} + C_{it}}\right)^2\right] \tag{7.5}$$

where N_f^+ and N_f^- are the number densities of positive and negative oxide fixed charges and N_{it}^+ and N_{it}^- are the number densities of positive and negative interface trap charges. Other symbols are

$$C_\lambda \equiv \frac{\epsilon_s + \epsilon_{ox}}{\lambda}$$

$$N_o = 4\pi \left(\frac{\epsilon_s + \epsilon_{ox}}{\beta q}\right)^2 \tag{7.6}$$

$$\approx 5.85 \times 10^{11}\, cm^{-2} \quad \text{at } 290°K$$

and λ is an adjustable parameter.

According to (7.5), σ_s^2 is gate bias dependent through N_{it}^+, N_{it}^-, and C_s. Temperature dependence is contained in N_o and C_s. Because C_s occurs logarithmically, its influence is weak. Similarly, the fitted parameter λ of (7.6) enters logarithmically and affects the magnitude of σ_s^2 but not its bias independence, temperature dependence, nor interface charge density dependence.

(a) Gate Bias Dependence

The early experimental work of Nicollian and Goetzberger[4] on wet oxides indicated that the variance of band bending was gate bias independent. This gate bias independence was confirmed for wet oxides by Fahrner and Goetzberger[15] (see Fig. 7.13a). Such a gate bias independence is expected from (7.5) provided that the interface charge density is not gate bias dependent, that is, provided that the change in interface trap charge number density with gate bias is small compared to the net number density of oxide fixed charges $N_f^+ + N_f^-$.

Fahrner and Goetzberger[15] also measured dry oxidized samples in which σ_s varied strongly with gate bias. However, this gate bias dependence can be accounted for as a gate bias dependent interface trap charge density because $Q_{it} \gtrsim Q_f$ in these samples, as can be inferred from D_{it} shown in Fig. 7.13b. The capture cross section behavior shown in Fig. 7.13c is discussed in Section 7.3.7.

Declerck et al.[16] also confirmed a rather weak gate bias dependence of σ_s in various p-type and n-type samples for which $N_{it} \ll N_f^+ + N_f^-$, as shown in Fig. 7.14. However, to obtain a reasonable fit to the *magnitude* of σ_s, a degree of compensation was hypothesized for their samples. That is, in addition to the usual positive oxide fixed charge density N_f^+, they assumed a negative oxide fixed charge density N_f^-. The degree of compensation needed is a function of the value assumed for λ, as is discussed in the next subsection. For $\lambda \approx 75\,Å$, $N_f^- \approx 1.4 \times 10^{11}\,cm^{-2}$, and for $\lambda \approx 5\,Å$, $N_f^- \approx$

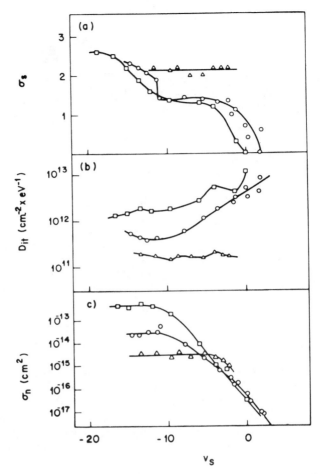

Fig. 7.13 Density of interface trap levels D_{it}, broadening parameter σ_s, and capture cross section σ_n plotted against band bending v_s. Three different oxides are illustrated: dry oxide (squares), dry annealed oxide (o), and wet oxide (Δ). The samples are n-type, (111) epitaxial material, $N_D = 5 \times 10^{15}$ cm^{-3} with Cr-Au contacts. After Fahrner and Goetzberger.[15]

4.4×10^{10} cm^{-2}. The fit shown in Fig. 7.14 does not vary much with the value of λ chosen. One sample, $n(111)$, does not fit well.

A possibility other than charge compensation is that a background contribution to the width of the conductance peak exists as a result of a distribution of individual capture cross sections.

The results due to Declerck et al.[16] raise the complication that charges of both signs may be needed to explain the data, at least on some samples at low levels of charge. This possibility means that *two* fitting parameters may be necessary: λ and the number density of negative oxide fixed charge. The investigation of such charge compensation was done for a p-type (100)

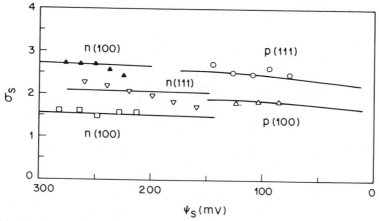

Fig. 7.14 Broadening parameter σ_s versus band bending for several different samples with values of D_{it} in the low $10^{10}\,\text{cm}^{-2}\,\text{eV}^{-1}$ range and oxide fixed charge densities in the low $10^{11}\,\text{cm}^{-2}$ range. The solid lines are predicted using (7.5) with $\lambda = 75\,\text{Å}$. Doping densities are in the low to middle $10^{15}\,\text{cm}^{-3}$ range. After Declerck et al.[16]

wet oxidized sample by Ziegler and Klausmann[17] and extended by Ziegler[18] to determine the donor or acceptor character of interface traps. This work is discussed next.

(b) Investigation of Presence of Two Signs of Interface Charge

According to (7.5), for $(N_f^+ + N_f^-) \gg (N_{it}^+ + N_{it}^-)$, the variance of silicon band bending is proportional to the sum $(N_f^+ + N_f^-)$, with each sign of charge contributing independently to the fluctuations. Flatband voltage shift V_{FB}, however, is a measure of the difference $(N_f^+ - N_f^-)$. Consequently, if both types of charge are present, it is possible to fit both σ_s^2 and V_{FB}, provided that σ_s^2 is not less than the value predicted taking V_{FB} as due to a single type of charge.

Experimentally, σ_s^2 and V_{FB} are known. Treating N_f^- and λ as adjustable parameters, we have

$$N_f^+ = N_f^- + \frac{C_{ox} V_{FB}}{q} \tag{7.7}$$

and, substituting (7.7) in (7.5)

$$\lambda = \lambda_o \left\{ \frac{\exp\left[\dfrac{N_o \sigma_s^2}{Q_f/q}\right] - 1}{\exp\left[\dfrac{N_o \sigma_s^2}{Q_f/q + 2N_f^-}\right] - 1} \right\}^{1/2} \tag{7.8a}$$

where $Q_f \equiv C_{ox} V_{FB}$ and λ_o is the value of λ that fits the known value of σ_s^2

when N_f^- is assumed to be zero. Solving (7.8a) for N_f^-, we obtain

$$N_f^- = \frac{N_o \sigma_s^2}{2 \ln\{1 + (\lambda_o/\lambda)^2 [\exp(N_o \sigma_s^2/(Q_f/q)) - 1]\}} - \frac{Q_f}{2q}. \tag{7.8b}$$

Using (7.8b), we can plot N_f^- against λ for any given σ_s^2 and Q_f; N_f^+ from (7.7) is then a parallel curve, shifted by Q_f/q. Such a set of curves is shown in Fig. 7.15 for $Q_f/q = 2 \times 10^{11}$ cm^{-2}, $\lambda_o = 75$ Å, and $\sigma_s^2 = 4$ (solid lines) and $\sigma_s^2 = 3$ (dashed lines). Thus λ is a strong function of N_f^- for these values of σ_s^2 and Q_f/q.

In principle, any pair of values (N_f^+, N_f^-) corresponding to the same choice of $\lambda \geq \lambda_o$ is an acceptable fit of (7.5) to experiment because any such pair reproduces the experimental σ_s^2 and Q_f/q. Ziegler and Klausmann[17] measured Q_f/q and measured σ_s^2 at a variety of positions of the Fermi level in the bandgap. Both temperature and bias were varied. Then, using (7.7) and (7.8), they computed N_f^+ and N_f^-. The results for a p-type (100) wet oxidized sample are shown in Fig. 7.16. Two choices of λ are illustrated: $\lambda = 75$ Å shown in Fig. 7.16a and $\lambda = 1200$ Å shown in Fig. 7.16b.

A comparison of these two fits shows that choosing $\lambda = 75$ Å provides N_f^+ and N_f^- values with much less scatter than $\lambda = 1200$ Å, thus indicating that the fit of (7.5) to σ_s^2 is better with $\lambda = 75$ Å. For values of λ less than 75 Å, a fit can be obtained only with a negative number of negative charges, so that values of $\lambda < 75$ Å are meaningless.

Figure 7.16a also shows that the resulting values of N_f^+ fit a straight line centered about Q_f/q. Because oxide fixed charge density is much larger

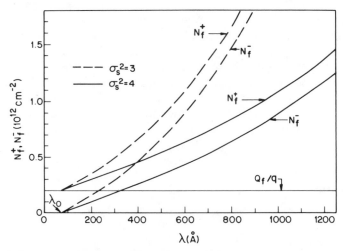

Fig. 7.15 Oxide fixed charge number densities N_f^+ and N_f^- versus λ from (7.7) and 7.8). Parameters are: $\lambda_o = 75$ Å, $Q_f/q = 2 \times 10^{11}$ cm^{-2}, and $\sigma_s^2 = 4$ (solid lines), $\sigma_s^2 = 3$ (dashed lines).

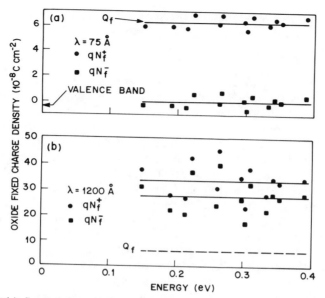

Fig. 7.16 Oxide fixed charge number densities N_f^+ and N_f^- for a p-type, (100), wet oxidized sample. (a) Parameters $\lambda = 75$ Å and N_f^+ and N_f^- computed using (7.7), and (7.8) with σ_s^2 determined by fitting conductance curves and Q_{it}/q determined from flatband voltage. It is assumed that interface trap charge density is negligible, as is observed at midgap. (b) Parameters $\lambda = 1200$ Å and N_f^+ and N_f^- computed as for (a), with the same values for σ_s^2 and Q_{it}/q. After Ziegler and Klausmann.[17]

than interface trap charge density in this sample (6.1×10^{-8} coul/cm^2 as compared to 0.2×10^{-8} coul/cm^2), this behavior of N_f^+ is expected but would not be reproduced if (7.5) failed to describe the gate bias and the temperature dependence of σ_s^2.

Finally, Fig. 7.16a shows there is no need to introduce negative oxide fixed charges to explain Ziegler and Klausmann's data: the N_f^- values are clustered closely about zero, again independent of gate bias or temperature. The difference between these data and those due to Declerck et al.[16] may be a result of different processing.

(c) Donor or Acceptor Nature of Interface Traps

Ziegler and Klausmann's[17] approach also can be applied to samples in which there is significant interface trap charge. Their method decides (in a limited portion of the bandgap) whether interface traps are donors, acceptors, or both. Donors follow the reactions

$$N_{it} \leftrightharpoons N_{it}^+ + e \qquad (7.9a)$$
$$N_{it} + h \leftrightharpoons N_{it}^+ \qquad (7.9b)$$

whereas acceptors follow the reactions

$$N_{it} \leftrightarrows N_{it}^- + h \tag{7.10a}$$

$$N_{it} + e \leftrightarrows N_{it}^- \tag{7.10b}$$

where e and h denote electrons and holes, respectively.

Interface traps that are donors can be distinguished from those that are acceptors. To do this, the change in net interface trap charge with band bending is computed and compared with the total *number* of charges, regardless of sign, inferred from the change of fluctuations in band bending as band bending is varied. To explain further, suppose that band bending is incremented an amount $d\psi_s$, thus moving the conduction band edge toward the Fermi level at the silicon surface. Then interface traps fill. Donor interface traps are neutralized, thereby decreasing both σ_s^2 and the interface trap charge density, whereas acceptor interface traps become negatively charged, thus increasing σ_s^2 and decreasing the interface trap charge density. Consequently, σ_s^2 may decrease, increase, or remain unchanged, according to the dominance of donor or acceptor interface traps or of neither. However, interface trap charge density can only decrease.

To be quantitative, we derive the change in interface trap charge density as band bending varies from a value ψ_{so} to a larger value ψ_s, for which the Fermi level at the silicon surface is closer to the conduction band. From (5.23), we find

$$\frac{dQ_{it}}{d\psi_s} = -q \int\limits_{E_v/q}^{E_c/q} d\zeta_s \left(\frac{kT}{q}\right) f_o(\zeta_s - \phi_s)[1 - f_o(\zeta_s - \phi_s)] D_{it}(\zeta_s) \tag{7.11}$$

where f_o is the Fermi distribution, ζ_s is the position of a typical interface trap energy level, and ϕ_s is the surface potential in electron volts. All energies are measured from the intrinsic level at the silicon *surface* (see Fig. 5.8). The density of interface trap levels D_{it} is the sum of the level densities for each type of interface trap, assuming no correlation between these level densities. For donor (superscript d) and acceptor (superscript a) interface traps, we obtain from (5.25)

$$D_{it}(\zeta_s) = D_{it}^d(\zeta_s) + D_{it}^a(\zeta_s). \tag{7.12}$$

Because $f_o(1 - f_o)$ is appreciable over only a few kT/q about zero argument, where $\zeta_s = \phi_s$, $D_{it}(\zeta_s)$ in (7.11) can be evaluated at $\zeta_s = \phi_s$ and removed from the integration provided that D_{it} does not vary appreciably over an energy interval of a few kT/q about ϕ_s. The observed D_{it} usually satisfies this condition. Therefore, (7.11) becomes

$$\begin{aligned} \frac{dQ_{it}}{d\psi_s} &= -q \, D_{it}(\phi_s) \\ &= -q \, D_{it}(\phi_B + \psi_s) \end{aligned} \tag{7.13}$$

where we have used (5.24) and introduced

$$\phi_s = \phi_B + \psi_s \tag{7.14}$$

and where ϕ_B locates the Fermi level relative to the intrinsic level in the silicon bulk [see Fig. 5.8]. Integrating (7.13) from ψ_{so} to ψ_s, the change in interface trap charge density over this change of band bending becomes

$$\Delta Q_{it} \approx -q \int_{\psi_{so}}^{\psi_s} d\psi\, D_{it}(\phi_B + \psi). \tag{7.15}$$

The derivation of (7.15) can be extended to include band-bending fluctuations, as in Section 8.3.2(a). There is no change in (7.15), but the restriction on the smoothness of D_{it} is more severe, requiring smoothness over an energy range in kT/q of $\pm\sqrt{2}\sigma_s$, rather than a few kT/q (see Section 5.6).

Consider experimental determination of the donor or acceptor nature of interface traps. The measured change in interface trap charge density given by (7.15) is the same regardless of whether $D_{it}(\phi_s)$ consists of donors or acceptors. That is, decreased donor occupancy increases interface charge by adding to the number of positive charges, whereas decreased acceptor occupancy increases interface charge by reducing the number of negative charges. The "blindness" of ΔQ_{it} to the donor or acceptor nature of interface traps means that a C-V measurement alone cannot distinguish donor from acceptor type interface traps. Therefore, measurement of σ_s^2 provides information not available from capacitance measurements alone.

Following Ziegler,[18] we consider three possibilities:

1 **Interface traps are all donors** In this case σ_s^2 will increase as interface traps are emptied, and a fit to σ_s^2 will provide a parameter $(N_f^+ + N_{it}^+)$ that will vary with gate bias just as (7.15).

2 **Interface traps are all acceptors** In this case σ_s^2 will decrease as interface traps are emptied, and a fit to σ_s^2 will provide a parameter $(N_f^- + N_{it}^-)$ that will vary with gate bias in just the opposite manner to (7.15).

3 **Interface traps are a mixture of donors and acceptors** For example, if the number of donor and acceptor interface traps were equal, σ_s^2 would not change and $(N_{it}^+ + N_{it}^-)$ from a fit would be constant with N_{it}^+ increasing as much as N_{it}^- decreases. Other quantities of donor and acceptor interface traps lead to $(N_{it}^+ + N_{it}^-)$ values that do not vary exactly like ΔQ_{it} but do vary in the same direction as ΔQ_{it} (donor interface traps dominant) or oppositely (acceptor interface traps dominant).

We conclude that we can determine whether the interface traps in the measured portion of the bandgap are *only* donors or *only* acceptors and

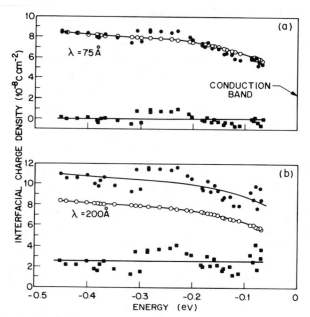

Fig. 7.17 Plot of $q(N_f^+ + N_{it}^+)$ and $q(N_f^- + N_{it}^-)$ as functions of bandgap energy for an n-type, (111), dry oxidized sample; $x_o = 1000$ Å, and $N_D = 4.9 \times 10^{15}$ cm^{-3}. Open circles, $Q_f + Q_{it}$; filled circles, $q(N_f^+ + N_{it}^+)$; filled squares, $q(N_f^- + N_{it}^-)$. Parameters: (a) $\lambda = 75$ Å; (b) $\lambda = 200$ Å. Equations (7.7) and (7.8) were used. After Ziegler.[18]

also whether or not both are present simultaneously, by comparing the behavior of $(N_f^+ + N_{it}^+ + N_f^- + N_{it}^-)$ determined from σ_s^2 with the behavior of ΔQ_{it} from (7.15).

An example for an n-type (111) dry oxidized sample is shown in Fig. 7.17. The values of $N_f^+ + N_{it}^+$ and $N_f^- + N_{it}^-$ determined from (7.7) and (7.8b) are computed from the measured value of σ_s^2 and the measured total interface charge density $(Q_f + \Delta Q_{it})/q$. The change ΔQ_{it} is computed from (7.15) using D_{it} obtained from measurement with $\psi_{so} = 0$ (flatbands).

In Fig. 7.17a $\lambda = 75$ Å. Scatter is small, and $N_f^+ + N_{it}^+$ decreases as the conduction band is approached in exactly the same manner as $Q_f + \Delta Q_{it}$ decreases. Because interface traps fill as the conduction band is approached, donor interface traps in this part of the bandgap far outnumber any acceptor interface traps.

In Fig. 7.17b $\lambda = 200$ Å. In this case scatter in the points is larger than for $\lambda = 75$ Å. Computed values of $N_f^- + N_{it}^-$ no longer cluster near zero, but they do not have any gate bias dependence. Therefore, although some acceptor interface traps may be present in the bandgap outside the region measured (i.e., near the conduction band edge or in the lower half of the gap), the conclusion drawn from Fig. 7.17b is again that donor interface traps far out number acceptor interface traps in the measured part of the bandgap.

Values of λ larger than 200 Å lead to scatter inconsistent with estimated measurement errors, suggesting that Fig. 7.17b shows *upper bounds* on the number of donor and acceptor interface traps in the measured device.

(d) Temperature Dependence

According to (7.5), σ_s^2 varies as T^{-2} apart from a weak dependence by way of C_s. In Fig. 7.18 σ_s computed from (7.5) (open symbols) is compared with experiment (filled symbols).

In Fig. 7.18a for the sample of Fig. 7.17, with an assumed value of $\lambda = 75$ Å and $N_{it}^- + N_f^- = 0$, the fit is very good. Variation of σ_s with energy at fixed temperature is due to the variation of interface trap charge.

In Fig. 7.18b similar results are shown for the sample of Fig. 7.16. Again, $\lambda = 75$ Å and $N_f^- = 0$. In this p-type sample, oxide fixed charge density is much larger than interface trap charge density, so that σ_s is independent of energy at fixed temperature.

These experimental results establish the band-bending fluctuation model, at least for charge densities of $\geq 2 \times 10^{11} \, \text{cm}^{-2}$ and temperatures from 132

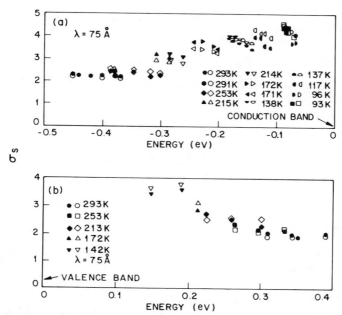

Fig. 7.18 Comparison of experimental (filled symbols) σ_s values with values computed from Brews's formula (7.5) (open symbols) at various temperatures and gate biases: (a) n-type dry oxidized sample in Fig. 7.17. After Ziegler[18]. (b) p-type wet oxidized sample in Fig. 7.16. After Ziegler and Klausmann[17].

to 293°K. At low charge densities, the contribution of capture cross section variations may become significant, but this possibility has not been resolved.

(e) Summary

From the comparisons of (7.5) with experiment in this section, it is clear that (7.5) fits experiment very well with regard to gate bias dependence for both the fixed charge dominated case (Figs. 7.14 and 7.16) and the case of significant interface trap charge (Fig. 7.17). In addition, temperature dependence is well described (Fig. 7.18). One parameter is required by the theory that must be obtained by fitting experiment, namely, $\lambda \approx 75$ Å.

In some cases a second parameter may be needed, the density of compensating fixed charge N_f^-. This parameter is not always needed, and its interpretation is ambiguous. It may represent a background distribution of cross sections, such as that of the Preier model, rather than compensating charge.

Using (7.5), the gate bias dependence of σ_s^2 can be used to infer the donor or acceptor nature of interface trap levels in the bandgap. For a (111) dry oxidized n-type sample, these interface trap levels have been shown to be donor levels in the upper half of the bandgap.

7.3.7 Overall Behavior of Conductance Measurements in Weak Inversion

(a) Conductance Measurements in Various Gate Bias Regimes

The observations discussed so far in this chapter are based on measurements made in *depletion*. This regime is the simplest to analyze; the equivalent circuit was derived in Chapters 5 and 6. In depletion only majority carrier capture is important, and the modulation of band-bending fluctuations by the ac gate voltage is negligible. A typical fit to experiment is shown in Fig. 7.19a. In weak inversion minority carriers play a role, leading to a conductance peak that follows single-level interface trap behavior as shown in Fig. 7.19b. The transition between these two types of behavior was studied by Cooper and Schwartz[19] and is discussed next.

(b) Weak Inversion

The gate bias range from midgap to weak inversion was examined by Cooper and Schwartz.[19] Some experimental results are shown in Fig. 5.21. The solid lines are theory and the points are experiment. At low frequencies corresponding to band bending in weak inversion, a single narrow peak, first reported by Nicollian and Goetzberger,[4] is observed. Near midgap, a very broad peak that is a mixture of weak-inversion-like and depletion-like responses is seen. In depletion, the broad single peak studied by Nicollian and Goetzberger[4] is found.

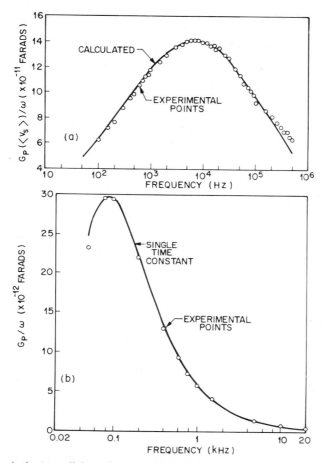

Fig. 7.19 Equivalent parallel conductance in depletion (a) and in weak inversion (b). After Nicollian and Goetzberger.[4] Copyright (1967), American Telephone and Telegraph Company. Reprinted with permission.

The theory from midgap to weak inversion is complicated by the need to include minority carriers (see Section 5.3.6). Therefore, minority carrier capture cross section becomes an additional fitting parameter in this gate bias region. However, the influence of this parameter is confined to biases that place the Fermi level within a few kT/q of midgap. The fit in weak inversion away from midgap depends only weakly on this parameter, and the fit in depletion away from midgap is independent of this parameter. Therefore, the fit of theory to experiment is convincing, and not attributable to the added flexibility introduced into the model by the additional parameter.

The successful explanation of the entire range of conductance behavior—from depletion, through midgap, to inversion—lends considerable support to the interface trap model used here.

7.3.8 Interface Trap Capture Cross Sections

Observations of capture cross sections are not so clear-cut as other aspects of conductance measurements. The magnitude of the capture cross section varies considerably from sample to sample and appears to depend on interface trap level density in some cases. Data in this area are fragmentary; however, no theory consistent with the observations has been presented.

(a) *Band Edge Behavior*

In Fig. 7.13*c* the data on capture cross section of Fahrner and Goetzberger are presented. Capture cross section dropped dramatically as the

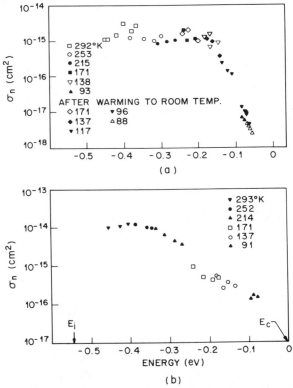

Fig. 7.20 Capture cross section near the conduction band edge for two samples: (a) dry oxidized, (111), *n*-type sample in Fig. 7.13; (*b*) a dual dielectric structure with 1000 Å of SiO$_2$ and 500 Å of Al$_2$O$_3$, $N_D = 4.4 \times 10^{14}$ cm^{-3}, and silicon orientation was (100). After Deuling et al.[13] Copyright (1972), Pergamon Press, Ltd. Reprinted with permission.

conduction band edge was approached. Measurement near the conduction band requires biases near flatbands or in accumulation where majority carrier density is large. These majority carriers screen any interfacial charges present, thus reducing the capture cross sections of any donor type interface traps on n-type silicon and of any acceptor type interface traps on p-type silicon. Screening effects on capture are inadequate to explain the observed drop of capture cross section. Moreover, Deuling et al.[13] find a sample dependence of this effect, shown in Fig. 7.20. Here the dropoff of the lower cross section device is markedly slower than that of the other device or that of Fahrner and Goetzberger's device.

One cause contributing to such effects is the failure of the Gaussian approximation to the Poisson distribution. That is, near flatbands the number of oxide fixed charges contributing to a fluctuation is curtailed, thus making a more careful approximation necessary. For positive oxide fixed charges, the tendency of such a correction is to cause a drop in cross section as shown. At the valence band edge, the effect would be reversed for p-type material because the oxide fixed charges would be repulsive. An additional complication may be present in the rapid increase of D_{it}, shown for these samples in Fig. 7.21.

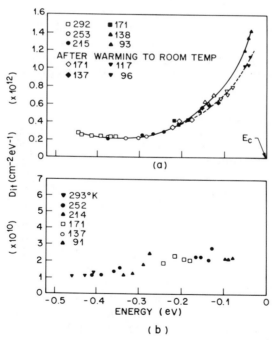

Fig. 7.21 Density of interface trap levels D_{it}, for the same two samples in Fig. 7.22. After Deuling et al.[13] Copyright (1972), Pergamon Press, Ltd. Reprinted with permission.

Three additional reservations in the cross section determination of Fig. 7.20 might be noted:

1   Near flatbands ac modulation of band-bending fluctuations can become important. Such modulation has not been included in the equivalent circuit used to extract the interface trap parameters.

2   Near flatbands rather high frequencies are needed to obtain a complete conductance peak. If only a portion of the G_P/ω versus log ω curve can be measured, fitting only this portion is prone to error, especially if the peak is not completely within the accessible frequency range.

3   When lower temperatures are used to bring the band edge interface trap response within the measurable frequency range, it is not known whether σ_n and D_{it} change with temperature. Some change in D_{it} with temperature is expected because the silicon bands move with temperature, and an interface trap energy level has a parentage that is a mixture of states from various bands.

A more sweeping reservation about our entire treatment holds near the band edges. Here, the wave functions of trapped electrons are spread out over large distances compared to the wave functions of electrons in deep interface traps. Therefore, the wave function of an electron captured in a shallow interface trap level may overlap the wave function for a similar level on a neighboring interface trap. As a result, interface trap levels near the band edges may be characterized by bandlike behavior, and the treatment in terms of isolated trap levels with capture cross sections may not be valid. All these reservations must be understood before interface traps can be characterized near the band edges.

(b) Capture Cross Sections and D_{it}

There appears to be some correlation between capture cross section falloff near the band edge and an increase in D_{it} near the band edge, as seen in Fig. 7.13 and in Figs. 7.20a,b. A similar correlation was found in the midgap region by Cooper and Schwartz[19] (see Fig. 7.22), who found that the minority carrier capture cross section tends to increase with D_{it} whereas the majority carrier capture cross section drops.

As discussed in Section 10.5, interface traps can contribute an error to the experimental value of V_{FB}, causing an error in band bending as well. This error stems from the inability to measure interface trap parameters near flatbands because of their very rapid response time. Because this error in V_{FB} becomes larger for larger D_{it}, the error in band bending also increases with D_{it}. Moreover, because hole and electron densities vary in opposite directions with band bending, there will be a tendency for such an error in band bending to cause σ_n and σ_p to appear to vary oppositely with D_{it}. Although such an error may not be the cause of Cooper and Schwartz's

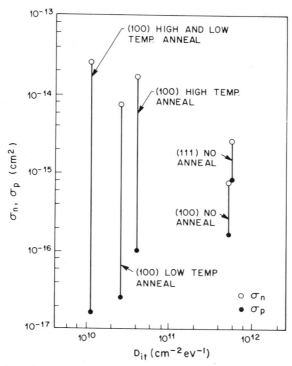

Fig. 7.22 Variation of majority and minority carrier capture cross sections with interface trap level density found near midgap. Samples were *n*-type. After Cooper and Schwartz.[19] Copyright (1974), Pergamon Press, Ltd. Reprinted with permission.

observations, it is too early to rule out such an explanation. Capture cross section dependence on D_{it} cannot be regarded as experimentally established as yet.

REFERENCES

1   B. E. Deal, M. Sklar, A. S. Grove, and E. H. Snow, *J. Electrochem. Soc.*, **114**, 266 (1967).

2   R. J. Powell and C. N. Berglund, *J. Appl. Phys.*, **42**, 4390 (1971).

3   J. R. Brews, *J. Appl. Phys.*, **44**, 379 (1973).

4   E. H. Nicollian and A. Goetzberger, *Bell Syst. Tech. J.*, **46**, 1055 (1967).

5   G. Declerck, R. Van Overstraeten, and G. Broux, *Solid-State Electron.*, **16**, 1451 (1973).

6   A Goetzberger, V. Heine, and E. H. Nicollian, *Appl. Phys. Lett.*, **12**, 95 (1968).

7   B. I. Halperin and M. Lax, *Phys. Rev.*, **148**, 722 (1966); *ibid.*, **153**, 802 (1967).

8   S. Pantelides, *Rev. Mod. Phys.*, **50**, 797 (1978).

9   H. J. Queisser, *Solid-State Electron.*, **21**, 1495 (1978).

10   K. Lehovec, *Appl. Phys. Lett.*, **8**, 48 (1966).

11   E. H. Nicollian and A. Goetzberger, *J. Appl. Phys.*, **40**, 899 (1969).
12   H. Preier, *Appl. Phys. Lett.*, **10**, 361 (1967).
13   H. Deuling, E. Klausmann, and A. Goetzberger, *Solid-State Electron.*, **15**, 559 (1972).
14   J. R. Brews, *J. Appl. Phys.*, **43**, 2306 (1972).
15   W. Fahrner and A. Goetzberger, *Appl. Phys. Lett.*, **17**, 16 (1970).
16   G. Declerck, R. Van Overstraeten, and G. Broux, *J. Appl. Phys.*, **45**, 2593 (1974).
17   K. Ziegler and E. Klausmann, *Appl. Phys. Lett.* **28**, 678 (1976).
18   K. Ziegler, *Appl. Phys. Lett.*, **32**, 249 (1978).
19   J. A. Cooper, Jr. and R. J. Schwartz, *Solid-State Electron.*, **17**, 641 (1974).

8

Extraction of Interface Trap Properties from the Capacitance

- INTRODUCTION, 319
- SMALL-SIGNAL STEADY-STATE CAPACITANCE METHODS, 321
- ERRORS IN EXTRACTING INTERFACE TRAP LEVEL DENSITY, 333
- GRAY-BROWN METHOD, 356
- INFORMATION FROM TRANSIENT CAPACITANCE MEASURE-MENTS, 359

8.1 INTRODUCTION

In this chapter we discuss how interface trap level density as a function of gate bias is extracted from the capacitive component of the measured admittance. There are three approaches to this problem: (1) comparison of the measured high frequency capacitance with a theoretical capacitance with no interface traps (Section 8.2.4), (2) comparison of a measured low frequency capacitance with a theoretical capacitance with no interface traps (Section 8.2.5), and (3) comparison of a measured high frequency capacitance with a measured low frequency capacitance (Section 8.2.6). Each approach is discussed in turn, and reasons are advanced for preferring the last method over the other two.

Following extraction of interface trap level density, the second step is to determine interface trap level density as a function of energy in the silicon bandgap. The extraction of the position of the Fermi level at the silicon surface as a function of gate bias is required for this step, as is described in Section 3.3.3.

To extract interface trap level density, capture cross section, and time constant dispersion, admittance must be measured as a function of frequency and gate bias. Both capacitance and equivalent parallel conductance contain the same interface trap information.* However, the measurement of equivalent parallel conductance is more sensitive and accurate because the conductance is related directly to the energy loss provided by the ac signal source during capture and emission of carriers by interface traps.† In extracting interface trap properties from the capacitance, *differences* in capacitance must be calculated because depletion layer capacitance is in parallel with interface trap capacitance. Taking differences leads to round-off error.

Because of its advantages, the conductance method is used when a complete characterization of interfacial properties is desired or when sensitivity and accuracy are paramount, as described in Section 5.5. However, in many applications only interface trap level density is of interest and the sensitivity and accuracy of the conductance method is not needed. Such applications are the study of the effects of stress or annealing on interface trap level density and the monitoring of interface trap level density during integrated circuit fabrication. For these applications, it is convenient to measure capacitance as a function of gate bias at only two frequencies, one so high that interface traps do not respond and one so low that they maintain equilibrium. As a consequence of using only these two frequencies, all dynamical information about interface traps is lost.

The errors in extracting interface trap level density using the capacitance methods are discussed in Section 8.3, and the sensitivity and the accuracy of the conductance and capacitance methods are compared in Section 8.3.6.

An extension of the capacitance methods is to measure capacitance transients, the emission time spectroscopy methods discussed in Section 8.5. By monitoring the charge emitted from filled interface traps in a narrow time interval, the density of interface trap levels with a specified emission constant can be determined by these methods. The possibility exists for obtaining information additional to that from the conductance method because different assumptions are needed to interpret the two measurements, as discussed in Section 8.5.2.

---

*These two components of the admittance are related to each other by the Kronig–Kramers relations.
†This ac loss is due to the phase shift of the capture and emission processes relative to the applied ac voltage, as discussed in Chapter 5.

8.2 SMALL-SIGNAL STEADY-STATE CAPACITANCE METHODS

Capacitance measurements reveal interface traps by way of (1) stretch-out of the C-V curves along the gate bias axis and (2) the contribution of interface traps to the capacitance. Consider these effects in turn.

8.2.1 Stretchout Due to Interface Traps

Interface traps change their charge state depending on whether they are filled or empty. *Acceptor* interface traps are negative when filled, and neutral when empty, whereas *donor* interface traps are neutral when filled and positive when empty. Both types of interface trap may exist, perhaps simultaneously in the same device.*

Because interface trap occupancy varies with gate bias, stretchout of C-V curves occurs. To understand stretchout, compare the change in band bending of two MOS capacitors, one without and one with interface traps, when the gate charge is changed by an amount δQ_G. Both MOS capacitors are allowed to reach equilibrium after this change in gate charge.

In the MOS capacitor without interface traps, overall charge neutrality requires the change in gate charge to be balanced by a change in silicon surface charge δQ_s. That is, $\delta Q_G + \delta Q_s = 0$. Therefore, band bending changes to bring about this balance.

In the MOS capacitor with interface traps, a change in interface trap charge density δQ_{it} also occurs with any change in band bending. Therefore, charge balance satisfies $\delta Q_G + \delta Q_{it} + \delta Q_s = 0$. Because the MOS capacitor with interface traps includes the additional change in charge density δQ_{it}, the required change in δQ_s is less. Thus the change in band bending is less in the MOS capacitor with interface traps than in the one without.† Consequently, to drive the MOS capacitor from accumulation to inversion requires a larger range of gate charge variation for the case with interface traps, than for that without interface traps. Because gate charge and gate bias are related $[dQ_G = C\,\delta V_G]$, a larger swing of gate bias also will be required. Thus the C-V curve is *stretched out* along the gate bias axis.

A quantitative treatment of stretchout can be obtained from Gauss's law

$$C_{ox}(V_G - \psi_s) = -Q_{it}(\psi_s) - Q_s(\psi_s) \tag{8.1}$$

where C_{ox} = oxide capacitance per unit area, V_G = gate bias, ψ_s = band bending, Q_{it} = interface trap charge per unit area, and Q_s = silicon surface charge per unit area. For a slow, infinitesimal change in gate bias dV_G, (8.1)

---

*A reliable method for determining which type is dominant in a limited region of the bandgap is discussed in Section 7.3.6(c), where it is shown that interface traps in thermally grown device-grade oxides are predominantly donor-type in the upper half of the silicon bandgap.
†This is true for n-type or p-type silicon and acceptor- or donor-type interface traps.

predicts a change in band bending $d\psi_s$ given by

$$C_{ox} \, dV_G = [C_{ox} + C_{it}(\psi_s) + C_s(\psi_s)] d\psi_s \qquad (8.2)$$

where

$$C_{it}(\psi_s) \equiv - \frac{dQ_{it}}{d\psi_s} \qquad (8.3)$$

and

$$C_s(\psi_s) = - \frac{dQ_s}{d\psi_s} \qquad (8.4)$$

are the interface trap and the silicon surface capacitances per unit area, respectively.

From (8.2), a given gate bias variation dV_G leads to a smaller band-bending variation when C_{it} is present than when C_{it} is zero. That is, (8.2) predicts stretchout of C-V curves.

8.2.2 Interface Trap Capacitance

The low frequency capacitance of the MOS capacitor is defined by

$$C_{LF} \equiv \frac{dQ_T}{dV_G} \qquad (8.5)$$

where the total charge per unit area is

$$Q_T = -(Q_s + Q_{it}). \qquad (8.6)$$

Combining (8.2) and (8.5) yields

$$C_{LF} = \frac{dQ_T}{d\psi_s} \frac{d\psi_s}{dV_G}$$

$$= (C_s + C_{it}) \frac{C_{ox}}{C_{ox} + C_s + C_{it}}. \qquad (8.7)$$

That is, (8.7) shows that in addition to stretchout, measured C_{LF} is increased as a result of C_{it}. The equivalent circuit corresponding to (8.7) is shown in Fig. 8.1.

The capacitance given by (8.7) is measured if the change in charge dQ_T is measured in response to a *very slow* change in gate bias. However, if a small-signal ac variation $dQ_T(\omega)$ occurs in response to a small-signal alternating component of gate bias $dV_G(\omega)$, interface traps will not respond fully. Only those interface traps that can capture electrons within the period of the ac gate voltage will respond. Consequently, C_{it} becomes $C_{it}(\omega)$, and for $\omega \to \infty$, $C_{it}(\psi) \to 0$. That is, for high frequencies, interface traps do not respond to the small-signal alternating component of gate voltage. Consequently, at high frequencies the capacitance measured at a

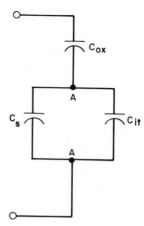

Fig. 8.1 Low frequency equivalent circuit of the MOS capacitor.

given *band bending* is identical in a comparison of MOS capacitors that differ only in interface trap level density. Stretchout is the only effect of interface traps at high frequencies.

For lower frequencies, interface traps respond to ac gate voltage changes. Hence the capacitance at a given band bending is not the same in comparison MOS capacitors but contains an additional *interface trap capacitance*, which is not present if there are no interface traps.

Thus several approaches to extracting interface trap level density from the capacitance exist. The three main methods are: (1) measurement of stretchout of the high frequency C-V curve by comparison with a theoretical curve, (2) measurement of interface trap capacitance by subtraction of calculated capacitance from a measured low frequency C-V curve, and (3) comparison of both high and low frequency C-V curves with no need for a calculated curve.

8.2.3 Relation Between Interface Trap Capacitance and D_{it}

To relate interface trap capacitance to interface trap level density at ac frequencies so low that interface traps maintain equilibrium with the ac voltage at every instant of time, we use the expression for the equilibrium interface trap charge density Q_{it} [see (5.22)]

$$Q_{it}(\psi_s) = q \int_{E_v/q}^{E_c/q} d\zeta_s \{- f_o(\zeta_s - \psi_s - \phi_B) D_{it}^a(\zeta_s) + [1 - f_o(\zeta_s - \psi_s - \phi_B)] D_{it}^d(\zeta_s)\}$$

$$(8.8)$$

where the superscripts a and d refer to acceptor- and donor-type interface traps, respectively.

In (8.8) ϕ_B locates the Fermi level relative to the intrinsic level in the bulk silicon (see Fig. 5.8). Now, under a very slow incremental change in band bending $\delta\psi_s$, the corresponding change in interface trap charge density δQ_{it} is, from (8.8)

$$\delta Q_{it}(\psi_s) = -q \int_{E_v/q}^{E_c/q} d\zeta_s \frac{\partial f_o}{\partial \psi_s} [D_{it}^a(\zeta_s) + D_{it}^d(\zeta_s)] \delta\psi_s. \tag{8.9}$$

Equation (8.9) shows that the *sum* of the acceptor and donor interface trap level densities determines δQ_{it}. *It is not possible* to distinquish donor-type from acceptor-type interface traps by measurement of δQ_{it}.

The definition of the low frequency interface trap capacitance C_{it} is given in (8.3). Therefore, from (8.9)

$$C_{it}(\psi_s) = -q \int_{E_v/q}^{E_c/q} d\zeta_s \frac{\partial f_o(\zeta_s - \psi_s - \psi_B)}{\partial \zeta_s} [D_{it}^a(\zeta_s) + D_{it}^d(\zeta_s)]. \tag{8.10}$$

Now, evaluate $\partial f_o/\partial \zeta_s$

$$\frac{\partial f_o}{\partial \zeta_s} \equiv \frac{\partial}{\partial \zeta_s} \left(\frac{1}{1 + \exp\left[\dfrac{q(\zeta_s + \psi_s - \phi_B)}{kT}\right]} \right)$$

$$= -\frac{q}{kT} f_o(1 - f_o). \tag{8.11}$$

As inferred from Fig. 2.2, $f_o(1 - f_o)$ is a function sharply peaked near zero argument, that is, near the value $\zeta = \phi_s \equiv \phi_B + \psi_s$, which corresponds to an energy level opposite the Fermi level. The width of this peak is approximately a range of energy $\Delta\zeta = kT/q$. Therefore, provided that $(D_{it}^a + D_{it}^d)$ is nearly constant in a range $\Delta\zeta$ about the value $\zeta = \phi_s$, we may evaluate the interface trap level density at $\zeta = \phi_s$ and remove it from the integration in (8.10). The integral is then an integral of $(\partial f_o/\partial \zeta)$

$$\int_{E_v/q}^{E_c/q} d\zeta_s \frac{\partial f_o(\zeta_s - \psi_s - \phi_B)}{\partial \zeta_s} = -1 \tag{8.12}$$

where we have used the fact that $f_o(E_c/q - \psi_s - \phi_B) = 0$ and $f_o(E_v/q - \psi_s - \phi_B) = 1$ for band bendings that do not place the Fermi level within a few kT/q of the band edges. Using (8.12) in (8.10), we obtain, for interface trap level densities that do not vary rapidly over a few kT/q

$$C_{it}(\psi_s) \approx q[D_{it}^a(\phi_s) + D_{it}^d(\phi_s)]$$

$$= q D_{it}(\phi_s) \tag{8.13}$$

where we have used the notation [see (5.25)]

$$D_{it}(\phi_s) \equiv D_{it}^a(\phi_s) + D_{it}^d(\phi_s) \tag{8.14}$$

for the sum of the two interface trap level densities. We refer to $D_{it}(\phi_s)$ in (8.14) as the *total* density of interface trap levels at a position in the bandgap ϕ_s from the intrinsic level at the silicon surface.

Using (8.13), we can relate a measured $C_{it}(\psi_s)$ to the total interface trap level density $D_{it}(\phi_s)$ at a position in the bandgap opposite the Fermi level. Then, knowledge of ψ_s and the bulk doping level allows computation of $\phi_s = (\phi_B + \psi_s)$, so D_{it} can be positioned correctly in the bandgap (see Section 3.3.3).

8.2.4 High Frequency Capacitance Method

Terman[1] developed and used the high frequency capacitance method for determining interface trap capacitance. In the high frequency capacitance method, capacitance is measured as a function of gate bias with frequency fixed at a high enough value so that interface traps do not respond.* The implementation of this measurement is described in Section 12.5.1.

Although interface traps do not follow the ac gate voltage in a high frequency C-V measurement, they *do* follow very slow changes in gate bias as the MOS capacitor is swept from accumulation to inversion. Because interface traps do not respond to the ac gate voltage, they contribute no capacitance to the high frequency C-V curve. However, as interface traps do follow changes in gate bias, they cause the high frequency C-V curve to stretch out along the gate bias axis because interface trap occupancy must be changed in addition to changing depletion layer charge. This stretchout is illustrated in Fig. 8.2, which shows a hypothetical high frequency C-V curve with interface traps compared to an ideal C-V curve. The ideal C-V curve is calculated for the same doping density and oxide thickness but without interface traps. The curve for the MOS capacitor with interface traps arbitrarily has been translated to cross the ideal curve at zero gate bias. Figure 8.2 shows that stretchout does *not* produce a parallel shift of the C-V curve, as do oxide fixed charge and work function differences. Rather, interface traps produce a distortion in the shape of the C-V curve.

As Fig. 8.2 shows, distortion in the shape of the C-V curve will be observed even if interface trap levels are uniformly distributed in energy over the silicon bandgap. The other extreme, an interface trap energy level distribution with pronounced structure, will be reflected in pronounced shape distortion of the high frequency C-V curve. For example, if interface trap level density increases abruptly somewhere in the silicon bandgap,

---

*In Chapter 4 "high frequency" meant a frequency high enough to rule out minority carrier response. Generally, minority carriers do not respond to frequencies much above 1 kHz in device grade silicon (see Fig. 4.1), whereas interface traps will respond to frequencies up to 100 MHz. Thus we broaden the term "high frequency" to mean a frequency sufficiently high that neither interface traps nor minority carriers follow the ac gate voltage.

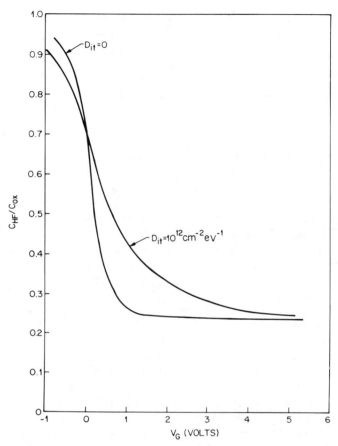

Fig. 8.2 A theoretical high frequency C-V curve with interface trap stretchout compared to a theoretical C-V curve, with no interface traps. Parameters = $N_D = 10^{15}$ cm^{-3}, $x_o =$ 1000 Å, and $D_{it} = 10^{12}$ cm^{-2} eV^{-1}.

capacitance will change much more slowly with gate bias (flatten out) as the abrupt increase in interface trap level density is swept past the Fermi level at the silicon surface by the gate bias.

At high frequencies the total capacitance is given by (8.7) with $C_{it}(\omega) = 0$ (because ω is too large for any ac response of interface traps). That is, the capacitance at high frequencies C_{HF} is given by

$$C_{HF} = \frac{C_s C_{ox}}{C_s + C_{ox}}. \tag{8.15}$$

This capacitance corresponds to the equivalent circuit shown in Fig. 8.3.

The circuit in Fig. 8.3 does not contain interface traps explicitly. Therefore, regardless of interface trap level density, the high frequency capaci-

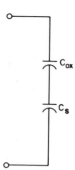

Fig. 8.3 High frequency equivalent circuit of the MOS capacitor.

tance of an MOS capacitor will be the same as that of an ideal one without interface traps, provided that C_s is the same. However, C_s varies with band bending ψ_s. Therefore, measured C_{HF} will be the same as the ideal if band bending is the same. Knowing ψ_s corresponding to a given C_{HF} in the ideal MOS capacitor and measuring V_G corresponding to the same C_{HF} in the real MOS capacitor, we can construct a ψ_s versus V_G curve for the MOS capacitor with interface traps. It is this ψ_s versus V_G relationship that contains all the information about interface trap level density in high frequency C-V measurements.

To obtain this ψ_s versus V_G relationship, we first find how C_s varies with band bending. For a given doping profile, C_s is a known function of band bending. For example, for uniform doping, C_s has been computed in Section 4.4. If biases in inversion are avoided, where the frequency response of the inversion layer introduces complications, C_s is frequency independent and easily calculated.

Now, we can make a theoretical plot of C_{HF} versus ψ_s. This plot can be compared with the measured plot of C_{HF} versus V_G. For any choice of C_{HF}, comparison of these two plots then determines ψ_s versus V_G. Figure 8.4 is a comparison of $C_{HF}(\psi_s)$ versus ψ_s with C_{HF} versus V_G for the same case illustrated in Fig. 8.2. To show how the ψ_s versus V_G curve is constructed, an example is given in Fig. 8.4. Corresponding to a gate bias $V_G = 1$ V, the capacitance ratio determined from the curve for the device with interface traps is $C_{HF}/C_{ox} = 0.431$. This capacitance ratio, when located on the curve for the ideal device, corresponds to the band bending $\psi_s = 5.8 \ kT/q$. Because high frequency capacitance is always the same at a given band bending regardless of interface trap level density, we now have determined for the MOS capacitor with $D_{it} = 10^{12} \ cm^{-2} \ eV^{-1}$ that the gate bias of 1 V corresponds to a band bending $5.8 \ kT/q$.

A full ψ_s versus V_G curve is shown in Fig. 8.5. The example point of Fig. 8.4 is shown by the box of dashed lines in Fig. 8.5. Figure 8.5 shows that interface traps stretch the ψ_s versus V_G curve along the gate bias axis. In fact, the amount of stretchout, as measured by $d\psi_s/dV_G$, determines D_{it}. Thus by graphical or numerical differentiation of the ψ_s versus V_G curve,

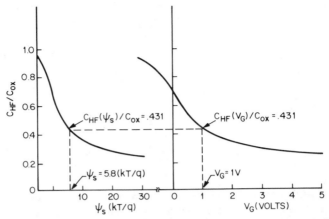

Fig. 8.4 A theoretical C_{HF} versus ψ_s plot compared with a hypothetical C_{HF} versus V_G plot for an MOS capacitor with $D_{it} = 10^{12} \, \text{cm}^{-2} \, \text{eV}^{-1}$. For any given C_{HF}, the same band bending ψ_s occurs so that ψ_s from the C_{HF} versus ψ_s curve corresponds to V_G from the C_{HF} versus V_G plot. Parameters are the same as those for Fig. 8.2.

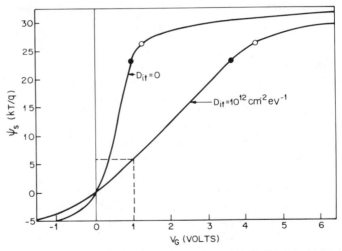

Fig. 8.5 Plot of ψ_s versus V_G for the example in Fig. 8.4. Solid dots indicate $\psi_s = 2\phi_B$, and open circles denote the Lindner band bending ψ_L.

the derivative $d\psi_s/dV_G$ is found. Then $C_{it}(\psi_s)$ can be determined from (8.2) as

$$C_{it}(\psi_s) = C_{ox} \left[\left(\frac{d\psi_s}{dV_G} \right)^{-1} - 1 \right] - C_s(\psi_s). \tag{8.16}$$

Once C_{it} is found from (8.16), D_{it} ordinarily is inferred using (8.13).

8.2.5 Low Frequency Capacitance Method

Berglund[2] developed and was first to use the low frequency C-V method. Again, a C-V curve is measured at a constant frequency, but now at a frequency so low that interface trap response is immediate.* Because interface traps immediately respond to the ac gate voltage, they will contribute an additional capacitance C_{it} to the measured low frequency C-V curve. In addition, interface traps follow changes in gate bias so that the measured low frequency C-V curve will be stretched out along the gate bias axis like the high frequency C-V curve.

At low frequencies the C-V characteristics are represented by the equivalent circuit in Fig. 8.1. This equivalent circuit shows that C_{it} can be extracted from the measured low frequency capacitance if C_s and C_{ox} are known. The relation between the capacitances in Fig. 8.1 can be rewritten from (8.7) as

$$\frac{1}{C_{LF}} = \frac{1}{C_{ox}} + \frac{1}{C_s + C_{it}} \tag{8.17}$$

where C_{LF} is the low frequency capacitance measured at gate bias V_G. Solving (8.17) for C_{it} yields

$$C_{it} = \left[\frac{1}{C_{LF}} - \frac{1}{C_{ox}}\right]^{-1} - C_s. \tag{8.18}$$

where C_{ox} can be measured in strong accumulation. Rather than use an ac method, in practice it has proven convenient to use a slow gate bias ramp[3-5] and to measure the resulting displacement current with an electrometer as described in Section 12.5.2.† This measurement gives $C_{LF}(V_G)$. Finally, we need $C_s(V_G)$ to obtain C_{it} from (8.18). The value $C_s(V_G)$ is obtained by calculation. There are two steps in this calculation. First, C_s is calculated as a function of ψ_s from (3.35) if the doping profile is uniform. If the doping profile is not uniform, it must be determined independently as described in Section 9.4 and then used to calculate $C_s(\psi_s)$. Second, $C_s(\psi_s)$ can be related to $C_s(V_G)$ if ψ_s versus V_G is known.

Two methods for experimentally determining ψ_s versus V_G are described in Section 3.3. However, these methods require measurements other than a low frequency C-V curve. We describe a method here of experimentally obtaining ψ_s versus V_G from a low frequency C-V curve alone. This

---

*In Chapter 4 "low frequency" meant a frequency low enough for immediate minority carrier response. Generally, it requires a lower frequency to elicit immediate minority carrier response than to elicit immediate interface trap response. Thus we broaden the term "low frequency" to mean a frequency low enough for immediate minority carrier *and* interface trap response.

†Alternatively, one can use the Q-V method with less convenience but considerably better accuracy [see Sections 3.3.1 and 12.5.2(c)].

method was used first by Berglund.[2] From (8.2) we obtain

$$\psi_s = \psi_{so} + \int_{V_{G0}}^{V_G} dV_G \frac{C_{ox}}{C_{ox} + C_{it} + C_s}. \tag{8.19}$$

Here V_{G0} is an initial gate bias, corresponding to ψ_{so}. Using (8.7), (8.19) becomes

$$\psi_s = \psi_{so} + \int_{V_{G0}}^{V_G} dV_G \left[1 - \frac{C_{it} + C_s}{C_{ox} + C_{it} + C_s} \right]$$

$$= \psi_{so} + \int_{V_{G0}}^{V_G} dV_G \left[1 - \frac{C_{LF}(V_G)}{C_{ox}} \right]. \tag{8.20}$$

The integrand in (8.20) is determined from a measured C_{LF} versus V_G curve. The value V_{G0} can be chosen arbitrarily. Because the corresponding ψ_{so} is not known, V_{G0} is chosen to minimize the uncertainty in ψ_{so}. For example, V_{G0} can be chosen in accumulation or in inversion, where band bending is only a weak function of gate bias.

With the experimental ψ_s versus V_G relation established, a derivative could be taken and (8.16) used again to obtain $C_{it}(\psi_s)$, just as with the high frequency C-V curve. However, the errors incurred by differentiation can be avoided by using (8.18).

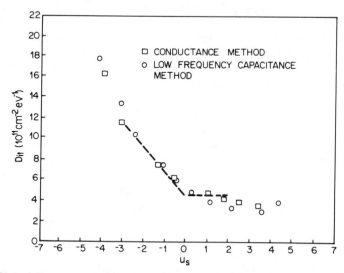

Fig. 8.6 Plot of D_{it} versus position in the bandgap measured from midgap u_s. Negative values of u_s correspond to weak inversion; positive values correspond to depletion. Also shown are the results of conductance measurements. Sample is an unannealed (111), n-type silicon substrate with $N_D = 8 \times 10^{16}$ cm^{-3}, dry oxidized to 878 Å at 1000°C. After Cooper.[6]

Just as with the high frequency capacitance method, a theoretical calculation of $C_s(\psi_s)$ is necessary. In addition, an error in ψ_s versus V_G due to the uncertain integration constant ψ_{so} of (8.20) can lead to a wrong value of C_s in (8.18) and a corresponding error in C_{it}. A full discussion of the errors in this method is deferred to Section 8.3.

An example of the interface trap level density obtained using this method is shown in Fig. 8.6, which also shows a comparison with results obtained using the conductance method (see Section 5.5) on the same sample. Agreement is satisfactory.

8.2.6 Combined High–Low Frequency Capacitance Method

Castagné and Vapaille[3] were the first to combine high and low frequency C-V curves* to obtain a *measured* C_s. The step eliminates the need for a theoretical computation of C_s and for measurement of the doping profile of

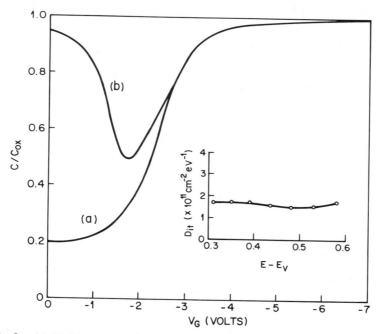

Fig. 8.7 Combined (a) high frequency (1 MHz) and (b) low frequency C-V curves for a Au—Cr—SiO$_2$—Si MOS capacitor; $N_A = 1.5 \times 10^{15}$ cm^{-3} and oxide thickness 500 Å. The resulting interface trap level distribution calculated from (8.22) is shown in the insert. After Kuhn.[4] Copyright (1970), Pergamon Press, Ltd. Reprinted with permission.

*A method for simultaneous measurement of a high and low frequency C-V curve is described in Section 12.5.4.

the device. Using (8.15) yields

$$C_s = \left(\frac{1}{C_{HF}} - \frac{1}{C_{ox}}\right)^{-1}. \tag{8.21}$$

Therefore, (8.18) becomes

$$C_{it} = \left(\frac{1}{C_{LF}} - \frac{1}{C_{ox}}\right)^{-1} - \left(\frac{1}{C_{HF}} - \frac{1}{C_{ox}}\right)^{-1}. \tag{8.22}$$

In this way C_{it} is obtained directly from the measured C-V curves, without the uncertainty introduced by a theoretical C_s and without uncertainty as to whether C_s has been calculated for the correct band bending. Examples of C_{HF} and C_{LF} versus V_G curves are shown in Fig. 8.7.

Of course, (8.22) yields C_{it} only as a function of gate bias. Interface trap level density is obtained as a function of position in the bandgap as described in Section 3.3.3.

Equation (8.21) and hence (8.22) is not valid for gate biases in or near inversion where minority carriers do not follow the high frequency gate voltage.

Instead of solving (8.22), we can obtain interface trap level density graphically by using Fig. 8.8.[7] Figure 8.8 is a plot of (8.22) rearranged for convenience as a function of C_{HF}. The low frequency capacitance C_{LF} is eliminated from (8.22) by the relation

$$C_{LF} = \Delta C + C_{HF} \tag{8.23}$$

where $\Delta C = C_{LF} - C_{HF}$. Substituting (8.23) into (8.22) yields

$$\begin{aligned}
D_{it} &= \frac{C_{ox}}{q}\left[\left(\frac{1}{\Delta C/C_{ox} + C_{HF}/C_{ox}} - 1\right)^{-1} - \left(\frac{1}{C_{HF}/C_{ox}} - 1\right)^{-1}\right] \\
&= \frac{\Delta C}{q}\left(1 - \frac{C_{HF} + \Delta C}{C_{ox}}\right)^{-1}\left(1 - \frac{C_{HF}}{C_{ox}}\right)^{-1} \tag{8.24}
\end{aligned}$$

where C_{it} is related to D_{it} by (8.13). Equation (8.24) from Wagner and Berglund[7] is plotted in Fig. 8.8 with

$$D_{it} = D_{ito}\frac{1000}{x_o} \tag{8.25}$$

where x_o is the oxide thickness in angstroms. To use Fig. 8.8, C_{HF}/C_{ox} is found for a given gate bias, say, V_1 in Fig. 8.9, $\Delta C(V_1)$ from (8.23) is found from the experimental curves as illustrated in Fig. 8.9 and D_{it} determined from (8.25). This procedure can be repeated for another value of gate bias to obtain D_{it} as a function of V_G. However, accurate values of D_{it} cannot be obtained over the entire gate bias range for a number of reasons explained in Section 8.3.5.

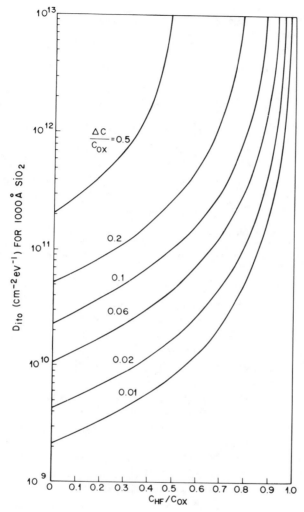

Fig. 8.8 Nomograph for obtaining interface trap level density versus gate bias using the high-low frequency capacitance method. Values of ΔC are obtained at each value of gate bias as shown in Fig. 8.9. After Wagner and Berglund.[7]

8.3 ERRORS IN EXTRACTING INTERFACE TRAP LEVEL DENSITY

In this section, some errors in the capacitance methods are discussed. For both the high and low frequency capacitance methods, one source of error is the theoretical $C_s(\psi_s)$ needed. To calculate $C_s(\psi_s)$, the doping profile must be known. Therefore, some discussion of errors due to doping profile inaccuracy is made in Section 8.3.1. Spatial nonuniformities of the

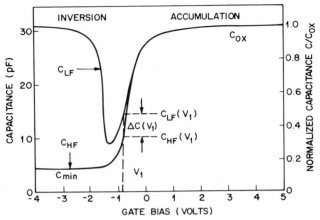

Fig. 8.9 High frequency capacitance C_{HF} and low frequency capacitance C_{LF} as functions of gate bias for an n-type sample illustrating the parameters needed to graphically extract interface trap level density using the high-low frequency capacitance method. After Wagner and Berglund.[7]

charge distribution in the oxide along the interfacial plane also can cause an error in the calculated $C_s(\psi_s)$, as discussed in Section 8.3.2. In the high frequency capacitance method, another error is due to differentiation of the ψ_s versus V_G curve. This error can be made acceptable by precision measurements. Another error is failure to measure a true high frequency C-V curve, particularly a problem near flatbands or in accumulation, as discussed in Section 8.3.3. In the low frequency capacitance method an accurate value of the integration constant used in extracting band bending as a function of gate bias as described in Section 8.3.4 is needed.

We also consider direct measurement of C_{it} using the high-low frequency capacitance comparison method. Here again, difficulties in measuring a true high frequency C-V curve arise near flatbands or in accumulation. Also, in weak-to-strong inversion the experimental evaluation of C_s as $(C_{HF}^{-1} - C_{ox}^{-1})^{-1}$, using (8.21) is unsatisfactory because the value of C_s obtained from a high frequency C-V curve does not incorporate minority carrier response. That is, the high-low frequency capacitance comparison is not valid in weak-to-strong inversion. This limitation is discussed in Section 8.3.5.

Errors due to roundoff are examined in Section 8.3.5. The capacitance methods all depend on subtraction of C_s from the measured $(C_s + C_{it})$. Regardless of whether C_s is measured or calculated, roundoff occurs when C_s becomes large compared to C_{it}. In contrast, the conductance method, described in Chapters 5–7, measures interface trap loss *directly*. Consequently roundoff error is not a problem there. A comparison of accuracy for the capacitance and conductance methods shows that for measurements done with comparable equipment, the accuracy of capacitance measurements is not sufficiently greater than that of conductance measurements to compensate for roundoff error. Particularly where the measurements become difficult, as for cases where (C_{it}/C_s) is small, the conductance method is more accurate than the capacitance methods. The capacitance and conductance methods are compared in Section 8.3.7.

8.3.1 Errors Due to Nonuniform Doping

For both the high and low frequency capacitance methods of determining D_{it}, a theoretical estimate of the semiconductor capacitance C_s is needed for either (8.16) or (8.18). To obtain this theoretical C_s, the doping profile must be known. Usually this profile will not be uniform, because of impurity redistribution during oxidation or due to implantation.

To estimate the error introduced in determining D_{it} where impurity redistribution is ignored, Baccarani et al.[8] calculated the *apparent* D_{it} deduced by using C_s based on uniform doping corresponding to the doping level before oxidation. To do this, they calculated C-V curves for the doping profiles that result from thermal redistribution. These profiles can be deduced from the theoretical model in Section 14.4.2. Then they solved Poisson's equation for the nonuniform profile and computed a C-V characteristic and used this characteristic as C_{LF} in (8.18) with C_s corresponding to uniform doping. The result was the *apparent* D_{it} that would distort the C-V curve for uniform doping to resemble that of the nonuniformly doped case.

Figure 8.10a shows a set of low frequency C-V curves calculated by Baccarani et al.[8] for boron doped silicon oxidized at 1100°C in dry O_2. The solid lines are the C-V characteristics calculated taking into account impurity redistribution, whereas the dashed lines are the corresponding ideal C-V curves calculated for the uniform doping concentration that existed before oxidation.

Figure 8.10b shows the C-V curves calculated for the high frequency case. In depletion, deviations from ideal are similar to those for the low frequency C-V curves in Fig. 8.10a.

The deviation of the C-V curves from the ideal increases with increasing initial uniform doping concentration. That is, as doping density increases, the nonuniformly doped region becomes a larger fraction of the depletion layer width, making the uniform doping approximation poorer. For a depth of the nonuniformly doped region greater than the maximum depletion layer width, the C-V curve will deviate from the ideal curve over its entire gate bias range. Thus deviations increase with doping level, as depletion layer widths are reduced.

For the boron doped samples (see Fig. 8.10), oxidation depletes boron near the surface. Therefore, the redistributed profile contains less boron than the initial uniform profile, leading to a larger depletion layer width at a given bias. As a result, the C-V curves for the redistributed ("experimentally observed") profile lie below those of the uniform ("theoretical")

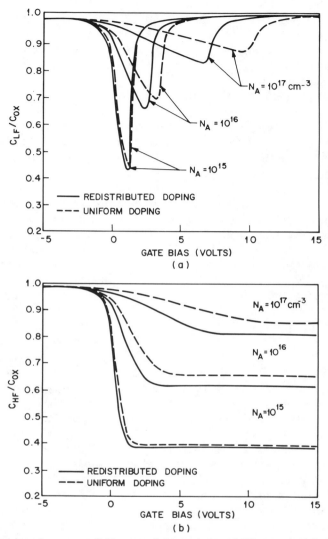

Fig. 8.10 (a) Low frequency C-V curves for boron doped silicon calculated for uniform doping from (3.35) and (3.11) and after impurity redistribution during oxidation at 1100°C to a thickness of 0.2 μm for 2.33 hr in dry oxygen. The doping profile was calculated using (14.4) and (14.5) with $D_o^{1/2} = 4 \times 10^3 \, \mu$m/hr$^{1/2}$, $D_s^{1/2} = 0.2 \, \mu$m/hr$^{1/2}$, $B = 0.131 \, \mu$m/hr$^{1/2}$, and $m = 0.3$, where D_o and D_s are the diffusion coefficients of boron in the oxide and the silicon, respectively; B is the parabolic rate constant; and m is the segregation coefficient. After Baccarani et al.[8] (b) High frequency C-V curves for the same boron doped sample as (a) calculated for uniform doping and after impurity redistribution during terminal oxidation. After Baccarani et al.[8]

profile for biases in depletion. Also, because there are fewer dopant ions, the redistributed structure will invert at a lower bias than the uniform structure, as seen in Fig. 8.10.

In Fig. 8.10 the C-V curves for uniform doping appear "stretched out" in comparison to C-V curves for nonuniform doping. Thus extraction of D_{it} using a theoretical C-V curve for uniform doping leads to an underestimate of D_{it} when the experimental structure has a nonuniform doping profile such as that used in Fig. 8.10. That is, any apparent stretchout, as a result of the use of a theoretical C-V curve based on uniform doping, will be subtracted from the stretchout of the experimental C-V curve due to D_{it}, resulting in too low a D_{it}.

However, for an impurity such as phosphorus, which piles up at the surface during oxidation, the opposite would be true: the redistributed profile with pileup would result in a larger capacitance and delayed inversion compared to the uniform structure. In this case extraction of D_{it} neglecting the redistributed profile would cause an overestimate of D_{it}.

To see the magnitude of the errors introduced in D_{it} when impurity redistribution is neglected, a low frequency analysis of the C-V curves in

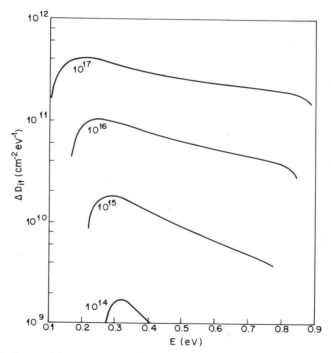

Fig. 8.11 The error ΔD_{it} in interface trap level density caused by neglecting boron redistribution during thermal oxidation. These curves were calculated by comparison of the C-V curves in Fig. 8.10a with curves computed for uniform doping at the bulk level. After Baccarani et al.[8]

Fig. 8.10a is made. Figure 8.11 shows the magnitude of the error ΔD_{it} as a function of bandgap energy with initially uniform doping concentration as the parameter for boron. The greatest error in D_{it} is for the largest doping concentrations. Figure 8.11 shows that the error made in extracting D_{it} is very significant for initially uniform doping concentrations above 10^{15} cm^{-3}, which is right in the range of device interest. Also, note from Fig. 8.11 that neglecting impurity redistribution will affect not only the magnitude of D_{it} extracted at each energy, but also the shape of the D_{it} versus bandgap energy curve.

A large error in D_{it} from neglecting impurity redistribution also would be obtained using the high frequency capacitance method. Moreover, there would be an additional error (not shown in Fig. 8.11) resulting from the differentiation required to obtain D_{it} from (8.16).

A common estimate of silicon surface doping density is the value of uniform doping that fits the measured ratio of the high frequency capacitance in accumulation (the maximum capacitance $\approx C_{ox}$) to its value in inversion (the minimum capacitance), as described in Section 9.4.6. When this average doping level is used in place of the bulk doping, Baccarani et al.[8] found the maximum error in D_{it} to be reduced to $\Delta D_{it} = 10^{11}$ cm^{-2} eV^{-1} at $N_A = 10^{17}$ cm^{-3} and remain the same as in Fig. 8.11 for $N_A = 10^{14}$ cm^{-3}.

8.3.2 Errors Due Lateral Nonuniformities in Interface Charge

Two kinds of interfacial charge nonuniformity have been discussed in Chapter 6. Gross interfacial charge nonuniformities are larger than a depletion layer width in size. They distort C-V curves in somewhat the same way as interface traps. Means to determine the origin of this distortion have been described in Section 6.2.5. In practice, gross interfacial charge nonuniformities are undesirable and would be eliminated from a device-grade interface. Interface trap level density determination on a sample with gross interfacial charge nonuniformities would rarely be attempted.

Apart from gross interfacial charge nonuniformities, small scale or microscopic nonuniformities due to localized oxide fixed and interface trap charges arise. As pointed out in Sections 6.3 and 7.3.5, these charges cause variations in band bending across the interface. In depletion and weak inversion these variations cause negligible changes in the silicon capacitance C_s. However, near flatbands their influence can be significant for the high frequency C_s and can be significant both near flatbands and near inversion for the low frequence C_s.

For bias regimes where spatial charge nonuniformities modify C_s, these nonuniformities prevent accurate determination of the doping profile. They also introduce a dependence of the apparent doping profile on interface trap level density (see Section 9.4).

Qualitatively, the presence of interfacial charge nonuniformities distorts the measured D_{it}. For example, a localized oxide fixed charge near an interface trap will alter the occupancy of this interface trap. Therefore, this interface trap will contribute to the measured capacitance at a gate bias different from the gate bias at which it would contribute if no oxide fixed charge were nearby. Conversely, a single oxide fixed charge causes a larger distortion of mobile carrier density when no interface traps are nearby than when nearby interface traps can change occupancy to screen out the oxide fixed charge. Thus it is not possible to calculate an ideal C-V curve with no interface traps that includes oxide fixed charge nonuniformities correctly. Nor is it possible to calculate a theoretical C-V curve including interface traps but not nonuniformity, because interface traps are themselves localized, causing nonuniformity. Thus wherever these effects are large, a combined approach is necessary in principle. Because all capacitive methods of determining D_{it} do not use such an approach, these methods are inaccurate outside depletion and weak inversion.

Even in depletion and weak inversion, the ability of the capacitance methods to resolve structure in D_{it} is limited by band-bending fluctuations, as discussed next.

(a) Mathematical Analysis

Suppose that the probability of a band bending ψ_s in the interval $[\psi_s, \psi_s + d\psi_s]$ is given by a Gaussian, $P(\psi_s - \langle \psi_s \rangle)$

$$P(\psi_s - \langle \psi_s \rangle) = (2\pi\sigma_s^2)^{-1/2} \exp\left\{ -\frac{[\beta(\psi_s - \langle \psi_s \rangle)]^2}{2\sigma_s^2} \right\} \tag{8.26}$$

where σ_s^2 is the variance of band-bending fluctuations and $\langle \psi_s \rangle$ is the average band bending. This distribution can be derived if the band-bending fluctuations are small enough to allow a linearized treatment of the Poisson equation (see Section 6.3). Such an approximation is valid for a Poisson distribution of localized interface charges for biases in depletion or weak inversion (see Section 6.3). With (8.26), the interface trap charge density, (8.8), becomes

$$\langle Q_{it}(\langle \psi_s \rangle) \rangle = q \int_{-\infty}^{\infty} d\psi_s \, P(\psi_s - \langle \psi_s \rangle) Q_{it}(\psi_s) \tag{8.27}$$

with $Q_{it}(\psi_s)$ from (8.8). Changing variables from ψ_s to $\xi = \psi_s - \langle \psi_s \rangle$, (8.27) becomes

$$\langle Q_{it}(\langle \psi_s \rangle) \rangle = q \int_{-\infty}^{\infty} d\xi \, P(\xi) Q_{it}(\xi + \langle \psi_s \rangle) \tag{8.28}$$

where angular brackets indicate averaging over band-bending fluctuations. In this case a very slow incremental change in mean band bending $\delta\langle \psi_s \rangle$

results in a variation in mean interface trap charge density $\delta\langle Q_{it}\rangle$ given by

$$\delta\langle Q_{it}\rangle = -\left\{ q \int_{-\infty}^{\infty} d\xi \, P(\xi) C_{it}(\xi + \langle\psi_s\rangle) + q \int_{-\infty}^{\infty} d\xi \, \frac{\partial P}{\partial\langle\psi_s\rangle} Q_{it}(\xi + \langle\psi_s\rangle) \right\} \delta\langle\psi_s\rangle. \tag{8.29}$$

The second term in (8.29) is zero only if σ_s^2 in (8.26) does not depend on $\langle\psi_s\rangle$.

For long wavelength nonuniformities, σ_s^2 always is dependent on $\langle\psi_s\rangle$. For example, σ_s^2 depends on $\langle\psi_s\rangle$ for gross charge nonuniformities of dimensions comparable to a depletion layer width or larger (see Fig. 6.4). If such nonuniformities are present, the extraction of D_{it} from a measurement of $\delta\langle Q_{it}\rangle$ is difficult.

For short wavelength nonuniformities due to localized interface charges, σ_s^2 is very nearly independent of $\langle\psi_s\rangle$ in depletion and weak inversion, as discussed in Section 6.3.2. However, in accumulation or inversion, even this case results in a σ_s^2 that varies with $\langle\psi_s\rangle$ (see Figs. 6.8 and 6.9).

Only in depletion and weak inversion and for a Poisson distribution of localized interface charges do we know for certain that the second term in (8.29) can be dropped. In this case, (8.29) shows that the capacitance $\delta\langle Q_{it}\rangle/\delta\langle\psi_s\rangle$ is

$$\frac{\delta\langle Q_{it}\rangle}{\delta\langle\psi_s\rangle} \approx -\langle C_{it}(\langle\psi_s\rangle)\rangle. \tag{8.30}$$

To relate $\delta\langle Q_{it}\rangle/\delta\langle\psi_s\rangle$ to the measured capacitance, $\delta\langle Q_{it}\rangle/\delta V_G$, we also must know $\delta\langle\psi_s\rangle/\delta V_G$. Under the same conditions imposed on (8.30), it is shown in Section 6.3 that $\delta\langle\psi_s\rangle/\delta V_G$ is given by

$$\frac{\delta\langle\psi_s\rangle}{\delta V_G} = \frac{C_{ox}}{C_{ox} + C_s(\bar{\psi}_s) + \langle C_{it}(\bar{\psi})\rangle} \tag{8.31}$$

where $\bar{\psi}_s$ is the band bending corresponding to the average interface charge density. Also, in the approximation of small band-bending fluctuations, we may replace $\bar{\psi}_s$ by $\langle\psi_s\rangle$ in (8.31).* Therefore, in depletion or weak inversion, the interface trap capacitance extracted in the presence of short-range charge nonuniformities is $\langle C_{it}(\langle\psi_s\rangle)\rangle$ related to D_{it} by

$$\langle C_{it}(\langle\psi_s\rangle)\rangle \approx q \int_{-\infty}^{\infty} d\psi_s P(\psi_s - \langle\psi_s\rangle)[D_{it}^a(\phi_B + \psi_s) + D_{it}^d(\phi_B + \psi_s)]. \tag{8.32}$$

From (8.32), band-bending fluctuations average interface trap level density over an energy interval of a few σ_s. Therefore, under the condition that

---

*Because of nonlinearities in the dependence of charge on band bending, the average band bending $\langle\psi_s\rangle$ differs from the band bending corresponding to the average charge density $\bar{\psi}_s$. For small fluctuations, this difference is of the order of the variance in band-bending fluctuations. Over such a small variation in band bending, C_{it} ordinarily does not vary much.

interface trap level density varies slowly over an energy interval of a few σ_s, D_{it} can be evaluated at $\psi_s = \langle \psi_s \rangle$ and removed from the integral. Because the probability $P(\psi_s - \langle \psi_s \rangle)$ integrates to unity, (8.32) becomes

$$\langle C_{it}(\langle \psi_s \rangle) \rangle \approx q D_{it}(\phi_B + \langle \psi_s \rangle) \tag{8.33}$$

where $D_{it} \equiv D_{it}^a + D_{it}^d$ [see (8.14)]. Equation (8.33) imposes a smoothness requirement on the variation of D_{it} more severe than (8.13), and this requirement will limit the resolution of structure in D_{it} when capacitive methods are used.

8.3.3   Failure to Obtain a True High Frequency C-V Curve

The high frequency C-V curve is basic to MOS capacitor measurements. It is necessary for the extraction of the silicon doping profile as explained in Section 9.4, for the extraction of interface trap level density using the high-low frequency capacitance method explained in Section 8.2.6, and for the determination of flatband voltage as explained in Section 10.5.

However, a measured high frequency C-V curve is rarely a good approximation to a true high frequency C-V curve over its entire bias range. As flatbands is approached, majority carrier density increases, making capture more rapid and interface trap time constants shorter. At practical frequencies, interface traps follow the ac gate voltage over part of the gate bias range. This interface trap response limits the bandgap energy range over which interface trap level density can be extracted from the high-low frequency capacitance method.

Figure 8.12 illustrates the preceding discussion at 1 MHz. Here the loss peak and interface trap capacitance normalized to interface trap level density $G_P(\omega)/(\omega q D_{it})$ and $C_{it}(\omega)/(q D_{it})$, are plotted against band bending for the values of the parameters listed in the caption. For band bendings below $3kT/q$, the loss drops with decreasing band bending as more and more interface traps are able to maintain equilibrium with the ac gate voltage. That is, with decreasing band bending, majority carrier density increases, decreasing interface trap response time. As a result of this approach to equilibrium, $C_{it}(\omega)$ approaches the low frequency interface trap capacitance, and $C_{it}(\omega)/(q D_{it}) \to 1$. In contrast, as band bending increases beyond $4kT/q$, fewer and fewer interface traps respond to the ac gate voltage and the loss drops. For the same reason, $C_{it}(\omega)$ drops below the low frequency interface trap capacitance and ultimately tends to zero.

Figure 8.12 also shows how increased broadening of the loss peak due to band-bending fluctuations delays the drop of $C_{it}(\omega)$ to larger band bendings, roughly by $\sqrt{2}\sigma_s kT/q$.

For various frequencies, Fig. 8.13 shows the minimum gate bias for which the C-V curve deviates less than 2% from a true high frequency C-V curve. This gate bias is calculated using (9.42) over the frequency

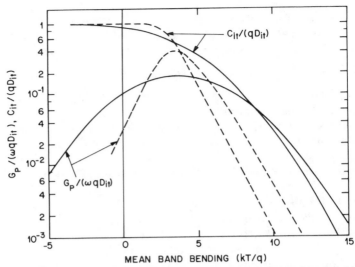

Fig. 8.12 Logarithmic plot of C_{it} and G_P/ω at 1 MHz, normalized to the dc interface trap capacitance qD_{it}, versus band bending. A capture probability of $10^{-8}\,\text{cm}^3\,\text{sec}^{-1}$ is assumed and two values for the standard deviation of band bending $\sigma_s = 0$ (dashed lines) and $\sigma_s = 3$ (solid lines). Doping density is $N_A = 10^{16}\,\text{cm}^{-3}$.

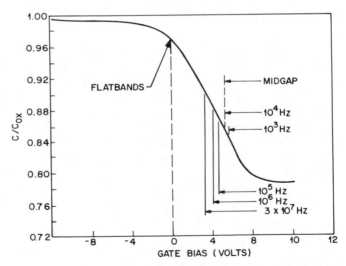

Fig. 8.13 Normalized capacitance calculated from (3.35), (5.73), and (3.11) as a function of gate bias for $N_A = 10^{18}\,\text{cm}^{-3}$, $x_o = 600\,\text{Å}$, $D_{it} = 10^{12}\,\text{cm}^{-2}\,\text{eV}^{-1}$, and $c_p = 2.2 \times 10^{-8}\,\text{cm}^3/\text{sec}$ at infinite frequency. The minimum gate bias above which a finite frequency C-V curve approximates this high frequency C-V curve within 2% is shown as a function of frequency.

range from 10^3 Hz to 30 MHz. Use of frequencies as high as 30 MHz presents experimental difficulties because of errors explained in Section 12.6.1. At 1 MHz none of these high frequency errors need be considered. Therefore, the most widely used frequency for measuring high frequency C-V curves is 1 MHz. However, Fig. 8.13 shows that with a reduction of frequency to 1 MHz, a much smaller fraction of the total bias range can be covered with less than 2% error.

The departure of the 1 MHz C-V curve from the ideal high frequency curve, caused by the increasing capacitive response of interface traps as flatbands is approached, leads to an apparent interface trap level density that is too small. That is, as flatbands is approached, the increasing interface trap capacitance contribution causes the 1 MHz capacitance to rise toward C_{ox} more rapidly than the true high frequency capacitance. A more rapid change of capacitance with gate bias is *opposite* to stretchout, leading to a fictitious reduction in interface trap level density when the high frequency capacitance method is used, as illustrated in Fig. 8.14. A fictitiously small interface trap level density also will result if the combined high-low frequency capacitance method is used because, compared to C_{HF} in (8.22), the 1 MHz capacitance is closer to C_{LF}.

8.3.4 Errors Due to Inaccurate Band Bending

To relate D_{it} as a function of gate bias to D_{it} as a function of position in the bandgap, band bending must be known at each gate bias. For this purpose, the integration of (8.20) may be used. Any error in the constant of integration leads only to a rigid translation of the D_{it} curve in energy, a tolerable error in most cases because D_{it} is only a weak function of energy.

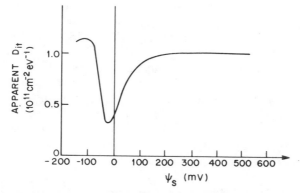

Fig. 8.14 Apparent interface trap level density D_{it} as a function of silicon bandgap energy for $N_A = 5 \times 10^{14}$ cm^{-3}, $x_o = 1500$ Å, $\sigma_p = 6 \times 10^{-16}$ cm^2, and an actual interface trap level density of 10^{11} cm^{-2} eV^{-1}. After Declerck et al.[9] Copyright (1973), Pergamon Press, Ltd. Reprinted with permission.

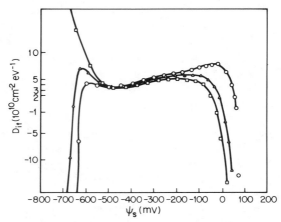

Fig. 8.15 Plot of D_{it} versus ψ_s calculated from an experimental low frequency C-V curve for different values of the integrating constant: $\psi_{so} = 185$ mV (O), $\psi_{so} = 195$ mV (Δ), $\psi_{so} = 205$ mV (□). After Declerck et al.[9] Copyright (1973), Pergamon Press, Ltd. Reprinted with permission.

A more serious error occurs in the low frequency capacitance method where a theoretical $C_s(\psi_s)$ is subtracted from the measured $(C_s + C_{it})$ [see (8.18)]. Here an error in ψ_s leads to subtraction of a value of C_s corresponding to the wrong band bending. This error will be most serious where (C_{it}/C_s) is small or where C_s is a rapid function of band bending.

To illustrate this error, Declerck et al.[9] calculated D_{it} for a given experimental C_{LF} versus V_G curve, using various values of the integration constant ψ_{so} in (8.20). In Fig. 8.15 the resulting D_{it} versus ψ_s is illustrated. In the midgap region, D_{it} is not sensitive to a 20 mV change in ψ_{so}. However, both near flatbands and near inversion where C_s is large and rapidly varying with ψ_s, the resulting D_{it} varies enormously with a 20 mV change in ψ_{so}. Inasmuch as ψ_{so} cannot be determined more accurately than this, the estimated D_{it} cannot be trusted where this estimate is sensitive to ψ_{so}.

The use of the high frequency capacitance method to obtain ψ_s versus V_G without inaccuracy due to integration has two disadvantages: (1) such an approach requires measurement of the doping profile, making this method more complex, and (2), because the profile cannot be measured all the way to the interface [see Section 9.4.3(b)], some inaccuracy in ψ_s results. This inaccuracy could be of the same order as that resulting from the integration technique.

8.3.5 Errors in the High-Low Frequency Capacitance Method

The advantage of the high-low frequency capacitance method is that C_s is determined experimentally from a high frequency C-V measurement for the *same gate bias* at which $(C_s + C_{it})$ is measured using a low frequency C-V measurement. Therefore, the subtracted C_s automatically corresponds to the correct band bending. Additional advantages are that a theoretical calculation is unnecessary and no doping profile measurement is needed.

Because of these advantages of the high-low frequency capacitance method, an extended discussion of its accuracy is presented. We begin with a discussion of round-off error, introduced by the subtraction of nearly equal quantities in (8.22). Then errors due to failure of a 1 MHz C-V curve to approximate a true high frequency C-V curve are considered. Finally, errors due to the onset of inversion are examined.

(a) Round-Off Errors

The first step in the high-low frequency capacitance method is to extract C_s from the high frequency C-V curve as explained in Section 8.2.4. Because we confine ourselves to depletion or weak inversion, we replace C_s by C_D, the depletion layer capacitance. The relative error in C_D caused by the relative measurement errors in C_{ox} and C_{HF} is found by differentiating (8.21) with $C_s = C_D$ and dividing the resulting expression by (8.21), which yields[25]

$$\frac{dC_D}{C_D} = \left(\frac{C_{HF}}{C_{ox} - C_{HF}}\right)\frac{dC_{ox}}{C_{ox}} + \left(\frac{C_{ox}}{C_{ox} - C_{HF}}\right)\frac{dC_{HF}}{C_{HF}} \qquad (8.34)$$

where dC_{ox}/C_{ox} is the relative error in measuring C_{ox} and dC_{HF}/C_{HF} is the relative error in measuring C_{HF}. The important thing to notice in (8.34) is that the relative error in C_{ox} is multiplied by a factor $C_{HF}/(C_{ox} - C_{HF})$ and the relative error in C_{HF} is multiplied by the factor $C_{ox}/(C_{ox} - C_{HF})$. These factors embody the round-off error incurred in subtracting two nearly equal quantities, namely, C_{ox}^{-1} and C_{HF}^{-1}. Both of these multiplying factors can become very large as C_{HF} approaches C_{ox}. It is convenient to express these multiplying factors in terms of a normalized capacitance $C_n = C_{HF}/C_{ox}$. The multiplying factor for dC_{ox}/C_{ox} in (8.34) now becomes $C_n/(1 - C_n)$, and for dC_{HF}/C_{HF} it becomes $1/(1 - C_n)$. Figure 5.26 illustrates how the two normalized multiplying factors increase as C_{HF} approaches C_{ox}.

Figure 5.26 shows that the value of C_D that can be extracted from a high frequency C-V curve will have a rapidly increasing error as C_n approaches unity regardless of the relative measurement error. Figure 8.16 illustrates the band bending v_s for which $(1 - C_{HF}/C_{ox})^{-1} = 10$ [or $(C_{HF}/C_{ox})(1 - C_{HF}/C_{ox})^{-1} = 9$]. Assuming the error in C_{ox} to be about 1%, these curves show the least band bending for which this error in C_{ox} introduces an error in C_D of about 10%.

The amplification of the error in C_{ox} and C_{HF} in (8.34) is due to rounding off. From Fig. 8.16, this round-off error is a serious limitation for thicker oxides and heavier doping, restricting how close to the majority carrier band edge an accurate C_D can be measured.

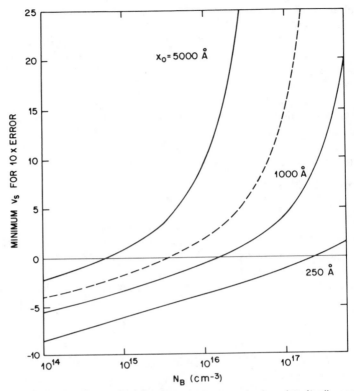

Fig. 8.16 Minimum band bending (in kT/q) v_s versus doping density (in cm^{-3}) N_B for several oxide thicknesses. For band bendings of less than v_s, the amplification factors in (8.34) are larger than 10. The dashed line shows the corresponding band bending for an amplification of 10 in (8.36) for $x_o = 1000 \, \text{Å}$ and $D_{it} = 10^{12} \, cm^{-2} \, eV^{-1}$.

A similar analysis can be made for the low frequency C-V curve. The equivalent parallel capacitance across terminals A–A in Fig. 8.1 at low frequency in depletion where $C_s = C_D$ is

$$C_P = C_{it} + C_D = \frac{C_{ox}C_{LF}}{C_{ox} - C_{LF}}. \tag{8.35}$$

The relative error in extracting C_P from the low frequency C-V curve is obtained by differentiating (8.35) and dividing the resulting expression by (8.35), which yields

$$\frac{dC_P}{C_P} = \left(\frac{C_{LF}}{C_{ox} - C_{LF}}\right)\frac{dC_{ox}}{C_{ox}} + \left(\frac{C_{ox}}{C_{ox} - C_{LF}}\right)\frac{dC_{LF}}{C_{LF}} \tag{8.36}$$

where dC_{LF}/C_{LF} is the relative error in measuring C_{LF}. Note that (8.36) has the same form as (8.34) so that Fig. 5.26 applies, showing how the multiplying factors vary with C_n. Now $C_n = C_{LF}/C_{ox}$. For $D_{it} = 0$, Fig. 8.16

also applies for $(1 - C_{LF}/C_{ox})^{-1} = 10$. For $D_{it} = 10^{12}\,\text{cm}^{-2}\,\text{eV}^{-1}$ and $x_o = 1000$ Å, the dashed line in Fig. 8.16 applies, showing that round-off error is more serious for (8.36) than for (8.34). The dashed line shows that measurement of accurate D_{it} values is limited to energies further and further from the band edge as D_{it} increases.

The quantity of real interest is C_{it}, the *difference* between $C_D + C_{it}$ measured at low frequencies and C_D measured at high frequencies. The relative error in C_{it} can be obtained from (8.22) by differentiating (8.22) and dividing the result by (8.22). We find

$$\frac{dC_{it}}{C_{it}} = \left(\frac{C_{HF}}{C_{ox} - C_{HF}} + \frac{C_{LF}}{C_{ox} - C_{LF}}\right)\frac{dC_{ox}}{C_{ox}} + \left(\frac{C_{ox} - C_{LF}}{C_{ox} - C_{HF}}\right)\left(\frac{C_{HF}/C_{LF}}{1 - C_{HF}/C_{LF}}\right)\frac{dC_{HF}}{C_{HF}}$$

$$+ \left(\frac{C_{ox} - C_{HF}}{C_{ox} - C_{LF}}\right)\left(\frac{1}{1 - C_{HF}/C_{LF}}\right)\frac{dC_{LF}}{C_{LF}}. \tag{8.37}$$

The first term in (8.37) contains the round-off error introduced when C_{LF} or C_{HF} approaches C_{ox}, as already discussed. This error limits the accuracy of D_{it} only for heavy doping, thick oxides, or large D_{it}. The remaining two terms contain factors involving C_{HF}/C_{LF}. These factors amplify the errors in C_{HF} and C_{LF} whenever C_{LF} and C_{HF} are comparable, expressing the round-off error introduced when C_D is subtracted from C_P [see (8.35)]. For $C_{HF} \approx C_{LF}$, the first factor in the error terms contributed by C_{HF} and C_{LF} becomes unity. Ignoring the error due to C_{ox}, (8.37) then becomes

$$\frac{dC_{it}}{C_{it}} \approx \left(\frac{C_{HF}/C_{LF}}{1 - C_{HF}/C_{LF}}\right)\frac{dC_{HF}}{C_{HF}} + \left(\frac{1}{1 - C_{HF}/C_{LF}}\right)\frac{dC_{LF}}{C_{LF}}. \tag{8.38}$$

When (8.38) is used, the relative errors in C_{HF} and C_{LF} again are amplified by the factors in Fig. 5.26, now with $C_n = C_{HF}/C_{LF}$. Thus the relative errors in C_{HF} and C_{LF} lead to greatly magnified errors in C_{it} when C_{it} is smaller than C_D. In or near accumulation, where C_D becomes large, this limitation will hold even for large values of C_{it}.

We now explore the errors in D_{it} due to errors in C_{HF} and C_{LF}.

(b) Error Due to Use of a 1 MHz C-V Curve

Rather than normal measurement errors, the principal error in C_{HF} is due to the use of a 1 MHz C-V curve for C_{HF}. At 1 MHz some interface trap response occurs, causing departure of the 1 MHz capacitance from the ideal C_{HF} that incorporates no interface trap response. The resulting relative error in D_{it} can be found from the second term in (8.37).

To estimate dC_{HF}/C_{HF}, suppose that this error is due entirely to the use of a 1 MHz C-V curve. Then in place of C_{HF}, given by

$$\frac{1}{C_{HF}} = \frac{1}{C_{ox}} + \frac{1}{C_D} \tag{8.39}$$

we use

$$\frac{1}{C_m(\omega)} = \frac{1}{C_{ox}} + \frac{1}{C_D + C_{it}(\omega)}. \tag{8.40}$$

The difference is the error

$$d\left(\frac{1}{C_{HF}}\right) = \frac{1}{C_D} - \frac{1}{C_D + C_{it}(\omega)} \approx \frac{C_{it}(\omega)}{C_D^2}. \tag{8.41}$$

Rewriting the second term in (8.37) in terms of C_D and C_{it}, we find the contribution to the relative error in C_{it} due to the use of a 1 MHz C-V curve

$$\begin{aligned}
\left(\frac{dC_{it}}{C_{it}}\right)_{HF} &= \frac{1/C_{LF} - 1/C_{ox}}{1/C_{HF} - 1/C_{ox}}\left(\frac{1}{1/C_{HF} - 1/C_{LF}}\right)\frac{dC_{HF}}{C_{HF}^2} \\
&= \frac{C_D^2}{C_{it}}d\left(\frac{1}{C_{HF}}\right) \\
&= \frac{C_{it}(\omega)}{C_{it}} \tag{8.42}
\end{aligned}$$

where (8.41) has been used for $d(1/C_{HF})$, $C_{it} = qD_{it}$ and where positive signs are used throughout because only the magnitude of the error is important.

Equation (8.42) shows that the error in C_{it} will be 10% if

$$C_{it}(\omega) \approx 0.1 C_{it}. \tag{8.43}$$

From curves of normalized C_{it} (1 MHz) versus mean band bending, such as those in Fig. 8.12, the minimum band bending for which (8.43) applies can be found. A plot of this minimum band bending $v_{s\,min}$ versus standard deviation of band-bending fluctuations σ_s is shown in Fig. 8.17. For values of $v_s < v_{s\,min}$, $C_{it}(1\,\text{MHz})$ exceeds $0.1C_{it}$. For band bendings greater than $v_{s\,min}$, $C_{it}(1\,\text{MHz})$ falls below $0.1C_{it}$ (see Fig. 8.12).

Figure 8.17 shows $v_{s\,min}$ to be a weak function of σ_s but a strong function of doping density. Because interface trap capture time is inversely proportional to $c_p N_B \exp(-v_s)$, the vertical spacing between curves in Fig. 8.17 is the difference in natural logarithms of the corresponding doping densities. A similar shift occurs if the capture probability is changed from the value chosen for Fig. 8.17, specifically, $c_p = 10^{-8}\,\text{cm}^3\,\text{sec}^{-1}$. Comparison of Figs. 8.16 and 8.17 shows that the error in C_{it} due to use of a 1 MHz C-V curve usually dominates the error due to C_{ox}.

(c) Error from C_{LF}

In (8.37) the last term determines the relative error in C_{it} due to the error in C_{LF}. At the low current levels necessitated by the slow voltage ramp rates used to ensure equilibrium, the relative error in C_{LF} is about 5%.* Using the Q-V method (see Section 3.3.1), much better accuracy is

---

*This error estimate assumes a 10 pA current level. For example, such a current results from using a 20 mV/sec voltage ramp rate for a capacitor with $x_o = 1000$ Å and a gate diameter of 1–2 mm.

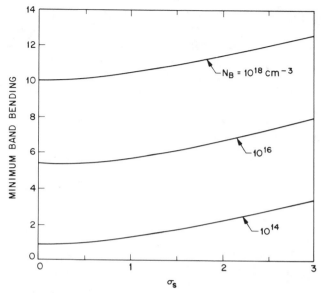

Fig. 8.17 Minimum band bending (in kT/q) above which $C_{it}(1\text{ MHz}) < 0.1\, qD_{it}$ versus standard deviation of band-bending fluctuations (in kT/q) σ_s. Capture probability is assumed to be $c_p = 10^{-8}\text{ cm}^3\text{ sec}^{-1}$.

possible. For this discussion, we assume the relative error in C_{LF} to be about 1%.

Rewriting the last term in (8.37) in terms of C_D and C_{it}, we find the contribution to the relative error in C_{it} due to that in C_{LF}

$$\left(\frac{dC_{it}}{C_{it}}\right)_{LF} = \left(\frac{1/C_{HF} - 1/C_{ox}}{1/C_{LF} - 1/C_{ox}}\right)\left(\frac{1}{1/C_{HF} - 1/C_{LF}}\right)\frac{dC_{LF}}{C_{LF}^2}$$

$$= \left(1 + \frac{C_D}{C_{it}}\right)\left(1 + \frac{C_P}{C_{ox}}\right)\frac{dC_{LF}}{C_{LF}}. \tag{8.44}$$

To keep the error in C_{it} of the order of 10%, we require (8.44) to be of this order. If the relative error in C_{LF} is 1%, approximately a 10% error in C_{it} will result if*

$$C_{it} \approx 0.1 C_D. \tag{8.45}$$

Using $v_{s\,min}$ from Fig. 8.17, we can calculate $C_D(v_{s\,min})$ for any choice of doping density. Substituting this value for C_D in (8.45) and using (8.33), we obtain the value of D_{it} for which the error due to C_{LF} and the error due to

---

*With the use of (8.45) in (8.44), the first factor in the error is $(1 + C_D/C_{it}) = 11$, and the second factor $(1 + C_P/C_{ox})$ is in the range 1–5, depending on doping density and oxide thickness. Therefore, the error in C_{it} is in the range 10–50%.

use of a 1 MHz C-V curve for C_{HF} are comparable. The resulting D_{it} is plotted in Fig. 8.18. For values of D_{it} lower than those in Fig. 8.18, the error in D_{it} due to that in C_{LF} is dominant, amplified by the first factor in (8.44). For values of D_{it} larger than those in Fig. 8.18, the error due to C_{LF} is reduced and that due to the use of a 1 MHz C-V curve is dominant. For these larger values of D_{it}, use of a 1 MHz C-V curve limits how close to the band edge D_{it} can be measured. Also, the larger D_{it} becomes, the larger the bias range becomes over which the dominant error in D_{it} is due to the use of a 1 MHz C-V curve.

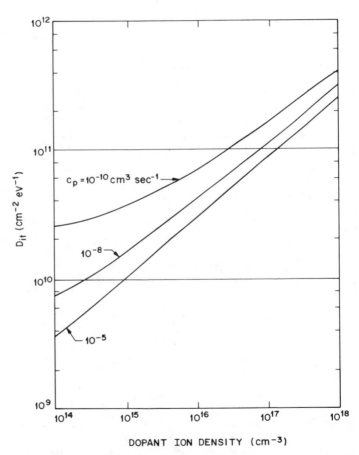

DOPANT ION DENSITY (cm⁻³)

Fig. 8.18 Plot of D_{it} versus doping density. For the plotted D_{it}, the error in D_{it} is caused equally by the error in the slow voltage ramp measurement of C_{LF} and the error due to the use of a 1 MHz C-V curve for C_{HF}. For lower D_{it}, the error due to C_{LF} dominates; for higher D_{it}, the error due to the 1 MHz C-V curve dominates. A relative measurement error in C_{LF} of 1% is assumed, and a relative error in D_{it} of 10% is contributed by both sources of error for D_{it} of these curves.

The error in C_{it} can be maintained of the order of 10% for values of D_{it} lower than shown in Fig. 8.18 by increasing the band bending beyond $v_{s\,min}$. This increase lowers C_D in (8.45), making this requirement less severe. However, minimum D_{it} values are reduced only about 25%, even for biases near midgap (see Table 8.1).

The assumed accuracy in C_{LF} of 1% is obtainable using the Q-V method but is optimistic for a slow ramp C_{LF} measurement. If this error is substantially greater, say 10%, criterion (8.45) becomes $C_{it} \approx C_D$ and the D_{it} values in Fig. 8.18 are 10 times larger. For such large values of D_{it}, the error in D_{it} due to that in C_{LF} will dominate the error due to the use of a 1 MHz C-V curve for all band bendings larger than $v_{s\,min}$. That is, whereas the error due to C_{LF} remains about 10% at all bend bendings, that due to the use of a 1 MHz C-V curve drops rapidly with band bending. In contrast, for a 1% error in C_{LF}, an increase in D_{it} by a factor of 10 reduces the error in D_{it} due to the error in C_{LF} to about 2% [see (8.44)], making the 1 MHz C-V curve the major source of error out to values of v_s sufficient to make $C_{it}(1\,\text{MHz}) = 0.02C_{it}$ [see (8.42)].

The data in Fig. 8.18 might suggest that lightly doped samples should be used to measure low values of D_{it}. However, series resistance then complicates the measurement, requiring use of epitaxial substrates or correction for series resistance.

The limitation of the bias range over which a 1 MHz C-V curve accurately approximates a high frequency C-V curve is not fundamental. The measured C-V curve can be corrected for interface trap response by using (5.71). However, this correction requires a knowledge of interface trap level density, capture probability, and standard deviation of band bending. These parameters can be obtained by using the conductance method, as explained in Section 5.5, but once all this has been done, there is nothing to be gained from the use of the high-low frequency capacitance method. Thus use of a 1 MHz C-V curve is limited by practical considerations to a fraction (typically less than half) of the silicon bandgap energy.

The limitation on the accuracy of measurement of C_{LF} also is not fundamental. By resorting to the Q-V method, very accurate values of C_{LF} can be obtained. However, convenience is lost when one abandons measurement by means of the displacement current caused by a slow gate bias ramp.

(d) Errors Due to Onset of Inversion

On the inversion side of depletion, as the minority carrier band edge is approached, D_{it} can be extracted accurately until inversion layer capacitance becomes a significant fraction of C_{it}. Only C_D, and not the sum of C_D and inversion layer capacitance C_I, can be extracted from the high frequency C-V curve. From the low frequency C-V curve, the value of

C_P extracted will be

$$C_P = C_{it} + C_D + C_I \qquad (8.46)$$

instead of (8.35). Because C_I is included in (8.46), $C_{it} + C_I$ will be the capacitance extracted from (8.22), thereby counting C_I as interface trap capacitance. The value of D_{it} extracted from this capacitance will be in error unless C_I is negligible compared to C_{it}. The capacitance C_I becomes significant when band bending is about $2\phi_B$, which is the onset of strong

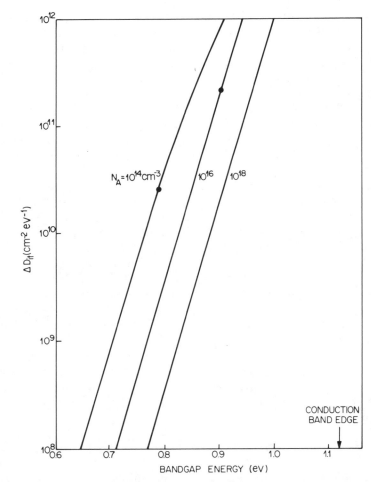

Fig. 8.19 Error in interface trap level density ΔD_{it} versus bandgap energy in the high-low frequency capacitance method due to the departure of the ideal low-frequency capacitance from the ideal high-frequency capacitance as minority carriers become important. p-type silicon is used for illustration. Heavy dots indicate the error at a band bending of $\psi_s = 2\phi_B$.

inversion, although there will be some error in weak inversion for $2\phi_B > \langle \psi_s \rangle < \phi_B$. Band bending greater than $2\phi_B$ will result in a rapidly increasing error in C_{it}.

Figure 8.19 is a plot of this error ΔD_{it} given by (3.35) and (4.152) for uniform acceptor doping by

$$q \, \Delta D_{it} = [C_s(\omega = 0) - C_s(\omega = \infty)]$$

$$= \frac{C_{FBS}}{\sqrt{2}} \left(\frac{1 - \exp(-v_s) + (n_i/N_A)^2[\exp(v_s) - 1]}{\{\exp(-v_s) + v_s - 1 + (n_i/N_A)^2[\exp(v_s) - v_s - 1]\}^{1/2}} \right.$$

$$\left. - \frac{1 - \exp(-v_s)}{\{\exp(-v_s) + v_s - 1\}^{1/2}} \right) \qquad (8.47)$$

where the flatband capacitance is $C_{FBS} = \epsilon_s/\lambda_p$ with $\lambda_p = (\epsilon_s/\beta q N_A)^{1/2}$ from (2.87) or (2.88). For $v_s > v_m$, the relation $v_s = v_m$ is used in the second term of (8.47), where v_m is given by (4.161). Figure 8.19 shows that the error rises exponentially and becomes intolerable before $\psi_s \approx 2\phi_B = 2(kT/q) \ln(N_A/n_i)$, indicated by heavy dots in Fig. 8.19. For example, if $N_A = 10^{14} \, \text{cm}^{-3}$, this error is greater than $\Delta D_{it} = 2 \times 10^{10} \, \text{cm}^{-2} \, \text{eV}^{-1}$ at $2\phi_B$ from Fig. 8.19. Particularly at low D_{it}, where $C_{it} < C_s$, this error in C_s will mask C_{it} entirely.

The relative error in C_{it} will increase toward the majority carrier band edge because of the multiplying factors in (8.37) and will increase again toward the minority carrier band edge because of C_l.

8.3.6 Sensitivity and Accuracy Comparison of the Capacitance and Conductance Methods

Interface trap properties can be extracted from the measured admittance with better accuracy and sensitivity by using the conductance method rather than the capacitance methods, even when the capacitance measurements are extended to include measurement of capacitance dispersion. The relative error of the conductance method depends on the relative error of measuring C_{ox}, C_m, G_m, and ω as shown in Section 5.8. In contrast, the relative error of the capacitance methods depends on the relative error of extracting C_P from the C-V curve at frequency ω and the relative error in extracting C_D from the high frequency C-V curve. At the general frequency ω, we find

$$C_{it}(\omega) = C_P(\omega) - C_D. \qquad (8.48)$$

Because this difference is not measured directly but must be calculated by subtracting the results of two separate capacitance measurements, a round-off error results in computing C_{it}. Moreover, because C_D is larger near flatbands than near midgap, this round-off error will be more severe near flatbands. This round-off error puts a limit on how close in energy to the majority carrier band edge D_{it} can be extracted. From (8.48), the relative error in $C_{it}(\omega)$ is

$$\frac{dC_{it}(\omega)}{C_{it}(\omega)} = \frac{dC_P(\omega) + dC_D}{C_P(\omega) - C_D}$$

$$= \left(\frac{1}{1 - C_D/C_P(\omega)} \frac{dC_P(\omega)}{C_P(\omega)}\right) + \left(\frac{1}{C_P(\omega)/C_D - 1} \frac{dC_D}{C_D}\right). \qquad (8.49)$$

The errors in (8.49) are added because only their magnitude is important. In (8.49), $dC_P(\omega)/C_P(\omega)$ and dC_D/C_D are the relative errors from all sources in measuring $C_P(\omega)$ and C_D, and the simplification of (8.42) is not invoked.

The multiplicative factors in (8.49) blow up as C_{it} becomes smaller and C_P approaches C_D. This blowup is due to round-off error in determining $C_{it}(\omega)$ as the difference (8.48) and is inherent in the capacitance methods. In contrast, no such subtraction arises in the conductance method, where the errors are given by (5.95). Therefore, when capacitance and equivalent parallel conductance are measured with the same relative error, interface trap properties can be extracted with greater sensitivity and accuracy by using the conductance method than by using any capacitance method, particularly for low values of D_{it}. However, the accuracy of the conductance method and the high-low frequency capacitance method are comparable as long as C_{it} is of the same order of magnitude as C_D.

Assuming that the limit of detectability of the high-low frequency capacitance method occurs when $C_{it} = 0.1C_D$ [see (8.45)], the smallest measurable interface trap level densities are as those listed in Table 8.1 for five different doping densities. Values of C_D corresponding to midgap were used in calculating this table. Midgap band bending is a reasonable choice

Table 8.1 Minimum Detectable Interface Trap Level Density $D_{it}(\text{min})$[a]

Doping Density (cm^{-3})	Minimum Detectable Interface Trap Level Density $(\text{cm}^{-2}\,\text{eV}^{-1})$
10^{14}	4×10^9
10^{15}	1×10^{10}
10^{16}	3×10^{10}
10^{17}	9×10^{10}
10^{18}	3×10^{11}

[a]Measurable by the high-low frequency capacitance method calculated from $D_{it}(\text{min}) \equiv 0.1C_D/q$, where C_D is calculated for a band bending corresponding to midgap from $C_D = [q^2\epsilon_s N_B/2kT \ln(N_B/n_i)]^{1/2}$, where N_B is the doping density.

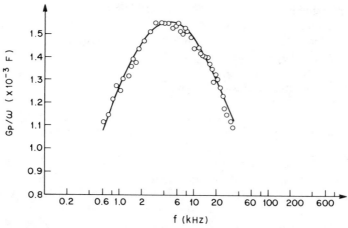

Fig. 8.20 Plot of $\langle G_P \rangle / \omega$ versus log frequency measured on an MOS capacitor having an n-type substrate with a (100) orientation, a donor density of $5.3 \times 10^{15}\,cm^{-3}$, an oxide thickness of 1500 Å, and a gate area of $2.07 \times 10^{-3}\,cm^2$. Gate bias was $-1.65\,V$, corresponding to a band bending of $-313\,mV$. After Declerck et al.[9] Copyright (1973), Pergamon Press, Ltd. Reprinted with permission.

for the purpose of comparison. Table 8.1 shows the dependence of the minimum detectable D_{it} on silicon doping density.

The lowest value of D_{it} measured with the conductance method was reported by Declerck et al.[9] Figure 8.20 shows their $G_P(\omega)/\omega$ versus log frequency curve. Extracting D_{it} from this curve by the method described in Section 5.5.1 yields an interface trap level density of $1.9 \times 10^9\,cm^{-2}\,eV^{-1}$, about an order of magnitude lower than the value from Table 8.1 for the high-low frequency capacitance method.

For the gate area of $2.07 \times 10^{-3}\,cm^2$ used in this sample, $D_{it} = 1.9 \times 10^9\,cm^{-2}\,eV^{-1}$ corresponds to 4×10^6 interface trap levels per electron volt. For an ac band bending of 20 mV peak to peak that is within the small-signal approximation, transitions involving only about 8×10^4 interface trap levels were detected, making the conductance method one of the most sensitive electrical measuring methods in the physical sciences. With an admittance bridge, it would be possible to detect interface trap level densities with the conductance method a decade lower, or in the low $10^8\,cm^{-2}\,eV^{-1}$ range, *independent* of silicon doping density. However, such low values of D_{it} never have been observed.

If C_{it} is very large, the conductance method again is more accurate than the capacitance methods. For the low frequency capacitance method and the high-low frequency capacitance method, round-off error amplifies the error contribution from C_{ox}, as discussed in Section 8.3.5(c). For the high frequency capacitance method, large values of D_{it} cause doping profile measurement errors, making the theoretical capacitance calculation inaccurate, particularly near flatbands [see Section 9.4.3(c)].

Series resistance, described in Section 5.7, limits sensitivity and accuracy of the conductance and capacitance methods. In some cases when D_{it} is small, series resistance loss can dominate the measured equivalent parallel conductance so that no interface trap information can be extracted. In the high-low frequency capacitance method, series resistance can cause the C-V curve measured at 1 MHz to deviate from a true high frequency C-V curve so that correction for series resistance must be made before D_{it} can be extracted. Bulk series resistance would be a limiting factor, particularly for 10^{14} and 10^{15} cm^{-3} doping densities unless an epitaxial layer is used. Series resistance can be made negligibly small by the methods described in Section 12.9 if the experimenter has complete control of the design and fabrication of his samples. However, if testers on an integrated circuit chip are to be measured, correction for series resistance may have to be made.

Bulk trap contributions to the admittance will not be an important factor in depletion for either the conductance or high-low frequency capacitance methods.

8.4   GRAY–BROWN METHOD[11, 12]

One application of the capacitance method yet to be considered is the Gray-Brown method. This method is aimed at measuring D_{it} near the majority carrier band edge, the region in which the conductance method becomes very difficult to use and interpret. The Gray-Brown method is a way of estimating D_{it} by measuring high frequency capacitance at various temperatures. Because the Fermi level moves with temperature, interface trap levels opposite the Fermi level at a given band bending are different at different temperatures. If it were possible to maintain a fixed band bending while cooling the device, the changing interface trap charge density due to the shifting Fermi level would be detectable through the changing gate bias needed to maintain the chosen band bending.

The idea is as follows. At a temperature $T_1 > T_2$ the Fermi level is closer to midgap than at T_2. For the sake of argument, we suppose that filled interface trap levels are neutral and empty ones are positive and assume n-type silicon. Then the field in the oxide is larger at T_1 than at T_2, and the difference is simply

$$F(T_1) - F(T_2) = \frac{q \displaystyle\int_{E_F(T_2)}^{E_F(T_1)} dE\, D_{it}(E)}{\epsilon_{ox}}. \tag{8.50}$$

Consequently, the change in gate bias needed to maintain the band bending at the lower temperature is

$$V_G(T_2) - V_G(T_1) = -\frac{q\displaystyle\int_{E_F(T_1)}^{E_F(T_2)} dE\, D_{it}(E)}{C_{ox}}. \qquad (8.51)$$

If we knew the temperature dependence of the Fermi level, how to maintain a fixed band bending as temperature was varied, and presumed D_{it} did not depend on temperature, we could determine D_{it} from

$$qD_{it} = \frac{d[V_G(T_1) - V_G(T_2)]}{dT_2}\left[\frac{dE_F(T_2)}{dT_2}\right]^{-1}. \qquad (8.52)$$

To determine the gate bias corresponding to a given band bending, the idea of Gray and Brown was to use the known capacitance corresponding to a given band bending. At high frequencies, where interface traps cannot follow the ac gate voltage, the measured capacitance is the depletion layer capacitance, which bears a known relation to band bending at any temperature. Therefore, the Gray-Brown method, in principle, consists of measuring a high frequency C-V curve at various temperatures, calculating the capacitance corresponding to a chosen band bending, picking the gate bias corresponding to this capacitance from the measured C-V curves (one bias for each temperature), and plotting the resulting gate bias versus temperature curve. Finally, the Fermi energy is computed at each temperature and the gate bias versus temperature curve converted to gate bias versus energy. Differentiation of the Q_{it} versus V_G curve provides D_{it} [see 8.52].

This method is rather involved, and we might ask why it should be used in place of the simpler high-low frequency capacitance method. Originally, the Gray-Brown method was proposed for use near the majority carrier band edge, where the other methods fail. Unfortunately, for this purpose, the method is not immune to some of the same basic problems:

1 The doping profile is not known right up to the interface. Consequently, the theoretical capacitance at flatbands is not known. Therefore, the capacitance corresponding to a given band bending near the majority carrier band edge, which means somewhere from accumulation to flatbands, is not known with confidence.

2 Interfacial charge nonuniformities alter the capacitance near flatbands, again making it difficult to find a given band bending.

3 Particularly near flatbands, a very high frequency is needed to ensure that a high frequency C-V curve actually is measured. If an insufficiently high frequency is used, a differing number of interface traps will respond at each temperature, leading to erroneous results. If a sufficiently high frequency is used, corrections for bulk series resistance and other precautions are demanded (see Section 5.7).

The effect of using too low a measurement frequency was illustrated very clearly by Boudry.[13] He used the results of Deuling et al.[14] and modeled D_{it} and σ_n to fit the results of Figs. 7.20a and 7.21a, as shown in Fig. 8.21. Application of the Gray-Brown method to the corresponding theoretical capacitance curves for temperatures between 80 and 300°K resulted in the plots of D_{it} shown in Fig. 8.22.

The peaks in Fig. 8.22 qualitatively can be understood as follows. At temperatures near 300°K and at a frequency of 1.5 MHz, some of the interface traps follow the ac gate voltage. The shift in gate bias for a capacitance corresponding to a given band bending will provide D_{it} through (8.52), but the experimental band bending will not correspond to the theoretical value because the theoretical capacitance does not incorporate interface trap capacitance. Therefore, at higher temperatures the D_{it} obtained will be translated closer to the band edge, on the energy scale, as seen in Fig. 8.22 for $\phi_B \approx -0.2$ V. As temperature is lowered, majority carrier density at the silicon surface is reduced, and the responding interface traps begin to freeze out, thus eliminating their contribution to the capacitance. Then the C-V curve moves closer to a true high frequency curve. The elimination of interface trap capacitance in itself causes a shift in gate bias corresponding to a given capacitance. This shift is additional to the shift of the C-V curve due to the temperature dependence of the Fermi level. Therefore, the derivative in (8.52) is too large, making D_{it} too

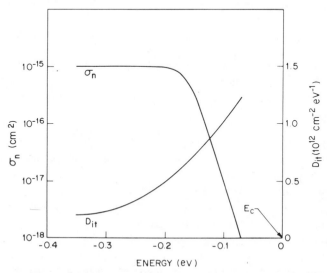

Fig. 8.21 Diagram for D_{it} and σ_n chosen to fit the results in Figs. 7.20a and 7.21a for an n-type MOS capacitor. The empirical formulas used were $\sigma_n = 10^{-15}\{1 + \exp[(\psi_s + 0.16)/0.013]\}^{-1}$ cm^2 and $D_{it} = 2 \times 10^{11}[1 + 65(\psi_s + 0.35)^2]$ cm^{-2} eV^{-1}. After Boudry.[13]

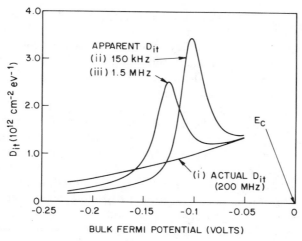

Fig. 8.22 Apparent D_{it} obtained by applying the Gray-Brown method at various frequencies to theoretical capacitance curves generated for an n-type sample with the parameters as in Fig. 8.21. Only for frequencies greater than 200 MHz is a realistic D_{it} obtained. After Boudry.[13]

big. Finally, at sufficiently low temperatures, a true high frequency curve is obtained and D_{it} becomes correct, as indicated in Fig. 8.22 for energies close to the band edge. This approach to the correct D_{it} near the band edge is aided substantially by the assumption of an exponential drop in capture cross section with approach to the band edge (see Fig. 8.21).

According to Boudry's analysis, only at frequencies of 200 MHz would a true D_{it} be obtained by using the Gray-Brown method. Such frequencies are not practical to use [see Section 12.6]. Of course, Boudry's analysis does not apply to every sample. Nonetheless, the conclusion that very high frequencies are needed is quite generally true. For these reasons, the Gray-Brown method is not quantitative, although it may be useful in qualitative comparisons of devices.

8.5 INFORMATION FROM TRANSIENT CAPACITANCE MEASUREMENTS

As pointed out in Section 4.5 in connection with low temperature MOS capacitor measurements, if gate bias is swept rapidly from accumulation toward inversion, mobile carriers are depleted from the interface and interface traps can change occupancy only by emission. Interface trap emission provides a promising opportunity to explore band edge interface trap levels by another method, complementary to steady-state measurements.

For example, a method for bulk traps developed by Lang[15] from a method of Sah et al.[16] called *emission time spectroscopy* can be used for interface traps as well. Preliminary results have been obtained by Schulz and others.[17-19] A closely related method employs a CCD to detect carriers emitted by interface traps, as discussed by Heald[20] and by Kriegler et al.[21]

The approach of emission time spectroscopy requires cooling the MOS capacitor to liquid nitrogen temperatures, to slow down the emission process. Cooling also makes the use of steady-state methods possible in the same time frame because capture and emission times are equal in steady-state. However, emission time spectroscopy permits the measurement of interface trap properties under conditions different from those in the steady-state methods. The first difference is that emission time is measured directly, allowing capture cross section to be determined without knowledge of the majority carrier density at the interface. The second difference is that emission time is independent of lateral nonuniformities, provided that an initial state in accumulation is chosen and that the measurement is completed long before capture by empty interface traps becomes significant. Because lateral interfacial charge nonuniformities are stronger at low temperatures, their avoidance should lead to better resolution than a low temperature, steady-state measurement. The third difference is that the method is sensitive to traps distributed into the oxide, opening the way to tests of the Preier model (see Section 7.3.5), and of capture cross section distributions in general. Finally, and perhaps of most importance, emission time spectroscopy does not require an equivalent circuit to relate interface trap response to the measured admittance. If these differences can be exploited, emission time spectroscopy offers the opportunity to check the presently accepted model of the interface described in Section 7.2. Information would be particularly interesting regarding the role of interface traps distributed into the oxide (Preier model) and of chemical inhomogeneity in general.

We begin with a brief discussion of nonequilibrium gate bias ramps in Section 8.5.1. This work, although qualitative, provides support for the more ambitious use of emission time spectroscopy discussed in Section 8.5.2.

8.5.1 Nonequilibrium Gate Bias Ramps

Consider measuring the high frequency capacitance as gate bias is varied. Two conditions can be imagined: (1) interface traps may remain in equilibrium with the gate bias, in which case the usual small-signal, steady-state, high frequency capacitance is measured; or (2) interface traps may not remain in equilibrium with the gate bias; that is, they may not change occupancy rapidly enough to follow the change in gate bias, in which case the measured capacitance will differ from the steady-state capacitance and contains information about interface trap dynamics.

Such *non-steady-state* *C-V* curves are most easily seen at low temperatures because the time constant for emission τ_e from an interface trap becomes very long. If electron emission is assumed, the emission time is given by

$$\tau_e = (\sigma_n \bar{v} M_c)^{-1} \exp\left(\frac{E_T}{kT}\right) \tag{8.53}$$

where E_T is the interface trap energy level measured from the band edge, σ_n is the capture cross section for electrons, $\bar{v}$ is the thermal velocity of electrons, and M_c is the effective conduction band density of states.

As a consequence of long emission times, if gate bias is swept from accumulation toward inversion in *n*-type, interface traps cannot emit electrons rapidly enough to maintain equilibrium with the gate bias. Because insufficient negative charge is emitted from interface traps, electrons at the depletion layer edge are forced to move instead, thus widening the depletion layer. Therefore, the capacitance drops below its steady-state value.

For *p*-type silicon, an analogous argument holds. In this case as gate bias swings toward inversion, interface traps tend to fill with electrons (i.e., emit holes). Because hole emission is too slow, extra holes at the depletion layer edge must move away from the gate, again widening the depletion layer and reducing the capacitance.

This effect is illustrated in Fig. 8.23. Here, measured *C-V* curves are shown for various sweep rates at 160°K. At the lowest sweep rate, an equilibrium *C-V* curve is measured out to negative biases. However, at

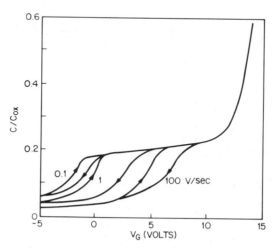

Fig. 8.23 Measured *C-V* curves at various sweep rates for $T = 160°$K. Insulator was silicon nitride, 800 Å thick, $N_D = 5 \times 10^{14}$ cm^{-3}, $D_{it} \approx 3 \times 10^{13}$ cm^{-2} eV^{-1}, and $\sigma_n \approx 10^{-17}$ cm^2. After Wei and Simmons.[22] Copyright (1974), Pergamon Press Ltd. Reprinted with permission.

faster sweep rates, interface traps cannot emit rapidly enough, and the capacitance drops below the steady-state value.

On a return sweep, hysteresis is apparent. This hysteresis occurs because interface traps have continued to emit during the forward sweep, leading to an effective "flatband shift" of the C-V curve. The return sweep is thus a translated version of the forward sweep in the initial period. As the return sweep continues, electron density builds up until, at some critical gate bias, interface traps regain equilibrium by electron capture. Within a narrow range of band bending (i.e., within a short time), interface traps regain equilibrium and then follow the steady-state C-V curve back to accumulation. The same effects were observed earlier on gated diodes by Goetzberger and Irvin.[23]

8.5.2 Emission Time Spectroscopy

To extract the interface trap dynamics contained in transient or non-steady-state C-V curves, one approach is to bias the capacitor to accumulation, filling all the interface traps. Then the capacitor is pulsed into depletion. The high frequency capacitance initially will be less than the steady-state value, as we have seen. As the capacitance relaxes toward steady-state, because interface traps are emitting their charge, a change in capacitance is observed between the two times t_1 and t_2. This change can

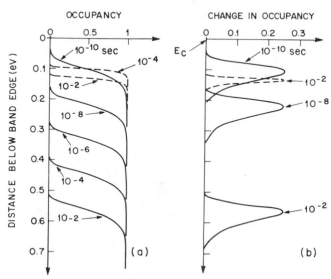

Fig. 8.24 Emptying of interface traps with time after a pulse into depletion: (a) occupancy from (8.55) for various times; (b) change in occupancy according to (8.56) during a time interval equal to the time indicated. Solid lines, $T = 290°K$; dashed lines, $T = 77°K$, $\sigma_n = 10^{-15}$ cm^2, and $\bar{v}M_c = 5.44 \times 10^{21}$ T^2 cm^{-2} sec^{-1}, $t_2 = 2t_1$, $g = 1$.

then be related to the charge emitted during this time interval and hence to the density of interface trap levels and their emission times.

To understand the method, we refer to Fig. 8.24. In Fig. 8.24a the occupancy of interface trap levels is shown at various times, supposing that all the interface trap levels were filled at time $t = 0$. Then the capacitor was pulsed into depletion and held there while the interface trap levels emptied. For this example, which assumes that all interface trap levels have the same capture cross section, interface trap levels steadily empty. Emission begins with levels near the conduction band edge E_c and moves down toward midgap as time goes on. As pointed out by Simmons and Wei,[24] the interface trap level occupancy function resembles an equilibrium Fermi function, with a "Fermi level" that is dropping steadily away from the band edge with time. The *change* in occupancy during the time interval $(t, 2t)$ is shown in Fig. 8.24b. This change in occupancy is strongly peaked near the half-occupancy point at time t. Dashed lines show the same results at 77°K. Here, emission very near the band edge occurs in times of the order of a millisecond.

To be more quantitative, the emission rate of an interface trap level e_n (we assume electron emission into the conduction band for our example) is given by*

$$e_n = g\bar{v}\sigma_n M_c \exp\left(-\frac{E_T}{kT}\right) \tag{8.54}$$

where E_T is measured from the conduction band edge and the degeneracy factor g is as defined in (5.1). If interface trap levels at energy E_T are filled at time $t = 0$, the fraction of interface trap levels occupied at time t when only emission occurs is

$$f(E_T, t) = \exp(-e_n t). \tag{8.55}$$

This function, $f(E_T, t)$, has been plotted in Fig. 8.24a.

The change in occupancy between times t_1 and t_2 is

$$S(E_T, t_1, t_2) \equiv \exp(-e_n t_2) - \exp(-e_n t_1). \tag{8.56}$$

This is the function plotted in Fig. 8.24b for $t_2 = 2t_1$.

The interface trap level corresponding to the maximum change in occupancy during the time interval (t_1, t_2) is found by setting the derivative of S with respect to E_T to zero. *We neglect any energy dependence of capture cross section.* This value is E_m

$$E_m = kT \ln\left[\frac{g\bar{v}\sigma_n M_c(t_2 - t_1)}{\ln(t_2/t_1)}\right]. \tag{8.57}$$

---

*The emission rate is taken to be the same as it is in equilibrium, where it equals the capture rate. Using (4.45), $e_n = c_n n \, (1 - f_o)/f_o$, which leads to (8.54) if we use (5.1) for f_o, (2.13) for n, and measure the trap level energy from the conduction band edge.

For the special case used by Lang[15] and Schulz and Johnson,[17] $t_2 = 2t_1$ and E_m is

$$E_m = kT \ln\left[\frac{g\bar{v}\sigma_n M_c t_1}{\ln 2}\right]. \qquad (8.58)$$

The non-steady-state occupancy of the interface traps is governed by E_m; interface trap levels above E_m tend to be empty, and those below E_m tend to be full.

If we measure E_T from E_m, rather than from the band edge, then S takes on a simple form. Let ξ be defined by

$$\xi = \frac{E_T - E_m}{kT}. \qquad (8.59)$$

Then S becomes

$$S = \exp\left[-\exp(-\xi)\left(\frac{t_1}{t_2 - t_1}\right)\ln\left(\frac{t_2}{t_1}\right)\right] - \exp\left[-\exp(-\xi)\left(\frac{t_2}{t_2 - t_1}\right)\ln\left(\frac{t_2}{t_1}\right)\right]. \qquad (8.60)$$

For the case $t_2 = 2t_1$, S becomes

$$S = \exp[-\exp(-\xi)\ln 2] - \exp[-\exp(-\xi)2\ln 2]. \qquad (8.61)$$

In particular, for $\xi = 0$, S takes on its maximum value S_m

$$S_m = 0.25. \qquad (8.62)$$

Two points can be made: (1) S (and S_m) do not depend on t_1, nor on the parameters ($g\bar{v}\sigma_n M_c$); and (2) when energy is measured in kT/q, as in (8.61), S is independent of temperature. (In particular, the width of S is always the same number of kT/q.)

In Fig. 8.25 E_m from (8.58) is plotted against capture cross section σ_n for two temperatures. Here again at 77°K, interface trap levels near the band edge are observable in reasonable times, and the smaller the capture cross section (the lower the emission rate), the closer to the band edge we can observe interface trap levels. This is particularly interesting because conductance measurements suggest that capture cross section may drop near the band edge. The conductance method is not reliable in this region, so this may not be true.

However, a complication in using the emission "spectroscopy" method is shown in Fig. 8.26. This figure is similar to Fig. 8.25, but now an exponential drop in cross section with energy has been assumed. The formula due to Boudry[13] has been used (see Fig. 8.21), which fits the results due to Deuling et al.[14] In this case, Fig. 8.26 shows that emission occurs from *several* regions of the gap simultaneously. The function S becomes *multiple-peaked* for some choices of time.

Shallow interface trap levels with low capture cross sections emit in the same time frame as deeper interface trap levels with larger capture cross

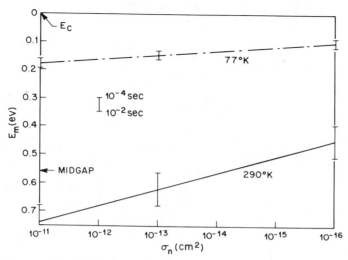

Fig. 8.25 Energy of half occupancy E_m from (8.58) versus capture cross section for the case $t_1 = 2t_2$. Vertical bars indicate the range of energies covered as t_1 ranges from 10^{-2} to 10^{-4} sec.

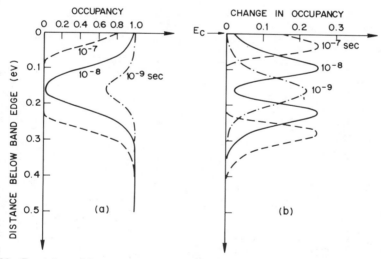

Fig. 8.26 Emptying of interface traps with time for a capture cross section variation given in Fig. 8.21 at $T = 290°K$: (a) occupancy at various times; (b) change in occupancy during the time interval indicated. The first interface trap levels to empty are about 0.15 eV from the band edge. By 10^{-8} sec, two regions of the gap are emptying simultaneously.

sections; therefore, care must be exercised in interpreting emission measurements under such circumstances.

Not to exaggerate this difficulty, the same model is used in Fig. 8.27 at 77°K. At this temperature only one range of energies is emitting. That is, cooling has separated the emission rates of the two sets of energy levels that have equal rates in Fig. 8.26. These two rates cannot remain equal at all temperatures because different values of E_T in (8.54) cause the rates to have different temperature dependencies. The emitting range of energies, indicated by the change in occupancy curves of Fig. 8.27b, is wider than for a constant capture cross section (dashed lines) and must be taken into account in estimating D_{it}.

If the actual interface trap dynamics are not too complex, and if observations are made at various times and temperatures, emission time spectroscopy may be able to provide new information near the band edge, not obtainable by the steady-state methods. Emission time spectroscopy measurements as yet have not yielded added insight into interface trap properties. Emission time spectroscopy has been described briefly in this section to point out its potential in studies of interface trap properties, but the instrumentation of this method is not included in Chapter 12.

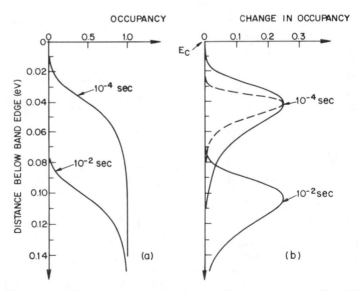

Fig. 8.27 Emptying of interface traps with time for the same case as Fig. 8.26, but at $T = 77°K$. In this case only one range of energies is emptying, and the only effect of rapid cross section variation is a widening of this region. The dashed-line curve in (b) shows the constant capture cross section change in occupancy, which is significantly narrower than the actual change in occupancy shown as a solid line.

(a) Extraction of D_{it} and σ_n

Our discussion has assumed that somehow interface trap emission in the window provided by S can be observed. How is this done? One procedure is to measure the relaxation of the depletion layer width when gate bias is held fixed. As interface traps emit, depletion layer width will shrink, increasing depletion layer capacitance. Therefore, the capacitance change between the two times (t_1, t_2) is a measure of the charge emitted by interface traps in this time interval provided the doping profile is known. Another procedure is to hold the depletion width constant by adjusting the gate bias during the relaxation process.[25] The change in gate charge then equals the change in interface trap charge.*

Because there are few mobile carriers near the interface in depletion, the capacitance can be measured at any frequency compatible with the sampling interval (t_1, t_2). That is, there is no opportunity for interface traps to follow the ac gate voltage used in determining capacitance.

One procedure to find the capture cross sections, suggested by Schulz and Johnson,[17] is to measure D_{it} using the time interval $(t_1, 2t_1)$ and to vary temperature to alter E_m. Then, the measurement is repeated, using another time, $(t_1', 2t_1')$. The two D_{it} curves then can be translated into coincidence, determining a temperature shift ΔT equivalent to the change in time t_1. Equating the E_m values at (t_1, T) and (t_1', T'), we have from (8.58) that

$$T \ln\left[\frac{g\bar{v}\sigma_n M_c t_1}{\ln 2}\right] = T' \ln\left[\frac{g\bar{v}\sigma_n M_c t_1'}{\ln 2}\right]. \tag{8.63}$$

Equation (8.63) determines σ_n, provided that the temperature dependence of σ_n is weak and that the temperature dependence of $(g\bar{v}M_c)$ is known. However, the accuracy of such a determination is poor, particularly because $(T - T')$ is not accurately known. An accurate value for $(T - T')$ is difficult to obtain for the nearly featureless D_{it} normally found because there is no identifiable feature to establish just when the D_{it} curves for T and T' have been brought into coincidence. The inadequacy of this procedure is illustrated by the conflicting conclusions of Schulz and Johnson,[17] further discussed by Yamasaki et al.[18] For emission time spectroscopy to be useful, capture cross section determination must be improved, perhaps by use of the prepulse method described later in Section (c) below.

(b) Problems in the Method

There are some problems in using emission time spectroscopy. One difficulty is the analysis of data where emission from several positions in

---

*Gate bias is adjusted automatically, using a feedback loop that senses the depletion layer capacitance. Because the depletion layer capacitance is measured in the feedback loop at 30 MHz, epitaxial samples with degenerately doped substrates are necessary to avoid series resistance errors.

the bandgap can occur, as shown in Fig. 8.26. This difficulty is addressed in Section (c) below. Also, emission from bulk traps must be distinguished from interface trap emission, requiring additional measurements. Finally, there is a limit on the time interval that can be studied. If one waits too long to observe emission, capture may play a significant role as steady state is approached. This complication is particularly severe near midgap, where time constants are long. Minority carriers also complicate the conductance method near midgap, as discussed in Sections 5.3.5 and 5.3.6. However, analysis has been taken farther for the conductance method.

Emission time spectroscopy requires knowledge of the temperature dependence of MOS parameters. In particular, the theoretical temperature dependence of g, $\bar{v}$, and M_c in (8.54) has not been compared with experiment in the temperature range of interest. Some temperature dependence of D_{it} and σ_n is expected, by analogy with bulk trap levels. At temperatures below 77°K, the approach to steady state of dopant atom levels may need consideration because some freezeout may occur, thus modifying depletion layer response.

An additional problem is in locating the emitting interface trap levels in the bandgap. Even in the simple case shown in Fig. 8.24, where only one narrow energy interval in the bandgap is emitting, the position E_m from (8.58) cannot be found unless the capture cross section is known. The capture cross section determination of (8.63) is too inaccurate for this purpose, and the prepulse approach must be used, as discussed next.

(c) Prepulse Approach

A better approach to locating the emitting energy levels is to vary the initial state. That is, rather than biasing into accumulation only, one may vary the initial or prepulse bias to fill interface trap levels only partially. If this prepulse fills the interface trap levels that emit in the chosen time frame, a signal will be seen. If the prepulse fails to fill these levels, no signal will be seen. By noting the band bending at which the signal drops, the energy level of the interface traps emitting in the chosen time frame can be deduced. This band bending is determined by using either the low frequency C-V method or the Q-V method discussed in Section 3.3. Data from the accumulated initial state (which is free from the effects of interfacial charge nonuniformities) can then be placed correctly in the bandgap by comparison with the results obtained from partially filled initial states (which are subject to the effects of lateral interfacial charge nonuniformities because screening is reduced). Work on such an approach has begun.[26,27]

Once the energy position of emitting levels has been found by using the prepulse, E_m in (8.57) is known, and capture cross section can be found by inverting (8.57)

$$\sigma_n = \exp\left(\frac{E_m}{kT}\right)\left[\frac{g\bar{v}M_c(t_2 - t_1)}{\ln(t_2/t_1)}\right]^{-1}. \qquad (8.64)$$

Unfortunately, the use of a prepulse requires a capacitive determination of band bending, introducing some of the same uncertainties found in the conductance method. For example, the prepulse establishes an initial state with an interface trap level occupancy governed by the Fermi distribution and by interfacial nonuniformities. Therefore, use of the prepulse degrades resolution. In addition, errors can arise in determining band bending. Unless the Q-V method is used, rather than the slow ramp C-V method, accuracy will be poor.

(d) Results

For electrons, the results from this method so far are consistent with an electron capture cross section such as that modeled by Boudry (see Fig. 8.21). However, as discussed earlier, uncertainty in the results in Ref. 17 is so large that they do little to confirm this behavior. Electron capture cross sections from Ref. 17 are about 3×10^{-15} cm^2 at room temperature, in agreement with the results of conductance measurements (see Figs. 7.8, 7.13, and 7.22).

More recent results of emission time spectroscopy reported in Ref. 26 for (111)-oriented n-type silicon with no post metallization anneal and midgap $D_{it} \approx 10^{11}$ cm^{-2} eV^{-1} indicate an energy independent capture cross section over the energy range from 0.15–0.4 eV, with a value of 4×10^{-15} cm^2. This behavior resembles that found with the conductance method, but no direct comparison of the two methods on comparable samples has been made. The range of energy explored in Ref. 24 does not reach close enough to the band edge to test the cross section behavior modeled by Boudry.

For holes, results depend strongly on processing. Emission time spectroscopy measurements were made on a p-type (100)-oriented epitaxial sample subject to a low temperature (450°C) postmetalization forming gas anneal for 30 min.[17] Resulting midgap D_{it} was about 10^{10} cm^{-2} eV^{-1}. The observed capture cross section was $\sigma_p \approx 5 \times 10^{-12}$ cm^2, independent of temperature and of energy in the range 0.2–0.5 eV from the valence band edge. This cross section is very much larger than the values found by using the conductance method ($\sigma_p \approx 10^{-16}$ cm^{-2}; see Figs. 7.8b and 7.22). If interface trap levels were present with σ_p in the range 10^{-14}–10^{-18} cm^2, they should have been observed. In addition, the interface trap level density disagreed with that found using the low frequency method. Whereas D_{it} from emission-time spectroscopy increased from the valence band edge toward midgap, D_{it} from the low frequency method increased in the usual manner, from midgap toward the valence band edge. No satisfactory explanation for these discrepancies is available.

For p-type samples oxidized at 1000°C to an oxide thickness of 1000 Å with a 30 min 1000°C argon anneal but without a postmetalization anneal, a peak in D_{it} about 0.3 eV from the valence band edge is reported.[28] This peak corresponds to $D_{it} \approx 5 \times 10^{11}$ cm^{-2} eV^{-1} and $\sigma_p \approx 1 \times 10^{-14}$ cm^2 on (100)-

and $D_{it} \simeq 5 \times 10^{12}\,\mathrm{cm}^{-2}\,\mathrm{eV}^{-1}$ and $\sigma_p \simeq 5 \times 10^{-14}\,\mathrm{cm}^2$ on (111)-oriented silicon. No data from conductance measurements are available for comparison, except Cooper's work (Fig. 7.22), which indicates a much smaller $\sigma_p \simeq 10^{-16}\,\mathrm{cm}^2$. Agreement with low frequency D_{it} is crude, but the spectrum is qualitatively similar, unlike the annealed case.

REFERENCES

1   L. M. Terman, *Solid-State Electron.*, **5**, 285 (1962).

2   C. N. Berglund, *IEEE Transact. Electron Devices*, **ED-13**, 701 (1966).

3   R. Castagne' and A. Vapaille, *Surface Sci.*, **28**, 557 (1971); *C. R. Acad. Sci. (Paris)*, **270**, 1347 (1970).

4   M. Kuhn, *Solid-State Electron.*, **13**, 873 (1970).

5   D. R. Kerr, Conference on Properties and Use of MIS Structures, Grenoble, France, June 17–21, 1969.

6   J. A. Cooper, Jr., "A Unified Treatment of the Conductance, Capacitance, and Noise Due to Surface States at the SiO₂–Si Interface," Ph.D. thesis, School of Electrical Engineering, Purdue University, West Lafayette, Indiana, 1973.

7   S. Wagner and C. N. Berglund, *Rev. Sci. Instrum.*, **43**, 1775 (1972).

8   G. Baccarani, S. Solmi, and G. Soncini, *Alta Frequenza*, **40**, 310E (1971).

9   G. Declerck, R. Van Overstraeten, and G. Broux, *Solid-State Electron.*, **16**, 1451 (1973).

10   A. Goetzberger, E. Klausmann, and M. J. Schulz, "Interface States on Semiconductor/Insulator Surfaces," *CRC Critical Reviews in Solid-State Science* (1976).

11   P. V. Gray and D. M. Brown, *Appl. Phys. Lett.*, **8**, 31 (1966).

12   D. M. Brown and P. V. Gray, *J. Electrochem. Soc.*, **115**, 760 (1968).

13   M. R. Boudry, *Appl. Phys. Lett.*, **22**, 530 (1973).

14   H. Deuling, E. Klausmann, and A. Goetzberger, *Solid-State Electron.*, **15**, 559 (1972).

15   D. V. Lang, *J. Appl. Phys.*, **45**, 3023 (1974).

16   C. T. Sah, W. W. Chan, H. S. Fu, and J. W. Walker, *Appl. Phys. Lett.*, **20**, 193 (1972).

17   M. Schulz and N. M. Johnson, *Appl. Phys. Lett.*, **31**, 622 (1977); *Solid-State Commun.*, **25**, 481 (1978); *Errata, Solid-State Commun.*, **26**(2), 126(i) (1978).

18   K. Yamasaki, M. Yoshida, and T. Sugano, *Jap. J. Appl. Phys.*, **18**, 113 (1979).

19   M. Schulz and E. Klausmann, *J. Appl. Phys.*, **18**, 16 (1979).

20   D. L. Heald, *Solid-State Electron.*, **20**, 657 (1977).

21   R. J. Kriegler, T. F. Devenyi, K. D. Chik, and J. Shapir, *J. Appl. Phys.*, **50**, 398 (1979).

22   L. S. Wei and J. G. Simmons, *Solid-State Electron.*, **17**, 1021 (1974).

23   A. Goetzberger and J. C. Irvin, *IEEE Transact. Electron Devices*, **ED-15**, 1009 (1968).

24   J. G. Simmons and L. S. Wei, *Solid-State Electron.*, **16**, 53 (1973).

25   N. M. Johnson, D. J. Bartelink, and M. Schulz, in *The Physics of SiO₂ and Its Interfaces*, S. T. Pantelides, Ed., Pergamon, New York, 1978, pp. 421–427; M. Schulz and E. Klausmann, *Appl. Phys.*, **18**, 169 (1979).

26   K. L. Wang, *IEEE Transact. Electron Devices*, **ED-26**, 819 (1979).

27   N. M. Johnson, *Appl. Phys. Lett.*, **34**, 802 (1979).

28   N. M. Johnson, D. J. Bartelink and J. P. McVittie, *J. Vac. Sci. Technol.*, **16**, 1407 (1979).

9

Measurement of Silicon Properties

9.1   INTRODUCTION

In Chapters 2, 3, and 4 we discussed the operating principles and derived the electrical characteristics of the MOS capacitor. In this chapter we view the MOS capacitor as a measurement tool for studying the properties of the silicon. The characteristics of this device are influenced by the electrical properties of the silicon particularly near its surface. The deviations of measured characteristics from the ideal characteristics calculated in Chapters 2, 3, and 4 are used to extract this information. The information about silicon properties we seek is important for device applications in integrated circuits (MOSFETs, bipolar transistors, and CCDs). That is, this information is useful in device design and in monitoring processing. As one part of this information, the interpretation of measurements of interfacial and oxide properties is needed.

We start with one of the most fundamental properties of silicon, namely, whether it is n-type or p-type. Section 9.2 discusses how conductivity type can be determined from MOS capacitor C-V measurements, Section 9.3 discusses the deep depletion condition, and Section 9.4 describes the most accurate way of determining the doping profile near the silicon surface from capacitance measurements. The advantages and limitations of using the MOS capacitor for this purpose are discussed in relation to other methods of measuring the doping profile at the silicon surface.

A review of the advantages, disadvantages, and limitations of the transient and steady-state methods of measuring lifetime with the MOS capacitor is given in Section 9.5. Measuring lifetime with the MOS capacitor is useful for characterizing starting material. Lifetime in the devices made on this material may be different because the subsequent processing could introduce trapping centers. However, a knowledge of the lifetime in the starting material gives a baseline for comparison and for accepting or rejecting the material.

In silicon, lifetime is much longer than the dielectric relaxation time (see Section 4.2). As a result, carriers injected into silicon in which a field is present will equilibrate with the field before they are trapped or recombine. Thermally grown SiO_2 is an example of a material where lifetime is much shorter than the dielectric relaxation time. In SiO_2, injected carriers are trapped before they can equilibrate with an applied field, particularly in the case of holes. Therefore, the electrical behavior of this material is fundamentally different from that of silicon. The details of these differences are beyond the scope of this book, but the interested reader can delve into these matters in Ref. 1.

9.2 CONDUCTIVITY TYPE

Historically, the conductivity type of a semiconductor was measured by Hall effect measurements. For silicon, observation of the sign of the Hall constant provided an unambiguous way of determining conductivity type[2-5]. In low-resistivity samples, the Hall constant would be inversely proportional to free carrier concentration and opposite in sign for n-type and p-type. For purer samples (high resistivity), contributions from both holes and electrons had to be taken into account. In this case the Hall constant fell to zero when $p = (\mu_n/\mu_p)^2 n$, where μ_n and μ_p are the electron and hole mobilities and p and n are the hole and electron densities. Therefore, ambiguities were possible for near-intrinsic samples. In all other cases the Hall constant provided a clear indication of conductivity type throughout the bulk of the semiconductor. The serious disadvantages of a Hall measurement were that a shaped sample had to be prepared and that the measurement itself was time consuming if perturbing effects were to be eliminated.

It proved preferable to use simpler, faster techniques that could be made on any sample geometry. One widely used method for conductivity type determination employed a thermoprobe.[4] With this method, conductivity type was determined by the sign of the thermoelectric electromotive force developed as the result of a thermal gradient established between a heated metal point in contact with the semiconductor surface and the back of the semiconductor wafer in contact with a heat sink. The heated point could be the tip of a miniature soldering iron. A thermal gradient was maintained by placing the semiconductor wafer on a large metal block, which acted as a heat sink, with the soldering iron tip contacting the upper surface of the wafer. A galvanometer or other millivoltmeter placed between the soldering iron tip and the metal block measured the electromotive force (emf). If the probe were positive with respect to the sink, the material was n-type. A negative emf would mean a p-type material.

Another simple method was to observe the I-V curve of a rectifier made by pressing a sharp metal whisker to the surface of the semiconductor.[5] For a p-type sample, forward current would be observed if the whisker were biased positively with respect to the substrate and conversely for an n-type sample.

Today, conductivity type is determined from a C-V curve measured on an MOS capacitor. This method is convenient, unambiguous, nondestructive, and effective even for nearly intrinsic (or heavily compensated) samples. The C-V method can be used without the necessity of removing the oxide layer, which the other methods would require. Moreover, measurements of the admittance of the MOS capacitor provide far more information than does conductivity type alone, as is seen later.

Although conductivity type can be determined from either a high frequency or a low frequency C-V curve, a high frequency C-V curve is more reliable because of its greater asymmetry. Figures 9.1a,b show high

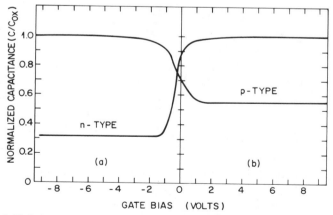

Fig. 9.1 (a) High frequency normalized capacitance measured at 1 MHz at 300°K as a function of gate bias for an n-type sample having a donor concentration of 1.5×10^{15} cm^{-3}, an oxide thickness of 1000 Å, and an interface trap level density in the 10^9 cm^{-2} eV^{-1} range. (b) High frequency normalized capacitance measured at 1 MHz and 300°K as a function of gate bias for a p-type sample having an acceptor concentration of 1.5×10^{16} cm^{-3}, an oxide thickness of 1000 Å, and an interface trap level density in the 10^9 cm^{-2} eV^{-1} range.

frequency C-V curves measured on n-type and p-type substrates, respectively. With the polarity convention described in Section 2.3.2(c), the C-V curve goes from accumulation to inversion with increasing gate bias for p-type while it goes from accumulation to inversion with decreasing gate bias for n-type, regardless of interface charge density or polarity.

Before interpreting any C-V curve, one must establish that accumulation has been achieved. Saturation of the maximum capacitance of a C-V curve is no guarantee that accumulation has been achieved. For example, the C-V curve may saturate at a high capacitance because of a large density of interface trap levels. Also, a large oxide fixed charge density may shift the C-V curve to such extreme gate biases that the oxide breaks down before accumulation is reached. Accumulation has been achieved only if the maximum saturated high frequency capacitance measured equals the calculated oxide layer capacitance. This calculation requires knowledge of the oxide thickness, which can be determined by one of the optical techniques described in Section 14.3 and of the gate area, which can be determined by a calibrated reticle on a metallurgical microscope.

Conductivity type also can be determined from the asymmetry of a low frequency C-V curve: the gate bias corresponding to the minimum capacitance is closer to inversion than accumulation. However, if there is a large interface trap level density, the low frequency curve may become symmetrical about the capacitance minimum, or structure may appear in the C-V curve making it impossible to determine conductivity type. Therefore, a high frequency C-V curve is better for determining conductivity type.

9.3 DEEP DEPLETION, AVALANCHE BREAKDOWN, AND TUNNELING

In this chapter the importance of deep depletion lies in (1) extending the depth over which doping profile in the silicon can be measured with the MOS capacitor (described in Section 9.4) and (2) simplifying the measurement of lifetime with the MOS capacitor described in Section 9.5. The term "deep depletion" (described in Section 4.5.1) refers to a nonequilibrium widening of the depletion layer beyond its thermal equilibrium width. When in equilibrium, the formation of an inversion layer limits the maximum depletion layer width to that given by a band bending of approximately $2(kT/q)\ln(N_B/n_i)$. However, when gate bias is changed too rapidly for thermal generation of minority carriers, overall charge neutrality of the MOS capacitor cannot be maintained by an increase in minority carrier density. Instead, charge balance is maintained by an increase in depletion layer charge and, therefore, in depletion layer width.

The depletion layer width continues to increase with gate bias until the onset of oxide breakdown or the onset of additional generation of minority carriers. Additional generation occurs when electron-hole pairs are created by avalanche or by tunneling between the silicon valence and conduction bands. At a gate bias such that avalanche or tunneling occurs, an inversion layer forms. Beyond this gate bias, depletion layer width no longer increases, as increased gate charge is balanced by increased inversion layer charge. Therefore, the high frequency capacitance saturates.

For very high doping densities ($> 10^{18} \, cm^{-3}$), tunneling—rather than avalanche breakdown—will determine when the high frequency capacitance saturates. In this event band bending cannot exceed the bandgap. Hence, the ultimate depth to which the profile can be measured will be virtually the same as for the equilibrium case. Therefore, the deep depletion condition is advantageous only for lighter doping densities ($< 10^{18} \, cm^{-3}$).

Deep depletion also can occur in thin oxide MOS capacitors, with oxides less than 50 Å thick. In such devices an inversion layer cannot form because minority carriers tunnel through the oxide to the gate electrode. Consequently, even for a time-independent gate bias, deep depletion occurs. However, we consider the usual thick oxide case here.

9.3.1 Deep Depletion

There are three principal types of C-V characteristic that can be measured on an MOS capacitor, as shown in Fig. 9.2. The deviations between the curves of Fig. 9.2 all occur in inversion where minority carrier density effects the capacitance. Curve a is the low frequency curve described in Section 3.2. Here, minority carriers are in equilibrium with both gate bias and ac gate voltage, contributing fully to the capacitance. Curve b is the high frequency curve described in Section 4.4. Here, minority carriers are in equilibrium with gate bias but do not follow the ac gate voltage and hence do not contribute to the capacitance. The only role of minority carriers is to pin the band bending, so the depletion layer width becomes fixed, independent of gate bias. Therefore, once inversion is reached, the high frequency capacitance saturates. Curve c is the deep depletion case described in Section 4.5.1. Here, minority carriers follow neither gate bias nor ac gate voltage. Now minority carriers play no role; in particular, band bending no longer is pinned, and the depletion layer width continues to increase with gate bias until avalanche breakdown occurs. Once avalanche breakdown occurs, minority carriers are produced rapidly,* and thus band bending becomes pinned. As a result, for gate biases beyond avalanche breakdown, depletion layer width and capacitance become fixed, independent of bias. All three curves coincide in accumulation and depletion (ignoring interface traps for clarity) because

---

*Typical times required to produce avalanche breakdown in silicon are less than a nanosecond.[6]

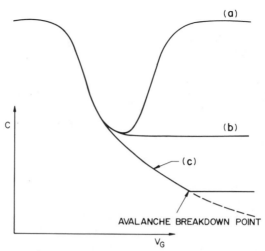

Fig. 9.2 Capacitance as a function of gate bias of the MOS capacitor for three different conditions: (a) low frequency (< 10 Hz for silicon) C-V curve when minority carriers contribute fully to the capacitance; (b) high frequency C-V curve when minority carriers do not contribute to the capacitance (capacitance saturates at a low value in inversion, as described in Section 4.4); (c) the deep depletion case obtained at high frequency when gate voltage is applied so rapidly that the buildup of an inversion layer is impossible. Capacitance decreases with gate bias until the avalanche breakdown point in the silicon is reached. Then the capacitance saturates, but at a lower value than in curve (b). After Goetzberger and Nicollian.[7]

equilibrium minority carrier density is so small in these regions that it does not matter whether it follows gate bias or ac gate voltage.

How is a deep depletion C-V curve produced experimentally? In a silicon MOS capacitor the time constants that govern the buildup of an inversion layer are very long. Times on the order of a fraction of a second to several seconds apply to the normal low-field process of thermal generation of minority carriers in the silicon surface depletion layer at room temperature (see Section 4.3). Therefore, one practical way to produce deep depletion in an MOS capacitor is to apply rectangular voltage pulses that have a duration short compared to the thermal generation time. Other ways of producing deep depletion are by applying a large amplitude sinusoidal voltage whose period is short compared to the thermal generation time constant or by applying a fast ramp voltage. Cooling the MOS capacitor below room temperature lengthens the thermal generation time constant, thereby enabling deep depletion to occur even when gate bias is changed slowly.

Perhaps the most convenient way to measure deep depletion C-V curves is by using repetitive rectangular voltage pulses so that gate voltage remains constant for each pulse duration. The superposition of a small-signal ac voltage enables the measurement of capacitance during each pulse, as described in Section 12.5.6. If pulse height is increased from one pulse to the next, silicon band bending is increased until avalanche breakdown occurs at the silicon surface. The capacitance curve then saturates as shown in curve c in Fig. 9.2.

In Fig. 9.3 an experimental deep depletion curve taken with pulsed voltage using the apparatus described in Section 12.5.6 is compared to an experimental high frequency C-V curve. Both curves were measured on the same MOS capacitor. The gate voltage at which avalanche breakdown in the silicon occurs is noted on the deep depletion curve. Capacitance saturates for pulses of higher voltage.

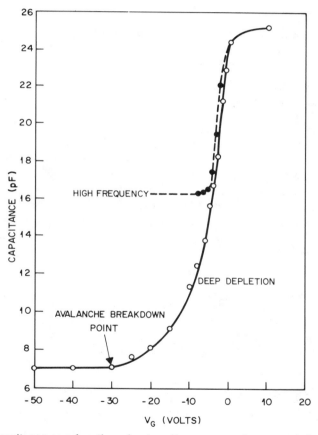

Fig. 9.3 Capacitance as a function of gate voltage on an n-type sample having a donor concentration of 6.8×10^{16} cm^{-3} and an oxide thickness of 1000 Å. The high frequency curve was measured at 500 kHz by manually changing gate bias. The deep depletion curve was measured at 500 kHz, by applying pulses of 6 msec duration and a duty cycle of 1% using the apparatus described in Section 12.5.6. After Goetzberger and Nicollian.[8]

9.3.2   Avalanche Breakdown and Tunneling

The MOS capacitor can be used to measure avalanche breakdown voltage and field in the silicon as described in Section 11.2 and in Ref. 8. However, our interest in avalanche breakdown in this chapter is that it limits the depth to which doping profile in the silicon can be measured with C-V methods. Avalanche breakdown and tunnel effects in silicon MOS capacitors can be divided into three regions, according to doping level. For an oxide thickness of 1000 Å, these are:

1 **High resistivity (doping density $< 10^{16}$ cm^{-3}).** Avalanche breakdown with pronounced lowering of avalanche breakdown voltage is observed as a result of field concentration at the edges of the metal gate.

2 **Medium resistivity (doping density from 5×10^{16} cm^{-3} to 10^{18} cm^{-3}).** Uniform avalanche breakdown is observed. The avalanche breakdown field is somewhat higher than in a step p-n junction.

3 **Tunnel range (doping density $> 10^{18}$ cm^{-3}).** Band bending has to be equal to the bandgap (~ 1.12 V).

It is important in measuring doping profiles that field concentration at the gate edges be avoided. Otherwise, the maximum depth limit set by avalanche breakdown will not be reached. Figure 9.4 illustrates field enhancement at the metal gate edges, showing the difference between high and low semiconductor doping density. Edge breakdown is expected to be more pronounced at low doping densities, (Fig. 9.4a) than in highly doped silicon where depletion layer width is only a small fraction of oxide thickness (Fig.

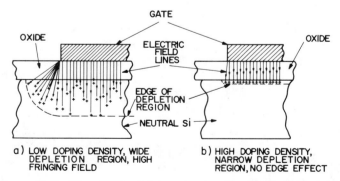

a) LOW DOPING DENSITY, WIDE DEPLETION REGION, HIGH FRINGING FIELD

b) HIGH DOPING DENSITY, NARROW DEPLETION REGION, NO EDGE EFFECT

Fig. 9.4   Cross section of an MOS capacitor showing the electric field at the edge of the gate for low and high doping density in the silicon. Field lines are terminated on ionized donors in the depletion region of the silicon. When edge avalanche breakdown occurs before avalanche breakdown near the center of the dot, the depletion layer width cannot be driven to its maximum value. After Goetzberger and Nicollian.[7]

9.4b). For the high doping density shown in Fig. 9.4b, bunching of the field lines occurs in the oxide only. There is no influence on the field distribution in the silicon for constant oxide thickness.

Rusu and Bulucea[9] have made two-dimensional computer calculations of step junction avalanche breakdown voltage for doping densities over the range 10^{14}–10^{18} cm^{-3} and oxide thicknesses of 0.05–5 μm, to estimate quantitatively the doping density and oxide thickness ranges for uniform and edge avalanche breakdown. Figure 9.5 shows the gate voltage V_{GB} at which avalanche breakdown occurs in the silicon as a function of doping density with oxide thickness as parameter. In Fig. 9.5 gate edge breakdown is dominant to the left of the minimum of each curve, whereas uniform breakdown under the gate is dominant to the right.

The curves in Fig. 9.5 are valid provided that the oxide does not break down. Under sinusoidal excitation, avalanche breakdown occurs in the silicon near the peak of the ac cycle, which drives the silicon into deep depletion as described in Sections 11.2 and 11.3. The other half of the ac

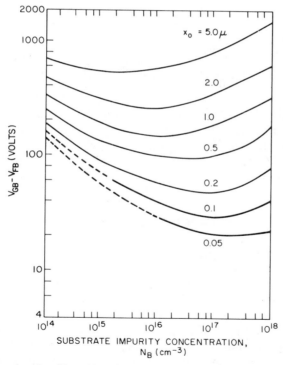

Fig. 9.5 Diagram showing $V_{GB} - V_{FB}$ as a function of silicon doping concentration with oxide thickness as parameter, where V_{GB} is the gate voltage corresponding to avalanche breakdown in the silicon and V_{FB} is the flatband voltage. These curves have been calculated for a field-induced step junction. After Rusu and Bulucea.[9] Copyright (1979), IEEE.

cycle is necessary to recombine the inversion layer carriers created during the avalanche half of the cycle. At the peak of the recombination half cycle, the silicon is driven into strong accumulation. The peak gate voltage must not exceed the dielectric breakdown strength of the oxide, or the oxide will break down while the silicon is in accumulation. Therefore, no calculated curves for x_o less than 0.05 μm are presented in Fig. 9.5. In the dashed portions of the curves calculated in Fig. 9.5 for $x_o = 0.05$ and 0.10 μm, the dielectric breakdown strength, assumed to be 6×10^6 V/cm,* is exceeded. No oxide breakdown occurs in the region of the solid curves.

To the left of the minima in the curves in Fig. 9.5, avalanche breakdown occurs in the silicon at the gate edge, making $V_{GB} - V_{FB}$ less than the value expected for uniform avalanche breakdown determined by the doping density. Edge avalanche breakdown values of $V_{GB} - V_{FB}$ increase with decreasing N_B, as shown on the left of the minima in Fig. 9.5. This behavior reflects the increase in avalanche breakdown voltage of the silicon. To the right of the minima in Fig. 9.5, where uniform breakdown occurs, $V_{GB} - V_{FB}$ increases with increasing N_B. This increase becomes more pronounced for thicker oxides and higher doping densities. To explain this increase,[9] we note that $V_{GB} - V_{FB} = (\epsilon_s/\epsilon_{ox})F_B x_o + \psi_B$, where F_B is the avalanche breakdown field at the silicon surface and ψ_B is the corresponding silicon band bending. For thick oxides and high doping densities, $(\epsilon_s/\epsilon_{ox})F_B x_o \gg \psi_B$ so that $V_{GB} - V_{FB} \approx (\epsilon_s/\epsilon_{ox})F_B x_o$. The field F_B increases with increasing doping density,[7, 10] causing $V_{GB} - V_{FB}$ to increase with N_B in this x_o and N_B range.

Although curves are shown in Fig. 9.5 for oxide thicknesses above 1 μm, it usually is not practical to use MOS capacitors that have such thick oxides. There are two difficulties with oxides thicker than 1 μm: (1) oxide thicknesses greater than 1 μm cannot be grown at 1 atm. (i.e., such thick oxides either must be deposited or grown under high pressure; see Section 14.6); and (2) high gate voltages must be applied.

9.4 IONIZED DOPANT IMPURITY DISTRIBUTION

The ionized dopant impurity distribution near the silicon surface must be known for an analysis of MOS device characteristics and for their interpretation in terms of the properties of the MOS system.

In practice, the ionized dopant impurity distribution of an MOS device is not the same at the silicon surface as in the bulk. One reason for nonuniformity in a direction perpendicular to the silicon surface is dopant impurity segregation during thermal oxidation, discussed in Section 14.4.2. Repeated growth and removal of the oxide layer during integrated circuit processing increases the nonuniformity caused by dopant impurity redistribution. Another source of nonuniform dopant impurity profiles per-

---

*This breakdown field is typical.

pendicular to the silicon surface is deliberate diffusion or implantation of dopant ions into the silicon surface region to control threshold voltage, channel depth, or punchthrough characteristics in a MOSFET. Regardless of the cause of nonuniformity, it is necessary to measure the final profile as a check on processing and analysis.

There are a number of ways to measure doping profiles. For example, one may etch off successive layers of the semiconductor and measure either the amount of dopant remaining in the semiconductor or the amount of dopant recovered. The four-point resistivity probe method[4] is one way of measuring the amount of dopant remaining after each etch-off step. This method measures the ionized dopant density, as does the capacitance method. Other measurement methods yield the sum of both ionized and un-ionized dopant impurity density. Some examples are helium ion back-scattering, secondary ion spectroscopy, and electron spectroscopy for chemical analysis. These techniques are useful only for rather heavy doping levels, greater than 10^{18} cm^{-3}. Other details of these methods, all of which are beyond the scope of this book, are described in Refs. 11 and 12.

The method treated here is based on C-V measurements of MOS capacitors, Schottky barrier diodes, and p-n junctions.

The three structures—MOS capacitor, Schottky diode, and p-n junction—are not equally practical or advantageous. The MOS capacitor has three important advantages over the Schottky diode and p-n junction. The first advantage is that the doping profile can be measured significantly closer to the semiconductor surface. The closest distance to the Si–SiO$_2$ interface at which an accurate profile can be obtained is a few extrinsic Debye lengths in the MOS capacitor, approximately the zero bias depletion layer width (which is longer than a few extrinsic Debye lengths) in the Schottky diode, and zero bias depletion layer width plus an additional width of neutral silicon between the edge of the zero bias depletion layer and the semiconductor surface in the p-n junction. Figure 9.6 compares the ranges of depth w over which doping profile can be accurately measured. Curve a refers to the MOS capacitor and curve b, to the abrupt junction diode.* Both structures are limited to the same maximum values of w to which doping density can be determined by avalanche breakdown shown in curve c.

The second advantage of the MOS capacitor is that its fabrication alters the doping profile less than does fabrication of the other structures. The abrupt junction diode requires a metal to be deposited and/or alloyed with the semiconductor, which can drastically alter the doping profile. The p-n junction requires a diffusion that certainly alters the doping profile, unless the profile of the diffusion itself is to be measured. The MOS capacitor affects the profile only through the redistribution of dopant ions that occurs during oxide growth. However, the MOS capacitor is not limited to

---

*The term "abrupt junction" means the Schottky diode or a step p-n junction.

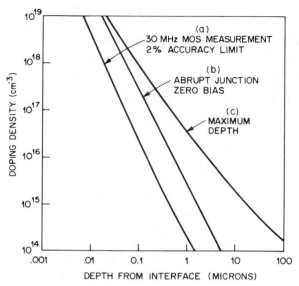

Fig. 9.6 Doping density versus depth. Curve a is the minimum depth measured from the Si-SiO₂ interface to which the doping profile can be accurately determined for an MOS capacitor calculated from (9.43) – (9.48) at 30 MHz, 300°K, and the interface trap parameters in Fig. 9.11. The 2% criterion is explained in Section 9.4.3(c). Curve b is the minimum depth to which the doping profile can be accurately determined for an abrupt junction diode. In curve b depth is measured from the edge of the abrupt doping discontinuity rather than from the Si-SiO₂ interface. Curve c is the maximum depth to which doping profile can be accurately determined limited by avalanche breakdown for both the MOS capacitor and abrupt junction diode. Comparison of the range for the MOS capacitor to the range for the abrupt junction diode shows that for each doping density shown, the range of depth is greater for the MOS capacitor because the doping profile can be measured closer to the Si-SiO₂ interface. After Nicollian et al.[20] Copyright (1973), IEEE.

structures fabricated with thermally grown oxide. By fabricating the MOS capacitor with low temperature-deposited oxide as described in Section 14.6.3, dopant impurity redistribution is avoided.

The third advantage of the MOS capacitor is that profiles can be measured in higher doping density ranges than with either the Schottky diode or *p-n* junction.* Therefore, the total doping density range that can be measured with the MOS capacitor is greater than with the Schottky diode or *p-n* junction.

The major drawback of the MOS capacitor is that interface traps can contribute a large error to the doping profile. A simple method for avoiding such errors is described in Section 9.4.1. Errors due to interface traps also

---

*At high doping densities the Schottky diode and *p-n* junction become ohmic, so that capacitance no longer can be measured reliably.

may arise with Schottky barrier diodes when profiles near the semiconductor surface are examined, unless precautions are taken to eliminate the native oxide layer between the metal and the semiconductor, which can transform a Schottky barrier diode into a thin oxide MOS capacitor.

Various capacitance methods have been used in conjunction with all three structures. Grove et al.[13] measured the redistribution of acceptor and donor impurities that had occurred during thermal oxidation by using an incremental C-V technique on shallow (0.2 μm) diffused p-n junctions. Theory and practice of the p-n junction and Schottky diode methods can be found in Refs. 14–17. A variation of the p-n junction method is described by Meyer and Gulbrandsen,[18] who apply a sine wave to the junction and then measure the ratio k_2 of the second harmonic voltage to the first harmonic voltage generated by the diode. They showed that doping concentration as a function of depth into the silicon is proportional to $C^2 \delta v_1/k_2$, where C is the measured capacitance and δv_1 is the applied ac voltage. Another elegant variation of the junction method, making direct plotting of the inverse doping profile possible, was described by Copeland.[19] Nicollian et al.[20] applied this second harmonic method to the MOS capacitor to make automatic plots of doping profile free of interface trap error over a certain range. Deal et al.[21] measured the redistribution of impurities using the MOS capacitor. They measured the maximum high frequency capacitance in strong accumulation and the minimum high frequency capacitance in strong inversion and calculated doping concentration at the silicon surface by applying the depletion approximation. Other workers have reported the same maximum-minimum high frequency capacitance method using essentially the same depletion approximation.[21-23] In an attempt to minimize interface trap error, Van Gelder and Nicollian[24] superposed a small high frequency ac voltage on a voltage pulse applied to the gate of an MOS capacitor and extracted the doping profile from the resulting C-V curve.

The basic theory of doping profile determination from high frequency capacitance measurements made on MOS capacitors is now described. Then a simple way of correcting the error introduced by interface traps is presented. Finally, second harmonic and pulsed MOS capacitance methods are compared to this approach.

9.4.1 Profiling with the MOS Capacitor

(a) Basic Analysis

Profiling with the MOS capacitor is based on the measurement of differential capacitance as a function of gate bias. However, differential capacitance is a measure of the differential change in charge produced by a differential change in voltage (see Section 3.2). Therefore, we can obtain the doping profile from a capacitance measurement *only* when the profile is closely related to the free carrier charge concentration flowing in response to the ac gate voltage. In accumulation and inversion, free carrier charge density is only *weakly* dependent on ionized dopant impurity concentration. For example, (2.42) and (2.44) show that the free carrier concentrations depend exponentially on band bending, with impurity density occurring only as a prefactor of the exponential term. Therefore, dopant impurity concentration cannot be obtained accurately in either accumulation or inversion. However, it can be obtained accurately in depletion, where the free carrier concentration depends strongly on dopant impurity concentration.

The ionized dopant impurity profile is obtained from the slope of a $(1/C_m^2)$ versus V_G curve, where C_m is the capacitance measured in depletion at gate bias V_G. Assuming that interface traps have a negligible influence, we derive an expression for the ionized dopant impurity concentration at the depletion layer edge.

Suppose that an incremental charge density dQ_G is added to the gate of the MOS capacitor. The result will be a movement of the depletion layer edge an amount dw just sufficient to change the depletion layer charge density to compensate dQ_G. If the ionized dopant impurity density at the depletion layer edge, a distance w from the silicon surface, is $N(w)$ and the depletion layer edge moves a distance dw, then

$$dQ_G = - qN(w)dw. \qquad (9.1)$$

Equation (9.1) is valid for both donors and acceptors. To derive (9.1), majority carrier density is assumed negligible in the depletion region up to the depletion layer edge and is assumed equal to the dopant ion density beyond the depletion layer edge. An improvement on this abrupt depletion layer edge approximation is discussed later [see (9.24)].

Equation (9.1) expresses overall charge neutrality, the balance between the change in charge density on the gate and in the silicon. Ionized dopant impurity density at position w can be calculated from (9.1) if dQ_G and dw are known. Now, dQ_G can be obtained from the measured capacitance C_m and the gate bias change dV_G; that is

$$dQ_G = C_m \, dV_G \qquad (9.2)$$

where C_m is the measured capacitance per unit area. To proceed, we need to obtain w and dw. Within the depletion approximation described in Section 2.3.4, the depletion layer capacitance is

$$C_D = \frac{\epsilon_s}{w} \qquad (9.3)$$

where C_D is the depletion layer capacitance per unit area and ϵ_s is the dielectric permittivity of silicon. Thus we can calculate w from (9.3) by extracting C_D from the measured capacitance using (3.11). Equation (9.3) also leads to the determination of dw. We simply determine the *change* in

$1/C_D$ when gate bias is changed by dV_G

$$dw = \epsilon_s d\left(\frac{1}{C_D}\right). \tag{9.4}$$

Because oxide layer capacitance per unit area C_{ox} is not voltage dependent, we can express (9.4) in terms of measured capacitance

$$dw = \epsilon_s d\left(\frac{1}{C_{ox}} + \frac{1}{C_D}\right). \tag{9.5}$$

From (3.11), (9.5) can be written

$$dw = \epsilon_s d\left(\frac{1}{C_m}\right). \tag{9.6}$$

We can express $N(w)$ in terms of a slope. First, eliminate dQ_G by equating (9.1) to (9.2) to obtain

$$dQ_G = -qN(w)dw = C_m\, dV_G. \tag{9.7}$$

Then, substitute (9.6) into (9.7) to eliminate dw; thus

$$-qN(w)\epsilon_s d\left(\frac{1}{C_m}\right) = C_m\, dV_G. \tag{9.8}$$

Solving (9.8) for $N(w)$ yields

$$N(w) = -\left[q\epsilon_s \frac{1}{C_m} \frac{d}{dV_G}\left(\frac{1}{C_m}\right)\right]^{-1}. \tag{9.9}$$

Applying the simple formula, $(d/dx)(1/x)^2 = (2/x)(d/dx)(1/x)$, for differentiation, we further simplify (9.9) to obtain

$$N(w) = -2\left[q\epsilon_s \frac{d}{dV_G}\left(\frac{1}{C_m^2}\right)\right]^{-1}. \tag{9.10}$$

Equation (9.10) originated with Van Gelder and Nicollian.[24] Equation (9.10) shows that $N(w)$ is related reciprocally to the slope of a $(1/C_m)^2$ versus V_G curve. A positive slope will give a negative $N(w)$ from (9.10) for *acceptors*, whereas a negative slope gives a positive $N(w)$ for *donors* using the sign convention for V_G given in Section 2.3.2(c).

To obtain the ionized dopant impurity profile as a function of depth w, we calculate $N(w)$ at each value of gate bias using (9.10). To get the corresponding value of w, solve (9.3) for w and use (3.11) for series capacitors to get w as a function of C_m; that is

$$w = \epsilon_s\left(\frac{1}{C_m} - \frac{1}{C_{ox}}\right). \tag{9.11}$$

(b) Analysis Including Interface Traps

We have treated the ideal case where interface trap level density is negligibly small. Although it is possible to fabricate MOS capacitors with a negligibly small interface trap level density, there are experiments and structures (e.g., using a deposited oxide) in which interface trap level densities are sufficiently high that their neglect would cause severe error in the extracted doping profile. Moreover, if a doping profile can be extracted even with large interface trap level densities present, the MOS capacitor becomes a more versatile tool. Also, although oxides can be grown where interface trap level density around the center of the silicon bandgap is in the low 10^{10} cm^{-2} eV^{-1} range, interface trap level density increases toward each band edge. This can be one of the limitations on how close the Si–SiO$_2$ interface profiling can be extended.

Taking interface trap response into account is most important for profiling near the Si–SiO$_2$ interface. In deep depletion the carrier concentration at the silicon surface becomes very small, making interface trap response times very long. Therefore, in deep depletion no correction for interface trap response to ac gate voltage is needed.

The next task is to modify (9.10) in a simple way to take interface trap effects into account. Interface traps influence doping profile measurements by affecting the slope of the measured C-V curve. Interface traps will affect this slope if they change occupancy in response to the ac gate voltage or to changes in gate bias.

The ac response of interface traps can be avoided by measuring capacitance at such a high frequency that interface traps do not respond over the gate bias range corresponding to depletion. How high a frequency is required will be treated quantitatively later in this section. Qualitatively, at such high frequencies interface traps cannot capture a carrier* rapidly enough during one phase of the ac gate voltage nor emit that carrier during the opposite phase; consequently, interface trap occupancy does not change with the ac gate voltage but lags behind. A C-V curve or a portion of a C-V curve over which interface traps do not respond to the ac gate voltage is called a *high frequency curve*.†

Provided that capacitance is measured at a high frequency, depletion layer width still can be obtained directly from (9.11). To emphasize this point, we introduce C_{HF} as the high frequency capacitance. Then (9.11) becomes

$$w = \epsilon_s \left(\frac{1}{C_{HF}} - \frac{1}{C_{ox}} \right). \qquad (9.12)$$

---

*In depletion only majority carriers are captured and emitted by interface traps because minority carrier density at the silicon surface is negligibly small in comparison. Therefore, for p-type silicon, only holes are captured and emitted whereas for n-type silicon, only electrons. See Chapter 5 for a more complete discussion.

†A C-V curve where minority carriers do not respond to the ac gate voltage also is called a *high frequency C-V curve*, and there can be a large difference in frequency between these two cases. Confusion can be avoided by keeping in mind the context in which a C-V curve is called a "high frequency" curve.

We measure C_{HF} and plot $(1/C_{HF})^2$ versus V_G. However, the *slope* of the resulting curve still will be affected by interface traps. Equation (9.10) must be modified further to take into account that, although interface traps do not respond to the ac gate voltage at high frequencies, they do follow changes in gate bias. Gate bias usually is changed so slowly that the system is in thermal equilibrium and interface traps change occupancy in harmony with changes in gate bias. Then, *slow* changes in gate charge are balanced not solely by a change in depletion layer charge density as in (9.1) but, in addition, by a change in interface trap charge density. The result is a "stretchout" of the high frequency C-V curve along the voltage axis, causing the slope of the high frequency C-V curve in depletion to decrease.

It is simple to modify (9.10) to express $N(w)$ in terms of the slope of a $(1/C_{HF})^2$ versus V_G curve, taking into account the voltage stretchout caused by interface traps. This modification is made by relating the change in gate bias when no interface traps are present, dV_{G0}, to the change when interface traps are present, dV_G. When no interface traps are present, we have

$$C_{ox}\, dV_{G0} = (C_{ox} + C_D)d\psi_s \qquad (9.13)$$

and when interface traps are present

$$C_{ox}\, dV_G = (C_{ox} + C_D + C_{it})d\psi_s \qquad (9.14)$$

where C_{it} is the interface trap capacitance per unit area. From (9.13) and (9.14), the same change in band bending $d\psi_s$ and hence the same change in depletion layer width occurs provided that

$$\frac{dV_{G0}}{dV_G} = \frac{C_{ox} + C_D}{C_{ox} + C_D + C_{it}}. \qquad (9.15)$$

To express (9.15) in terms of measured capacitances, C_D can be obtained from a high frequency C-V curve; thus

$$\frac{C_{HF}}{C_{ox}} = \frac{C_D}{C_{ox} + C_D}, \qquad (9.16)$$

and C_{it} can be obtained from a low frequency C-V curve; thus

$$\frac{C_{LF}}{C_{ox}} = \frac{C_D + C_{it}}{C_{ox} + C_D + C_{it}} \qquad (9.17)$$

where C_{LF} is the measured low frequency capacitance. A low frequency C-V curve is one in which interface traps follow the ac gate voltage.* When (9.16) is solved for C_D and (9.17) for $C_D + C_{it}$ and the result

---

*Again, a low frequency C-V curve also means one in which minority carriers follow the ac gate voltage. The distinction can be made as for the high frequency C-V curve by keeping the context in mind.

substituted into (9.15), (9.15) becomes

$$\frac{dV_{G0}}{dV_G} = \left(1 - \frac{C_{LF}}{C_{ox}}\right)\left(1 - \frac{C_{HF}}{C_{ox}}\right)^{-1}. \tag{9.18}$$

Rewriting (9.10) and replacing C by C_{HF} yields

$$N(w) = -2\left[q\epsilon_s \frac{d}{dV_{G0}}\left(\frac{1}{C_{HF}^2}\right)\right]^{-1}. \tag{9.19}$$

If voltage stretchout is taken into account, (9.19) becomes

$$N(w) = -2\left[q\epsilon_s \frac{d}{dV_G}\left(\frac{1}{C_{HF}^2}\right)\right]^{-1}\frac{dV_{G0}}{dV_G}. \tag{9.20}$$

Finally, substituting (9.18) into (9.20), yields

$$N(w) = -2\left(\frac{1 - C_{LF}/C_{ox}}{1 - C_{HF}/C_{ox}}\right)\left[q\epsilon_s \frac{d}{dV_G}\left(\frac{1}{C_{HF}^2}\right)\right]^{-1}. \tag{9.21}$$

If there is no stretchout due to interface traps, $(1 - C_{LF}/C_{ox})/(1 - C_{HF}/C_{ox}) = 1$, $V_G = V_{G0}$, and (9.21) is identical to (9.19). Equation (9.21) first was obtained by Brews,[25] and a generalization of (9.21) to arbitrary frequencies was made by Zohta.[26]

9.4.2 Combined High and Low Frequency Capacitance Method

To obtain $N(w)$ experimentally using (9.21), a high and a low frequency C-V curve must be measured on the sample as described in Section 12.5. Equation (9.21) will be accurate as long as the high and low frequency C-V curves are measured with a sufficiently slow gate bias sweep rate. Otherwise, the contribution of interface traps to voltage stretchout will not correspond to the low frequency interface trap capacitance. In many cases a gate bias sweep rate of 10–50 mV/sec will be adequate. The optimal gate bias sweep rate in this case is the highest sweep rate for which no hysteresis is observed in sweeping back and forth between accumulation and inversion. Then $N(w)$ is obtained from (9.21) by calculating the slope of the high frequency C-V curve, whereas w is calculated from (9.12). Both C_{LF} and C_{HF} are measured at the same value of gate bias. The doping profile is obtained by repeating this procedure for different values of gate bias in depletion.

The slope of the high frequency C-V curve can be calculated directly. However, the derivative of the square of the reciprocal capacitance is related to the doping. Therefore, it is advantageous to compute the slope of $(1/C_{HF})^2$ versus V_G [see (9.21)] rather than that of C_{HF} versus V_G. For example, for uniform doping, $(1/C_{HF})^2$ versus V_G in depletion will approximate a straight line, making determination of the slope particularly simple.

In general, the slope of the high frequency $(1/C_{HF})^2$ versus V_G curve can be calculated most easily by drawing chords between adjacent measured points on either side of the chosen gate bias point. If the length of the chord is too short, there will be a round-off error. That is, the minimal chord length depends on the number of significant figures to which capacitance and gate bias are measured, as well as the location on the $(1/C_{HF})^2$ versus V_G curve where the chord is drawn. If the length of the chord is too long, the slope will be in error. Again, the longest chord that can be drawn without significant error depends on where on the $(1/C_{HF})^2$ versus V_G curve the chord is chosen. Therefore, the optimal chord length depends on the accuracy with which capacitance and gate bias are measured and the bias point at which the slope is calculated. The optimal chord length will vary over the depletion region, and finding it is a matter of good judgment. This method of finding the slope can be implemented graphically or by digital computer.

9.4.3 Limitations and Errors

The doping profile obtained using (9.21) is subject to a number of limitations. We list these limitations and then discuss quantitatively the basic extrinsic Debye length limitation and the limitation imposed by a failure to measure a high frequency C-V curve over the entire depletion region:

1 In cases where doping density changes abruptly in a distance comparable to an extrinsic Debye length or less, the profile given by (9.21) will be an average of the true profile[27, 28] over an extrinsic Debye length or so. The reason, as is shown shortly, is that (9.21) measures the majority carrier density in the vicinity of the depletion layer edge, rather than the dopant ion density itself. Changes in majority carrier density occur over distances of a few extrinsic Debye lengths. Therefore, changes of doping density over a distance comparable to an extrinsic Debye length are not resolved. The extrinsic *Debye length* is a measure of the limiting *resolution* in obtaining a doping profile from a capacitance measurement.

2 Equation (9.21) gives a good approximation to the doping profile if w is greater than several extrinsic Debye lengths. For distances within a few extrinsic Debye lengths of the Si–SiO$_2$ interface, majority carrier density is modified by the presence of the interface itself. That is, the majority carrier density at distance w is no longer equal to $N(w)$ but begins to increase exponentially as the interface is approached (see Section 2.3). Hence (9.21) becomes dominated by the interface, rather than by the doping density, leading to spurious results close to the interface.[20, 29] We call this the *interface proximity* limitation.

3 Another limitation on the accuracy of profiling near the Si–SiO$_2$

interface is the difficulty in obtaining a true high frequency capacitance C_{HF} to use in (9.21), especially at small depletion layer widths. The small-signal measurement frequency must be increased at smaller depletion layer widths to avoid an interface trap contribution to C_{HF}.[20] Briefly, majority carrier density at the silicon surface increases as w is reduced, thus increasing the interface trap capture rate and shortening interface trap response times.

4   The maximum depth to which doping profile can be measured is limited by the onset of inversion when gate bias is applied slowly enough for minority carriers to follow. This depth can be extended significantly by applying gate voltage rapidly enough so that minority carriers cannot follow and the silicon is driven into deep depletion (see Section 4.5.1). Doping density then can be profiled to a maximum depth limited by the onset of avalanche breakdown.

However, when interface traps cannot be neglected, deep depletion must be used with care to extend the depth of profile measured, as is discussed later.

(a) Relation of Doping Profile Measurement to Majority Carrier Density: Debye Length Limitation

Kennedy et al.[30] proposed that (9.1) was not as accurate an expression of the charge variation as

$$dQ_G = qp(w)dw \qquad (9.22)$$

where $p(w)$ = majority carrier density per unit volume (taken here as holes, for illustration). This relation, when substituted for (9.1), leads to the conclusion that (9.10) should be replaced by

$$p(w) = 2\left[\frac{q\epsilon_s d(1/C_m^2)}{dV_G}\right]^{-1}. \qquad (9.23)$$

That is, *majority carrier* density at the depletion layer edge is measured, not ionized dopant density. Extensive computer calculations[27, 28] showed that (9.22) was a better approximation than (9.1). A derivation of (9.22) is now presented.[31]

In place of the abrupt depletion layer edge approximation used to obtain (9.1), namely

$$p(x) = 0; \qquad 0 < x < w$$
$$p(x) = N_A(x); \qquad x > w$$

a more realistic approximation is

$$p(x) = 0; \qquad 0 < x < w \qquad (9.24)$$
$$p(x) = p_o(x); \qquad x > w$$

where $p_o(x)$ is the majority carrier density in the region unaffected by the gate bias variation. The carrier density $p_o(x)$ differs from $N_A(x)$ because a nonuniform doping profile implies a nonzero band bending at point x [to counteract the influence of differing doping densities on the Fermi levels at adjacent points] and, therefore, a nonzero field. That is, charge neutrality cannot prevail in the presence of a nonuniform doping profile, and so $p_o(x) \neq N_A(x)$.

Using the approximation (9.24), we can give the net charge per unit area in the semiconductor Q_s by

$$
Q_s = -\epsilon_s \frac{d\psi}{dx}\bigg|_{x=0}
$$

$$
\approx -\epsilon_s \frac{d\psi_o}{dx}\bigg|_{x=w} - \frac{\epsilon_s}{\beta\lambda_p^2} \int_0^w dx_1 \frac{N_A(x_1)}{N_B} \qquad (9.25)
$$

where N_B is the acceptor density in the bulk and $\psi_o(x)$ is the band-bending potential in the region $x > w$ corresponding to $p_o(x)$; that is

$$
\frac{d\psi_o}{dx}\bigg|_{x=w} = \frac{1}{\beta\lambda_p^2} \int_w^\infty dx_1 \frac{N_A(x_1) - p_o(x_1)}{N_B} \qquad (9.26)
$$

and $\beta = q/kT$, where k is Boltzmann's constant and T the absolute temperature. If a small bias change is made, causing a change in depletion width δw, then (9.25) and (9.26) provide

$$
\begin{aligned}
\delta Q_s &= -\epsilon_s \delta\left(\frac{d\psi_o}{dx}\right) - \epsilon_s \frac{1}{\beta N_B \lambda_p^2} N_A(w)\delta w \\
&= q[N_A(w) - p_o(w)]\delta w - qN_A(w)\delta w \\
&= -qp_o(w)\delta w.
\end{aligned} \qquad (9.27)
$$

From charge neutrality, we have

$$
\delta Q_G + \delta Q_s = 0. \qquad (9.28)
$$

Combining (9.27) and (9.28) yields

$$
\delta Q_G = qp_o(w)\delta w \qquad (9.29)
$$

which agrees with the Kennedy-Murley-Kleinfelder conjecture, with $p_o(w)$ the majority carrier density *undisturbed* by external bias.

A complication of this derivation is that (9.25) defines a depletion layer width that may not agree with the experimental depletion layer width given by (9.11). That is, (9.25) is a dc definition of w, whereas (9.11) is an ac definition. Because both values of w lie in the transition region near the depletion layer edge, they cannot differ by more than a Debye length or so. As now shown, the Debye length is the limit of resolution of the method, anyway, so a careful examination of the relation between these two estimates of w is not warranted.

If (9.29) is used in place of (9.1) in the derivation of (9.10), the "doping" profile obtained using (9.10) or (9.21) is more accurately a majority carrier density profile. That is, a more accurate equation than (9.21) is

$$p_o(w) = 2\left(\frac{1 - C_{LF}/C_{ox}}{1 - C_{HF}/C_{ox}}\right)\left[q\epsilon_s \frac{d}{dV_G}\left(\frac{1}{C_{HF}^2}\right)\right]^{-1}. \tag{9.30}$$

Because majority carrier density is a direct function of band bending, not dopant ion density, $p_o(w)$ cannot change as abruptly as the doping profile itself. For example, in a region of near neutrality, in the vicinity of a point $x = x'$

$$p_o(x) = n_i \exp[-v(x) + u_F] \cong p_o(x')[1 - v(x) + v(x')],$$

the Poisson equation becomes

$$\frac{d^2v}{dx^2} = \frac{p_o(x')}{N_A}[v(x) - v(x')]\lambda_p^{-2}$$

which has the solution

$$v(x) \cong v(x') \exp\left[-\frac{(x - x')}{\lambda(x')}\right]$$

where the "local" Debye length $\lambda(x')$ is

$$\lambda(x') = \left[\frac{\beta q p_o(x')}{\epsilon_s}\right]^{1/2}.$$

Therefore, band bending varies significantly only over distances of the order of $\lambda(x')$, and this will be approximately true of $p_o(x) \equiv p_o[v(x)]$ as well.

Johnson and Panousis[27] made extensive computer calculations for abruptly varying profiles. They computed the profile obtained from (9.10) and also majority carrier density. Figure 9.7 illustrates their results for two profiles.

Figure 9.7a shows that for a step that occurs in one Debye length, majority carrier density (solid line) and apparent profile (broken lines) agree fairly well and differ considerably from the true profile.

Figure 9.7b shows that for a more gradual transition, occurring over five Debye lengths, majority carrier density and the two profiles agree very closely. The true profile also resembles the apparent profile, but is shifted a Debye length or so along the x-direction. The corners in the true profile are not at all well represented by the apparent profile, again because of the Debye length limitation on resolution inherent in capacitive methods of profiling.

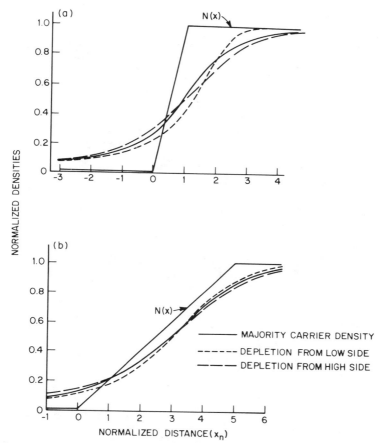

Fig. 9.7 Comparison of apparent doping profile, deduced from a capacitance measurement using (9.10), with majority carrier density. Dopant concentration was assumed to rise linearly from a low value to a high value. The vertical scale is doping density normalized to the highest value of doping density. The horizontal scale is distance in Debye lengths of the heavily doped region: (a) 100:1 change of doping density in one Debye length; (b) 100:1 change of doping density in five Debye lengths. The solid lines represent the assumed profile (straight-line segments) and the majority carrier density (smooth line). The broken lines represent the apparent profile of (9.10) for the two cases of a contact on the lightly doped or on the heavily doped side of the jump in dopant concentration. After Johnson and Panousis.[27] Copyright (1971), IEEE.

(b) Interface Proximity Limitation

Consider first the case with no interface traps, no oxide fixed charge, and a uniform doping profile to illustrate the basic limitation on the accuracy with which doping density can be determined as the depletion layer edge approaches the silicon surface. For this purpose, we evaluate (9.10). From

Appendix VI, we find that

$$\frac{d}{dV_G}\left(\frac{1}{C^2}\right) = \frac{d}{d\psi_s}\left(\frac{1}{C_D^2}\right)$$

so that (9.10) becomes

$$N(w) = -2\left[q\epsilon_s\frac{d}{d\psi_s}\left(\frac{1}{C_D^2}\right)\right]^{-1}. \tag{9.31}$$

For p-type and positive values of band bending (depletion), when the Fermi level is more than several kT/q volts from the intrinsic level the depletion layer capacitance is, from (3.35)

$$C_D = \left(\frac{q\epsilon_s\beta N_A}{2[\exp(-\beta\psi_s) + \beta\psi_s - 1]}\right)^{1/2}[1 - \exp(-\beta\psi_s)] \tag{9.32}$$

where ψ_s is the silicon surface band bending.

Substituting (9.32) into (9.31) yields

$$N(w) = N_A\frac{[1 - \exp(-\beta\psi_s)]^3}{1 - 2\beta\psi_s\exp(-\beta\psi_s) - \exp(-2\beta\psi_s)}. \tag{9.33}$$

If (9.32) is substituted into (9.3), w becomes

$$w = \frac{\epsilon_s}{C_D}$$

$$= \lambda_p\sqrt{2}\frac{[\exp(-\beta\psi_s) + \beta\psi_s - 1]^{1/2}}{1 - \exp(-\beta\psi_s)}. \tag{9.34}$$

For (9.33) to produce the true doping density N_A to within 2%, we find $\beta\psi_s \geq 6$. Then, $w \geq \sqrt{10}\lambda_p \approx 3\lambda_p$. Figure 9.8, calculated from (9.33), shows the departure of $N(w)$ from N_A graphically. This figure is independent of N_A and λ_p when plotted as shown, in terms of dimensionless quantities. The interface proximity limitation is seen in the region of the curve of Fig. 9.8, where there is a sharp increase of $N(w)$ from its constant value at low values of w.

It is also of interest to note that for $\beta\psi_s$ large and negative, $N(w) \rightarrow N_A\exp(-\beta\psi_s)$, which is the majority carrier density at the silicon surface corresponding to the band-bending ψ_s.

In profiling near the interface, mobile carrier density departs widely from the dopant ion density because of the exponential variation of $N(w)$ with ψ_s. The capacitance also has an exponential dependence on ψ_s. The doping profile information contained in the capacitance is small compared to the overwhelming exponential. Consequently, very accurate determination of capacitance and band bending are needed to obtain accurate profiles near the interface. The required accuracy may be so difficult to attain that only sophisticated measurement techniques would suffice, and elaborate error analysis would be needed to establish confidence in the resulting profile.

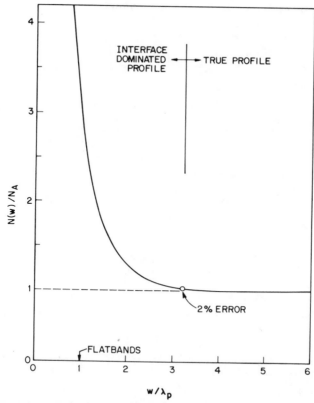

Fig. 9.8 Apparent doping density as a function of depth from the Si-SiO₂ interface calculated from (9.33) and (9.34). The 2% error point in the apparent profile occurs for $\psi_s \sim 6.2\,kT/q$ and $w \sim 3.2\,\lambda_p$.

Note that the interface proximity limitation in Fig. 9.8 cannot be attained in practice, even if *no* interface traps are present. The reason is that the localization of oxide fixed charge causes nonuniformities in band bending near the interface. If the depletion layer edge is made to approach the interface to profile close to the interface these band-bending nonuniformities will alter the bias dependence of the depletion layer capacitance. Consequently, an error in doping density will occur even before the interface proximity limitation occurs. Calculation of this nonuniformity effect on the capacitance is difficult (see Chapter 6).

(c) *Measurement Frequency Limitation*

Interface traps add a frequency dependent capacitance and conductance $C_{it}(\omega)$ and $G_P(\omega)$ in parallel with the semiconductor capacitance, as shown in Section 5.4. Therefore, at intermediate frequencies the measured capacitance C_m becomes frequency dependent and is no longer the high frequency capacitance C_{HF}.

Using $C_{it}(\omega)$, we can determine quantitative limitations under which a 1 MHz C-V curve approximates a true high frequency C-V curve. Interface trap capacitance and silicon capacitance are in parallel, suggesting that analysis be based on

$$C_P(\omega) = C_s + C_{it}(\omega). \tag{9.35}$$

At high frequencies $C_P(\omega) = C_s$, and at very low frequencies $C_P(0) = C_s + qD_{it}$, where D_{it} is the interface trap level density. Therefore, we adopt as a criterion for realizing a true high frequency C-V curve the condition that $C_P(\omega)$ departs no more than 10% from its high frequency value C_s. That is, we require $C_{it}(\omega)$ to satisfy

$$C_{it}(\omega) \le 0.1 C_s \tag{9.36}$$

where ω corresponds to 1 MHz. Consequently

$$D_{it} \le 0.1 C_s \left[\frac{qD_{it}}{C_{it}(\omega)} \right]. \tag{9.37}$$

From curves such as those in Fig. 8.12, we can determine at any band bending the value of $C_{it}(\omega)/(qD_{it})$ at 1 MHz. For a given doping level, C_s is known. Hence (9.37) can be evaluated.

Equation (9.37) is plotted against band bending in Fig. 9.9 with doping density as parameter. To illustrate the use of Fig. 9.9, consider the solid line curve for a doping density of 10^{14} cm^{-3}. At a band bending of $8kT/q$, Fig. 9.9 shows that a high frequency C-V curve will be obtained at 1 MHz if D_{it} is less than 8×10^{10} cm^{-2} eV^{-1}. For smaller band bendings, a high frequency C-V curve will be obtained only for smaller D_{it}. Similarly, using this same curve, for $D_{it} = 8 \times 10^{10}$ cm^{-2} eV^{-1}, a 1 MHz C-V curve will be a high frequency C-V curve only for band bendings larger than $8kT/q$.

The curves in Fig. 9.9 can be understood qualitatively as follows. The plotted value of D_{it} increases with band bending because a smaller and smaller fraction of D_{it} contributes to $C_{it}(\omega)$ as band bending increases, as seen in Fig. 8.12. However, at small band bendings a minimum in the plotted D_{it} occurs. Near this minimum, nearly all interface traps respond, but C_s is large enough to make (9.36) true, even for the low frequency value of $C_{it}(0) = qD_{it}$. Reduction of band bending below the value at the minimum causes C_s to increase, so that D_{it} also increases with decreasing band bending to the left of the minimum.

In the range of low band bendings, increased doping density increases the maximum acceptable D_{it} because C_s increases with doping density. However, in the range of large band bendings, D_{it} is reduced by increased doping density because interface trap response time is shortened, thus increasing the fraction of D_{it} contributing to the loss.

Figure 9.9 shows results for two capture probabilities. As expected, the

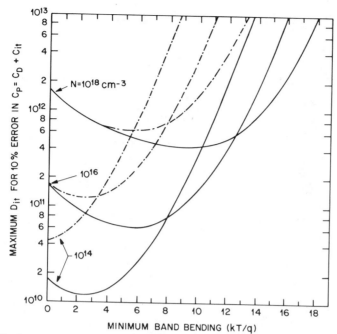

Fig. 9.9 Maximum allowed D_{it} versus minimum band bending for several doping densities. Solid line, $c_p = 10^{-6}$ cm^3/sec; broken line, $c_p = 10^{-8}$ cm^3/sec. For a 1 MHz C-V curve to be a high frequency C-V curve at a given band bending, D_{it} must be below the curve. For a given D_{it}, the band bending must lie to the right of the curve. The standard deviation of band bending in kT/q has the value $\sigma_s = 2.2$.

smaller capture probability (broken lines) leads to slower interface trap response and a larger acceptable D_{it} at all band bendings.

The presence of minima in the curves in Fig. 9.9 shows that a value of D_{it} exists such that the entire C-V curve at 1 MHz will be within 10% or less of a high frequency C-V curve. This maximum D_{it} for which the entire 1 MHz C-V curve is close to a high frequency C-V curve is plotted against capture probability in Fig. 9.10 with doping density as parameter. Larger capture probabilities reduce the allowed D_{it} because the minimum in the curves in Fig. 9.9 is pushed to larger band bendings, where C_s is lower.

Finally, the curves in Figs. 9.9 and 9.10 have been calculated for $\sigma_s = 2.2$. Curves for other values of σ_s are not necessary because D_{it} is not a strong function of σ_s in the usual range $0 < \sigma_s < 3$.

To illustrate how high a frequency is required to avoid interface trap response to the ac gate voltage, an expression is derived for doping density taking interface trap response to the ac gate voltage into account. In this case (9.21) becomes

$$N_\omega(w) = -2\left(\frac{1 - C_{LF}/C_{ox}}{1 - C_m(\omega)/C_{ox}}\right)\left(q\epsilon_s \frac{d}{dV_G} C_m^{-2}(\omega)\right)^{-1} \qquad (9.38)$$

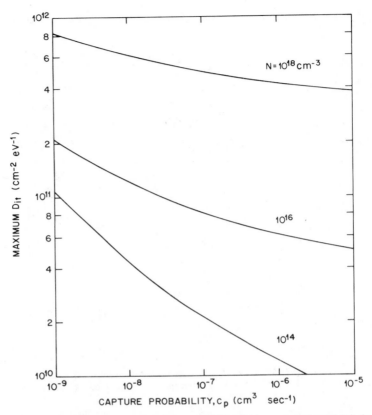

Fig. 9.10 Maximum allowed D_{it} for the entire 1 MHz C-V curve to be a high frequency C-V curve versus capture probability. The standard deviation of band bending in kT/q is $\sigma_s = 2.2$.

and (9.12) is replaced by

$$w = \epsilon_s \left[\frac{1}{C_m(\omega)} - \frac{1}{C_{ox}} \right]. \tag{9.39}$$

Equation (9.38) becomes (9.21) at large ω. The capacitance $C_m(\omega)$ is related to the measured admittance $Y_m(\omega) = G_m(\omega) + j\omega C_m(\omega)$, where Y_m is the series combination of $j\omega C_{ox}$ and the semiconductor admittance Y_s given by (5.48)

$$Y_s = j\omega C_P + G_P \tag{9.40}$$

where

$$C_P(\omega) = C_D + C_{it}(\omega). \tag{9.41}$$

Solving for C_m, we find

$$C_m(\omega) = \frac{C_{ox}\{C_P(\omega)[C_{ox} + C_P(\omega)] + [G_P(\omega)/\omega]^2\}}{[C_{ox} + C_P(\omega)]^2 + [G_P(\omega)/\omega]^2}. \tag{9.42}$$

Using (9.42), for any *given* profile we can compute $C_m(\omega)$ and use the result in (9.38) to find $N_\omega(w)$. By comparing this result with the given profile, we can determine, for any angular frequency ω, the error introduced by using (9.38). Such a comparison now will be made for the case of uniform doping.

Equation (9.42) can be simplified by noting that G_P/ω occurs only as a squared term, while C_{it} occurs only in the combination $C_P = C_D + C_{it}$. Consequently, both *linear* and *quadratic* terms in C_{it} occur in (9.42), but only *quadratic* terms in G_P/ω occur. Because C_{it} and G_P/ω both are small, the quadratic terms can be dropped; that is, G_P/ω can be neglected altogether. With this approximation, (9.42) becomes

$$C_m^{-1}(\omega) = C_{ox}^{-1} + C_P^{-1}(\omega) \tag{9.43}$$

and (9.38) can be simplified, as shown in Appendix X, to read

$$N_\omega(w) = -\left\{ C_P^{-1}(\omega) q\epsilon_s \frac{d}{d\psi_s} C_P^{-1}(\omega) \right\}^{-1} \tag{9.44}$$

and now

$$w = \epsilon_s C_P^{-1}(\omega). \tag{9.45}$$

Equation (9.44) does not depend on C_{ox}, so the profile will not depend on C_{ox} as long as the neglect of G_P/ω is justified.

To use (9.44), we ignore interfacial charge nonuniformity effects on depletion layer capacitance but include them for interface traps. In practice, usually interface traps limit the accuracy of profiling, not the interfacial charge nonuniformity effects on depletion layer capacitance.* The effect of microscopic interfacial charge nonuniformities on depletion layer capacitance becomes substantial only near flatbands, whereas the effect on interface traps is prominent even in depletion (see introductions to Sections 6.3 and 6.4).

From (5.73), we have

$$C_{it}(\psi_s, \omega) = q\beta D_{it} \int_{-\infty}^{\infty} \theta^{-1} \tan^{-1}(\theta) P(\psi_s - \psi_s', \sigma_s) d\psi_s' \tag{9.46}$$

where

$$\theta = \frac{\omega \exp(\beta\psi_s')}{c_p N_A} \tag{9.47}$$

where c_p is the majority carrier capture probability and $P(\psi_s - \psi_s', \sigma_s)$ is the Gaussian probability function (5.69), with σ_s the standard deviation of

---

*Gross interfacial charge nonuniformities discussed in Section 6.2 will influence the capacitance in depletion. However, gross nonuniformities are seldom encountered in device-grade oxides.

band bending. Equation (9.46) is based on (5.49d), and thus assumes that both D_{it} and c_p are such slowly varying functions of ψ_s that they can be considered to be constants. These assumptions are in accord with experimental observations around the center of the bandgap in thermally oxidized silicon as described in Section 7.3. Near the band edges (high frequencies) there might be some error introduced by these assumptions, but this error is not significant for our purpose here. Equation (9.46) takes the experimentally observed interface trap time constant dispersion into account and has been found in Section 7.3 to give a good fit to the measured frequency dependence of interface trap capacitance. Equation (9.46) also can be derived theoretically, as done in Section 6.3 from a model in which interface trap time constant dispersion is caused by band-bending fluctuations arising from a random distribution of oxide charge over the interfacial plane. However, even without a theoretical basis, (9.46) is a convenient analytical expression that fits experiment. With the use of (9.46), the derivative of $C_{it}(\omega)$ with respect to ψ_s is

$$\frac{dC_{it}(\psi_s, \omega)}{d\psi_s} = qD_{it}\beta^2 \int\limits_{-\infty}^{\infty} \left(\frac{P(\psi_s - \psi'_s, \sigma_s)d\psi'_s}{1 + \theta^2}\right) - \beta C_{it}(\psi_s, \omega). \tag{9.48}$$

Equations (9.46) and (9.48) have been numerically integrated, the limits of integration being from minus to plus four times the standard deviation of band bending. Integrating between the $4\sigma_s$ limits results in a very good approximation to the integrals. Then, $N_\omega(w)$ is calculated from (9.44) using (9.43), (9.46), and (9.48), for uniform N_A and D_{it}.

Figure 9.11 shows the results of this calculation. The curve marked infinite frequency in Fig. 9.11 is, in effect, the case $D_{it} = 0$. This curve shows the interface proximity limit for $N_A = 10^{18}\,cm^{-3}$. Also, Fig. 9.11 shows that doping density is accurately determined down to $0.013\,\mu m$, but that for shallower depths there will be a rapidly increasing error. The extrinsic Debye length from (3.28) for $N_A = 10^{18}\,cm^{-3}$ and $T = 290°K$ is $0.0040\,\mu m$. Therefore, the error due to the interface proximity limitation becomes significant for w less than about three extrinsic Debye lengths.

Figure 9.11 shows that interface traps produce fictitious structure in the profile when too low a frequency is used. Both peaks and dips can occur, even for uniform doping.[20, 26] The peaks occur at depletion layer widths corresponding to the onset of the transition of C_{it} from its low frequency value to zero. In this region the capacitance $C_P = C_s + C_{it}$ can display a shoulder if C_{it} is large enough compared to C_s. In this event the derivative $dC_P/d\psi_s$ displays a maximum, and the apparent profile displays a peak. The minimum in the apparent profile is caused by the minimum in $dC_{it}/d\psi_s$, which occurs close to the depletion layer width corresponding to the maximum in G_P/ω.

Figure 9.11 also shows that at any given frequency, interface trap error is insignificant for all values of w beyond some minimum value. The reason is

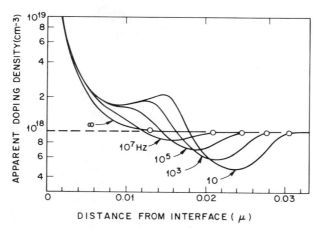

DISTANCE FROM INTERFACE (μ)

Fig. 9.11 Apparent doping density versus apparent depth from the Si-SiO$_2$ interface w as determined from (9.38) and (9.39). The curves are calculated from (9.44) and (9.45) with frequency as parameter, using (9.46)–(9.48), $N_A = 10^{18}\,\text{cm}^{-3}$, $D_{it} = 10^{12}\,\text{cm}^{-2}\,\text{eV}^{-1}$, $c_p = 10^{-8}\,\text{cm}^3/\text{sec}$, $\sigma_s = 2.6$, and $T = 290°\text{K}$; D_{it}, c_p, and σ_s are assumed to be independent of N_A. Open dots indicate 2% accuracy points limited by interface traps, except for the infinite frequency curve, for which the 2% accuracy point is due to the interface proximity limitation.

that longer interface trap time constants occur for larger values of w (larger values of band bending ψ_s) as predicted by (9.47). This minimum value of w decreases as the measurement frequency increases. Consequently, at higher frequencies one can profile to smaller values of w before interface traps begin to respond to the ac gate voltage. At very high frequencies interface traps cease to limit profiling, and only the inherent extrinsic Debye length limitation and interface proximity limitation remain.

To permit a quantitative assessment of the interface proximity and interface trap limitations, we define the value of w at which doping density is in error by 2% as the smallest depth to which doping density can be accurately determined. Not only is this a reasonable and conservative definition, but by varying D_{it} and c_p over a decade, it was found by calculation that this value of w is insensitive to D_{it}, c_p, and σ_s. In Fig. 9.11 the point of 2% accuracy is indicated by open circles. In all cases interface traps are the dominant limiting factor, except for the infinite frequency curve, limited only by interface proximity.

Figure 9.12 illustrates the features of Fig. 9.11 over a useful range of doping densities that have been obtained from families of curves, such as those in Fig. 9.11, using the 2% accuracy limit for the ordinate. In Fig. 9.12 the interface proximity limit on minimum depth for 2% accuracy decreases with increasing doping density because the extrinsic Debye length decreases as doping density increases, as seen from (3.28). The interface trap limit on minimum depth for 2% accuracy at a given frequency also

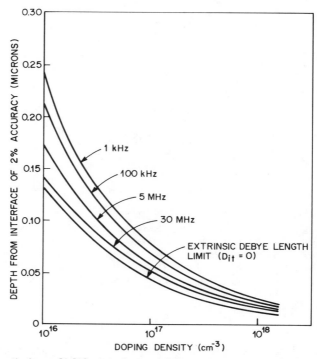

Fig. 9.12 Depth from Si-SiO₂ interface of 2% accuracy versus doping density with frequency as parameter obtained from calculated sets of curves such as those in Fig. 9.11. Interface trap parameters are the same as in Fig. 9.11, but $c_p = 10^{-9}$ cm/sec. After Nicollian et al.[20] Copyright (1973), IEEE.

decreases with increasing doping density. The explanation of this statement follows from (9.47). At any value of ψ_s, interface trap time constants are shorter at larger doping densities. That is, at a given frequency the value of θ at which interface traps begin to respond to the ac gate voltage will correspond to a slightly larger band bending, the larger the doping density. For this given value of θ, ψ_s increases by $(kT/q)\ln 10 \approx 2.3(kT/q)$ for every decade increase in doping density. The value of w corresponding to the given value of θ decreases because the Debye length decreases as the square root of doping density. Despite the increase in ψ_s, the large decrease in Debye length causes w to decrease by a factor of 2–3 for every decade increase in doping density for $\psi_s > kT/q$.

Figure 9.12 shows, as does Fig. 9.11, that the interface trap limit approaches the extrinsic Debye length limit as frequency increases. Also, at high doping densities all the curves in Fig. 9.12 tend to converge to a common minimum depth for 2% accuracy. Thus interface traps have a much smaller influence on the minimum depth at which doping density can be accurately determined at high doping densities and high frequencies. At lower doping densities, frequency must be significantly higher to profile accurately close to the interface.

The frequency that should be used depends on the doping density range to be measured, the value of D_{it}, and how close to the interface an accurate profile is needed. The frequency can be 100 MHz or higher if a profile as near to the interface as allowed by the interface proximity limitation is desired. Such a measurement would be difficult (see Section 12.6.1). However, for many applications, a frequency of 1 MHz, which poses no difficulties, is quite adequate. Frequencies below 1 MHz are to be avoided in most cases because the range of w over which an accurate profile can be extracted becomes too limited, with no compensating advantage.

All the arguments presented for an illustrative p-type sample are general and apply to n-type samples as well. Because the curves shown in Figs. 9.11 and 9.12 are insensitive to capture cross section, they will apply quite well to n-type samples.

Making measurements below room temperature allows profiling only marginally closer to the interface. Figure 9.13 shows the influence of temperature on the minimum depth at which doping density can be determined. Figure 9.13 shows that the depth of 2% accuracy varies less rapidly than the Debye length with temperature over the range shown. Over this

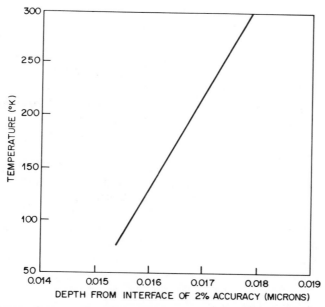

Fig. 9.13 Temperature versus depth from the Si-SiO₂ interface of 2% accuracy calculated from sets of curves such as those in Fig. 9.11 for $N_A = 10^{18}$ cm^{-3}, $D_{it} = 10^{12}$ cm^{-2} eV^{-1}, $c_p = 10^{-9}$ cm³/sec, $\sigma_s/\beta = 67.5$ mV, and $f = 30$ MHz. It is assumed that c_p is independent of temperature. After Nicollian et al.[20] Copyright (1973), IEEE.

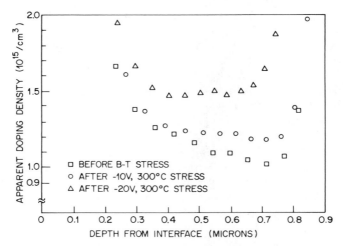

Fig. 9.14 Apparent doping density versus depth from the interface obtained from high frequency C-V curves without correction for interface trap effects. Bias-temperature stress was used to change interface trap level density. Failure to correct for interface trap errors leads to a wide disparity in the three profiles. After Brews.[25]

temperature range and for such heavy doping, freeze out of the dopant ion charge is important. Figure 9.13 does not include freeze out, which may make accurate profiling impractical at low temperatures with this method.

Figure 9.14 shows the magnitude of the error in doping profile if the voltage stretchout caused by interface traps is not taken into account. Doping density was estimated by using (9.19) and three C-V curves measured at 1 MHz. The first C-V curve was measured after fabrication (squares). The second and third curves (circles and triangles) were measured after interface trap level density was increased in the sample by negative bias-temperature stress at successively higher stress bias. A discussion of negative bias-temperature stress aging is given in Section 15.6. Figure 9.14 shows that failure to correct for interface trap voltage stretchout leads to a wide disparity in doping profile that depends on interface trap level density. The sharp increase in apparent doping density at small values of w in Fig 9.14 is the interface proximity limitation due to majority carriers (accumulation). Because gate bias was varied in the measurement slowly enough for minority carriers to follow, the profile appears to rise sharply again at large values of w due to the Debye length limitation of minority carriers (inversion).

9.4.4 The Pulse Method

The doping profile obtained using (9.21) and measured high and low frequency C-V curves is limited to a maximum depth given by the depletion layer width in inversion. This depth is the largest value of w obtainable when minority carriers remain in equilibrium with the gate bias.

In principle, doping profile can be extended over a wider range of depth by applying a voltage pulse to the gate of such short width that neither minority carriers nor interface traps follow. Then there will be no interface trap stretchout, only a high frequency C-V curve need be measured, and doping profile can be extracted from (9.10) and (9.12). However, pulse widths of microseconds or less have to be used to avoid interface trap stretchout over the entire bias range. Such short pulse widths require very high frequencies for a sufficient number of cycles to be obtained during a pulse to measure capacitance. As explained in Section 12.6.1, the measurement of capacitance at such high frequencies becomes difficult. Therefore, it is more practical to apply pulses of $10\ \mu\text{sec}$ widths that preclude both interface trap and minority carrier response only for sufficiently large depletion layer widths. For profiling closer to the surface, the method based on (9.21) is best.

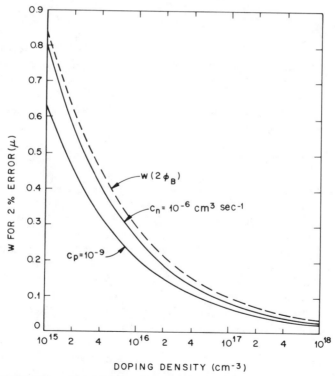

Fig. 9.15 Depth below which interface traps begin to respond to a 100 kHz signal versus doping density. Two capture probabilities are illustrated, spanning the range of observed values (see Section 7.3.8). The broken line indicates the depletion layer width corresponding to a band bending of $2\phi_B$, which is approximately the onset of inversion.

In Fig. 9.15 the depletion layer width for 2% error at 100 kHz is plotted against doping level for uniformly doped substrates and $D_{it} = 10^{12} \, \text{cm}^{-2} \, \text{eV}^{-1}$. This depletion layer width is the smallest consistent with negligible interface trap response at this frequency. Also shown is the depletion layer width corresponding to $2\phi_B$, that is, the depletion layer width beyond which (9.21) becomes impractical because inversion sets in. Figure 9.15 shows that $w(2\phi_B) > w$ (2% error), even for the extreme case of a capture probability of $10^{-6} \, \text{cm}^3 \, \text{sec}^{-1}$. Therefore, the method of (9.21) and the pulse method complement each other.

Figure 9.15 leads to the conclusion that 10 μsec pulses not only will deep deplete the MOS capacitor but will be too fast to allow interface trap response for biases beyond the normal onset of inversion. As a result, one can profile in deep depletion by pulsing the MOS capacitor from a bias near inversion into deep depletion. To eliminate interface trap capacitance and stretchout, a 10 μsec pulse should be used and, say, a 1 MHz small-signal ac voltage superposed to allow capacitance measurement during the pulse. The doping profile then is given by (9.10). Comparison with the profile of (9.21) in the overlapping range of depletion layer widths for which both methods are valid provides an experimental check that the pulse width is short enough to validate (9.10).

In summary, for profiling close to the interface, (9.21) is recommended, at a frequency greater than 1 MHz. For profiling in deep depletion, (9.10) is recommended with a pulse method. The capacitor should be pulsed from a bias just short of inversion into deep depletion, with a pulse width of about 10 μsec and the capacitance measured during the pulse by superposition of a small ac voltage at 1 MHz.

9.4.5 Second Harmonic Methods

The second harmonic methods[20] described briefly in the introduction to Section 9.4 also permit profiling out to depths limited by avalanche breakdown and are free of interface trap stretchout effects. There are spurious second harmonic problems associated with this method[32] that make it more difficult to use than the high-low frequency capacitance method. In addition, small gate area devices are necessary, to increase the amplitude of the second harmonic, which varies as the inverse square of this area.[20] However, this gate area cannot be so small as to introduce edge effects; a radius of many depletion layer widths is needed. As a result, the second harmonic method requires the test capacitor to be made with a gate area related to the doping. This restriction does not apply to other profiling methods. Finally, the second harmonic method is sensitive to stray capacitance and cannot be used to measure a sample at the end of a long coaxial cable.[20, 32]

9.4.6 Maximum-Minimum Capacitance Method

The simplest MOS capacitor measurement method for getting doping density is from a measurement of the maximum high frequency capacitance at a gate bias in strong accumulation and the minimum high frequency capacitance at a gate bias in strong inversion.[21] This method is independent of interface trap effects because the maximum high frequency capacitance measured in accumulation is just C_{ox}, and the minimum high frequency capacitance is measured in strong inversion where interface traps do not contribute any capacitance. There is no C_{it} because interface trap response times are orders of magnitude longer than the period of the applied ac signal in strong inversion. This method is most accurate for a uniform doping profile. Because of its simplicity, the maximum-minimum high frequency capacitance method is widely used for obtaining doping density, particularly for routine monitoring where accuracy may not be the paramount consideration.

The basic idea behind this method is as follows. The depletion layer width is a maximum when high frequency capacitance is at its lowest value in strong inversion. Therefore, the maximum depletion layer width w_{max} can be obtained from the minimum high frequency capacitance using (9.12)

$$w_{max} = \epsilon_s \left(\frac{1}{C_{HF}(\min)} - \frac{1}{C_{ox}} \right). \qquad (9.49)$$

This maximum depletion layer width is approximately the sum of the inversion and depletion layer widths for a band-bending $\psi_s(\text{inv}) \approx \phi_m$, where ϕ_m is the maximum band bending from (4.161). If we neglect inversion layer width and simply use the depletion approximation width calculated from (2.82) evaluated at ϕ_m for the minimum capacitance, for *uniform* doping density, we obtain

$$w_{max}^2 = \frac{2\epsilon_s \phi_m}{qN_A} \qquad (9.50)$$

where N_A is the uniform acceptor concentration (we have taken p-type as an illustrative example). For our purposes, ϕ_m can be approximated as $2\phi_B + (kT/q)\ln[2q\phi_B/(kT) - 1]$, which is very nearly the Lindner potential of (4.160). The correction of $0.75 \, kT/q$ in (4.161) is negligible here. Using $\phi_B = (kT/q)\ln(N_A/n_i)$ from (2.45) and combining (9.49) and (9.50), we obtain

$$\frac{N_A}{\ln(N_A/n_i) + \frac{1}{2}\ln[2\ln(N_A/n_i) - 1]} = \frac{4kT\epsilon_{ox}^2}{q^2\epsilon_s x_o^2} \left(\frac{C_{ox}}{C_{HF}(\min)} - 1 \right)^{-2} \qquad (9.51)$$

where ϵ_{ox} is the dielectric permittivity of SiO_2, n_i is the intrinsic carrier concentration, and x_o is the oxide layer thickness. Equation (9.51) is a transcendental equation in N_A that can be solved by iteration. Figure 9.16 shows a plot of N_A as a function of $C_{HF}(\min)/C_{ox}$ with oxide thickness as

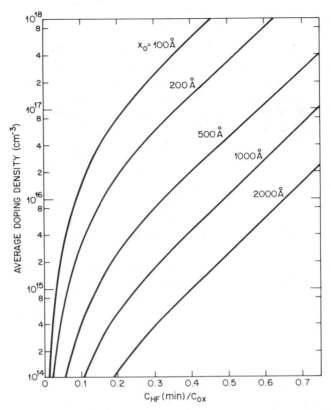

Fig. 9.16 Acceptor doping concentration N_A versus $C_{HF}(min)/C_{ox}$ with oxide thickness as parameter based on (9.51).

parameter based on (9.51). Figure 9.16 can be used to obtain N_A graphically, avoiding an iterative solution of (9.51).

For a nonuniform doping profile, the value of N_A obtained from (9.51) or Fig. 9.16 is an *average** of the profile over w_{max}. That is, (9.51) leads to an "effective" doping density that is only an approximation of the true doping density. Even for uniform doping density, (9.51) is only approximate because it approximates polarization effects in the inversion layer [see Section (4.4)] and neglects inversion layer thickness. Therefore, (9.51) will be accurate to within about 3% for uniform doping density. In the usual case where the doping density is not uniform, this method yields only the average doping density so it is not worthwhile to correct for these errors.

---

*The max-min doping is related to the first moment of the profile in the depletion layer, not to a simple spatial average.

9.5 LIFETIME

The free carrier lifetime in a semiconductor is the time required for the free carrier concentration in the bulk semiconductor to return to thermal equilibrium after a disturbance of the equilibrium state has occurred. Such a disturbance, for example, would be a sudden change of electric field in the semiconductor. Once equilibrium has been disturbed, both majority and minority carrier concentrations must change to reestablish thermal equilibrium. At room temperature, minority carrier concentration will reach thermal equilibrium by generation or recombination through bulk traps. If the disturbance has reduced minority carrier concentration below its equilibrium value, then thermal equilibrium will be reestablished by generation, as described in Section 4.3.2(c). If the disturbance has increased minority carrier concentration above its equilibrium value, then thermal equilibrium will be reestablished by recombination, also described in Section 4.3.2(c).

Generation requires that traps alternately emit holes and electrons. Lifetime thus depends on both minority and majority carrier processes. If majority carrier processes take longer than minority carrier processes, for example, lifetime will be determined by majority carrier processes.

There are various ways for measuring lifetime in semiconductors. Some of these methods not discussed here are (1) Haynes-Shockley,[33] (2) photoconductive decay,[34] (3) surface photovoltage,[35] (4) the Many bridge,[5] and (5) charge pumping.[36] A discussion of all these methods is beyond the scope of this book. The interested reader can find a brief discussion of the first four methods in Ref. 5 and charge pumping in Ref. 36. The first four methods require correction for generation and recombination through interface traps.

In integrated circuit work, it is convenient to measure lifetime using an MOS capacitor structure, which usually is included during circuit manufacture as a tester on the silicon wafer. We concentrate here on ways of determining lifetime from MOS capacitor measurements.

There are two major MOS capacitor methods for determining lifetime: steady-state methods and transient methods. Because they are better understood, it is preferable to use the steady-state methods rather than the transient methods. Although the small-signal steady-state method described in Sections 4.3 and 9.5.1 is more accurate, requiring no perimeter corrections, the steady-state ramp method described in Sections 9.5.3 and 12.12 is easier to use and to implement.

9.5.1 The Small-Signal Steady-State Method

In Section 4.3.7 the bulk hole lifetime τ_{Tp} and the electron lifetime τ_{Tn} were defined by (4.104)

$$\tau_{Tp} = (c_p n_T)^{-1} \tag{9.52a}$$

$$\tau_{Tn} = (c_n n_T)^{-1} \tag{9.52b}$$

(where n_T is the volume density of SRH[37] centers and c_p and c_n are the capture rates). These lifetimes were related to the small-signal admittance of the MOS capacitor biased in strong inversion. It was shown that the minority carrier lifetime could be obtained from measurements at temperatures high enough that *diffusion* dominated the minority carrier response, whereas the product of majority and minority carrier lifetimes could be obtained at lower temperatures where *generation-recombination* dominated minority carrier response. .

The small-signal determination of lifetimes provides the geometric mean of majority and minority carrier lifetimes. It is a rather involved method, requiring measurement at several frequencies as a function of temperature, and requiring some computation. Consequently, some faster, more convenient approaches have been suggested that are based on measurement of the response of the MOS high frequency capacitance to gate bias pulses or to a gate bias ramp. In Section 9.5.2 a discussion of several of these methods is given.

9.5.2 Capacitance Transient Methods

If the gate bias of an inverted MOS capacitor is changed in the direction of stronger inversion, inversion layer occupancy tends to increase with time. However, the minority carriers needed to increase occupancy of the inversion layer cannot be provided instantly but either are generated in the depletion layer or are provided by diffusion through the depletion layer from the quasi-neutral bulk. If gate bias is changed slowly compared to this supply rate, the inversion layer will receive all the carriers that it requires and the inversion layer will remain in equilibrium throughout the change of gate bias. If gate bias is changed rapidly, minority carriers will not be generated quickly enough to satisfy the demand for inversion layer charge. In this case depletion layer width will increase to maintain overall charge neutrality, compensating gate charge with ionized dopant ions until such time as sufficient minority carriers can be supplied. While inversion layer carrier density is increasing, depletion layer width will shrink until equilibrium is obtained.

This behavior of the depletion layer width can be seen in the behavior of the high frequency capacitance, shown in Fig. 9.17. Figure 9.17 illustrates an MOS capacitor biased in inversion at times $t < 0$, so the capacitance C_{HF} is the equilibrium capacitance C_∞. At $t = 0$ a depleting voltage step is applied to the gate. The capacitance drops to a value C_0 as the depletion layer width expands to its maximum value. Then, as time progresses, C_{HF} increases as depletion layer width contracts as a result of generation of inversion layer carriers by bulk traps, returning to the equilibrium value C_∞ as $t \to \infty$.

The idea behind a pulse measurement of lifetime now can be explained: gate bias is pulsed to a new value, depletion layer width expands, and then

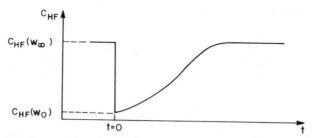

Fig. 9.17 Transient response of capacitance after a voltage pulse on the gate (schematic).

the decay of depletion layer width toward its equilibrium value is measured. To relate the decay of depletion layer width to lifetime, two steps are used: (1) the change in depletion layer width is related to the change in inversion layer carrier density (this was done by Zerbst[38]); and (2) the change in inversion layer carrier density is related to lifetime. This is where some difficulty arises because of the complex nature of generation and recombination in a depletion layer that is not in steady state.

(a) Zerbst Relation

At any time t, by Gauss's law, the oxide field is proportional to the net charge per unit area in the semiconductor; that is

$$C_{ox}[V_G - \psi_s(t)] = q\left[N_I(t) + \int_0^{w(t)} dx\, N_B(x)\right] \qquad (9.53)$$

where $\psi_s(t)$ is the instantaneous band bending, $N_I(t)$ is the instantaneous inversion layer carrier density, and $N_B(x)$ is the dopant density at position x. The instantaneous value of depletion layer width is $w(t)$.

Gate bias is held fixed, so differentiating (9.53) with respect to time and rearranging terms, we find

$$\frac{dN_I}{dt} = -\frac{C_{ox}}{q}\left(\frac{d\psi_s}{dt}\right) - N_B[w(t)]\frac{dw}{dt}. \qquad (9.54)$$

The depletion layer width is experimentally observable if it is defined in terms of C_{HF}; that is

$$w(t) = \epsilon_s\left(\frac{1}{C_{HF}(t)} - \frac{1}{C_{ox}}\right). \qquad (9.55)$$

Hence (dN_I/dt) is measurable, provided that $d\psi_s/dt$ can be related to w and that the doping profile $N_B(x)$ is measured.

Zerbst related ψ_s to w by neglecting the change in voltage drop across the inversion layer. Under these conditions, within the depletion approximation

$$\psi_s(t) = \frac{q}{\epsilon_s} \int\limits_{0}^{w(t)} dx \, x N_B(x). \tag{9.56}$$

That is

$$\frac{d\psi_s}{dt} = \frac{qw(t)}{\epsilon_s} N_B[w(t)] \frac{dw}{dt}. \tag{9.57}$$

Combining (9.54) and (9.57) yields

$$\frac{dN_I}{dt} = -\left(\frac{C_{ox}}{\epsilon_s} w + 1\right) N_B(w) \frac{dw}{dt}. \tag{9.58}$$

Using (9.55), we obtain

$$\begin{aligned}\frac{dN_I}{dt} &= -\left(\frac{C_{ox}}{C_{HF}}\right) N_B(w)\epsilon_s \frac{d}{dt}\left(\frac{1}{C_{HF}} - \frac{1}{C_{ox}}\right)\\ &= -\frac{N_B(w)\epsilon_s}{2C_{ox}} \frac{d}{dt}\left(\frac{C_{ox}}{C_{HF}}\right)^2.\end{aligned} \tag{9.59}$$

Equation (9.59) is the Zerbst relation between rate of change of inversion layer carrier density and the rate of change of depletion layer width (or reciprocal high frequency capacitance). This relation is quite general; the major approximation is the treatment in (9.56) of the band bending. An additional requirement is that the density of recombination-generation centers controlling lifetime be much smaller than the bulk dopant level. If this were not the case, the charge in these traps would affect the depletion layer width, thereby modifying (9.56).

(b) Generation-Recombination Away from Steady State

Most analyses of the pulsed MOS capacitor propose to approximate generation-recombination by a constant generation rate in the depletion region. That is, they adopt

$$\frac{dN_I}{dt} = \frac{n_i}{\tau} [w - w(\infty)] \tag{9.60}$$

where

$$\tau = \tau_{Tn} \exp(-v_T) + \tau_{Tp} \exp(v_T). \tag{9.61}$$

In (9.61) v_T is the energy in units of kT/q of the bulk trap level measured with respect to midgap, as shown in Fig. 4.3. In (9.60), $[w - w(\infty)]$ is the excess of the nonequilibrium depletion layer width over the equilibrium portion. The factor $[w - w(\infty)]$ on the right of (9.60) tends to zero as w tends to the equilibrium value $w(\infty)$. This factor ensures that dN_I/dt tends to zero as $t \to \infty$. The rate (n_i/τ) is taken from the steady-state analysis of a reverse biased p-n junction due to Sah et al.[39] However, Collins and

Churchill[40] have shown that this rate is inappropriate, except during the final stages of the approach to equilibrium, and even then is not accurate in the limit $t \to \infty$. We discuss their work shortly. For now, assuming that (9.60) is adequate, we rewrite (9.60) using (9.55) as

$$\frac{dN_I}{dt} = \frac{n_i}{\tau} \epsilon_s \left[\frac{1}{C_{HF}(t)} - \frac{1}{C_{HF}(\infty)} \right]. \tag{9.62}$$

Combining (9.62) with (9.59), we see that for *uniform* doping, a plot of $-d(1/C_{HF})^2/dt$ versus $[1/C_{HF}(t)]$ should provide a straight line with slope $[2n_i/(\tau C_{ox} N_B)]$ and intercept $[1/C_{HF}(\infty)]$.

An example of a Zerbst plot is shown in Fig. 9.18. A straight-line region is apparent for times that are not too short (far right of figure) nor too long (near origin of figure). Equation (9.62) applies during this time interval, and the slope of this straight-line region determines τ.

We examine (9.60) further. Collins and Churchill[40] considered the relaxation of an MOS capacitor pulsed from flatbands to strong inversion. They found that the relaxation could be divided into three main regimes: (1) the dielectric relaxation regime ($t < 10^{-9}$ sec) during which depletion of majority carriers occurs, (2) the depletion regime (10^{-9} sec $< t < 10^{-2}$ sec) during which band bending remains at the deep-depleted value but minority carriers diffuse in from the neutral bulk to fill some SRH centers while other SRH centers emit *majority* carriers at the steady state rate, and (3) the equilibration regime ($t > 10^{-2}$ sec) during which bulk traps generate minority *and* majority carriers at the steady-state rate, causing the inversion layer to build up and the depletion layer width to relax to its equilibrium value. The times mentioned here are illustrative but will vary with bulk trap parameters and with the importance of diffusion (i.e., with sample thickness and temperature). The main point is that if bulk trap

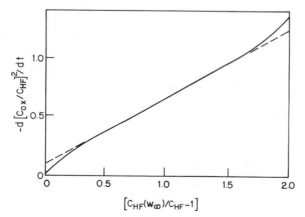

Fig. 9.18 Zerbst plot ·from transient capacitance data. After Schroder and Guldberg.[46] Copyright (1971), Pergamon Press, Ltd. Reprinted with permission.

density is low compared to bulk dopant density, the depletion layer width response occurs during the equilibration period (period 3), which is dominated by the steady-state generation expression. The initial phases of bulk trap response, which are not steady state, cause little change in depletion or inversion layer charge density. Therefore, the work of Collins and Churchill[40] supports the traditional use of steady-state theory for the analysis of depletion and inversion layer response. However, as we shall see, in the limit of very long times, the accurate steady-state generation expression [see (9.64)] must be used, rather than (9.60).

(c) Steady-State Theory

Sah et al.[39] point out that in steady state the capture rate of holes must equal that of electrons for the bulk trap occupancy to be time independent. As a result, we have (see Section 4.3.3)

$$r_a - r_b = \frac{nf_{Tp} - n_1 f_T}{\tau_{Tn}} = r_c - r_d = \frac{pf_T - p_1 f_{Tp}}{\tau_{Tp}}. \tag{9.63}$$

where $r_a - r_b$ and $r_c - r_d$ are net capture rates for electrons and holes, f_T is the fraction of bulk traps occupied by electrons, $f_{Tp} = 1 - f_T$, and n_1 and p_1 are the electron and hole densities when the Fermi level is located at the bulk trap level. Manipulation of (9.63) to eliminate f_T and f_{Tp} leads to

$$r_a - r_b = r_c - r_d \equiv U = \frac{pn - n_i^2}{(n + n_1)\tau_{Tp} + (p + p_1)\tau_{Tn}}. \tag{9.64}$$

In the depletion layer both p and n are small. Consequently, in the depletion layer

$$U = -\frac{n_i^2}{n_1 \tau_{Tp} + p_1 \tau_{Tn}} = -\frac{n_i}{\exp(v_T)\tau_{Tp} + \exp(-v_T)\tau_{Tn}}. \tag{9.65}$$

A careful examination of (9.64) shows that (9.65) is a good approximation in the part of the depletion layer between the points where the bulk trap level crosses the minority and majority carrier quasi-Fermi levels.[41] As the capacitor relaxes toward equilibrium, the "reverse bias" across the depletion layer is reduced, and the width of this generation region also is reduced. This contraction of the generation region is illustrated in Fig. 9.19 from Collins and Churchill,[40] who found the generation region to be about $[w - w(\infty)]$ in size initially but tending to become smaller than $[w - w(\infty)]$ as time progresses—that is, dN_I/dt falls below $(n_i/\tau)[w - w(\infty)]$. This effect can be seen in Fig. 9.20, which shows calculated values of dN_I/dt versus $[(w - w(\infty))]$. It is apparent that proportionality is only approximate, with dN_I/dt tending to fall more rapidly as $[w - w(\infty)]$ is reduced. The explanation for this more rapid drop of dN_I/dt is that mobile carriers encroach on the depletion region at smaller $[w - w(\infty)]$. Consequently, generation is reduced, and (9.64) for U must be used in place of (9.65).

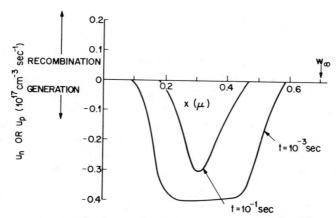

Fig. 9.19 Electron and hole net recombination versus x for several short times using the steady-state approximations. After Collins and Churchill.[40] Copyright (1975), IEEE.

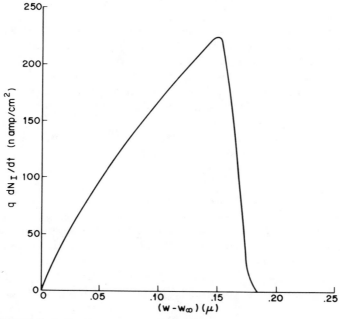

Fig. 9.20 Rate of change of inversion layer carrier density per unit area versus $(w - w_\infty)$. After Collins and Churchill.[40] Copyright (1975), IEEE.

Therefore, we conclude that a measurement of the time derivative (dN_I/dt) using the Zerbst relation (9.59) will be approximately a straight-line function of $[w - w(\infty)]$ or, according to (9.62), of $\epsilon_s(1/C_{HF} - 1/C_{HF}(\infty))$ provided that (1) the doping profile is uniform and (2) the time interval chosen is not too early and not too late. In practice, the appropriate time interval is found experimentally as the interval over which a linear Zerbst plot is observed.

9.5.3 The Steady-State Ramp Method

Pierret[42] and Pierret and Small[43] have suggested a variant of the afore-mentioned method that uses a gate bias ramp to measure lifetime. Assuming that (9.60) is valid, if a gate bias ramp is applied to an MOS capacitor so as to maintain a constant w greater than the equilibrium width, then dN_I/dt is constant. Returning to (9.53) and assuming that w is held constant, independent of time, we find, by differentiating (9.53), that

$$C_{ox} \frac{dV_G}{dt} = q \frac{dN_I}{dt}. \tag{9.66}$$

Using (9.60) yields

$$\left(\frac{C_{ox}}{q}\right) \frac{dV_G}{dt} = \left(\frac{n_i}{\tau}\right) [w - w(\infty)]. \tag{9.67}$$

Consequently, a plot of the measured ramp rate (dV_G/dt) versus depletion layer width, measured by using (9.55), should provide a straight line with slope $qn_i/(\tau C_{ox})$. This method appears to be simpler to use than the pulse method because (1) virtually no computation is involved, (2) uniform doping need not be assumed, (3) steady-state analysis is applicable, and (4) failure to obtain a straight-line plot provides a warning that the sample is not obeying the model analysis.

Examples of experimental data obtained from this method are given in the next section. The instrumentation for this method is described in Section 12.12.

9.5.4 Effects of Surface Generation

When interface traps are present, surface generation of minority carriers occurs in addition to generation by bulk traps considered already. The contribution of interface traps is minimized by pulsing from inversion to stronger inversion because the interface trap levels are filled to a level well above midgap throughout the pulse, thereby reducing their generation rate to near zero.

However, as pointed out by Schroder and Nathanson[44], an unexpected contribution of interface traps still exists from the *perimeter* of the MOS capacitor. As illustrated in Fig. 9.21 from Kano and Shibata,[45] the depletion layer under the gate bends up to the interface near the edge of the gate. Consequently, as the capacitor is pulsed, a surface generation component from interface traps or other surface defects within this deep-depleted perimeter portion of the interface will contribute to inversion layer response. References 44–47 discuss various ways to correct for this effect.

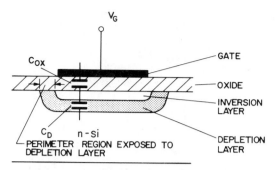

Fig. 9.21 Cross section of an MOS capacitor illustrating the perimeter region exposed to the depletion layer when generation by interface traps must be considered in determining inversion layer response. After Kano and Shibata.[45]

Here, we describe the approach of Small and Pierret,[47] whose method requires a dual-ring dot structure such as that shown in Fig. 9.22 and involves three measurements of the total generation rate: (1) measurement under the central dot, with both rings biased in accumulation (Fig. 9.22b), (2) measurement under the center ring, with outer ring and dot biased to accumulation (Fig. 9.22c), and (3) measurement under center ring and dot, with outer ring biased to accumulation (Fig. 9.22d). The depletion regions in each case are shown in Fig. 9.22. The region *between* electrodes is held in accumulation in case 1 by introducing ionic charges in a wet N_2 environment by using an applied bias and locking them in place by changing to a dry ambient. The measurement is then performed in a dry N_2 ambient (see Section 4.3.12). In cases 2 and 3 ions are used to invert the interelectrode regions, thus eliminating interface generation in these regions (see Section 5.3.7).

Examination of Fig. 9.22 shows that in case 1 the total generation rate G_1 is composed of depletion layer generation under the dot G_{D1} and surface generation from the perimeter of this dot G_{S1}

$$G_1 = G_{D1} + G_{S1}. \tag{9.68}$$

Similarly, in case 2 the total generation rate G_2 is

$$G_2 = G_{D2} + G'_{S1} + G_{S2} \tag{9.69}$$

where G'_{S1} is the surface generation from the periphery of the central dot, which may differ somewhat from G_{S1} in (9.68) because the interelectrode surface region contributing to G_{S1} and G'_{S1} may behave somewhat differently under the differing bias conditions in cases 1 and 2. Finally, in case 3 we obtain

$$G_3 = G_{D1} + G_{D2} + G_{S2}. \tag{9.70}$$

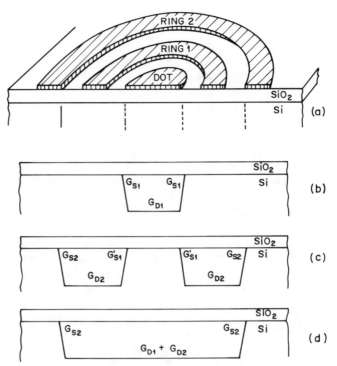

Fig. 9.22 The dual ring-dot structure according to Small and Pierret.[47] (a) A cross section of the structure in (b)–(d) shows the depletion regions for measurements of cases 1–3. Copyright (1976), Pergamon Press, Ltd. Reprinted with permission.

Combining (9.69) and (9.70) yields

$$G_3 - G_2 = G_{D1} - G'_{S1}. \tag{9.71}$$

Comparing (9.71) and (9.68), we see that if G_{S1} and G'_{S1} are small compared to G_D, then G_1 and $G_3 - G_2$ will be equal. Hence comparison of experimental results for G_1 and $G_3 - G_2$ indicates immediately whether G_{S1} and G'_{S1} can be neglected. In unannealed samples (large surface generation rate samples) G_{S1} and G'_{S1} could not be neglected but had to be estimated on the assumption

$$G_{S1} \approx G'_{S1}. \tag{9.72}$$

Figure 9.23 shows typical results for generation rate versus $\Delta w/w$ from steady-state ramp measurements for an annealed structure, where Δw is the excess of the deep-depleted depletion layer width beyond the equilibrium value w. The three generation rates G_1, G_2, and G_3 are shown, and $G_3 - G_2$ is compared with G_1. The agreement of these last two shows that the neglect of G_{S1} and G'_{S1} is justified in this case.

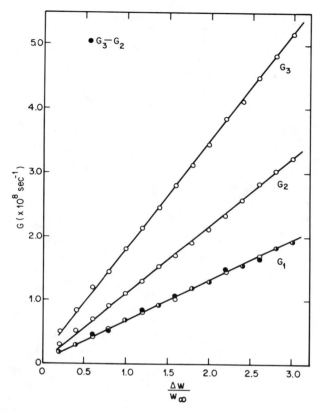

Fig. 9.23 Typical generation rates versus $\Delta w/w_\infty$ for annealed structures. After Small and Pierret.[47] Copyright (1976), Pergamon Press, Ltd. Reprinted with permission.

Figure 9.24 shows typical results for an unannealed structure, where G_{S1} was determined by assuming $G'_{S1} \approx G_{S1}$ so that subtraction of (9.71) from (9.68) yields

$$G_{S1} = \frac{G_1 + G_2 - G_3}{2}. \tag{9.73}$$

These results show that the surface generation rate is appreciable in unannealed samples, and it can be estimated from dual-ring dot structures. More quantitative estimates must rely on gated diode measurements,[48–50] which are beyond the scope of our discussion.

Perimeter effects usually will be negligible in the small-signal steady-state method described in Sections 4.3 and 9.5.1 because generation from interface traps at the silicon surface does not occur in the entire perimeter region shown in Fig. 9.21. For a small-signal measurement, generation occurs only for those interface trap levels located near midgap, as discussed in Section 4.3.1c.

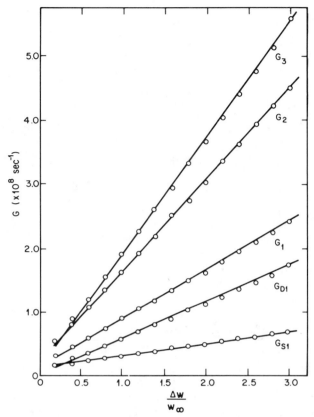

Fig. 9.24 Typical generation rates versus $\Delta w / w_\infty$ for unannealed structures. After Small and Pierret.[47] Copyright (1976), Pergamon Press, Ltd. Reprinted with permission.

Midgap levels are completely empty at one side of the perimeter region and completely full at the other. Only in a narrow annulus where the band bending is within kT/q of midgap are the midgap interface trap levels near the Fermi level. Hence, the density of interface traps contributing to generation-recombination is given by (see (4.64)) $(kT/q)D_{it}(\partial v_s/\partial R)^{-1}2\pi R$, where R is the radius of the annulus and $\partial v_s/\partial R$ is the surface field in the annulus in kT/q per cm. Comparison with the bulk trap contribution to generation, $n_T(\partial v/\partial x)^{-1}\pi R^2$, shows for comparable fields that interface trap generation is a small effect. For example, for a 0.1 cm radius dot, $n_T = 10^{12}$ cm^{-3}, and $D_{it} = 10^{11}$ cm^{-2} eV^{-1}, the interface trap contribution to generation is 5% of the bulk trap contribution.

REFERENCES

1 H. J. Queisser, *Semiconductors in the Relaxation Regime*, 1972 European Solid-State Devices Research Conference, Lancaster, England, Institute of Physics, London, paper 8, pp. 145–168.

2 O. Lindberg, *Proc. Inst. Radio Eng.*, **40**, 1414 (1962).

3 A. J. Dekker, *Solid-State Physics*, Prentice-Hall, New York, 1957, Chapter 13.

4 E. H. Putley, *The Hall Effect and Related Phenomena*, Butterworths, London, 1960.

5 J. Evans, *Fundamental Principles of Transistors*, 2nd ed., Van Nostrand, Princeton, N.J., 1962.

6 H. Melchior, A. Goetzberger, E. H. Nicollian, and W. T. Lynch, *Solid-State Electron.*, **12**, 449 (1969).

7 A. Goetzberger and E. H. Nicollian, *J. Appl. Phys.*, **38**, 4582 (1967).

8 A. Goetzberger and E. H. Nicollian, *Appl. Phys. Lett.*, **9**, 444 (1966).

9 A. Rusu and C. Bulucea, *IEEE Transact. Electron Devices*, **ED-26**, 201 (1979).

10 S. M. Sze and G. Gibbons, *Appl. Phys. Lett.*, **8**, 111 (1966).

11 P. F. Kane and G. B. Larrabec, Eds., *Characterization of Solid Surfaces*, Plenum, New York, 1974.

12 *Proceedings of ARPA/NBS Workshop IV, Surface Analysis for Silicon Devices*, NBS, Gaithersburg, Maryland, April 23–24, 1975 (issued March 1976).

13 A. S. Grove, O. Leistiko, Jr., and C. T. Sah, *J. Appl. Phys.*, **35**, 2695 (1964).

14 J. Hilibrand and R. D. Gold, *RCA Rev.*, **21**, 245 (1960).

15 C. O. Thomas, D. Kahng, and R. C. Manz, *J. Electrochem. Soc.*, **109**, 1055 (1962).

16 I. Amron, *Electrochem. Technol.*, **12**, 327 (1964).

17 D. R. Decker, *J. Electrochem. Soc.*, **115**, 1085 (1968).

18 N. I. Meyer and T. Guldbrandsen, *Proc. IEEE*, **51**, 1631 (1963).

19 J. A. Copeland, *IEEE Transact. Electron Devices*, **ED-16**, 445 (1969).

20 E. H. Nicollian, M. H. Hanes, and J. R. Brews, *IEEE Transact. Electron Devices*, **ED-20**, 380 (1973).

21 B. E. Deal, A. S. Grove, E. H. Snow, and C. T. Sah, *J. Electrochem. Soc.*, **112**, 308 (1965).

22 C. Jund and R. Poirier, *Solid-State Electron.*, **9**, 315 (1966).

23 D. C. Gupta and N. G. Anantha, *Proc. IEEE*, **55**, 1108 (1967).

24 W. Van Gelder and E. H. Nicollian, *J. Electrochem. Soc.*, **118**, 138 (1971).

25 J. R. Brews, *J. Appl. Phys.* **44**, 3228 (1973).

26 Y. Zohta, *Solid-State Electron.*, **17**, 1299 (1974).

27 W. C. Johnson and P. T. Panousis, *IEEE Transact. Electron Devices*, **ED-18**, 965 (1971).

28 W. E. Carter, H. K. Gunmel, and B. R. Chawla, *Solid-State Electron.*, **15**, 195 (1972).

29 A. R. LeBlanc, D. D. Kleppinger, and J. P. Walsh, *J. Electrochem. Soc.*, **119**, 1068 (1972).

30 P. P. Kennedy, R. C. Murley, and W. Kleinfelder, *IBM J. Res. Dev.*, **12**, 399 (1968).

31 J. R. Brews, *IEEE Transact. Electron Devices*, **ED-26**, 1696 (1979).

32 D. C. Gupta and J. Y. Chan, *J. Appl. Phys.*, **43**, 515 (1972).

33 J. R. Haynes and W. Shockley, *Phys. Rev.*, **81**, 835 (1951).

34 D. T. Stevenson and R. J. Keyes, *J. Appl. Phys.*, **26**, 190 (1955).

35   E. O. Johnson, *J. Appl. Phys.*, **28**, 1349 (1957).

36   J. S. Brugler and P. G. A. Jespers, *IEEE Transact. Electron Devices*, **ED-16**, 297 (1969); J. Golder and E. Baldinger, *Helv. Phys. Acta*, **44**, 387 (1971); M. Declercq and P. Jespers, *Revue HF (Belg.)*, **9**, 244 (1974).

37   W. Shockley and W. T. Read, Jr., *Phys. Rev.*, **87**, 835 (1952).

38   M. Zerbst, *Z. Angew. Phys.*, **22**, 30 (1966).

39   C. T. Sah, R. N. Noyce, and W. Shockley, *Proc. IRE*, **45**, 1228 (1957).

40   T. W. Collins and J. N. Churchill, *IEEE Transact. Electron Devices*, **ED-22**, 90 (1975).

41   P. U. Calzolari and S. Graffi, *Solid-State Electron.*, **15**, 1003 (1972).

42   R. F. Pierret, *IEEE Transact. Electron Devices*, **ED-19**, 869 (1972).

43   R. F. Pierret and D. W. Small, *IEEE Transact. Electron Devices*, **ED-22**, 1051 (1975).

44   D. K. Schroder and H. C. Nathanson, *Solid-State Electron.*, **13**, 577 (1970).

45   Y. Kano and A. Shibata, *Jap. J. Appl. Phys.*, **11**, 1161 (1972).

46   D. K. Schroder and J. Guldberg, *Solid-State Electron.*, **14**, 1285 (1971).

47   D. W. Small and R. F. Pierret, *Solid-State Electron.*, **19**, 505 (1976).

48   M. V. Whelan, IEDM Washington, D.C., October 1969 and *Philips Research Reports*, Supplement No. 6 (1970).

49   R. F. Pierret, *Solid-State Electron.*, **17**, 1257 (1974).

50   R. F. Pierret and R. J. Grossman, *Solid-State Electron.*, **20**, 373 (1977).

10

Charges, Barrier Heights, and Flatband Voltage

10.1 INTRODUCTION

In this chapter we discuss (1) the contribution of charges and of work function differences to the flatband voltage, (2) how these contributions are determined, and (3) how flatband voltage is measured. This chapter lays the groundwork for parts of Chapters 11 and 15.

Section 10.2 describes how oxide charge affects MOS capacitor characteristics and how its polarity and magnitude can be determined. Section 10.3 describes how internal photoemission can be used to determine the energy barrier height between silicon and SiO_2 and between the gate and the SiO_2. Work function differences can be determined from these barrier heights. Section 10.4 describes how the work function difference between gate material and silicon affects C-V characteristics and how it is measured. Section 10.5 describes how flatband voltage is measured when interface traps are present and the uncertainties involved in the extraction of oxide charge from flatband voltage caused by charged interface traps and doping nonuniformities.

10.2 OXIDE CHARGE

There are two basic types of charge in the SiO_2 layer: (1) interface trap charge, discussed in Chapters 5–7, and (2) oxide charge. The feature that distinguishes interface trap charge from oxide charge is that interface trap charge varies with gate bias, whereas oxide charge is independent of gate bias. We ignore interface trap charge in this section and concentrate on oxide charge.

There are three types of oxide charge Q_o, that are technologically important. The first type, oxide fixed charge Q_f, is the charge density remaining after interface trap charge is annealed out (see Section 15.4.3). Oxide fixed charge is located at or very near the $Si–SiO_2$ interface (see Chapters 11 and 15). In electrical measurements Q_f can be regarded as a charge sheet located at the $Si–SiO_2$ interface.

The second type of oxide charge, oxide trapped charge Q_{ot}, usually is located either at the metal-SiO_2 interface or at the $Si–SiO_2$ interface. One exception is when the oxide traps are introduced by ion implantation. In this case Q_{ot} can be distributed within the oxide layer. Oxide trapped charge is commonly produced by the injection of hot electrons or holes from an avalanche plasma in a high field region in the silicon, injection of carriers by photoemission or by exposure to ionizing radiation. These effects are discussed in Chapter 11.

The third type, mobile ionic charge Q_m, most commonly is caused by the presence of ionized alkali metal atoms such as sodium or potassium. This type of charge is located either at the metal-SiO_2 interface, where it originally entered the oxide layer, or at the $Si–SiO_2$ interface, where it has drifted under an applied field. Drift can occur because such ions are mobile in SiO_2 at relatively low temperatures.

Immobile oxide charge can be distinguished from mobile ionic charge by a bias-temperature aging experiment. Gate bias is applied for a given length of time while the sample is held at a moderately elevated temperature. The density of oxide fixed charge may change under this treatment, but the centers responsible for this charge do not move. However, the mobile ionic charge can be cycled back and forth between the metal-SiO_2 interface and the $Si–SiO_2$ interface by this treatment without discharge, and the resulting ionic current can be detected.

10.2.1 Measurement of Oxide Charge

Oxide charge is an important parameter in devices. For example, it can alter the threshold voltage of a MOSFET, alter the silicon surface potential and thereby change reverse surface leakage current in a p-n junction, alter the avalanche breakdown voltage of a p-n junction, and alter common emitter current gain at low collector current in a bipolar transistor. It can invert the silicon, resulting in unwanted current paths between elements in

an integrated circuit. It also must be understood for the correct interpretation of C-V curves measured on an MOS capacitor.

The simplest and most widely used method for measuring oxide charge
density is to infer this density from the voltage shift of a C-V curve. We
start by calculating the effect of oxide charge on the C-V curve. Let $n_o(x)$
be the *volume* density of oxide charge. This charge induces an *image*
charge in the silicon surface. Therefore, the capacitance at any given gate
bias will be different from what it would be if $n_o(x) = 0$. The gate bias
required to compensate the image charge produced in the silicon surface by
this oxide charge and produce the flatband condition ($\psi_s = 0$) is given by
integration of Poisson's equation in the oxide. Poisson's equation is

$$\frac{d^2\psi}{dx^2} = -\frac{qn_o(x)}{\epsilon_{ox}} \tag{10.1}$$

where x is measured from the metal-SiO$_2$ interface, $n_o(x)$ is the volume
density of positive oxide charge, and ψ is the band bending in the oxide.
Integrating (10.1) from a point x in the oxide to the silicon surface, $x = x_o$,
where $d\psi/dx$ vanishes in the flatband condition,* (10.1) becomes

$$-\frac{d\psi}{dx} = -\frac{q}{\epsilon_{ox}} \int_x^{x_o} dx'\, n_o(x'). \tag{10.2}$$

Integrating (10.1) a second time, from the gate $x = 0$ to the silicon surface,
(10.2) becomes

$$\psi(0) = -\frac{q}{\epsilon_{ox}} \int_0^{x_o} dx \int_x^{x_o} dx'\, n_o(x'). \tag{10.3}$$

At the gate $\psi(0)$ is related to the gate bias at flatbands V_{FB} by

$$\psi(0) = V_{FB} - W_{ms} \tag{10.4}$$

where W_{ms} represents the *work function difference* discussed in detail in
Section 10.4; W_{ms} is a nonelectrical contribution to the gate bias, resulting
from work done against chemical forces in moving an electron from the
silicon to the gate electrode. The double integral in (10.3) can be reduced to
a single integral by interchanging the order of integration

$$\int_0^{x_o} dx \int_x^{x_o} dx'\, n_o(x') = \int_0^{x_o} dx'\, n_o(x') \int_0^{x'} dx = \int_0^{x_o} dx'\, x'\, n_o(x'). \tag{10.5}$$

---

*For nonuniform doping, $(d\psi/dx)$ does not vanish at the silicon surface when ψ_s vanishes. In
this case the left-hand-side of (10.2) becomes $Q_s(0)/\epsilon_{ox} - d\psi/dx$, where $Q_s(0)$ is the silicon
surface charge per unit area at zero band bending (see Section 10.5).

When (10.5) and (10.4) are used in (10.3), (10.3) becomes

$$V_{FB} - W_{ms} = -\frac{q}{\epsilon_{ox}} \int_0^{x_o} dx' \, x' \, n_o(x').$$ (10.6)

From (10.6), $V_{FB} - W_{ms}$ is the portion of the flatband voltage due to oxide charge.

The relationship (10.6) between V_{FB} and the charge distributed in the oxide is clarified by introducing an average distance $\bar{x}$ called the *centroid* of the charge distribution. The charge centroid is defined by

$$\bar{x} = \frac{\displaystyle\int_0^{x_o} x' n_o(x') \, dx'}{\displaystyle\int_0^{x_o} n_o(x') \, dx'}.$$ (10.7)

Because $q \int_0^{x_0} n_o(x') \, dx' = Q_o$, (10.7) becomes

$$\bar{x} Q_o = q \int_0^{x_o} x' n_o(x') \, dx'.$$ (10.8)

From (10.6) and (10.8), we obtain

$$V_{FB} - W_{ms} = \frac{\bar{x} Q_o}{\epsilon_{ox}}.$$ (10.9)

The quantity $\bar{x} Q_o$ is called the *first moment* of the charge distribution. Equation (10.9) shows that $V_{FB} - W_{ms}$ is a measure of the first moment of the charge distribution. Thus the oxide charge distribution cannot be obtained from a measurement of V_{FB} alone. In fact, not even $\bar{x}$ and Q_o can be obtained from such a measurement. To separate $\bar{x}$ and Q_o, either the oxide charge density or the centroid must be known independently. For the common and important case of thermally grown oxides, where oxide charge is located in a sheet at the Si–SiO$_2$ interface, Q_o is obtained directly from a measurement of V_{FB} because the centroid is known to be the oxide thickness (within about 30 Å or less). For this case, (10.8) or (10.9) becomes

$$V_{FB} - W_{ms} = -\frac{x_o Q_o}{\epsilon_{ox}} = -\frac{Q_o}{C_{ox}}$$ (10.10)

where $\epsilon_{ox}/x_o = C_{ox}$, the oxide layer capacitance per unit area. Equation (10.10) still shows that $V_{FB} - W_{ms}$ is proportional to the first moment of the charge distribution. But now we can calculate Q_o from (10.10) because $\bar{x}$ is independently known from C_{ox}.

Apart from a discussion of mobile ionic charge, the remainder of this chapter is restricted to the case of a charge sheet located at the Si–SiO$_2$ interface where V_{FB} and oxide charge density are related by (10.10). This case applies to oxide charge density characteristic of thermal oxidation. The case of bulk charge distributed in the oxide is discussed in Section 11.4, where photo I-V measurements are used to separate Q_o and $\bar{x}$.

Equation (10.6) or (10.10) shows that the entire C-V curve is shifted along the voltage axis with respect to the ideal C-V curve [$n_o(x) = 0$] by the amount $V_{FB} - W_{ms}$. The capacitance at each value of bias is shifted by $V_{FB} - W_{ms}$, so that the shape of the C-V curve is unaltered, and the C-V curve is parallel to the ideal C-V curve.

Figures 10.1a,b illustrate the shift along the voltage axis of a high frequency C-V curve when positive Q_o is present in the oxide. The voltage shift is measured with respect to an ideal high frequency C-V curve where $Q_o = 0$ and W_{ms} corresponds to the gate material of the experimental device. The C-V curve for a p-type substrate is shown in Fig. 10.1a and for an n-type substrate, in Fig. 10.1b. In both cases positive Q_o causes the C-V curve to shift to more negative values of gate bias with respect to the ideal C-V curve. Positive Q_o is the most commonly observed process

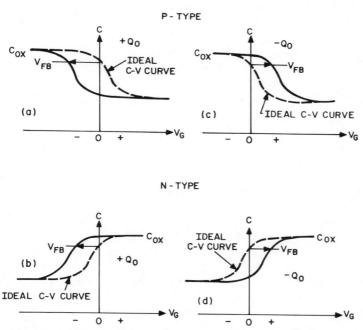

Fig. 10.1 High frequency capacitance as a function of gate bias showing the effect of oxide charge. The C-V curves marked "ideal" have no oxide charge. (a) Positive oxide charge, p-type; (b) positive oxide charge, n-type; (c) negative oxide charge, p-type; (d) negative oxide charge, n-type.

related charge. Although negative Q_o is not as common, it can be produced deliberately by the injection of electrons into the SiO_2, for example. To show that the C-V curves can be used to determine the polarity as well as the magnitude of oxide charge, Figs. 10.1c, d show the shift of a high frequency C-V curve along the voltage axis with respect to the ideal C-V curve for negative Q_o. Figure 10.1c shows a p-type substrate and Fig. 10.1d, an n-type substrate. The C-V curve is shifted to more positive bias values for both p-type and n-type substrates for negative oxide charge. Thus, from the direction of the voltage shift of the C-V curve, the polarity of Q_o can be determined, provided that Q_o is the sole cause of the shift. From the magnitude of the voltage shift of the C-V curve, the value of Q_o can be determined from (10.10). A low frequency C-V curve also could be used for this purpose.

The voltage shift of the C-V curve caused by Q_o is explained as follows, using an n-type substrate for illustration. For a given gate bias and $Q_o = 0$, the depletion layer width is such that negative charge on the gate is balanced by positive dopant ion charge in the depletion layer. If positive Q_o is introduced into the Si–SiO_2 interface, this charge balance is upset. Image charge is introduced in the gate and in the silicon. For the gate, this image charge increases the negative gate charge. For the silicon, additional electrons are added at the depletion layer edge, and the depletion layer width is reduced. Thus the capacitance for n-type is greater than for the ideal capacitor with $Q_o = 0$. The capacitance is greater for all values of gate bias, except strong accumulation where the capacitance in both cases equals C_{ox}, and in strong inversion where the capacitance reaches the same saturation value at high frequency or reaches C_{ox} again at low frequency. The result is a shift of the C-V curve to more negative gate bias as seen in Fig. 10.1b.

10.2.2 Measurement of Mobile and Oxide Fixed Charge

Consider how to distinguish between flatband voltage shifts due to mobile ionic charge and those due to oxide fixed charge.

Consider an experiment where the only oxide charge is oxide fixed charge. The initial high frequency C-V curve measured is labeled (i) in Fig. 10.2a. After heating at 180°C for half an hour with a positive gate bias producing a field of a few million volts per centimeter across the oxide, and cooling back to room temperature, the curve labeled (f_+) in Fig. 10.2a is obtained. Repeating the bias-temperature aging with negative gate bias yields curve (f_-) in Fig. 10.2a. Figure 10.2a shows that no shift in the C-V curve was produced. Therefore, from (10.9), the first moment of the oxide fixed charge density did not change under this treatment. Because it is improbable that the density of oxide charge and its centroid both changed in a manner to keep the first moment constant, we conclude that oxide fixed charge centers are immobile.

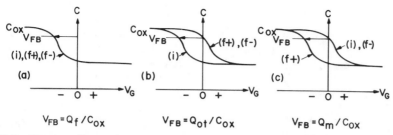

Fig. 10.2 Diagram illustrating how oxide fixed charge, oxide trapped charge, and mobile ionic charge can be distinguished from a bias-temperature aging experiment. The Symbols (i) denotes the initial C-V curve; (f_+), after positive bias-temperature aging, and (f_-), after negative bias-temperature aging. (a) Oxide fixed charge; (b) oxide trapped charge; (c) mobile ionic charge.

Repeating this bias-temperature aging experiment with oxide trapped charge of the type that anneals out at low temperatures, curves (i), (f_+), and (f_-) in Fig. 10.2b show that the first moment of oxide trapped charge is reduced. Gate bias polarity has no effect. Again, it is most likely that the oxide trapped charge centers are immobile.

Finally, Fig. 10.2c shows the results of repeating this experiment on an oxide contaminated by mobile ions. Initially V_{FB} is low, and after positive bias aging it increases. With negative bias aging, V_{FB} returns to its original value. These results are attributed to mobile ion movement that alters the centroid of the mobile ion distribution $\bar{x}$ and so alters V_{FB} according to (10.9). The alternative explanation, that the charge density also varies, is ruled out by other measurements, described in Section 10.2.3.

Assuming the ions are mobile, Fig. 10.2c is explained as follows. Initially, the ions were concentrated at the metal-SiO_2 interface where they entered the oxide. Because most of the ionic charge is imaged in the gate rather than the silicon, the initial C-V curve has a low V_{FB}. During positive gate bias aging, ions drift from the metal-SiO_2 interface to the Si-SiO_2 interface so that almost all the ionic charge is imaged in the silicon producing the shift shown by curve (f_+) in Fig. 10.2b. Aging with negative gate bias aging, ions drift from the metal-SiO_2 interface to the Si–SiO_2 condition; that is, curves (f_-) and (i) are identical.

10.2.3 Measurement of Mobile Ions

The most important mobile ion in SiO_2 is sodium for reasons given in Section 15.3. Potassium ions also are mobile. Because they are metallic, these ions are positive in SiO_2. This section describes the most practical methods of measuring mobile ion density in SiO_2. The technology for minimizing the effects of sodium on device characteristics is described in Section 15.3.4.

(a) The Sodium Profile

To better understand the ion density measurement techniques, it is necessary to know how sodium and sodium ions are distributed in SiO_2. In a classic paper, Yon et al.[1] determined the total and mobile ion distribution in SiO_2. Figure 10.3 illustrates the sodium profile they found in an as-grown oxide where no pains were taken to minimize sodium contamination. This curve was obtained by neutron activation analysis described in Section 10.4.1(a). The sodium contamination is greatest near the air-oxide interface where sodium entered the oxide.

The total sodium profile is shown in Fig. 10.3. However, not all the sodium is incorporated in the oxide as mobile ions. Some of the sodium is chemically bound in the silica lattice. This sodium is electrically inactive, as it is uncharged and immobile. Several other workers[2-4] have reported electrically inactive sodium. It remains unclear how the conditions of oxide growth or the method of contamination affects the amount of sodium that remains electrically inactive.* However, our interest is in the mobile ionized sodium that affects device stability.

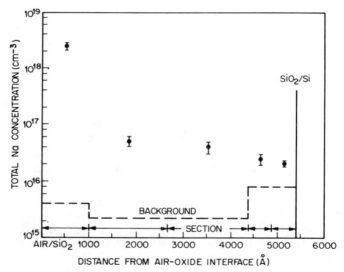

Fig. 10.3 Residual sodium concentration in thermally grown SiO_2 versus distance into the oxide measured from the air-oxide interface. These results are typical of nine oxides grown in wet oxygen and two oxides in "dry" oxygen. There was no measurable sodium in the silicon. The dashed curve is background activity shown as equivalent concentration. Arrows show oxide steps successively etched off to obtain the profile. Initial oxide thickness was 5400 Å and acceptor density was $5 \times 10^{14}\,cm^{-3}$. After Yon et al.[1] Copyright (1966), IEEE.

---

*It has been established that sodium originating from sodium halides evaporated onto the oxide before metallization all becomes mobile. This is a common way of deliberately introducing sodium contamination.[5]

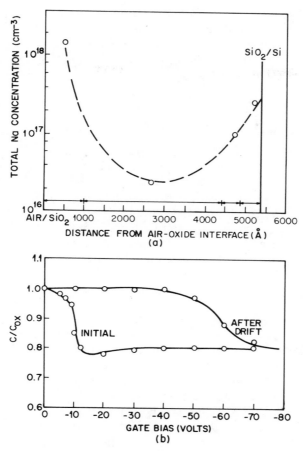

Fig. 10.4 (a) Sodium profile obtained after bias-temperature drift. Arrows show the size of the oxide etch-off steps. Initial oxide thickness was 5400 Å and acceptor density was $5 \times 10^{14} \, cm^{-3}$. After Yon et al.[1] Copyright (1966), IEEE. (b) Diagram illustrating C-V characteristics measured at 1 MHz normalized to the capacitance at $V_G = 0$ before and after bias-temperature drift at 200°C for 10 min with an applied oxide field of 5×10^5 V/cm. The initial oxide fixed charge density was $4 \times 10^{11} \, cm^{-2}$ and the change after drift was 1.9×10^{12} sodium ions/cm². The sample is the same as in (a) before etch-off. After Yon et al.[1] Copyright (1966), IEEE.

The mobile ionized sodium in the profile of Fig. 10.3 was drifted toward the Si–SiO₂ interface by heating the sample to 200°C for 10 min with a positive gate bias applied corresponding to an electric field of 5×10^5 V/cm. Figure 10.4a shows the sodium profile after this treatment. Notice the pile up of sodium at the Si–SiO₂ interface as a result of the drift of mobile sodium under the applied field. Assuming that all the charged sodium is in the last etch-off step, that is, within 1000 Å of the Si–SiO₂ interface, reasonable agreement is obtained between the sodium density in this etch-off step, $N_m = 9.7 \times 10^{11} \, cm^{-2}$, and the charge density as calculated from the flatband voltage shift of the C-V curves measured before and after high temperature drift shown in Fig. 10.4b, $N_m = 1.9 \times 10^{12}$ cm^{-2}.

The drifted sodium ions are actually much less than 1000 Å from the Si–SiO$_2$ interface, Williams[6] performed a photoemission experiment (see Section 11.4.3) in which sodium ions were drifted to the Si–SiO$_2$ interface, inverting the p-type silicon substrate, so that photoemission from the conduction band could be observed. From this experiment, an order of magnitude estimate can be made, placing the sodium $\approx$ 10–100 Å from the Si–SiO$_2$ interface. Using the photo I-V method described in Sections 11.4.2 and 11.4.3, DiMaria[7] found the drifted sodium ions to be within about 50 Å of the Si–SiO$_2$ interface. In this experiment sodium ions in concentrations up to 2.6×10^{12} cm^{-2} were drifted alternately to the Si–SiO$_2$ interface and to the Al–SiO$_2$ interface with oxide fields between 2×10^6 and 4.5×10^6 V/cm at temperatures between 20 and 40°C. Photo I-V measurements then were done at 77°K to avoid ionic motion under gate bias. Ionized sodium was found to pile up within about 50 Å of both interfaces. Therefore, the sodium ions after drift can be considered to be a sheet of charge at the interface to which they were drifted.

(b) C-V Method

There are four steps in the sodium story: (1) the discovery that mobile sodium ions in the oxide were responsible for the drift in device characteristics universally experienced in the early days of the MOS technology, (2) the identification of the most common sources of sodium and the steps taken to minimize or eliminate sodium contamination during manufacture, (3) the introduction of monitoring at various steps in the fabrication to maintain control of the processing, and (4) the development of protective coatings and encapsulants to keep sodium contamination out of the oxide during device operation. All these steps are more fully described in Section 15.3.

In all four historical steps, the C-V method was used to measure mobile sodium ion density. Snow et al.[8] were the first to use the C-V method for studying sodium contamination in SiO$_2$. The C-V method still is the most widely used method.

In the C-V method, a high frequency (usually 1 MHz) C-V curve of an MOS capacitor is measured. Then the MOS capacitor is heated to various temperatures up to 300°C and held there for up to 30 min, which is long enough to ensure that all of the available sodium ions at the given temperature drift completely across the oxide. A positive gate bias is applied sufficient to cause an oxide field of a few MV/cm. After the given time period at elevated temperature and high field, the MOS capacitor is cooled back to room temperature and another C-V curve measured. The flatband voltage shift between the C-V curve before and after bias-temperature drift is a measure of the mobile ion concentration drifted at the given temperature.

Because mobile ions pile up in a charge sheet at the Si–SiO$_2$ interface, (10.10) can be used to calculate charge density from the flatband voltage shift. This calculation will give an accurate value of sodium ion density because the error in flatband voltage, discussed in Section 10.5, is unaffected by bias-temperature drift provided the change in interface trap level density is negligible, and gross nonuniformities in interfacial charge are not produced (see Chapter 6). Therefore, these errors cancel when the flatband voltage of the curve after drift is subtracted from the flatband voltage of the curve before drift to obtain the flatband voltage shift. In many applications bias-temperature treatment causes negligible changes in either interface trap level density or interfacial charge uniformity. In these cases the C-V curves before and after bias-temperature drift are parallel to each other. Then the flatband voltage shift is simply the parallel voltage shift. For those cases where significant changes occur in interface trap level density or interfacial charge uniformity, a more accurate method for obtaining sodium ion density is described in Section (c) below.

Figure 10.5 illustrates the effect of sodium ion drift on the C-V characteristics of an MOS capacitor with a negligible change of interface trap level density and no gross interfacial charge nonuniformity. Figure 10.5a shows the shift of the C-V curve along the voltage axis before and after bias-temperature aging, and Fig. 10.5b shows the charged sodium distribution in the oxide corresponding to each C-V curve; C-V curve 1 is the original C-V curve measured at 1 MHz, and C-V curve 2 is measured after heating for 5 min at 150°C with -10 V applied to the gate. For both cases, ionized sodium is piled up at the metal-oxide interface so that the image charge is induced primarily in the metal and there is virtually no effect on the C-V curve. Then C-V curve 3 is measured after heating for 5 min at 150°C with $+10$ V on the gate. Now, the ionized sodium has drifted to the Si–SiO$_2$ interface, where it induces a large image charge in the silicon, causing a large shift of the C-V curve toward more negative bias values. Finally, C-V curve 4 is measured after heating for 5 min at 150°C with the gate shorted to the substrate. The C-V curve has returned to its initial bias range because the sodium ions have drifted back to the metal electrode. The sodium ions end up at the metal-oxide interface under a zero applied field because the activation energy there is greater than at the Si–SiO$_2$ interface.

A very important observation is that the same number of sodium ions can be cycled back and forth between the silicon and metal electrodes a number of times.* Therefore, sodium ions do not exchange charge with either electrode and do not become discharged even though they pile up to within tunneling distance at each electrode.† As a consequence, the ion distribution for moderate concentrations is frozen in at room temperature

---

*An exception is in oxides grown in an ambient containing chlorine (see Section 15.3).
†An electrode at which ions do not discharge is called a *blocking electrode*.

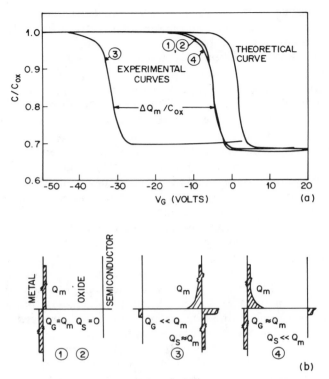

Fig. 10.5 (a) Plot of C/C_{ox} versus gate bias measured before and after sodium drift. Oxide is thermally grown in "dry" oxygen to a thickness of 2000 Å. Acceptor density was 1.5×10^{16} cm^{-3}. After Snow et al.[8] (b) Charged sodium distribution in the oxide corresponding to the various C-V curves in (a); Q_m is the ionized sodium charge density, Q_G is the image charge density induced in the metal by Q_m, and Q_s is the image charge density induced in the silicon by Q_m. After Snow et al.[8]

after drift at higher temperatures. This fact is the basis for the C-V method.

This failure to discharge, along with the high mobility of sodium ions in SiO$_2$, causes large instabilities in silicon devices. In contrast, mobile ions that become neutralized at either electrode can contribute only a transient instability during transit from one interface to the other and once discharged at an electrode, contribute no further instability. The observation that sodium ions do not discharge at the electrodes is limited to temperatures below 390°C. Chou[9] has shown experimentally that sodium ions will begin to discharge at temperatures above 390°C.

The number of sodium ions that can drift in the SiO$_2$ is limited by emission from the interface where the sodium initially was located (see Section 15.3 and Chapter 16). The sodium ions must overcome energy barriers at this interface before they can drift across the oxide. Therefore, there is a distribution of activation energies associated with ion drift.[10] The fraction of the total sodium concentration that becomes ionized and drifts is temperature dependent.

After measuring the amount of ionizable sodium in the oxide, the next problem is to relate this measurement to device stability. Sodium ion densities in the low 10^{10} cm^{-2} range drifted at 200°C are considered tolerable for most integrated circuit applications, and such low sodium ion densities routinely are achieved in manufacture as described in Section 15.3.4.

The sodium ion concentration usually is measured at a temperature above the operating temperature of the device. This is done to detect smaller sodium ion densities than are detectable at operating temperature. It is difficult to predict from high temperature data the amount of sodium that would drift at operating temperature because there is a distribution of activation energies for sodium ion emission from the interfaces. For example, extrapolation using an Arrhenius plot of the drifted amount of sodium against the reciprocal of aging temperature is unjustified. In practice, the processing is adjusted until the amount drifted at aging temperature is sufficiently small that even if that amount drifted at operating temperature, device characteristics would stay within specifications. Because at operating temperature the amount that drifts is significantly lower than at aging temperature, this approach is conservative.

(c) *Triangular Voltage Sweep Method*

In the triangular voltage sweep (TVS) method a triangular voltage ramp is applied to the gate and ionic displacement current measured at an elevated temperature. This method has four advantages over the C-V method:

1 The mobile ion density is accurately obtained even if interface trap level density changes significantly by heating or if gross nonuniformity in interfacial charge occurs (e.g., as a result of defect decoration). An example where significant changes in interface trap level density occur is an MOS capacitor having an aluminum gate deposited by electron-gun evaporation. X-Rays generated when electrons strike the molten aluminum target cause ionizing radiation damage in the SiO_2 as described in Section 11.6. Although the electrical effects of this damage can be annealed out, subsequent bias-temperature treatments will cause large increases in interface trap level density.

2 Different mobile ionic species such as sodium and potassium can be separated at a given temperature because the peak in gate current occurs at a different gate bias. The density of each can be determined.[11] The C-V method would measure the sum total of all mobile ionic species in the oxide.

3 Ion densities as low as $10^9 \, cm^{-2}$ can be detected, making this method more sensitive than the C-V method. The greater sensitivity of the TVS method makes it possible to measure sodium ion drift at lower temperatures than possible with the C-V method.

4 The TVS method is faster than the C-V method because only one curve is needed and the sample can remain heated for many measurements.

The one disadvantage of the TVS method is that the analysis is more complicated than that of the C-V method. For the major application of monitoring the fabrication process, the C-V method, because of its simpler analysis, is used rather than the TVS method. However, the advantages of the TVS method warrant its further discussion.

Yamin[12] was the first to use the TVS method for detecting mobile ions in SiO_2. Kuhn and Silversmith[13] and Chou[9] discussed the application of the TVS method for determining mobile ion charge density in SiO_2.

In the TVS method the MOS capacitor is held at a constant temperature while a linear voltage ramp is applied to the gate and the resulting gate current measured against gate bias as the mobile ions drift from one electrode to the other. The triangular ramp voltage then sweeps the ions back to the other electrode and the measured gate current changes direction. The circuit shown in Fig. 12.5 and described in Section 12.5.2 can be used to measure gate current, and the pedestal design shown in Fig. 12.18 can be used to heat and contact the sample.

The idealized gate current as a function of applied bias with and without mobile ionic charge is shown in Fig. 10.6. At room temperature, gate current is

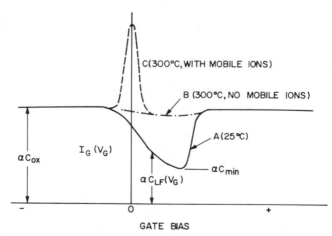

Fig. 10.6 Gate current as a function of gate bias in response to a linear voltage ramp in an MOS capacitor having a p-type substrate. Curves A (25°C) and B (300°C) would be measured when there are no mobile ions in the oxide. Curve C is the displacement current due to mobile ion drift from the oxide bulk to an interface. After Kuhn and Silversmith.[13] Reprinted by permission of the publisher, The Electrochemical Society.

proportional to the low frequency differential capacitance shown in curve A. An increase in temperature results in an increase of C_{min}/C_{ox} as described in Section 3.2.4. At temperatures typically about 300°C, gate current of an uncontaminated device becomes nearly constant and equal to C_{ox}, as shown in curve B. This constant capacitance results from the large value of n_i at 300°C, which makes the silicon surface capacitance large compared to C_{ox}. However, at 300°C with mobile ions present, a gate current peak near $V_G = 0$ is superposed on the C-V characteristic, as shown in curve C.

The peak in current of curve C arises as follows. At large negative gate bias all the mobile ions are at the metal-oxide interface, and the gate current that flows is proportional to C_{ox}. As gate voltage increases, mobile ions begin drifting toward the Si–SiO$_2$ interface, attracting an increasing number of electrons to the silicon surface. That is, the ionic movement causes extra electrons to flow from the gate to the silicon through the external circuit, thereby increasing gate current. This excess current peaks when the largest number of mobile ions is traversing the oxide layer. As gate voltage is increased further, mobile ions pile up at the Si–SiO$_2$ interface. Fewer ions flow, and the excess gate current falls. Ultimately, all mobile ions that will drift at the given temperature have piled up at an interface. Then the gate current again becomes proportional to C_{ox}.

The mobile ion density drifted at a given temperature is proportional to the area under the peak in the gate current caused by the ionic motion. To show this, we determine the gate current per unit area I_G defined as

$$I_G \equiv \frac{dQ_G}{dt} \tag{10.11}$$

where Q_G is the gate charge, given by charge neutrality as·

$$Q_G = -Q_m - Q_{ot} - Q_f - Q_{it} - Q_s. \tag{10.12}$$

Although the mobile charge Q_m moves, the total amount is time independent. Therefore, taking time derivatives of (10.12), we find

$$I_G = -\frac{dQ_{it}}{dt} - \frac{dQ_s}{dt}. \tag{10.13}$$

If V_G varies slowly enough, the equilibrium expressions relate Q_{it} and Q_s to the band bending $\psi_s(t)$. Under these conditions

$$-\frac{dQ_{it}}{dt} = C_{it}(\psi_s)\frac{d\psi_s}{dt} \tag{10.14a}$$

and

$$-\frac{dQ_s}{dt} = C_s(\psi_s)\frac{d\psi_s}{dt}. \tag{10.14b}$$

To find $(d\psi_s/dt)$, we use Gauss's law in the form

$$C_{ox}(V_G - V_{FB} - \psi_s) = -Q_s(\psi_s) - Q_{it}(\psi_s) \tag{10.15}$$

where V_G, V_{FB}, and ψ_s all are time varying. Taking time derivatives of (10.15), we find

$$\frac{d\psi_s}{dt} = \frac{d(V_G - V_{FB})}{dt} \frac{C_{ox}}{C_{ox} + C_s + C_{it}}. \tag{10.16}$$

Substituting (10.14) in (10.13) and using (10.16), we obtain I_G

$$I_G = C_{LF}(V_G) \frac{d(V_G - V_{FB})}{dt} \tag{10.17}$$

where, as usual, the low frequency MOS capacitance per unit area C_{LF} is defined by

$$C_{LF} \equiv C_{ox} \frac{C_s + C_{it}}{C_{ox} + C_s + C_{it}}. \tag{10.18}$$

As shown in Fig. 10.6, to a fair approximation at elevated temperatures

$$C_{LF} \approx C_{ox}. \tag{10.19}$$

Also, we use a linear voltage ramp, so

$$\frac{dV_G}{dt} = \alpha \tag{10.20}$$

where α is the constant voltage sweep rate. If we substitute (10.19) and (10.20) in (10.17), I_G becomes

$$I_G \approx C_{ox} \left(\alpha - \frac{dV_{FB}}{dt} \right). \tag{10.21}$$

If we take αC_{ox} to the left of (10.21) and integrate from a gate bias $(-V_G)$ to a gate bias V_G, (10.21) becomes

$$\int_{-V_G}^{V_G} dV_G [I_G - \alpha C_{ox}] = -C_{ox} \int_{-V_G}^{V_G} dV_G \left(\frac{dV_{FB}}{dt} \right)$$

$$= -\alpha C_{ox} \int_{t(-V_G)}^{t(V_G)} dt \left(\frac{dV_{FB}}{dt} \right)$$

$$= -\alpha C_{ox} \{ V_{FB}[t(V_G)] - V_{FB}[t(-V_G)] \}. \tag{10.22}$$

The integral over gate bias on the right of (10.22) is made into an integral over time using (10.20). The integral on the left of (10.22) is the area between the I_G-V_G curve and the straight line $I_G = \alpha C_{ox}$, representing the gate current of the MOS capacitor when no ions move. The right side of (10.22) can be evaluated by using (10.9). We suppose that at the time when the gate bias was $-V_G$, at time $t(-V_G)$, the mobile charge centroid was $\bar{x}(-V_G)$. Similarly, when the gate bias was V_G, at time $t(V_G)$, the centroid was $\bar{x}(V_G)$. Then (10.9) provides

$$V_{FB}[t(V_G)] - V_{FB}[t(-V_G)] = \frac{qN_m}{\epsilon_{ox}} [\bar{x}(V_G) - \bar{x}(-V_G)] \qquad (10.23)$$

where N_m is the mobile ion density per unit area. Therefore, (10.22) becomes

$$\int_{-V_G}^{V_G} dV_G[I_G - \alpha C_{ox}] = \alpha q N_m \left[\frac{\bar{x}(V_G)}{x_o} - \frac{\bar{x}(-V_G)}{x_o} \right]. \qquad (10.24)$$

For sufficient biases and times, the centroids in (10.24) are bias and time independent. Consequently, (10.24) shows that for a linear voltage ramp at temperatures sufficient to make $C_{LF} \approx C_{ox}$, the area under an I_G-V_G curve in excess of $I_G = \alpha C_{ox}$ is proportional to the mobile ion density per unit area.

To use (10.24) to estimate N_m, one ordinarily assumes $\bar{x}(V_G) = x_o$ (all positive ions drifted to the silicon surface) and $\bar{x}(-V_G) = 0$ (all positive ions drifted to the gate). These assumptions probably are valid to within 100 Å.

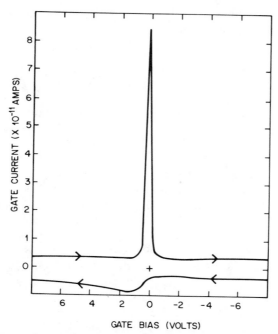

Fig. 10.7 Gate current as a function of gate bias measured at 202°C on a heavily contaminated oxide. The substrate of the MOS capacitor was an n-type silicon epitaxial layer 5 μm thick with a resistivity of 1 Ω-cm and a (111) orientation. The oxide was grown in an oxygen plasma to a thickness of 1000 Å. The gate was Cr–Au, with an area of 1.3 × 10^{-3} cm^2. Voltage sweep rate was 77.5mV/sec. After Kuhn and Silversmith.[13] Reprinted by permission of the publisher, The Electrochemical Society.

Figure 10.7 shows an I_G-V_G curve measured at 202°C on an oxide heavily contaminated with sodium. The sweep from positive to negative gate bias (left to right) where the ions drift from the Si–SiO$_2$ interface to the metal-oxide interface is shown in the upper curve. The area under this curve corresponds to 2.5×10^{12} ions/cm^2. The sweep from negative to positive gate bias (right to left) where the ions drift from the metal-oxide interface to the Si–SiO$_2$ interface is shown in the lower curve.

The most striking feature of these curves is the strong asymmetry of the I_G-V_G characteristics in the two sweep directions. Reasons for this asymmetry are described briefly in Section (e) below. The areas under both peaks in Fig. 10.7 are identical so that all ions transported toward the metal-oxide interface can be returned to the Si–SiO$_2$ interface. The areas under the peaks are independent of voltage sweep rate if the range of bias variation is sufficient and if the time spent at extreme bias is long enough to achieve a time independent pile up at each interface.

Our derivation of (10.24) does not require a detailed understanding of the time dependence of V_{FB}. This is fortunate because ionic transport alone is not sufficient to model V_{FB}. As discussed later, mobile ion motion is dominated by interfacial barrier effects.

(d)   Other Measurement Methods

The electrical methods described in Sections (b) and (c) above are the most sensitive of the methods used to detect sodium. However, the electrical methods do not measure the neutral sodium profile or the sodium in chemical reagents. Analytic methods used for sodium profiling are the radioactive tracer method,[4, 14] neutron activation analysis[1, 3, 14] described briefly in Section 11.4.1(a), flame photometry,[15, 16] and secondary ion mass spectroscopy[17] (SIMS).

In SIMS the mass spectroscopic analysis of secondary ions proceeds simultaneously with the sputter etching of the oxide by primary ions. However, sodium becomes mobile[17] in the electric field created by the primary ions, so that only the total amount of sodium and not its profile can be measured. Secondary ion mass spectroscopy is the only analytical method whose limit of sensitivity is comparable to the electrical methods, but as work with this method has been confined to small sample areas, its full sensitivity has yet to be exploited. This method requires complex instrumentation and is destructive to the sample. Therefore, it is not used as widely as the other analytical methods.

Among the methods that require a large sample area, the radioactive tracer method combines sensitivity with simplicity. Neutron activation analysis is possible because sodium incorporated "in process" can be activated and detected. However, this method requires access to a nuclear pile.

The least sensitive but simplest analytical method is flame photometry, which is used in monitoring sodium contamination of chemical reagents

Table 10.1 Commonly Used Techniques for Detection of Sodium in SiO$_2$ Films.[a]

Method	Limiting Factors	Sensitivity		Ref.
		Na Atoms	Na Ions	
Neutron activation analysis Isotope, ^{24}Na; half-life, 15 hr; γ-ray emission at 1.4 MeV and 2.75 MeV	1. Neutron flux 2. Background radiation	10^{11}–10^{12}		1, 3, 14
Radioactive tracer Isotope, ^{22}Na; half-life, 2.6 yr, γ-ray emission at 0.51 MeV	1. ^{22}Na/^{23}Na $< 10^{-2}$ (usually) 2. Background radiation	10^{11}–5×10^{11}		4, 14
Flame photometry Sodium emission 5890–5896 Å	1. Minimum volume of liquid sample required 2. Sodium contamination of HF solvent	8×10^{11}		15, 16
Secondary ion mass spectroscopy ^{23}Na; O$_2^+$ primary beam	1. Ionization efficiency of Na 2. Usually small sample area ($\sim 10^{-5}$ cm^{-2})	$\sim 10^6$		17
C-V method			10^{10} cm^{-2}	8
Triangular voltage sweep			10^9 cm^{-2}	13

[a] After Kriegler.[18]

used in integrated circuit processing. In this method sodium atoms in solution are atomized into a flame* that excites the characteristic sodium emission line. The intensity of the emitted light is directly proportional to the number of sodium atoms excited by the flame and thus to the sodium concentration in the solution. The sodium emission line is resolved by a monochrometer and detected by a photomultiplier tube. The instrument is calibrated with standard solutions containing a known amount of sodium to establish a relationship between the emission intensity of a known spectral line and the concentration of sodium in solution. Similar intensity measurements are made on the sample solution and converted into concentration by using the calibration curve.

These analytical methods are summarized in Table 10.1, which also shows, for comparison, the two electrical methods. The sensitivity of the C-V method in Table 10.1 is based on $x_o = 1000$ Å and a detectable voltage shift of 0.1 V assuming no change in interface trap level density. Although a smaller voltage shift could be measured, the extra effort required would nullify the simplicity of this method; thus the TVS method might as well be used.

(e) Kinetics of Ion Drift

Hofstein[4] was the first to observe that the bias-temperature drift of sodium ions from the silicon to the metal was much faster than the drift from the metal to the silicon. The mobility of sodium ions in the silica bulk has been estimated to be about 4×10^{-12} cm^2/V-sec at room temperature.[18, 19] In the MOS capacitor, the rate limiting step appears to be emission over an energy barrier from the interface. The energy barrier appears to be higher at the metal-oxide interface than at the Si–SiO$_2$ interface, accounting for the asymmetry observed in Fig. 10.7.

Ionic transport through the oxide can be studied without the complications of interfacial detrapping.[18, 19] Such studies are based on observation of the time derivative of flatband voltage, as observed using the current given by (10.21), for example. As explained earlier, as ions detrap and drift across the oxide, the current rises until all the available ions detrap, and then saturates while all the ions are in transit across the oxide. When the ions begin to collect at the opposite electrode, the current drops toward the baseline value again. The time period between the current's initial rise and its drop is the time for the leading ions to traverse the oxide.

On the basis of crude assumptions of constant drift velocity and negligible influence of the ionic charges themselves on the field, Stagg[19] found that his transit time measurements implied an activated bulk mobility in dry oxides of the form $\mu = \mu_o \exp(-qE_A/kT)$. Here, for sodium, $\mu_o \approx$

---

*An oxyhydrogen flame is commonly used.

$1 \, cm^2/V \, sec$ and $E_A = 0.66 \, eV$ and for potassium, $\mu_o \approx 0.03 \, cm^2/V \, sec$ and $E_A = 1.1 \, eV.$*

Sodium ion emission from the electrodes has been studied using two methods:[10, 20, 21] the isothermal transient ionic current (ITIC) method and the thermally stimulated ionic current (TSIC) method. In the ITIC method ionic displacement current is measured as a function of time while the MOS capacitor is held at an elevated temperature with a field applied across the oxide. In the TSIC method ionic displacement current is measured as a function of temperature at a given applied oxide field. These two methods are useful mainly for research studies of ion emission from the silicon and the gate electrodes. The understanding of ion emission from the electrodes gained from these methods is at an early stage. These methods are not discussed further.

10.3 INTERNAL PHOTOEMISSION

Internal photoemission is used to determine the energy barrier between silicon and the conduction band of SiO_2† and the energy barrier between various gate metals and the SiO_2 conduction band. The barrier heights between the gate metal and SiO_2 and between silicon and SiO_2 can be used to estimate the work function difference between the gate and the silicon.

This measurement, described in Sections 11.4.2 and 11.4.3, provides a check of the directly measured work function differences described in Section 10.4. The measurement of the SiO_2 bandgap energy is described in Section 11.6.5(a).

In this section the use of internal photoemission to determine electrode energy barriers is described. Of the many approaches to the use of photoemission,[22] we restrict ourselves to those used on the MOS system. Sections 11.2, 11.3, and 11.4 discuss the use of avalanche injection and internal photoemission to study trapping phenomena in SiO_2.

No appreciable current flows between the metal and the silicon in the MOS system when fields below the dielectric breakdown strength of the oxide are applied because (1) SiO_2 is an insulator with a wide bandgap $(8.8 \, eV)$[23] (therefore, free hole and electron densities intrinsic to SiO_2 are negligible at all temperatures of interest) and (2) electrons (or holes) in each electrode must overcome large energy barriers (several electron volts) to enter the SiO_2 as free carriers (therefore, injection of holes and electrons into the SiO_2 is difficult, limiting current flow).

---

*These measurements were made using a voltage step method rather than the TVS method. When a voltage step is used, the oxide field in the absence of ionic charges is constant, rather than linearly increasing as with the ramp method.

†The term "conduction band" is applied loosely to SiO_2. In an energy band an electron is not localized at a particular lattice site but is free to move about the entire crystal. The band assumption may not apply to thermally grown SiO_2.

One way to get electrons (or holes) from either electrode into the SiO_2 over the large energy barriers is to heat them to high energies. There are two methods of heating electrons sufficiently. The first method is internal photoemission, where photons excite electrons over the energy barrier. The second method, described in Section 11.2, injects electrons or holes from an avalanche plasma in the silicon substrate into the SiO_2. In this method electrons (or holes) are heated by acceleration in an electric field at the silicon surface.*

10.3.1 Implementation

In a pioneering paper, Williams[24] was first to measure the barrier height between the valence band of silicon and the conduction band of SiO_2 in the MOS system. He used MOS capacitors with semitransparent gold electrodes, thermally grown SiO_2 layers 2μm thick, and both p-type 18 and 20 Ω-cm and n-type 10 and 15 Ω-cm silicon substrates. Upon illumination of such a structure, light passes through the semitransparent gold electrode and the oxide but is absorbed in the silicon if the wavelength is sufficiently short (infrared light). For nondegenerate silicon, light is absorbed primarily by electrons in the silicon valence band, rather than in the silicon conduction band because the concentration of valence band electrons is several orders of magnitude greater than the concentration of conduction band electrons.

Absorption of the light, through interaction of the electric field of the lightwave with a valence electron, causes the electron to make a transition to the silicon conduction band. The result is a hole-electron pair. With the gate biased positively with respect to the silicon, electrons drift toward the SiO_2 and holes toward the bulk silicon. In the bulk silicon, holes recombine with electrons. As the frequency of the incident light is increased, a critical frequency is reached at which some valence electrons have sufficient energy to surmount the barrier and enter the SiO_2. These injected electrons drift to the gate in the electric field applied across the oxide.† The resulting photocurrent can be detected in an external circuit as described in Section 12.10.

Figure 10.8 shows the conduction band as a function of distance in the SiO_2 under an applied bias. To the left of the vertical axis in Fig. 10.8 is plotted the density of electrons in the electrode, heated by incident light of energy $h\nu$. In Fig. 10.8 $h\nu$ is sufficient to excite electrons in the shaded portion of the distribution over the interfacial energy barrier.

Williams[24] determined a photoemission *threshold* of 4.25 eV, independent

---

*Another way to get electrons into the SiO_2 is by tunneling. This approach is not discussed here.

†Electrons move freely in SiO_2, with a mobility of 20–30 cm^2/V sec.[25, 26] Hole transport in SiO_2 is much slower (see Section 11.6.5).

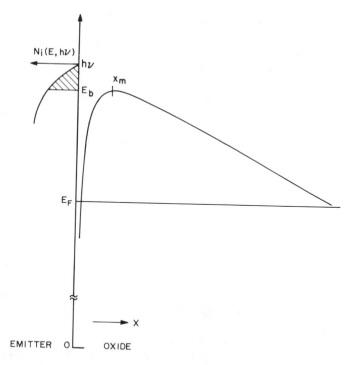

Fig. 10.8 Energy-band diagram showing the energy barrier between an electrode (metal or silicon) and the conduction band of the SiO_2 under an applied electric field. The barrier height E_b is shown as well as the hot electron distribution at incident light energy $h\nu$ just above E_b. After Powell and Berglund.[27]

of whether the silicon was p-type or n-type. This threshold energy is interpreted as the barrier height or energy difference between the valence band of silicon and the conduction band of SiO_2. Figure 10.9 shows the resulting energy-band diagram of the Si–SiO_2 system. The barrier height shown in Fig. 10.9 was measured by the more accurate method described in Section 10.3.3(d).

The barrier height between SiO_2 and various metals deposited as the gate electrode can be determined in the same way, by noting the critical light frequency above which electrons contribute a photocurrent when the gate is *negatively* biased.

In such barrier determinations it is critical that hole injection from the opposite contact be negligible. Hole photocurrent from the metal would flow in the same direction as electron current from the silicon, making interpretation of threshold ambiguous. Thus, the usefulness of internal photoemission depends on ascertaining the contribution of hole emission to the measured photocurrent.

Williams[24] showed in his experiments that hole emission from his gold

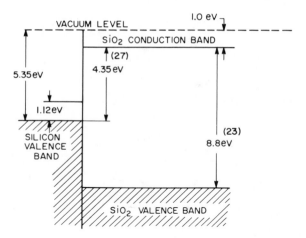

Fig. 10.9 Energy-band diagram of the Si–SiO₂ system obtained from photoemission measurements. After Williams.[24]

gate electrode was negligible. This was done by measuring spectral response with a gold gate electrode and then repeating the measurement with a gate electrode of de-ionized water on the same area of oxide. He found that the threshold energies differed by no more than 0.1 eV, within the scatter obtained for different gold electrodes. More significantly, the quantum yield [see Section 10.3.3(a)] of photocurrent for the water electrode was the same as for gold electrodes (5×10^{-5} electrons per quantum at a wavelength of 2300 Å). Because the optical properties and the energy-band structures of gold and water are vastly different, the agreement in threshold energy and quantum yield for the two different electrodes meant that hole emission was negligible, and the photocurrent measured was due to electrons from the silicon. In Section 10.3.2 it can be seen that optical interference in thin oxides also can be used to rule out hole emission.

10.3.2 Optical Interference and Hole Injection

Not all of the incident light on an MOS structure is absorbed in the silicon or metal electrodes. Some is reflected at the silicon surface towards the metal electrode. This reflected light travels through the oxide (the oxide is transparent to a good approximation at the frequencies of interest in photoemission experiments), and some of it again is reflected at the metal surface back toward the silicon. The reflected light changes phase as it travels through the oxide. If the oxide thickness is comparable to the wavelength of the light, then constructive and destructive interference of the light will occur. The wavelengths for intensity maxima at one electrode surface will correspond approximately to intensity minima at the other electrode surface. Now, the electron and hole contributions to the measured photocurrent originate from opposite electrodes. Therefore, hole current will be a maximum for wavelengths where the electron current is a minimum and vice versa.*

Powell[28] used optical interference to show that hole emission was negligible. Figure 10.10 shows the photocurrent spectra of an Au–SiO$_2$–Si structure for the two bias directions measured on samples with an oxide thickness of 3000 Å using the techniques described in Section 12.10. Shown are the actual photocurrents as a function of wavelength for an applied electric field across the oxide of about 3×10^6 V/cm. These photocurrents are not corrected for the wavelength dependence of the intensity of the incident light source. The relative intensity of the incident light is shown as a dashed curve in Fig. 10.10 for comparison with the spectra. Most important in Fig. 10.10 is the approximate coincidence of the minima for one gate bias polarity with the maxima for the other polarity and vice versa.

The optical behavior of the three-layered MOS structure must be understood to interpret the photoelectric threshold and to determine the electrode energy barriers correctly. The steps followed by Powell[28] were:

1 The reflectance of the structure was calculated using the known optical constants of Au, SiO$_2$, and Si and the measured SiO$_2$ and Au electrode thicknesses.

2 A comparison was made between the calculated and measured reflectance to determine the energy range over which the degree of

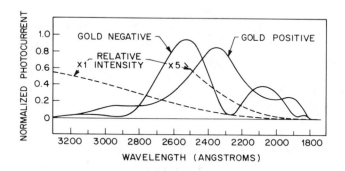

Fig. 10.10 Photocurrent spectra of an Au–SiO$_2$–Si structure. The oxide was grown in steam at 1050°C to a thickness of 1920 Å on the (100) surface of a 0.1 Ω-cm. p-type silicon substrate. Gold thickness was 90 Å. Full scale photocurrents are 10^{-9} and 10^{-10} A for positive and negative biases, respectively. The dashed curve is the relative light intensity incident on the sample. After Powell.[28]

---

*Optical interference becomes important at oxide thicknesses in the 2000–3000 Å range in photoemission experiments and can be avoided by using very thick (several micrometers) or very thin (≤ 500 Å) oxide thicknesses.

agreement is sufficient to ensure reasonable accuracy of the solution for correction of the photoelectric spectra.

3 Using the absolute value of the incident light power and the fraction absorbed in the two electrodes, the absolute quantum yields (the quantum yield of a photoemitter is defined as the ratio of the number of electrons emitted per unit time to the number of photons absorbed in the same period) are determined under the assumption that only electron emission makes a significant contribution to the photocurrent. For example, the photocurrent spectrum for positive gate bias is corrected for intensity fluctuation only at the silicon surface.

4 The resulting yield spectra are examined for evidence of hole emission. If such contributions are significant, they will be seen as undulations in the yield because no corrections have been made for the interference photocurrent of hole emission.

To correct the photocurrent spectra for variations in absorbed light intensity with photon energy, the reflectance R, the fraction of the incident light power reflected from the MOS structure, and the transmittance T, the fraction transmitted into and absorbed in the substrate must be known. Then, under the assumption that no light is absorbed in the oxide, the absorptance A, the fraction of the incident light power absorbed in the gold film, is

$$A = 1 - R - T. \tag{10.25}$$

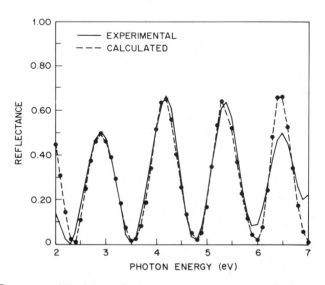

Fig. 10.11 Computer plot of the reflectance spectrum of an Au-SiO₂–Si structure. The solid curve is the experimental reflectance, and the dashed curve is the reflectance calculated for a gold thickness of 90 Å and an oxide thickness of 2920 Å. After Powell.[28]

The calculation of R and T in terms of the optical constants is described in Appendix XI.

The reflectance spectrum of an Au–SiO$_2$–Si structure is shown in Fig. 10.11. The solid curve* is the experimental reflectance, and the dashed curve is the reflectance calculated as described in Appendix XI. Agreement between calculated and measured reflectance is quite good, particularly in the 2.5–6.0 eV photon energy range. This is the energy range encompassing the electrode barrier energies and is the important energy range for our purposes. The discrepancies below 2.5 eV probably are caused by errors in the optical constants of gold in this range.[28] The discrepancies above 6.0 eV probably are caused by absorption in the SiO$_2$, and absorption is assumed negligible in the calculation.[28] The values of T and A used to correct the photocurrent spectra were calculated by using the actual oxide thickness measured with an interferometer, because the photocurrent was measured at normal incidence.

The calculated transmittance and absorption of the samples in Fig. 10.11 are shown as functions of photon energy in Fig. 10.12. Figure 10.12 shows that the absorptance has a much larger variation with photon energy than the transmittance. Therefore, one might expect hole photocurrent from the gold electrode to be detected readily. However, no hole current is detected.

The experimental evidence against a hole current hinges on the depen-

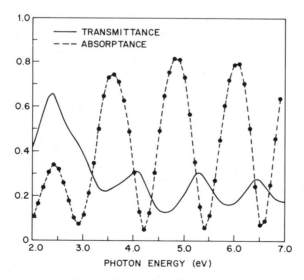

Fig. 10.12 Computer plot of transmittance T (solid curve) and the absorptance A for an Au–SiO$_2$–Si structure with a gold thickness of 90 Å and an oxide thickness of 2870 Å. After Powell.[28]

*These curves are computer plots. They are not smooth because the calculated points simply have been joined by straight-line segments.

dence of photocurrent on A and T. This dependence is derived as follows. Not every photon absorbed in the silicon that has an energy greater than the barrier height results in an electron injected into the SiO$_2$. The *quantum yield* discussed in Section 10.3.3(a) is the number of injected electrons per absorbed photon. The expression for the quantum yield Y in terms of photocurrent I and photon flux F is

$$Y = \frac{I}{qF} \quad \text{(electrons/absorbed photon)} \tag{10.26}$$

where I is in amperes,* F is in photons/sec, and q is the electronic charge in coulombs. It is convenient to express Y in terms of absorbed power P, which is related to F and the photon energy $h\nu$ (in joules) by

$$P = Fh\nu \tag{10.27}$$

(in watts). If we substitute (10.27) into (10.26), Y is

$$Y = \frac{Ih\nu}{Pq}. \tag{10.28}$$

In (10.28) Y is the electron yield from the silicon under positive gate bias or the electron yield from the metal under negative gate bias.

Consider the photocurrent in an MOS structure with positive gate bias. If the photocurrent under positive gate bias I_f consists of both electrons and holes, we can state, using (10.28), that

$$I_f = I_{nf} + I_{pf} = \frac{qY_{nf}P_iT}{h\nu} + \frac{qY_{pf}P_iA}{h\nu} \tag{10.29}$$

where the subscripts n and p refer to electrons and holes, respectively, and f refers to positive (forward) gate bias. For positive gate bias, electrons are emitted from the silicon. Light enters with transmittance T, so that the absorbed power is $P = P_iT$. Holes are emitted from the metal; hence the absorptance A in the second term in (10.29). Recombination in the oxide is assumed to be negligible in (10.29).

We first correct I_f only for intensity variations at the silicon surface. This correction is done by substituting (10.29) into (10.28) to obtain

$$Y'_f = Y_{nf} + Y_{pf}\frac{A}{T} \tag{10.30}$$

where Y'_f is the quantum yield determined from the measured photocurrent from (10.28) assuming only electron emission from the silicon surface. Next, consider correcting the photocurrent under negative gate bias only

---

*Equation (10.26) applies when I is a steady-state photocurrent. Therefore, trapping of photoemitted electrons in the oxide is assumed to be negligible, which is a reasonable assumption for the conditions of this experiment.

for intensity variations at the metal surface. In a similar way, we obtain

$$Y'_r = Y_{nr} + Y_{pr}\frac{T}{A} \tag{10.31}$$

where the subscript r denotes negative (reverse) gate bias and Y'_r is the quantum yield determined from the measured photocurrent assuming only electron emission from the metal surface.

The factors A/T and T/A modulating the hole quantum yields in (10.30) and (10.31), respectively, can be large because the maxima of A correspond to the minima of T and vice versa, as shown in Fig. 10.12. These factors can be larger than 5. Therefore, if hole photocurrents make a significant contribution to the quantum yield, undulations will appear in Y_{nf} and Y_{nr} as functions of photon energy corresponding to maxima and minima in A/T and T/A. Figure 10.13 shows Y_{nf} and A/T as functions of photon energy, and Fig. 10.14 shows Y_{nr} and T/A as functions of photon energy. Because there is no structure in the yield spectra at the appropriate locations, hole photocurrents make no significant contribution to the measured quantum yields.

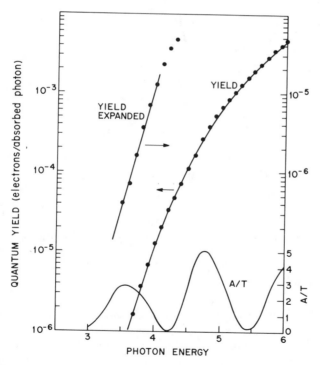

Fig. 10.13 Absolute quantum yield as a function of photon energy. Positive gate bias corresponding to an electric field of 3×10^6 V/cm was applied to the Au–SiO$_2$–Si structure. The quantity A/T (linear scale) is plotted at the lower right. The yield is primarily electron emission from the silicon. After Powell.[28]

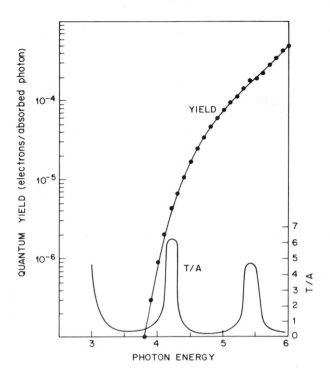

Fig. 10.14 Absolute quantum yield as a function of photon energy. Negative gate bias corresponding to an electric field of 3×10^6 V/cm was applied to the Au–SiO$_2$–Si structure. The quantity T/A (linear scale) is plotted at the lower right. The yield is primarily electron emission from the gold. After Powell.[28]

To make this conclusion more quantitative, consider the smallest hole quantum yields that could be detected by this method. Detection of small undulations in the yield curve of Fig. 10.13 is difficult because Y_f' increases over 3 orders of magnitude for photon energies between 4 and 6 eV. Nonetheless, undulations of 25% of the average yield could be detected. As the change in modulating factor A/T is about 5, an average quantum yield for hole emission over this photon energy range must be less than 5% of the total yield.

10.3.3 Interface Barrier Height Determination

Having eliminated hole injection, let us now examine how the photo-emission experiment is used to determine barrier heights. Interpretation of this method is based on a functional form for the photoelectric quantum yield, such as that given by the Fowler model.[29] Given the form of the yield, determination of the barrier heights conventionally is found following these steps:[30-32] (1) photocurrent is measured as a function of wavelength with oxide field as parameter; (2) the quantum yield is found from the photocurrent spectrum (to do this, corrections must be made for light source intensity variations with photon energy, for variations of the optical constants of the materials with photon energy, and for the effects of optical interference as described in Section 10.3.2); (3) the quantum yield is plotted in the functional form of the theoretical model as a function of photon energy and extrapolated to zero yield to obtain the barrier height corresponding to the chosen oxide field; and (4) finally, the barrier heights determined this way are plotted as functions of oxide field and extrapolated to zero oxide field to get the zero field barrier height.

We also describe an alternative method developed by Powell[33] in which oxide field is taken as the variable with photon energy as parameter. Powell's method permits direct barrier height measurement, independent of absorbed light intensity. There are two steps necessary in this method to determine barrier height and its gate bias dependence: (1) the photocurrent V-I characteristics are plotted directly (electronically) for different photon energies; and (2) the barrier height and its field dependence are determined by the extrapolation of $V^{1/2} - I^{1/p}$ characteristics to zero photocurrent,* or the barrier height alone can be determined simply by examining the V-I characteristics for a sign change in the second derivative. These methods are described in detail because they are more sensitive and accurate than the Fowler method, and the barrier energy can be determined regardless of the functional form of the quantum yield.

(a) Quantum yield

We start by describing the dependence of the quantum yield on photon energy and gate bias. Quantum yield depends on three basic processes: (1) electron photoexcitation, (2) transport to the emitter surfaces, and (3) escape over the energy barrier. The first two processes are taken into account by the energy distribution of photoexcited electrons at the emitter surface $N_i(E, h\nu)$ in units of electrons per photon per electron volt as shown in Fig. 10.8 for photon energies just above the barrier energy. The escape process can be described by an energy-dependent surface transmission probability $P(E)$ that accounts for the group velocity and directional dependence of the electron escape probability. The quantum yield can be written in terms of these two functions as

$$Y(h\nu, E_b) = \int_{E_b}^{h\nu} N_i(E, h\nu)P(E)\, dE \qquad (10.32)$$

where E_b is the barrier height in electron volts.

*Note: $p = 2$ or $p = 3$, such as in (10.43).

The internal energy distribution of electrons at the surface $N_i(E, h\nu)$ depends on the selection rules governing optical transition probabilities.* In the usual band model of a solid, optical transitions are either direct or indirect. For direct transitions, both energy and crystal momentum are conserved, and $N_i(E, h\nu)$ depends on the joint density of initial and final states.[34,38–40] For indirect transitions, both energy and crystal momentum are conserved, but phonon emission or annihilation is involved, and $N_i(E, h\nu)$ depends on a convolution of the initial and final density of states.[41] When the initial or final state wave functions are inadequately represented by independent electron Bloch functions, a third type of transition becomes more likely, the nondirect transition.[34] Now, conservation of crystal momentum is not important as a selection rule. In such transitions, $N_i(E, h\nu)$ will depend on the product of the initial and final density of states.

It is not worthwhile to dwell at length on the form of the distribution of $N_i(E, h\nu)$ because it depends not only on the density of states, but also on the transport of the excited electrons to the emitting surface. Thus $N_i(E, h\nu)$ best may be inferred from experiment.

Powell[33] has found experimentally that for emission from the silicon valence band, $N_i(E, h\nu)$ is well approximated as a ramp proportional to $E - h\nu$ and for the silicon conduction band, is approximately given by a step function. For our development, we shall not specify $N_i(E, h\nu)$ but assume that varying the photon energy simply translates this distribution in energy; that is

$$N_i(E, h\nu) = N_i(E - h\nu). \tag{10.33}$$

The transmission probability will be calculated using a model.[29] This model assumes that the necessary and sufficient condition for electron escape is that the electrons have a component of group velocity normal to the surface greater than some critical value. In this model the energy dependence of the transmission probability is given by

$$P(E) = \frac{1}{2}\left[1 - \left(\frac{E_b - E_o}{E + E_o}\right)^{1/2}\right]$$
$$\approx \frac{1}{2}\frac{E - E_b}{E + E_o + [(E + E_o)(E_b + E_o)]^{1/2}}. \tag{10.34}$$

In (10.34) the energy E and the barrier height E_b are measured with respect to the top of the silicon valence band, and E_o is the energy zero for a free electron model that best fits the actual group velocity, which in silicon is $E_o \approx 12$ eV.[42] The denominator in the approximate form of $P(E)$ in (10.34) varies only slightly if E or E_b is varied a few tenths of an electron volt because E_o and E_b are several electron volts in magnitude. Therefore, the transmission probability is a function of $E - E_b$ to a good approximation,

---

*The situation can become quite complex, depending on the photon energies involved.[34–39]

and we can write

$$P(E) \approx C(E - E_b) \tag{10.35}$$

where C is a constant. Substituting (10.33) and (10.35) into (10.32) yields

$$Y(h\nu - E_b) = C \int_{E_b}^{h\nu} N_i(E - h\nu)(E - E_b)\, dE. \tag{10.36}$$

Figure 10.15 shows how (10.36) can be interpreted. Figure 10.15a shows separate plots of the two factors in the integrand of (10.36) as functions of photon energy. The product of these components is shown as a function of photon energy in Fig. 10.15b. The quantum yield is proportional to the area beneath this curve.

The photocurrent is bias dependent because the barrier height depends on electric field (see Section 11.4.2 for a derivation of the dependence of barrier height on electric field). Barrier height decreases with increasing electric field. Only oxide fields in the intermediate to high regime ($> 10^6$ V/cm), where the bias dependence of the photocurrent is dominated by field dependent barrier lowering, is considered. Barrier lowering for high electric fields increases $P(E)$. This effect is shown in Fig. 10.15a, where E_{b1} is the barrier at a lower field than E_{b2}. The new product is shown as the dashed curve in Fig. 10.15b. The area under this dashed curve is greater than under the solid curve. Therefore, the quantum yield and the photocurrent will increase.

(b) Photocurrent

It is convenient to transform (10.36) with the relation $E = h\nu - E'$ to obtain

$$Y(h\nu - E_b) = -C \int_0^{h\nu - E_b} N_i(-E')(h\nu - E_b - E')\, dE'. \tag{10.37}$$

When (10.37) is combined with (10.28), the photocurrent is

$$I = A(h\nu) \int_0^{h\nu - E_b} N_i(-E')(h\nu - E_b - E')\, dE' \tag{10.38}$$

where $A(h\nu)$ accounts for the constant factor and the photon energy dependences of P and $h\nu$ in (10.28).

Next, we express the bias dependence of the photocurrent explicitly. The mechanisms responsible for field dependent barrier lowering are image-force lowering and field penetration into the electrode (band bending). In nondegenerate silicon the latter effect is large and can dominate the

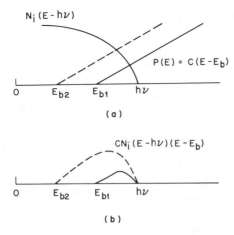

Fig. 10.15 Illustration of the dependence of photocurrent on bias. (a) The product $N_i(E - h\nu)$ is a hypothetical energy distribution of electrons at the surface of a photoemitter. The transmission probability is the straight line $P(E) = C(E - E_b)$. The dashed line $P(E)$ represents a higher field than the solid line. After Powell.[33] (b) The product $CN_i(E - h\nu)(E - E_b)$ as a function of photon energy. The area under the curve is the yield. The dashed curve is calculated from the dashed line in (a). After Powell.[33]

bias dependence of the photocurrent. For degenerate silicon and metals, the effect is significant only at very high fields and the major contribution to barrier lowering is the image-force effect. Although the theoretical expressions presented here can be extended to include the effect of field penetration, this effect is neglected here because all measurements are done on degenerate silicon.

Including image-force lowering and assuming no oxide charge,* the barrier height for degenerate silicon ($\psi_s \approx 0$) is given by [see Section 11.4.2(a)][42]

$$E_b = E_{bo} - qK(V_G - W_{ms})^{1/2} \qquad (10.39)$$

where V_G is the applied gate bias, W_{ms} is the work function difference between the metal and the silicon (see Section 10.4), E_{bo} is the zero field barrier height, $K = (q/4\pi\epsilon_i x_o)^{1/2}$, ϵ_i is the high frequency dielectric permittivity in the image-force region,† and x_o is the interelectrode separation or oxide thickness. Equation (10.39) is derived in Section 10.4. Substituting (10.39) into (10.38) makes the photocurrent an explicit function of gate bias

$$I(h\nu, V_G) = A(h\nu) \int_0^{h\nu - E_{bo} - K(V_G - W_{ms})^{1/2}} N_i(-E')$$

$$\times [h\nu - E_{bo} - K(V_G - W_{ms})^{1/2} - E']\, dE'. \qquad (10.40)$$

Two cases that apply to photoinjection from silicon and metals are

---

*In device-grade thermally grown SiO_2, only oxide fixed charge and charged interface traps are present. It is shown in Section 11.4 that these charges are located so close to the Si–SiO₂ interface that they do not affect the photocurrent.

†Because electron transit time through the image-force region is very short compared to the period of the highest frequency lattice vibrational mode in the oxide, ϵ_i is the electronic or optical permittivity.

considered. In the first case $N_i(-E')$ is a ramp, that is, $N_i(-E') = E'$. This is approximately the case for photoemission of electrons from the silicon valence band. Integrating (10.40) gives

$$I(h\nu, V_G) = A(h\nu)[h\nu - E_{bo} + K(V_G - W_{ms})^{1/2}]^3. \tag{10.41}$$

In the second case $N_i(-E')$ is a step and $N_i(-E') = N(E')$. The second case applies approximately to photoemission from the conduction band of degenerate silicon and some metals. Integrating (10.40) gives

$$I(h\nu, V_G) = A(h\nu)[h\nu - E_{bo} + K(V_G - W_{ms})^{1/2}]^2. \tag{10.42}$$

Equations (10.41) and (10.42) show that the dependence of photocurrent on photon energy is complicated by $A(h\nu)$. However, the gate bias dependence is entirely in the second factor that permits the direct determination of E_{bo} and K.

(c) Determination of Barrier Height

Considering photoemission from the valence band of the silicon, we take the cube root of (10.41) to obtain

$$I^{1/3} = [A(h\nu)]^{1/3}[h\nu - E_{bo} + K(V_G - W_{ms})^{1/2}]. \tag{10.43}$$

Therefore, a plot of $I^{1/3}$ versus $(V_G - W_{ms})^{1/2}$ for a given $h\nu$ will be a straight line, and its intercept on the $(V_G - W_{ms})^{1/2}$ axis at $I^{1/3} = 0$ is

$$(V_G - W_{ms})^{1/2}|_{I=0} = \frac{E_{bo} - h\nu}{K}. \tag{10.44}$$

As $h\nu$ is varied, the intercept is still given by (10.44) regardless of the value of $A(h\nu)$. Equation (10.44) shows that for $h\nu = E_{bo}$, the intercept of the line is zero; that is, the line passes through the origin. Furthermore, the change of the intercept with $h\nu$ will determine K and, hence, ϵ_i in the image force. The determination of E_{bo} and K in this way is independent of $A(h\nu)$, so $I^{1/3}\text{-}V^{1/2}$ plots can be scaled by adjusting the value of $A(h\nu)$, by varying either light intensity or recorder sensitivity to make the lines have the same ordinate at the maximum gate bias. Doing this results in a family of curves, such as those in Fig. 10.16. In Fig. 10.16 the line for $h\nu > E_{bo}$ intersects the $I^{1/3}$ axis, which would not happen in a measured curve. In reality, as shown by the dashed curve, a deviation from linearity occurs because barrier lowering no longer has a simple bias dependence at low fields.

Figure 10.17 is a family of $I^{1/3}\text{-}V^{1/2}$ curves for photoemission from the valence band measured on a degenerate silicon sample with photon energy as parameter. For each photon energy, the gain of the recorder was adjusted to obtain the same ordinate at maximum bias. The data in Fig. 10.17 fit (10.41) quite well. The characteristics are linear over about 85% of the bias range 16–100 V. The deviation from linearity for $h\nu \geq 4.2 \, \text{eV}$ results from low field effects, and the deviations for $h\nu \leq 4.16 \, \text{eV}$ result

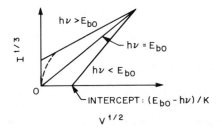

Fig. 10.16 Family of $I^{1/3}$-$V^{1/2}$ curves illustratin[g] a method of determining E_{bo} and K. Aft[er] Powell.[33]

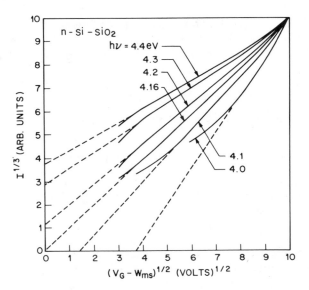

Fig. 10.17 $I^{1/3}$-$V^{1/2}$ characteristics for the emission of electrons from the valence band of degenerate n-type silicon under positive gate bias. The oxide was steam grown at 1050°C to a thickness of 3240 Å on a (111) surface of 0.001 Ω-cm n-type silicon. The semitransparent gold gate electrode was ~ 100 Å thick. After Powell.[33]

from the contribution of conduction band emission. Applying (10.43) to Fig. 10.17, $E_{bo} = 4.16$ eV in reasonable agreement with Williams[24] and $K = 0.044$ (V)$^{1/2}$. This value of K corresponds to an image-force dielectric permittivity of 1.95×10^{-13} F/cm. The large change of intercept in Fig. 10.17 between $h\nu = 4.16$ and 4.1 eV means that the accuracy of this method is about 1%.

Figure 10.18 is a family of $I^{1/2}$-$V^{1/2}$ curves for photoemission from the conduction band measured on a degenerate silicon sample with photon energy as parameter. Figure 10.18 shows that the experimental data obey (10.42). An electron distribution approximated by a delta function rather than a step function might be expected for photoemission from a narrow band of filled states such as the conduction band in degenerate silicon. A

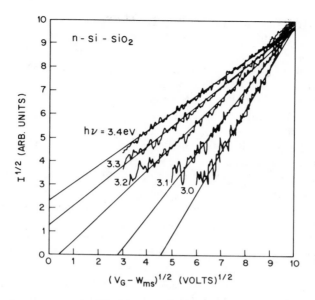

Fig. 10.18 Diagram illustrating $I^{1/2}$-$V^{1/2}$ characteristics for emission from the conduction band under positive gate bias of the sample in Fig. 10.17. After Powell.[33]

delta function electron distribution would result in a linear $I^{1/2}$-$V^{1/2}$ characteristic. However, broadening of the electron distribution by scattering may account for a step function distribution fitting the data. The barrier height from Fig. 10.18 is about 3.2 eV. The departure from linearity at high bias values corresponding to fields near 3×10^6 V/cm may result from deviation of the barrier height field dependence from the classical image-force model used in deriving (10.42). Such a deviation could result from a failure of the classical image-force model, or tunneling, or field penetration into the degenerate silicon at high electric fields.

(d) Barrier Height from Shape of I-V Curves

Barrier height also can be obtained by direct examination of the shape of the measured I-V characteristics. A functional form for N_i that gives a set of linear characteristics need not be found. The only requirement is that the quantum yield must be a function of $(h\nu - E_b)$ as in (10.37).

The method based on the shape of I-V characteristics requires the second derivative of the photocurrent with respect to $V_G - W_{ms}$. It is shown in Appendix XII that

$$4(V_G - W_{ms})^{3/2} [KAN_i\xi]^{-1} \frac{d^2I}{d(V_G - W_{ms})^2} = K(V_G - W_{ms})^{1/2} \left(\frac{1}{\xi} - 1\right) - (h\nu - E_{bo})$$

$$(10.45)$$

where ξ is defined in Appendix XII. For practical cases, $(1/\xi - 1) \geq 0$. Equation (10.45) shows that: (1) for $h\nu \leq E_{bo}$, $d^2I/d(V_G - W_{ms})^2 > 0$ for all $V_G - W_{ms}$. That is, the I-V characteristic is concave upward CU for all biases; (2) for $h\nu$ somewhat greater than E_{bo}, the I-V characteristic will be concave downward CD at low biases and change to CU at higher biases; and (3) for $h\nu$ sufficiently greater than E_{bo}, the I-V characteristic will be CD for all biases, assuming that bias is limited by breakdown or other considerations.

The barrier height is determined by increasing $h\nu$ until a change from CU to CD is observed at low bias values. For metals, where ξ is nearly 1, the curve corresponding to the transition between CU to CD will be nearly linear at low biases. The photon energy for this curve equals the barrier height energy, and the I-V characteristics will change abruptly from CU to CD at $h\nu = E_{bo}$ independent of bias.

Figure 10.19 shows measured I-V characteristics for photoemission from the valence band of a degenerate silicon sample. For each photon energy, the recorder gain was adjusted to about the same ordinate at the maximum bias. The curves are CD for $h\nu > E_{bo}$ and CU for $h\nu < E_{bo}$. At $h\nu = 4.4\,\text{eV}$ the curve changes from CD at low bias to CU at higher biases. The curve separating CU and CD curves at low biases corresponds to $h\nu = 4.2\,\text{eV}$. This curve is almost linear, so ξ is near unity, making $4.2\,\text{eV}$ a good estimate of the zero field barrier height.

For $h\nu$ less than E_{bo} (e.g., the $h\nu = 4.0\,\text{eV}$ curve in Fig. 10.19), the photocurrent should be zero for biases less than a critical value necessary to reduce the barrier height to equal $h\nu$. However, the $h\nu = 4.0\,\text{eV}$ curve in Fig. 10.19 is not zero at low biases because of emission from the conduction band. For $h\nu > E_{bo}$, the photocurrent results from emission from both

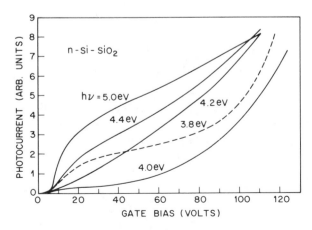

Fig. 10.19 Experimental I-V characteristics for photoemission under positive gate bias primarily from the valence band for the sample in Fig. 10.17. After Powell.[33]

the valence and conduction bands, but emission from the conduction band is negligible compared to emission from the valence band because the density of electrons in the conduction band ($\sim 10^{20}\,\text{cm}^{-3}$) is about 1% that of the valence band. As $h\nu$ is reduced below E_{bo}, emission from the conduction band becomes dominant. An example is the $h\nu = 4.0\,\text{eV}$ curve at low biases.

When $h\nu$ is reduced below 4.0 eV, the photocurrent is nearly all emission from the conduction band. An example of this is the $h\nu = 3.8\,\text{eV}$ curve in Fig. 10.19 below 80 V. Above 80 V, barrier lowering is sufficient to allow significant emission from the valence band, as shown in the rapidly rising portion of the dashed curve. Because ξ in (10.45) can vary with both photon energy and gate bias, some judgment is required in interpreting I-V curves. It is advisable to avoid estimating barrier height by this method unless ξ is near unity over an extended bias range.

Measurement of photoemission for photon energies far below the barrier for valence band emission gives the curves shown in Fig. 10.20. The photocurrents here result entirely from conduction band emission and their absolute values are about 1% of the photocurrents in Fig. 10.19. Additional evidence for the interpretation that the curves in Fig. 10.20 result from emission from the conduction band was that no photocurrent was detected for the same photon energies from a degenerate p-type sample.

Figures 10.19 and 10.20 are similar in shape. The crossover from CU to CD occurs in Fig. 10.20 at about 3.2 eV, which is the barrier height for the silicon conduction band. For photon energies below the conduction band barrier height in Fig. 10.20 (e.g., $h\nu = 3.0\,\text{eV}$), the photocurrent is zero until a critical bias is reached where the barrier height equals $h\nu$, as expected.

When the gate is negatively biased, photoemission from the gate metal will occur. Figure 10.21 shows the I-V characteristics for photoemission from the gold gate electrode. Like the previous data, the recorder gain is adjusted to give all the curves the same ordinate at maximum bias. The

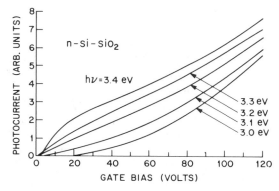

Fig. 10.20 Experimental I-V characteristics for photoemission under positive gate bias from the conduction band for the sample in Fig. 10.17. After Powell.[33]

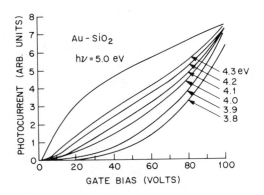

Fig. 10.21 Experimental *I-V* characteristics for emission under negative gate bias from the gold for the sample in Fig. 10.17. After Powell.[33]

curves for $h\nu = 4.1$–4.2 eV are nearly linear in the 20–60 V range of gate bias, so that the barrier height for gold is 4.1–4.2 eV.

In attempting to fit the data in Fig. 10.21, it was found that the square law case given by (10.42) fits well at low biases whereas the cube law case given by (10.41) fits at higher biases. This lack of agreement with a single model for N_i suggests that the electron energy distribution may approximate a superposition of step and ramp distributions. Because ξ is the ratio of the average of N_i to its value at $(h\nu - E_b)$, ξ also varies with bias in this case. Therefore, getting E_{bo} from the shape of the *I-V* characteristics or by the method of intercepts is somewhat uncertain in this case.

10.4 WORK FUNCTION DIFFERENCE

In this section we describe work function differences, and their relation to the barrier height determinations of the previous section. First, we discuss the origin of work function differences.

The gate material used in an MOS capacitor can be quite different from the silicon substrate. Consequently, an electron in the gate can be in a very different chemical environment than in the substrate. For example, if the gate were aluminum, any particular electron would move among other electrons and the aluminum ions at a density appropriate for this metal. Therefore, to remove an electron from this environment would be a task involving readjustment of the remaining electrons, a task requiring an amount of work peculiar to aluminum and different from the work necessary to remove an electron from another metal or from the silicon substrate.

If the MOS capacitor is placed in a circuit, or even if the capacitor is connected so that the gate is shorted to the substrate, electrons are free to travel from gate to substrate or vice versa. If, for example, electrons can leave the gate doing less work than when they leave the substrate, electrons will leave the gate and enter the substrate until they charge the substrate up to such a negative potential that there is no further advantage in this charge transfer. That is, for our example, electrons do less work leaving the gate than the substrate but must do so much work to overcome the repulsive force of the negatively charged substrate that no advantage is gained. If the work to leave the metal is W_m (in electron volts) and to leave the silicon is W_s, the substrate charges up until it reaches a potential relative to the gate of ΔV

$$\Delta V = -(W_s - W_m) \equiv W_{ms} \tag{10.46}$$

where W_{ms} is called the *work function difference*.

A simple example is the case of a degenerately (very heavily) doped n-type polysilicon gate on an MOS capacitor with a p-type substrate. In this case the chemical environments of gate and substrate are the same except for the dopant ions. Consequently, the difference $(W_m - W_s)$ is the same as the difference between the Fermi level of the n-type gate ϕ_{Fn} and that of the p-type substrate ϕ_{Fp}. This difference drives electrons from the gate to the substrate. In the substrate these electrons recombine with holes. As the holes disappear by recombination, the silicon depletes, and we end up with a positively charged gate, balanced by a negative acceptor ion charge in a surface depletion layer. There is now a potential rise across the oxide and the silicon depletion layer, placing the gate at a potential $(\phi_{Fp} - \phi_{Fn})$ above the substrate. This situation occurs spontaneously, without external bias. Consequently, to obtain a flatband condition, a bias on the gate of $V_{FB} = -(\phi_{Fp} - \phi_{Fn})$ must be applied relative to the substrate. More generally, as a result of work function difference, we have

$$V_{FB} = W_{ms} = W_m - W_s. \tag{10.47}$$

A band-bending diagram is shown in Fig. 10.22a for the case of an MOS capacitor with gate short-circuited to substrate. In this example the work function of the metal is less than that of the silicon, so electrons have been transferred to the silicon from the gate. These electrons recombine with holes, forming a depletion layer of acceptor ions. To accommodate this depletion layer, the conduction band of the silicon has bent downward, toward an inversion condition for this p-type substrate. If enough electrons are transferred to the silicon, an inversion layer will form.

As a result of this electron exchange between metal and silicon, an electric dipole is formed, as shown in the lower part of Fig. 10.22a. This dipole layer consists of positive metal ions on the gate and a balancing depletion layer of negative acceptor ions in the silicon. These two equal and opposite charge layers are separated by the gate oxide that, therefore, is subject to a uniform electric field, resulting in the voltage drop V_{ox} in Fig. 10.22a.

Clearly, W_{ms} is a property of the bulk silicon and the gate metal; W_{ms} is

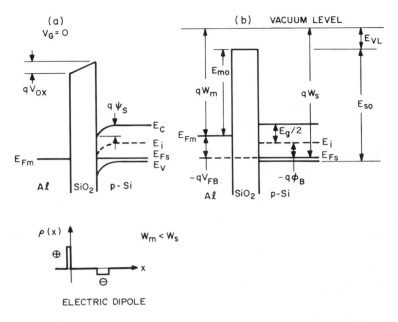

Fig. 10.22 Energy-band diagram showing the effect of work function difference between aluminum and p-type silicon on the potential distribution in an MOS capacitor. (a) Conditions for $V_G = 0$, where E_c and E_v are the conduction and valence band edges, respectively; E_i is the intrinsic Fermi level; and E_{FS} and E_{FM} are the Fermi levels in the silicon and metal, respectively. (b) Flatband condition, where W_m is the work function of the aluminum and W_s is the work function of the silicon; $E_{VL}/q = 1$ V is the vacuum level measured with respect to the conduction band edge of the oxide; E_{so} and E_{mo} are the zero field barrier heights from the silicon valence band and the aluminum Fermi level to the SiO₂ conduction band. After Grove.[48]

not affected by interfacial effects, such as the silicon doping profile near the silicon surface, nor by charges in the SiO₂. However, these factors do influence the amount of charge transfer needed to bring the two bulk Fermi levels into coincidence. For example, if positive oxide charge is moved from the vicinity of the gate to the vicinity of an n-type silicon substrate, the overall potential drop W_{ms} becomes distributed differently between the oxide and the silicon depletion layer. For this example, a larger portion of W_{ms} appears across the oxide and less across the depletion layer as the oxide charge is moved toward the silicon. Thus fewer electrons are transfered from the silicon to the gate. Consequently, a smaller gate bias V_{FB} is needed to bring the silicon to a flatband condition. This effect of oxide charge is included in (10.6) for V_{FB}.*

*Equation (10.6) also can be used to find the effect of oxide dipole layers, by placing a double layer of closely spaced positive and negative charges in $n_o(x)$.

10.4.1 Effect of Work Function Difference on MOS Capacitor Characteristics

To see the effect of the work function difference alone on the C-V characteristics of the MOS capacitor, consider applying just enough gate bias to counterbalance the built-in potential W_{ms}. A counterbalancing gate bias will produce the flatband condition ($V_{ox} = \psi_s = 0$), where V_{ox} and ψ_s are the potentials across the SiO$_2$ and silicon, respectively, produced by the work function difference. Flatbands is illustrated in Fig. 10.22b. In the example in Fig. 10.22, V_{FB} is negative. Hence $(-qV_{FB})$ is the energy interval separating E_{Fm} and E_{Fs}. The sum of the potentials going from the Fermi level in the metal to the Fermi level in the silicon must be zero. From Fig. 10.22b, we can rederive (10.47).

The work function difference W_{ms} is positive when $W_m > W_s$ and negative when $W_m < W_s$. Therefore, the C-V curve will be shifted to more positive voltage in the first case and to more negative in the second. Figure 10.22b shows that for p-type

$$W_s = \frac{E_{so} + E_{VL} - E_g/2 - q\phi_B}{q}. \tag{10.48a}$$

For the metal, W_m is given by

$$W_m = \frac{E_{mo} + E_{VL}}{q}. \tag{10.48b}$$

In (10.48) E_{so} is the barrier height between SiO$_2$ and the silicon valence band; E_{mo} is the barrier height between SiO$_2$ and the Fermi level in the metal gate electrode, both of which can be measured by photoinjection; E_{VL} is the energy difference between the vacuum level and the SiO$_2$ conduction band (E_{VL} need not be known to find the work function difference because it always cancels); $(E_g/2 + q\phi_B)$ is the energy interval between the valence band and the silicon Fermi level, with E_g the silicon band gap; and ϕ_B is from (2.43) or (2.45). For n-type, ϕ_B is positive; for p-type, ϕ_B is negative. Thus

$$\phi_B = \frac{kT}{q} \ln \frac{N_D}{n_i}; \qquad n\text{-type} \tag{10.49a}$$

$$= \frac{kT}{q} \ln \frac{n_i}{N_A}; \qquad p\text{-type.} \tag{10.49b}$$

Combining (10.47) and (10.48), the work function difference is

$$W_{ms} = \frac{E_{mo} - E_{so} + E_g/2 + q\phi_B}{q}. \tag{10.50}$$

If we include the voltage shift contributed by $Q_f + Q_{it}$ and W_{ms}, (10.47) becomes*

$$V_{FB} = -\frac{Q_f + Q_{it}}{C_{ox}} + W_{ms} \tag{10.51}$$

where Q_{it} is the interface trap charge density evaluated at $\psi_s = 0$ or at flatbands. Thus Q_f and W_{ms} cause a parallel shift of the C-V curve along the voltage axis with respect to the ideal by an additive constant equal to $-Q_f/C_{ox} + W_{ms}$. Interface traps add $Q_{it}(\psi_s = 0)/C_{ox}$ to the flatband point and also cause the shape of the C-V curve to change compared to the ideal.

Figure 10.23 is a plot showing how W_{ms} for aluminum gate electrodes varies with donor and acceptor concentrations. Curve a is for acceptors, and curve b is for donors. Figure 10.23 is based on the measurements by Deal et al.[43] of the flatband voltage shift of a C-V curve as a function of oxide thickness. The details of this method are discussed in Section 10.4.2. Values of W_{ms} from Fig. 10.23 are found to agree with (10.50) using values of E_{mo} and E_{so} obtained from photoemission measurements.

The arguments and relationships just derived for metal gates also apply for the polysilicon gate commonly used in integrated circuits. The work function difference is now between the Fermi level in the polysilicon gate and that of the more lightly doped silicon substrate. Figure 10.24 shows W_{ms} for the polysilicon-SiO_2-Si system as a function of doping density. The dopant impurity usually is diffused into the polysilicon gate simul-

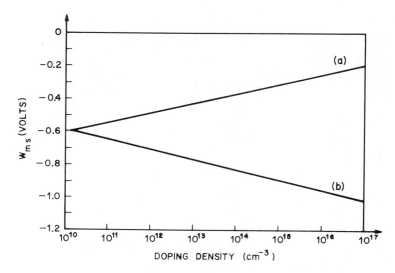

Fig. 10.23 Work function difference W_{ms} as a function of silicon doping density for aluminum gate electrodes. Curve a is for n-type; curve b, for p-type. After Deal et al.[43]

*Equation (10.51) can be derived from (10.6) by setting $n_o(x') = (Q_f + Q_{it})\delta(x' - x_o)$, that is, by modeling Q_f and Q_{it} as charge sheets at the oxide-silicon interface.

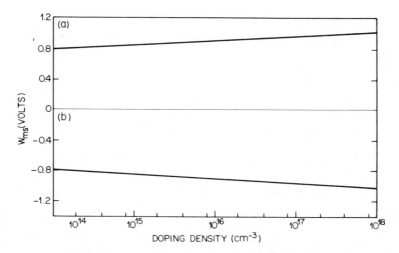

Fig. 10.24 Work function difference between a degenerately doped polysilicon gate and silicon as a function of doping density in the silicon substrate. (a) The polysilicon is doped with boron to a concentration of about 10^{20} cm^{-3}, and the silicon substrate is n-type. (b) The polysilicon gate is doped with phosphorous to a concentration of about 10^{20} cm^{-3}, and the silicon substrate is p-type.

taneously with the source and drain diffusion. Therefore, in an n-channel MOSFET, the polysilicon will be n-type and the substrate p-type, whereas for a p-channel MOSFET, the polysilicon will be p-type and the substrate n-type.

In Fig. 10.24 the polysilicon is degenerately doped with boron in curve a and with phosphorous in curve b. The curves in Fig. 10.24 were calculated assuming the Fermi level in the gate to be at the appropriate band edge, located 0.56 eV from midgap. The Fermi level of the substrate was taken as $(kT/q)\ln(N_D/n_i)$ for n-type and as $(kT/q)\ln(n_i/N_A)$ for p-type. Then W_{ms} is computed as

$$W_{ms} = -0.56\,\text{eV} + \frac{kT}{q}\ln\frac{N_D}{n_i} \qquad (10.52a)$$

for n-type and as

$$W_{ms} = -0.56 + \frac{kT}{q}\ln\frac{n_i}{N_A} \qquad (10.52b)$$

for p-type.

10.4.2   Measurement of Work Function Differences

The value W_{ms} can be obtained from the flatband voltage in a C-V measurement. Alternatively, the zero field barrier heights between SiO$_2$ and both the metal E_{mo} and the silicon E_{so} can be measured by electron photoemission, and W_{ms} can be calculated from (10.50). These approaches are now discussed.

(a) W_{ms} from Two C-V Methods

At flatbands $V_G = V_{FB}$, and for uniform doping, $Q_s = \psi_s = 0$. For simplicity, we also assume $Q_{it}(0) = 0$. Then (10.51) becomes $V_{FB} = -Q_f/C_{ox} + W_{ms}$. Thus V_{FB} depends on both W_{ms} and Q_f. The problem is to separate the term $-Q_f/C_{ox}$ from W_{ms}. Two methods of doing this are described next.

In the first method relative values of W_{ms} are obtained by comparing the flatband voltage shifts of MOS capacitors made using different gate metals. To obtain absolute values of W_{ms}, the value of W_{ms} for one metal must be independently known, from a photoemission measurement for example. In the second method V_{FB} is measured as a function of oxide thickness, varying oxide thickness by etching it off in small steps. The second method of obtaining W_{ms} is independent of other experiments and is useful for comparison. Which method is best depends on the motivation and convenience of the experimenter.

The first method is illustrated in Fig. 10.25, which shows two high frequency C-V curves taken using two different gate metals having work functions W_{m1} and W_{m2}. The difference in flatband voltage is

$$(V_{FB})_1 - (V_{FB})_2 = (W_{ms})_1 - (W_{ms})_2 \tag{10.53}$$

where the terms in (10.51) due to Q_f cancel out in (10.53) because Q_f is the

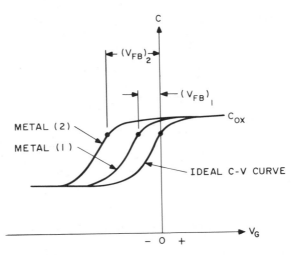

Fig. 10.25 High frequency capacitance as a function of bias with two different metals used as gate electrodes and an ideal C-V curve assuming no work function difference.

same for each capacitor. This cancellation of the Q_f term* is the reason we take differences and use two or more different metals. If W_{ms} is known for one metal, it can be found for all the others from (10.53).

If the work function difference using one metal is known (it can be calculated from (10.50) using E_{mo} and E_{so} measured by photoemission), all the others can be determined from (10.53). The only requirement is that all the MOS capacitors be prepared under the same conditions with the same doping density. Then Q_f and ϕ_B will be the same in all the capacitors and will cancel out when the difference in flatband voltage is taken.

Interface traps and impurity redistribution resulting from thermal oxidation both affect the shape of the C-V curve. If preparation conditions are the same, these effects will be the same for each C-V curve. Therefore, the C-V curves will be parallel, only shifted along the voltage axis, and interface traps and impurity redistribution introduce no error.

The second method of separating $-Q_f/C_{ox}$ from W_{ms} is illustrated in Fig. 10.26, where V_{FB} is plotted as a function of oxide thickness. Oxide thickness is varied by etching off decremental steps of oxide. The flatband voltage V_{FB} is measured after each etch-off step using a C-V curve measured at 1 MHz. The flatband voltage will be given by (10.51). Assuming $Q_{it} \ll Q_f$, V_{FB} becomes

$$V_{FB} = W_{ms} - \frac{Q_f}{C_{ox}} = W_{ms} - \frac{Q_f x_o}{\epsilon_{ox}}. \qquad (10.54)$$

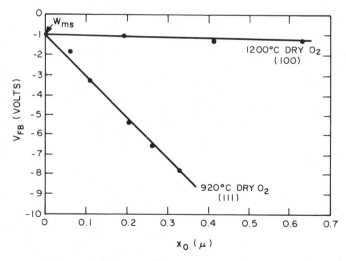

Fig. 10.26 Flatband voltage shift as a function of oxide thickness. This figure shows the determination of metal-semiconductor work function difference from MOS flatband voltage measurements at 1 MHz. The gate metal was aluminum, and acceptor concentration in the silicon was 10^{16} cm^{-3}. After Grove.[43]

*In addition to Q_f, there may be other constant terms contributing to V_{FB} as described in Section 10.5.1.

From (10.54), a straight line plot of V_{FB} versus x_o has a slope equal to $-Q_f/\epsilon_{ox}$ and intercept, at $x_o = 0$, equal to W_{ms}. Figure 10.26 is such a plot. The samples were aluminum-SiO$_2$-p-type silicon with $N_A = 10^{16}$ cm^{-3}. Notice that in this case oxidation conditions and silicon orientation are different for the two samples measured in Fig. 10.26 so that Q_f and, hence, the slope is different. However, the intercept $W_{ms} = -1$ V is the same for each sample.

The key point is that V_{FB} is measured after each etch-off step on the *same* sample. Etching the oxide has no effect on Q_f, interface trap level density, and impurity redistribution, which are not located in the removed oxide and, consequently, are the same throughout the measurement. Once the oxide has been etched to about 200 Å, the gate will become shorted to the substrate and the measurement concluded. Short-circuiting occurs long before the distributions of Q_f and interface traps at the Si–SiO$_2$ interface are reached.[44]

Before V_{FB} is plotted using this method, the correct flatband point is located on each C-V curve for each oxide thickness. Determination of V_{FB} is discussed in Section 10.5.

In (10.54) V_{FB} has been written as a function of W_{ms} and Q_f only. This idealization is made to illustrate the concepts. In practice, V_{FB} also depends on interface trap charge density and, for the usual case of a nonuniform doping profile, on doping profile. Therefore, the slopes of the curves in Fig. 10.26 depend on these factors in addition to Q_f, as explained in more detail in Section 10.5.1.

Another complication can arise in the etch-off method when charges are present in the oxide. For example, if (10.54) is replaced by (10.9), we find

$$V_{FB} - W_{ms} = -x_o \frac{Q_o}{\epsilon_{ox}} + \frac{x_o - \bar{x}}{\epsilon_{ox}} Q_o. \qquad (10.55)$$

The distance of the charge centroid from the silicon surface $(x_o - \bar{x})$, is unaffected by etchoff, as long as the etch-off step does not remove any of the oxide charge. Under these conditions a plot of V_{FB} versus x_o will have an intercept that is not W_{ms}, but $W_{ms} + (x_o - \bar{x})/\epsilon_{ox} Q_o$. Therefore, it is important that the etch-off steps be continued as close to the silicon surface as possible, or that other checks be made to ensure that no charge is located within the oxide layer other than Q_f.

Another method of obtaining W_{ms} from an etch-off experiment is to plot the change in flatband voltage ΔV_{FB} rather than V_{FB} as a function of x_o. Then because W_{ms} is the same at all x_o, it cancels and we get $\Delta V_{FB} + Q_f \Delta x_o/\epsilon_{ox} = 0$. A plot of ΔV_{FB} versus x_o is a straight line passing through the origin with a slope proportional to Q_f. Then W_{ms} is determined from (10.54).

The C-V methods can be used for MOS capacitors with any doping level. The theory of these methods is simple. However, only relative values of W_{ms} can be found from (10.53) using the first C-V method. To find W_{ms} for any particular choice of metal and substrate, a known value of W_{ms} is needed as a benchmark. Such a value can be found from the second C-V method or from photoemission, discussed next.

(b) W_{ms} by Photoemission

Now we turn to the optical determination of work function difference. The value W_{ms} is not directly measured by photoemission. The quantity measured is the zero field metal-SiO_2 barrier height, E_{mo}.

If we make the assumption that the barrier heights for injection into SiO_2 are determined by the distance from the Fermi level to the SiO_2 conduction band, then by reference to (10.50) or to Fig. 10.22b, we see that for a degenerately doped p-type silicon substrate $[q\phi_B = -E_g/2]$

$$W_{ms} = \frac{E_{mo} - E_{so}}{q} \qquad (10.56)$$

where E_{mo} and E_{so} are the zero field barrier heights in joules for injection from the metal and from the silicon. Equation (10.56) is based on the assumption that the interface does not modify the barrier to injection into SiO_2 differently for different metals or for silicon itself. In particular, we assume that high charge densities (due to ionized sodium, for example) are not present near the injecting electrodes and that any interfacial strain or disorder is slight.*

Figure 10.27 illustrates how the barrier height for injection into SiO_2, E_{mo}/q, can be found from a photoemission measurement. Figure 10.27a shows the cube root of photoresponse, which is directly proportional to photoemitted electron current, as a function of incident photon energy for MOS capacitors by using various gate metals.[43] The extrapolated intercept on the photon energy axis is equal to the barrier energy E_m/q for the chosen gate bias. By repeating these measurements at various gate biases, E_m/q can be plotted against oxide field, and E_{mo}/q can be determined as the zero field intercept of such a plot. Table 10.2 lists values of E_{mo}/q obtained from photoemission measurements for these metals. The value of E_{so}/q for silicon is also listed.

(c) Results

The values of E_{mo}/q obtained from C-V measurements, using various gate metals, agree quite well with the values obtained from photoemission measurements as shown in Table 10.2. The values of W_{ms} for aluminum measured by Werner[45] with the C-V technique illustrated in Fig. 10.26 are questionable because they differ from the photoemission measurements by more than 0.3 eV. These discrepancies are large and possibly would be reduced if the procedure described in Section 10.5 were used. In particular,

---

*The effect of charges near the interface on barrier heights measured by photoinjection is discussed in Section 11.4.2(c). Interfacial charges may make the oxide field dependence of the barrier height complex, interfering with the determination of the zero field barrier height. Interfacial disorder also could alter this oxide field dependence, by altering the density of states near the SiO_2 conduction band edge for example.

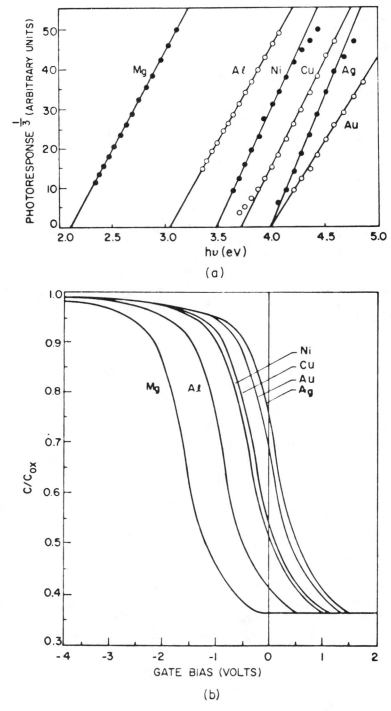

Fig. 10.27 (a) Cube root of photoresponse as a function of incident photon energy for MOS capacitors using various gate metals. After Deal et al.[43] (b) Normalized capacitance as a function of gate bias for a variety of gate metals. After Deal et al.[43]

Table 10.2 Metal-SiO$_2$ and Si–SiO$_2$
Barrier Height Energies in Electron
Volts as Inferred from MOS Capacitor Flatband Voltages and Measured
by Internal Photoemission[a]

Metal	Photon	MOS
Mg	2.25	2.4
Al	3.2	3.2[c]
Ni	3.7	3.6
Cu	3.8	3.8
Ag	4.15	4.2
Au	4.1	4.1
Si	4.35[b]	

[a] After Grove.[48]

[b] Arbitrarily chosen as reference from
which the remaining barrier heights in
this column are calculated. This barrier
height is measured from the top of the
silicon valence band.

[c] Energy E_{mo} after Deal et al.,[43]
measured by photoemission, is used as
reference in determining E_{mo} by C-V
measurements using various gate
metals.

the effect of oxide charge, interface traps, and nonuniformity of the doping
profile all must be evaluated.

The C-V data in Table 10.2 were obtained using curves such as those in
Fig. 10.27b. Figure 10.27b shows a family of C-V curves measured at
1 MHz for the same metals used in Fig. 10.27a. Silicon substrate doping is
the same for all samples. Also, processing is the same for all samples to
ensure the same Q_f, interface trap level density, and impurity redistribution. Figure 10.27b shows the parallel shift of the C-V curves along the
voltage axis as a result of the different W_{ms} for each metal.

Using (10.50) and (10.51), we can relate the shifts in the C-V curves in
Fig. 10.27b to the barrier heights determined from photoemission by

$$(V_{FB})_1 - (V_{FB})_2 = \frac{(E_{mo})_1 - (E_{mo})_2}{q}. \qquad (10.57)$$

Here, as in (10.53), subscripts "one" and "two" refer respectively to a first
and a second choice of gate metal.

For values of $(Q_f + Q_{it})/q$ $(>10^{11}\,\text{cm}^{-2})$ sufficient to make $Q_f + Q_{it}$ the
dominant factor in V_{FB}, $Q_f + Q_{it}$ is found to be positive. However, for small

values of $(Q_f + Q_{it})/q$ (in the 10^{10} cm^{-2} range typical in integrated circuit manufacture) where V_{FB} is dominated by W_{ms}, negative $Q_f + Q_{it}$ is found on n-type samples, using the value of W_{ms} for aluminum from Fig. 10.23. Although there is no reason why $Q_f + Q_{it}$ cannot be negative, more accurate values of W_{ms} are needed before negative values of $Q_f + Q_{it}$ can be established.

The agreement between the W_{ms} values of the C-V methods and those of the photoemission measurements indicates that, despite the many theoretical assumptions needed to interpret photoemission measurements, reasonable results can be obtained. Photoemission measurements are sensitive to small changes in barrier height, suggesting that photoemission can provide benchmark values of W_{ms} for use with the C-V measurements. However, photoemission is limited to degenerately doped silicon substrates,* it can be affected by layers of interfacial or oxide charge (which can make a zero field barrier height unmeasurable), and it is theoretically much more complex than the C-V methods. Thus, although photoemission is sensitive to the barrier height, it also depends upon various experimental and theoretical complications. The accuracy of the barrier heights obtained from photoemission is unknown, although W_{ms} calculated from these values may be more accurate than W_{ms} measured using the C-V methods.

10.4.3 Application to Integrated Circuits

In practice, the combination of work function difference and positive oxide fixed charge will result in some surface band bending in the silicon, even with no applied gate bias. For simplicity, consider the idealized case where $Q_{it} = 0$. For an aluminum gate, $W_m < W_s$ so that electrons are transferred from the aluminum to the silicon, and the bands at the silicon surface bend down. The bands in the silicon also are bent down as a result of positive oxide fixed charge. Therefore, an oxidized n-type silicon surface, with an aluminum gate, is in accumulation at zero gate bias. For p-type silicon at zero gate bias, the surface can be either depleted or inverted, depending on the doping concentration in the silicon, oxide thickness, and oxide fixed charge density. The charge density Q_f can be varied over a wide range by processing as described in Section 15.4.3.

To evaluate the effects of Q_f and W_{ms} on ψ_s, we find the values of Q_f/q required to make the surface of p-type silicon intrinsic as a function of acceptor concentration for an aluminum gate electrode shorted to the silicon substrate.

As described in Section 10.4, when gate is short-circuited to substrate, the gate loses electrons to the substrate until it reaches a positive potential $(-W_{ms})$ relative to the grounded substrate. This gate potential is the sum of

---

*See Section 10.3.3(b). In nondegenerate silicon the band bending in the depletion layer affects the photoemission barrier height in a complicated way.

the potential drops across the oxide and the silicon depletion layer; that is

$$-W_{ms} = -\frac{Q_f + Q_s}{C_{ox}} + \psi_s. \qquad (10.58)$$

For a p-type substrate, from (2.45) the surface is intrinsic for $\psi_s = (kT/q)\ln(N_A/n_i) \equiv -\phi_B$. At this band bending, Q_s is the depletion layer charge from (2.93), namely, $Q_s = -[2\epsilon_s q N_A(-\phi_B)]^{1/2}$. Substituting these results into (10.58), solving for Q_f and using (10.50), we find

$$Q_f = C_{ox}(W_{ms} - \phi_B) - Q_s$$

$$= C_{ox}\frac{(E_{mo} - E_{so} + E_g/2)}{q} + [2\epsilon_s q N_A(-\phi_B)]^{1/2}. \qquad (10.59)$$

Figure 10.28 is a plot of Q_f/q from (10.59) as a function of N_A with $E_{mo} = 3.2$ eV for aluminum (see Table 10.2) and several oxide thicknesses. For $x_o = 1000$ Å, Fig. 10.28 shows that the silicon surface will be inverted for acceptor concentrations less than 4×10^{15} cm^{-3}. Regardless of the value of Q_f/q, W_{ms} inverts the silicon surface. For a given x_o, Fig. 10.28 also shows that the heavier the doping, the larger Q_f must be to produce an intrinsic surface or the onset of inversion. For a typical value of $Q_f/q = 2 \times 10^{11}$ cm^{-2} and $x_o = 1000$ Å, Fig. 10.28 shows that: (1) for $N_A <$

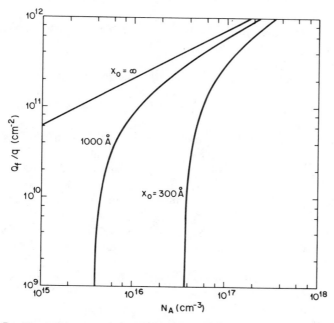

Fig. 10.28 Positive oxide charge density per unit area versus acceptor density per unit volume for several choices of oxide thickness. This value of Q_f/q will make the silicon surface intrinsic according to (10.59).

2.2×10^{16} cm^{-3}, the silicon surface will be on the n-type side of intrinsic; (2) for $N_A = 2.2 \times 10^{16}$ cm^{-3}, the silicon surface will be intrinsic; and (3) for $N_A > 2.2 \times 10^{16}$ cm^{-3}, the silicon surface will be on the p-type side of intrinsic. The curves in Fig. 10.28 differ for different values of E_{mo}.

The curve for infinitely thick oxide in Fig. 10.28 approximates a field oxide. Typically, such a field oxide is 0.5–1 μm thick. Comparison of the curves in Fig. 10.28 shows that as acceptor concentration increases, the curves approach one another at large values of Q_f and that the importance of x_o diminishes.

The tendency of thermally oxidized surfaces to be n-type has a number of consequences in integrated circuit applications and MOS capacitor measurements of interfacial properties. For an n-type substrate, used in making an integrated circuit using n-p-n bipolar transistors and/or p-channel MOSFETs, the oxidized surface is in accumulation in the emitter and collector regions of the bipolar transistor and in accumulation under the gate of the MOSFET at zero gate bias. Accumulation causes no problem in isolating one device from another, but the threshold voltage of the MOSFET is larger than for a depleted surface. The base of the n-p-n bipolar transistor could be inverted. However, base region doping is usually greater than 10^{17} cm^{-3}, so that even with $Q_f/q = 2 \times 10^{11}$ cm^{-2}, base layer inversion will not be a problem. Collector avalanche breakdown voltage is decreased. No problems arise from an accumulated surface in MOS capacitor measurements.

However, there is a problem for a p-type substrate used in making p-n-p bipolar transistors and/or n-channel MOSFETs. For thick oxides, Fig. 10.28 shows that for $Q_f/q = 2 \times 10^{11}$ cm^{-2}, for example, and $N_A < 10^{16}$ cm^{-3}, the entire surface is inverted. Even with diffused n-type regions to isolate one device from another, an n^+p junction such as a drain or a collector junction with an inverted surface tends to be more leaky than one with a surface that is not inverted.

Suppose that an aluminum lead interconnecting two devices in an integrated circuit overlaid a 1000 Å oxide. Again, for $Q_f/q = 2 \times 10^{11}$ cm^{-2}, $N_A < 2.2 \times 10^{16}$ cm^{-3}, the surface under such an aluminum lead is inverted. For two interconnected n^+p junctions, the inversion layer under the metal lead short-circuits them, even with an n^+ isolation region between them. The bias applied to metal leads evaporated onto the oxide surface also can invert the silicon substrate underneath. Inversion is avoided by using thick SiO$_2$ in regions where metal leads pass over the silicon and by using heavier substrate doping.

Consider the gate region of an n-channel MOSFET. The effect of Q_f and W_{ms} is to reduce threshold voltage, and for low N_A and large Q_f in the gate region, the MOSFET could be switched on without the application of a gate bias. To avoid this problem, threshold voltage usually is controlled with an ion implant in the channel region. For an enhancement mode MOSFET, the implanted impurity is of the same type as in the substrate.

Generally, Q_f and W_{ms} influence the performance of a MOSFET more than that of a bipolar transistor or that of metal interconnections. An increase in Q_f during the life of the device or integrated circuit is treated in Section 15.4.3(d).

10.5 DETERMINATION OF FLATBAND VOLTAGE

Real MOS devices are not doped uniformly. For a nonuniformly doped structure, the gate bias corresponding to $\psi_s = 0$ need not correspond to the bias at which the silicon charge density per unit area vanishes. Consequently, for determining oxide charge, the gate bias corresponding to $\psi_s = 0$ has no basic significance. Rather, it is necessary to know how much a measured C-V curve has *shifted* relative to a theoretical C-V curve. It is the *shift* in V_{FB} rather than the gate bias corresponding to $\psi_s = 0$ that matters. However, to maintain continuity with common terminology, we continue to call this shift the flatband voltage.

A common method of measuring flatband voltage is from the gate bias corresponding to a calculated flatband capacitance on a C-V curve. This method is described in Section 10.5.3 for completeness. However, this method has large uncertainty, as described in Section 10.5.4. Therefore, the preferred method is to obtain flatband voltage by comparing a measured C-V curve to an ideal theoretical C-V curve. The theoretical curve is calculated for a device without oxide charges or work function difference, but with the same oxide thickness and doping profile as the experimental device.

When interface trap level density is small (low $10^{10}\,\mathrm{cm}^{-2}\,\mathrm{eV}^{-1}$ range), the shift in the measured C-V curve relative to a theoretical C-V curve is found directly. That is, the doping profile is measured, and the theoretical C-V curve is calculated for this profile assuming zero W_{ms}, zero D_{it}, and zero Q_f. Then this C-V curve is plotted on the same axes as the measured C-V curve, and the separation of the two C-V curves is V_{FB}.

When D_{it} is not small, the doping profile still can be measured using (9.21). Both a high and a low frequency C-V curve are needed for this purpose. For the measured profile, a theoretical C-V curve again can be plotted on the same axes as the experimental curve. Now the two curves are not simply translated, because of the stretchout introduced by D_{it} (see Section 8.2.1). Moreover, both the theoretical and the measured C-V curves are liable to error in the flatband-to-accumulation region. The errors in this bias range are due to uncertainty in the measured profile near the interface (see Section 9.4) and the likelihood of some interface trap capacitance contribution to the measured high frequency curve [see Section 9.4.1(b)]. Therefore, a more reliable estimate of flatband voltage is obtained from the portion of the high frequency C-V curve corresponding to depletion.

To find V_{FB} when D_{it} is not small, we choose a capacitance in depletion. On the theoretical C-V curve this capacitance corresponds to a gate bias we denote as V_{G0}. On the measured high frequency C-V curve the chosen capacitance corresponds to a gate bias V_G. Then, because a given high frequency capacitance value corresponds to the same band bending on both curves (see Section 8.2.4), we find from Gauss's law [see (2.94) and (2.91)] for the theoretical curve

$$C_{ox}(V_{G0} - \psi_s) = -Q_s(\psi_s) \tag{10.60a}$$

and for the experimental curve

$$C_{ox}(V_G - W_{ms} - \psi_s) = -Q_s(\psi_s) - Q_{it}(\psi_s) - Q_f. \tag{10.60b}$$

Subtracting these two equations, we find

$$C_{ox}(V_G - V_{G0} - W_{ms}) = -Q_{it}(\psi_s) - Q_f. \tag{10.61}$$

By definition, V_{FB} is given by

$$V_{FB} = W_{ms} - \frac{Q_f}{C_{ox}} - \frac{Q_{it}(0)}{C_{ox}} - \frac{Q_s(0)}{C_{ox}}. \tag{10.62}$$

Some clarification of (10.62) is needed. For nonuniform profiles, the condition $\psi_s = 0$ does not necessarily imply that $Q_s = 0$. Therefore, the term in $Q_s(0)/C_{ox}$ is subtracted from V_{FB}.* Alternatively, we could choose V_{FB} to correspond to the condition $Q_s = 0$ and introduce ψ_{so} as the band bending corresponding to this zero field condition. Either approach can be used, but the choice of $\psi_s = 0$ leads to somewhat simpler formulas. In particular, $Q_s(0)/C_{ox}$ is easier to calculate than ψ_{so} because it requires one less integration of Poisson's equation. Regardless of the choice made, V_{FB} no longer is independent of the silicon when $Q_s(0)$ is nonzero.

Using (10.61), we can rewrite (10.62) as

$$V_{FB} = V_G - V_{G0} + \frac{Q_{it}(\psi_s) - Q_{it}(0)}{C_{ox}} + \frac{Q_s(\psi_s) - Q_s(0)}{C_{ox}}. \tag{10.63}$$

The terms involving interface trap charge density in (10.63) can be rewritten using the interface trap level density D_{it}. From (5.22) we obtain

$$Q_{it}(\psi_s) - Q_{it}(0) = q \int_{E_v/q}^{E_c/q} d\zeta_s[f_o(\zeta_s - \phi_B) - f_o(\zeta_s - \phi_B - \psi_s)]D_{it}(\zeta_s). \tag{10.64}$$

To obtain V_{FB} from (10.63), both V_G and D_{it} must be measured.† Then

---

*This term arises as described in the footnote to (10.2).
†Unfortunately, D_{it} cannot be measured all the way to flatbands because of difficulty in measuring a true high frequency C-V curve near flatbands in the high-low frequency capacitance method (see Section 8.3.5) or because the conductance method is not accurate near flatbands (see Section 6.3). Therefore, the estimate (10.64) will be in error.

(10.64) is evaluated, using ψ_s as obtained from a comparison of the theoretical and experimental high frequency C-V curves at the chosen value of capacitance. The term $Q_s(0)/C_{ox}$ is computed using the measured doping profile.

An alternative, simpler procedure is to *correct* the measured C-V curve for interface trap stretchout. This correction requires the calculation of a revised gate bias V_{G0} from (10.60a). Let ψ_s be incremented by $d\psi_s$, and let the corresponding increments in gate bias in (10.60) be dV_{G0} and dV_G. Then from (10.60) we have

$$C_{ox}\, dV_{G0} = (C_{ox} + C_s)\, d\psi_s \qquad (10.65a)$$

and

$$C_{ox}\, dV_G = (C_{ox} + C_{it} + C_s)\, d\psi_s. \qquad (10.65b)$$

Taking the ratio of these two equations, we obtain

$$\frac{dV_{G0}}{dV_G} = \frac{C_{ox} + C_s}{C_{ox} + C_s + C_{it}}$$

$$= \left(\frac{C_{ox}}{C_{ox} + C_s + C_{it}}\right)\left(\frac{C_{ox} + C_s}{C_{ox}}\right)$$

$$\approx \left(1 - \frac{C_{LF}}{C_{ox}}\right)\left(1 - \frac{C_{HF}}{C_{ox}}\right)^{-1}. \qquad (10.66)$$

The result in (10.66) agrees with that in (9.18), which applies in depletion. Integrating (10.66), we obtain V_{G0} from V_G as

$$V_{G0} = \int_{V_{FB}}^{V_G} dV \frac{\left(1 - \dfrac{C_{LF}}{C_{ox}}\right)}{\left(1 - \dfrac{C_{HF}}{C_{ox}}\right)} - \frac{Q_s(0)}{C_{ox}}. \qquad (10.67)$$

The limits in (10.67) imply that $V_{G0} = -Q_s(0)/C_{ox}$ corresponds to $V_G = V_{FB}$. This result is correct, as can be seen from (10.60). Choosing $\psi_s = 0$ results in $V_{G0} = -Q_s(0)/C_{ox}$ from (10.60a) and $V_G = V_{FB}$ from (10.60b) and (10.62). Given V_{FB} and the theoretical value of $Q_s(0)$, (10.67) determines V_{G0} in terms of measured quantities for all biases short of inversion.* Hence the measured C_{HF} versus V_G curve can be plotted instead as a measured C_{HF} versus V_{G0} curve. For the correct choice of V_{FB}, the measured C_{HF} versus V_{G0} curve should coincide with the theoretically calculated C_{HF} versus V_{G0} curve, at least in depletion, where measurement errors are smallest.

Different choices of V_{FB} in (10.67) simply translate the measured C_{HF} versus V_{G0} curve. Therefore, V_{FB} can be determined by choosing $V_{FB} = 0$

---

*Equation (10.66) shows that the integrand in (10.67) is unity if $C_{it} = 0$ or far enough into into accumulation that $C_s \gg C_{it}$. In inversion, $(C_{ox} + C_s)/C_{ox}$ cannot be approximated as $(1 - C_{HF}/C_{ox})^{-1}$, so this replacement in (10.66) is not valid in inversion.

in (10.67), plotting measured C_{HF} versus V_{G0}, and translating this curve to coincide with the theoretical curve. Let the amount of translation be ΔV_{G0}. Then V_{FB} is found by subtracting V_{G0} obtained from (10.67) from V_{G0} obtained from (10.67) with $V_{FB} = 0$. The result is

$$\Delta V_{G0} = \int_0^{V_{FB}} dV \, \frac{\left(1 - \dfrac{C_{LF}}{C_{ox}}\right)}{\left(1 - \dfrac{C_{HF}}{C_{ox}}\right)}. \tag{10.68}$$

A simple way to solve (10.68) is to plot the integral in (10.68) versus V_{FB} and then choose the value of V_{FB} that corresponds to the measured ΔV_{G0}.

Equations (10.67) and (10.68) can be solved numerically using measured values of C_{HF} and C_{LF} versus gate bias. C_{HF} and C_{LF} should not be measured by the high-low frequency capacitance method described in Section 12.5.4, nor C_{LF} by the voltage ramp method described in Section 12.5.2, because the voltage ramp method is too inaccurate. For this application, a better way to measure C_{LF} is the Q-V method described in Section 12.5.2(c). This method requires precision voltage measurements to avoid round off error when C_{LF} is obtained by differentiating the measured Q-V relationship.

Because of the complexity of (10.68), a better way to find V_{FB} in practice is the direct use of the Q-V method discussed in Section 10.5.6. Our discussion here shows that the determination of V_{FB} from the shift of C-V curves is too complicated to be practical when interface traps are present. However, such a discussion is necessary to illustrate the pitfalls that underlie the customary use of C-V curves.

Another problem in using (10.67) and (10.68) arises because, in practice, the doping profile is not known accurately. As a result, within experimental error, there will be a range of acceptable profiles, each profile leading to its own theoretical C_{HF} versus V_{G0} curve. In depletion, these curves have the same shape, but are displaced from one another along the V_{G0} axis. Consequently, uncertainty in the doping profile leads to uncertainty in ΔV_{G0}, and finally in V_{FB}. This uncertainty is not large. If the doping profile is measured carefully, this error might be only a few kT/q. However, for work function determination using the etch off method of Section 10.4.2(a), this error may exceed that of the photoemission method.

The extraction of V_{FB} using (10.68) can be divided conveniently into two cases. Case 1, the case of a fully depleted region of nonuniform doping, results in a C-V curve for the nonuniformly doped structure that is a simple translation of a uniformly doped C-V curve, over that part of the bias range where the depletion layer edge is in the uniformly doped region. For case 1, shifts in V_{FB} can be deduced from the shift relative to a uniformly doped C-V curve. Case 2, the case of a nonuniformly doped region that never is fully depleted, results in a C-V curve that differs from the uniformly doped C-V curve over the entire bias range. In this case a theoretical C-V curve based on the actual doping profile must be calculated to find the shift in V_{FB}.

10.5.1 Case 1: Nonuniformly Doped Region Completely Within the Depletion Layer

In case 1 the corrected experimental C-V curve will be parallel to the calculated ideal C-V curve somewhere in depletion, and the voltage shift between the two will provide the shift in V_{FB} apart from a correction for the difference in profiles, (10.71) or (10.72), derived later. To obtain the shift in V_{FB}, the squared reciprocal of C_{HF} is plotted because, for uniform doping, this quantity is a straight line in depletion, with a slope related to the dopant ion density [see (9.10)]. The shift in the two C-V curves can be obtained more accurately as the shift in the straight-line portion of $1/C_{HF}^2$ versus V_{G0} curves.

Curve a in Fig. 10.29 is a plot of $(C_{ox}/C_{HF})^2$ as a function of V_{G0}, where (10.67) was used to obtain V_{G0} versus V_G. The measured curve corresponds to a boron doped sample that has a nonuniform doping profile caused by impurity redistribution during thermal oxidation and is representative of case 1. Curve b is the calculated $(C_{ox}/C)^2$ versus V_{G0} curve in depletion assuming *uniform* doping. This curve is calculated by first obtaining ψ_s versus V_{G0}. Use of (3.29) results in $V_{G0} = -Q_D/C_{ox} + \psi_s$, where $Q_D = qN_Aw$ in depletion. Eliminating the depletion layer width w, using (2.82) with $N_A \gg N_D$ and $w = (2\epsilon_s\psi_s/qN_A)^{1/2}$, yields a quadratic equation in ψ_s with the

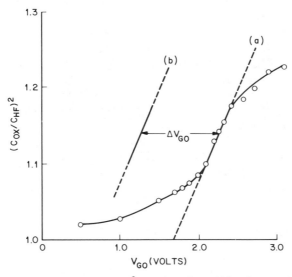

Fig. 10.29 Diagram illustrating $(C_{ox}/C_{HF})^2$ as a function of V_{G0}. Curve a is determined as described in the text, and curve b is calculated for $N_A = 1.5 \times 10^{16}$ cm^{-3} and $x_0 = 1000$ Å.

solution

$$\psi_s = V_{G0} + \frac{\epsilon_s q N_A}{C_{ox}^2}\left[1 - \left(1 + \frac{2C_{ox}^2 V_{G0}}{q N_A \epsilon_s}\right)^{1/2}\right]. \tag{10.69}$$

Using the values of ψ_s versus V_{G0}, we obtain C_D versus V_{G0}, where $C_D = \epsilon_s/w = (q N_A \epsilon_s/2\psi_s)^{1/2}$. Finally, $(C_{ox}/C)^2$ versus V_{G0} is obtained from the relation for series capacitance (3.11), $(C_{ox}/C)^2 = (C_{ox} + C_D)^2 C_D^{-2}$. The voltage shift between the linear parts of curves a and b (in depletion) in Fig. 10.29 is related to the shift in V_{FB} and the difference in profiles, as discussed below.

To calculate the shift in flatband voltage ΔV_{FB} after an experiment or treatment in which oxide charge density is changed, the linear portions of the C_{HF}^{-2} versus V_{G0} curves before and after the treatment are shifted until they become superimposed on the theoretical C-V curve. The shift in flatband voltage due to the experiment is found by determining the solution to (10.68) before and after the treatment.

There is one practical case that does not require taking interface trap stretchout into account. This is the case when only the difference in flatband voltage ΔV_{FB} between two curves that have the *same* interface trap level density is required. In this case ΔV_{FB} equals the parallel voltage shift along the voltage axis. An example of this case is in finding the work function difference using two or more different metals as described in Section 10.4.2.

(a) *Deficit Charge Density Concept*

In case 1 above, a theoretical curve for uniform doping was used to find the shift ΔV_{FB}. This value of ΔV_{FB} does not correspond to the shift in the C-V curves because a voltage shift corresponding to the depletion of ionized boron atoms near the silicon surface was not included in the theoretical curve. To illustrate this *deficit charge density* concept, Fig. 10.30 shows $N_s(x)/N_B$ as a function of depth into the silicon x, where $N_s(x)$ is the doping density at x and N_B is the doping density deep in the silicon, where the doping is uniform. Figure 10.30 shows boron redistribution after oxidation in dry O_2 at 1000°C from Fig. 14.6 with the maximum depletion width for $N_A = 10^{15}\ \mathrm{cm}^{-3}$ shown. The shaded area in Fig. 10.30 resulting from boron depletion is proportional to the deficit charge density Q_N. For an impurity that piles up at the silicon surface such as phosphorus, Q_N would be negative. The deficit charge density is

$$Q_N = q \int_0^\infty [N_B - N_s(x)]\, dx \tag{10.70}$$

and the contribution to the total flatband voltage is approximately

$$V_N = \frac{Q_N}{C_{ox}} \tag{10.71}$$

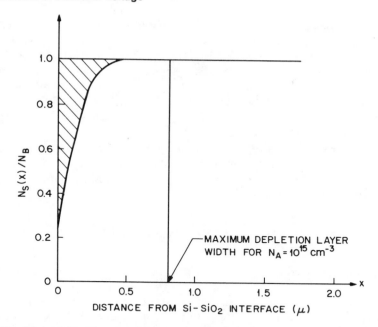

Fig. 10.30 Diagram illustrating $N_s(x)/N_B$ as a function of depth into the silicon from Fig. 14.6 after oxidation at 1000°C. The depleted impurity is boron, and the shaded area is proportional to the deficit charge density Q_N.

where Q_N is positive for an impurity that depletes like boron and negative for an impurity that piles up like phosphorus.* The contribution made to V_{FB} by the deficit charge makes V_{FB} dependent on oxidation conditions, the dopant impurity species, and the doping density.

10.5.2 Case 2: Nonuniformly Doped Region Extending Beyond the Depletion Layer

For case 2, the experimental $(C_{ox}/C_{HF})^2$ versus V_{G0} curve is obtained in the same way and the doping profile is extracted as explained in Section 9.4.1. In this case comparison of experiment with a theoretical curve for uniform doping is inadequate, and the entire doping profile must be used to calculate an ideal C-V curve.† However, the doping profile cannot be measured all the way to the silicon surface because of various limitations

---

*A correction to the estimate $V_N = Q_N/C_{ox}$ is an adjustment in the depletion layer width due to the nonuniform profile. For redistribution due to a single oxidation, this correction is minor.[46]
†Oxide charge density can be obtained more easily from the shift in threshold voltage of an MOSFET than from the shift of a C-V curve in this case.

explained in Section 9.4.3. This uncertainty in the profile is reflected in an uncertainty in the ideal C-V curve. The best approach to reduce this uncertainty is to get values for the segregation and diffusion coefficients and the parabolic rate constant. These values can be found from a fit of (14.4) and (14.5) to the measured profile over the range of depths where the profile is known accurately. Then, using these values and (14.4) and (14.5), the doping profile can be calculated all the way to the silicon surface by modeling the effects of oxidation.

Thus, instead of a uniform profile, a theoretical profile can be used that approximates the real nonuniform profile. Nonetheless, a discrepancy between the profiles will exist, leading to an apparent flatband contribution V_N given by (10.71) where, in place of (10.70)

$$Q_N = q \int_0^\infty [N_{theory}(x) - N_s(x)] \, dx \qquad (10.72)$$

where $N_{theory}(x)$ is the theoretical profile calculated right up to the silicon surface.

In the most extreme case, no effort is made to model the nonuniform profile, and comparison is made with a theoretical C-V curve for uniform doping. Using such an approach. Table 10.3 shows calculated shifts in V_{FB}[47] due to nonuniform doping densities resulting from the redistribution of boron and phosphorus during thermal oxidation. Case 1 applies to the lightest doping densities and case 2, to all the rest in Table 10.3. Table 10.3 shows that the effect of nonuniform doping is to produce a negative shift in V_{FB} relative to the uniformly doped case, a shift that makes a significant contribution to the overall flatband voltage.

Table 10.3 illustrates the error that can result if nonuniform doping is ignored. However, Table 10.3 does not list the actual shift in C-V curves because the tabulated ΔV_{FB} was determined as the difference in gate bias corresponding to $\psi_s = 0$. For case 1, this ΔV_{FB} does not correspond to the shift at a given capacitance. For case 2, the uniform profile C-V curve has little relation to the correct C-V curve.

This discussion applies for doping densities that vary slowly over a distance of a few extrinsic Debye lengths. If doping density changes rapidly over this distance, the method for determining V_{FB} described in this section does not work. The methods for finding V_{FB} also fail when gross charge nonuniformities are present. Gross nonuniformities are discussed in Section 6.2.

10.5.3 Charged Interface Trap Contribution

Even if interface trap stretchout is taken into account in determining V_{FB}, there still are charged interface traps that contribute to V_{FB} at the flatband point. Interface traps that are unoccupied donors above the Fermi

Table 10.3 Estimate of Contribution of Nonuniform Doping Profile to V_{FB}[a]

Boron		Phosphorus	
N_A (cm^{-3})	ΔV_{FB} (V)	N_D (cm^{-3})	ΔV_{FB} (V)
10^{15}	-0.08	10^{15}	-0.05
10^{16}	-0.33	10^{16}	-0.25
10^{17}	-1.14	10^{17}	-0.96

[a] After Baccarani et al.[47] Values of ΔV_{FB} in Table 10.3 were obtained from C-V curves calculated using (14.4) and (14.5) for the nonuniform doping profile resulting from impurity redistribution after thermal oxidation. The uniform doping density prior to oxidation is listed in the table. Then, V_{FB} is the voltage at the flatband point defined by the condition $\psi_s = 0$. Finally, $\Delta V_{FB} = V_{FB}$ (nonuniformly doped) $- V_{FB}$ (uniformly doped). The oxidation conditions were dry oxygen at 1100°C for 2.33 hr, $m = 0.3$, $D_{ox} = 4.4 \times 10^{-17}$ cm^2/sec, $D_s = 1.1 \times 10^{-13}$ cm^2/sec, and $B = 2.2 \times 10^{-7}$ cm^2/sec$^{1/2}$. For phosphorus, oxidation conditions were the same—$m = 10$, $D_{ox} = 3.4 \times 10^{-14}$ cm^2/sec, $D_s = 2 \times 10^{-13}$ cm^2/sec, and $B = 2.2 \times 10^{-7}$ cm^2/sec$^{1/2}$, where m is the segregation coefficient; D_{ox} and D_s are the diffusion coefficients in the oxide and silicon respectively; and B is the parabolic rate constant.

level or occupied acceptors below the Fermi level make such a contribution at flatbands (see Section 7.3). The interface trap charge density at $\psi_s = 0$ (flatbands) is [using (5.22)]

$$Q_{it}(\psi_s = 0) = -q \int_{E_v/q}^{E_c/q} d\zeta_s f_o(\zeta_s - \phi_B) D_{it}^a(\zeta_s)$$

$$+ q \int_{E_v/q}^{E_c/q} d\zeta_s [1 - f(\zeta_s - \phi_B)] D_{it}^d(\zeta_s) \qquad (10.73)$$

where D_{it}^a is the density of acceptor interface trap levels per unit area per electron volt and D_{it}^d is the corresponding density of donor interface trap levels. In (10.73) E_v/q and E_c/q are the energies of the valence and conduction band edges, ζ_s is the position of a typical interface trap energy level, ϕ_B is the Fermi level measured with respect to the intrinsic level, and f_o is the Fermi distribution. All energies (E_v/q, E_c/q, ζ_s, and ϕ_B) are measured from the intrinsic level, which at flatbands is the same at the silicon surface as in the silicon bulk. Equation (10.73) shows that the contribution made by $Q_{it}(\psi_s = 0)$ to V_{FB} makes V_{FB} a function of interface trap level density, temperature, and doping density. These last two dependencies arise primarily from the temperature and doping dependence of the Fermi energy.

In general, it is not possible to measure $Q_{it}(\psi_s = 0)$ accurately right up to the majority carrier band edge. However, it is possible to estimate the order of magnitude contribution that charged interface traps make to V_{FB}. Assuming a uniform interface trap level density over the entire silicon bandgap for simplicity, an oxide thickness of 1000 Å, and $N_A = 10^{15} \, \text{cm}^{-3}$, V_{FB} was calculated using the relation $V_{FB} = qD_{it}(\phi_B + E_g/2q)/C_{ox}$ and listed in Table 10.4. Because D_{it} invariably rises near the band edges as shown in Section 7.3, values of D_{it} were chosen on the high side for Table 10.4. This table shows that charged interface traps can make a significant contribution to V_{FB}.

In summary, the overall flatband voltage will be

$$V_{FB} = W_{ms} + V_N - C_{ox}^{-1}[Q_f + Q_{it}(0) + Q_s(0)] \qquad (10.74)$$

where V_N is the voltage shift caused by inaccuracy in calculating the nonuniform doping profile, that is, as illustrated in Table 10.3.

10.5.4 Applications

To extract Q_f from a measurement of V_{FB}, we use (10.74) to obtain

$$Q_f = -C_{ox}[V_{FB} - W_{ms} - V_N] - Q_{it}(0) - Q_s(0). \qquad (10.75)$$

In (10.75) V_{FB} is measured, and W_{ms} can be obtained from Table 10.2 for the gate metal used. With only these two quantities available, the charge

Table 10.4 Estimate of Contribution of Charged Interface Traps to $V_{FB}{}^a$

D_{it} (cm^{-2} eV^{-1})	V_{FB} (V)
10^{11}	0.12
10^{12}	1.18
10^{13}	11.81

[a]This estimate was obtained using the relation $V_{FB} = qD_{it}[\phi_B + E_g/2q]/C_{ox}$ for acceptor-type interface traps, $N_A = 10^{15} \, \text{cm}^{-3}$ and $C_{ox} = 3.4 \times 10^{-8} \, \text{F/cm}^2$ ($x_0 = 1000$ Å).

density that can be extracted is

$$Q_T = -C_{ox}[V_{FB} - W_{ms}] = Q_f + Q_{it}(0) + Q_s(0) - C_{ox}V_N. \qquad (10.76)$$

The exact values of Q_f, characteristic of thermally grown oxides, cannot be determined exactly from a measurement of V_{FB} unless V_N, $Q_{it}(0)$, and $Q_s(0)$ are known. It should be borne in mind that values of Q_f in the literature are in error by $[Q_{it}(0) + Q_s(0) - C_{ox}V_N]$, an unknown error in most cases. However, changes in Q_f produced by aging can be accurately measured provided that the remaining terms in (10.76) do not change.

Other consequences of the inability to accurately determine $Q_{it}(0)$, $Q_s(0)$, and V_N are: (1) the energy scale for interface trap level density and capture probability, calculated as described in Section 3.3.3, uses V_{FB}, and hence this energy scale will have a fictitious dependence on temperature, interface trap level density, doping profile, and oxidation conditions; and (2) interface trap capture probabilities, calculated as described in Section 5.5.3, will have the same fictitious dependences as the energy scale if ψ_s is determined using V_{FB}.

These effects are minimal if D_{it} near the band edges is small and doping nonuniformity mild. In modern integrated circuit technology D_{it} around the center of the silicon bandgap and Q_f are both small, but highly nonuniform doping profiles are purposely introduced by an implant. Thus the uncertainty introduced by V_N can be large. Small-signal steady-state measurements should be interpreted with these limitations in mind. Finally, these limitations restrict our ability to model the interface.

10.5.5 Flatband Capacitance Method

A widely used method for determining V_{FB} uses the silicon flatband capacitance for uniform doping, calculated from the relation $C_{FBS} = \epsilon_s/\lambda_p$ (using p-type as example), and the corresponding total high frequency flatband capacitance calculated from the relation $C_{FB} = C_{FBS}C_{ox}/(C_{ox} + C_{FBS})$. Finding this value of C_{FB} on the measured high frequency C-V curve is taken to locate the flatband point and the value of V_{FB}. This method is widely used because of its simplicity. However, this method is in error whenever interface traps and charge nonuniformities cannot be ignored and when the doping is not uniform all the way to the interface. These errors cannot be assessed with certainty.

Interface traps cannot be ignored at flatbands because they respond to a frequency of 1 MHz, the frequency used almost universally to measure high frequency C-V curves. Therefore, $C_{it}(\psi_s = 0)$ must be added to C_{FBS} to calculate C_{FB}. This correction results in a voltage shift of the flatband point. An estimate of this shift in V_{FB} is shown in Table 10.4. To correct for interface trap capacitance, one must know both interface trap level density and capture probability around flatbands. These quantities are measurable with some error. However, the flatband point is not known *a priori* for this measurement; hence these quantities must be slowly varying with bandgap energy to be useful in such a correction. No reliable information is available, but the available measurements indicate a failure of this assumption.

The doping profile in the silicon is seldom uniform, because of either an impurity redistribution during thermal oxidation or an implant. As mentioned above, doping density at the silicon surface cannot be measured accurately because of various limitations. Unfortunately, the flatband capacitance method requires that the doping density be known near the silicon surface to calculate C_{FB}.

The effect of nonuniform doping is to produce a negative V_{FB}, as shown in Table 10.3, whereas interface trap capacitance can produce either a positive or negative V_{FB}. Shown in Table 10.4 are V_{FB} values calculated for acceptor type interface traps. For acceptor type interface traps, the two effects tend to cancel. However, the cancellation is not exact. For donor type interface traps, the two effects are additive.

We have discussed the errors inherent in the flatband capacitance method and have estimated the magnitude of these errors. The flatband capacitance method is deceptively simple but is not practical for determining accurate values of V_{FB}.

10.5.6 The *Q-V* Method

In Section 3.3.1 it was shown that band bending versus gate bias can be obtained to within an additive constant using the *Q-V* method. We now explain why the *shape* of this ψ_s versus V_G curve should be more accurate than that obtained from either the high frequency method (Section 8.2.4) or the low frequency method (Section 3.3.2).

(a) *Comparison with the High Frequency Method*

The high frequency ψ_s versus V_G curve (see Section 8.2.4) is in error for two reasons: (1) a true high frequency *C-V* curve seldom can be measured over the entire bias range because ac interface trap response cannot be eliminated, particularly near flatbands (see Section 8.3.3). Hence the experimental high frequency *C-V* curve is in error; and (2) the theoretical high frequency *C*-ψ_s curve also is in error. This error is a result of uncertainty in the doping profile, particularly near the silicon surface (see Section 9.4.3). Because both the experimental *C-V* curve and the theoretical *C*-ψ_s curve are in error, so is the ψ_s versus V_G extracted from the comparison of these capacitance curves.

It is apparent that the ψ_s versus V_G curve from the *Q-V* method is not subject to either of these errors: (1) it is a static measurement, thus there is no need to eliminate interface trap response; and (2) no theoretical input is required, and the doping profile need not be known. The only uncertainty in the Q-V method is uncertainty in the additive constant $\psi_s(0)$ (see (3.63))]. This uncertainty results in an uncertain rigid shift of the ψ_s versus V_G curve along the ψ_s-axis, but does not lead to any uncertainty in the shape of the curve.

(b) Comparison with the Low Frequency Method

The low frequency method depends upon measurement of a low frequency C-V curve. If this C-V curve is measured using the voltage ramp method, an admittance bridge, or a lock-in amplifier, current levels are very low, noise levels are high, requiring a filter and accuracy can be poor unless a very large gate area is used (see Section 12.5.2). This C-V curve then must be integrated, leading to ψ_s versus V_G to within an additive constant of integration.

The Q-V method is superior to the low frequency method because it is a precision voltage measurement with good noise immunity, does not require a large gate area, and leads to ψ_s versus V_G without the labor of an integration.

(c) Finding the Flatband Voltage

If ψ_s versus V_G is known, the flatband voltage V_{FB} is found as the value of V_G corresponding to $\psi_s = 0$. Because the Q-V method leads to the best *shape* of the ψ_s versus V_G relation, especially near flatbands, it is desirable to use the Q-V method to find V_{FB}. To do this, the additive constant of the method $\psi_s(0)$ must be found.

A crude determination of $\psi_s(0)$ is based upon strong inversion. It is presumed that ψ_s becomes more or less independent of V_G in strong inversion, and is clamped near the value $\psi_s = (kT/q)\ln(2u_B)$. In fact, ψ_s continues to increase appreciably with V_G, even in strong inversion. Therefore a more accurate procedure is needed.

One such procedure is to compute ψ_s versus V_G from a high frequency C-V curve and compare this curve with that from the Q-V method. A ψ_s versus V_G curve can be obtained from a high frequency C-V curve as described in Section 8.2.4. The ψ_s versus V_G curve obtained from the high frequency C-V curve is most accurate in depletion and weak inversion (see Section 8.3). Therefore the ψ_s versus V_G curve from the Q-V method is shifted vertically to obtain the best fit to the ψ_s versus V_G curve from the high frequency C-V curve in depletion and weak inversion. V_{FB} is determined as the intercept on the V_G axis at $\psi_s = 0$ of this shifted ψ_s versus V_G curve from the Q-V method.

This method of finding V_{FB} involves the same measurements as the method of stretch-out corrected C-V curves based on (10.68). However, the direct use of the Q-V method of this section is simpler. To use (10.68) requires both C_{LF} (and thus differentiation of the measured Q-V relation), and also evaluation of the integral in (10.68). None of these steps is needed here.

REFERENCES

1   E. Yon, W. H. Ko, and A. B. Kuper, *IEEE Transact. Electron Devices*, **ED-13**, 276 (1966).

2   E. Kooi and M. V. Whelan, *Appl. Phys. Lett.*, **9**, 314 (1966).

3   H. G. Carlson, G.A. Brown, C. R. Fuller, and J. Osborne, *Phys. Fail. Electron.*, **4**, 390 (1966).

4   S. R. Hofstein, *IEEE Transact. Electron Devices*, **ED-14**, 794 (1967).

5   J. S. Logan and D. R. Kerr, "Migration Rates of Alkali Ions in SiO_2 Films," paper presented at IEEE Solid-State Device Research Conference, Princeton, N.J., 1965.

6   R. Williams, *J. Appl. Phys.*, **37**, 1491 (1966).

7   D. J. DiMaria, *J. Appl. Phys.*, **48**, 5149 (1977).

8   E. H. Snow, A. S. Grove, B. E. Deal, and C. T. Sah, *J. Appl. Phys.*, **36**, 1664 (1965).

9   N. J. Chou, *J. Electrochem. Soc.*, **118**, 601 (1971).

10   M. R. Boudry and J. P. Stagg, *J. Appl. Phys.*, **50**, 942 (1979).

11   G. F. Derbenwick, *J. Appl. Phys.*, **48**, 1127 (1977).

12   M. Yamin, *IEEE Transact. Electron Devices*, **ED-12**, 88 (1965).

13   M. Kuhn and D. J. Silversmith, *J. Electrochem. Soc.*, **118**, 966 (1971).

14   T. M. Buck, F. G. Allen, J. V. Dalton, and J. D. Struthers, *J. Electrochem. Soc.*, **114**, 862 (1967).

15   B. Yurash and B. E. Deal, *J. Electrochem. Soc.*, **115**, 1191 (1968).

16   J. E. Barry, H. M. Donega, and T. E. Burgess, *J. Electrochem. Soc.*, **116**, 257 (1969).

17   H. L. Hughes and R. D. Baxter, *IEEE Transact. Nuclear Sci.*, **NS-19**, 256 (1972).

18   R. J. Kriegler, *12th Annual Proceedings Reliability Physics*, 1974, pp. 250–258, IEEE Catalog No. 74CH0839-1 Phy; R. J. Kriegler and T. F. Devenyi, *Thin Solid Films*, **36**, 435 (1976).

19   J. P. Stagg, *Appl. Phys. Lett.*, **31**, 532 (1977).

20   T. W. Hickmott, *J. Appl. Phys.*, **46**, 2583 (1975).

21   C. B. Viswanathan, Technical Report ECOM-0283-F on Contract DAABO 7-70-0283, July 1973, pp. 2–26; R. J. Roedel, "Transport on Mobile Ions in Metal-Silicon Dioxide-Silicon Structures," Ph.D. dissertation, Department of Electrical Engineering, University of California, Los Angeles, California, 1976.

22   M. Cardona and L. Ley, *Photoemission in Solids*, Vol. 26 of *Topics in Applied Physics*, Springer-Verlag, Berlin, 1978.

23   R. J. Powell and C. G. Derbenwick, *IEEE Transact. Nucl. Sci.*, **NS-18**, 99 (1971).

24   R. Williams, *Phys. Rev.*, **140**, A569 (1965).

25   A. M. Goodman, *Phys. Rev.*, **164**, 1145 (1967).

26   W. T. Lynch, "Theory, Experiment, Results, and Analysis of Three Terminal Electron Hall Mobility Measurements in Thermally Grown Silicon Dioxide," Ph.D. thesis, Department of Electrical Engineering, Princeton University, Princeton, N. J., 1971.

27   R. J. Powell and C. N. Berglund, *J. Appl. Phys.*, **42**, 4390 (1971).

28   R. J. Powell, *J. Appl. Phys.*, **40**, 5093 (1969).

29   R. H. Fowler, *Phys. Rev.*, **38**, 45 (1931).

30   A. M. Goodman, *Phys. Rev.*, **144**, 588 (1966).

31   C. A. Mead, E. H. Snow, and B. E. Deal, *Appl. Phys. Lett.*, **9**, 53 (1966).

32   W. Ludwig and B. Kornfield, *Phys. Status Solidi*, **24**, K137 (1967).

33   R. J. Powell, *J. Appl. Phys.*, **41**, 2424 (1970).

34 G. F. Derbenwick, D. T. Pierce and W. E. Spicer, in *Methods of Experimental Physics*, Vol. 11, R. V. Coleman, Ed. Academic, New York, 1974, p. 67.

35 P. M. Williams, *Optical and Electrical Properties*, Vol. 4 of *Physics and Chemistry of Materials with Layered Structures*, P. A. Lee, Ed., Riedel, Boston, Mass., 1976, p. 273.

36 B. Feuerbacker and B. Fitton, "Photoemission Spectroscopy," in *Electron Spectroscopy for Surface Analysis*, Vol. 4 of *Topics in Current Physics*, H. Ibach, Ed., Springer-Verlag, New York, 1977, p. 151.

37 W. E. Spicer, in *Electron and Ion Spectroscopy of Solids*, L. Fiermans et al., Eds., Plenum, New York, 1978, p. 54.

38 D. Brust, M. L. Cohen, and J. C. Phillips, *Phys. Rev. Lett.*, **9**, 389 (1962).

39 J. C. Phillips, *Phys. Rev.*, **133**, A452 (1964).

40 D. Brust, *Phys. Rev.*, **134**, A1337 (1964).

41 R. A. Smith, *Semiconductors*, 2nd ed., Cambridge University Press, London, 1978, pp. 309–326.

42 E. O. Kane, *J. Phys. Soc. Jap.*, **21** (supplement), S37 (1966).

43 B. E. Deal, E. H. Snow, and C. A. Mead, *J. Phys. Chem. Solids*, **27**, 1873 (1966).

44 A. D. Lopez, *J. Electrochem. Soc.*, **113**, 89 (1966).

45 M. Werner, *Solid-State Electron.*, **17**, 769 (1974).

46 J. R. Brews, *IEEE Transact. Electron Devices*, **ED-26**, 1696 (1979).

47 G. Baccarani, S. Solmi, and G. Soncini, *Alta Frequenza*, **40**, 301E (1971).

48 A. S. Grove, *Physics and Technology of Semiconductor Devices*, Wiley, New York, 1967.

11

Charge Trapping in the Oxide

- INTRODUCTION, 492
- AVALANCHE INJECTION OF ELECTRONS, 495
- AVALANCHE INJECTION OF HOLES, 505
- DETERMINATION OF THE TRAP DISTRIBUTION, 508
- MEASUREMENT OF BULK OXIDE TRAP PROPERTIES, 531
- RADIATION EFFECTS IN SiO$_2$, 549

11.1 INTRODUCTION

In this chapter we discuss bulk oxide traps in thermally grown SiO$_2$. These traps are distinguished from the interface traps located at the Si–SiO$_2$ interface, discussed in Chapters 5–8. Bulk oxide traps can be located not only at the metal-oxide or Si–SiO$_2$ interfaces, but deep in the oxide. Bulk oxide traps are associated with defects in the SiO$_2$ such as impurities and broken bonds. They can be reduced, sometimes to negligible proportions, by the processing described in Section 15.4.3. Bulk oxide traps, usually uncharged, become charged when electrons and holes are introduced into the oxide.

Bulk oxide traps are important in: (1) nonvolatile memory applications. where bulk oxide traps are used for charge storage, and (2) device stability. Hot carriers generated during device operation can enter the oxide and become trapped, altering device characteristics. Also, either in manufacture or in operation, electron-hole pairs can be created in the SiO$_2$ by ionizing radiation, with subsequent trapping and deterioration of device characteristics.

Commonly used methods for injecting electrons and holes into SiO$_2$ and for finding bulk oxide charge properties are described in this chapter. Two methods for injecting carriers into SiO_2 are internal photoemission, described in Section 10.3, and avalanche injection, described in Sections 11.2 and 11.3. Section 11.4 discusses determination of the trap distribution in the oxide. Section 11.5 discusses bulk oxide trap measurements and the properties found. Section 11.6 deals with ionizing radiation in SiO_2. Sections 11.2, 11.4, and 11.5 deal with electron injection and trapping, and Sections 11.3 and 11.6 treat hole injection and trapping. Discussion mainly refers to the MOS capacitor, the structure preferred for these studies.

To study the properties of bulk traps in SiO_2, free carriers must be introduced into the SiO_2 that then can be trapped at impurity sites. Free carrier injection is difficult because there are large energy barriers between the silicon and metal electrodes and the SiO_2, as discussed in Section 10.3.3. Alternatively, carriers can be created within the SiO_2 layer by electron-hole pair production using ionizing radiation that has energy greater than the SiO_2 bandgap (8.8 eV).[1]

There are various ways of injecting carriers over the electrode energy barriers. These methods are (1) internal photoemission from either the silicon or the gate electrode as described in Section 10.3, (2) avalanche injection from the silicon[2-6] as described in Section 11.2, (3) Fowler-Nordheim tunneling[7-13] from either the silicon or gate electrode into thick oxides or direct tunneling into thin oxides, and (4) injection using a low energy electron beam in vacuum.

Fowler-Nordheim tunneling is not widely used in oxide trapping studies because the injected current densities are small and very high oxide fields ($\leq 7 \times 10^6$ V/cm) are required, favoring field ionization of the trapped charge and breakdown of the SiO_2. Direct tunneling has been used to measure interface trap properties, as described in Section 5.10. Low energy electron injection in a vacuum also is not widely used because it is more difficult than internal photoemission and avalanche injection.

Our understanding of those bulk oxide trap properties important in integrated circuit technology comes from internal photoemission and avalanche injection measurements. Internal photoemission is used in obtaining the barrier height at each electrode and the oxide trap charge distribution, neither of which can be found by avalanche injection.

Avalanche injection is used for charging oxide traps. Photoinjection also can be used, but avalanche injection has three advantages for this purpose:

1 Relatively large current densities* (≤ 10 mA/cm^2 in a thermally grown oxide about 1000 Å thick[4]) can be controllably injected into the SiO_2. High injected current densities are desirable for detection of oxide traps with small capture probabilities (such as the dominant water related electron traps) in a reasonable time. To obtain similar injected current

*The maximum current density is limited by oxide breakdown.

densities with photoinjection, a light source in the kilowatt range is required.

2   Avalanche injection uses a simpler apparatus than photoinjection.

3   Because no photons are involved,* only one type of carrier can be injected.

4   There are no problems with free carrier recombination or photoionization of bulk oxide traps as there might be with photoinjection.

The primary disadvantage of avalanche injection is that electrons entering the oxide have a rather wide energy distribution; thus it is not possible to study the energy dependence of trapping in the oxide by avalanche injection. The primary disadvantage of the MOS capacitor in avalanche injection studies is that the field in the oxide cannot be adjusted independent of injected current; thus it is not possible to study the field dependence of the capture cross section with the MOS capacitor.

The combination of avalanche injection for charging the oxide traps and studying charging kinetics and the photo I-V method for determining the trap distribution in the oxide is the most effective way to study bulk oxide traps.

The practical impact of hot carrier injection into SiO_2 lies in very large scale integration (VLSI), multilayer gate memory,[14] and high voltage diodes.

Very large scale integration incorporates MOSFETs with very small channel lengths. For n-channel MOSFETs, usually found in VLSI circuits, hot electrons are emitted from the silicon into the gate oxide when applied voltages are sufficiently large. Hot electrons can originate from either the surface channel[15] or the silicon substrate.[16] Subsequent trapping of the electrons injected into the oxide can cause instability in the form of transconductance degradation and threshold voltage drift with time.[15-20] These effects also can result from hot hole injection, but hot hole injection occurs only when the drain is in strong avalanche breakdown, which is a rare condition in practice. The device aspects of this problem are discussed in Ref. 21.

In the multilayer gate memory cell information is written into the cell by electron transport from the silicon through an SiO_2 layer to charge storage sites. The cell is erased by electron transport back through the thin oxide to the silicon. Although the electrons moving through the thin SiO_2 layer during the write and erase cycles need not be hot, they still can become trapped in the thin SiO_2 layer. Buildup of oxide charge and interface traps causes the current flowing across the oxide to decrease to such a low level that the cell becomes inoperative.

---

*There is a very low intensity light emission from the avalanche plasma in the silicon during avalanche injection that is too weak and not energetic enough (its spectrum is peaked in the visible) to produce any detectable effects on bulk oxide traps.

Finally, high voltage planar silicon diodes frequently must be operated in avalanche breakdown. The fringing field across the junction is sufficient for injection into the SiO_2 of hot carriers from the avalanche plasma in the junction depletion layer. Subsequent trapping in the SiO_2 causes the avalanche breakdown voltage to change with time. In an n^+p diode, electrons are injected, resulting in negative oxide charge. This negative oxide charge attracts an accumulation layer of holes over the p-region in the silicon. The resulting n^+p^+ junction has a lower avalanche breakdown voltage. In a p^+n diode, holes are injected, resulting in positive oxide charge. This positive oxide charge attracts an inversion layer of electrons over the n-region in the silicon. The resulting p^+n^+ junction has a lower avalanche breakdown voltage. The oxide trapping phenomena are similar in all these cases. In this chapter we discuss what is known about trapping of electrons and holes introduced into the oxide.

Hot carrier injection has been observed in the MOS capacitor[2-6] as well as MOSFET and gate controlled diode structures.[22-32] Although the MOSFET and gate controlled diode have been used in oxide trap studies,[27,29] injection current and therefore oxide charge in these structures is not uniform over the gate area. Thus quantitative interpretation of charging phenomena in these structures is difficult. In the MOS capacitor, injection current is uniform over the gate area, making the MOS capacitor the preferred structure for quantitative oxide trap studies. We concentrate on the MOS capacitor in this chapter.

A periodic waveform is required to produce avalanche injection in an MOS capacitor. Sinusoidal, square, and sawtooth waveforms can be used. The avalanche injection measurements described in this chapter all have been done with sinusoidal excitation.

11.2 AVALANCHE INJECTION OF ELECTRONS

In avalanche injection, carriers are accelerated by an applied electric field. When the field at the silicon surface reaches the avalanche breakdown field ($\geq 3 \times 10^5$ V/cm, depending on doping density), carriers generated in the surface depletion layer are accelerated to sufficient energy for impact ionization to occur. A plasma of energetic electron-hole pairs is created in the silicon surface depletion layer. Some of the electrons or holes have enough energy to surmount the interfacial energy barrier and enter the SiO_2. Generally, electrons are injected when the substrate is p-type and holes when the substrate is n-type.[2,3]

In many devices, hot carrier or avalanche injection into the oxide layer occurs during normal operation. Control of injection as well as of subsequent trapping is important in limiting undesirable effects on device characteristics. Briefly, some instances where device characteristics are affected by hot carrier or avalanche injection are:

1   In a MOSFET, channel carriers can be accelerated by the source-to-drain electric field to sufficient energy to surmount the Si–SiO$_2$ interfacial energy barrier,[22,23] particularly when the channel length is very small. Once in the oxide, some of the carriers may be trapped, creating both oxide fixed charge and interface traps. These effects cause undesirable changes in device characteristics during operation. In addition, the gate current resulting from hot carrier injection is undesirable, particularly in memory applications.

2   The junction breakdown voltage, gain factor, and threshold voltage of a MOSFET can vary as a result of trapped oxide charge and increased interface trap density introduced by avalanche breakdown of the source and drain junctions.[33]

3   If an emitter-base junction of a planar bipolar transistor is driven to avalanche breakdown during operation, degradation of the low current common emitter current gain will occur.[2,34,35] This degradation is caused by avalanche injection of carriers into the oxide and subsequent trapping, creating both oxide charge and interface traps.

4   Avalanche injection and trapping in the oxide occurring after avalanche breakdown of a high voltage planar diode causes an undesirable reduction in the avalanche breakdown voltage of the diode.[33,36,37]

5   Avalanche injection from the drain junction of a MOSFET across an oxide layer (~ 1000 Å thick) to charge up a polysilicon gate has been used as a memory element.[14,38,39]

11.2.1   Mechanism

(a)   *Physical Description*

Hot carriers in an avalanche plasma in the silicon can surmount the energy barrier between the silicon and the SiO$_2$. The first step toward injection is to produce an avalanche plasma at the silicon surface of the MOS capacitor. Avalanche breakdown in the silicon cannot occur in thermal equilibrium because the maximum band bending in thermal equilibrium is approximately the bandgap or 1 V, an insufficient voltage. To produce sufficient band bending, the silicon must be driven into deep depletion (see Section 4.5.1). Deep depletion is produced by applying a large amplitude sinusoidal ac signal of a frequency too high for minority carriers to follow. As the ac voltage is increased, band bending at the silicon surface increases, until the field at the silicon surface becomes high enough for avalanche breakdown. An avalanche plasma then forms once per cycle. A small fraction of the electrons in this plasma have sufficient energy to surmount the interfacial energy barrier. Once in the oxide, these electrons drift to the gate, driven by the electric field across the oxide. They then flow through the external circuit where they are detected. Thus an injected current pulse is detected once per cycle. We describe the frequency and the electric field dependence of the injected gate current based on this picture, following the work of Nicollian and Berglund.[6]

To study the injection mechanism, the injected current must be *emission limited*. That is, it is essential that trapping in the oxide be negligible. Otherwise, the injected current will decay with time because traps charge up, thus altering the electric field in the oxide. Electron trapping, being better understood, is much easier to control than hole trapping in the oxide. Oxide charging resulting from electron injection can be reduced to an insignificant level by using a dry oxide that is sodium ion free.* Therefore, we confine attention to electron injection from p-type silicon in this section. Hole injection results when an n-type silicon substrate is used as described in Section 11.3.

The injection process is illustrated in Fig. 11.1 for p-type silicon. Figure 11.1a shows one cycle of the electric field across the oxide produced by the large amplitude ac drive. Figures 11.1b–d show band bending in both oxide and silicon at various times during the ac cycle. At time $t = 0$ (Fig. 11.1b), the bands are assumed to be flat.† Then the bands begin to bend as

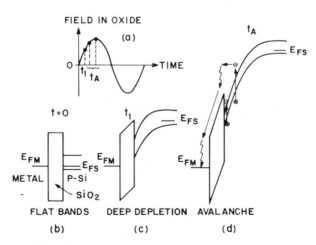

Fig. 11.1 Diagram illustrating the principles of avalanche injection of electrons: (a) one cycle of the oxide field is produced by the external ac drive; (b), (c), and (d) show band bending in the silicon at various times during the ac cycle. The energy E_{FM} is the Fermi level in the metal gate, and E_{FS} is the Fermi level in the silicon. After Nicollian and Berglund.[6]

* A sodium ion free dry oxide is: (1) an oxide that contains no more than 10^{10} sodium ions/cm^2, (2) an oxide that has been grown in dry oxygen containing no more than 1 ppm water vapor, and (3) an oxide in which any water that has diffused in from the room ambient after oxidation has been removed either by etching off or baking out.
† Residual interfacial oxide charge and work function differences are ignored for simplicity and in any case are not important for understanding the injection process.

the silicon surface tends towards inversion.* Deep depletion is reached at time $t = t_1$ (Fig. 11.1c). Band bending increases with further increase in field until avalanche breakdown, initiated by thermally generated carriers, occurs at the silicon surface. The silicon remains in the avalanche breakdown condition during the time interval t_A shown in Fig. 11.1a.

Figure 11.1d shows band bending at some time during the avalanche interval t_A. Electrons are excited from the valence band to the conduction band of the silicon by impact ionization. These electrons drift toward the Si–SiO$_2$ interface under the influence of the electric field in the depletion layer. Most electrons arriving at the interface have energies below the interfacial barrier height E_b, and these form an inversion layer. However, a few electrons have energies above E_b. Those not scattered back into the silicon at the barrier will enter the SiO$_2$. Once in the SiO$_2$, they become thermalized after several collisions with the silica lattice [the electron scattering mean free path at thermal energy in SiO$_2$ is about 34 Å;[40] see Section 11.4.2(b)] and drift toward the gate driven by the electric field across the oxide. Avalanche breakdown in the silicon is extinguished when the oxide field passes its peak value.† Thus a pulse of electrons is injected into the oxide once per cycle during the interval t_A.

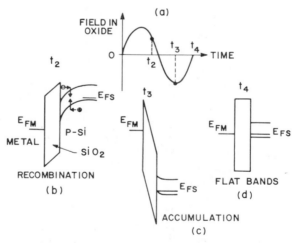

Fig. 11.2 Diagram illustrating band bending in the silicon and in the oxide after the avalanche portion of the cycle has been completed. After Nicollian and Berglund.[6]

*When the silicon bands are bent rapidly enough so that thermal equilibrium is not maintained, buildup of an inversion layer by thermal processes does not occur. Time constants for the buildup of an inversion layer by thermal generation in the silicon surface depletion layer are usually 0.1–1 sec in silicon, but the frequencies of the large signal ac drive are in the megahertz range, so that the silicon surface is driven into deep depletion (see Section 4.5.1).
†As soon as avalanche is initiated in the silicon, an inversion layer begins to form. This inversion layer charge reduces the electric field in the silicon. Therefore, to maintain the field in the silicon at the avalanche value, the applied field must increase.

Figure 11.2 illustrates the remainder of the cycle. At time $t = t_2$, when band bending is decreasing toward the flatband condition (Fig. 11.2b), electrons in the inversion layer generated during the interval t_A are injected back into the silicon, where they recombine. At time $t = t_3$ (Fig. 11.2c), the silicon is in accumulation. No carriers are injected from the silicon into the SiO$_2$ during this part of the cycle because the holes at the silicon surface, which are at lattice temperature, see a barrier greater than 4 eV. Thus the injected gate current is unidirectional. Finally, at $t = t_4$ (Fig. 11.2d) the initial flatband condition is reached and the cycle repeats.

(b) Analysis

We assume that: (1) the MOS capacitor is driven by a sinusoidal current source and (2) when an electron current is flowing in the silicon, electrons have a Maxwellian energy distribution with a characteristic temperature T_n depending only on the instantaneous electric field. Referring to Fig. 11.3 and ignoring diffusion, the instantaneous electron current density I_n at any point is given by

$$I_n = \int_0^\infty qv(E, F)M_c(E) \exp\left[-\frac{E - E_o}{kT_n(F)}\right] dE \qquad (11.1)$$

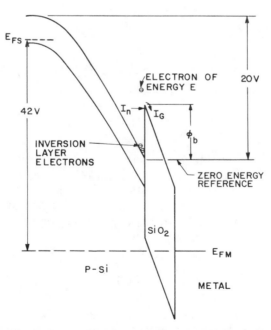

Fig. 11.3 Diagram illustrating magnitudes of the band bending during avalanche for an acceptor density of 5.6×10^{16} cm^{-3} and an oxide thickness of 1330 Å. After Nicollian and Berglund.[6]

where F is the time-dependent electric field, $v(E, F)$ is the drift velocity of an electron of energy E in a field F, $M_c(E)$ is the density of states in the silicon conduction band, E_o is a constant energy adjusted to give the proper density of electrons in the silicon, q is the electronic charge, and k is Boltzmann's constant. If $P(E)$ is the probability that an electron of energy E striking the Si–SiO$_2$ interface will enter the SiO$_2$, the injected current density I_i will be

$$I_i = \int_0^\infty qP(E)v(E, F)M_c(E) \exp\left[-\frac{E - E_o}{kT_n(F)}\right] dE \qquad (11.2)$$

where F is now defined as the time-varying electric field at the Si–SiO$_2$ interface. We assume that $P(E) = 0$ for E less than the (image force lowered) interfacial barrier height E_b and $P(E) = P_o$ for $E > E_b$. When variables have been changed and this condition applied, (11.2) becomes

$$I_i = P_o \exp\left(-\frac{E_b}{kT_n}\right) \int_0^\infty qv[(E + E_b), F]M_c(E + E_b) \exp\left(-\frac{E - E_o}{kT_n}\right) dE. \qquad (11.3)$$

Equation (11.3) can be written

$$I_i = I_n s \exp\left(-\frac{q\phi_b}{kT_n}\right) \qquad (11.4)$$

where $\phi_b = E_b/q$ and

$$s = P_o \frac{\displaystyle\int_0^\infty v[(E + E_b), F]M_c(E + E_b) \exp\left(-\frac{E}{kT_n}\right) dE}{\displaystyle\int_0^\infty v(E, F)M_c(E) \exp\left(-\frac{E}{kT_n}\right) dE}. \qquad (11.5)$$

If $P_o = 1$ and if both the drift velocity and the density of states are approximately energy independent, or if kT_n is much greater than the barrier height energy, s will be unity. In general, s will be a complicated function of E_b, T_n, and F, the type of carrier injected, and will be less than unity. Using (11.4), we can express the average or dc component of the current density flowing through the SiO$_2$ as

$$I_G = \frac{\omega}{2\pi} \int_0^{\pi/2\omega} I_i \, dt = \frac{\omega}{2\pi} \int_0^{\pi/2\omega} I_n s \exp\left(-\frac{E_b}{kT_n}\right) dt \qquad (11.6)$$

where ω is the angular frequency of the sinusoidal displacement current.

If the field dependencies of s, E_b, T_n, and I_n were known, the solution of (11.6) could be done numerically. For example, Bulucea[30,31] has considered the scattering factor s in more detail. However, it is more useful to make reasonable assumptions that allow a closed-form approximate solution to (11.6). This solution is valid when the ac voltage is of sufficient amplitude to produce avalanche breakdown in the silicon during a significant part of each cycle. It is assumed that I_n is nonzero only when the silicon is in avalanche breakdown. In the time interval when the silicon is in avalanche breakdown, its band bending is approximately constant at the avalanche breakdown voltage.* Therefore, during this time interval I_n is approximately equal to the displacement current density I_D in the oxide. Because silicon band bending is constant, s and T_n are constants that have the values s_o and T_{no} determined by the magnitude of the avalanche breakdown field in the silicon. The electric field dependence, and thus the time dependence of E_b, is small because the maximum change in barrier height expected for the oxide fields applied over the interval t_A is about 0.08 eV whereas the previously reported value of kT_n/q at avalanche is about 0.5 eV.[41] Therefore, ϕ_b is assumed constant at its value when I_n is a maximum. Then (11.6) becomes

$$I_G = s_o \exp\left(\frac{-q\phi_{bo}}{kT_{no}}\right)\frac{\omega}{2\pi}\int_{\pi/2\omega-t_A}^{\pi/2\omega} I_D\, dt \tag{11.7}$$

where ϕ_{bo} is the barrier height at the peak value of I_n (at the onset of avalanche breakdown in the silicon).

With the assumption of a sinusoidal displacement current, the electric field F in the oxide is

$$F = F_o \sin \omega t. \tag{11.8}$$

Then

$$I_D = \omega\epsilon_{ox}F_o \cos \omega t \tag{11.9}$$

where ϵ_{ox} is the dielectric permittivity of SiO$_2$ and F_o is the peak electric field in the oxide resulting from the ac excitation. If F_B is defined as the electric field in the oxide at the onset of avalanche breakdown in the silicon, (11.7) becomes

$$I_G = s_o \exp\left(-\frac{q\phi_{bo}}{kT_{no}}\right)\frac{\omega}{2\pi}\int_{[\arc\sin(F_B/F_o)]\omega^{-1}}^{\pi/2\omega} \omega\epsilon_{ox}F_o \cos \omega t\, dt$$

$$= \epsilon_{ox}(F_o - F_B)fs_o \exp\left(-\frac{q\phi_{bo}}{kT_{no}}\right) \tag{11.10}$$

where f is the frequency corresponding to ω.

*To keep displacement continuous across the interface, the field at the silicon surface increases slightly as the inversion layer charge density generated by the avalanche increases. However, for typical inversion layer charge densities (10^7 coul/cm^2) and typical inversion layer thickness (20 Å), the increased band bending required is only 20 mV. This is negligible compared to the total band bending of several volts at avalanche breakdown.

Intuitively, (11.10) can be derived as follows. From Gauss's law, $\epsilon_{ox}(F_o - F_B)$ is the total electron charge density arriving at the interface each cycle. A fraction $\exp(-q\phi_{bo}/kT_{no})$ of this charge has energy above the barrier ϕ_{bo}. Of this fraction, there is a probability s_o that the electrons will enter the SiO_2. The average current density I_G is just this injected charge per cycle multiplied by the frequency f.

In the previous discussion F_B was constant, equal to the field just required to produce avalanche breakdown in the silicon. However, a more accurate definition would set F_B equal to the electric field in the oxide at which electron current in the depletion layer equals displacement current in the oxide. At low frequencies, when all electrons recombine between avalanche periods, the two definitions coincide. This situation applies to the experiments to be described later in this section.

Equation (11.10) is expected to be accurate only over a limited range of frequency and oxide field. At higher frequencies, some electrons may not recombine before the next avalanche period begins. In this case F_B will become dependent on both frequency and F_o. In addition, significant dc current may flow at fields below F_B, current caused by electron multiplication in the silicon surface depletion layer, not described by (11.10). Although a quantitative treatment of such effects is not given here, it is evident that at a given F_o the frequency and electric field dependence of I_G predicted by (11.10) may become inaccurate when the frequency becomes comparable to or less than the reciprocal minority carrier lifetime. That is, if the frequency becomes too low, interface traps and bulk minority carrier effects will become important.

The use of avalanche injection for investigating oxide trapping effects depends on operating the MOS capacitor within the range where the simple relation (11.10) applies. Next, we find the range of frequency and oxide field over which (11.10) is accurate.

(c) Experiment

In this section we establish (11.10). To test (11.10) by experiment, I_G must be limited by hot carrier emission over the interfacial barrier, not by space charge in the oxide. The following two independent observations show that I_G is emission limited: (1) I_G at a given value of F_o is independent of oxide thickness in the range 500–2000 Å; and (2) I_G can be varied by shining light from a tungsten lamp on the capacitor.*

Figure 11.4 is a plot of I_G as a function of F_o measured at 0.95 MHz as

---

*Normally, the avalanche is initiated in the silicon by thermally generated carriers. To make I_G light sensitive, it is necessary to initiate the avalanche by light-generated carriers. This can be accomplished by either (1) using a frequency sufficiently low that interface traps, usually the dominant generation centers, are in equilibrium with the silicon at all times so that they do not generate or (2) to prevent the interface traps from filling with holes. This is done by superposing a bias that allows the ac to drive the silicon into depletion but not as far as flatbands or accumulation.

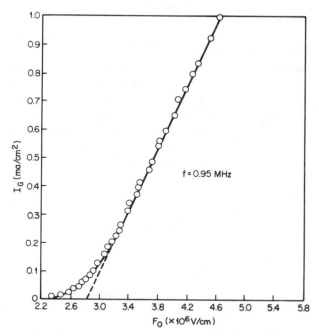

Fig. 11.4 Plot of I_G versus F_o measured at 0.95 MHz. The sample measured had an acceptor density of 5.6×10^{16} cm^{-3} and an oxide thickness of 1330 Å. Frequency was measured on an oscilloscope. After Nicollian and Berglund.[6]

described in Section 12.11.1(a). The current measured is in the direction corresponding to electron flow from the silicon to the gate. To quantitatively fit (11.10), I_G must be linear in F_o and linear in f, extrapolate to the expected value of F_B when $I_G = 0$, and have the expected slope. Figure 11.4 shows that in the high field region, I_G varies linearly with F_o. At $I_G = 0$, the linear position of the curve extrapolates to a field of 2.83×10^6 V/cm. The expected value of F_B is $F_B = (\epsilon_s/\epsilon_{ox})F_s$, where ϵ_s is the dielectric permittivity of silicon and F_s is the avalanche breakdown field in the silicon. To match the extrapolated field, the quantity V_{FB}/x_o must be added to F_B, where V_{FB} is the flatband voltage and x_o is the oxide thickness. The quantities F_s and V_{FB} are obtained from a measurement of capacitance versus pulse bias on the same capacitor by the method described in Section 12.5.6. For the capacitor in Fig. 11.4, the values found this way were $F_s = 1.65 \times 10^6$ V/cm and $F_B + V_{FB}/x_o = 2.18 \times 10^6$ V/cm. This result is within 30% of the extrapolated field value of 2.83×10^6 V/cm shown in Fig. 11.4 and is within the precision expected of the measurements.

In Fig. 11.4 an abrupt current threshold is not observed; rather, there is a current tail extending to fields well below F_B. This current tail arises from electron multiplication in the silicon surface depletion layer as electrons are heated by the applied electric field from lattice temperature to the temperature of the avalanche plasma. The current tail in the low field region is not described by (11.10).

Figure 11.5, which is a plot of I_G/f versus F_o measured at 0.95 and 4.6 MHz, shows that the frequency dependence is linear in the high field region as predicted by (11.10). The current density in the low field region has a nonlinear frequency dependence because of generation and recombination in interface traps and bulk centers in the silicon surface depletion layer and by minority carrier diffusion rates.

The slope of the linear portion of the characteristic in Fig. 11.5 predicted by (11.10) is

$$m = \epsilon_{ox}s_o \exp\left(-\frac{q\phi_{bo}}{kT_{no}}\right). \tag{11.11}$$

Equation (11.11) shows that the slope depends on T_{no}. Solving (11.11) for T_{no}, we obtain

$$T_{no} = \frac{q\phi_{bo}}{k \ln(\epsilon_{ox}s_o/m)}. \tag{11.12}$$

The value of ϕ_{bo} in (11.12) is the zero field barrier height of 3.10 eV between the silicon and SiO_2 conduction bands as determined from internal photoemission measurements described in Section 10.3 less the maximum barrier height lowering of 0.2 eV to obtain $\phi_{bo} = 2.90$ eV. Assuming as a reasonable first approximation that s_o in (11.10) is unity, $T_{no} = 5200°K$, which is in reasonably good agreement with the previously published value of 6000°K.[41] Thus this slope is the value expected.

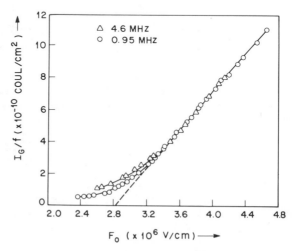

Fig. 11.5 Plot of I_G/f versus F_o measured at 0.95 and 4.6 MHz on the same sample as in Fig. 11.4. After Nicollian and Berglund.[6]

(d) Doping Density Range

Avalanche injection experiments are practical over only a limited range of silicon doping density. The upper end of this range is about 10^{18} cm^{-3}. Beyond this upper limit, interband tunneling rather than avalanche breakdown occurs in the silicon. Two factors determine the lower end of the range: (1) avalanche breakdown at the gate edge and (2) breakdown of the oxide during the accumulation portion of the ac cycle ($t = t_3$ in Fig. 11.2c).

Breakdown over the entire gate area is necessary to the analysis of avalanche breakdown experiments. Edge breakdown should be avoided. Edge breakdown depends on the relative sizes of the oxide and depletion layers as discussed in Section 9.3.2. Figure 9.5 shows the lowest doping density for which edge breakdown can be avoided for a given oxide thickness.

The field across the oxide is highest during the accumulation part of the cycle, when all the applied voltage appears across the oxide. During the avalanche portion of the cycle, the field in the oxide is lower because the ac voltage divides between the oxide and silicon depletion layers. As doping density is decreased, avalanche breakdown voltage increases. For sufficiently low doping density, at voltages sufficiently high to cause avalanche, the dielectric breakdown strength of the oxide can be exceeded during the accumulation portion of the cycle. Dielectric breakdown destroys the sample.

Dielectric breakdown cannot be avoided by eliminating the accumulation portion of the cycle, because this portion of the cycle is needed to recombine the inversion layer carriers created during the avalanche portion of the cycle. If the inversion layer charge is not completely recombined on one cycle, a higher ac voltage will be required to initiate avalanche on the next cycle. Dielectric breakdown can be ameliorated somewhat by (1) using a thick oxide and (2) superposing the ac on a bias chosen to reduce the maximum voltage across the oxide during the accumulation portion of the cycle.* However, this bias must not be so high as to prevent complete recombination of the inversion layer carriers.

Doping density should be made high, or a thin epitaxial layer on a thick degenerate substrate should be used, to minimize series resistance. Series resistance, described in Section 5.7, should be low to obtain an accurate high frequency C-V curve, used in the measurement of flatband voltage (see Section 10.5) and interface trap density (see Section 8.2).

11.3 AVALANCHE INJECTION OF HOLES

For an n-type silicon substrate, avalanche results in hole injection into the SiO$_2$. The mechanism is that described for electrons in Section 11.2.1,

---

*This superposition makes the ac voltage asymmetric about zero.

except holes are injected during the negative half of the cycle. One major problem in interpreting hole injection measurements is that the injected gate current is not entirely a hole current. Besides the flow of holes from the silicon to the gate, a large electron current flows from the gate to the silicon. This electron current is not important in the avalanche injection of electrons, for reasons explained later.

11.3.1 Edge Effect

When the substrate is n-type, electron current is caused by field emission from the edges of the gate electrode. Because of the sharpness of the gate edge, electric fields at the edge can be much higher ($\sim 10^7$–10^8 V/cm over a few tens of angstroms from the gate edge) than over the gate area ($\geq 10^6$ V/cm) at a given bias.

The electric field at the gate edge will be determined by the gate edge geometry (it is independent of x_o) only if the potential applied to the gate does not build up along the oxide surface beyond the gate. Potential buildup beyond the gate can be caused by ion migration if the ambient is humid.* Even in a humid ambient the time constant for potential buildup along the oxide surface is the order of several seconds to several minutes. Thus, for frequencies above 500 kHz, normally used in avalanche injection experiments, there will be no time for potential spreading along the oxide surface during either half of the ac cycle. Therefore, edge geometry will determine the field, and field emission will occur.

To measure the electron component of current emitted from the gate edge independent of the hole current emitted by the silicon, degenerate n-type or p-type silicon can be used to prevent avalanche breakdown at the high voltages required for field emission. Figure 11.6 demonstrates that electrons are injected from the gate edges under ac excitation (no current is detectable under dc excitation at voltages equal to the peak ac voltage). The method of measuring injected current under ac excitation is described in Section 12.11.

To demonstrate that this current is emitted from the gate edges, several gates can be used, each with a different perimeter and area. For example, the gate can be deposited in the form of a ring. Each ring has the same outer diameter, but the inner diameter is made successively smaller, thereby increasing the area and decreasing the perimeter. Figure 11.6 shows plots of injected current as a function of peak oxide field with gate perimeter and area as parameter. (The peak field is just the peak ac voltage divided by x_o.) At a given peak oxide field, the injected current is greatest for the largest perimeter and smallest area, demonstrating that the current is injected from the gate edges.

---

*This potential buildup invariably will occur with dc excitation, except when the experiment is done in a high vacuum and all the adsorbed water has been driven off the oxide surface.

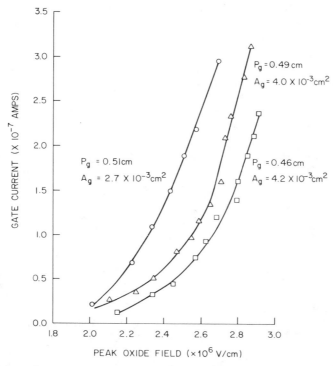

Fig. 11.6 Injected gate current as a function of peak oxide field with gate perimeter (P_g) and gate area (A_g) as parameters measured at 500 kHz on degenerate n-type silicon. Oxide thickness $x_o = 1000$ Å.

To add to the injected hole current, the gate edge current must be carried by electrons and, therefore, be injected during the negative half of the cycle. Figure 11.7 shows that gate edge current emission occurs during the negative half cycle of the ac drive. This curve is obtained by superposing an ac voltage at a fixed peak amplitude on a bias that is varied. As the bias is made more positive, the injected current is extinguished; that is, the injected current drops to zero as the peak ac field drops below a threshold value on each negative half-cycle. We conclude that edge current is injected once per cycle when the ac drives the gate negative. Electrons must be the carrier because injection occurs when the gate is negative and the measured current was found to be in the direction expected for negative charge flowing from the gate to the silicon. Thus edge injected electron current is a substantial fraction of the hole current observed in avalanche injection experiments because the peak oxide fields at which edge injection occurs are identical to the peak oxide fields during avalanche injection of holes. Also, edge injected currents are comparable to the injected currents observed during avalanche injection of holes.

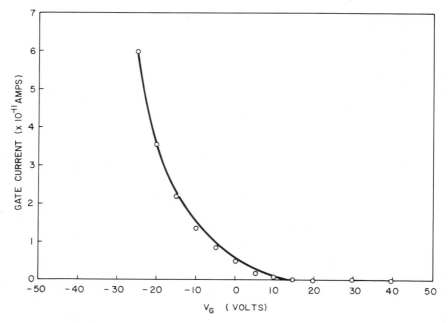

Fig. 11.7 Injected gate edge current as a function of V_G superposed on a peak ac voltage of 26 V measured at 500 kHz on the same sample as in Fig. 11.6.

Electron injection from the gate edges does not play a significant role in electron avalanche injection measurements because the electron current injected from the gate edges during the negative half cycle of the ac drive is more than 3 orders of magnitude smaller than the electron current injected from the silicon during the positive half of the ac drive at all values of peak oxide field. It is not feasible to correct for edge injection in a hole avalanche injection experiment because electron injection from the gate edge is not controllable. Despite the electron current, net positive charging is observed, thus indicating hole trapping. An increase in interface trap density also is seen.[4]

11.4 DETERMINATION OF THE TRAP DISTRIBUTION

Either the density of traps as a function of distance in the oxide or the trap density and its centroid must be known before the effects of oxide traps on device characteristics can be quantitatively determined. Two of the most commonly used methods for finding the trap distribution are the etch-off method and the photo I-V method, which is described in Section 11.4.2. A number of other profiling methods have been applied, mainly to profiling chlorine in the oxide and to determining the extent of the region between silicon and its thermally grown oxide in which the chemical composition changes from silicon to SiO_2. A detailed description of these methods is beyond the scope of this book.

11.4.1 Etch-Off Methods

The spatial distribution of traps or impurities $n(x)/cm^3$ in the oxide can be determined by the etch-off method. The first step is to etch off a thickness decrement of oxide small compared to the spatial extent of the trap or impurity distribution in the oxide and to measure either the number of traps in the etched-off oxide decrement or the number remaining in the oxide. The trap or impurity distribution is profiled by repeatedly etching off oxide decrements of known thickness and measuring the amount of traps or impurities etched off or remaining until the entire distribution is etched away.

(a) The Radioactive Methods

Radioactive methods for obtaining impurity or trap profiles in SiO_2 employ radioactive tracers or neutron activation.

In the radioactive tracer method, the radioactive form of the impurity must be incorporated into the oxide, usually by diffusion. Either the radioactivity in the etching solution is counted after each increment of oxide thickness is etched off, or the radioactivity left in the remaining oxide is counted. This method has been used for measuring sodium, chlorine, and water in SiO_2. The count is proportional to the volume density of the radioactive impurity. The oxide thickness corresponding to this measured impurity density needed to obtain the distribution must be determined from the known etch rate and etch-off time or from some other measurement such as from C-V measurements.

In neutron activation analyses, the oxide is exposed to thermal neutrons in a pile. Normal ^{23}Na in the oxide is activated to ^{24}Na by neutron capture. The activated ^{23}Na can be detected because it emits γ-rays at 1.4 and 2.75 MeV. The sodium profile is measured by etching-off the oxide in steps and measuring the ^{24}Na concentration either in the etching solution or in the remaining oxide with a scintillation counter.

The advantages of the radioactive methods are: (1) the impurity species is known; (2) it does not matter whether the impurity is charged or neutral; and (3) the impurity distribution at both the air-SiO_2 and Si–SiO_2 interfaces can be determined. A disadvantage of the radioactive tracer method is that additional impurity must be incorporated into the oxide beyond what may be introduced during a particular fabrication step. Neutron activation analysis does not suffer from this disadvantage because the impurity already in the oxide is made radioactive. However, neutron activation analysis has been used primarily on sodium and may not work as well for other impurities.

(b) The C-V Method

The C-V method, in which flatband voltage is measured after each successive increment of oxide is etched off, is useful for impurity profiling only when the impurity is charged and located near the air-SiO$_2$ interface. Examples where the C-V method is suitable are profiling alkali metals and traps charged by avalanche injection or photoinjection located near the air-SiO$_2$ interface.

Only charge distributions near the air-SiO$_2$ interface can be profiled by the C-V method because the gate short-circuits to the silicon substrate once the oxide is etched to within about 200 Å of the interface.[42] The reason for this short-circuiting is that certain localized regions, related to contamination, etch faster than the surrounding regions. After the oxide is etched to within 200 Å of the silicon, some of these localized regions etch all the way to the silicon, allowing the gate metal to fill the resulting hollow pipe or pinhole and contact the silicon.

Another problem with the C-V method is the need for a metal gate. A new gate can be deposited after every oxide etch-off step and then etched off after each measurement. However, metal atoms can penetrate 20 Å or more into the oxide, possibly perturbing the oxide charge distribution. This perturbation could be significant because 20 Å is a significant fraction of the incremental oxide thickness (50–100 Å) etched off in each step. To avoid this problem, the mercury probe described in Section 12.8.3 can be used as the gate electrode.

To obtain the charge profile, the oxide thickness and the charge density must be measured after each etch-off step. First, consider measurement of the oxide thickness remaining after each etch-off step. The gate area of a mercury probe is unknown, so oxide thickness cannot be determined simply from a measurement of the oxide layer capacitance. Oxide thickness can be obtained from the slope of the linear portion of a $(C_{ox}/C_{HF})^2$ versus V_{G0} curve, where C_{HF} is the high frequency capacitance and V_{G0} is the gate bias corrected for interface trap voltage stretchout as described in Sections 9.4.1(b) and 10.5.* The slope m of the linear portion of this curve depends on x_o and the uniform doping density N_B according to the relation[43]

$$x_o = \left[\frac{2\epsilon_{ox}^2}{q\epsilon_s} \right]^{1/2} (mN_B)^{-1/2} = 1.17 \times 10^3 \, (mN_B)^{-1/2} \qquad (11.13)$$

obtained by solving (9.19)† for x_o. The value of N_B must be known or determined independently, as described in Section 9.4, to determine x_o from (11.13). For this determination, a metal gate of known area is

---

*The ratio (C_{ox}/C_{HF}) is the same as the ratio of the measured device capacitances because the device area cancels.

†Equation (9.19) first must be divided by C_{ox}^2.

evaporated on the oxide prior to etch off in a region of the wafer that does not disturb the properties under study.

This procedure requires measuring both a high frequency and a low frequency C-V curve to obtain a value of m independent of interface trap voltage stretchout. Because the etch-off method is a relatively coarse profiling method, some error can be tolerated in m in exchange for a reduced amount of measurement labor. For low interface trap level densities ($\leq 10^{10}$ cm^{-2}eV^{-1}), m can be obtained with tolerable error from the linear part of a $(C_{ox}/C_{HF})^2$ versus V_G curve because $V_G \approx V_{G0}$.

Where charge trapping was locally induced by avalanche injection or by photoinjection using the initially evaporated metal gate, it is important to place the mercury probe on the same charged area after each oxide etch-off step.* Because the charged area will etch at a different rate than the adjacent uncharged areas, the charged area can be distinguished visually. A negatively charged area will etch more slowly than the surrounding area,[43] whereas a positively charged area will etch more rapidly.[44] As an example of the magnitude of the differences in etch rate between charged and uncharged regions, if 400 Å of uncharged oxide is etched away, the negatively charged region will be about 50 Å thicker than the uncharged region. The negative charge being located near the air-oxide interface is itself etched away and has a diminishing influence on the etch rate as etching proceeds.

The volume charge density can be obtained from a measurement of flatband voltage as a function of x using (10.6). The measurement of flatband voltage is described in Section 10.5. Given an experimental set of V_{FB} versus x data, (10.6) can be inverted to yield $n_o(x)$. This procedure is accurate if the oxide charge is uniformly distributed or varies slowly over the decremental oxide thickness etched off in each step. However, if the oxide charge distribution varies rapidly over the decremental oxide thickness, this procedure is unreliable. The size of the decremental oxide thickness that can be etched off reproducibly is limited by the degree of control that can be exercised over the etch rate. Chemical etching in buffered hydrogen flouride is the etching method least likely to perturb the oxide charge distribution. Oxide decrements in the 50–100 Å range are about the smallest possible for easy control with chemical etching.

Oxide charge located near the Si–SiO$_2$ interface and not removed by etching will be measured as a constant charge density. If no charge is removed before shorting occurs, the oxide charge must be closer to the Si–SiO$_2$ interface than the remaining oxide thickness just prior to the last etch-off step when the gate shorts to the silicon.

In such an etch-off experiment, Deal et al.[45] found that oxide fixed charge and charged interface traps were within 200 Å of the Si–SiO$_2$ interface.

---

*Any perturbation of the oxide charge will be restricted to the first increment (~ 50–100 Å thick) of oxide etched off.

11.4.2   Photo *I-V* Method—Basics

For studying trapping phenomena or charged impurity centers in the oxide, the photo *I-V* method is both nondestructive and more accurate than the etch-off method. The photo *I-V* method is based on internal photoemission from the metal and the silicon. An electron excited by the absorption of ultraviolet light in either the metal gate or the silicon of an MOS capacitor will be injected into the SiO_2, provided that it can *reach* and *surmount* the interfacial energy barrier.

The ability of an electron to reach the barrier depends on the probability of scattering along the way, which, in turn, depends on the distance of the energy barrier from the injecting interface. The further the barrier maximum is from the interface, the greater the probability of scattering. The ability of an electron to surmount the barrier depends on the barrier height and the energy of arriving electrons.

Both the barrier position and the barrier height are affected by oxide charge and the gate bias. Therefore, oxide charge information is contained in two ways in the photo *I-V* curves: (1) through the position of the barrier maximum and (2) through the maximum barrier height. The photoinjection can be performed so that one or the other effect is dominant.

The use of photoemission to determine barrier height was discussed in Section 10.3. In barrier height determination, the only gate bias dependence to be considered was image force barrier lowering. To apply photoemission to the determination of oxide charge distributions, it is necessary to consider, in addition, the bias dependence of the photocurrent caused by electron scattering. We consider electron scattering and derive an expression for the photocurrent that takes into account the gate bias dependence caused by both scattering and image-force lowering. Then, in Section (*b*), we use this expression to fit an experimental photo *I-V* curve. This curve is measured in such a way that scattering rather than image-force barrier lowering dominates the gate bias dependence and determines the electron scattering length.

For clarity, the expression for the photo *I-V* characteristic is derived first in the absence of oxide charge. The characteristic derived assuming no oxide charge accurately describes a photo *I-V* characteristic measured on an MOS capacitor having oxide charge, provided that the oxide charge is located very near an interface. Such is the case for device grade thermally grown oxides. For oxide charge very near an interface, the centroid and the maximum extent of this charge can be estimated.

Deeper oxide charge distributions can be measured directly and profiled in some cases. To extract information about deeply distributed oxide charge from photo *I-V* characteristics, the expression for the bias dependence of the photocurrent derived in Section (a) below must be extended to include oxide charge. This extension is made in Section (c).

Powell and Berglund,[46] who introduced the photo *I-V* method for oxide charge determination, used it to find how close oxide fixed charge and charged interface traps were to the Si–SiO$_2$ interface. They also used the photo I-V method to profile the oxide charge distribution near an Au–SiO$_2$ interface produced by the capture of photoinjected electrons by the water-related traps described in Section 11.5.1. Sixt et al.[47] attempted to determine the profile of positively charged cesium ions implanted into the SiO$_2$ layer of an MOS capacitor. DiMaria[48] used photoinjection from the metal and the silicon to determine the density and centroid of a thin sheet of oxide charge located far beyond the interfaces. Brews[49] discussed the limitations of the photo I-V method for measuring charge distributions. This section follows the work of Berglund and Powell,[50] Powell and Berglund,[46] Powell,[51] Brews,[49] and DiMaria.[48]

(a) Photo I-V Characteristics—No Oxide Charge[50]

To calculate the gate bias dependence of the photocurrent taking both image force barrier lowering and scattering into account, it is necessary to obtain the barrier height and the position of the barrier maximum with respect to the interfacial plane as functions of gate bias in the absence of oxide charge.

Figure 11.8 is an energy-distance diagram at the metal-oxide interface. The energy-distance diagram at the Si–SiO$_2$ interface is identical, but the metal-oxide interface is shown because distance will be measured from this interface. Figure 11.8 shows the oxide conduction band or barrier height, the energy resulting from an applied electric field, and the image force energy, all as functions of distance into the oxide in the absence of oxide charge. Referring to Fig. 11.8, the barrier height at zero applied field is

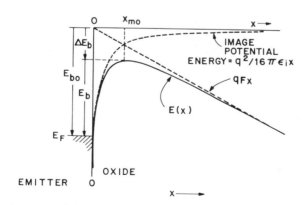

Fig. 11.8 Energy-band diagram between the emitter surface and SiO$_2$. The zero field barrier height at $x = 0$ is E_{bo}. Barrier height is lowered by the presence of oxide charge and an applied field, making the effective barrier height E_b. The barrier height lowering ΔE_b is due to the combined effects of the electric field, consisting of the contributions from oxide charge at $x > x_{mo}$ and an applied bias and the image force. After Sze.[52]

$\phi_{bo} = E_{bo}/q$. This zero field barrier height includes the work done by an electron against the attraction of its image charge in the injecting electrode. Suppose that we are to move an electron to infinity from a point located a distance x from the injecting interface. The work remaining to be done against the image charge is then $q^2/(16\pi\epsilon_i x)$.* Therefore, the energy needed by an electron in the injecting electrode to reach point x is $\phi_{bi}(x)$, given by $q\phi_{bi}(x) = q\phi_{bo} - q^2/(16\pi\epsilon_i x)$. As $x \to \infty$, $\phi_{bi} \to \phi_{bo}$.

If a spatially uniform field F is applied to assist electrons in leaving the injecting electrode, then at a distance x the electron has gained an energy qFx from the field, also shown in Fig. 11.8. If this energy, plus the energy acquired from a photon of energy $h\nu$, equals $q\phi_{bi}(x)$, the electron can travel from the injecting electrode as far as x, provided that no scattering intervenes. Thus, for the electron to reach x, the photon energy must satisfy the relation

$$h\nu + qFx = q\phi_{bi}(x)$$

or, rearranging terms

$$h\nu = q\left[\phi_{bo} - \frac{q}{(16\pi\epsilon_i x)} - Fx\right]$$
$$\equiv E(x). \tag{11.14}$$

The barrier energy $E(x)$ from (11.14) is plotted as the solid line in Fig. 11.8. This curve has a maximum at a distance x_{mo}. This maximum means that once an electron reaches x_{mo}, it can escape into the oxide from the injecting electrode, assuming that it is not scattered back.

The value of x_{mo} is obtained by setting the derivative of $E(x)$ at $x = x_{mo}$ to zero, to find

$$x_{mo} = \left(\frac{q}{16\pi\epsilon_i F}\right)^{1/2}. \tag{11.15}$$

The least photon energy needed for injection is obtained by substituting (11.15) into (11.14) to obtain

$$h\nu = q\left[\phi_{bo} - \left(\frac{qF}{4\pi\epsilon_i}\right)^{1/2}\right]. \tag{11.16}$$

Thus, with an applied field, the zero field barrier height is reduced by an amount $\Delta E_b/q = \Delta\phi$, where

$$\Delta\phi = \left(\frac{qF}{4\pi\epsilon_i}\right)^{1/2}. \tag{11.17}$$

---

*The image force permittivity ϵ_i is determined experimentally by fitting the oxide field dependence of the photocurrent [see Section 10.3.3(c)]. The experimental value found for SiO$_2$ in Chapter 10 corresponds to the optical dielectric constant of 2.2, lower than the dc dielectric constant of 3.9. Presumably, ϵ_i is the optical value of the permittivity because an emitted electron reaches the barrier maximum at x_{mo} too quickly for the SiO$_2$ to fully screen the image-force interaction.

The image-force barrier lowering $\Delta\phi$ represents the saving in escape energy made possible by the applied field because the image force must be combatted only to a distance x_{mo}, and not to infinity.

The field F can be related to gate bias by solving Poisson's equation. Consider the simple case of no oxide charge; then

$$F = \frac{V_G - W_{ms} - \psi_s}{x_o} \tag{11.18}$$

where W_{ms} is the work function difference between the gate metal and the silicon described in Section 10.4. We assume degenerately doped silicon, so $\psi_s \approx 0$. Substituting (11.18) with $\psi_s = 0$ into (11.15) gives the gate bias dependence of x_{mo}, which is

$$x_{mo} = \left(\frac{qx_o}{16\pi\epsilon_i(V_G - W_{ms})}\right)^{1/2}. \tag{11.19}$$

Equation (11.19) shows that the position of the barrier maximum moves closer to the interface as V_G is increased.

Taking electron scattering into account, it is shown in Appendix XIII that the photocurrent is given by

$$I_o = A(h\nu)[h\nu - E_{bo} - q\Delta\phi]^p \, \exp\left(-\frac{x_{mo}}{l}\right) \tag{11.20}$$

where the form of the preexponential factor was derived in Section 10.3, l is the electron scattering length in SiO_2, and p depends on the hot electron energy distribution in the emitter. From (10.41), the exponent of the energy dependent terms of the photocurrent is 3 for electron emission from the silicon valence band, and from (10.42), the exponent is 2 for electron emission from the conduction band of degenerate n-type silicon or from the metal. From (11.17), (11.18), with $\psi_s = 0$, and (11.19), (11.20) becomes the following explicit function of gate bias

$$I_o = A(h\nu)[h\nu - E_{bo} + K(V_G - W_{ms})^{1/2}]^p \, \exp\left(-\frac{V_l}{V_G - W_{ms}}\right)^{1/2} \tag{11.21}$$

where $K = (q^3/4\pi\epsilon_i x_o)^{1/2}$ and $V_l = qx_o/16\pi\epsilon_i l^2$.

(b) Electron Scattering Length[50]

The electron scattering length l in SiO_2 can be extracted from a measured photo I-V curve. This is done by fitting (11.21) to a measured photo I-V curve. The shape of this curve is determined by the simple parameter l because $A(h\nu)$ in (11.21) is just a scale factor. To see how sensitive a photo I-V curve is to l, we calculate I_o as a function of $V_G - W_{ms}$ from (11.21) with l as parameter. Figure 11.9 shows a family of photo I-V curves calculated from (11.21) for electron emission from the valence band of n-type degenerate silicon. The value of $A(h\nu)$ was selected arbitrarily to make all the curves coincide at $V_G - W_{ms} = 30$ V. The photo

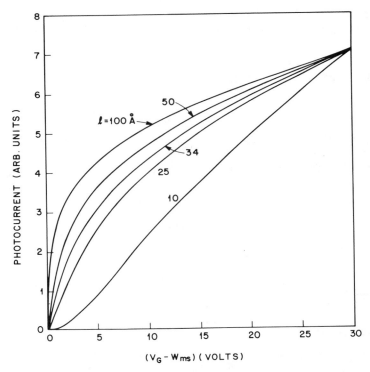

Fig. 11.9 Photocurrent in arbitrary units as a function of V_G-W_{ms} with electron scattering length l as parameter calculated from (11.21) for $x_o = 2450\,\text{Å}$, $p = 2$, $h\nu = 5\,\text{eV}$, $\epsilon_i = 1.97 \times 10^{-13}\,\text{F/cm}$, $A = 7.8\,\text{V}^{-1}\,\text{sec}^{-1}$, and no oxide charge.

I-V curves are most sensitive to l when the exponential term due to scattering in (11.21) dominates the bias dependence of I_o. Scattering dominates the bias dependence of I_o in (11.21) when $(h\nu - E_{bo}) > K(V_G - W_{ms})^{1/2}$. A photon energy of 5 eV is sufficient for this to occur because it is well above the barrier height of $\phi_{bo} = 4.16\,\text{eV}$ (see Section 10.3) for electron emission from the silicon valence band, and the barrier lowering is $K(V_G - W_{ms})^{1/2} = 0.052\,\text{eV}$ for $V_G - W_{ms} = 0.9\,\text{V}$ and 0.25 eV for $V_G - W_{ms} = 18.9\,\text{V}$ in our illustrative example. Figure 11.9 shows that the photo I-V characteristics are sensitive enough to l under these conditions to permit its extraction reliably, particularly at low values of $V_G (< 10\,\text{V})$.

Figure 11.10 shows a measured (points) and a calculated (solid curve) photo I-V characteristic. The experimental points were fitted to the theoretical curve with $V_l = 3.24\,\text{V}$. A value of $A(h\nu) = 8.4\,\text{V}^{-1}\,\text{sec}^{-1}$ was used to scale the calculated curve to coincide with the experimental points. Thus using (11.21) to calculate l from the value of V_l and to fit the shape of the experimental curve, a value of $l = 34\,\text{Å}$ was found to be the scattering length in SiO_2 for thermal electrons.

The deviation of the experimental points from the theoretical curve in Fig. 11.10 at about 29 V is due to expected deviations from the image force theory at high values of V_G because x_{mo} becomes small (< 10 Å). Also, other effects such as tunneling and electrode field penetration (band bending) that have not been taken into account in the theory may come into play at high values of V_G. However, this deviation at high values of V_G is of no consequence because the low V_G region is where the fit must be obtained.

In the determination of barrier height described in Section 10.3.3, electron photoemission is measured at high values of V_G, so that scattering is unimportant. With zero charge (see Fig. 11.11), x_{mo} varies slowly with V_G and x_{mo} is small compared to l ($l = 34$ Å in SiO$_2$) in the high gate bias range. Therefore, $\exp(-x_{mo}/l) \approx 1$, and electron scattering can be neglected, making (11.20) or (11.21) assume the same form as (10.41) or (10.42) depending on p.

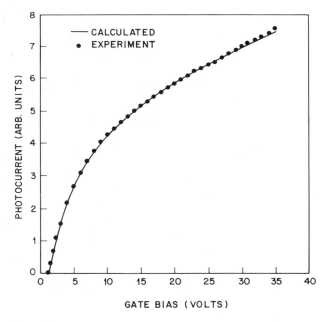

Fig. 11.10 Photocurrent in arbitrary units as a function of gate bias comparing the calculated (solid lines) and measured (points) gate bias dependence of electron current photoinjected from the silicon into a 2450 Å thick SiO$_2$ layer. The experimental points were measured on a degenerate n-type silicon sample with a photon energy of 5.0 eV. The structure was Au–SiO$_2$–Si. The solid curve was calculated from (11.21) using $p = 2$, $W_{ms} = 1.1$ V for the Au–Si work function difference $\epsilon_i = 1.97 \times 10^{-13}$ F/cm, and $E_{bo} = 4.16$ eV. The remaining parameters needed to fit the experimental points are described in the text. After Berglund and Powell.[50]

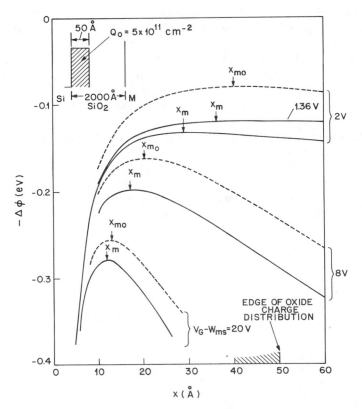

Fig. 11.11 Diagram illustrating $\Delta\phi$ as a function of x calculated from (11.26) with V_G-W_{ms} as parameter for no oxide charge (dashed lines), and also for the charge distribution shown in the insert (solid lines). The insert shows a rectangular positive oxide charge of 5×10^{11} cm^{-2} extending 50 Å into the oxide.

(c) *Photo I-V Characteristics—With Oxide Charge*[49,50]

We derive an expression for the photocurrent as a function of gate bias taking oxide charge into account. Consider the injecting electrode to be located at $x = 0$ and the oxide to lie in the region $0 \le x \le x_o$. Thus the Poisson equation in the oxide when a charge density of $n_o(x)$/cm^3 is present is

$$\frac{d^2V}{dx^2} = -\frac{qn_{ot}(x)}{\epsilon_{ox}} \tag{11.22}$$

subject to the boundary conditions

$$V(0) = 0 \quad and \quad V(x_o) = V_G - W_{ms}. \tag{11.23}$$

The solution of (11.22) at the emitter with the boundary condition of (11.23) and degenerate n-type silicon is (see, e.g., Appendix in Ref. 49)

$$V(x) = (V_G - W_{ms})\frac{x}{x_o} + \frac{q}{\epsilon_{ox}}\left[x\int_x^{x_o} n_o(x')dx' + \int_0^x x'n_o(x')dx' - \frac{x}{x_o}\int_0^{x_o} x'n_o(x')dx'\right].$$

(11.24)

Differentiating (11.24), the electric field at $x = x_m$ is

$$F(x_m) = \frac{V_G - W_{ms}}{x_o} + \frac{q}{\epsilon_{ox}}\left[\int_{x_m}^{x_o} n_o(x')dx' - \int_0^{x_o} \frac{x'}{x_o} n_o(x')dx'\right]$$

(11.25)

where x_m is the barrier height position with respect to the emitter surface in the presence of oxide charge. The image-force barrier lowering at the emitter is [see (11.14) for a uniform field]

$$\Delta\phi \equiv \phi_{bo} - \phi_b = V(x_m) + \frac{q}{16\pi\epsilon_i x_m}$$

(11.26)

where $V(x_m)$ is (11.24) evaluated at $x = x_m$. Finally, the barrier position in the presence of oxide charge, obtained the same way as (11.15), is

$$x_m = \left[\frac{q}{16\pi\epsilon_i F(x_m)}\right]^{1/2}.$$

(11.27)

To obtain the bias dependence of the photocurrent, we first must evaluate (11.26). Solving (11.25) for $V_G - W_{ms}/x_o$ using $F(x_m) = q/16\pi\epsilon_i x_m^2$ from (11.27), we obtain

$$\frac{V_G - W_{ms}}{x_o} = \frac{q}{16\pi\epsilon_i x_m^2} - \frac{q}{\epsilon_{ox}}\left[\int_{x_m}^{x_o} n_o(x')dx' - \frac{1}{x_o}\int_0^{x_o} x'n_o(x')dx'\right].$$

(11.28)

Substituting (11.28) into (11.24) for $x = x_m$ yields

$$V(x_m) = \frac{q}{16\pi\epsilon_i x_m} + \frac{q}{\epsilon_{ox}}\int_0^{x_m} x'n_o(x')dx'.$$

(11.29)

Substituting (11.29) into (11.26) gives

$$\Delta\phi = \frac{2q}{16\pi\epsilon_i x_m} + \frac{q}{\epsilon_{ox}}\int_0^{x_m} x'n_o(x')dx'$$

(11.30)

or

$$\Delta\phi = \frac{2q}{16\pi\epsilon_i x_m} + \frac{Q_o}{\epsilon_{ox}}[1 - f(x_m)]\bar{x}$$

(11.31)

where $Q_o = q\int_0^{x_o} n_o(x')dx'$ is the total oxide charge density, $f(x_m) =$

$q \int_{x_m}^{x_o} n_o(x')dx'/Q_o$ is the fraction of the total charge that lies beyond x_m, and $\bar{x} = \int_0^{x_m} x' n_o(x')dx'/\int_0^{x_m} n_o(x')dx'$ is the centroid of the charge distribution. Using (11.31), we can express the photocurrent [see (11.20)] as

$$ I = A(h\nu)\left\{ h\nu - E_{bo} - \frac{2q^2}{16\pi\epsilon_i x_m} - qQ_o[1 - f(x_m)]\frac{\bar{x}}{\epsilon_{ox}} \right\}^p \exp\left(-\frac{x_m}{l}\right). \quad (11.32) $$

Because the photocurrent in (11.32) is a function of x_m, which, in turn, is a function of $F(x_m)$ through (11.27), the influence of oxide charge on the bias dependence of I depends on $F(x_m)$. Therefore, the extraction of oxide charge distributions from photo I-V measurements is based on three properties of (11.25). First, if the oxide charge is confined very near $x = 0$, the integrals in (11.25) are small, so that the photo I-V curve may be experimentally indistinguishable from the case with no oxide charge. However, the density of oxide charge in a sheet at the Si–SiO$_2$ interface can be measured from the flatband voltage shift as explained in Section 10.5. Combining C-V measurements of this density and photo I-V measurements, a limit can be placed on how far the charge extends into the oxide. Second, if the barrier maximum is within the charge distribution, the charge distribution can be profiled. By varying gate bias, the barrier maximum is swept through the oxide charge distribution, affecting the last term in (11.25). Thus the resulting photo I-V curve is distorted in shape compared to the case where the barrier maximum is never in the oxide charge distribution. Charge profile information is contained in this shape distortion of the photo I-V curve. Third, if there is oxide charge only beyond the barrier maximum, the integrals in (11.25) are independent of x_m and of gate bias. Then the photo I-V curve will be translated along the voltage axis, remaining parallel to the photo I-V curve for no oxide charge or for oxide charge located only in a charge sheet at the injecting interface. The charge density and its centroid can be extracted from the voltage shift measured first with the metal as emitter (gate negative) and then with the silicon as emitter (gate positive). These three applications are described in Section 11.4.3.

To provide some insight into the charge profile information contained in the photocurrent, the shape and the position of the barrier to injection are explored in Figs. 11.11 and 11.12. Figure 11.11 is a plot of $\Delta\phi$ as a function of x with $V_G - W_{ms}$ as parameter showing the influence of oxide charge on barrier height and position. The curves in Fig. 11.11 were calculated from (11.26) using (11.24) evaluated at $x = x_m$ for $V(x_m)$. The gate bias and oxide charge dependence of x_m in (11.26) was calculated by substituting (11.25) into (11.27). A rectangular positive oxide charge density of 5×10^{11} charges/cm^2 extending 50 Å into the oxide, illustrated in the insert of Fig. 11.11, was used in calculating the solid curves, and the dotted curves represent no oxide charge. In the gate bias range selected in Fig. 11.11, $x_m < 50$ Å so that the barrier maximum never passes beyond the oxide charge distribution.

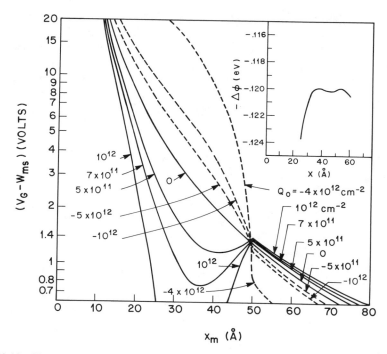

Fig. 11.12 $V_G\text{-}W_{ms}$ as a function of x_m with oxide charge density as parameter calculated from (11.28) for the oxide charge distribution in the insert of Fig. 11.11, with $x_o = 2000$ Å. *Insert:* $\Delta\phi$ as a function of x calculated from (11.26) for $V_G\text{-}W_{ms} = 1.23$ V.

Figure 11.11 shows that the barrier changes shape with gate bias and becomes lower and closer to the interface when oxide charge is present. Oxide charge affects the position of the barrier maximum much less than it affects the barrier height. As gate bias is decreased, the position of the barrier maximum becomes more sensitive to the oxide charge distribution. However, the barrier height is larger, so the photocurrent is small and becomes progressively more difficult to detect as gate bias is decreased. Also, the barrier maximum becomes very broad, degrading the resolution of the profile.

The behavior of the barrier described above shows that the best way to measure oxide charge near an interface is to make lowering of the barrier height dominate the bias dependence by using a photon energy just barely greater than the barrier height. However, at photon energies very close to the barrier energy, the photocurrent is small. Therefore, the lowest photon energy that can be used is limited by how small a photocurrent can be detected.

The photocurrent also cannot be too large because the injected electrons cause oxide charging, as discussed in Section 11.5. Photocurrent increases with both light intensity and photon energy. Hence, to keep oxide charging low, light intensity and photon energy also must be low.

Figure 11.12 illustrates the bias dependence of x_m with oxide charge density as parameter. The same oxide charge distribution shown in the insert of Fig. 11.11 is used, but Q_o is varied from a positive charge density (solid lines) of 10^{12} cm^{-2} to a negative charge density (dotted lines) of 5×10^{12} cm^{-2}. The curves in Fig. 11.12 were calculated from (11.27) using (11.25) for $F(x_m)$.

At small values of x_m (high gate bias), all the curves in Fig. 11.12 tend toward the same zero charge density curve, making barrier position independent of oxide charge density and polarity.

The curves in Fig. 11.12 are more complicated for positive oxide charge than for negative oxide charge. Consider sweeping the barrier maximum through the $Q_o = 5 \times 10^{11}$ cm^{-2} charge distribution by reducing gate bias. The barrier maximum moves deeper into the oxide until gate bias is reduced to $V_G - W_{ms} = 1.4$ V. At around $V_G - W_{ms} = 1.4$ V, there are *two* possible values for x_m. The barrier potential goes through a broad maximum. For a range of V_G values below $V_G - W_{ms} = 1.4$ V, there are *three* possible values of x_m. The barrier goes through two maxima as a function of x. For example, at $V_G - W_{ms} = 1.2$ V, barrier potential goes through a maximum at 35.7 Å, a minimum at 46 Å, and another maximum at 53 Å, as shown in the insert of Fig. 11.12. This insert also shows that the variation in barrier height through the two maxima is quite small. Thus, for practical purposes, the barrier is approximately flat over the range of depths between the various x_m values corresponding to the chosen V_G. As gate bias is reduced further, the two maxima move deeper into the oxide. Ultimately, again only one value of x_m exists. For example, at $V_G - W_{ms} = 1$ V, the barrier will have only one maximum at 59 Å, past the oxide charge distribution. The curve in Fig. 11.12 for $Q_o = 10^{12}$ cm^{-2} has a gap because the calculation was not extended to negative values of $V_G - W_{ms}$.

Consider sweeping the barrier maximum through any of the negative oxide charge distributions in Fig. 11.12. There is only one barrier maximum, and it moves deeper into the oxide with decreasing gate bias. Notice that for negative oxide charge densities, as gate bias is reduced, the position of maximum barrier height moves least rapidly as the outer edge of the charge distribution is approached.

The range of depth over which oxide charge density can be accurately determined is limited. At high gate bias the minimum x_m will be about 10 Å from the interface because the classical image force theory used is no longer very accurate and other effects such as tunneling can occur. At low gate bias the maximum x_m is limited by the sensitivity with which photocurrent can be measured, by oxide charge polarity, and by its density. Using the examples in Fig. 11.12, a quantitative estimate of the maximum x_m can be given. For $Q_o = 10^{12}$ cm^{-2}, photocurrents would be too small to measure reliably below about $V_G - W_{ms} = 0.7$ V so that the maximum x_m is about 25 Å. Using the same criterion for $Q_o = 5 \times 10^{11}\,\mathrm{cm}^{-2}$ to $Q_o = -4 \times 10^{12}\,\mathrm{cm}^{-2}$, the entire charge distribution out to 50 Å can be determined. For more realistic charge distributions, the low bias range limit on depth will be different.

11.4.3   Photo *I-V* Method—Distribution of Oxide Charge

Although photo I-V curves do contain information about the distribution of charges in the oxide, this information is limited by an experimental problem. This problem is that trapping of injected electrons alters the oxide charge distribution under investigation. Charging is due to water-related traps, described in Section 11.5.1.

Water-related traps are present in all oxides. By heating the sample in vacuum at temperatures above 600°C water can be driven out. However, this heating also may disturb the oxide charge distribution under investigation. Another approach is to use an oxide grown in dry oxygen and etching off the first 500 Å or so of oxide where the water entered from the ambient before making a photo I-V measurement. The most practical approach to charge trapping, which is least disturbing to the sample and thus is most commonly used, is to make photo I-V measurements at sufficiently low light intensity and over a short enough time period to ensure no significant change in oxide charge density during measurement. The price paid for using low light intensity is reduced sensitivity and, therefore, a more severe limitation on the maximum depth to which oxide charge can be profiled.

(a)   *Location of Oxide Fixed Charge and Charged Interface Traps*[46,49]

How close to the Si–SiO$_2$ interface are oxide fixed charge and charged interface traps? In an experiment by Powell and Berglund,[46] C-V and photo I-V curves (emission from the silicon) were measured on an MOS capacitor having a nondegenerate silicon substrate.* The sample then was annealed in oxygen for 3 hr at 550°C and the C-V and photo I-V measurements repeated. This annealing produced a flatband voltage shift of 5 V in the C-V curve corresponding to an increase of total positive oxide charge density of $5 \times 10^{11}\,\mathrm{cm}^{-2}$ at $x = x_o$. Figure 11.13 shows that the photo I-V curve was the same after annealing as before. Therefore, the oxide fixed charge and charged interface traps must be located between the Si–SiO$_2$ interface and the barrier height position. If this position and the minimum detectable change in photocurrent are known, a limit can be placed on how far from the interface charge can extend into the oxide. To

---

*Nondegenerate silicon must be used so that a C-V curve can be measured on the same sample as the photo I-V curves are measured. For nondegenerate silicon, the first term in (11.25) becomes $(V_G - W_{ms} - \psi_s)/x_o$, where ψ_s is the silicon surface band bending.

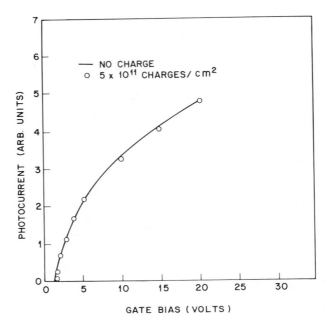

Fig. 11.13 Experimental photo *I-V* curve. Solid curve, before annealing; points, after annealing, which increased the oxide charge to 5×10^{11} cm^{-2} as measured by the flatband shift of a *C-V* curve. Electron emission was from the valence band of a silicon substrate, with $N_A = 1.5 \times 10^{16}$ cm^{-3}, into an oxide layer 2230 Å thick grown in dry oxygen at 1100°C. After Powell and Berglund.[46]

do this, the relative change in photocurrent due to the presence of oxide charge $(I - I_o)/I_o$ must be calculated as a function of oxide charge density from (11.20) and (11.32). We start by subtracting (11.28) from $(V_G - W_{ms})/x_o$ as given by (11.19) and find for large x_o that

$$\frac{q}{16\pi\epsilon_i}\left(\frac{1}{x_m^2} - \frac{1}{x_{mo}^2}\right) = \frac{q}{\epsilon_{ox}} \int_{x_m}^{x_o} n_o(x')dx' = \frac{Q_o}{\epsilon_{ox}} f(x_m). \quad (11.33)$$

Using (11.33), the image-force lowering term in (11.32) can be reexpressed as

$$\frac{2q}{16\pi\epsilon_i x_m} = \frac{2q}{16\pi\epsilon_i x_{mo}} + \left(\frac{2x_{mo}x_m}{x_m + x_{mo}}\right)\frac{Q_o}{\epsilon_{ox}} f(x_m). \quad (11.34)$$

It follows from (11.32), (11.34), (11.19), and (11.20) that

$$\frac{I}{I_o} = \left[1 + \left(\frac{qQ_o}{\epsilon_{ox}}\right)\frac{2x_m x_{mo}f/(x_m + x_{mo}) - \bar{x}(1-f)}{h\upsilon - E_{bo} + K(V_G - W_{ms})^{1/2}}\right]^p \exp\left(-\frac{x_m - x_{mo}}{l}\right). \quad (11.35)$$

For small oxide charge density, we can make a Taylor series expansion of (11.35) about $Q_o = 0$. Neglect of second-order terms yields[49]

$$\left|\frac{I - I_o}{I_o}\right| \approx p\left|(1 - f)\frac{\bar{x}}{l_c} + f\frac{x_{mo}}{l_c}\right| + \left|\left(\frac{x_{mo}}{l}\right)^3 8\pi\frac{Q_o}{q}l^2\left(\frac{\epsilon_i}{\epsilon_{ox}}\right)f\right| \quad (11.36)$$

where $l_c = \dfrac{\epsilon_{ox}}{qQ_o}[h\nu - E_{bo} + K(V_G - W_{ms})^{1/2}]$.

Two extreme modes can be identified from (11.36). In the barrier position mode, photon energy is large compared to the barrier height. As a result, l_c is large, making the second term in (11.36) dominant. In the barrier height mode, photon energy is only slightly greater than the barrier height and x_{mo} is very small, making the first term in (11.36) dominant. We analyze the data of Fig. 11.14 in terms of the barrier position mode to determine how far from the Si–SiO$_2$ interface, oxide fixed charge, and charged interface traps can extend into the oxide. Brews[49] has also found the centroid of this charge distribution to be 18 Å or less by analyzing the data in Fig. 11.13 in terms of the barrier height mode, which we do not consider.

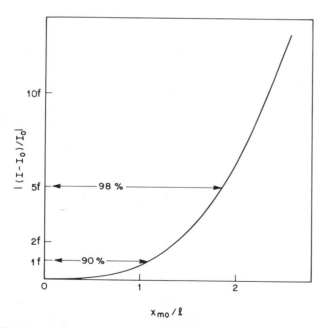

Fig. 11.14 Relative change in photocurrent caused by oxide charge as a function of barrier height position for photon energies well above the barrier maximum (barrier position mode) calculated from the second term of (11.36) for the same parameters as in Fig. 11.13. The estimates of charge confinement indicated by arrows assume relative photocurrent changes of 10% are experimentally observable. After Brews.[49]

In the barrier position mode, photon energy is large compared to the barrier height, so that (11.35) simplifies to

$$I = I_o \exp\left(-\frac{x_m - x_{mo}}{l}\right). \qquad (11.37)$$

For the limiting case where the oxide charge Q_o produces only a small effect on the barrier position x_m, (11.37) predicts

$$\left|\frac{I - I_o}{I_o}\right| = \left|\frac{x_m - x_{mo}}{l}\right|. \qquad (11.38)$$

That is, the relative change in photocurrent due to the presence of oxide charge is given by the change in barrier position measured in units of the scattering length. Suppose that the minimum observable change in photocurrent is 10%. Then the condition for observing the charge is $|(x_m - x_{mo})/l| \geq 0.1$ using (11.38). The relation between $(I - I_o)/I_o$ and Q_o is given by the second term in (11.36) (barrier position mode dominant). Figure 11.14 illustrates how the limit of the oxide charge distribution can be estimated. Figure 11.14 is a plot of $|(I - I_o)/I_o|$ in multiples of the fraction f of charge beyond the barrier maximum, which is the detectable charge, as a function of x_{mo}/l calculated from the second term in (11.36) using $Q_o = 5 \times 10^{11}$ cm^{-2}, $l = 34$ Å, $\epsilon_i = 1.97 \times 10^{-13}$ F/cm, and $\epsilon_{ox} = 3.4 \times 10^{-13}$ F/cm. To use Fig. 11.14, an ordinate say $|(I - I_o)/I_o| = 5f$ is chosen. At this point on the curve the detectable charge must lie beyond a distance $x_{mo} \approx 2l$. To find the percentage of the charge distribution lying beyond $2l$, we must stipulate the minimum observable change in photocurrent. Assuming the minimum detectable change to be 10%, $|(I - I_o)/I_o| \geq 0.1$. Therefore, for the ordinate equal to $5f$ in Fig. 11.14, $5f \geq 0.1$, $f \geq \frac{1}{50}$, or 2% of the charge must lie beyond a distance equal to $2l$. Or, the undetectability of any charge when $x_{mo} \approx 2l$ implies that 98% or so of the charge lies within two scattering lengths of the Si–SiO$_2$ interface. Similar reasoning provides the 90% limit also indicated in Fig. 11.14.

Applying Fig. 11.14 to the data in Fig. 11.13, the inobservability of any effect of the charge at $V_G \approx 4$ V corresponding to $x_{mo} \approx l$ [calculated from (11.19) for $W_{ms} = 1.1$ V for the Au–Si work function difference] means that no more than 10% of the oxide charge can extend beyond one scattering length or 34 Å from the Si–SiO$_2$ interface. The continued inobservability of charge at higher gate bias (x_{mo} closer to the interface) is expected because the photocurrent becomes progressively more insensitive to oxide charge, as seen in Fig. 11.12. For example, the inobservability of charge at $V_G = 14$ V corresponding to $x_{mo} = l/2$ tells us nothing because all the charge could be beyond $l/2$ and would not be detected. Inobservability at V_G less than 4 V gives information only about the tail 10% of the charge distribution, which is not terribly interesting. However, the estimate that oxide fixed charge and charged interface traps must lie closer than 34 Å to the Si–SiO$_2$ interface with a centroid below 18 Å is a great improvement over the earlier bound of 200 Å found from an etch-off experiment.[45]

It should be noted that $l = 34$ Å was determined from photo I-V measurements made at $h\nu = 5.0$ eV. The dependence of l on photon and electron energy is not known.

(b) Profiling [46,51]

An oxide charge distribution can be profiled by the photo I-V method provided that the barrier maximum lies within the charge distribution. Then the photo I-V curve must be measured before and after the introduction of the oxide charge distribution. These requirements make the photo I-V profiling method suitable for studying oxide trapping and charged impurity centers such as sodium that can be deliberately introduced into the oxide. The photo I-V curve measured after the charge distribution has been produced will be distorted in shape compared to the initial photo I-V curve. The profile is extracted by measuring the voltage difference between the initial and final photo I-V curves at a given photocurrent.

Figure 11.15 shows photo I-V curves measured before and after electron trapping by the water-related centers described in Section 11.5.1. Electrons photoinjected from the silicon are trapped near the metal electrode, producing negative oxide charge there. The curves in Fig. 11.15 were

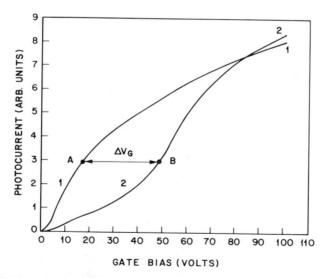

Fig. 11.15 Photocurrent as a function of gate bias for electron injection from the metal. Curve 1 is before charging water related traps located near the metal-SiO$_2$ interface, and curve 2 is after trapping electrons photoinjected from the silicon at photocurrents in the 10^{-8} A range, producing a negative oxide charge distribution. Full scale photocurrent is less than 10^{-11} A, so that negligible charging occurred during the measurement of these curves. After Powell and Berglund.[46]

obtained by photoinjecting electrons from the metal into the SiO_2 at such a low photocurrent that no further trapping occurred. The voltage difference ΔV_G is obtained between the points A and B. An expression for ΔV_G in terms of oxide charge is obtained by evaluating (11.28) with and without oxide charge and taking the difference with the result

$$\Delta V_G = \frac{qx_o}{16\pi\epsilon_i}\left(\frac{1}{x_m^2} - \frac{1}{x_{mo}^2}\right) - \frac{qx_o}{\epsilon_{ox}}\left[\int_{x_m}^{x_o} n_o(x')dx' - \frac{1}{x_o}\int_0^{x_o} x'n_o(x')dx'\right]. \quad (11.39)$$

Because x_m and x_{mo} are implicit functions of I, (11.39) can be differentiated with respect to I, yielding

$$\frac{d\,\Delta V_G}{dI} = -\frac{2qx_o}{16\pi\epsilon_i}\left[\frac{1}{x_m^3}\left(\frac{dx_m}{dI}\right) - \frac{1}{x_{mo}^3}\left(\frac{dx_{mo}}{dI}\right)\right] + \frac{qn_o(x)x_o}{\epsilon_{ox}}\left(\frac{dx_m}{dI}\right). \quad (11.40)$$

From experimental photo I-V curves such as those in Fig. 11.15, ΔV_G can be found at successive values of I, each differing by ΔI. Then the oxide charge profile can be extracted by using (11.40). To use (11.40), dI/dx_m and dI/dx_{mo} must be determined. Differentiating (11.32) yields

$$\frac{dI}{dx_m} = -\left(\frac{pI}{h\nu - E_{bo} + q\,\Delta\phi}\right)\frac{d\,\Delta\phi}{d\,x_m} - \frac{I}{l}. \quad (11.41)$$

Differentiating (11.30) yields

$$\frac{d\,\Delta\phi}{d\,x_m} = -\frac{2q}{16\pi\epsilon_i x_m^2} + \frac{qx_m n_o(x)}{\epsilon_{ox}} \quad (11.42)$$

and

$$\frac{d\,\Delta\phi}{dx_{mo}} = -\frac{2q}{16\pi\epsilon_i x_{mo}^2}, \quad n_o(x) = 0. \quad (11.43)$$

For negative oxide charge, substitution of (11.42) and (11.43) into (11.41) yields

$$\frac{dI}{dx_m} = I\left[\left(\frac{p}{h\nu - E_{bo} + q\,\Delta\phi}\right)\left(\frac{2q}{16\pi\epsilon_i x_m^2} + \frac{qx_m n_o(x)}{\epsilon_{ox}}\right) + \frac{1}{l}\right] \quad (11.44)$$

and for $Q_o = 0$

$$\frac{dI}{dx_m} = I\left[\left(\frac{p}{h\nu - E_{bo} + q\Delta\phi}\right)\left(\frac{2q}{16\pi\epsilon_i x_{mo}^2}\right) + \frac{1}{l}\right]. \quad (11.45)$$

For photon energies much greater than the barrier height and low gate bias (x_m large), the barrier position mode will be dominant, and the second term in both (11.44) and (11.45) will be dominant, so that (11.44) and (11.45) become

$$\frac{dI}{dx_m} = \frac{dI}{dx_{mo}} = \frac{I}{l}. \quad (11.46)$$

For low gate bias, x_m and x_{mo} will be large, making the second term in (11.40) dominant, and if (11.46) is used, (11.40) becomes

$$\frac{d\,\Delta V_G}{dI} = \frac{q x_o n_o(x)}{\epsilon_{ox}} \left| \frac{l}{I} \right|. \tag{11.47}$$

The profile can be determined using (11.47) far from the emitting surface. As gate bias is increased and the charge distribution is sampled closer to the interface, it will not be possible to neglect the first term in (11.40). Because the first term in (11.40) cannot be evaluated if the oxide charge distribution is unknown, it is necessary to determine the charge profile by iteration. As no method of iteration has been developed for this problem, oxide charge profiling by the photo I-V method is limited to distances that are large compared to l (i.e., 60–100 Å).

(c) *Oxide Charge Density and Centroid*[48]

When the oxide charge distribution lies beyond the barrier maximum, the charge density and its centroid can be determined with the photo I-V method. Applications for this case are the multilayer gate memory element[53-55] and oxide charge distributions produced by ion implantation into the oxide.[47,56] Unless the location of the charge is known, a C-V measurement of such an oxide charge distribution will give only the first moment of the charge distribution as described in Section 10.2.1.

Because there are two unknowns, namely, the oxide charge density and its centroid, two separate measurements are required to extract both of these parameters. The two measurements are a set of photo I-V curves with electron emission from the metal before and after charging and a set of photo I-V curves with electron emission from the silicon before and after charging. The charge density and centroid are extracted from the parallel voltage shift of each set. This method is valid provided that the voltage drop across the charge distribution is small compared to the voltage across the oxide and the spatial extent of the charge distribution is small compared to oxide thickness.

Taking the origin of coordinates at the metal-oxide interface, we can find $F(x_m)$ at the metal from (11.25) and at the silicon, from

$$F(x_m) = -\frac{(V_G - W_{ms})}{x_o} + \frac{q}{\epsilon_{ox}} \left[\int_0^{x_o} \frac{x'}{x_o} n_o(x')dx' - \int_{x_m}^{x_o} n_o(x')dx' \right]. \tag{11.48}$$

For oxide charge distributions that are beyond the barrier maximum at each interface and that are large compared to the oxide charge near the interface, (11.25) becomes

$$F(x_m) = \frac{V_G - W_{ms}}{x_o} - \frac{1}{\epsilon_{ox}} \left(1 - \frac{\bar{x}}{x_o} \right) Q_o \tag{11.49}$$

and (11.48) becomes

$$F(x_m) = -\frac{(V_G - W_{ms})}{x_o} + \frac{1}{\epsilon_{ox}}\left(\frac{\bar{x}}{x_o}\right)Q_o. \tag{11.50}$$

At a given photocurrent, $F(x_m)$ at the emitter is the same before and after charging. Therefore, if (11.49) is used, the voltage shift ΔV_{GM} with the

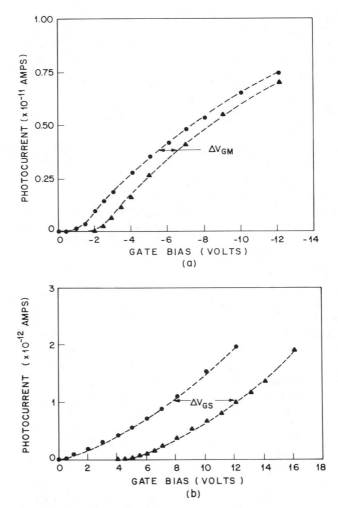

Fig. 11.16 (a) Photocurrent measured as a function of V_G for electron emission from the gate metal before (solid circles) and after (solid triangles) partial charging of the tungsten layer by electron capture (negative trapped charge). After DiMaria.[48] (b) Photocurrent measured as a function of gate bias for electron emission from the silicon before (solid circles) and after (solid triangles) partial charging of the tungsten layer. The sample was the same as in (a). After DiMaria.[48]

metal as emitter (negative gate bias) is

$$\Delta V_{GM} = \left(\frac{x_o - \bar{x}}{\epsilon_{ox}}\right) Q_o \tag{11.51}$$

and if (11.50) is used, the voltage shift ΔV_{GS} with the silicon as emitter (positive gate bias) is

$$\Delta V_{GS} = -\frac{\bar{x}}{\epsilon_{ox}} Q_o. \tag{11.52}$$

Equations (11.51) and (11.52) are two simultaneous equations in two unknowns that can be solved to obtain

$$Q_o = \frac{\epsilon_{ox}}{x_o} (\Delta V_{GM} - \Delta V_{GS}) \tag{11.53}$$

and

$$\bar{x} = x_o \left(1 - \frac{\Delta V_{GM}}{\Delta V_{GS}}\right)^{-1}. \tag{11.54}$$

Figure 11.16a shows a set of before-after charging photo I-V curves with the metal as emitter, and Fig. 11.16b shows a set of before-after photo I-V curves on the same sample before and after charging with the silicon as emitter. The sample was a metal–SiO_2–W–SiO_2–Si structure with evaporated tungsten atoms 80 Å from the Si–SiO_2 interface acting as a sheet of electron traps. Figure 11.16 illustrates the parallel shift of the photo I-V curves as a result of positively charging the tungsten layer and the way in which ΔV_{GM} and ΔV_{GS} are determined.

11.5 MEASUREMENT OF BULK OXIDE TRAP PROPERTIES

We discuss how bulk oxide trap properties can be measured and the results of these measurements. Avalanche injection is the method most widely used in bulk oxide trapping studies because it is the most convenient method for getting electrons and holes into SiO_2. Emphasis is on this method; electron trapping is discussed in Sections 11.5.1 and 11.5.2 and hole trapping, in Section 11.5.3.

Our development requires that the fraction of injected free carriers that are trapped always is small, and the dielectric breakdown field of the oxide is high ($> 5 \times 10^6$ V/cm). In nearly all cases of interest, only a small fraction of the injected carriers become trapped. Therefore, no other case need be considered. Oxides with high dielectric breakdown strength can be made routinely, to avoid any problem of dielectric breakdown.

The electron capture cross sections for bulk oxide traps in SiO_2, measured mainly by the study of charging kinetics using the method of avalanche injection, are listed for comparison in Table 11.1. The im-

Table 11.1   Electron Trap Capture Cross Sections

Trap	σ_n (cm^2)	Ref.
Water related	1.5×10^{-17}	43
Sodium related	10^{-12}	78
Deposited W	10^{-14}–5×10^{-14}	79
Implanted W	10^{-15}	79
Aluminum	10^{-15}–10^{-18}	79
Arsenic	10^{-15}	79
Phosphorus	3×10^{-17}	79

portance of the capture cross section is that it determines how long it takes to fill traps at a given injected current density.

There are two different types of electron trapping center in as-grown device-grade SiO_2. The dominant one is the water-related trap first observed by Nicollian et al.[43] The second type is the sodium-related trap. The water-related electron trap is described in Section 11.5.1 and the second trap, in Section 11.5.2.

11.5.1   Water-Related Traps

All thermally grown SiO_2 films contain electron traps related to the presence of "water" in the film.[4,43,56-63] These traps are called *water-related* traps because their density is proportional to the amount of water incorporated in the oxide, but the exact chemical nature of the center is ᵤnknown. In an as-grown oxide, water-related traps are electrically neutral and cannot be detected until charged by electron injection into the SiO_2. Electron trapping by the water-related traps is different from the conventional trapping common in semiconductors. In conventional trapping an electron is trapped and later emitted without changing the chemical composition of the trap. In a water-related trap, trapping of an electron initiates a *chemical* reaction. The products of this chemical reaction are a negatively charged center and hydrogen. This chemical reaction consumes the water-related traps until no further trapping occurs.

We follow Nicollian et al.,[43] who avalanche injected electrons into the SiO_2 as described in Section 11.2. Figures 11.17a,b illustrate what happens after avalanche injection of electrons into wet and dry oxides. Figure 11.17a shows that in a steam-grown oxide, the C-V curve is shifted to more positive voltage after avalanche injection. Therefore, negative oxide charge has been produced. Figure 11.17a also shows that the peak value of the equivalent parallel conductance has increased after avalanche injection. Therefore, interface traps have been produced. Figure 11.17b

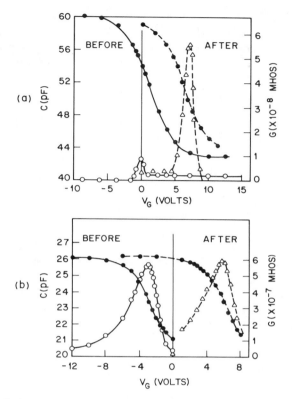

Fig. 11.17 (a) Capacitance and equivalent parallel conductance before and after the avalanche injection of electrons as functions of gate bias for an oxide grown in steam in an electric field of 1500 V/cm (silicon positive) to minimize sodium contamination[64] before and after avalanche injection of electrons. Measuring frequency was 10 kHz, gate diameter was 125 μm, acceptor density was 2×10^{17} cm^{-3}, and oxide thickness was 700 Å. After Nicollian et al.[43] (b) Capacitance and equivalent parallel conductance as functions of gate bias before and after avalanche injection of electrons for an oxide that was first grown in dry oxygen and then exposed to water vapor before the evaporation of the gate contact. Measuring frequency was 100 kHz, gate diameter was 250 μm, acceptor density was 2×10^{17} cm^{-3}, and oxide thickness was 660 Å. After Nicollian et al.[43]

shows that in an oxide first grown in dry oxygen and then exposed to water vapor before gate contact evaporation, negative oxide charge was produced and there was no detectable change in interface trap level density. The important difference between Figs. 11.17a and 11.17b is that interface trap level density near midgap, determined by the conductance method described in Section 5.5, has increased from 1.2 to 7.9×10^{11} cm^{-2} eV^{-1} as a result of avalanche injection of electrons into steam-grown oxide. However, interface trap level density has remained unchanged at 1.7×10^{12} cm^{-2} eV^{-1} (changes of 10^{11} cm^{-2} eV^{-1} could easily have been detected over this background) in the oxide in Fig. 11.17b, where water has not been permitted to diffuse all the way to the Si–SiO$_2$ interface. There is a marked increase of interface trap level density after avalanche injection of electrons only when water-related traps extend to the Si–SiO$_2$ interface.

(a) Measurement Methods

The increase of interface trap level density after avalanche injection unnecessarily complicates the interpretation of the measurements in oxide charge trapping studies.

To avoid significant increases of interface trap level density by avalanche injection, the oxides discussed in this section were hydrated in a carefully controlled manner. Oxides with a minimum water content were required. Dry oxides were grown in a radio frequency (RF) heated furnace as described by Goetzberger.[64] By using an RF heated furnace, the fused quartz furnace tube walls remained relatively cool, thus minimizing the diffusion of water through them from the ambient. The oxygen was dried by passage through a liquid nitrogen trap. Then, to control the penetration depth of the water into the oxide, water was diffused in at temperatures below 200°C. The apparatus used to diffuse controlled amounts of water into the oxide is described in Section 12.11.2.

Before gate electrodes were evaporated over the oxide surface, the sample was held at diffusion temperature for a given time (≥ 10 min) while a stream of nitrogen gas saturated with water vapor (25–55 Torr) flowed over it.

The next step was to charge the oxide by avalanche injection. Because water was not diffused all the way to the Si–SiO$_2$ interface, avalanche injection produced no negative charge or interface traps there. Thus avalanche injection produced a voltage shift of the C-V curve without altering its shape. This voltage shift, called the *flatband voltage shift* for convenience, is a measure of the negative oxide charge density produced. The final step was to measure this flatband voltage shift as a function of time. Flatband voltage shift was measured continuously while an avalanche injected dc current flowed through the oxide by using the feedback circuit described in Section 12.11.1(b).

(b) Charging Kinetics

The kinetics of the negative oxide charging effect can be studied by measuring flatband voltage shift ΔV_{FB} as a function of time while an injected electron current flows. Figure 11.18a shows a typical set of curves of ΔV_{FB} versus t with partial pressure of water p_{H_2O} as parameter. The wafer was sequentially exposed to the increasing values of p_{H_2O} shown. Exposure was always for 10 min at 150°C. All curves were measured with a constant injected dc electron current density of 4.4×10^{-5} A/cm^2 flowing through the oxide. Each curve tends to saturate at a value of V_{FB} that we call ΔV_{FBS} and that increases with increasing p_{H_2O}. Figure 11.18b shows a

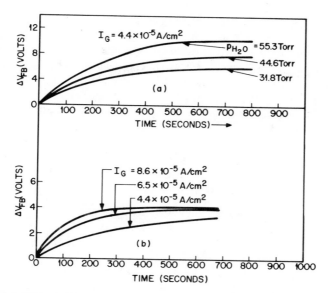

Fig. 11.18 (a) Plot of ΔV_{FB} versus t measured with constant I_G flowing with p_{H_2O} as parameter, with $N_A = 1.5 \times 10^{17}$ cm^{-3} and $x_o = 1500$ Å. The wafer was sequentially exposed to the increasing values of p_{H_2O} shown. After each exposure, measurement was made at a different position on the oxide surface with a mercury probe. The area of the probe was calculated from C_{ox} and x_o. After Nicollian et al.[43] (b) Plot of ΔV_{FB} versus t with I_G as parameter, with $N_A = 1.5 \times 10^{17}$ cm^{-3} and $x_o = 1500$ Å. Diffusion conditions are given in the text. Each measurement was made on a different metal gate (Cr-Au) evaporated after the diffusion. Gate diameter was 250 μm. After Nicollian et al.[43]

set of curves of ΔV_{FB} versus t for an oxide diffused at $p_{H_2O} = 25$ Torr for 10 min at 150°C with injected electron current density as parameter. All the curves tend to saturate at the same value of ΔV_{FB} for a given amount of water in the oxide, but the rate at which they tend to saturate increases with increasing injected electron current density.

To interpret these experimental results quantitatively, we let n_{ot} be the density of water-related traps per unit volume in the oxide that can capture an electron with probability c_n. Capture of an electron initiates a chemical reaction that has as one of its products an immobile negatively charged center that causes the flatband voltage shift measured in Fig. 11.18. Because all the curves in Fig. 11.18b tend to saturate at about the same value of ΔV_{FB} regardless of the magnitude of I_G, it is concluded that all the water-related traps have become charged. That is, the chemical reaction goes to completion and all the water-related traps are used up. Because electron capture determines the chemical reaction kinetics, and the emission probability of an electron is zero after the reaction goes to completion, we can write from Nicollian et al.[43]

$$\frac{dn_{ot}(t)}{dt} = -c_n[n_{ot}(0) - n_{ot}(t)]n_e \qquad (11.55)$$

where $c_n = \bar{v}\sigma_n$, $\bar{v}$ is the mean thermal velocity of electrons in SiO_2 at room temperature (cm/sec), σ_n is the electron capture cross section (cm²), $n_{ot}(t)$ is the instantaneous density of negative charged centers (cm⁻³), and n_e is the average injected electron density (cm⁻³). The quantity $[n_{ot}(0) - n_{ot}(t)]$ is the instantaneous density of water-related traps available for electron capture. Because only about one electron in 10^4 at most is captured, n_e is approximately constant. The solution of (11.55), with the boundary condition $n_{ot}(t) = 0$ when $t = 0$, is

$$n_{ot}(t) = n_{ot}(0)\left[1 - \exp\left(-\frac{t}{\tau}\right)\right] \tag{11.56}$$

where

$$\tau = (c_n n_e)^{-1} = (\bar{v}\sigma_n n_e)^{-1}. \tag{11.57}$$

To rewrite (11.56) in terms of the measured quantity ΔV_{FB}, we note that ΔV_{FB} must have the same proportionality to $n_{ot}(t)$ as ΔV_{FBS} has to $n_{ot}(0)$. Thus (11.56) becomes

$$\Delta V_{FB} = \Delta V_{FBS}\left[1 - \exp\left(-\frac{t}{\tau}\right)\right]. \tag{11.58}$$

Figure 11.19 shows that (11.58) accurately predicts the experimental results. The points in Fig. 11.19 represent two values of I_G taken from Fig.

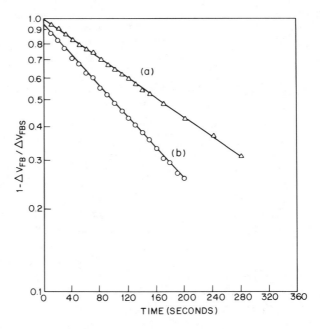

Fig. 11.19 Diagram illustrating log $(1 - \Delta V_{FB}/\Delta V_{FBS})$ as a function of t. Curve a is for $I_G = 4.4 \times 10^{-5}$ A/cm², and curve b is for $I_G = 6.5 \times 10^{-5}$ A/cm². The data for these curves are taken from the data in Fig. 11.18 b. After Nicollian et al.[43]

11.18*b*, and the solid lines were calculated from (11.58), where a value of τ was chosen for best fit.

According to (11.57), τ should vary inversely with n provided that c_n is a constant; c_n can be considered a constant even though it may be field dependent and oxide fields of the order of several million volts per centimeter were used in obtaining the data in Fig. 11.19. This assumption is supported experimentally, as is now discussed. Although I_G increases linearly with peak oxide field as shown in Fig. 11.4, the peak value of the injected current pulses always occurs at about the same value of oxide field, namely, that field that slightly exceeds F_B, the oxide field at the onset of avalanche breakdown in the silicon. Thus n_e is approximately proportional to I_G. Figure 11.20 shows values of τ, obtained by fitting the curves in Fig. 11.19, as a function of $(I_G)^{-1}$. The plot in Fig. 11.20 is linear. Therefore, c_n is constant over the range of parameters measured.

If it is assumed that $\bar{v}$ in (11.57) is approximately equal to the electron drift velocity, (11.57) can be written as

$$\tau = q(\sigma_n I_G)^{-1}. \tag{11.59}$$

With this interpretation, (11.59) shows that σ_n is related to the slope of the line in Fig. 11.20. From (11.59), σ_n for the water-related trap is estimated to be 1.5×10^{-17} cm². This value is of atomic dimensions as expected for a neutral trap. Therefore, this value of σ_n appears to be reasonable.

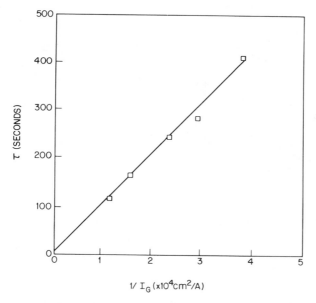

Fig. 11.20 Diagram illustrating τ as a function of $1/I_G$. Data for this curve are taken from the best-fit values of τ to the data of Fig. 11.19. After Nicollian et al.[43]

(c)   *The Role of Water*

Figure 11.21 shows the dependence of ΔV_{FBS}, obtained from Fig. 11.18*a*, on p_{H_2O}. Figure 11.21 shows that the dependence of ΔV_{FBS} on p_{H_2O} is linear and that no charging would occur if the sample had not been exposed to water vapor. Additional evidence that the electron traps are water related is that after heating the sample in vacuum at temperatures above 600°C, a process known to drive water out of SiO_2, no negative oxide charging is observed at any of the injected current densities in Fig. 11.18*b*.

In Fig. 11.21, ΔV_{FBS} varies linearly with water vapor pressure and hence so does the concentration of negative oxide charge. However, the water-related trap concentration in the silica is not the concentration in equilibrium with the water vapor pressure in Fig. 11.21. Nonequilibrium is present, as the water-related traps are not uniformly distributed through the oxide as shown in Fig. 11.22. Work by Holmberg et al.[65] suggests that equilibrium would be achieved only over many hours, whereas the data of Fig. 11.21 represent diffusion for only 10 min. Moreover, flatband voltage is proportional to the centroid of the charged centers as measured from the gate electrode. Consequently, even if the surface of the sample were in equilibrium with the ambient, the measurement would emphasize the nonequilibrium tail region.

There is indirect evidence that equilibrium with the ambient H_2O leads to the incorporation of SiOH centers in SiO_2. At elevated temperatures (800–1200°C) H_2O introduces hydroxyl groups into fused silica. Infrared absorption measurements show that the saturation concentration of

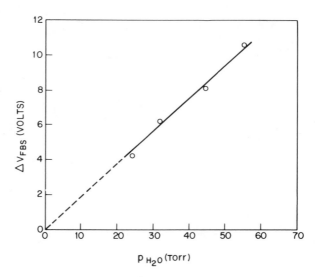

Fig. 11.21   Diagram illustrating ΔV_{FBS} as a function of p_{H_2O}. The data for this curve are taken from those in Fig. 11.18a. After Nicollian et al.[43]

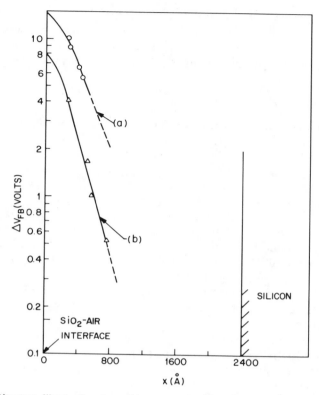

Fig. 11.22 Diagram illustrating log ΔV_{FB} as a function of x obtained from an oxide etch-off experiment. Curve a is for a diffusion time of 40 min, and curve b for a diffusion time of 10 min ($N_A = 1.5 \times 10^{17}$ cm^{-3}). After Nicollian et al.[43]

hydroxyl groups varies as the square root of the water vapor pressure.[66-69] Holmberg et al.[65] diffused tritiated water into thermal oxides at the low temperature, low water vapor pressure range used in Fig. 11.18. A square root dependence of tritium (hydrogen tracer) concentration on water vapor pressure was found. From the law of mass action, such a square root dependence of hydrogen or hydroxyl is expected if the following reaction takes place between water vapor and silica[70]

$$H_2O(\text{vapor}) + (\equiv Si-O-Si\equiv)(\text{solid}) \rightleftarrows 2(\equiv Si-OH)(\text{solid}) \qquad (1)$$

This reaction is the simplest one that leads to a square root dependence.

The likelihood is that such SiOH centers can capture an electron becoming negatively charged (a nonbridging oxygen* or SiO$^-$) and releasing atomic hydrogen, thus explaining the observations on water related traps.

---

*A nonbridging oxygen is a silica lattice defect in which an oxygen atom is bonded to only one silicon atom.

The situation is more complicated near the Si–SiO$_2$ interface, where, according to Beckmann and Harrick,[71] not only SiOH, but large concentrations of SiH may be expected. Thus the tritium profiles in SiO$_2$ films previously described in the literature[72,73] are not necessarily identical to the actual SiOH profile.

(d) Chemical Reaction

It is clear that negative oxide charge produced by injecting an electron current into SiO$_2$ is caused by electron capture at a water-related trap in the oxide. It has to be shown that the saturation of the curves in Fig. 11.18 results because traps are consumed in a chemical reaction. Negative charge density cannot be changed by exposure to photons of energy up to 4 eV but can be removed completely by heating in room air at 200°C for 15 min. It is unlikely that thermal detrapping is occurring where optical detrapping is so ineffective. Rather, this behavior suggests that electron capture by a neutral water-related trap initiates a chemical reaction. That is, the capture of an electron is accompanied by the evolution of some chemical species. As the evidence suggests that chemical decomposition at a water-related trap occurs as a result of electron capture, and in SiO$_2$ hydrogen is the most rapidly diffusing water related species, it is most likely that hydrogen is the species evolving.[74]

To test this hypothesis, two samples of SiO$_2$ grown in dry oxygen were exposed to tritiated water at room temperature for 24 hrs. One of the samples was then subjected to avalanche injection of electrons. This was done by placing a mercury probe on the oxide surface at several hundred points in succession and charging the oxide to saturation at each point. In this way only about 0.1–0.5 cm^2 of the total sample area of 1 cm^2 was charged. Because tritium is a β or electron emitter, β activity of the avalanche injected sample could be compared to that of the control to determine whether any tritium had been lost as a result of avalanche injection. A statistically significant reduction of β count compared to the control was observed on the avalanche injected sample, and the amount of tritium lost was found to have about a one-to-one correspondence with the amount of negative charge produced.* This result proves that hydrogen evolution occurs when an electron current flows through hydrated SiO$_2$. Thus a hydrogen atom bound to the water-related trap captures an electron. Electron capture breaks the bond between the hydrogen and the trap permitting the resulting neutral hydrogen atom to escape. Thus a reasonable conclusion is that atomic hydrogen rather than molecular hydrogen is the immediate product of the reaction initiated by electron capture at a water-related trap.

---

*The charged area was not determined accurately enough to correlate conclusively the amount of tritium introduced into the oxide with the amount of charge produced.

(e) Negative Charge Distribution

To determine the negative oxide charge distribution in the oxide after a hydrated oxide was charged to saturation, an oxide etch-off experiment was performed as described in Section 11.4.1(b). Two samples were heated to 150°C with $p_{H_2O} = 55.3$ Torr, one for 10 min and one for 40 min. Following charging, each oxide was etched off in approximately 100 Å steps, and flatband voltage was obtained from a C-V measurement at 1 MHz after each etch-off step using a mercury probe. The voltage difference ΔV_{FB} was then calculated taking $Q_f + Q_{it}$ into account. Values of ΔV_{FB} obtained this way were plotted as a function of distance x measured from the air-SiO_2 interface in Fig. 11.22. Figure 11.22 shows that all the negative charges are located very near the original air-SiO_2 interface. After 400 Å of oxide was etched away, no detectable oxide charging occured for the same electron currents as in unetched oxides. This observation and the fact that all the water-related traps are charged at saturation, means that the negative charge profile must be the same as that of the originally neutral water-related traps.

The profiles in Fig. 11.22 were obtained from (10.1). In principle, given an experimental plot of ΔV_{FB} versus x as obtained from an etch-off experiment, (10.1) can be inverted to yield $n_{ot}(x)$. However, a simpler way is to assume a distribution and see if it fits the experimental data. The exponential distribution

$$n_{ot}(x) = n_{ot}(0) \exp\left(-\frac{x}{x_c}\right) \tag{11.60}$$

where $n_{ot}(0)$ is the negative charge density at $x = 0$ and x_c is the characteristic decay length of the charge distribution, was found to fit. To plot ΔV_{FB} against x, we substitute (11.60) into (10.1), neglecting W_{ms}, which is small compared to ΔV_{FB}, and integrate from 0 to x

$$\Delta V_{FB} = \left[q n_{ot}(0) \frac{x_c^2}{\epsilon_{ox}} \right]\left(1 + \frac{x}{x_c}\right) \exp\left(-\frac{x}{x_c}\right). \tag{11.61}$$

In Fig. 11.22 the solid curves are calculated from (11.61) with $n_{ot}(0)$ and x_c chosen for best fit. For the 10 min diffusion (curve b in Fig. 11.22), $x_c \approx 160$ Å and $n_{ot}(0) \approx 6.7 \times 10^{18}$ cm^{-3}, whereas for the 40 min diffusion (curve a in Fig. 11.22), $x_c \approx 225$ Å and $n_{ot}(0) = 6.0 \times 10^{19}$ cm^{-3}. Thus $n_{ot}(0)$ is about the same at fixed temperature and pressure for two different diffusion times, and x_c increases with increasing diffusion time, as predicted by the hypothesis that a water-related trap is diffusing into the SiO_2.

(f) Diffusion of Water

There are two models for the diffusion of water into SiO_2 that may be valid at temperatures below 200°C: (1) uniform diffusion, which occurs in many single-crystal materials, and (2) a pore model. In model 1, x_c in

(11.60) can be compared approximately to $(4Dt)^{1/2}$, where t is diffusion time and D is the diffusion coefficient, and the profile should follow the complementary error function. Model 2 is diffusion through pores of molecular or larger dimensions that have a larger diffusion coefficient than the surrounding material. A model applicable to this case has been mathematically analyzed by Fisher.[75] In the case of micropore diffusion, x_c should vary as $t^{1/4}$ and the profile should be exponential.

The data in Fig. 11.22 are consistent with the micropore model because an exponential profile fits the data and x_c varies nearly as $t^{1/4}$. The 100 Å etch-off steps are comparable to x_c. Therefore, this experiment was too coarse to show conclusively that water diffuses through micropores in the oxide at low temperatures. However, Holmberg et al.[65] find a micropore model consistent with low temperature water-diffusion measurements. Also, evidence summarized by Revez and Zaininger[76] suggests the existence of a micropore structure in SiO_2. The strongest experimental evidence in favor of a micropore diffusion model is the low activation energy (~ 0.35 eV) measured in both Ref. 71 and in the isochronal annealing of the negative oxide charge described in Section (g). As pointed out in Ref. 65, the activation energy measured at low temperatures, where diffusion through micropores would be dominant, is about half the bulk SiO_2 value measured at high temperatures, where bulk diffusion would be dominant. Not only the activation energy but the distribution of diffusant, as previously explained, is different for micropore diffusion and bulk diffusion. This may explain the difference in the profiles near the Si–SiO_2 interface reported by Burkhardt,[72] who diffused water at high temperatures and Holmberg et al.,[65] who diffused water at low temperatures.

(g) Annealing

To determine the conditions required to discharge the negatively charged enters, a hydrated oxide was first charged to saturation and then annealed in various ambients. Of the ambients tried, it was found that the negative charge could be discharged only in a wet ambient. Again by comparison with the ineffectiveness of optical discharge, a chemical reaction is needed between the water diffusing into the oxide from the ambient and the negatively charged center. It also was found that the oxide annealed in a wet ambient could be charged again. In other words, the chemical reaction initiated by electron capture is reversible.

Two annealing experiments show that water is needed to discharge the negative centers. Annealing in either vacuum or dry H_2 at 200°C (H_2 does not chemically react with SiO_2 at temperatures below 350°C) for 30 min will produce no detectable change in the negative charge density. But annealing at 200°C for 15 min in an ambient that has a partial pressure of water vapor of 11 Torr [room ambient at 50% relative humidity (RH)] will result in the complete discharge of the oxide.

The activation energy for discharging the negative oxide charge in a water vapor ambient is found from an isochronal annealing experiment. In a typical isochronal annealing experiment, the sample is annealed at $p_{H_2O} = 11$ Torr at temperatures of 80–200°C with a constant annealing time of 10 min at each temperature. The results of this experiment are shown in Fig. 11.23. Because the curve in Fig. 11.23 is linear, the discharging mechanism can be represented by a single activation energy of approximately 0.35 eV. This value is about the same as that reported for the diffusion of water in SiO_2 at low temperatures in Ref. 65.

With a low activation energy of 0.35 eV, water should diffuse readily into the oxide at room temperature forming water-related traps. This is observed to be the case. Most of the water-related electron traps are removed by etching off about 400 Å of oxide. No negative oxide charge is detected after avalanche injection using a mercury electrode. However, after exposure to

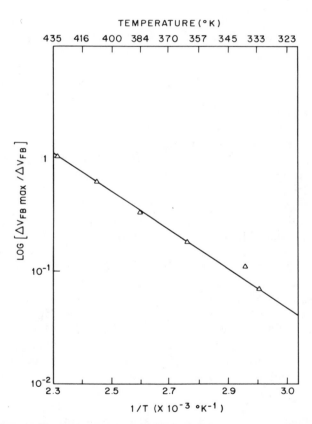

Fig. 11.23 Diagram illustrating $\log \log(\Delta V_{FBM}/\Delta V_{FB})$ as a function of T for an isochronal annealing experiment where ΔV_{FBM} is the initial saturation flatband voltage shift after annealing at a given temperature for 10 min in air having $p_{H_2O} = 11$ Torr ($N_A = 1.5 \times 10^{17}$ cm^{-3}). After Nicollian et al.[43]

room air in which $p_{H_2O} \approx 11$ Torr for about 1 week, negative oxide charge again is observed after avalanche injection.

A reasonable hypothesis is that water from the ambient diffuses into the oxide to a negatively charged center and chemically reacts, rendering the center electrically neutral and capable of capturing an electron again. Because the activation energy of this process is found to be 0.35 eV, which is identical to the activation energy of water diffusion at low temperatures, the overall process probably is diffusion limited; that is, all the water arriving at a negatively charged center reacts, but the amount arriving is limited by diffusion. The charge of the center can be repeatedly cycled from neutral to negative and back to neutral by starting with a hydrated oxide, charging the water-related traps to saturation by avalanche injection of electrons and restoring the original conditions by annealing in a wet ambient. The fact that negative charge can be removed by diffusing additional water into the oxide could mean that a hydrogen ion is recaptured by each negatively charged center from the diffusing water species. However, details of this reaction are not understood.

(h) Role of Sodium

In Refs. 61 and 62 it has been argued that neutral sodium also might be involved in the water-related trap. This argument is based on (1) the similarity between the neutral sodium profile and the profile of water diffused into the oxide at low temperatures* and (2) the increase in negative charge density with increasing sodium concentration in the oxide. It is not clear whether sodium is an integral part of the trap along with water or whether it plays an indirect role by making it easier for water to enter the oxide.

To show convincingly that sodium is an integral part of the trap, it would be necessary to relate the negative charge density observed to the amount of neutral sodium. This could be accomplished by varying the sodium content while the water content is kept fixed. Until such a controlled experiment is done, the role of sodium will remain unclear.

It is likely that in sodium-contaminated samples, neutral sodium enhances the diffusion of water. Both neutral sodium and water usually are incorporated into SiO_2 on nonbridging oxygen sites. Therefore, water can diffuse by an exchange reaction with sodium, and no Si–O bonds have to be broken. This exchange reaction is a lower energy process than creation of nonbridging oxygen sites, so that water enters the oxide more easily when neutral sodium is present.

(i) Positive Oxide Charge

At very high total injected electron densities, positive oxide charge is produced,[59,77] as evidenced by Fig. 11.24. Figure 11.24 shows that the ΔV_{FB}

---

*As shown in Section (e) above, the water profile is identical to the negative charge and, therefore, the water-related trap profile.

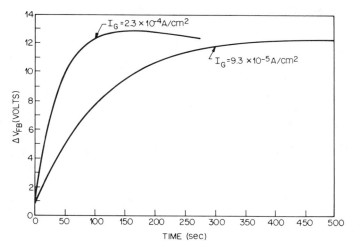

Fig. 11.24 Plot of ΔV_{FB} versus time for two values of I_G. The oxide was grown in dry oxygen to a thickness of 1250 Å and then contaminated with water at 150°C ($N_A = 1.5 \times 10^{17}$ cm^{-3}; $A_g = 1.07 \times 10^{-3}$ cm^2).

versus t curve saturates for injected current densities below 9.3×10^{-5} A/cm^2. This current density results in too small a total injected charge density on the time scale shown in Fig. 11.24 to produce positive charge. However, for $I_G = 2.3 \times 10^{-4}$ A/cm^2, the curve no longer saturates but goes through a maximum and then declines. This behavior results because, initially, the dominant oxide charge in this curve is negative. At later times, when the production of negative oxide charge saturates, the production of positive oxide charge becomes significant, causing ΔV_{FB} to decline. Photo I-V measurements confirm that this decline is caused by the buildup of positive charge at the Si–SiO$_2$ interface.[77] When the water content of the oxide is high, positive oxide charge will be observed at a lower total injected electron density than in Fig. 11.24.

The mechanism for the production of positive oxide charge is not understood. However, according to Young et al.,[77] this positive charge is caused by donor-type interface traps that have a high density near the silicon conduction band edge.

According to Young et al.,[77] this positive oxide charge (1) increases after avalanche injection ceases if a positive bias corresponding to an oxide field of 4–6×10^6 V/cm is applied and decreases when polarity is reversed, (2) is not generated if electron injection is at 77°K, and (3) is not detected if electrons are injected at elevated temperatures (373–423°K).

In summary, positive oxide charge is observed only when total injected electron density and the amount of water in the oxide are both large. In the samples measured in this section, positive oxide charge has been avoided by keeping the total injected electron charge density below $\sim 10^{-1}$ coul/cm^2 [$(10^{-4}$ A/cm^2) $(10^3$ sec$)$] and the amount of water diffused into the oxide below the amount diffused in at a water vapor pressure of 55 Torr for 40 min at 150°C.

(j) Summary

In summary, the phenomena observed when electrons are injected into hydrated SiO_2 are complex and only partially understood. By restricting water-related traps to the vicinity of the air-SiO_2 interface, the generation of significant interface trap density is avoided. By keeping the total injected electron density moderate and the amount of water incorporated into the oxide also moderate (nonequilibrium condition), positive oxide charging is avoided. Without these complicating phenomena, the mechanism for negative oxide charge generation has been found to be an electrochemical reaction initiated by the capture of an injected electron by a water-related trap. The products of this reaction are a negatively charged center and hydrogen.

11.5.2 Sodium-Related Centers

Williams[78] first observed an electron trap about 2 eV* below the conduction band of the SiO_2. This electron trap appears to be related to sodium in the oxide, particularly in oxides grown in steam at 1000–1200°C, because it disappears when sodium levels are reduced. This trap along with the water-related trap described in Section 11.5.1 are the only electron† traps detected at room temperature that are not deliberately introduced into thermally grown SiO_2. A fundamental difference between these two traps is that the sodium-related trap remains chemically unaltered by electron capture whereas the water-related trap undergoes a chemical reaction initiated by electron capture as described in Section 11.5.1.

Williams[78] detected the sodium-related electron traps by initially filling them with electrons by photoemission from the silicon and then photodepopulating them with photons at energies too low for electron emission from either electrode. As the traps empty, the measured photocurrent decays with time. The energy level of the trap is found from the peak photocurrent measured at $t = 0$, when the light that depopulates the traps is just turned on, as a function of photon energy. The density, centroid, and capture cross section of this trap can be obtained from the area under the photocurrent-time curves measured as a function of gate bias during photodepopulation. Because the photoionization cross section is small ($\sim 10^{-18}$–10^{-17} cm²),[79] fairly thick oxide layers must be used to depopulate the traps in a reasonable time. However, if the oxide is made too thick, the probability of retrapping increases. A reasonable compromise between

---

* Several workers[61,79] have subsequently found that the energy of this trap has a distribution about 0.5 eV wide centered at about 2.4 eV below the conduction band of the SiO_2.

† The sodium-related trap also can capture holes.

these two contradictory requirements is to use an oxide thickness in the range 1–2 μm.

This method of measuring the properties of bulk oxide traps is called the photodepopulation method. Determination of the density and centroid of the trap distribution using this method is discussed in the literature.[79] However, we do not discuss this method because the photo I-V method discussed in Section 11.4 is better for determining the oxide trap distribution.

The photodepopulation method is limited to traps that are no deeper than about 4 eV, because at greater photon energies the photocurrent will be dominated by photoemission from the electrodes, rather than by oxide trap depopulation. There appear to be few other oxide traps of interest in thermally grown SiO_2 that have energy levels in the range suitable for this method. The photo I-V method does not suffer from this limitation. To date, the photodepopulation method has been used only for measuring the sodium-related trap for which it is admirably suited.

Williams[78] found a density of about 3×10^{14} cm^{-3} and a capture cross section of 10^{-12} cm^2 for the sodium-related trap. It was possible to detect this electron trap because its capture cross section is about 5 orders of magnitude larger than the capture cross section of the dominant water-related electron trap (see Table 11.1), and the photoinjected current densities used were 4–6 orders of magnitude smaller than the avalanche injected current densities used for charging the water-related traps. Therefore, the total amount of electronic charge photoinjected by Williams during his experiments was so small that charging of the water-related traps was negligible compared to the sodium-related traps.

Although the sodium-related trap is initially neutral, its relatively large capture cross section (see Table 11.1) is generally interpreted as being due to a coulombic center. Such a center has a more pronounced electric field dependence than a center like the water-related trap, which has a capture cross section of atomic dimensions.

Although this coulombic trap appears to be related to the sodium content in the oxide, sodium ions do not trap electrons,[57,79] except at 77°K.[58] Furthermore, this trap is found to be uniformly distributed throughout the SiO_2[63,79] whereas the sodium is primarily concentrated at the air-oxide or metal-oxide interface as described in Section 10.2.3(a). The exact chemical nature of this trap and how it is related to the sodium are not known.

11.5.3 Deliberately Introduced Traps

In this section we consider traps in the oxide produced by tungsten, phosphorus, arsenic, and aluminum that have been deliberately introduced by either diffusion or ion implantation. These traps are chemically unaltered by electron capture. They have been studied using the powerful combination of avalanche injection to charge the traps and study charging kinetics and the photo I-V method to determine the resulting charged trap distribution.

(a) Tungsten

Tungsten is of interest as a trapping center in applications such as in a nonvolatile memory element[54,55] or for reducing leakage and improving dielectric breakdown characteristics of MOS structures.[80]

In these studies[54,79] tungsten atoms are deposited less than a monolayer thick ($\leq 10^{15}$ atoms/cm^2) between a thermally grown SiO_2 layer and a chemical vapor deposited (CVD) Al_2O_3 layer in a capacitor structure. The tungsten traps are initially neutral. The electron capture cross section of this trap is listed in Table 11.1. It depends on the number of deposited tungsten atoms.[79] Ion implanted tungsten, after annealing out radiation damage, was found to have a smaller capture cross section, also listed in Table 11.1. Trap density is nearly constant regardless of small variations in the number of deposited tungsten atoms, which is a factor of about 10–40 less than the number of implanted tungsten atoms.

The application of tungsten traps is based on the fact that they are stable and not easily discharged after electron capture. Tungsten traps charged by electron capture cannot be discharged with photon energies up to 6 eV[79] but can be discharged thermally at 200–400°C in nitrogen with an activation energy of 0.8 eV.[76] They also can be discharged electrically under fields sufficiently high to allow tunneling. A chemical reaction may be involved in the thermal discharge, but this possibility has not been explored.

(b) Ion Implanted Impurities[47,79]

The ion implanted impurities whose trapping properties have been studied are boron, aluminum, arsenic, and phosphorus. None of these impurities will act as a trap if it is thermally incorporated into the oxide or if the sample is oxidized further after implantation. If these impurities are ion implanted, some of them can act as traps. However, ion implantation creates numerous traps resulting from damage that results as energetic ions loose energy to the silica lattice. Usually, this damage is so extensive that any carrier injected into the SiO_2 is trapped. Annealing at high temperatures (~ 1000°C) in an *inert* ambient after implantation removes most of the traps caused by the damage. Therefore, trapping associated with the impurities themselves can be seen. The exact nature of the traps related to implanted impurity atoms is not known.

Ion implanted boron does not act as a trap. Therefore, if boron is implanted through the gate oxide of an n-channel MOSFET to control threshold voltage, any boron remaining in the oxide will not trap electrons injected into the oxide during device operation.

Traps related to implanted aluminum are initially neutral and have the electron capture cross section listed in Table 11.1. Trap densities are proportional to the implantation energy for a given number of implanted ions or proportional to the number of implanted ions for a given ion energy. For a fluence of 10^{13} ions/cm², the total oxide trap density varies from 10^{12} cm^{-2} to 10^{13} cm^{-2} depending on ion energy. The trapped charge and the implanted ions have the same spatial distribution. Charged traps cannot be discharged at photon energies up to 6 eV, so that aluminum related traps are stable.

The electron capture cross section of the dominant oxide traps related to implanted arsenic and phosphorus are listed in Table 11.1. The densities of these oxide traps are not functions of implantation energy as for aluminum. Instead, these densities are functions of the number of implanted ions and have a value of about half this number. The charged oxide traps have the same centroid as the implanted ions. These oxide traps can be discharged at photon energies above 3–4 eV. Thermal detrapping at temperatures of 100–350°C in a nitrogen ambient has been observed. The activation energy for thermal discharge is 0.15–0.25 eV, suggesting that discharge is by a chemical reaction involving water that diffuses to the trap site. Arsenic and phosphorus related oxide traps also can trap holes.

(c) Trapping at Low Temperature

Ning[81] has found shallow trapping levels in SiO₂ that can be observed only at 77°K. The peak of the energy distribution of these traps is found to be 300 ± 50 MeV below the SiO₂ conduction band, and the half-width is about 200 MeV. The physical origin of these traps is unknown. However, the density of these traps is about the same for wet, dry, and HCl oxides, suggesting that they are not related to the oxide growth ambient. These traps are not of major importance in device applications at room temperature because they retain charge only at low temperatures.

11.6 RADIATION EFFECTS IN SiO₂

This section covers the subject of radiation damage in SiO₂ and how it affects semiconductor device characteristics. First, the experimentally established facts are summarized, and then the models and physical mechanisms that appear to account for these facts are discussed.

The importance of radiation arises from interest in (1) operating semiconductor devices in industrial, space, and military applications in a radiation environment, (2) device fabrication processes such as electron beam deposition, sputtering, ion etching, and ion implantation that expose the device to radiation, and (3) devices such as the silicon vidicon target that contain an inherent radiation source. Radiation consists of high energy particles such as electrons, neutrons and protons, and energetic X-ray and γ-ray photons.

The principal cause of radiation damage varies with the application. For industrial applications associated with instrumentation in nuclear reactors and high energy X-ray machines, as well as military applications where the devices must survive exposure to a nuclear weapon, energetic neutrons and photons are of primary importance. In processing such as electron deposition and sputtering, photons and ions are of primary importance. In a device such as the vidicon target, electrons and photons are of primary importance. For space applications, electrons and protons produce most of the damage in semiconductor devices.

In a bipolar transistor and a MOSFET, radiation damage in either the silicon or the SiO_2 will affect device characteristics. In either material, radiation effects can be placed in two broad classes: (1) effects arising from hole-electron pairs produced in the material by ionizing radiation and (2) effects resulting from defects in the lattice produced by high energy particles. These two phenomena often occur together, but they result from quite different interactions of the radiation with the solid. They have different consequences in the semiconductor and in the oxide.

Ionization or the creation of hole-electron pairs in a solid can be produced by either charged particles or photons. Figure 11.25 illustrates electron-hole pair production in a crystalline solid. Any charged particle passing through the crystal produces ionization by collisions with electrons bound to lattice atoms. These collisions excite electrons into the conduction band, leaving behind holes in the valence band. Thus electron-hole pairs are produced as the incident charged particle loses energy to electrons bound to the lattice, in exact analogy with the production of hole-electron pairs by photons. Neutrons and γ-rays also cause ionization of the lattice through intermediate reactions that produce charged particles. As far as effects that depend on ionization or electron-hole pair production are concerned, the originating particle's identity is incidental. All that matters is how much energy is lost in the solid by the particle. The number of electron-hole pairs produced is then proportional to this energy loss.

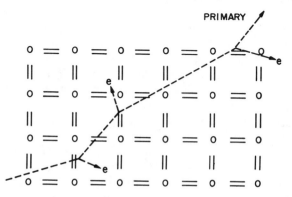

Fig. 11.25 Production of hole-electron pairs by the collision of a charged particle with valence electrons in silicon. After Peck et al.[82] Copyright (1963), American Telephone and Telegraph Company. Reprinted with permission.

Figure 11.26 illustrates the ionization processes in crystalline SiO_2 caused by photons. In order of increasing energy, these ionization processes include (1) the photoelectric effect, (2) Compton scattering, and (3) electron-positron pair production. At photon energies of less than the bandgap of SiO_2 (8.8 eV), no electron-hole pairs can be generated. However, if the photon energy is greater than the barrier energy at either the metal-oxide or oxide-silicon interface, photocurrent will occur by carrier injection from one of the electrodes. Which electrode depends on the polarity of the bias applied across the MOS structure. Such currents due to electron emission by UV light have been studied in detail.[78,80-87] In general, the injected photocurrent saturates at a relatively low applied bias, which means that practically all of the injected electrons reach the opposite electrode.

Photons with energy greater than the bandgap of silicon or SiO_2 can cause photoconductivity by the generation of electron-hole pairs. In SiO_2, this would require 8–9 eV photons, whereas 3–5 eV photons cause only photoemission of electrons from the electrodes. The energy of the incident photon is completely absorbed, and the valence electron injected into the conduction band has an energy that is equal to the initial photon energy minus the binding energy of the electron to the lattice atom. This energy is carried away as kinetic energy by the electron injected into the conduction band. If the injected electron has sufficient kinetic energy, it can produce

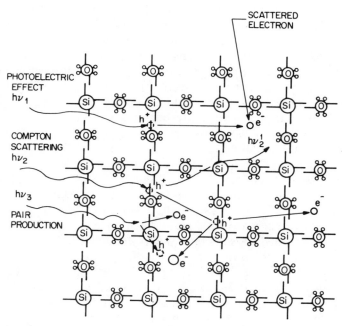

Fig. 11.26 Schematic representation of ionization processes in SiO₂. After Gwyn.[83]

secondary electrons as it moves through the lattice. For even higher energy photons, electrons can be ejected from the K and L shells of the lattice atoms. The subsequent transition of electrons from the outer electron shells to the K or L shell is accomplished by the emission of additional X-ray photons, which in turn can produce further ionization in the SiO_2.

In a Compton scattering collision, which occurs at still higher photon energies, an incident photon creates a hole-electron pair. The photon energy is large compared to the binding energy of an electron to a lattice atom. Only some of the energy of the incident photon is transferred to the electron, in contrast to the photoelectric effect, where all the photon energy is absorbed. The scattered photon in the Compton effect can continue to produce more electron-hole pairs until it is completely absorbed. The kinetic energy of the scattered electron can be a significant fraction ($> \frac{1}{2}$) of the incident photon energy so that a large amount of secondary ionization can be produced as the electron is decelerated by the lattice. If the scattered electron energy is of the order of several hundred kilo-electron volts, displacement damage also can be produced.

When the incident photon energy exceeds 1.02 MeV, the interaction of the photon with the coulomb field of the nuclei of the lattice atoms may result in the creation of an electron-positron pair. The photon is completely annihilated, and the energy in excess of that required to produce the electron-positron pair is about evenly distributed between the two particles as kinetic energy. Photon energies high enough to produce electron-positron pairs are not usually encountered in radiation environments in most applications. Electron-hole pair production by lower energy photons and charged particles is the usual type of radiation effect encountered.

In silicon the generated electron-hole pairs tend to recombine with a time constant that is the conventional lifetime. Therefore, all effects in this class in silicon are transient and persist only for the order of a lifetime after the excitation is removed. The electron-hole pairs produced in the silicon alter its conductivity and also contribute currents in p-n junction diodes and transistors. In fact, p-n junctions can be used as radiation detectors.

Pulsed ionizing radiation can alter conductivities by many orders of magnitude. On the other hand, in response to a single energetic particle, these effects can be very small.

Figure 11.27 illustrates the other type of radiation damage, displacement damage, which can occur in a crystalline solid as a result of collisions of energetic particles such as neutrons with the nuclei of the lattice atoms. If such a collision transfers sufficient energy to the struck lattice atom, this atom can be dislodged from its normal lattice site and subsequently strike and dislodge other lattice atoms before coming to rest in an interstitial site. The cascade of displaced atoms form disordered regions or damage clusters containing interstitial atoms, lattice vacancies, and more complex intrinsic defects. In addition to displacement damage, a large amount of ionization also can be produced as valence electron bonds are broken in the displacement process.

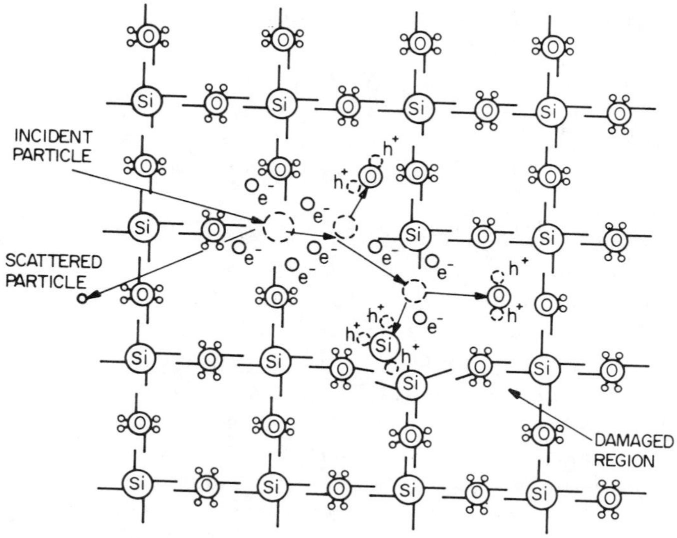

Fig. 11.27 Schematic representation of displacement damage in SiO$_2$. After Gwyn.[83]

Displacement damage is rare compared to the ionization events produced by charged particles and photons discussed previously, but displacement damage creates permanent or at least semipermanent defects in the lattice structure. Defects and defect clusters can act as recombination-generation centers in silicon. These centers reduce minority carrier lifetime in the silicon, thereby decreasing the current gain of a bipolar transistor at low collector currents. The reason for the reduction in current gain is that now a larger fraction of the minority carriers injected by the emitter can recombine in the emitter-base depletion region and in the base region through these defects. As a result fewer injected carriers reach the collector junction. At sufficiently high injection levels, there is no penalty in current gain because the recombination current depends on the finite number of recombination centers and majority carriers available. However, degradation of low current gain by displacement damage is important because many applications are small-signal applications.

Increasing the density of generation-recombination centers in silicon by displacement damage also will increase diode reverse current due to increased carrier generation.

In contrast to electron-hole pair production by ionizing radiation, which depends only on energy but not the type of incident particle, radiation damage that produces displacements of lattice atoms is extremely dependent on the particular particle involved. Energetic protons, for example, are much more effective in producing lattice damage than energetic electrons. This type of radiation effect is of major importance for long term power conversion using solar cells in space. This problem is beyond the scope of this book.

11.6.1 Telstar Experiment

Most of the early studies of the effects of radiation on semiconductor devices were concerned exclusively with the semiconductor bulk. The Telstar communications satellite experiment in 1962 showed for the first time that in a radiation environment, surface effects can be as important if not more important than bulk effects. In the Telstar satellite, mesa-type bipolar transistors (see Section 15.3.1) encapsulated in a metal can backfilled with an inert gas were used in the command circuitry.

The failure of Telstar was explained successfully by Peck et al.[82] in terms of surface effects caused by the radiation received by the satellite during transit through the Van Allen belt. Figure 11.28 illustrates how failure occurred. On exposure to radiation, the gas in the can became ionized. The electric field between the metal can and the surface of the base region of the transistor served to attract positive ions to the surface of the base region. The positive ions then came under the influence of the fringing field caused by the collector junction reverse bias, which concentrated them on the surface of the base region. As a result, the surface of the base region became charged by either adsorption of ions or charge transfer from ions to surface impurities already present. The surface charge layer thus created caused an inversion layer to form in the base region, leading first to increased collector leakage current and finally to an emitter-base short-circuit when the base inverted along its entire length.

This work sparked interest in the radiation sensitivity of semiconductor surfaces. About this time, thermal oxide passivation technology leading to planar bipolar transistors was rapidly replacing mesa bipolar transistors

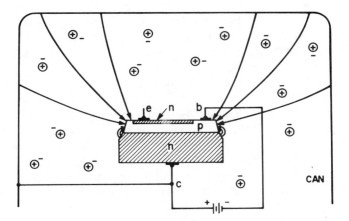

Fig. 11.28 Enhancement of ion collection at a mesa-type bipolar transistor surface due to an electric field between the transistor and its metal encapsulating case. After Peck et al.[82] Copyright (1963), American Telephone and Telegraph Company. Reprinted with permission.

(see Section 15.3.1). Also, MOSFET devices were coming into use in integrated circuit applications. It was felt that the MOSFET would be inherently resistant to radiation because it is a majority carrier device and degradation of lifetime caused by radiation would have negligible effect, unlike a bipolar transistor. If this immunity to radiation really existed, it would make the MOSFET ideal for many space applications.

As a result of the discovery of radiation-induced surface effects on transistors in the Telstar experiment, MOSFETs were tested in a radiation environment before being sent into space. Experiments by Hughes and Giroux[88] in 1964 on both n-channel and p-channel MOSFETs showed that large changes in transconductance occurred during exposure to ionizing radiation. Thus MOSFETs were not radiation insensitive as first supposed.

Szedon and Sandor[89] were among the first to recognize that the change in MOSFET characteristics was caused by the formation of positive charge in the oxide. This led to an intensive study of radiation effects in SiO₂. In many cases radiation damage in the SiO₂ was the dominant effect.

In present-day semiconductor technology, radiation damage in the semiconductor bulk has been reduced in bipolar transistors by using very shallow junctions. Shallow junctions are effective because the amount of ionization or lattice damage produced depends on the amount of energy absorbed from the incident particle by the silicon lattice. The incident particle loses most of its energy to the silicon lattice at some depth into the silicon. Therefore, making the junction shallow removes it from the region of maximum damage, which lies much deeper.

Another reason why radiation damage in the semiconductor is less important than in the oxide is that the incident radiation may lose so much energy creating damage in the oxide that it has insufficient energy to create any damage in the semiconductor. This is true mainly for lower energy particles.

One device that is inherently radiation resistant is the JFET, a majority carrier device that does not suffer intolerable degradation if lifetime decreases. Moreover, the JFET does not require a radiation-sensitive oxide. Nevertheless, efforts have continued to devise radiation-hardened SiO₂ and other insulators to permit use of integrated circuits, which must employ planar bipolar transistors and MOSFETs, in a radiation environment.

Table 11.2 summarizes the vulnerability of several types of semiconductor device to a radiation environment. Four degradation mechanisms occur in semiconductor devices. The three failure modes listed in Table 11.2 are lifetime degradation, carrier removal and trapping, and mobility degradation. These are the primary failure modes in all devices except the MOSFET. They are caused by displacement-type damage in the semiconductor bulk resulting mainly from exposure to neutrons. The primary failure mode of the MOSFET in a radiation environment is that listed in Table 11.2 under the column headed "oxide charge and surface effects."

Table 11.2 Radiation-Induced Degradation of Semiconductor Devices[a]

Device Type[b]	Lifetime Degradation	Carrier Removal and Trapping	Mobility Degradation	Oxide Charge and Surface Effects
		Failure Mechanisms		
Semiconductor resistors		P	S	
Diodes (breakdown voltage, depletion capacitance)	S	P		
Diodes (leakage, forward current, photocurrent)	P	S		S
Microwave diode sources (IMPATT, TRAPATT, BARITT)	S	P		
Junction field effect devices	S	P		
MOS field effect devices and capacitors				P
Microwave bulk oscillators (Gunn, LSA)		P	S	
Bipolar transistors	P	S		
Switching devices (SCR, UJT)	P	S		S
Optoelectronic devices (LED, junction laser)	P			

[a] After Gwyn.[83]

[b] Acronyms: IMPATT, impact avalanche transit time; TRAPATT, trapped plasma avalanche triggered transit; BARITT, barrier injection transit time; LSA, limited space charge accumulation; SCR, semiconductor controlled rectifier; UJT, unipolar junction transistor or JFET.

[c] Code: P, primary; S, secondary.

This failure is caused by ionizing radiation. Table 11.2 shows that displacement damage, which is rare in most radiation applications, is the primary failure mechanism in all semiconductor devices listed except the MOSFET.

The difference between ionizing radiation damage in SiO_2 and in semiconductors is that in semiconductors, ionizing radiation produces a transient effect, whereas in SiO_2, ionizing radiation produces a transient followed by a permanent or semipermanent effect. We concentrate on ionizing radiation effects in SiO_2 for the remainder of this section.

11.6.2 Definition of Units and Dosimetry

Before describing the experimental results of radiation damage in SiO_2, it is necessary to define the unit for measuring the total amount of radiation absorbed by a material, the *rad* (radiation-absorbed dose). One rad absorbed means that each gram of material has received 100 ergs of energy. A transistor in an earth satellite passing repeatedly through the Van Allen belts could absorb a total dose of 10^6 rads in one year. Outside the Van Allen belts or in interplanetary space, the dose absorbed in one year would be several orders of magnitude less.

The *roentgen* measures the total exposure to radiation, that is, the incident energy, not the amount absorbed. Therefore, this term is used to describe the magnitude of the radiation source rather than the extent of interaction of the radiation with a solid material.

(a) Dose Rate for X-Rays

The dose rate for X-rays can be estimated in two ways. First, a steady-state photocurrent flows in an MOS capacitor when X-rays create hole-electron pairs in the oxide. This photocurrent increases at first with increasing gate bias and finally saturates as gate bias is increased. Assuming that the saturation value of the photocurrent corresponds to the collection of all the carriers generated in the oxide, the absorbed dose rate R can be obtained from the relation

$$R = \frac{E_{pair}J}{q\rho x_o} \tag{11.62}$$

where E_{pair} is the energy absorbed per electron-hole pair created, J is the saturated photocurrent density, and ρ is the density of SiO_2 (about $2.3\ g/cm^3$). A typical X-ray source used in this type of work is a tungsten target tube operating at an anode voltage of 25 kV and an anode current of 35 mA. This will produce a saturated photocurrent density in an oxide 4000 Å thick of $4 \times 10^{-8}\ A/cm^2$. To calculate the absorbed dose rate from (11.62), we take E_{pair} to be about two times the bandgap of SiO_2 ($\sim 18\ eV$).[90] From (11.62), $R = 8 \times 10^4\ erg/g\text{-}sec = 800\ rads/sec$. The second way of estimating dose is obtained from the reverse current in a silicon *p-n* junction, again using (11.62) but with $E_{pair} = 3$ eV, with x_o replaced by the minority carrier diffusion length in silicon. The estimates of R obtained in both these ways are in fair agreement.

(b) Dose Rate for Low Energy Electrons

For low energy electron radiation, (≤ 20 keV), the flux of incident primary electrons can be measured by using a Faraday cup. The average energy ($E_o - E_p$) dissipated in the oxide by each primary electron can be determined by assuming the Thomson-Whiddington square law relationship[91] $E_o^2 - E_p^2 = bx_o$, where E_o is the energy of the incident electrons, E_p is the most probable energy of the transmitted electrons, and x_o is the oxide thickness. The constant b is 1.2×10^{12} eV^2/cm obtained by correcting the available empirical values reported for aluminum[91] for the difference in density between aluminum and SiO$_2$. Thus, for 20 keV electrons incident on a 2000 Å thick oxide, the dose in rads is given approximately by 2×10^{-7} times the total number of incident electrons per square centimeter. For high energy electrons (1 MeV), the dose can be calculated from the integrated electron current density by assuming that one electron per cubic centimeter corresponds to 3×10^{-8} rads.[92] Within the accuracy of this dosimetry, a given absorbed dose of either high or low energy electrons, X-rays, or γ-rays will have the same effect. The choice is then a matter of convenience. Low energy electron radiation is usually the most convenient to implement and gives a much wider range of dose rates than is possible easily with X-rays or γ-rays.

11.6.3 The MOS Capacitor

The MOS capacitor is widely used in the study of radiation effects in SiO$_2$. Figure 11.29 illustrates the effect of ionizing radiation on the C-V characteristics of an MOS capacitor. Figure 11.29 shows a set of high frequency C-V curves before radiation and after exposure to a dose of 6×10^4 and 2×10^6 rads with $+2.5$ V on the gate. The C-V curves shift along the voltage axis to higher negative voltages, thus indicating that positive charge is being induced in the oxide as a result of exposure to ionizing radiation. Also, the induced positive oxide charge density increases with increasing radiation dose. In addition to the shift of the C-V curve along the voltage axis, the shape of the C-V curve has been altered. After exposure to ionizing radiation, the C-V curves are stretched out along the voltage axis. This voltage stretchout can be caused either by an increase of gross charge nonuniformities (see Section 6.2) or by an increase in interface trap level density. As discussed later, measurement of surface recombination velocity using a gate controlled diode structure shows that voltage stretchout results from an increase in interface trap level density, and that surface recombination velocity increases with dose.

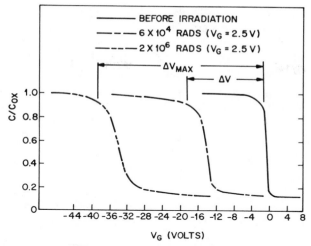

Fig. 11.29 Shift along the voltage axis in MOS capacitor C-V characteristics as a result of irradiation at two different dose levels under the same positive gate bias. After Gwyn.[83]

11.6.4 Properties of Radiation-Induced Oxide Charge

(a) *Electrical Properties*

Figure 11.30 is the voltage shift ΔV of the C-V characteristic of an MOS capacitor as a function of radiation dose at various dose rates. The flatband voltage shift ΔV was measured at the flatband point of the C-V curve with the preradiation curve taken as reference. A different capacitor from the same run was used for each indicated dose rate. A noticeable voltage shift

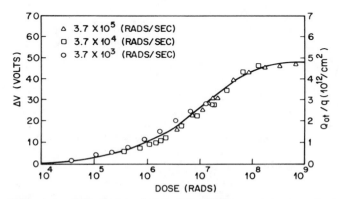

Fig. 11.30 Shift of the C-V characteristics of an MOS capacitor along the voltage axis ΔV and the corresponding charge $Q_{ot} = \Delta V C_{ox}$ as a function of dose with dose rate as parameter. A gate bias of $+2\,V$ was applied during irradiation. Oxide thickness was 2000 Å. After Snow et al.[93] Copyright (1967), IEEE.

begins to appear at about 10^4 rads, and ΔV saturates above a dose of 10^8 rads. The two important points in Fig. 11.30 are (1) the saturation of ΔV with dose and (2) the dependence of ΔV not on dose rate, but only on total dose $D = Rt$, where t = time. The density per unit area Q_{ot}, of the radiation induced charge is given approximately by $Q_{ot} = C_{ox} \Delta V$ because all the charge for positive gate bias during radiation will reside at or very near the Si–SiO$_2$ interface. A scale for Q_{ot} also is shown in Fig. 11.30.

Figure 11.31a shows ΔV as a function of dose with positive gate bias applied during radiation as parameter. Figure 11.31b shows ΔV as a function of dose with negative gate bias applied during radiation as parameter. Note that at high doses ΔV saturates and that ΔV at saturation

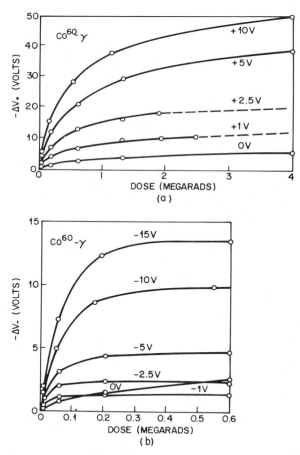

Fig. 11.31 (a) Shift of the C-V curve measured between the gate and the substrate of a MOSFET along the voltage axis ΔV_+ for positive aging gate bias as a function of Co^{60} γ-ray dose. (b) Same as (a), but for negative gate bias during irradiation. The sample has a 1600 Å deposited oxide. After Mitchell.[94] Copyright (1967), IEEE.

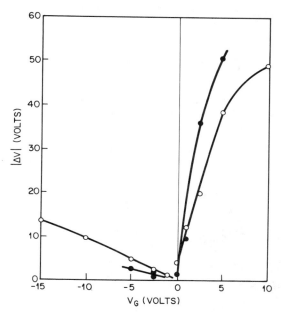

Fig. 11.32 Change in the threshold voltage of a MOSFET as a function of the gate bias applied during radiation after a dose of 4×10^6 rads of ^{60}Co γ-rays. Results for two different *n*-type silicon MOS capacitors are shown. Open circles denote a 1600 Å deposited oxide with a Cr–Au gate. Filled circles denote a 1200 Å dry thermal oxide with an aluminum gate. After Mitchell.[94] Copyright (1967), IEEE.

depends on the polarity and the magnitude of the gate bias applied during exposure.

The values of ΔV at a dose of 4×10^6 rads, denoted $|\Delta V|$ from curves such as those in Figs. 11.31*a,b* are plotted as a function of gate bias applied during radiation for both polarities in Fig. 11.32.[94] For negative bias, $|\Delta V|$ is the saturated value of ΔV, whereas for positive bias, it is about two-thirds of saturation. The value of $|\Delta V|$ at a given gate bias depends strongly on processing. Figure 11.32 shows that the saturation voltage shift due to positive oxide charge buildup is much greater for positive than for negative gate bias applied during radiation. A model that qualitatively explains this behavior is described in Section 11.6.5.

(b) Location in the Oxide

Etching experiments on samples that have been radiated with positive gate bias show that the radiation-induced positive oxide charge and charged interface traps are located within a few hundred angstroms of the Si–SiO₂ interface, in the case of X-ray, γ-ray, or high energy electron radiation. There is no charge distributed through the bulk of the oxide. Etch-off experiments and barrier height reduction measurements by inter-

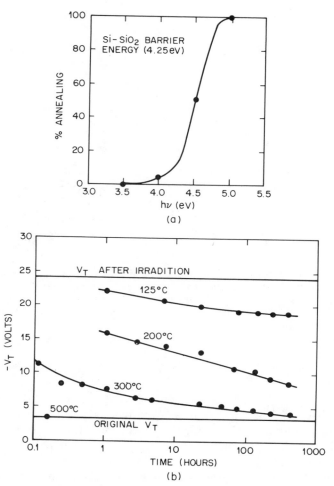

Fig. 11.33 (a) Percent annealing of the radiation induced oxide charge as a function of photon energy. The data points correspond to four different devices exposed to equal numbers of photons of the indicated energies. After Snow et al.[93] Copyright (1967), IEEE. (b) Thermal annealing of the radiation induced oxide charge as a function of time and temperature. A gate bias of $+2\,V$ was applied during irradiation. Samples were p-channel MOSFETs with a gate oxide thickness of 1200 Å and a substrate donor density of $10^{16}\,cm^{-3}$. After Snow et al.[93] Copyright (1967), IEEE.

nal photoemission of electrons from the gate metal[95] show that after radiation with negative gate bias, some of the positive oxide charge produced by radiation is localized at the metal-SiO₂ interface.* Barrier height lowering at the metal-SiO₂ interface measured by photoemission of electrons from the metal gate is conclusive evidence that some radiation induced charge is located very near the metal-SiO₂ interface. Barrier lowering of the Al–SiO₂ barrier by 0.65 eV for a dose of 10^6 rads and 0.90 eV for a dose of 10^7 rads is observed. Because the image potential at the interface, which accounts for the lowering, is prominent in the first 20 Å or so from the interface, a portion of the charge must be located very near the interface. Otherwise, barrier lowering of the magnitude observed would not be possible. However, the charge may extend beyond this depth (see Section 11.4.3).

(c) Annealing

Figure 11.33 shows how oxide charge can be annealed. Figure 11.33a shows annealing of oxide charge by illumination with UV light. To investigate this effect quantitatively, MOS capacitors with 200 Å thick gate electrodes, semitransparent to UV light, were prepared. Four units were then radiated with a gate bias of + 2 V. Then, using a UV light source and monochrometer, the units were exposed to equal numbers of 3.5, 4.0, 4.5, and 5.0 eV photons. The percent annealing for the four units is shown in Fig. 11.33a. There is a rather sharp threshold energy for annealing, which is consistent with the Si–SiO₂ barrier energy of 4.25 eV. Therefore, electrons are injected from the silicon into the oxide conduction band and subsequently neutralize the positive charge.

The results of thermal annealing using a group of p-channel MOSFETs with 1200 Å gate oxides that had been radiated with X-rays with + 2 V on the gate are shown in Fig. 11.33b. During radiation, the threshold voltage V_T of the MOSFETs increased from 3.4 to 24 V, corresponding to a radiation-induced positive charge density of 3.8×10^{12} cm⁻². The devices were then annealed at various temperatures and threshold voltage measured as a function of time as shown in Fig. 11.33b. An Arrhenius plot of the data in Fig. 11.33b at any given time is not linear. Therefore, a single activation energy characteristic of the annealing process cannot be extracted from these data.

11.6.5 Model for Radiation-Induced Oxide Charge

Our next task is to explain the experimental observations with a model. This model will explain features of the radiation induced positive oxide charge.

---

*The density of hole traps at the metal-SiO₂ interface may depend on contaminants introduced during processing.[96] Therefore, these remarks are pertinent only to these particular samples.

Zaininger,[97] followed independently by Grove and Snow[98] soon after, proposed the following simple model that quantitatively explains most of the observed features of positive oxide charge buildup. Ionizing radiation passing through the oxide creates hole-electron pairs in the oxide by breaking Si—O bonds as shown in Fig. 11.34. Some of the generated carriers recombine, but most are driven toward the electrodes by the oxide field applied during irradiation. Electrons rapidly drift toward the positive electrode, where most flow out into the external circuit, as shown in Fig. 11.34. Very few electrons become trapped in the oxide. Holes drift much more slowly toward the negative electrode. Once holes reach the Si–SiO₂ interface, a fraction becomes trapped in this interfacial region, as shown in Fig. 11.34.*

There is very little permanent hole trapping anywhere in the oxide except at the Si–SiO₂ interface and possibly at the metal-SiO₂ interface, where hole traps are located. Trapped holes at the Si–SiO₂ interface constitute the radiation induced positive oxide charge observed. These trapped holes also may be responsible for the increased interface trap level

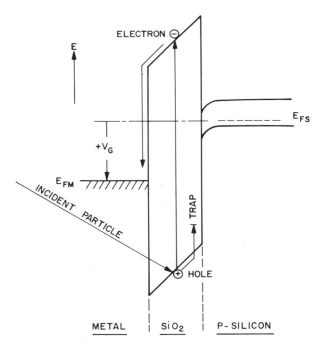

Fig. 11.34 Energy-band diagram of the metal–SiO₂–Si system illustrating the creation of a hole-electron pair in the SiO₂ by ionizing radiation, the subsequent drift under positive gate bias of the electron to the gate and the drift of the hole to the Si-SiO₂ interface, where it is trapped to form positive oxide charge.

*This fraction varies with oxide fabrication and may be as small as a few percent or less.[99]

density usually associated with ionizing radiation at large doses, but exactly how these interface traps are created is not known (see Section 11.6.7).

This model requires hole transport through the SiO₂. Hole transport in SiO₂ could be controlled by bulk oxide traps.* If so, the times that holes spend in these traps are spread over a wide range and are comparable with the transit time of the holes across the oxide. As a result, the transit time t_T increases more rapidly than linearly with oxide thickness x_o. The customary mobility $\mu = x_o/(t_T F_{ox})$ (where F_{ox} is the oxide field) is then oxide thickness dependent, field dependent, and time dependent. Hole transport cannot be described by drift alone, but must include capture and release of holes from bulk traps.[100] A number of models of this type have been fitted to hole transport. One of these[99,101] describes transport as phonon assisted hopping between bulk oxide trap sites. Theoretical objections to this model have yet to be settled.[102] Another model describes transport as drift in the valence band interrupted by residence in bulk oxide traps.[103] These two models are formally equivalent,[100] thus making it difficult to identify the microscopic transport process.

Experimentally, hole transport is observed to be very slow. Using measurement times as short as 10 nsec, where bulk oxide trapping is assumed to have only a small effect on transport, Hughes[104] found the intrinsic mobility of holes in SiO₂ to be $\mu = 1.2 \times 10^{-2} \exp(-0.16q/kT)$ cm²/V-sec, or about 2×10^{-5} cm²/V-sec at room temperature. The defect-controlled transport also is thermally activated, becomes faster as temperature increases, and depends very strongly on field and oxide growth conditions. Wet oxides show much lower effective hole mobilities than do dry oxides.[105]

The fraction of electron-hole pairs that escape recombination is a function of the oxide field applied during irradiation and of the energy of the incident radiation. This fractional yield is shown in Fig. 11.35 for several types of radiation.[106] At low oxide fields many pairs recombine before the field sweeps the electrons away from the holes. The fractional yield increases almost linearly with field in the region below 5×10^5 V/cm. At fields above 2×10^6 V/cm, recombination is suppressed and nearly all the generated carriers are transported to the interfaces.

Using this model and the fractional yield dependence on field of Fig. 11.35, the MOSFET threshold voltage shift versus gate bias during irradiation shown in Fig. 11.32 can be understood qualitatively. First, for the low oxide fields shown in Fig. 11.32, the fractional yield is proportional to oxide field (from Fig. 11.35) and hence to gate bias; thus a linear dependence of threshold voltage shift on gate bias is expected. Second, the steeper slope of threshold voltage shift versus gate bias for positive gate bias compared to the slope for negative gate bias is explained by the different number of

---

*If such bulk oxide traps exist, they do not trap holes permanently. They differ in this respect from hole traps at the Si–SiO₂ interface, which trap holes for a long time.

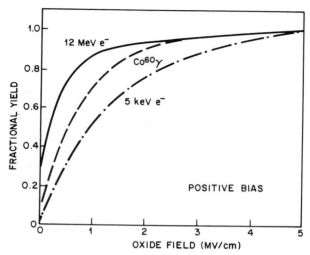

Fig. 11.35 Field and energy dependence of the fraction of generated electron-hole pairs that escape recombination in MOS capacitors with thermally grown SiO_2. After McGarrity.[106] Copyright (1980), IEEE.

holes trapped near the Si–SiO_2 interface for these two polarities. Thus, for a given oxide field applied during radiation, the same number of holes may be generated regardless of field polarity. However, for positive gate bias, most holes drift to the Si–SiO_2 interface, and a fraction of all these holes become trapped. For negative gate bias, most holes drift to the metal gate, and only a fraction of those holes actually generated near the Si–SiO_2 interface become trapped there—a small number compared to the positive bias case.

Holes trapped near the metal-SiO_2 interface are not very germane to this discussion because it is the product of trapped charge and its centroid with respect to the gate that determines threshold [see (10.9)], and the centroid of charge trapped near the gate is near zero.

(a) Experimental Verification of the Model

The experiments of Powell and Derbenwick[107] support the model just presented for the creation of radiation-induced oxide charge. The key feature of these experiments is the use of vacuum ultraviolet (VUV) photons as the ionizing particles. The advantages of VUV are that the depth of light absorption in the oxide can be controlled by varying photon energy and that only hole-electron pairs are produced without displacement damage or other effects.*

---

*Lattice relaxation effects associated with hole-electron pair production may occur.

These experiments were done on MOS capacitors having a semitrans-parent gold electrode about 100 Å thick deposited on a 2000 Å thick oxide thermally grown in dry oxygen on a 0.8 Ω-cm p-type silicon substrate. The electrode configuration and optical apparatus are described in Section 12.10. In these experiments (1) there is an energy threshold for the effect near the fundamental absorption edge of SiO₂ (8.8 eV), (2) light penetration was on the order of 100 Å, thus localizing hole-electron pair production to the vicinity of the metal-oxide interface, and (3) the positive oxide charge produced was found to be localized at the Si–SiO₂ interface by an etch-off experiment.

Figure 11.36 shows the optical absorption coefficient and absorption depth as a function of photon energy. Extrapolation of the upper portion of the curve gives a value of approximately 9 eV for the fundamental ab-sorption edge of SiO₂. Thus the radiation charging threshold should occur at about this energy. The absorption depth in the oxide is about 100 Å for photon energies greater than about 10 eV, as seen from the right-hand scale in Fig. 11.36. Therefore, for exposure of reasonably thick samples (say, 2000 Å) to such radiation, virtually no light reaches the Si–SiO₂ interface and essentially all the hole-electron pairs are produced in the outermost few hundred angstroms of the SiO₂.

It is difficult to define an absorption edge from optical transmission measurements in SiO₂ because of the long, low energy tail in the absorption.

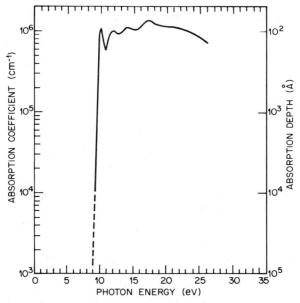

Fig. 11.36 Silica optical absorption coefficient and absorption depth as a function of photon energy. After Powell and Derbenwick.[107] Copyright (1971), IEEE.

Therefore, the optical properties of SiO_2 shown in Fig. 11.36 have been obtained from reflectance measurements that are insensitive to the small absorption coefficients normally measured in transmission. Powell and Derbenwick[107] found that the reflectance of their SiO_2 samples agreed quite well with that of Phillipp[108] for fused silica. Figure 11.36 is obtained by a Kramers-Kronig analysis of Phillipp's reflectance data and is in agreement with the work of DiStefano and Eastman.[109]

Figure 11.37 shows that the threshold energy for positive oxide charge production coincides with the bandgap energy of SiO_2. Figure 11.37 was obtained by radiating samples with photon energies from well below to well above the optical absorption edge or bandgap energy. Figures 11.37a and 11.37b show the initial and post-irradiation high and low frequency C-V curves, respectively, with photon energy as parameter. A positive gate bias of 20 V was applied during radiation at a constant dose of approximately 6.7×10^{14} photons/cm^2.* The C-V curves are shifted to more negative gate

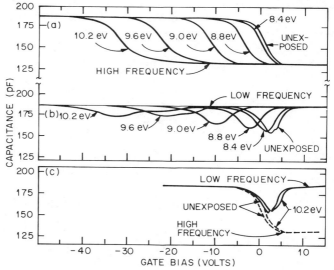

Fig. 11.37 (a) High frequency MOS capacitor C-V curves for irradiation at constant dose under positive gate bias with photon energy as parameter. After Powell and Derbenwick.[107] Copyright (1971), IEEE. (b) Low frequency MOS capacitor C-V curves for irradiation at constant dose under positive gate bias with photon energy as parameter. After Powell and Derbenwick.[107] Copyright (1971), IEEE. (c) High and low frequency MOS capacitor C-V curves with constant dose at a photon energy of 10.2 eV under negative gate bias. After Powell and Derbenwick.[107] Copyright (1971), IEEE.

*This dose corresponds to 2.4×10^8 rads at 10.2 eV. However, because the amount of positive charging varies directly with the number of hole-electron pairs produced, the photon dose is more meaningful than rads as a measure of the incident radiation.

bias corresponding to the creation of positive oxide charge. Figures 11.37a, b show that the shift along the voltage axis is a strong function of photon energy varying from essentially zero for 8.4 eV photon exposure to over 30 V for 10.2 eV photon exposure. The charging threshold is about 8.8 eV, quite close to the value of 9 eV predicted from the optical data in Fig. 11.36.

Figure 11.37c shows C-V curves before and after 10.2 eV photon exposure at the same dose as in Figs. 11.37a,b, but with a negative gate bias of 20 V applied. No net positive oxide charge was produced. Figure 11.37 is interpreted as follows. Consider first positive gate bias and large (> 10 eV) photon energies. There are two processes to consider under these conditions: hole-electron pair production in the oxide near the gate and hole injection from the gold gate into the oxide. Hole injection from the gold is minor because very little light is absorbed by the semitransparent gold. There is no electron injection from the silicon* because no light reaches the silicon at the higher photon energies. The applied field rapidly sweeps electrons to the gate, thereby avoiding significant recombination. There is negligible electron trapping in the oxide because electron trapping is a very inefficient process, as described in Section 11.5.1, and most of the holes arriving at the Si–SiO₂ interface become trapped there. Therefore, every photon that creates a hole-electron pair contributes a potential positive oxide charge.

Next, consider positive gate bias at lower photon energies. Whereas 9.6, 9.0, and 8.8 eV photons produce hole-electron pairs in SiO₂, progressively more light reaches the silicon, thus producing electron injection from the silicon into the oxide. These injected electrons can annihilate both trapped positive charge at the Si–SiO₂ interface and mobile holes. Therefore, as photon energy is lowered, charging decreases for two reasons: (1) fewer hole-electron pairs are produced in the SiO₂ as more light penetrates all the way to the silicon; and (2) the competing process of electron injection from the silicon annihilates some of the positive charge. Thus the amount of charge produced at a given photon energy depends on the detailed balance between pair production and electron injection and the capture cross sections of both mobile and trapped holes.

For irradiation under negative gate bias, only pair production in the oxide near the gate and electron injection from the gate can occur. There will be little or no oxide charging because holes will move toward the metal gate and, even if trapped, can be annihilated by electrons injected from the gate. At lower photon energies, some positive oxide charging might occur

---

*The accumulation of a large positive oxide charge can result in sufficient field at the Si–SiO₂ interface to produce significant electron tunneling from the silicon into the oxide. This is one mechanism that limits oxide charge accumulation at large doses and is believed to be responsible for at least part of some small relaxation effects observed following cessation of irradiation.

because hole-electron pairs are created near the Si–SiO$_2$ interface and hole injection from the silicon may become significant as more light reaches the silicon. This positive charging depends on the magnitude of electron injection from the gate and may be enhanced at a photon energy for which interference produces a light intensity minimum at the gate (see Section 10.3.2) and little electron injection. At a sufficiently large photon energy, electron injection will annihilate all the holes and no charging will be observed. Figure 11.37c shows a slight negative oxide charging that may result from electron trapping by water-related traps in the oxide discussed in Section 11.5.1.

Positive charging is not observed for photon energies less than SiO$_2$ bandgap energy, even though holes can be injected from the electrodes. Hole injection is smaller than electron injection from the opposite electrode, probably because of a very short scattering length for holes in SiO$_2$. Most holes scatter before they reach the image-force potential maximum and are then returned to the injecting electrode.

The positive oxide charge was located by an oxide etch-off experiment. After irradiation with 10.2 eV photons and positive gate bias, the sample was successively etched. Measurement of the C-V curve between etching steps showed that within the sensitivity of the measurement, no oxide charge was removed to within 300 Å of the Si–SiO$_2$ interface. The increase of interface trap level density also observed in these experiments suggests that hole trapping occurs very near the Si–SiO$_2$ interface. For 10.2 eV photon irradiation, virtually all hole-electron pairs were created within the outer few hundred angstroms of the SiO$_2$ layer. Thus hole transport from the region of creation to the region of trapping at the Si–SiO$_2$ interface must have occurred. If holes were immobile, the positive oxide charge would have been distributed exponentially in the region of light absorption with virtually no charge near the Si–SiO$_2$ interface.

With positive gate bias during irradiation and a light intensity of about 2×10^{11} photons sec^{-1} cm^{-2}, decaying photocurrents of $\sim 10^{-11}$ A at a photon energy of 10.2 eV were observed to have the sign corresponding to the flow of holes from the gate toward the silicon. The decay rate depended on applied gate bias, but a typical decay time to 50% of initial value was about 2 min. A similar decay was observed when illumination was terminated, thus suggesting that hole mobility is very small and perhaps dominated by the effects of capture and emission from hole traps. However, these results suggest that although hole mobility in SiO$_2$ is quite small, it is not zero. Some holes drift to the vicinity of the Si–SiO$_2$ interface, where a large fraction of them become trapped, resulting in a net positive oxide charge.

To show that oxide charging by irradiation is independent of the gate metal, an electrodeless radiation experiment was done. The initial C-V curve was measured with a mercury probe described in Section 12.8.3. The sample then was irradiated in a vacuum with 10.2 eV photons. To provide a field in the oxide to drift holes to the Si–SiO$_2$ interface, a metal collector plate with a hole in it to admit light was placed near the SiO₂ surface. A bias was applied to this collector plate, positive with respect to the silicon substrate. The SiO₂ surface became positively charged when electrons were photoemitted out of the SiO₂ by the 10.2 eV photons and collected by the positively biased collector plate. After irradiation, the C-V curve measured on the irradiated area showed a 15-V shift corresponding to positive oxide charge, proving that positive oxide charge buildup is independent of a metal gate electrode.*

In the electrodeless experiment, continuity of current was maintained in the sample by hole transport across the oxide. If holes were absolutely immobile, the oxide surface would quickly charge to a sufficiently positive potential to quench the photocurrent. Lack of quenching of the photocurrent provides further evidence for hole transport across the SiO₂.

These experiments are interpreted in terms of hole transport and trapping, although there is only indirect evidence that it is holes that are transported. Instead of holes, could some positive ionic species be responsible for positive charge transport? One species present in the oxide that could become positively charged is the hydrogen atom, which is bonded to an oxygen atom in a water-related trap. The direct breaking of oxygen-hydrogen bonds by photons is much less likely than hole-electron pair production because the density of valence electrons in the oxide is 10^3–10^4 times greater than the density of water-related traps in the first 400–500 Å of oxide. In addition, the hydrogen-oxygen bond energy (~ 4.8 eV[110]) is less than the SiO₂ bandgap energy, so that positive oxide charging by protons would not be likely to have a threshold at 8.8 eV. Thus direct generation of protons by irradiation is unlikely.

Another possibility is that protons are produced indirectly through the free carriers generated by incoming radiation. For electrons, compelling evidence against this comes from the avalanche injection experiments described in Section 11.5.1. When electrons are injected into the oxide, some of them are captured by water-related traps initiating an electrochemical reaction whose products are neutral free hydrogen and an immobile *negatively* charged center. Therefore, electrons produced by ionizing radiation do not create free protons. When holes are avalanche injected into the oxide, no residual mobile charge is found. A buildup of fixed positive oxide charge and an increase in interface trap level density are observed. These results suggest that, like electrons, holes produced by ionizing radiation do not produce free protons, either.

Because alkali metal atoms also bond to oxygen atoms in the silica lattice, they might be the mobile positive species. The arguments given to eliminate free protons apply equally to free alkali metal ions. In addition,

---

*No charging was observed under zero bias irradiation, that is, with the metal electrode short-circuited to the silicon substrate.

alkali metals move too slowly to account for the observed transit times of the mobile positive charge.[90]

These arguments eliminate free protons and free alkali metal ions as the cause of mobile positive oxide charge resulting from exposure to ionizing radiation. However, the role of free hydrogen produced by the capture of electrons at water-related traps during irradiation has yet to be established.

(b) Oxide Thickness Dependence of Flatband Voltage Shift

The radiation-induced flatband voltage shift in oxides grown in dry oxygen varies linearly with oxide thickness[111] (or oxidation time at a given oxidation temperature) for low energy (10.2 eV) photon radiation that does not penetrate deeply into the oxide. Because radiation-induced positive oxide charge is located at the Si–SiO₂ interface, the charge centroid is equal to the oxide thickness. Provided that the radiation-induced oxide charge density is independent of oxide thickness, the flatband voltage shift will vary linearly with the charge centroid or oxide thickness as predicted by (10.9). This is the case for low energy radiation where the energy is almost entirely (95%) absorbed within the outer 300 Å of oxide.[111] Therefore, the net electron-hole pair generation is always confined to a thin surface layer of oxide and Q_{ot} is virtually independent of oxide thickness[111] ($x_o \geq 500$ Å).

A linear variation of flatband voltage with x_o is not observed at high energies (10^6 eV). Because high energy radiation penetrates the oxide deeply, electron-hole pair generation can occur throughout the oxide layer. Assuming uniform generation in the oxide and the trapping of a constant fraction of the generated holes, Q_{ot} will vary linearly with thickness, and (10.9) predicts a square-law dependence.[111]

A square-law dependence is not commonly observed. The dependence on oxide thickness varies with oxide quality and with oxide thickness. For example, Derbenwick and Gregory[112] have found that the flatband voltage shift varies as the cube of oxide thickness for dry oxides for high energy radiation. A distribution of hole traps into the oxide from the Si–SiO₂ interface, the possible dependence of hole trap density on oxide thickness, and interface traps generated at high doses could alter the simple square-law dependence. It is not known why the oxide thickness dependence of the flatband voltage shift is so varied. The main point is that the radiation induced charge density increases with increasing oxide thickness for high energy radiation.

11.6.6 Properties of Intrinsic Hole Traps at the Si–SiO₂ Interface

Because ionizing radiation does not create traps, hole traps are intrinsic to the Si–SiO₂ interface. They can be studied by avalanche injection as well as exposure to ionizing radiation. From avalanche injection measurements, intrinsic hole trap densities of up to 10^{13} cm⁻² have been found.[4,113] Intrinsic hole traps do not undergo a chemical change when a hole is trapped because these traps can be completely discharged by electrons photoemitted from either the metal or the silicon. The energy levels of these intrinsic hole traps must be more than 4 eV above the SiO_2 valence band because photon energies below the energy required for photoemission from either electrode does not discharge them.[114] However, they can be discharged thermally at temperatures of 150–400°C, with an activation energy of about 0.35 eV.[4] This activation energy suggests that thermal discharge occurs as a result of a diffusion limited chemical reaction with water. Intrinsic hole traps are initially neutral, and the capture cross section for holes is between 10^{-13} cm² and 10^{-14} cm²,[113] suggesting a Coulombic center. Thus, for a given injected current density, hole charging is approximately 10^3–10^4 times more rapid than electron charging (see Table 11.1 for the water-related trap). Therefore, the two intrinsic traps most important in SiO_2 are water-related traps that capture electrons and intrinsic hole traps.

Because intrinsic hole traps have energy levels more than 4 eV above the SiO_2 valence band, hole emission after capture at room temperature is unlikely. Therefore, (11.58) describes Figs. 11.30 and 11.31.

The nature of intrinsic hole traps is discussed in Section 16.5 and the method for preparation of radiation-insensitive or hardened oxides is described in Section 15.7.4. Radiation-induced bulk oxide traps, discussed in Ref. 79, are not considered.

11.6.7 Radiation-Induced Interface Traps

The major effect of ionizing radiation on MOS devices in most industrial and space applications is the generation of positive oxide charge resulting from hole trapping at the Si–SiO_2 interface, as discussed in the previous sections. Another effect of importance mainly at larger doses ($\gtrsim 10^5$ rads) is the generation of interface trap levels within the silicon bandgap.

(a) Evidence for Interface Trap Generation

Evidence for interface trap levels is the radiation-induced distortion of C-V curves of MOS capacitors and of I-V curves of MOSFETs. However, as discussed in Section 6.2.5, gross lateral charge nonuniformities and interface traps can produce similar distortions in the C-V curves of an MOS capacitor and in the I-V curves of an MOSFET. To distinguish between these two sources of distortion, one procedure is to measure the apparent doping profile, as discussed in Section 6.2.5. An alternative method is to measure surface recombination velocity. If interface traps are produced uniformly across the interface, surface recombination velocity is proportional to interface trap level density and the square root of the product of the hole and electron capture cross sections. These parameters can be measured independently using the conductance method to ensure that interface traps are produced by ionizing radiation.

To measure surface recombination velocity S_o, a small reverse bias ($V_R = 0.3$ V) is applied to the junction of a gate controlled diode, and reverse current is measured as a function of gate bias. A peak occurs in the I-V characteristic when the silicon surface under the gate is depleted because depletion allows current to be generated through interface trap levels near midgap. The current peak is due to the surface component of generation current $I_{gen\,s}$ and is related to S_o by[93]

$$I_{gen\,s} = q n_i S_o A_g \tag{11.63}$$

where A_g is the gate area.

Figure 11.38 shows S_o as a function of radiation dose exposed with $V_G = 0$. The value of S_o prior to radiation exposure was 5 cm/sec. Figure 11.38 shows that S_o increases with dose and finally saturates above 10^8 rads. The increase in S_o takes place over the same dose range as the distortion of the C-V and I_D versus V_G curves, suggesting that S_o increases because radiation causes D_{it} to increase.

Conductance measurements confirm that D_{it} is uniformly distributed in energy near the center of the bandgap.* Therefore, S_o and D_{it} are related as[93]

$$S_o = \tfrac{1}{2}(\sigma_n \sigma_p)^{1/2} \bar{v} \pi k T\, D_{it} \tag{11.64}$$

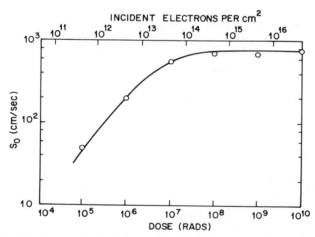

Fig. 11.38 Surface recombination velocity S_o of a depleted silicon surface as a function of radiation dose. The preirradiation value of S_o was 5 cm/sec. Measurements were made on an n^+p gate controlled diode with $x_o = 2000$ Å and $N_A = 5 \times 10^{15}$ cm³. After Snow et al.[93] Copyright (1967), IEEE.

---

*This conclusion is based on unpublished data by E. H. Nicollian and A. Goetzberger on samples prepared and exposed to X-rays by A. S. Grove and D. J. Fitzgerald. This work was done in 1966 and 1967.

where σ_n and σ_p are electron and hole capture cross sections, respectively; $\bar{v}$ is the thermal velocity, assumed to be the same for holes and electrons; k is Boltzmann's constant; and T is the absolute temperature. An order of magnitude estimate of the product $\sigma_n \sigma_p$ for interface traps created by radiation is found from (11.64) using S_o obtained by measurements on the gate controlled diode and D_{it} from conductance measurements. Such an estimate leads to $(\sigma_n \sigma_p)^{1/2}$ of the order of 10^{-16}–10^{-15} cm². These values are in good agreement with σ_n and σ_p obtained directly from conductance measurements of interface traps on the same thermal oxide after radiation.* The conductance measurements also show that σ_n and σ_p do not

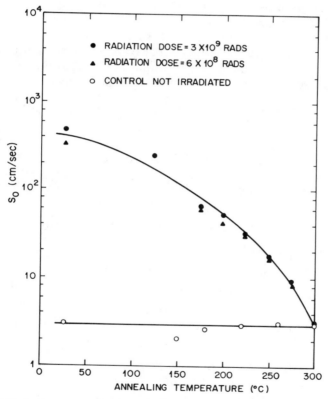

Fig. 11.39 Surface recombination velocity S_o of an electron irradiated oxide as a function of annealing temperature. The devices were annealed for 5 min at each temperature, with temperature increased in 24°C steps. Substrate acceptor density was 5×10^{15} cm⁻³, and gate oxide thickness was 2000 Å. After Snow et al.[93] Copyright (1967), IEEE.

*See previous footnote.

change with irradiation. Consequently, the increase in S_o with dose in Fig. 11.38 is due to an increase in D_{it} with dose, according to (11.64).

(b) Annealing

The increase in surface recombination velocity caused by irradiation can be annealed thermally in the same temperature range as discussed previously for oxide charge. Isochronal annealing curves are shown in Fig. 11.39. Devices were radiated with $\sim 10^9$ rads of 20 keV electrons at zero gate bias. Then temperature was increased in 25°C steps and the devices heated for 5 min at each temperature. Figure 11.39 shows that some annealing took place over the entire temperature range and that the devices were completely annealed at 300°C. Because S_o is proportional to interface trap level density, Fig. 11.39 also shows the thermal annealing of radiation induced interface traps.

For doses between 2×10^5 and 2×10^7 rads (below the saturation dose given in Fig. 11.38), Winokur et al.[115,116] observe a distortion in C-V curves, possibly due to interface trap generation. This distortion continues to increase for several hours following radiation exposure.

11.6.8 MOSFET Threshold Voltage Degradation

Figure 11.40 shows threshold voltage as a function of radiation dose for n-channel and p-channel MOSFETs. The curves in this figure are representative of hardened oxides (see caption to Fig. 11.40 and Section 15.7.4). In curve a in Fig. 11.40 for n-channel, the formation of positive

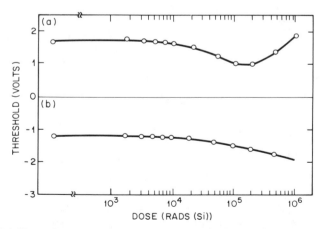

Fig. 11.40 (a) Threshold voltage of an n-channel MOSFET versus total radiation dose. (b) Threshold voltage of a p-channel MOSFET versus total radiation dose. Oxides in both transistors were grown to a thickness of 700 Å at 850°C in HCl-steam followed by a nitrogen anneal at 850°C. Gates were aluminum, and the devices were irradiated with Co^{60} γ-rays under a bias of +5 V. After King and Martin.[117] Copyright (1977), IEEE.

oxide charge with increasing dose causes threshold voltage to decrease. This decrease becomes significant above a dose of about 10^4 rads. Above 10^5 rads, threshold voltage increases, presumably as a result of formation of negatively charged interface traps. In curve b for p-channel, the formation of positive oxide charge with increasing dose causes threshold voltage to increase. This increase becomes significant above a dose of about 10^4 rads, just as for n-channel. However, unlike n-channel, threshold voltage increases monotonically without the appearance of any compensating charge. This difference between n-channel and p-channel devices is not understood.

REFERENCES

1   R. J. Powell and G. F. Derbenwick, *IEEE Transact. Nucl. Sci.*, NS-18 (6), 99 (1971).
2   E. H. Nicollian and A. Goetzberger, *Proceedings of Sixth Annual Reliability Physics Symposium*, 1967, pp. 66–67. IEEE catalog No. 7–15C58.
3   E. H. Nicollian and A. Goetzberger, *IEEE Transact. Electron Devices*, ED-15, 686 (1968).
4   E. H. Nicollian, A. Goetzberger, and C. N. Berglund, *Appl. Phys. Lett.*, 15, 174 (1969).
5   R. Poirier and J. Olivier, *Appl. Phys. Lett.*, 15, 364 (1969).
6   E. H. Nicollian and C. N. Berglund, *J. Appl. Phys.*, 41, 3052 (1970).
7   E. H. Snow, *Solid-State Commun.*, 5, 813 (1967).
8   M. Lenzlinger and E. H. Snow, *J. Appl. Phys.*, 40, 278 (1969).
9   C. M. Osburn and E. J. Weitzman, *J. Electrochem. Soc.*, 119, 603 (1972).
10   Z. A. Weinberg and A. Hartstein, *Solid-State Commun.*, 20, 179 (1976).
11   Z. A. Weinberg, *Solid-State Electron.*, 20, 11 (1972).
12   G. Lewicki and J. Maserjian, *J. Appl. Phys.*, 46, 3032 (1975).
13   G. Lewicki, *J. Appl. Phys.*, 47, 1552 (1976).
14   D. Kahng and E. H. Nicollian, "Physics of Multilayer-Gate IGFET Memories," *Applied Solid State Science*, Vol. 3, R. Wolfe, Ed., Academic, New York, 1972, pp. 2–70.
15   S. A. Abbas and R. C. Dockerty, *Appl. Phys. Lett.* 27, 147 (1975).
16   T. H. Ning, C. M. Osburn, and H. N. Yu, *Appl. Phys. Lett.*, 29, 198 (1976).
17   T. H. Ning, C. M. Osburn, and H. N. Yu, *J. Electron. Mater.*, 6, 65 (1977).
18   P. E. Cottrell and E. M. Buturla, *International Electron Devices Meeting Technical Digest*, New York, 1975, pp. 51–54.
19   S. A. Abbas and R. C. Dockerty, *International Electron Devices Meeting Technical Digest*, New York, 1975, pp. 35–38.
20   H. Hara, Y. Okamoto, and H. Ohnuma, *Jap. J. Appl. Phys.*, 9, 1103 (1970).
21   T. H. Ning, P. W. Cook, R. H. Dennard, C. M. Osburn, S. E. Schuster, and H. N. Yu, *IEEE Transact. Electron Devices*, ED-26, 346 (1979); *IEEE J. Solid-State Circuits*, SC-14, 268 (1979).
22   H. G. Dill, R. W. Bower, K. G. Aubuchon, and T. N. Toombs, *Digest of Technical Papers, International Solid-State Circuits Conference*, 1969, pp. 44–45.
23   D. M. Erb, H. G. Dill, and T. N. Toombs, *IEEE Transact. Electron Devices*, ED-18, 105 (1971).

24 M. Pepper, *J. Phys.*, **D6**, 2124 (1973).

25 M. Pepper, *IEEE Transact. Electron Devices*, **ED-21**, 174 (1974).

26 C. A. Neugebauer, J. F. Burgess, R. E. Joynson, and J. L. Mundy, *Thin Solid Films*, **13**, 5 (1972).

27 J. F. Verwey, *J. Appl. Phys.*, **44**, 2681 (1973).

28 C. A. Bosselaar, *Solid-State Electron.*, **16**, 648 (1973).

29 T. H. Ning and H. N. Yu, *J. Appl. Phys.*, **45**, 5373 (1974).

30 C. Bulucea, *Solid-State Electron.*, **18**, 363 (1975).

31 C. Bulucea, *Solid-State Electron.*, **18**, 381 (1975).

32 D. R. Young, *J. Appl. Phys.*, **47**, 2098 (1976).

33 P. J. Dunn and P. J. T. Mellor, *Microelectron. Reliab.*, **11**, 369 (1972).

34 B. McDonald, *International Electron Devices Meeting Technical Digest*, 1968, p. 70.

35 B. McDonald, *IEEE Transact. Electron Devices*, **ED-17**, 871 (1970).

36 J. F. Schenck, *Proceedings of the Sixth Annual Reliability Physics Symposium*, 1967, pp. 31–39. IEEE Catalog No. 7–15C58.

37 F. Conti and M. Conti, *Solid-State Electron.*, **15**, 93 (1972).

38 D. Frohman-Bentchkowski, *ISSCC Dig. Tech. Papers*, **15**, 80 (1971).

39 D. Frohman-Bentchkowski, *Appl. Phys. Lett.*, **18**, 332 (1971).

40 C. M. Berglund and R. J. Powell, *J. Appl. Phys.*, **42**, 573 (1971).

41 D. J. Bartelink, J. L. Moll, and N. I. Meyer, *Phys. Rev.*, **130**, 972 (1963).

42 A. D. Lopez, *J. Electrochem. Soc.*, **113**, 89 (1966).

43 E. H. Nicollian, C. N. Berglund, P. F. Schmidt, and J. M. Andrews, *J. Appl. Phys.*, **42**, 5654 (1971).

44 T. W. O'Keefe and R. M. Handy, *Solid-State Electron.*, **11**, 261 (1968).

45 B. E. Deal, M. Sklar, A. S. Grove, and E. H. Snow, *J. Electrochem. Soc.*, **114**, 266 (1967).

46 R. J. Powell and C. N. Berglund, *J. Appl. Phys.*, **42**, 4390 (1971).

47 G. Sixt, M. Schulz, and A. Goetzberger, *Appl. Phys.*, **4**, 217 (1974).

48 D. J. DiMaria, *J. Appl. Phys.*, **47**, 4073 (1976).

49 J. R. Brews, *J. Appl. Phys.*, **44**, 379 (1973).

50 C. N. Berglund and R. J. Powell, *J. Appl. Phys.*, **42**, 573 (1971).

51 R. J. Powell, *IEEE Transact. Nucl. Sci.*, **NS-17** (6), 41 (1970).

52 S. M. Sze, *Physics of Semiconductor Devices*, Wiley, New York, 1969, Chapter 8.

53 D. Kahng and E. H. Nicollian, in *Applied Solid State Science; Advances in Materials and Device Research*, Vol. 3, R. Wolfe, Ed., Academic, New York, 1972, Chapter 1.

54 D. Kahng, W. J. Sundberg, D. M. Boulin, and J. R. Ligenza, *Bell Syst. Tech. J.*, **53**, 1723 (1976).

55 K. K. Thornber, D. Kahng, and C. T. Neppell, *Bell Syst. Tech. J.*, **53**, 1741 (1974).

56 D. J. DiMaria, D. R. Young, R. F. DeKeersmaecker, W. R. Hunter, and C. M. Serrano, *J. Appl. Phys.*, **49**, 5441 (1978).

57 J. M. Aitken and D. R. Young, *J. Appl. Phys.*, **47**, 1196 (1976).

58 D. J. DiMaria, J. M. Aitken, and D. R. Young, *J. Appl. Phys.*, **47**, 2740 (1976).

59 R. A. Gdula, *J. Electrochem. Soc.*, **123**, 42 (1976).

60 A. Vshirokawa, E. Suzuki, and M. Warashina, *Jap. J. Appl. Phys.*, **12**, 398 (1973).

61 V. J. Kapoor, F. J. Feigl, and S. R. Butler, *J. Appl. Phys.*, **48**, 739 (1977).

62 S. R. Butler, F. J. Feigl, Y. Ota, and D. J. DiMaria, in *Thermal and Photostimulated Currents in Insulators*, D. M. Smyth, Ed., Electrochemical Society, Princeton, N. J., 1976, pp. 149–161.

63 J. M. Aitken, D. R. Young, and K. Pan, *J. Appl. Phys.*, **49**, 3386 (1978).

64 A. Goetzberger, *J. Electrochem. Soc.*, **113**, 138 (1966).

65 G. L. Holmberg, A. B. Kuper, and D. F. Miraldi, *J. Electrochem. Soc.*, **117**, 677 (1970).

66 A. J. Moulson and J. P. Roberts, *Transact. Br. Ceramics Soc.*, **59**, 388 (1960).

67 G. Stephenson and K. H. Jack, *Transact. Br. Ceramics Soc.*, **59**, 397 (1960).

68 A. J. Moulson and J. P. Roberts, *Transact. Faraday Soc.*, **57**, 1208 (1961).

69 G. Hetherington and K. H. Jack, *Phys. Chem. Glasses*, **3**, 129 (1970).

70 T. Drury and J. P. Roberts, *Phys. Chem. Glasses*, **4**, 79 (1963).

71 K. H. Beckmann and N. J. Harrick, *J. Electrochem. Soc.*, **118**, 614 (1971).

72 P. J. Burkhardt, *J. Electrochem. Soc.*, **114**, 196 (1967).

73 F. M. Fowkes and T. E. Burgess, *Surface Sci.*, **13**, 184 (1969).

74 P. H. Emmet, R. Livingston, H. Zeldes, and R. J. Kokes, *J. Phys. Chem.*, **66**, 921 (1962).

75 J. C. Fisher, *J. Appl. Phys.*, **22**, 74 (1951).

76 A. G. Revez and K. H. Zaininger, *RCA Rev.*, **29**, 22 (1968).

77 D. R. Young, E. A. Irene, D. J. DiMaria, and R. F. DeKeersmaecker, *J. Appl. Phys.*, **50**, 6366 (1979).

78 R. Williams, *Phys. Rev.*, **140**, A569 (1965).

79 D. J. DiMaria, in *The Physics of SiO₂ and its Interface*, S. T. Pantelides, Ed., Pergamon, New York, 1978, Chapter 4, pp. 160–178.

80 D. J. DiMaria, D. R. Young, and D. W. Ormond, *Appl. Phys. Lett.*, **31**, 680 (1977).

81 T. H. Ning, *J. Appl. Phys.*, **49**, 5997 (1978).

82 D. S. Peck, R. R. Blair, W. L. Brown, and F. M. Smits, *Bell Syst. Tech. J.*, **42**, 95 (1963).

83 C. W. Gwyn, Sandia Report SLA-73-0013, "Ionizing Radiation Effects in the Insulator-Region of MOS Devices," January 1973.

84 A. M. Goodman, *Phys. Rev.*, **144**, 588 (1966).

85 C. A. Mead, E. H. Snow, and B. E. Deal, *Appl. Phys. Lett.*, **9**, 53 (1966).

86 A. M. Goodman and J. J. O'Neill, Sr., *J. Appl. Phys.*, **37**, 3580 (1966).

87 B. E. Deal, E. H. Snow, and C. A. Mead, *J. Phys. Chem. Solids*, **27**, 1873 (1966).

88 H. L. Hughes and R. R. Giroux, *Electronics*, December 28, **37**, 58–60 (1964).

89 J. R. Szedon and J. E. Sandor, *Appl. Phys. Lett.*, **6**, 18 (1965).

90 J. R. Srour, O. L. Curtis, Jr., and K. Y. Chiu, *IEEE Transact. Nucl. Sci.*, **NS-21** (6), 73, (1974).

91 V. E. Corslett and R. N. Thomas, *Br. J. Appl. Phys.*, **15**, 1283 (1964).

92 J. P. Mitchell and D. K. Wilson, *Bell Syst. Tech. J.*, **46**, 1 (1967).

93 E. H. Snow, A. S. Grove, and D. J. Fitzgerald, *Proc. IEEE*, **55**, 1168 (1967).

94 J. P. Mitchell, *IEEE Transact. Electron Devices*, **ED-14**, 764 (1967).

95 J. L. Peel and R. C. Eden, *IEEE Transact. Nucl. Sci.*, **NS-18** (6), 84 (1971).

96 J. M. Aitken, private communication.

97 K. H. Zaininger, *Appl. Phys. Lett.*, **8**, 140 (1966).

98 A. S. Grove and E. H. Snow, *Proc. IEEE*, **54**, 894 (1966).

99 F. B. McLean, H. E. Boesch, and J. M. McGarrity, *IEEE Transact. Nucl. Sci.*, **NS-23** (6), 1506 (1976).

100 G. Pfister and H. Scher, *Adv. Phys.*, **27**, 747 (1978).

101 F. B. McLean and G. A. Ausman, Jr., *Phys. Rev.*, **15**, 1052 (1977).

102 J. M. Marshall, *Phil. Mag.*, **B-38**, 335 (1978).

103 O. L. Curtis, Jr. and J. R. Srour, *J. Appl. Phys.*, **48**, 3819 (1977).

104 R. C. Hughes, *Phys. Rev.*, **B-15**, 2012 (1977).

105 R. C. Hughes, E. P. Eer Nisse, and H. J. Stein, *IEEE Transact. Nucl. Sci.*, **NS-22** (6), 2227 (1975).

106 J. M. McGarrity, *IEEE Transact. Nucl. Sci.*, **NS-27** (6), 1739 (1980).

107 R. J. Powell and G. F. Derbenwick, *IEEE Transact. Nucl. Sci.*, **NS-18** (6), 99 (1971).

108 H. R. Phillipp, *Solid-State Commun.*, **4**, 73 (1966).

109 T. H. DiStefano and D. E. Eastman, *Solid-State Commun.*, **9**, 2259 (1971).

110 R. T. Sanderson, *Chemical Bonds and Bond Energy*, Academic, New York, 1971.

111 G. W. Hughes, R. J. Powell, and M. H. Woods, *Appl. Phys. Lett.*, **29**, 377 (1976).

112 G. F. Derbenwick and B. L. Gregory, *IEEE Transact. Nucl. Sci.*, **NS-22** (6), 2151 (1975).

113 J. M. Aitken and D. R. Young, *IEEE Transact. Nuclear Sci.*, **NS-24** (6), 2128 (1977).

114 E. Harari, S. Wang, and B. S. H. Royce, *J. Appl. Phys.*, **46**, 1310 (1975).

115 P. S. Winokur, J. M. McGarrity, and H. E. Boesch, Jr., *IEEE Trans. Nucl. Sci.*, **NS-23** (6), 1580 (1976).

116 P. S. Winokur, H. E. Boesch, Jr., J. M. McGarrity, and F. B. McLean, *J. Appl. Phys.*, **50**, 3492 (1979).

117 E. E. King and R. L. Martin, *IEEE Transact. Nucl. Sci.*, **NS-24** (6), 2172 (1977).

12

Instrumentation for Measuring Capacitor Characteristics

12.1 INTRODUCTION

In this chapter we discuss instrumentation of the measurements described in detail in Chapters 3–11. The purpose is to enable the reader to make measurements, whether for production monitoring, developmental studies, or research.

Nearly all the instrumentation described is commercially available. To enable the reader to assemble instrumentation, specific instruments are mentioned in this chapter that the authors have used and found to do a satisfactory job. However, the principles behind these instruments are described so that with ingenuity the reader can use equivalent or improved instruments.

Most of the instrumentation described in this chapter uses type BNC connectors permitting interconnection by 50–75 Ω coaxial cable. The experimenter can assemble and wire the circuits described in this chapter with tools common in an electronic wiring shop. Construction of a probe station and associated hot stage or cold finger will require access to a machine shop as well. It is assumed that the reader has access to the instruction manuals of the selected instruments and can make the adjustments necessary for optimal performance of these instruments.

The data taken with the instrumentation described in this chapter usually must be further processed to extract desired quantities such as interface trap properties, doping profile, and so on. With the dramatic decline in the cost of minicomputers in recent years, it is becoming feasible to feed data directly from the measurement instrumentation into a minicomputer. The minicomputer can be programmed to operate the instrumentation acquiring the data and to process it to yield the desired results.

Once the basic ideas behind the measurement methods described in this book and the instrumentation described in this chapter are understood, it is straightforward to digitalize the measured data and feed them directly into the computer if desired.

12.2 MOS CAPACITOR CHARACTERISTICS

First consider measurement of the small-signal *admittance* of the MOS capacitor as a function of *gate bias* and *frequency*. The environment for measurement varies from laboratory to production line. Each environment has different requirements and budgets. There is no one type of apparatus that will fulfill all measurement needs. We describe the types of apparatus available and indicate, in general, their major application. The physics behind these measurements has been described in the previous chapters.

The two basic instruments for measuring the admittance of an MOS capacitor are (1) the *admittance bridge* and (2) the *lock-in amplifier*. Both types of instrument measure the differential capacitance and equivalent parallel conductance of the MOS capacitor, independent of stray capacitance to ground. The admittance bridge is more accurate, but the lock-in amplifier is much easier to automate. Because it is slow, the admittance bridge is used primarily in basic studies. Because of its greater speed, the automated lock-in amplifier is used in both basic studies and production monitoring.

First, we describe admittance bridge measurements in Section 12.3 to illustrate the various problems involved. Next, in Section 12.4 we describe the lock-in amplifier methods, and in Sections 12.5 and 12.6, some of the limitations and precautions necessary in admittance measurements. In

Section 12.7 we discuss a linear voltage ramp generator for automatic measurements; in Section 12.8, the construction of a probe station for three terminal measurements; and in Section 12.9, the fabrication of MOS capacitor arrays. Finally, in Section 12.10 we discuss instrumentation for internal photoemission measurements; in Section 12.11, for hot carrier injection into SiO_2; and in Section 12.12, for lifetime measurements.

12.3 THE ADMITTANCE BRIDGE

The admittance bridge has the highest sensitivity and is the most accurate instrument for measuring interface trap properties. Therefore, it remains the most important instrument for such studies.

An admittance bridge suitable for studying interface trap properties, as described in Chapters 5 and 8, must satisfy three criteria: (1) it must measure differential capacitance and equivalent parallel conductance accurately over ranges of these parameters usually encountered in experimental samples (1–1000 pF and 0.1 nmho–1 mho); (2) it must have adequate sensitivity for measuring equivalent parallel conductance accurately when the loss angle is very small (typically <20°) in the small-signal range (<50 mV peak) needed for MOS capacitor measurements; and (3) it must cover a wide frequency range over which it is easy to extract the desired information without need to correct for lead inductance. Corrections for

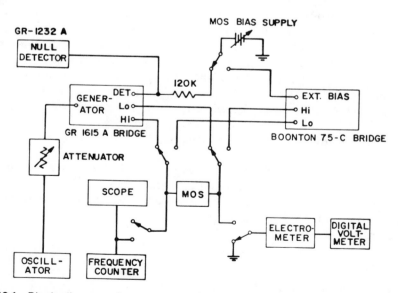

Fig. 12.1 Block diagram of conductance measuring apparatus. After Nicollian and Goetzberger.[2] Copyright (1967), American Telephone and Telegraph Company. Reprinted with permission.

series resistance and lead inductance become so severe that they limit our discussion to frequencies below a few megahertz.

Two admittance bridges that fulfill these criteria are: (1) the Boonton 75-C, which can measure *small-signal differential capacitance* over the range 1–1000 pF and *small-signal equivalent parallel conductance* from about 1 nmho–1 mho over the frequency range 5–500 kHz, and (2) the General Radio 1615-A, which measures small-signal capacitance over an even larger range than the Boonton 75-C and small-signal equivalent parallel conductance in the range 0.1 nmho–0.1 mho over the frequency range 50–100 kHz. Combined as shown in Fig. 12.1, these two bridges cover the frequency range 50 Hz–500 kHz. Both measure smaller values of C and G than stated above, but to do so requires a signal voltage higher than the small-signal range of the MOS capacitor. These two particular bridges are described here because they have been used by the authors. Other bridges also would be adequate if they fulfilled the criteria.

12.3.1 Three-Terminal Capacitance Measurements

One of the most important features of an admittance bridge is that *three-terminal* or *direct* differential capacitance measurements can be made with it. In a three-terminal capacitance measurement, only the direct capacitance of the MOS capacitor is measured by the bridge, independent of unavoidable terminal, cable, and stray capacitances to ground.[1] For example, capacitance to ground can be on the order of 50–100 pF, depending mainly on cable length.* In many cases such capacitance to ground is comparable to the MOS capacitor itself. Stray capacitances between leads, typically less than 1 pF, can be zeroed out by adjustments on the bridge. In a three-terminal measurement, *two* of the terminals are from the MOS capacitor and the *third* terminal must be the grounded metal enclosure surrounding the MOS capacitor. The probe station described in Section 12.8 is designed specifically for three-terminal capacitance measurements.

12.3.2 Bridge Accuracy and Sensitivity

Depending on the bridge used, accuracy is typically 0.01–0.25%, or four to five significant figures, for measuring capacitance, and 1–10%, or two to three significant figures, for measuring equivalent parallel conductance. These levels of accuracy are the best possible today in admittance measurements.

---

*The coaxial cable between the MOS capacitor and the bridge terminals can add capacitance to ground and series inductance. To avoid corrections due to these effects given in the bridge manufacturer's instruction manual, cable lengths should be kept short, no more than a foot or two.

The sensitivity of these bridges is exploited most fully by the conductance method, as described in Section 5.5. With these bridges combined as shown in Fig. 12.1, interface trap densities in the $10^9 \, cm^{-2} \, eV^{-1}$ range near midgap have been measured using the conductance method. Although this density range is not the smallest detectable with an admittance bridge, samples that have interface trap level densities below the $10^9 \, cm^{-2} \, eV^{-1}$ range, today's state of the art, would be necessary to put these instruments to the test.

Three parameters require particular attention in admittance bridge measurements: (1) frequency, (2) signal amplitude, and (3) gate bias. The following discussion of these parameters also applies in general to the automatic admittance measuring instrumentation described in Sections 12.4 and 12.5.

12.3.3 Frequency

When measuring admittance as a function of gate bias and frequency, frequency must be measured with a counter to ensure that the precision of the extracted interface trap properties is limited by the precision of the bridges. The frequency calibration marks on the apparatus usually are not accurate enough.

12.3.4 Signal Amplitude: Small-Signal Range

To keep harmonics of the signal frequency from giving rise to spurious conductance and capacitance values, only ac voltages of small amplitude can be applied. Capacitance, conductance, and all the equivalent circuits in this book have been calculated for the small-signal range. Moreover, the admittance bridge does not give accurate values of the admittance of an MOS capacitor except in the small-signal range. Thus it is essential to employ the small-signal range to allow interpretation of the experimental results. The small-signal range is defined as the range of signal amplitude in which capacitance and equivalent parallel conductance are independent of the ac gate voltage amplitude.

The small-signal range for an MOS capacitor depends on oxide thickness, silicon doping concentration, gate bias, interface trap density, and oxide charge nonuniformities. All these parameters influence the voltage dependence of equivalent parallel capacitance and conductance. In view of such complexity, it is practical to determine the small-signal range by experiment.

To find the small-signal range experimentally, set the bridge null amplifier at maximum sensitivity, apply a gate bias to the MOS capacitor, and null the bridge with the smallest signal amplitude for which a null can be obtained. At the maximum sensitivity setting of the bridge null amplifier, null is found in the small-signal range at a null meter reading of $\approx \frac{1}{2} \mu A$ on the Boonton bridge and $\approx \frac{3}{4}\,\mu V$ on the General Radio bridge. These null meter readings are not zero because of system noise.

As signal amplitude is increased, initially there will be no change in null meter readings in the small-signal range. As signal amplitude is increased beyond the small-signal range, nulls will be obtained at successively higher null meter readings. Changes in admittance produced by the signal are no longer linear, and the generated harmonics have enough amplitude to get through the tuned null amplifier, become rectified, and cause a corresponding increase in null meter reading. These readings no longer are true nulls because only the fundamental frequency has the correct phase for a true null.

The small-signal range is then the range of signal amplitudes over which null meter readings remain constant. Because the small-signal range is gate bias dependent, this range must be determined at various gate bias settings.

Figure 12.2 shows that there is a range of signal amplitude at which null readings remain constant. The values C_{max} and G_{max} are equivalent parallel capacitance and conductance measured in the small-signal range, and C_m and G_m are measured by obtaining nulls at successively higher null meter readings. The small-signal range in Fig. 12.2 is the range of amplitudes for which *both* C_m/C_{max} and G_m/G_{max} remain unity. The true values of C_m and G_m are measured in this range. At larger signal amplitudes the reduction of C_m and G_m reflects variation from true null.

The lower the small-signal amplitude, the lower the bridge sensitivity. Therefore, optimum ac signal amplitude is at the high amplitude end of the

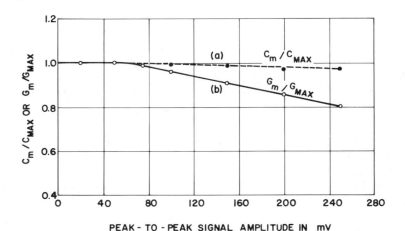

PEAK - TO - PEAK SIGNAL AMPLITUDE IN mV

Fig. 12.2 (a) Normalized capacitance and (b) normalized equivalent parallel conductance at fixed gate bias. The maximum capacitance was 367.8 pF, and equivalent parallel conductance was 58.8 nmho. Frequency was 100 Hz. Gate bias corresponds to peak conductance at this frequency. After Nicollian and Goetzberger.[2] Copyright (1967), American Telephone and Telegraph Company. Reprinted with permission.

small-signal range. If the small-signal range has been found at different values of gate bias, the largest signal amplitude in the narrowest range should be chosen for measuring the entire admittance-voltage characteristics. The small-signal range extends to higher amplitudes as the oxide is made thicker, silicon doping concentration is increased, or at large interface trap densities. For oxide thicknesses of 500–700 Å and silicon doping concentrations around 10^{16} cm^{-3}, for a typical sample such as in Fig. 12.2, maximum peak-to-peak signal voltage still in the small-signal range is about 50 mV for measurements in depletion.

To maintain the optimum small-signal level throughout the frequency range,* it is necessary to remeasure and readjust the small-signal voltage amplitude whenever frequency is changed. During readjustment, signal amplitude is monitored by an oscilloscope as shown in Fig. 12.1. Such monitoring is a necessary precaution because the oscillators feeding each bridge have an output voltage that varies with frequency.

12.3.5 Gate Bias

Because interface trap time constants are an exponential function of silicon band bending (see Section 7.3), gate bias must be measured to at least three decimal places to obtain values of admittance versus gate bias accurate enough to extract interface trap time constants. A very high input resistance digital voltmeter, such as the Cimron 6900A, must be used because of the manner in which bias is applied to the MOS capacitor through the bridges in Fig. 12.1. Gate bias supplied to the MOS capacitor through a voltage divider in each bridge draws current from the bias supply. Therefore, gate bias must be measured across the MOS capacitor rather than across the supply terminals. The voltmeter puts an additional capacitance and conductance across the MOS capacitor that would be read by the bridge. To avoid this correction, the voltmeter is used first to read the gate bias setting and then is switched out of the circuit, as shown in Fig. 12.1, before the bridge is balanced. Because the input resistance of the voltmeter no longer shunts the MOS capacitor, current through the voltage divider in the bridge decreases, making the bias across the MOS capacitor slightly larger than that measured. To minimize this change of bias, a high impedance voltmeter is required. No correction is needed with a voltmeter such as the Cimron 6900A, which has an input resistance of $10^9 \, \Omega$ when balanced. However, for a digital voltmeter that has a low input resistance of a few megohms, voltage changes across the MOS capacitor become unacceptably large. These voltage changes occur when switching the conductance multiplier on the Boonton bridge, with the conductance loss factor switch on the General Radio bridge, or in switching from one bridge to another. Switching is necessitated because frequency is varied during

---

*Small-signal range may be dependent on frequency.

the measurement. An alternative to direct use of a low input resistance digital voltmeter is its use in conjunction with an electrometer that has a high input resistance ($10^{15}\,\Omega$). The output voltage of the electrometer is read by the low input impedance digital voltmeter. The electrometer, in this arrangement, acts as an impedance transformer. The electrometer-digital voltmeter combination is shown in Fig. 12.1.

To permit control of the bias within a fraction of a millivolt, a voltage divider consisting of a bank of fixed resistors in series with a 10-turn helipot is used in conjunction with a heavy-duty 45 V battery in a shielded box. A battery rather than an electronic power supply is used to avoid problems of ac ripple and ground loops. Ground loops in particular must be avoided in operating the General Radio bridge.

12.3.6 Automatic Instrumentation

In an admittance bridge, differential capacitance and equivalent parallel conductance are read off two separate dials or indicators adjusted manually to null the bridge. The need for manual adjustment is the major drawback of conventional admittance bridge measurements. Automated measurements are an answer to more rapid data acquisition. Automated instruments that operate over a wide frequency range are commercially available such as HP models 4274A and 4275A. These automatic instruments, designed specifically for semiconductor measurements, have been recently introduced. They operate at signal voltages as low as 1 mV root mean square (rms), which is well within the small-signal range, and provide digital readout of capacitance of 1 pF–1 F full scale and resistance of 10^{-1}–$10^{7}\,\Omega$ full scale over a frequency range of 100 Hz–100 kHz in model 4274A and 1 pF–100 μF and resistance of 1–$10^{7}\,\Omega$ over a frequency range of 10 kHz–10 MHz in model 4275 A. These instruments interface readily with a computer. Model 4274A covers the same frequency, capacitance, and equivalent parallel conductance ranges as the manually nulled admittance bridges described earlier. However, model 4275A covers a much higher frequency range. Another instrument, the HP model 4192A LF Impedance Analyzer, has an automatic frequency sweep capability—an advantage over the other two HP instruments for conductance measurements—but does not measure as low a conductance.

12.4 THE LOCK-IN AMPLIFIER

In this section we describe application of the lock-in amplifier to the measurement of the admittance of an MOS capacitor. We deal with this application, and not with the details of how lock-in amplifiers work. The interested reader can find a discussion of how lock-in amplifiers work in Refs. 3 and 4.

The major advantage of the lock-in amplifier over the admittance bridge is the ease with which the lock-in amplifier can be automated over a range of frequencies. To automate, all that is required is a linear voltage ramp generator to supply a time varying gate bias to the MOS capacitor and an x-y recorder or a two-pen recorder to plot capacitance and equivalent parallel conductance simultaneously as functions of gate bias.

The major disadvantage of the lock-in amplifier is that it is neither as accurate nor as sensitive as the admittance bridge. Capacitance readings to three significant figures and conductance readings to two or three significant figures depending on the size of the phase angle are possible with a lock-in amplifier provided that its output voltage is measured with a digital voltmeter.

12.4.1 Admittance

Figure 12.3 shows a circuit for making admittance measurements as a function of gate bias and frequency using a lock-in amplifier. Direct capacitance measurements can be made using the lock-in amplifier in this circuit configuration. A linear voltage ramp generator, such as the one described in Section 12.7, applies a slowly varying gate bias to the MOS capacitor. Superposed on this ramp voltage is a small ac voltage, generated inside the lock-in amplifier used in the internal reference mode. The resulting displacement current flowing through the MOS capacitor flows into a current preamplifier.

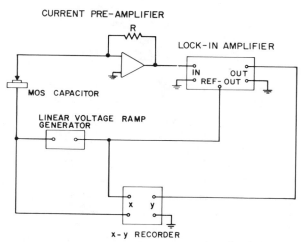

Fig. 12.3 Block diagram of apparatus for making automatic admittance measurements of an MOS capacitor with a lock-in amplifier. Lock-in amplifier can be a PAR HR-8 with a type A preamplifier and a model 181 current amplifier or an Ithaco model 193 with model 164 current amplifier. Connecting the input of the preamplifier to the gate rather than the substrate of the MOS capacitor minimizes noise.

There are three reasons for using a current preamplifier: (1) the input impedance of the current preamplifier is much lower than the impedance of the MOS capacitor over the entire frequency range covered by the lock-in amplifier, and thus virtually all of the ac voltage appears across the MOS capacitor; (2) the output of the current preamplifier is unaffected by input shunt capacitance to ground, and its frequency response is essentially flat over the frequency range of the lock-in amplifier; and (3) the current preamplifier converts the ac current flowing through the MOS capacitor to an ac voltage measured by the lock-in amplifier, which is an ac voltage measuring instrument.

The lock-in amplifier separates the 90° component from the 0° component of the input voltage. The 90° component of the current through the MOS capacitor is $\omega C_m \eta v_R$, and the 0° component is $G_m \eta v_R$, where v_R is the rms reference voltage, η is an attenuation factor introduced by the voltage ramp generator because the ac voltage is fed in series with the ramp voltage, ω is the angular frequency, C_m is the capacitance, and G_m is the equivalent parallel conductance of the MOS capacitor. The capacitance is

$$C_m = \frac{v_1 S}{\omega \eta v_R} \tag{12.1}$$

and the equivalent parallel conductance is

$$G_m = \frac{v_2 S}{\eta v_R} \tag{12.2}$$

where v_1 and v_2 are the 90 and 0° components of the rms input voltage, respectively, read by the lock-in amplifier and S is the sensitivity of the current preamplifier in units of A/V. Output of the lock-in amplifier, proportional to v_1 and v_2, can be applied to the y-axis of an x-y recorder and gate bias to the x-axis to plot C_m and G_m as functions of V_G at a given frequency.

The lock-in amplifier is a measuring tool used in a wide variety of measurements. Because it is a general purpose tool, it requires calibration and some skill to adjust and operate it properly.

(a) Calibration

There are two ways that this system can be calibrated. First, ω and v_R can be measured, and with the known values of η and S, C_m and G_m can be calculated from the values of v_1 and v_2 measured by the lock-in amplifier using (12.1) and (12.2). To maintain accuracy, frequency should be measured with a counter and the peak value of v_R measured with a Tektronix type W comparator in a 545 oscilloscope or equivalent. The second method of calibration is to use a test circuit consisting of a parallel combination of known and voltage independent C and G. The values of C and G of the test circuit should be similar to those of C_m and G_m of the MOS capacitor to be measured. The test circuit can be measured with an admittance bridge to ensure that its admittance is accurately known. Then (12.1) and (12.2) become

$$C_m = \frac{K_1 v_1}{\omega}$$ (12.3)

and

$$G_m = K_2 v_2$$ (12.4)

where K_1 and K_2 are determined using the test circuit. This second method of calibration is a good check on the first method.

(b) Adjustment of Signal Amplitude

To adjust v_R for the small-signal range, v_2 will be more sensitive than v_1, as seen from Fig. 12.2. In the small-signal range v_2 will scale linearly with v_R. Outside the small-signal range, v_2 will increase more slowly with v_R. If η is less than unity, v_R must be correspondingly larger. For an ac voltage across the MOS capacitor of 10 mV rms, $v_R = 0.5$ V rms for $\eta = \frac{1}{50}$, which is the case if the ramp generator of the PAR model 410 C-V plotter, discussed later in this section, is used. The value of η, if unknown, can be ascertained by using the test circuit if S is known.

(c) Frequency

The lock-in amplifiers of interest to us operate in the 10 Hz–200 kHz frequency range. As frequency is lowered, the equivalent parallel conductance for a given interface trap level density becomes smaller until noise makes measurement impractical. For an interface trap level density in the 10^{11} cm^{-2} eV^{-1} range near midgap, 100 Hz is about the lowest frequency at which equivalent parallel conductance can be measured with the lock-in amplifier system.

(d) C-V Plotter

A lock-in amplifier system specialized for measuring the capacitance and equivalent parallel inductance of the MOS capacitor is the PAR model 410 C–V plotter. It is suitable for both production monitoring and basic studies, where convenience is more important than the accuracy provided by a bridge. This lock-in amplifier is designed specifically for measurements on the MOS capacitor as a function of bias at 1 MHz. This detector measures direct capacitance of 0.5–2000 pF full scale and equivalent parallel conductance of 0.5 μmho–2 mmho full scale.

This instrument has a built-in ramp generator that applies a bias (± 100 V maximum), linearly varying with time, to the MOS capacitor to permit automatic plotting of the C-V and G-V characteristics. Application of output voltage of the instrument, (corresponding to either capacitance or equivalent parallel conductance) to the y-axis and the ramp voltage across the MOS capacitor to the x-axis of an x-y recorder permits the automatic plotting of C_m or G_m versus V_G. The PAR model 410 C-V plotter, used in conjunction with a commercially available heated probe stage, can be programmed to measure C-V characteristics before and after a predetermined bias-temperature aging cycle.

Accuracy is 2% for capacitance with this instrument. By using a standard capacitance, the accuracy of this instrument can be increased to 0.1% for capacitance. Signal amplitude is 15 mV rms, which should be within the small-signal range for most samples.

The model 410 C-V plotter can be used in conjunction with a separate lock-in amplifier to measured admittance over a range of frequencies and to avoid the calibration procedure of the lock-in amplifier just described for capacitance. First, a 1 MHz C-V curve is plotted with the model 410 C-V plotter. Then, with the separate lock-in set at the desired frequency, the amplitude of the reference voltage is adjusted until the capacitance coincides with C_{ox} measured at 1 MHz. If the zero point on the capacitance scale is the same, the capacitance scale determined for the 1 MHz curve is the same. To avoid error, it is essential that series resistance be negligible. Because G_m is frequency dependent, a similar procedure cannot be used to calibrate G_m, and the procedure embodied in (12.4) must be used.

A suitable lock-in amplifier for measuring admittance at frequencies below 1 MHz is the Ithaco model 393 Dynatrac 3, which provides output for both C_m and G_m. This feature, with a two-pen recorder such as the HP 7046A x-y recorder, permits the simultaneous plotting of C_m and G_m versus V_G with frequency as parameter. The Ithaco lock-in amplifier can be combined with the PAR model 410 C-V plotter, which provides a voltage ramp, to cover the frequency range up to 1 MHz.

12.5 DIFFERENTIAL CAPACITANCE

Considerable simplification in the measuring instrumentation can be achieved by measuring only the capacitive part of the admittance. Further simplification is possible by limiting C-V measurements to the high and low frequency ranges. Chapter 8 describes how interface trap properties are extracted from high and low frequency C-V curves. High and low frequency C-V curves are used for extracting many other properties of interest in the MOS system, as described in Chapters 6 and 9–11.

12.5.1 High Frequency

The most widely used instrument for high frequency C-V measurements, described in Section 4.4, in the Boonton L-C Meter (model 71-A or 72 AD). This L-C meter, operating at 1 MHz, measures the direct or three-terminal effective parallel capacitance of the MOS capacitor from 1 to 2000 pF. Figure 12.4 shows a low-cost, automatic C-V plotter widely

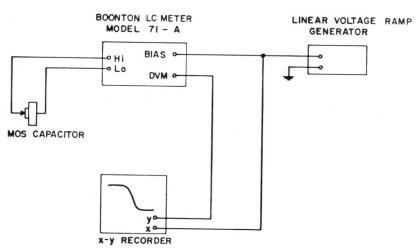

Fig. 12.4 Block diagram for measuring the high frequency C-V curve of an MOS capacitor using a Boonton L-C meter model 71-A or 72 AD.

used for production monitoring using an L-C meter to measure differential capacitance, a linear voltage ramp generator to apply a time varying gate bias to the MOS capacitor, and an x-y recorder to display the resulting C-V characteristics.

The Boonton L-C meter also works on the lock-in amplifier principle. The signal level it puts out is 15 mV rms, which is well within the small-signal range for most samples. When the Q of the MOS capacitor ($Q = \omega C_m G_m^{-1}$) exceeds 5, accuracy on the low capacitance scales is 0.25%, decreasing to 0.4% on the highest scale. Accuracy is degraded for lower values of Q. For Q between 1 and 5, accuracy is 0.5% on the low capacitance scales decreasing to 2% on the highest scale. For good quality oxides, the Q of the MOS capacitor will be 20 or higher at 1 MHz. Therefore, in some ranges the L-C meter will have an accuracy comparable to a bridge. To minimize errors introduced into the capacitance readings by stray capacitance to ground and series inductance, the coaxial cables connecting the MOS capacitor to the L-C meter should be kept short as possible. Corrections for these errors are given in the instruction manual.

A PAR model 410 C-V plotter also can be used to measure a high frequency C-V curve at 1 MHz. The sensitivity and the accuracy of this instrument are comparable to the L-C meter as both are lock-in amplifiers. However, the model 410 C-V plotter has additional features, which are described in Section 12.4.1(d), so the choice of instrument depends on the needs and budget of the experimenter.

12.5.2 Low Frequency

A C-V curve can be considered to be a low frequency curve (see Chapter 3 for a description of low frequency C-V curves) when the system is in thermal equilibrium during measurement. That is, both minority carriers and interface traps respond to, or follow, both the ac gate voltage and changes in gate bias. Frequencies of 1 Hz or below are usually adequate for measuring a low frequency C-V curve. The major problem of measuring capacitance at such low frequencies is that very small displacement currents (in the picoampere range) must be measured so that noise is a major problem. Special precautions, described in this Section, must be taken to ensure success.

A low frequency C-V curve can be measured using an admittance bridge or a lock-in amplifier. However, admittance bridge and lock-in amplifier measurements below 50 Hz require noise filters with long response times. At low frequencies, the response time of the filter can be the dominant factor determining sweep rate.

Another method for obtaining low frequency C-V curves is the voltage ramp method. This method is more widely used for measuring low frequency C-V curves than the admittance bridge or lock-in amplifier because equivalent frequencies of a fraction of a Hz are more easily attainable. However, like the other methods, displacement currents are in the picoampere range. Therefore similar noise problems are encountered requiring a filter with a long response time.

In the voltage ramp method, a voltage ramp generator is used to apply a time varying gate bias. The resulting displacement current flowing through the MOS capacitor is measured with an electrometer. This displacement current is directly proportional to the differential capacitance.[5-7] To quantitatively describe this method, we need the relationship between linear ramp voltage, displacement current through the MOS capacitor, and differential capacitance. The displacement current I_G flowing through the MOS capacitor in response to a time varying voltage V is related to the differential capacitance* defined in section 3.2 by

$$\frac{dQ}{dt} = I_G = \left(\frac{dQ}{dV}\right)\left(\frac{dV}{dt}\right) = C\left(\frac{dV}{dt}\right). \qquad (12.5)$$

Choosing a linear voltage ramp $V = \alpha t$, $dV/dt = \alpha$, where t is time, and using (12.5), we relate differential capacitance to displacement current by $C = I_G/\alpha$.

A circuit for measuring the displacement current through the MOS capacitor in response to a linear voltage ramp is shown in Fig. 12.5. The circuit shown in Fig. 12.5 is essentially an analogue differentiator that

---

*From the definition given in Section 3.2, the relationship between the differential and static capacitances under a time varying voltage is $C = C_{stat} + V(dC_{stat}/dV)$.

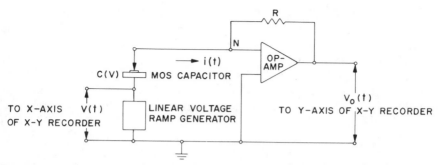

Fig. 12.5 Block diagram for measuring the low frequency C-V curve of an MOS capacitor using an operational amplifier to detect displacement current through the MOS capacitor produced by a linear voltage ramp. The operational amplifier can be a Keithley model 602 electrometer operated in the fast mode at unity gain. More important than in Fig. 12.3, the gate rather than the substrate of the MOS capacitor should be connected to the input of the electrometer for low noise. After Kuhn.[6] Copyright (1970), Pergamon Press, Ltd. Reprinted with permission.

incorporates the MOS device as the capacitive element. The amplifier is a high gain, high impedance, operational amplifier that maintains point N of the circuit at ground potential. When a voltage $V(t)$ is applied, the output voltage $V_o(t)$ is

$$V_o(t) = -R\,C(t)\frac{dV(t)}{dt}. \tag{12.6}$$

Thus when $V(t)$ is a linear ramp of the form $V(t) = V_1 \pm \alpha t$, the output voltage is directly proportional to the differential capacitance

$$V_o(t) = \mp \alpha R\,C(t). \tag{12.7}$$

A simultaneous display of $V_0(t)$ and $C(t)$ on an x-y recorder will yield a directly scaled $C(V)$ versus V_G curve with time as an independent parameter.

The circuit shown in Fig. 12.5 can be conveniently realized using an electrometer such as the Keithley model 602 operated in the "fast" mode as an operational amplifier. A simple operational amplifier integrator, described in Section 12.7, is used to obtain a linear ramp voltage with a linearity of less than 0.5%.

To obtain a true low frequency C-V curve (system in thermal equilibrium), the gate bias sweep rate must be slow enough for both interface traps and minority carriers to follow over the entire gate bias range. Sweep rates in the range of 50 mV/sec* or below, depending on minority carrier lifetime, will be slow enough in most cases. As a check on the voltage

---

*The corresponding frequency of a sinewave with an amplitude of 50 mV would be about 0.16 Hz.

sweep rate, C-V curves should be independent of both sweep direction and sweep rate. If so, and if the capacitance rises to C_{ox} in inversion, a true low frequency C-V curve has been measured.

If the arrangement shown in Figure 12.5 is used, a sweep rate of 50 mV/sec applied to an MOS capacitor with $C_{ox} = 150\,pF$ will result in displacement currents in the picoampere range. The absolute accuracy of most electrometers in this current range is about 5% of the current reading. To measure such low currents, the response time of the electrometer will be inordinately long if operated in the "normal" mode. With $C_{ox} = 150\,pF$, response time of the electrometer can be as long as 15 sec, in the picoampere range. Therefore, the electrometer must be operated in the "fast" mode to neutralize effects of input capacitance. For example, with 50 pF to ground, the rise time of the electrometer will be less than 1 sec in the "fast" mode. With such a rise time, the electrometer easily can follow sweep rates of 50 mV/sec or less, sweeping a total voltage span of several volts. An added advantage of operating in the "fast" mode is that the offset voltage across the electrometer input terminals is less than 1 mV in all ranges, so practically all the applied voltage appears across the MOS capacitor.

Stray capacitance to ground should be minimized by (1) connecting the input terminal of the electrometer to the probe wire contacting the gate rather than the pedestal, (2) making this interconnection with rigid General Radio coaxial connectors rather than coaxial cable, and (3) making this interconnection as short as possible. Clamping these connectors firmly to the laboratory bench suppresses spurious currents (noise) caused by vibrations. If stray capacitance has not been reduced below detectable limits by these procedures, it must be taken into account and measured. To do this, first lift the probe wire a short distance off the gate of the MOS capacitor and zero the electrometer according to the instruction manual. Then stray capacitance can be measured by applying the voltage ramp at the desired sweep rate and noting the resulting displacement current. This value of stray capacitance can either be subtracted from the readings measured with the MOS capacitor connected or compensated.

The best way to compensate for stray capacitance is to apply a current in parallel with the MOS capacitor using a constant current source operating in the picoampere range. The constant current source is adjusted, with the MOS capacitor disconnected, to just cancel the displacement current through the stray capacitance.

A commercially available instrument that measures a low frequency C-V curve is the HP model 4140A pA meter/dc voltage source. It has a built-in ramp generator, has a zero offset control to cancel displacement current through stray capacitance and any leakage current through test leads or test fixtures, presents capacitance versus voltage or normalized capacitance versus voltage, has digital readout, and can be interfaced with a computer. It has an accuracy of 5% for long integration times (1–2 sec) at current levels of 10 pA. This instrument operates as described in this section.

(a) Leaky Oxide[8]

Any shunting resistance across the MOS capacitor will show up as an ohmic current added to the displacement current flowing through the MOS capacitor. Shunting resistance can originate from a leaky oxide, leakage current paths along the oxide between gate and back contact, shunt paths from the probe wire to ground, or from shunt resistance within the measuring instrumentation itself. Shunt resistances of the order of $10^{12}\,\Omega$ have visible effects. Shunt paths along the oxide of the MOS capacitor or from the probe wire to ground, usually caused by moisture, can be avoided by placing the MOS capacitor in a probe box that has a dry ambient as described in Section 12.8.

Leakage current through a shunt resistance can be detected by stopping the voltage ramp at some high value of voltage. If the current read by the electrometer drops back to zero, there is no shunt resistance provided the current compensating circuit has been turned off. The effect of current flowing through a shunt resistance on the low frequency C-V curve is to skew the C-V curve. An example of this skewing is shown in Fig. 12.6 for a p-type sample swept from accumulation to inversion (dashed curve, with shunt resistance; solid curve, without shunt resistance). The skewing effect occurs because at negative values of gate bias in accumulation, current through the shunt resistance is opposite to displacement current through the capacitance. Shunt current, which is smaller than the displacement current in the example shown in Fig. 12.6, subtracts from the displacement current. As bias becomes less negative, this opposing shunt current decreases, causing an upward tilt of the C-V curve. Then, as bias becomes

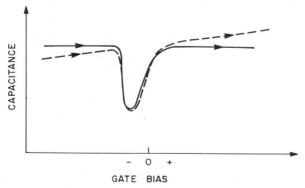

Fig. 12.6 Effect of a shunting resistance current component on the low frequency C-V curve (dashed line) compared to an MOS capacitor with no shunting resistance (solid line) measured with the circuit in Fig. 12.5. Bias sweep direction indicated by the arrows is from accumulation to inversion. After Derbenwick.[8]

positive, shunt current adds to the displacement current, causing a further upward tilt as shown. If bias is swept from inversion to accumulation, the resulting C-V curve will have a downward tilt because the displacement current flows in the direction opposite to that shown in Fig. 12.6.

(b) Deviations from Thermal Equilibrium[9]

Deviation from thermal equilibrium will lead to large errors in the application of low frequency C-V curves. Such deviations are caused by too fast a sweep rate or by light. A typical set of experimental response characteristics is shown in Fig. 12.7 for both thermal equilibrium and

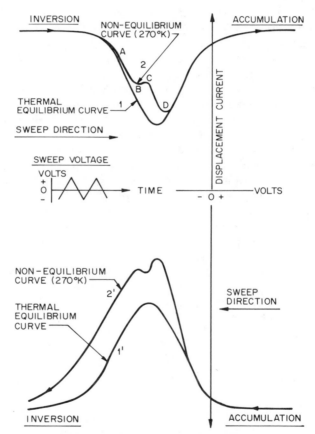

Fig. 12.7 Temperature dependence of the response of an MOS capacitor to a triangular voltage sweep. The silicon substrate was 5 Ω-cm, n-type. The oxide was grown in steam in an electric field of 1500 V/cm (silicon positive) to a thickness of 800 Å. The gate was Cr–Au, and the sweep rate was 100 mV/sec. Similar curves are obtained if, in place of decreasing temperature, a faster sweep rate is applied. After Kuhn and Nicollian.[9] Copyright (1971), the Electrochemical Society. Reprinted with permission.

nonequilibrium conditions and both sweep directions.* Curve 1 is the thermal equilibrium response when sweeping from inversion towards accumulation, and curve 1' is the thermal equilibrium response for the opposite sweep direction.† For comparison, curves 2 and 2' show the typical nonequilibrium response measured at 270°K to exaggerate the effect. The nonequilibrium characteristic, when sweeping from inversion toward accumulation (curve 2), shows a small voltage shift in strong inversion and a peak near weak inversion. The nonequilibrium characteristic obtained when sweeping from accumulation to inversion (curve 2') is similar.

Deviation from thermal equilibrium is due to those physical processes characterized by long time constants, that is, inversion layer and interface trap response. Inversion layer response time and interface trap response time both increase exponentially with decreasing temperature; therefore, effects on ramp response become dominant at low temperatures.

Consider first the sweep direction from accumulation toward inversion. Because inversion layer formation is governed by minority carrier generation, the system can remain in thermal equilibrium only if the maximum displacement current required by the sweep rate is much smaller than the generation current; that is, $C_{ox}(dV_G/dt) \ll qn_iw/\tau$, where w is the depletion layer width, τ is the minority carrier lifetime, and n_i is the intrinsic carrier concentration. At low temperatures, n_i and hence the minority carrier generation rate become so small that the preceding inequality no longer can be satisfied at normal sweep rates. Consequently, an equilibrium inversion layer cannot form, and the depletion layer expands more deeply into the bulk to satisfy charge neutrality. The increasing depletion layer width results in a reduced displacement current and an increased generation current. Both these effects tend to restore thermal equilibrium †† The peak in curve 2' in Fig. 12.7 is due to interface trap generation as the quasi-Fermi level passes through midgap at the silicon surface. At lower temperatures, interface trap generation would become so low that no peak would be observed when sweeping toward accumulation.

Sweeping from accumulation toward inversion, a small forward bias V_F develops across the field induced junction. Initially, net recombination of

---

*Edge effects can be ruled out in these measurements because displacement currents scale linearly with gate area and response characteristics are similar on both n- and p-type samples.
†The ideal thermal equilibrium response would be symmetrical in sweep direction; the difference in scales of measured curve 1 and 1' is due to a small difference in the magnitude of the sweep rate caused by the instrumentation.
††Response in this regime is determined by a complex dynamic equilibrium between increasing gate bias that tends to increase depletion layer width and minority carrier generation that tends to reduce depletion layer width and restore equilibrium. At temperatures as low as 77°K when the generation rate becomes negligible compared to displacement current, response is determined by oxide capacitance and yields the depletion characteristics. These characteristics can be obtained at room temperature, but a considerably faster sweep rate would be required.

minority carriers from the inversion layer with majority carriers from the bulk silicon occurs in the depletion layer corresponding to region A of the characteristic in Fig. 12.7. This forward bias is necessary to sustain displacement current and leads to a voltage shift between curves 1 and 2 in Fig. 12.7, given by $\Delta V = V_F + \Delta Q_s/C_{ox}$, where ΔQ_s is the excess minority carrier silicon surface charge in the inversion layer and in interface traps. In region A, ΔV is of the order of several tenths of a volt and minority carrier surface charge and silicon band bending are being slowly reduced while V_F and ΔQ_s remain nearly constant as ramp voltage decreases. At point B in Fig. 12.7 band bending has been reduced sufficiently to make diffusion current dominant.* Inversion layer charge is now rapidly injected into the bulk silicon, thereby dissipating the excess minority carrier charge at the silicon surface. The discharging of excess minority carrier silicon surface charge ΔQ_s by injection produces the current peak in the region BCD in Fig. 12.7. This observed peak is a nonequilibrium effect and cannot be attributed to any particular interface trap distribution. This peak would be observed in an ideal MOS capacitor without interface traps. The magnitude of the peak, however, depends on interface trap density only through the excess minority carrier charge stored in interface traps.

Under some conditions nonequilibrium effects can be encountered in low frequency C-V measurements that can cause distortion of the response even at the lowest practical sweep rates of 5–10 mV/sec. These conditions can occur even at room temperature for long lifetime silicon ($\tau >$ 100 μsec), at low temperatures, and under illumination. These nonequilibrium effects result in a small voltage shift of the inversion response, in a peak near weak inversion when sweeping from inversion to accumulation, and in a partial deep depletion condition when sweeping from accumulation to inversion. These nonequilibrium effects must be minimized to obtain meaningful results in extracting silicon band bending and interface trap density. They can be made negligible by using short lifetime silicon ($\tau \leq 1 \mu$sec), making measurements in the dark, and using sweep rates smaller than 50 mV/sec. Measurements on longer lifetime silicon can be performed by using elevated temperatures, which are normally held below 100°C to avoid ionic drift. In the high-low frequency capacitance method for determining interface trap level density both the low and the high frequency C-V curve should be measured at the same temperature.

To obtain a large displacement current, which is easy to measure, a fast ramp speed is required. However, ramp speed cannot be made too large without driving the system out of thermal equilibrium as we have seen. Also ramp speed must not be so fast that the noise filter does not respond immediately. A reasonable compromise is to use as large a gate area as

---

*Forward bias at which diffusion current becomes dominant is less in a field induced junction than in a metallurgical junction because in a field induced junction, inversion layer charge density is a functon of V_F.

possible to allow a reasonably large displacement current to be measured at a low ramp speed. For example, a maximum displacement current of 340 pA will flow through an MOS capacitor that has a gate area of $10^{-1}\,cm^2$ and an oxide thickness of 500 Å at a ramp speed of 50 mV/sec.

(c) *Charge Measurement Method*[10]

Low frequency capacitance can be obtained from a measurement of the charge Q on an MOS capacitor as a function of gate bias as well as by direct measurement as previously described in this section. In the Q-V method, described in Section 3.3.1, a relatively large voltage is measured. Therefore, this method does not suffer from the severe noise of the admittance bridge, lock-in amplifier, and voltage ramp methods. No filtering is needed in the Q-V method. Also, any gate area can be used. However, low frequency capacitance must be obtained by numerical differentiation of the measured Q-V data, necessitating precision voltage measurements.

In the Q-V method a bias is applied across the series combination of the MOS capacitor and a voltage independent capacitor consisting of the parallel combination of a variable air capacitor with ceramic insulation (a 250 pF capacitor is suitable), the input cable capacitance, and the input capacitance of the voltmeter. The voltage across the voltage-independent capacitor is measured and converted into a charge by the relation $Q = CV$. For a given bias across two capacitors in series, the voltage divides between the two, and each charges up to the same value of charge. Therefore, the charge measured across the voltage-independent capacitor is the same as the charge on the MOS capacitor.

Bias across the series combination can be varied point by point or with a slow voltage ramp so that the MOS capacitor is always in thermal equilibrium. Pointwise measurement allows digital measurement of voltage and, therefore, of the charge density necessary for differentiation to obtain C_{LF}.

The basic circuit of the Q-V method is shown in Fig. 12.8. A step voltage for point by point measurement is shown in Fig. 12.8 although a slow voltage ramp ($\sim$0.1 V/sec) also can be used for display on an x-y recorder. The voltage independent capacitor in series with the MOS capacitor is C_i. An adequate value of C_i is about 300 pF for many applications. The stray capacitance across the MOS capacitor is C_w.

The ratio C_i/C_{ox} needed in (3.63) to solve for ψ_s is obtained by biasing the MOS capacitor into strong accumulation, and measuring V_a and V_i. Then $C_i/C_{ox} = V_a/V_i - 1$. Similarly C_w/C_{ox} needed in (3.63) is obtained by lifting the probe, and measuring V_a and V_i at a convenient value of V_a. Then $C_w/C_{ox} = (V_a/V_i - 1)(V_a'/V_i' - 1)^{-1}$ where V_a and V_i have been measured with the MOS capacitor in strong accumulation as just described, and V_a' and V_i' have been measured with the probe lifted. All voltage measurements must be started with $V_a = V_i = 0$. For best results both V_a and V_i should be measured to within a millivolt.

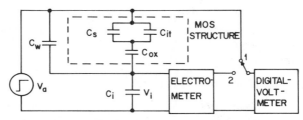

Fig. 12.8 Block diagram of a circuit for measuring the Q-V characteristics of an MOS capacitor. The electrometer is a Keithley 642 operated as a voltmeter. These electrometers are accurate to about 0.05%. This voltmeter does not need an electrometer buffer because it has sufficiently high input impedance and sufficiently low offset current. Capacitance C_i is measured with an admittance bridge, and a typical value is 300 pF. Before applying a new voltage, C_i and the MOS capacitor must be short-circuited to eliminate extraneous charge arising from the off-set current of the electrometer and the finite time constants of the capacitors. Voltage V_a should be applied slowly to avoid instabilities in the electrometer. After Ziegler and Klausmann.[10]

The electrometer, operated in the voltage mode, has a high input resistance ($>10^{14}\,\Omega$) and acts as an impedance transformer of unity gain between C_i and the digital voltmeter. Such an impedance transformer is necessary to avoid discharging C_i through the voltmeter input resistance. Using an impedance transformer results in another problem, namely charging of the voltmeter input capacitance (part of C_i) by the voltmeter offset current. Consequently, a voltmeter with a low input offset current ($>5 \times 10^{-15}$ A) and a low input capacitance (<20 pF) must be used. The switch in Fig. 12.8 is necessary only if the same voltmeter is used for measuring both V_a and V_i.

The voltage across the MOS capacitor is

$$V_G = (V_a - V_i). \tag{12.8}$$

The measured charge on the MOS capacitor is

$$\Delta Q_G = C_i V_i - C_w V_G. \tag{12.9}$$

By varying the applied bias V_a the charge ΔQ_G can be obtained as a function of V_G. Provided that the system is always in thermal equilibrium, differentiating ΔQ_G in (12.9) with respect to V_G yields the low frequency capacitance

$$C_{LF} = \frac{d\Delta Q_G}{dV_G}. \tag{12.10}$$

The low frequency capacitance C_{LF} can be obtained from (12.10) quite accurately because the accuracy with which voltage can be measured is very high, although the differentiation has to be performed numerically.

12.5.3 Silicon Band Bending

Silicon surface band bending ψ_s can be found as a function of gate bias by the Q-V method[10] from (3.63) (see Section 10.5.6) or from a low frequency or thermal equilibrium C-V curve where both interface traps and minority carriers follow the ac signal and gate bias sweep[11] as described in Section 3.3. Figure 4.1 shows that frequencies below 10 Hz will be inadequate for some samples. Therefore, a low frequency C-V curve over the entire gate bias range must be measured at frequencies below 10 Hz for these samples.

For samples that have very long minority carrier response time (seconds or more), it will be very difficult to measure ψ_s versus V_G for reasons given in section 12.5.2(b). One way to avoid this problem is to measure at 100°C. The problem with measuring a low frequency C-V curve at 100°C is that the unknown error introduced into the ψ_s versus V_G relationship by charged interface traps is different than at room temperature. For low interface trap level densities, this difference may be small. However, it is preferable to use the Q-V method for long response time samples.

To find ψ_s over most of the silicon bandgap from a low frequency C-V measurement, (3.65) is integrated from V_{FB} to values of V_G in strong inversion and from V_{FB} to values of V_G in strong accumulation. The integration of (3.65) can be done with a planimeter from a plot of $(1 - C_{LF}/C_{ox})$ versus V_G or by a digital computer. To evaluate (3.65), the remaining unknown required is V_{FB}. Determination of V_{FB} is explained in Section 10.5.

A way of automatically measuring ψ_s versus V_G to within an additive constant is with a lock-in amplifier tuned to a very low frequency. The instrumentation required is an x-y recorder, a linear voltage ramp generator, and a lock-in amplifier with an integrating feature such as the PAR model 126 with a type 184 current sensitive preamplifier. Details of this arrangement are described in Ref. 12. A lock-in amplifier without a built-in integrating capability also can be used, but then an integrating circuit must be used.

Tonner and Simmons[13] describe simple instrumentation consisting of a slow voltage ramp generator, an operational amplifier integrator, and an operational amplifier adder for measuring ψ_s versus V_G to within an additive constant.

12.5.4 High–Low Frequency C-V Technique[5]

(a) Sequential Method

A high and low frequency C-V curve can be combined to measure interface trap level density as described in Section 8.2.6. Rather than the high frequency or low frequency capacitance methods, the high–low frequency capacitance method should be used for measuring interface trap level density, for reasons described in Section 8.2. This method is effective only with a large gate area. To instrument this method in a convenient way, the MOS capacitor is switched from the low frequency circuit shown in Fig. 12.5, after plotting a low frequency C-V curve, to the high frequency circuit shown in Fig. 12.4. The y-axis gain of the x-y recorder is then adjusted to make the high frequency capacitance in very strong accumulation coincide with the low frequency capacitance in very strong accumulation. This high frequency capacitance measured in strong accumulation is equal to C_{ox} provided that *series resistance is negligible*. To complete the calibration, zero capacitance is set at the bottom of the chart. If the zero is offset, another value of capacitance, say, the high frequency capacitance in strong inversion, must be recorded. This pair of calibrated high and low frequency C-V curves constitute the raw data of the high–low frequency capacitance method.

(b) *Simultaneous Method*

It is more convenient to automate and more accurate to measure high and low frequency C-V curves simultaneously rather than sequentially.[8] Simultaneous measurement of high and low frequency C-V curves also ensures that there is no voltage shift between curves, which may be present when the two curves have to be aligned in the sequential method. For these reasons, the simultaneous method is preferred over the sequential method. Figure 12.9 shows a simple block diagram for the simultaneous measurement of high and low frequency C-V curves. Output is shown fed into a computer instead of an x-y recorder. The Boonton L-C meter and the Keithley electrometer still are used to measure the high and low frequency C-V curves, respectively, as described in Sections 12.5.1 and 12.5.2. But now the L-C meter and the electrometer have a common input to which the MOS capacitor is connected.

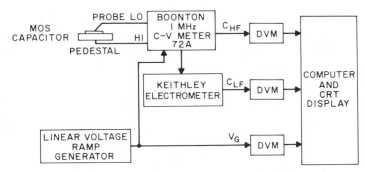

Fig. 12.9 Block diagram for simultaneously measuring a high and a low frequency C-V curve. Output is to a computer and cathode ray display rather than an x-y recorder. The digital voltmeters act as analogue-to-digital converters. After Derbenwick.[8]

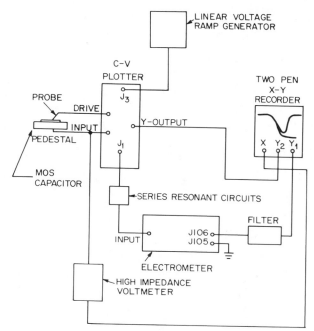

Fig. 12.10 Block diagram for simultaneously measuring a high and a low frequency *C-V* curve using the PAR model 410 *C-V* plotter. The electrometer can be a Keithley 602 operated in the fast mode. Connectors J_1 and J_3 are the same as those in Fig. 12.12*b*.

The PAR model 410 *C-V* plotter can be used instead of the *L-C* meter to measure the high frequency *C-V* curve at 1 MHz. Figure 12.10 shows a simple block diagram for using the *C-V* plotter. The voltage ramp generator is internal to the *C-V* plotter. Output is shown into a two-pen *x-y* recorder instead of a computer. A computer such as that shown in Fig. 12.9 can be substituted for the two-pen *x-y* recorder in Fig. 12.10, and the two-pen *x-y* recorder shown in Fig. 12.10 can be substituted for the computer in Fig. 12.9. The computer, processing the data digitally, is more accurate than analogue graphical methods for quantitative analysis and more convenient because the analogue results must eventually be processed by computer.

Modification of the *L-C* Meter To make a common input for the electrometer and the *L-C* meter, one must modify the circuitry of the *L-C* meter. The need for this modification arises because the electrometer operating in the picoampere range is very sensitive to shunt capacitance and resistance. The effect of shunt resistance on a low frequency *C-V* curve was described in Section 12.5.2. Shunting capacitance will increase electrometer noise substantially if it is very large. A shunting capacitance

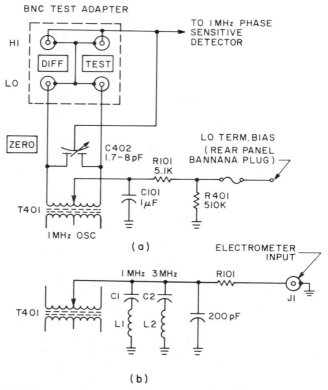

Fig. 12.11 Critical portions at the Boonton model 72A circuit (a) before and (b) after modification. Coils L_1 and L_2 can be wound on torroidal cores, or they can be variable inductors. Nominally, $L_1 = 63\,\mu H$ and $L_2 = 28.1\,\mu H$. For the torroidal core inductors, $C_1 = 450\,pF$ and $C_2 = 150\,pF$ are air or ceramic tuning capacitors. For the variable inductors, $C_1 = 402\,pF$ and $C_2 = 100\,pF$ mica capacitors. Connector J_1 is a UHF connector shielded where the cable attaches to it behind the panel. After Derbenwick.[8]

of $1\,\mu F$, as in the circuit of the model 72A *L-C* meter, will result in nanoamperes of noise in the electrometer for example. Figure 12.11 shows the critical portion of the *L-C* meter circuitry that must be modified. Figure 12.11a is the critical part of the Boonton model 72A circuitry before modification and Fig. 12.11b after modification. The aim of this modification is to minimize shunting capacitance and maximize shunting resistance across the input of the electrometer. The required modifications are[8]:

1   Reduced shunting capacitance.

   (a)   Remove $1\,\mu F$ bypass capacitor $C101$.

   (b)   Connect series resonant circuits tuned to 1 and 3 MHz and a 200 pF shunt capacitance from the center tap of the transformer to ground.

(c)   The values of C_1 and C_2 should not exceed several hundred picofarads, and L_1 and L_2 can be wound on torroidal cores. Suggested values of these quantities are given in Fig. 12.11.

2   Increased shunting resistance.

(a)   Remove the 510 kΩ protection resistor $R401$ between "dc low" and ground. This resistor can be left in or removed without affecting performance.

(b)   Replace all critical phenolic and fiber washers with Teflon or ceramic on transformer T401.

(c)   Replace printed circuit board wiring to zeroing capacitor $C402$ with shielded wiring.

3   Pickup noise reduction.

(a)   Shield the ac line to the power switch as well as the power switch itself in the front section of the instrument.

(b)   Shield the "dc low" lead to the series resonant circuits.

(c)   Use a UHF connector for J_1 to connect to the electrometer input. Use a "bonnet" shield where the cable connects to the connector behind the panel.

(d)   The same rigid coaxial connectors described in Section 12.5.2 should be used between the terminals of the electrometer and J_1 on the L–C meter shown in Fig. 12.11b.

These modifications require recalibration of the L-C meter. First, the center tapped transformer T401 may require retuning. Replacement of the phenolic and fiber washers on this transformer also may affect the center tap, which should then be repositioned. With an oscilloscope connected to J_1, the series resonant circuits L_1, C_1 and L_2, C_2 should be tuned by minimizing the 1 and 3 MHz ac voltages seen on the oscilloscope. These tuned circuits are required to shunt the ac to ground from the transformer center tap without introducing large shunt capacitance. The 1 MHz circuit shunts the fundamental oscillator frequency. However, the L-C meter will not read the correct capacitance unless the odd harmonics are also shunted to ground. The dominant odd harmonic is the third that is shunted by the series resonant circuit tuned to 3 MHz. The 200 pF capacitance shunts the higher harmonics. Calibration is completed by calibrating the L-C meter according to the instruction manual. Bias is supplied to the MOS capacitor in the modified circuit through the "Hi" terminal to ground on the back panel of the L-C meter.

Modification of the older Boonton model 71-A, which has a slightly different input circuit, proceeds in the same way. The center tap of input transformer T101 is lifted off ground and the two series resonant circuits connected to it as shown in Fig. 12.11b. The electrometer can be connected through J_1 as before. Steps 1–3 are the same, except 1(a) and 2(a) are not necessary, and the bias terminals are unaffected.

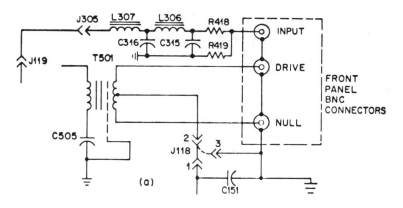

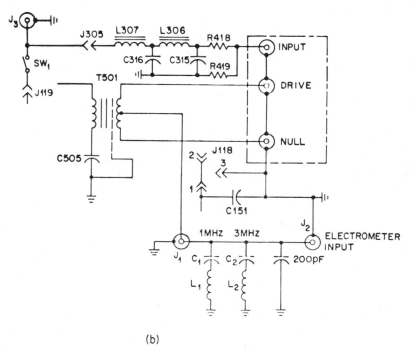

(b)

Fig. 12.12 Critical portions of the PAR model 410 C-V plotter (a) before and (b) after modification; L_1, L_2, C_1, and C_2 are the same as in Fig. 12.11b.

Modification of the C-V Plotter The PAR model 410 C-V plotter can be modified in a similar way so that a high and a low frequency C-V curve can be measured simultaneously. Figure 12.12a shows the critical part of the input circuitry before modification, and Fig. 12.12b shows it after modification. The required modifications are:

1 Increased shunting resistance.

(a) Remove the center tap connection between the secondary of transformer T501 and J118 from the printed circuit board and connect the two center tap leads to a shielded cable connected to a VHF connector mounted on the cover. The holes in the VHF connector can be threaded and the connector mounted so that it protrudes from the cover to facilitate removal of the cover.

(b) Remove the wires from the secondary of transformer T501 from the printed circuit board and the wires from the BNC connectors labeled "drive" and "null" from the printed circuit board. Connect the secondary leads of the transformer directly to the "drive" and "null" BNC connectors with shielded wire.

2 Reduced shunting capacitance.

(a) Connect the center tap of transformer T501 to series resonant circuits tuned to 1 and 3 MHz and the 200 pF shunt capacitor to ground as shown in Fig. 12.12b. The resonant circuits are tuned as described in the preceding section (under "modification of the L–C meter").

3 Pickup noise reduction.

(a) "Drive" on the C-V plotter is connected to the probe wire as shown in Fig. 12.10 rather than the pedestal to minimize noise pickup by the electrometer. However, if a temperature controller is connected to heaters in the pedestal, its power plug should be disconnected from the ac power line, or there will be excessive 60 Hz pickup by the C-V plotter, making the high frequency C-V curve noisy. To measure simultaneous high and low frequency C-V curves at temperatures above room temperature, dc should be used to power the heaters in the pedestal.

(b) The series resonant circuits and 200 pF shunt capacitor in Fig. 12.12b can be mounted in a minibox equipped with a male UHF connector connected to the female UHF connector J_1 on the cover by means of rigid coaxial connectors described in Section 12.5. Rigid coaxial connectors should be used between the UHF connector J_2 on the minibox and the input of the electrometer and between the probe lead coming out of the probe station (see Section 12.8) through a UHF connector and the BNC connector labeled "drive" on the front panel of the C-V plotter.

(c) Place a shield cap over the BNC connector labeled "null" on the front panel of the C-V plotter to shield it from the environment.

4 Accurate low frequency C-V curve.

(a) Use of the internal ramp generator of the C-V plotter will result in a distorted low frequency C-V curve. An external voltage ramp generator that has better linearity, such as the ramp generator described in Section 12.7, must be used. To make the modification required for an external ramp generator, the lead going to the BNC connector labeled "input" on the front panel of the C-V plotter is unplugged from J119 and wired to a BNC connector J_3 as shown in Fig. 12.12b. The BNC connector J_3 can be mounted on the top cover of the C-V plotter. A toggle switch SW_1 also is mounted on the top cover of the C-V plotter and wired between the BNC connector J_3 and the pin on J119 as shown in Fig. 12.12b. This switch permits use of the internal ramp generator for other measurements.

(b) The resistors $R418 = 2\,k\Omega$ and $R419 = 100\,k\Omega$ in Fig. 12.12b constitute a voltage divider. Therefore, gate bias must be measured at the input BNC on the C-V plotter rather than at J_3. Because the C-V plotter does not measure capacitance to ground, capacitance to ground of the voltmeter reading gate bias introduces negligible error if it is small. A high impedance voltmeter with unity gain, such as the Keithley model 602 digital electrometer, can be used between the input terminal of the C-V plotter and the x-axis input of the x-y recorder as shown in Fig. 12.10 to measure gate bias.

(c) The accuracy of the C-V plotter for measuring capacitance is unaffected by these modifications. However, the accuracy for measuring equivalent parallel conductance is adversely affected. To measure equivalent parallel conductance, the input of the electrometer should be short-circuited to ground to directly ground the center tap of transformer T501.

Recorder   To make a simultaneous plot of a high and low frequency C-V curve, a two-pen x-y recorder, such as the HP 7046A, can be used. To minimize 60 Hz noise, a low-pass filter, such as the HP model 17175A, which provides -55 to -70 dB of rejection of ac signals 50 Hz and higher, must be inserted between the output of the electrometer and one of the y-inputs of the x-y recorder if the HP 7046A two-pen recorder is used. The other y-input of the recorder is connected to the output of the L-C meter or the C-V plotter as the case may be.

To expand the y-scales, the zero settings will be below the chart paper. It is essential that both y-axes have the same zero. This can be accomplished by connecting a small dc voltage (a potentiometer and a battery) to each y-axis in turn, bringing the pens back onto the chart. With the range setting of both y-axes the same and a given external dc voltage applied, the zero adjust knobs are turned to bring each pen to the same arbitrarily chosen point on the chart paper. Ranges on the y-axes can be changed without affecting the zero.

To calibrate, the gain of the pen recording the high frequency C-V curve can be adjusted to make the high frequency capacitance coincide with the low frequency capacitance in strong accumulation. Then the high frequency capacitance in strong accumulation (C_{ox}) and in strong inversion is read and marked on the chart paper.

12.5.5   Other Techniques

Differential capacitance can be measured by superposing a small ac voltage on the bias and measuring the ac displacement current flowing through and the voltage across the MOS capacitor.[13] Also see Section 12.3.6.

12.5.6   Pulse C-V Technique[15]

In the preceding sections three-terminal capacitance measurement methods were described with a small ac voltage superposed on the bias or slowly varying ramp voltage. In this section three-terminal capacitance measurement methods are described with a small ac voltage superposed on a rectangular voltage pulse. Such measurements are made for determining doping profile or avalanche breakdown voltage in the semiconductor as described in Chapters 9 and 11.

An admittance bridge such as the Boonton 75-C can be used to measure three-terminal capacitance. Figure 12.13 shows a circuit for measuring pulse C-V curves. Capacitance and equivalent parallel conductance readings are made by nulling the ac output of the bridge null detector during the pulse. This nulling is accomplished by observing the ac null voltage from the output of the null amplifier on an oscilloscope synchronized with the pulse generator. When the pulse turns on and off, a brief transient is seen on the oscilloscope delineating the duration of the pulse. To get a null while pulse voltage exists across the MOS capacitor, the bridge oscillator frequency must be high enough to allow at least four or five complete cycles to be seen on the oscilloscope. At 200 kHz and a pulse width of 6 msec, there are 300 cycles during the pulse, which amply fulfills this requirement.

Pulse voltage is fed in parallel with the MOS capacitor and the bridge through the parallel feed network shown in Fig. 12.13a. Because the bridge looks like a short circuit to the pulse source in parallel with it, the pulse source is isolated from the bridge by the two 5000 pF capacitors C_1 and C_2 in Fig. 12.13a. The current transient through the bridge needed to charge C_1 and C_2 is partially reduced by the two 500 pF capacitors C_3 and C_4 connected from each bridge terminal to ground. The two 100 kΩ resistors R_1 and R_2 connect the pulse source to the MOS capacitor and isolate the bridge terminals from ground. The internal impedance of the pulse source is 2.5 kΩ, which is small compared to R_1 and R_2. By using an emitter follower stage in the output of the pulse source as shown in Fig. 12.13b, the output impedance is kept much smaller than 100 kΩ. Changing the position of the helipot R_6 to vary pulse voltage does not significantly change the

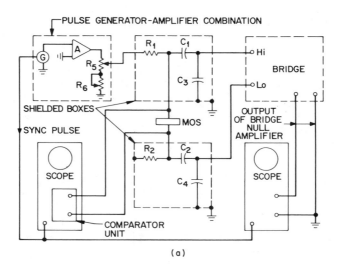

PULSE GENERATOR-AMPLIFIER COMBINATION

(a)

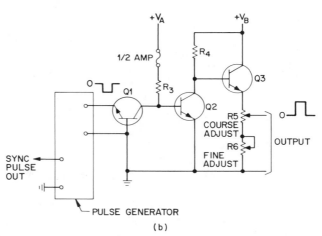

(b)

Fig. 12.13 (a) Circuit for applying voltage pulses to an MOS capacitor and measuring MOS capacitance as a function of pulse height during the pulse with a Boonton 75-C admittance bridge. Typically, the output impedance of the pulse generator G is 50 Ω. After Goetzberger and Nicollian.[15] (b) Wiring diagram of a pulse amplifier. The component values are: $R_3 = 1$ kΩ, 0.5 W; $R_4 = 6.8$ kΩ, 10 W; $R_5 = 5$ kΩ, 15 W General Radio potentiometer; $R_6 = 500$ Ω, 2 W, 25 turn Helipot; $Q_1 =$ any low voltage n-p-n transistor, Q_2 and $Q_3 = $ 2N3584 n-p-n transistors, $V_A = 22.5$ V from a heavy-duty 45 V B battery, and $V_B = 230$ V at 40 mA. A suitable power supply that can be used for V_B would have an internal impedance of 0.01 Ω and a ripple of 100 μV rms.

conductance to ground or the conductance seen at the bridge terminals. No current flows from the pulse source through R_1 and R_2 except during the transient at the beginning and at the end of the pulse. Therefore, full pulse voltage appears across the MOS capacitor.

The high and low terminal circuits R_1, C_1, C_3 and R_2, C_2, C_4 are placed in separate grounded metal boxes, and all connecting cables are coaxial to minimize stray capacitance between these elements, which would appear across the bridge terminals. This minimized stray capacitance will be small enough to be zeroed out by the bridge zero settings. If stray line to line capacitance or conductance is too large, measured capacitance of the MOS capacitor can be reduced 10% or more. In addition, the conductance measured by the bridge with the MOS capacitor in the circuit is less than the zero setting of the bridge with the MOS capacitor out of the circuit. When the MOS capacitor is being measured, some of the stray susceptance is short-circuited out. With the series isolating capacitors (C_1 and C_2) mixing the actual capacitance and conductance, shorting shows itself as a decrease in the effective conductance seen by the bridge. This shorting effect can be eliminated by reducing stray capacitance other than to ground, by increasing resistance to ground, by working at higher bridge oscillator frequencies, and by keeping the isolating capacitors as large as possible. It is preferable to eliminate this short-circuit effect rather than to correct for it because it can vary with the capacitance of the MOS capacitor and would thus have to be determined at every voltage and for every sample. With the circuit suggested in Fig. 12.13a, it was necessary to work above 150 kHz to eliminate this effect.

There are two known and constant errors introduced into the measurement of capacitance by the network consisting of R_1, C_1, C_3 and R_2, C_2, C_4. The first error arises from the conductance to ground provided by R_1, R_2 and the internal resistance of the pulse generator. The bridge operated in the three-terminal mode is relatively insensitive to any capacitances to ground, but conductance to ground will cause a reduction in the value of the measured capacitance of the MOS capacitor. The conductance of the circuit consisting of R_1, R_2 and the pulse generator is small, and at frequencies above 150 kHz, R_1 and R_2 are effectively short-circuited by C_3 and C_4 for the values of these components used in Fig. 12.13a. Therefore, negligible error is introduced. However, C_3 and C_4 are sufficiently large capacitances to ground to introduce a small but constant error into the measured capacitance. This error depends on frequency, the magnitude of C_3 and C_4, and can be obtained from a graph provided by the bridge manufacturer in his instruction manual. Capacitance measured by the bridge is reduced by a constant 2% at 200 kHz by C_3 and C_4; that is

$$C_c = 1.02C_m \qquad (12.11)$$

where C_c is the correct value and C_m is the measured value of capacitance.

The second error arises because the MOS capacitor is in series with C_1 and C_2. Therefore, the capacitance of the MOS capacitor C is calculated from

$$C = \frac{C_c C_1 C_2}{C_1 C_2 - C_c (C_1 - C_2)} . \tag{12.12}$$

Values of C calculated from (12.12), using (12.11), will be accurate within 0.25%, the accuracy of the Boonton 75-C bridge, provided C_1, C_2, C_3, and C_4 are known to this accuracy by measuring them on this bridge.

Rise time of the pulse voltage is determined by the RC time constant $R_1 C_1$ or $R_2 C_2$ in the circuit shown in Fig. 12.13a. This RC time constant is 0.5 msec. To reduce this rise time, it is better to reduce the size of C_1 and C_2 rather than R_1 and R_2, which cannot be reduced much below 100 kΩ without affecting the accuracy of the bridge measurement. Because sensitivity is lost when the isolating capacitators C_1 and C_2 are not large compared to the capacitance measured, the lower limit of the rise time is about 0.5 msec.

A signal amplitude of 300 mV peak to peak can be applied to the network containing the MOS capacitor to obtain high bridge sensitivity and still remain within the small-signal range. Pulse voltage can be measured within 0.1% with a Tektronix 5A13N differential capacitor unit in a 5440 oscilloscope or equivalent as shown in Fig. 12.13a. In the circuit shown in Fig. 12.13a measurements must be taken manually point by point. This manual measurement is done by varying pulse voltage with R_5 and R_6 and nulling the bridge at each pulse voltage.

Because many commercial pulse generators do not deliver high voltage pulses, a pulse amplifier fed by a commercial pulse generator such as that shown in Fig. 12.13b is used. This circuit delivers rectangular pulses with rise times of 0.4 μsec and an amplitude that is variable from 0 to 200 V and a duty cycle as low as 1%. A drop in pulse height with time of less than 0.01% can be obtained if V_B is supplied by a constant voltage source.

Transistor Q_3 is an emitter follower stage allowing the maximum output impedance from the wiper on R_5 to ground to be as low as 2.5 kΩ. The collector of transistor Q_2 swings between the saturation voltage of 0.3 V and the supply voltage V_B reduced by the voltage divider consisting of R_4 and $\beta_3 R_5$, where β_3 is the large-signal beta (or common emitter current gain) of transistor Q_3. A negative pulse from the pulse generator pulls the base of transistor Q_2 below ground taking Q_2 from saturation to cutoff. The output of this amplifier is a string of positive voltage pulses whose width, rise time, and duty cycle are set by the controls of the pulse generator.

12.6   EXTENDING THE ENERGY RANGE OF THE ADMITTANCE METHODS

There are two ways of extending the energy range over which interface trap properties can be extracted in any of the admittance (capacitance or conductance) methods described in Chapters 5 and 8: (1) to vary frequency while keeping temperature fixed and (2) to vary temperature while keeping frequency fixed.

12.6.1 Frequency

At fixed temperature the frequency range over which admittance measurements are made determines the energy range in the silicon bandgap over which interface trap properties can be extracted.

At 300°K admittance measurements can be made over the frequency range 50 Hz–500 kHz. Over this frequency range, interface trap properties can be extracted from about $2kT/q$ V from midgap to about $2kT/q$ V from flatbands.[2] However, an important segment of the interface trap level density distribution from flatbands to the majority carrier band edge is not measured in this frequency range. Because small-signal steady-state methods yield useful interface trap information between midgap and the majority carrier band edge, both n-type and p-type samples must be measured to cover both halves of the silicon bandgap. One way of extending the range of energy toward the majority carrier band edge is to make measurements at frequencies above 500 kHz. There are few reliable measurements available in the literature at frequencies much above a few megahertz because such measurements are difficult to make. There are two major problems that make very high frequency admittance measurements difficult.

First, series resistance, no matter how small, must be taken into account as described in Section 5.7.1. To give the reader an idea of how limiting series resistance can be, assume that the only series resistance is caused by the uniformly doped bulk silicon. For example, the series resistance of a 10Ω-cm epitaxial layer 10 μm thick will be 10 Ω for a typical gate area of 10^{-3} cm^2. At very high frequencies (100 MHz) this series resistance will dominate the loss and no information on interfacial properties can be extracted from either measured capacitance or equivalent parallel conductance. Thus as frequency is increased up to 100 MHz, an increasingly large correction for series resistance must be made. If an epitaxial layer is not used or if there is additional series resistance caused by other factors and total series resistance is more than 10 Ω, this series resistance will dominate the loss at frequencies well below 100 MHz. Series resistance is manifest most obviously by a frequency-dependent value of capacitance in strong accumulation.

Second, the inductance of the probe wire, which is typically more than $\frac{1}{2}$ inch long and 5–10 mils in diameter (see Section 12.8), introduces a frequency-dependent inductive reactance that must be taken into account. At sufficiently high frequency, depending on the inductance of the probe wire, inductive reactance will dominate the measured admittance and no information on interface trap properties can be extracted from either capacitance or equivalent parallel conductance. The inductance problem can be minimized by mounting the MOS capacitor in a low inductance, high frequency package such as used for varactor or Impatt diodes. However, this packaging entails additional sample preparation and heat treatment in bonding the MOS capacitor to the package and bonding a low inductance lead to the gate electrode. These heat treatments can alter the electrical properties of the MOS system sufficiently that the properties characteristic of the packaging process rather than the process of interest are measured. Another problem with packaging is pin to pin leakage currents in low frequency measurements with the ramp voltage method.

Once these factors have been dealt with, the best way to measure the admittance of the MOS capacitor up to frequencies of 100 MHz or higher is with an admittance bridge, which is commercially available in this frequency range. Another way is by reflection measurements using a vector voltmeter.[16] However, this method is inaccurate compared to the admittance bridge.

12.6.2 Temperature

As discussed in Section 2.2, the Fermi level at the silicon surface can be made to move from midgap to the majority carrier band edge by varying temperature. Therefore, interface trap properties can be extracted over half of the bandgap by varying temperature with frequency fixed. In general, it is easier to scan the silicon bandgap by varying frequency with temperature fixed. At room temperature high frequencies are required to extract interface trap properties near the majority carrier band edge with attendant problems as described in Section 12.6.1. The energy range can be extended toward the majority carrier band edge by measuring admittance versus gate bias and frequency at low temperatures.

One major problem is to maintain an ohmic back contact over a wide temperature range even though it is ohmic at room temperature. If the back contact is no longer ohmic at some temperature, a significant error will appear in the measured admittance. The Cr–Au or Ti–Au contact described in Section 12.9 should be ohmic down to liquid nitrogen temperature (77°K).

At low temperatures, interface trap response times become very long at all bandgap energies except those within a few kT/q volts of flatbands. Response times of minutes, days, and even years will be the rule over most of the gate bias range. Therefore, as temperature is decreased below room temperature, admittance measurements with the instrumentation described in Sections 12.3–12.5 can be made only over a progressively narrower energy range. To extend the energy range closer to the majority carrier band edge, measurements must be taken at a large number of different temperatures. Scanning the energy range by varying temperature requires very accurate temperature control and measurement because interface trap response times depend exponentially on temperature. Thus slight variations of temperature during the course of measurement or errors in the measurement of temperature will have a large effect on the admittance, and lead to erroneous interface trap properties. Because of the many difficulties in low temperature admittance measurements, few have been made to date.

The MOS capacitor can be heated above room temperature to extend the energy range toward midgap. The problem in this case is to activate or introduce ionic contaminants into the oxide, thereby changing interface trap level density and/or oxide fixed charge density and causing the MOS capacitor characteristics to drift during measurement. As a result, there are few measurements of interface trap properties at temperatures above room temperature.

An accurate method of measuring minority carrier lifetime involves measuring the admittance of the MOS capacitor at various temperatures above room temperature, is described in Sections 4.3 and 9.5. Instrumentation for the measurement of MOS capacitor characteristics both above and below room temperature are described in Section 12.8.

12.7 LINEAR VOLTAGE RAMP GENERATOR

The linear voltage ramp generator shown as a black box in Figs. 12.3, 12.4, 12.5, and 12.9 can be a commercial function generator operated to put out a triangular waveform. A circuit shown in Fig. 12.14 is specially

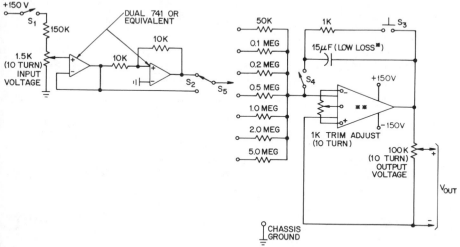

Fig. 12.14 Circuit for a linear voltage ramp generator having a linearity of 0.5%. The operational amplifier labeled ** is a Philbrick/Nexus Op-Amp 1022 or equivalent. The integrating capacitor labeled * is a low loss capacitor, 15 μF, ±10%; 400 V dc, made by EAI, West Long Branch, N. J., PSR 16004 BGP-G905. The switches are S_1 (on-hold-off), S_2 (positive-negative), S_3 (zero reset), S_4 (zero-hold), and S_5 sets the sweep rate. All resistors are 0.5 W, 10%. The dual 714 operational amplifiers can be powered from the ±150 V power supply through a voltage divider.

designed for MOS capacitor measurements. This circuit produces a voltage ramp that has a linearity better than 0.5%, and it produces an output voltage in range from 0 to ±150 V. This voltage range and linearity is adequate for most MOS capacitor measurements. Voltage sweep rates can be varied from 20 mV/sec to 2 V/sec in the circuit shown in Fig. 12.14, using S_5 and the 1.5 kΩ input potentiometer. Maximum output voltage can be varied from 0 to 150 V by varying the 100 kΩ output potentiometer. This circuit is an operational amplifier integrator with an input dc voltage, provided by the 1.5 kΩ potentiometer, which generates a linear voltage ramp.

To take advantage of the performance possibilities of an operational amplifier as an integrator, a feedback capacitor must be selected with a dielectric leakage current that is less than the bias current of the amplifier. Such a capacitor is suggested in the caption to Fig. 12.14. A more detailed description of the principles of operation of an operational amplifier used as an integrator can be found in Ref. 17. Sweep rate is determined by the input voltage divided by the RC product between the resistor selected by S_5 and the 15 μF integrating capacitor. Sweep rate can be further slowed by reducing input voltage with the 1.5 kΩ potentiometer.

12.8 PROBE STATION*

With both admittance bridges and lock-in amplifiers, *three-terminal* or *direct* differential capacitance measurements should be made. An MOS capacitor is made into a three-terminal capacitor by surrounding it by a grounded metal enclosure. Two terminals of the capacitor are now the gate electrode and the back contact, and the third terminal is the grounded metal enclosure. A grounded metal enclosure around the MOS capacitor also provides effective shielding for low frequency capacitance measurements, using an electrometer-operational amplifier described in Section 12.5.2.

There are two ways of putting the MOS capacitor in a grounded metal enclosure for three-terminal measurements. The first method is to encapsulate each individual MOS capacitor in a suitable package. The disadvantages of encapsulation are discussed in Section 12.6.1. It is almost always more convenient to probe the MOS capacitor in a probe station such as illustrated, in cross section, in Fig. 12.15. This probe station is suited for measuring MOS capacitors made in an array on a silicon wafer or MOS capacitors incorporated in a test pattern in a wafer containing an array of integrated circuits. Such an MOS capacitor array is shown in Fig. 12.16.

---

*The authors are indebted to E. Labate for assistance in constructing the probe stations described in this section.

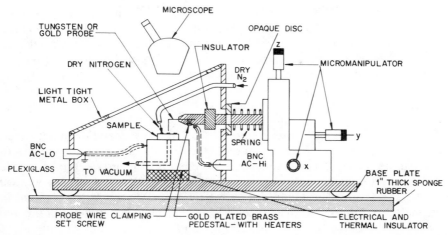

Fig. 12.15 Cross section through a metal probe station suitable for three terminal measurements of the admittance of an MOS capacitor in an array made on a wafer. The ac low and ac high terminals are BNC connectors that go to the terminals marked "Hi" and "Lo" in Figs. 12.1, 12.3, 12.4, 12.5 and 12.9. The micromanipulator permits the probe wire to be moved in the three mutually perpendicular directions, x, y, and z.

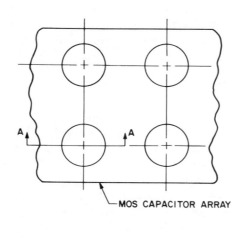

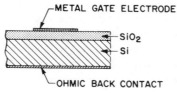

SECTION A–A

Fig. 12.16 Typical MOS capacitor array on a wafer for monitoring or experimental studies. An MOS capacitor is shown in cross section in section A-A.

In sections 12.8.1–12.8.5 two different designs for a probe station and its components are described. The purpose of discussing these particular designs is to illustrate the general principles involved. If these principles are followed, the actual design used is limited only by the imagination, needs, and resources of the experimenter.

12.8.1 Probe Box

There are three basic requirements that the probe station must meet, which apply equally to basic studies and production monitoring: (1) to make three-terminal capacitance measurements, the probe box must be made of metal maintained at ground potential as previously explained; (2) the metal box must be lightproof (see Section 4.3); and (3) the ambient in the metal box must be dry to avoid leakage currents flowing along the oxide surface between the gate electrode and the back contact of the MOS capacitor as described in Section 4.3.12(b). Leakage currents between terminals on the probe box to ground are also avoided by maintaining a dry ambient in the box. An adequately dry ambient can be maintained in the box by directing a gentle stream of dry nitrogen at the wafer. A small pipe for this purpose is shown in Fig. 12.15. The probe station shown in Fig. 12.15 meets all three of these requirements. Other probe box designs are possible. For example, the micromanipulator and the microscope both can be mounted inside the probe box, making the box design simpler but bulkier.

12.8.2 Probe Wires

A wire probe mounted on a micromanipulator and insulated from ground is used to contact the gate of an individual MOS capacitor on a wafer as shown in Fig. 12.15. To avoid high contact resistance to aluminum gates, a tungsten wire 5 or 10 mils in diameter, $\frac{1}{2}$–1 inch long, and sharpened electrolytically should be used.

Tungsten can be electrolytically sharpened by making it the anode in a solution of 10% NaOH or KOH by volume. The cathode can be a platinum wire. Only a few volts is required for the sharpening process. Only the tip of the wire to be sharpened is immersed into the solution. The length of time required for sharpening can be judged by examining the point with a magnifying glass or low power microscope after a minute or two in the solution with voltage applied. If the point is made too sharp, it becomes weak and bends easily when the slightest pressure is applied. A compromise between an oversharp weak point and a blunt point must be made.

The tungsten wire is mounted vertically to permit application of pressure to the contact pad. The tungsten wire should have an S-shaped bend in its shank to provide some springiness to take up the pressure applied by the micromanipulator. With the application of pressure, produced by vertical movement of the micromanipulator, the sharp, hard tungsten point can be made to penetrate the protective native oxide covering the aluminum to make an ohmic contact without scratching the aluminum.

An alternative method is to mount the tungsten wire horizontally and bend it down at a 15–20° angle to the vertical. However, this method is more prone to scratching the gate metal surface, thereby reducing its area or making poor contact.

The contact between tungsten probe wires and aluminum or gold pads tends to produce noise. This problem can be alleviated by plating the tungsten point after forming, first with nickel and then with rhodium.[18] The nickel is for better adherence and the rhodium for hardness and immunity to corrosion. In contacting a p-n junction, the problem of breaking through the oxide film can be solved by passing a large forward current through the junction to form or weld the probe wire before making the desired measurements.

For gate or contact metals such as gold (see Section 12.9.2) that do not form a protective oxide layer, a soft gold wire 5 or 10 mils in diameter and $\frac{1}{2}$–1 inch long can be used. Tungsten will scratch such soft metals. Sharpening is not required; simply cutting the wire at a 45° angle is sufficient.

Several micromanipulators can be used to simultaneously contact more than one MOS capacitor in the array. If repetitive measurements are to be made on a given MOS capacitor array, a fixture holding several pre-positioned probe wires can be used to simultaneously contact the array, dispensing with the need for separate micromanipulators.

Because the probe wires are very fine and delicate, any vibration transmitted to them can result in a loss of contact during measurement. To minimize transmission of vibrations to the probe wires, the entire probe station should be placed on a $\frac{1}{4}$ inch thick plexiglass sheet, which, in turn, is placed on a 1 inch thick sponge rubber sheet.

Micromanipulators are used because they permit precise motion of the probe wire in three mutually perpendicular directions, labeled x, y, and z in Fig. 12.15, making it easy to position the probe wire on a small gate or contact pad area. There are several micromanipulator designs commercially available. One commonly used design is manufactured by the Line Tool Company and another, by Electroglass. Both are very convenient to use. Other useful micromanipulator designs are described in Ref. 19.

The only major problem is to maintain the probe box lightproof while providing a mechanism for moving the probe wire in the box in three mutually perpendicular directions when the micromanipulator is mounted outside the box. To provide for ample motion in the probe box design shown in Fig. 12.15, a large hole must be cut into the side of the box. With such a hole, the box is not lightproof. One simple way to solve this problem is to use a spring loaded opaque disk to cover the large hole in the metal box, as shown in Fig. 12.15, through which a cylindrical bar, connecting the probe mechanically to the micromanipulator, passes with a tight but sliding fit. A sliding panel on the sloping top of the box can be opened to permit observation of the sample with a microscope to facilitate contacting with the probe. Closing this panel makes the box light tight during measurement.

12.8.3 Mercury Probe

In most applications, an evaporated metal gate electrode contacted by a thin probe wire described in Section 12.8.2 is preferred. However, there are certain experiments where evaporation of a metal electrode would affect the measured results. For example, in experiments where an exchange of some species takes place between the oxide and the ambient, an evaporated metal gate can act as a barrier against this exchange. In many cases etching-off the metal electrode to permit the exchange and reevaporating the metal to electrically evaluate the results of the exchange may disturb the process under study. Also, there are experiments where a charge distribution in the oxide layer is to be profiled. In such experiments the oxide layer is etched off in steps requiring reevaporation of the metal gate after each etch-off step. Again, reevaporation of the metal may disturb the charge distribution being measured. To avoid these problems and provide a more rapid experimental technique, a mercury probe can be used.

A mercury probe is a small ball of mercury held at the end of a glass capillary tube that makes contact with the oxide surface of an oxidized silicon wafer and acts as a gate electrode. Figure 12.17a shows a simple mercury probe. A cylindrical glass capillary tube about 2 inches long with a 0.508 mm (20 mil) inner diameter is clamped to a micromanipulator of the type made by the Line Tool Company. The lower end is ground to a taper. The capillary is filled with clean triple distilled mercury about one-half to three-fourths full using a hypodermic needle.

A 0.508 mm (20 mil) inner diameter capillary is used because it is small enough for surface tension to prevent the mercury from flowing out of the end by gravity when the tube is held vertically. Insertion of a 0.254 mm (10 mil) diameter tungsten wire into the mercury column causes a small ball of mercury to protrude from the capillary. A ball forms because mercury has a convex meniscus. This mercury ball is pressed against the oxide surface by the micromanipulator to form the gate electrode, and the tungsten wire makes electrical contact to the mercury. This probe is safe to use if its temperature does not exceed 30°C.

The major problem with the mercury probe is that the area of mercury contact to the oxide is unknown and is, in fact, different every time contact is made. The contact area A_g is most easily obtained from the relation

$$A_g = C_{ox}\left(\frac{x_o}{\epsilon_{ox}}\right) \tag{12.13}$$

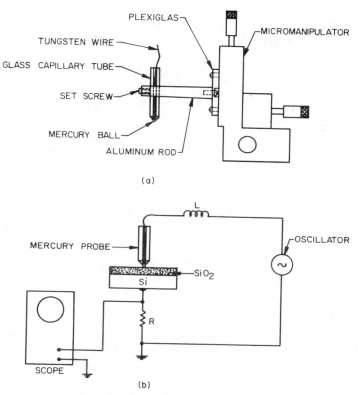

Fig. 12.17 (a) Simple mercury probe. (b) Series resonant circuit to reestablish constant contact area using the probe of part (a). The values of the components are: R 10 Ω, and L 1 mH. This combination will give a resonant frequency of about 1 MHz with a typical MOS capacitor with $C_{ox} \approx 25$ pF.

where x_o is the oxide thickness, ϵ_{ox} is the dielectric permittivity of SiO$_2$, and C_{ox} is the measured oxide layer capacitance. To use (12.13), x_o also must be known. Oxide layer thickness can be determined from any of the methods described in Section 14.3. However, an easy way is to deposit a metal gate somewhere on the wafer whose area is known [measured as described in Section 12.9.2(a)] and determine x_o from (12.13). Then (12.13) can be used to determine A_g for the mercury probe.

Normally, the area of contact between the mercury ball and the oxide surface is different every time contact is made. A very simple way to ensure that contact area is the same every time is to make a series resonant circuit with the MOS capacitor formed by the mercury gate and an inductance in series as shown in Fig. 12.17b. A small resistor R is placed in series to permit detection of displacement current on an oscilloscope, and the entire circuit is fed by a tuneable oscillator. The initial contact area is established by tuning the oscillator for maximum displacement current and noting the resonant frequency. To reestablish the same contact area, the contact area is adjusted with the micromanipulator until the same resonant frequency is obtained. Alternatively, the contact area could be adjusted for the same C_{ox}, but this would involve the application of an accurate value of bias.

Another solution to the variable contact area can be accomplished by a different mechanical design of a mercury probe more elaborate than that in Fig.12.17a.[20] In Ref. 20, the sample is positioned on a rubber O-ring into which mercury is pressed from below. A contact surface area precision of 5% is claimed. With Teflon replacing the rubber, surface area precision of 0.5% is claimed.

A commercial mercury probe available from the Materials Development Corporation (Santa Monica, California; model 754) or from MSI Electronics (Woodside, New York) uses vacuum to hold the mercury against the oxide surface within a predetermined area. This probe entails more difficulty in repeated contact to a given area than does the probe illustrated in Fig. 12.17a, but its use requires less skill.

12.8.4 Pedestal

The back contact of the wafer is electrically and thermally contacted by the gold plated brass plate or pedestal shown in Fig. 12.15. There are two basic requirements for a pedestal for three-terminal capacitance measurements. First, good electrical and thermal contact must be made between this pedestal and the back contact of the MOS capacitor. To ensure a good electrical and thermal contact, the pedestal should be gold plated to avoid oxide layers, and the wafer should be held tightly against the pedestal. Vacuum can be used to hold the wafer. This can be done by placing the sample over a small hole and grooves in the pedestal connected to a vacuum line as shown in Fig. 12.18. Second, the pedestal must be well insulated from ground to permit three-terminal capacitance measurements. To keep the pedestal above ground, it must be supported in the probe station on insulators, with Teflon tubing used in the vacuum line.

If three-terminal capacitance measurements are made at temperatures above room temperature (300–600°K), the design of the pedestal becomes more complicated. This temperature range is adequate for most measurements and for aging. At higher temperatures, electrochemical reactions can occur that are not of prime interest in integrated circuit technology (see Section 15.3).

One pedestal design, which illustrates the principles, is shown in Fig. 12.18. Dimensions of the pedestal can be made to accommodate the largest wafer diameter used. Heat is provided by heaters placed into holes drilled in the pedestal as shown in Fig. 12.18. The pedestal is thermally insulated from the probe box by ceramic spacers to confine heat to the pedestal. The

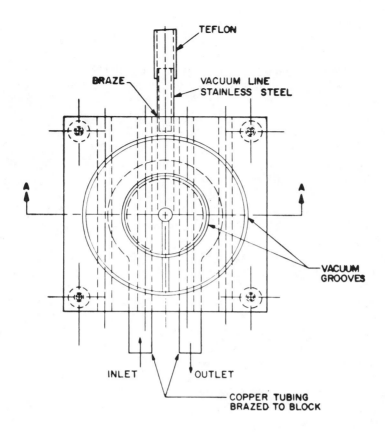

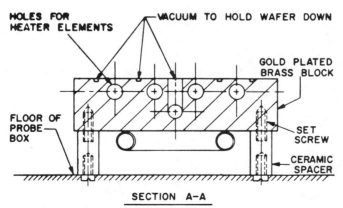

SECTION A-A

Fig. 12.18 Pedestal design suitable for use in the probe box shown in Fig. 12.15. Grooves on the surface of the pedestal distribute the vacuum underneath the wafer, making the pressure on the wafer against the pedestal more uniform. Copper tubing is brazed underneath the block through which a coolant such as tap water flows to quickly cool the block back to room temperature after heating. Dimensions are left to the discretion of the user.

heaters are connected to a temperature controller to regulate or vary the temperature of the pedestal. The pedestal must be electrically isolated from but in good thermal contact with the heat source. These requirements can be fulfilled by electrically insulating the heater wires from the pedestal. This insulation can be provided in the design of the heating elements.

In one heater design suitable for use in Fig. 12.18, a spiral groove is cut into a cylindrical ceramic rod so that the depth of the groove is slightly greater than the diameter of the heater wire. The outer diameter of the ceramic rod just fits into the holes drilled into the pedestal so that the heater wire wound in the spiral groove does not touch the pedestal. Another design suitable for use in Fig. 12.18 is a commercial heater such as a Hotwatt miniature stainless steel sheathed cartridge heating element. The cartridge can be snugly fitted into holes in the pedestal for good thermal contact while internal ceramic electrically insulates the heater wire from the stainless steel sheath. Another way to electrically isolate the pedestal is with a beryllia (beryllium oxide ceramic) spacer from another brass block containing the heaters. Beryllia has a low electrical conductivity typical of a ceramic and a high thermal conductivity comparable to aluminum. Thus the wafer in these designs is electrically isolated from ground and is in good thermal contact with the heat source.

Teflon tubing should be used in the vacuum line because it can withstand higher temperatures than most other plastics. Care should be taken not to heat the Teflon above 260°C, as it decomposes above this temperature. One of the decomposition products is phosgene, which is a highly toxic gas (used in World War I). The metal part of the vacuum line should be brazed to the pedestal and the probe wires mounted by mechanical clamping to the micromanipulator to withstand the elevated sample temperatures. Also, the leads connecting the probe to BNC connectors on the wall of the probe box should be soldered with a high temperature solder.

Bias-temperature aging also can be done with this pedestal design. However, if high temperature measurements are not contemplated, the sample design described previously can be used with bias-temperature aging done on a separate heating station.

A commercially available heated (and cooled) pedestal and temperature controller designed for use with the PAR model 410 $C-V$ plotter is the model TP36 Thermochuck System made by the Temptronic Corporation.

12.8.5 Low Temperatures

For measurements at temperatures below room temperature (77–300°K), additional complications arise. At low temperatures moisture condensation will occur on the surface of the MOS capacitor. Such condensation usually provides a current leakage path between gate and substrate or other electrodes in the structure interfering with the reliability and accuracy of the measurements. This condensation is avoided by keeping the probe

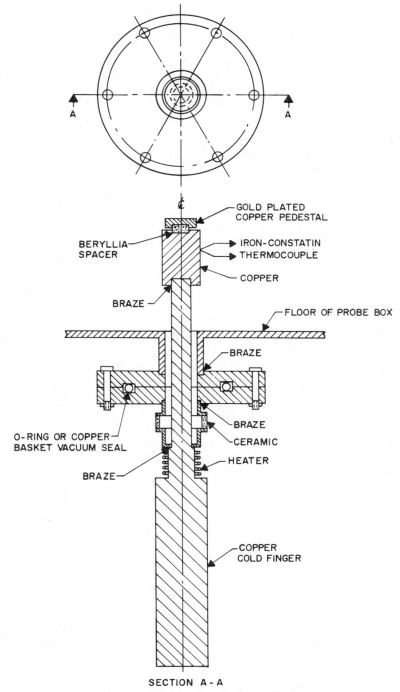

Fig. 12.19 Low temperature pedestal design suitable for a vacuum tight probe box. This pedestal does not feature a vacuum chuck and thus is most suitable for small wafers (<1 inch diameter).

station dry, which is done best by putting it under vacuum. The probe station in Fig. 12.15 is not adequate and must be redesigned to be vacuum-tight. The metal access cover must be made vacuum-tight by using a gasket or an O-ring seal. Motion of the probe wires must be brought into the station from each external micromanipulator by a vacuum-tight (and lightproof) metal bellows, and the box and cover walls must be strong enough to withstand atmospheric pressure. Ultrahigh vacuum in the box is not required. Vacuum provided by a mechanical vacuum pump or a molecular adsorption pump usually will suffice. Details of vacuum practice can be found in Ref. 21.

The pedestal also must be redesigned. The pedestal design, shown in Fig. 12.19, has the pedestal of Fig. 12.18 replaced by a long copper or a brass cylinder called a *cooling finger*, which is immersed in a Dewar flask containing either liquid nitrogen, a mixture of dry ice and acetone, or a mixture of ice and water, depending on the lowest temperature desired. The cooling finger and pedestal assembly are attached to the probe box through a vacuum-tight seal as shown in Fig. 12.19 and are thermally isolated from the probe station by a ceramic insulator. Temperatures above bath temperature in the Dewar can be obtained by wrapping a heater around the throat of the cooling finger. Energy is supplied to the heater to produce a steady-state temperature in the pedestal. Electrical isolation and good thermal contact between pedestal and cold finger is provided by a beryllia spacer. A vacuum chuck within the pedestal, as shown in Fig. 12.18, no longer will be effective for holding the wafer against the pedestal. The pressure of the probe wire must suffice.

An alternative to such a probe station is encapsulation of the device in a hermetically sealed package, either under vacuum or backfilled with a dry gas. The packaged device can be immersed in liquid nitrogen or other coolant solution to obtain low temperature. To obtain temperatures higher than the coolant bath, the encapsulated device must be embedded in a metal block with good thermal contact and a heated cold finger used.

If rapid cooling and heating over a limited temperature range ($-10°C$ to $100°C$) is required, a commercially available thermoelectric element can be used.

12.9 FABRICATION OF THE MOS CAPACITOR

The MOS capacitor consists of a silicon substrate, an ohmic back contact to the silicon substrate, a thin oxide layer, and a metal gate electrode. Such an array of MOS capacitors on a silicon wafer is shown in Fig. 12.16. The requirements of the silicon substrate have been discussed in Chapters 2 and 3 and of the oxide layer in Chapters 13 and 14. In this section we consider the ohmic back contact to the silicon substrate and the gate deposited on the oxide surface. In the case where the MOS capacitors are part of a test pattern on an integrated circuit chip, contact to the substrate and fabrication of the gate, usually by vacuum deposition and photolithography, are part of the overall processing of the integrated circuit, and no further discussion is necessary. However, if an array of MOS capacitors are to be separately made for the study of the effect of a particular processing step or other basic study, a few considerations (described in the following sections) may be helpful.

12.9.1 Back Contact

In most cases of interest the first step in the fabrication of an MOS capacitor array is the highest temperature step, the growth of an oxide layer. Then the polished side of the wafer is waxed onto a platinum holder and the oxide layer is etched off the back side in buffered HF. An *ohmic* back contact (defined in Chapter 3) must then be made to this surface. The $Si–SiO_2$ interfacial properties can be changed by heating the wafer at this stage, making the interface characteristic of the back contact process rather than of the process under study. An ohmic back contact can be made without heating by contacting an abraded region on the back surface of the wafer. Emery cloth is used to create a damaged region in the surface of the silicon. This damaged region is electrically equivalent to a diffused degenerate layer from the standpoint of an ohmic contact. There are two ways of contacting this damaged region to make an ohmic contact:

1 Initially, about 200 Å of chromium or titanium can be deposited by evaporation in a vacuum system held at 10^{-6}–10^{-7} Torr, followed, without breaking the vacuum, by the deposition by evaporation of at least 3000 Å of gold. To accomplish these depositions, two tungsten filaments or boats, one loaded with titanium or chromium and the other loaded with gold, placed side by side either above or below the wafer are required. Each metal is deposited by first heating one filament and then the other. A shutter kept closed prior to each deposition permits impurities, which evaporate first to coat the shutter rather than the wafer. During evaporation the wafer can become heated to several hundred degrees Celsius by the hot tungsten filaments used to vaporize the metals. Temperature rise can be held to a few degrees Celsius by placing the tungsten filaments 12–15 inches from the wafer and placing the wafer on a massive copper block that acts as a heat sink at room temperature during evaporation.

2 When sample size is small, the damaged area can be contacted by applying a paste to it made by mixing indium and gallium together. Because indium–gallium paste will attack the gold plating on the pedestal, a thin copper sheet should be inserted between the wafer and the pedestal.

An alternative way of making an ohmic back contact is to sinter after the deposition of the chromium and gold or the titanium and gold. Creation of a damaged region prior to deposition is not required. To avoid changing the interfacial properties, the sintering step should be done at a low temperature (300°C) for a short time (5 min). The choice of back contact process depends on the experiment contemplated. The back contact made by damaging the back surface of the wafer is the most useful if the properties of the as-grown oxide are to be measured. The sintered back contact usually is automatically provided on a test MOS capacitor on an integrated circuit chip. This back contact is useful if oxide properties are to be measured after all processing or in experiments such as avalanche injection, where the initial as-grown conditions are not important.

12.9.2 Gate Electrode

(a) *Thick Gate Electrode*

To make the gate electrode, aluminum can be deposited about 3000 Å thick in a high vacuum system by evaporation. It also can be deposited by electron gun evaporation but this method affects interfacial properties because ionizing radiation, described in Section 11.6, is produced in this process.

It is difficult to contact an aluminum gate with a probe without significant series resistance because of the protective oxide layer that forms on it on exposure to room air. It is much easier to make a low resistance contact to a metal such as gold, which does not form a protective oxide. Unfortunately, because gold is not an active metal, it does not bond strongly to the oxide surface.

To obtain reproducible results, it is desirable that the gate electrode adhere strongly to the oxide surface. This problem is solved by using a bimetallic gate. An active metal such as chromium or titanium is first deposited on the oxide surface. These metals chemically react with SiO_2 and strongly adhere. Next, gold is immediately deposited over the active metal. Because the active metal is in a vacuum, no protective oxide layer forms. Thus gold forms an adherent ohmic metal-to-metal contact but does not form a protective oxide layer when exposed to room air. Then Cr–Au or Ti–Au can be deposited by evaporation to the same thickness and in the same manner as described in Section 12.9.1 for the deposition of this combination of metals for the back contact.

Keeping the substrate cool during metal deposition is particularly important in measurements on thin oxides. Any heating will cause the gate metal to diffuse into the oxide, and if the oxide is thin, metal atoms can penetrate all the way to the Si–SiO_2 interface, altering the electrical properties there. The substrate can be kept cool by using a heat sink and keeping the evaporation filaments far away as described in Section 12.9.1.

Photolithography can be used to make the dot pattern, but this process requires some heating. Instead, a shadow mask, which is a thin sheet of metal* can be used to define the gate area during evaporation. The thin metal shadow mask is held tightly against the oxide surface of the wafer by means of a jig and exposed to the stream of evaporating metal. The shadow mask must be as thin as possible to provide sharp definition of the pattern. Such a mask is most easily made by clamping the thin sheet between two brass blocks and drilling an array of holes of the desired diameter. For evaporating Cr–Au or Ti–Au gate electrodes, the two filaments holding the chromium or the titanium and the gold must be as close together as possible and the shadow mask as thin as possible to ensure that gold entirely covers the initially deposited chromium or titanium dot. The deposited gate electrode area can be accurately measured by using a calibrated reticle in a metallurgical microscope.

The major drawback to making two-layer metal contacts with a shadow mask is that the two metals are unlikely to register perfectly. Thus the oxide will be contacted mostly by chromium or titanium, but there may be a small area of gold contacting the oxide where the registry is off. This presents no problems in the adhesion of the contact, but in certain situations there can be an electrical effect. This electrical effect comes about because the work function difference is not the same between (1) chromium or titanium and silicon and (2) gold and silicon. Therefore, band bending under the tiny gold area is slightly different than under the chromium or the titanium. Conductance measurements are the most sensitive to this small effect. Off-registry can be minimized when using a shadow mask by rotating the wafer so that it receives a more uniform exposure from each filamentary source.

Generally, the larger the dot diameter, the larger the admittance, if everything else is the same. Because measurement is easier with larger admittance, larger dot diameters should be used. However, making dot diameter much larger than 60 mils lowers yield because the dot is more likely to encompass a bad spot in the oxide that breaks down catastrophically at low voltages.

(b) Semitransparent Gate Electrode

For admitting light into the MOS capacitor structure for internal photoemission measurements (see Sections 10.3 and 11.4) or for seeing emitted light in avalanche breakdown measurements (see Sections 11.2 and 11.3), the gate electrode must be made semitransparent. A suitable semitransparent gate electrode can be made by depositing a gold layer about 50 Å thick by vacuum evaporation through a shadow mask. Because gold sticks to SiO_2 by very weak Van der Waal forces, such a transparent gate electrode is

---

*This mask is usually molybdenum a few mils thick with an array of holes generally 5, 10, 15, 20, or up to 60 mils in diameter spaced 100 mils apart in several rows.

difficult to contact electrically with a probe wire because the probe wire scratches the gold off. This problem is solved by first depositing a 5–10 mil diameter thick Cr–Au or Ti–Au contact as previously described. Then, the 50 Å thick gate electrode, which has a 20 mil or larger diameter, is deposited over the Cr–Au or Ti–Au contact and aligned so that the Cr–Au or Ti–Au contact is either in the center or to one side of the semitransparent gate. With such large diameters, the shadow mask for the thin gold gate can be aligned visually by using a microscope over the previously deposited Cr–Au dot. One electrode configuration of an MOS capacitor structure suitable for internal photoemission measurements is shown in Fig. 12.22. Details of metal evaporating techniques can be found in Ref. 22.

Aluminum also can be used to make a semitransparent gate electrode. Deposition of 75–100 Å of aluminum will suffice. Because aluminum adheres well to SiO$_2$, a thick bonding pad is not required as with gold. The semitransparent aluminum gate can be contacted with a 5 mil diameter tungsten probe wire.

(c) Processing

Figure 12.20 shows the steps necessary to fabricate an MOS capacitor array on a p-type substrate that has a negligible series resistance. Fabri-

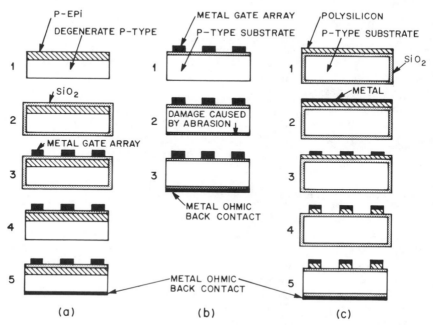

Fig. 12.20 (a) Steps in fabricating an MOS capacitor array on a p-type epitaxial layer. (b) Steps in fabricating an MOS capacitor array on a p-type silicon substrate. (c) Steps in fabricating a silicon gate MOS capacitor array.

cation of MOS capacitor arrays on an n-type substrate is similar. Figures 12.20a,b emphasize fabrication of an MOS capacitor array with no exposure to elevated temperatures after thermal oxidation. Figure 12.20c shows fabrication of a silicon gate MOS capacitor array.

Figure 12.20a shows the steps involved in fabricating an MOS capacitor array on an epitaxial layer. To minimize bulk series resistance, the epitaxial layer should be 5–10 μm thick, doped to a density of 10^{15}–10^{16} cm^{-3} and be deposited on a degenerate silicon substrate of the same conductivity type. Step 1 shows the epitaxial layer grown on the degenerate substrate. Step 2 shows the wafer after thermal oxidation. Step 3 shows the wafer after gate metal has been deposited through a shadow mask. The gate metal array is deposited immediately after oxidation so as to minimize the possibility of contaminating the oxide underneath the gate areas. Step 4 shows the oxide etched away except over the epitaxial layer.* Step 5 shows the completed wafer with a deposited ohmic back contact.

Figure 12.20b shows fabrication of an MOS capacitor array on a silicon wafer. Step 1 in Fig. 12.20b corresponds to step 4 in Fig. 12.20a. The substrate should be doped to about 10^{17} cm^{-3}, and silicon thickness should be no more than 5–6 mils to keep bulk series resistance to a minimum. Because the silicon substrate is not degenerate, its back surface must be abraided or scratched to create a damaged region to which an ohmic contact can be made by depositing a metal. Step 3 shows the completed MOS capacitor array with the back contact metal deposited over the damaged region.

The fabrication steps in Figs. 12.20a,b avoid any thermal treatment. However, there are three advantages of thermal treatment after the final fabrication step: (1) when an active metal is used for the gate, a low temperature thermal anneal will reduce interface trap density as described in Section 15.4.3(b); (2) the back contact need not be abraided as in Fig. 12.20b to produce an ohmic contact because the back metal can alloy with the silicon to produce an ohmic contact; and (3) photolithographic techniques can be used to define the gate areas instead of a shadow mask. Aluminum is one example of a gate and back contact metal with which all these benefits can be obtained.

There are two disadvantages of such heat treatment, however: (1) the interfacial properties of the oxidation process are not measured—rather, the interfacial properties produced by a low temperature anneal are measured; and (2) gate metal can diffuse into the oxide, which may be

---

*This can be accomplished by waxing the wafer, gate face down, onto a platinum holder and exposing to buffered HF. Apiezon wax, soluble in trichloroethelene, or biwax, soluble in acetone, can be applied and removed by melting or by first dissolving in the solvent and then painting onto the gate face. Commercially available tapes can be used instead of wax. Photoresist over the gate face also can be used. The simplest and cleanest approach is to hold the wafer, gate face down, and apply buffered HF to the back with a swab.

undesirable in thin oxides (≤ 100 Å) because small quantities of gate metal atoms diffusing all the way to the Si–SiO$_2$ interface alter the interfacial properties. Very small numbers of metal atoms at the interface can produce interface traps and oxide fixed charge, as discussed in Section 12.9.2(a).

Figure 12.20c shows how to make an MOS capacitor array with silicon gates. Step 1 shows polysilicon gate material deposited all over the oxidized silicon wafer. To make polysilicon a good electrical conductor, it is doped to degeneracy by diffusing either phosphorus or boron into it, depending on the desired work function difference (see Section 10.4). Step 2 shows metal deposited over the polysilicon to facilitate contact. Step 3 shows the gate areas delineated in the metal film by photolithographic techniques. Step 4 shows the gate areas delineated in the polysilicon using the metal layer as an etch mask. Step 5 shows the completed MOS capacitor array with the oxide etched away as described in the fabrication of the array in Fig. 12.20a. Thermal treatments cannot be avoided in fabricating a silicon gate array; hence this process is useful for studying the electrical properties of the interface characteristic of the silicon gate process or studying processes that do not depend on the fabrication process used.

12.10   INTERNAL PHOTOEMISSION[23]

In this section we describe the apparatus required to make the internal photoemission measurements discussed in Sections 10.3, 11.4.2, and 11.4.3. The apparatus consists of a monochromatic source of light whose wavelength can be varied, an optical system for focusing light onto the sample, and a detection system for measuring photocurrent. Because the energy barriers between silicon and SiO$_2$ and between SiO$_2$ and a variety of metals is of the order of several electron volts, short wavelength (UV) light in the approximate range 1000–3300 Å is required. As light at these wavelengths is readily absorbed by air, a vacuum monochromator such as the McPherson model 235 is suitable. A source of light must be used to produce the wavelengths required and be of intensity sufficient to provide a detectable photocurrent. A light source meeting these two requirements is an Oriel Optics 1000 W xenon arc lamp.

Because the monochromatic beam divergence is large ($f/11$) in the horizontal plane, a cylindrical lens is positioned near the exit slit of the monochromator to collimate the beam before it is focused onto the sample by a spherical lens. Such an optical system makes the image of the monochromator exit slit as small as practical at the sample plane. Figure 12.21a shows a cross-sectional view of an evacuable light focusing assembly and sample holder.

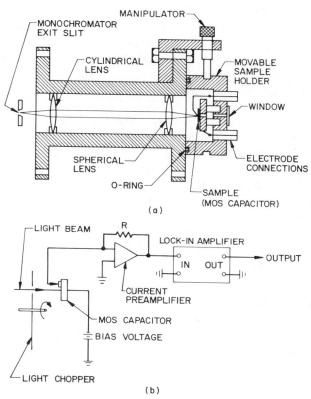

Fig. 12.21 (a) Cross-sectional view of an evacuable focusing assembly and sample holder. After Powell.[23] (b) Circuit for synchronous detection of photocurrent. The considerations are similar to those for measuring admittance.

To focus light onto the sample plane, the sample is replaced with a ground glass of the same thickness and the lens position adjusted to obtain the smallest image viewed through the window. Focusing is required only once because subsequent sample changes require sample positioning only in a plane perpendicular to the light beam. Optimal sample position is obtained by positioning the sample holder for maximum photocurrent. Three manipulators spaced 120° apart slide the sample holder along the face of the lens holder. The O-ring seal between the sample holder and the lens holder prevents metal-to-metal contact, facilitating movement as well as providing a vacuum-tight seal. With the optimal sample position and a 1000 W xenon light source, monochromatic resolution of 0.05 eV is possible with photocurrents as large as 2.5×10^{-8} Å in silicon to SiO_2 emission.

Figure 12.21b illustrates detection of photocurrent. The light beam is chopped by a rotating disk containing a series of holes near the rim. The chopped light produces an ac photocurrent that is detected by a lock-in amplifier. The lock-in amplifier is synchronized with the light chopper by tuning the frequency of the oscillator in the lock-in amplifier to produce maximum photocurrent. To find the field dependence of the photoinjection barrier, gate bias is applied to the sample with a battery or dc power supply to vary the electric field across the oxide. With the polarity shown in Fig. 12.21b, emission will be from the silicon into the SiO$_2$. For emission from the metal gate into the SiO$_2$, the polarity of the battery must be reversed.

A synchronous detection system shown in Fig. 12.21b is used to reduce background noise because relatively small photocurrents are measured. However, this system is suitable only when the silicon sample is degenerate. When nondegenerate silicon is used, dc photocurrent must be measured. Otherwise, the large ac surface photovoltage generated by the use of chopped light produces a displacement current in the oxide, generally much greater than the photocurrent. For dc photocurrent measurements, no chopper is used and the lock-in amplifier is replaced as detector by an electrometer such as the Keithley model 602.

Figure 12.22 shows a typical electrode configuration of an MOS capacitor suitable for photoemission work. A semitransparent gate electrode allows light to enter the structure while an electric field is applied. Suitable semitransparent gate electrodes can be made as described in Section 12.9.2(b). The dimensions shown in Fig. 12.22 are small enough to obtain high dielectric breakdown fields in most samples yet large enough to contain the image of the monochromatic exit slit.

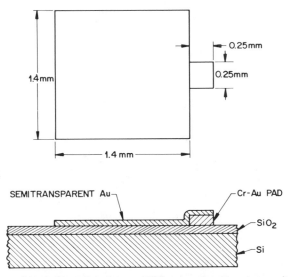

Fig. 12.22 Electrode configuration for an MOS capacitor structure suitable for internal photoemission measurements. The oxide is usually thermally grown. After Powell.[23]

To determine the light power absorbed, it is necessary to measure the Au–SiO$_2$–Si reflectance spectrum and the absolute incident light intensity. Reflectance measurements can be made using an RCA No. 935 photodiode for light detector. The window surface of the photodiode should be coated with a diffuse layer of sodium salicylate to reduce the positional sensitivity of the response and to act as a fluorescent wavelength converter for the shorter wavelengths. The window area of the tube should be marked to ensure that incident and reflected beams illuminate the same region on the detector. Reflectance measurements should be repeated twice to determine reproducibility. Filters should be used to eliminate second-order responses and scattered light.

To interpret reflectance measurements, it is essential to know both oxide and semitransparent gold film thicknesses. Therefore, after reflectance measurements on a sample are concluded, a step is etched in the oxide using photoresist masking. A film of silver is evaporated over the step and on both sides to make the entire region highly reflecting. Then thickness of the oxide and semitransparent gold electrode can be measured with an interferometer microscope as described in Section 14.3.3.

Relative intensity of the light incident on the sample can be determined by measuring the output of a vacuum ultraviolet thermopile with chopped light and synchronous detection. Absolute intensity can then be measured at several photon energies with a calibrated thermopile and these data used to scale the relative intensity data to provide absolute intensity over the range of interest. Details of reflectance measurements and how they are used to interpret photocurrent measurements are discussed in Section 10.3.1.

12.11 HOT CARRIER INJECTION INTO SiO$_2$

Hot carrier injection into SiO$_2$ can be studied using the MOS capacitor. Hot carrier injection into SiO$_2$ is discussed in Sections 11.2 and 11.3. In Section 12.11.1 we discuss instrumentation for studying hot carrier injection using the MOS capacitor.

12.11.1 Avalanche Injection in the MOS Capacitor Structure[24,25]

The two quantities that must be measured during hot carrier injection are the dc injected current density I_G and the electric field across the oxide layer F_o. It is easiest to measure the electric field across the oxide layer if the periodic voltage applied to the gate is sinusoidal. The frequency must be sufficiently high to prevent minority carrier response. Then the silicon surface can be driven into deep depletion (see Section 4.5.1) or to avalanche breakdown.

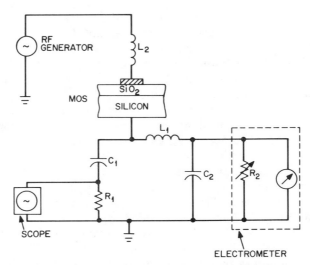

Fig. 12.23 Circuit for measuring injected dc electron current and peak oxide field. The electrometer can be a Keithley 602 operated in the normal mode. The values of the components are: $L_1 = 10$ mH, $L_2 = 1$ mH, $C_1 = 6000$ pF, $C_2 = 6000$ pF, $R_1 = 10 \, \Omega$, and R_2 is varied from 10^4 to $10^8 \, \Omega$ by the electrometer scale setting. Components L_2, C_1, C_2, and R_1 are each put in a General Radio-type connector. For an oxide layer 500 Å thick or more and a typical gate area of 5×10^{-4} cm^2, the oxide layer capacitance is 34 pF or less, making C_{ox} much less than C_1. After Nicollian and Berglund.[24]

(a) *Measurement of Injected Current and Oxide Field*

Figure 12.23 shows a circuit used for measuring I_G and F_o. Along with I_G, there is a displacement current density I_D more than a thousand times larger than I_G. These two currents are separated by shunting I_D through the series circuit C_1, R_1 shown in Fig. 12.23. The capacitor C_1 is much larger than C_{ox} and the resistor R_1 is small, so that most of the applied ac field appears across the MOS capacitor. The injected or dc component of the current flows through the input resistor R_2 of a Keithley model 602 electrometer. The time constant R_2C_2 is always very long (60 μsec–0.6 sec) compared to the period of the sinusoidal gate voltage (0.2–1 μsec), so that the electrometer measures the injected average direct current. Coil L_1 and capacitors C_1 and C_2 serve to keep ac entering the electrometer at a low level to prevent overdriving its input stage. To avoid significant biasing of the MOS capacitor by direct current flowing through R_2, the electrometer is operated in the 30 mV full-scale range or below.

To obtain F_o, peak displacement current through R_1 is measured with a Tektronix 5A13N differential comparator unit in a 5440 oscilloscope or equivalent. Other types of ac voltmeter of appropriate sensitivity also could be used. If displacement current is sinusoidal, F_o is calculated from

$$F_o = \epsilon_{ox}^{-1} \int_0^{\pi/2\omega} I_p \cos \omega t \, dt = \frac{I_p}{\epsilon_{ox}\omega} \qquad (12.14)$$

where I_p is the measured peak displacement current density $I_p = (v_p/A_g R_1)$, where v_p is the peak ac voltage measured by the comparator.

Displacement current can be made sinusoidal by driving the MOS capacitor, which is nonlinear under large ac voltage excitation, by a *constant current* source. In Fig. 12.23 a good approximation to a constant current source is achieved by using a series resonant circuit made up of the MOS capacitor and L_2. In this circuit the resonant or fundamental frequency sees a minimum impedance whereas higher harmonics are sharply attenuated as they see a much higher impedance. Resonance is obtained by tuning the oscillator for maximum displacement current through R_1. An additional benefit of using a series resonant circuit is that the voltage across the MOS capacitor is the Q of coil L_2 multiplied by the ac generator voltage. For example, a Q of 20 will result in 200 V rms across the MOS capacitor when using an ac generator producing only 10 V rms. This feature is useful because many ac generators do not produce the high output voltage needed for hot electron injection work.

(b) Measurement of Flatband Voltage Shift as a Function of Time

To study trapping and charging effects in SiO₂ resulting from hot carrier injection, it is necessary to measure flatband voltage shift as a function of time. Interpretation of such a measurement is simple only when interface trap level density is unchanged by hot carrier injection. Figure 12.24 shows a feedback circuit added to the circuit in Fig. 12.23 to measure flatband voltage shift as a function of time while an injected dc current flows through the MOS capacitor. The operational amplifier in the circuit in Fig. 12.24, designed to operate with *p*-type substrates for electron injection, senses the injected dc current from the output of the electrometer. The injected dc current is kept at a constant value determined by the reference voltage $-V_R$ by producing a bias of the right magnitude that is applied to the gate of the MOS capacitor in parallel with the ac drive. To keep injected dc current constant, the bias applied by the feedback circuit to the MOS capacitor must be equal to the flatband voltage shift because injected dc current is directly proportional to the sum of the peak ac field across the oxide and the field associated with the flatband voltage shift, as seen from Fig. 11.4. Equality of the voltage produced by the feedback circuit to flatband voltage shift, based on the linearity of the curve in Fig. 11.4, depends on there being no change in interface trap density produced by hot carrier injection. Conditions necessary for this requirement are described in detail in Section 11.5.

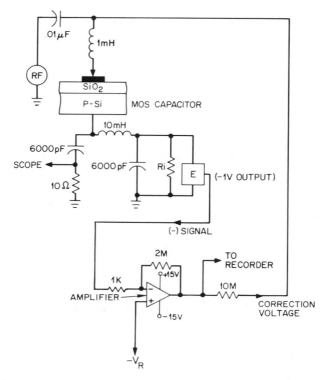

Fig. 12.24 Circuit for measuring the parallel shift of the C-V curve along the voltage axis or flatband voltage shift while a constant injected dc electron current flows through the oxide layer. The electrometer can be a Keithley 602 operated in the normal mode. Input resistance R_i of the electrometer is $10^7 \Omega$. Overall time constant is 0.1 sec, overall gain of the amplifier stage is 2000, and the output voltage range is ± 14.7 V. The amplifier is AD 540 or equivalent. Voltage $-V_R$ is an adjustable reference voltage from a low impedance source, used to set the value of I_G. After Nicollian et al.[25]

If interface trap level density increases during avalanche injection, the voltage produced by the feedback circuit will not be equal to the flatband voltage shift. Rather, it will have a component produced by the change in interface trap level occupancy in addition to the change in oxide trapped charge required to keep the injected current constant.

The feedback circuit voltage with reference voltage $-V_R$ initially will be negative and have a magnitude dependent on peak ac voltage. Thus for a given peak ac voltage, the zero on the voltage scale of the x-t recorder can be adjusted to make the flatband voltage shift equal to zero when $t = 0$. An example of the correction voltage produced by the feedback circuit in Fig. 12.24 and plotted on an x-t recorder is shown in Fig. 11.18.

12.11.2   Diffusion of Water into SiO₂

Figure 12.25 is a schematic diagram of the apparatus used to diffuse controlled amounts of water into the oxide for electron trapping studies (see Section 11.5.1). The amount of water incorporated into the oxide is controlled by exposing the sample at a given temperature (<200°C) for a given time (≥ 10 min) to a given partial pressure of water vapor (25–55 Torr). A temperature controller controls furnace temperature to within ±5°C.

Partial pressure of water vapor over the sample is established by flowing nitrogen saturated with water vapor through the diffusion furnace. Nitrogen saturated with water vapor is produced by bubbling nitrogen gas through a water bath heated by an immersion heater. Total pressure of nitrogen plus water vapor is kept at just slightly above atmospheric pressure to prevent back flow of room air into the system and to produce a flow rate sufficiently slow to keep the partial pressure of water vapor in the nitrogen stream at its equilibrium value determined by the water bath temperature.

Water bath temperature is maintained within ±0.1°C by a proportional temperature controller. If water bath temperature is known within ±0.1°C,

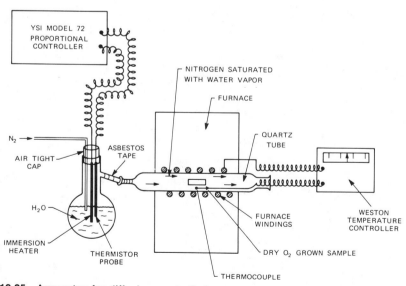

Fig. 12.25   Apparatus for diffusing controlled amounts of water into the oxide. Diffusion furnace temperatures are held within ±5°C by the Weston temperature controller. The asbestos tape on the tube leading from the water bath to the furnace tube is maintained at 100°C to prevent condensation of water before it reaches the furnace tube. Nitrogen and water vapor escape out the end of the furnace tube so that the system operates against atmospheric pressure. After Nicollian et al.[25]

water vapor pressure is obtained within ±0.1 Torr over the range of partial pressure measured (25–55 Torr) from a table of temperature versus water vapor pressure.[26]

12.12 LIFETIME

The MOS capacitor can be used to measure lifetime in the silicon as described in Section 9.5. Measurement should be made in the dark. The circuit shown in Fig. 12.26 can be used[27] for lifetime in the microsecond range or longer. A block diagram of this circuit is shown in Fig. 12.26a. This circuit functions by keeping the capacitance of the MOS capacitor at a predetermined value by applying a time-varying gate bias to balance the charge per second added to the inversion layer forming by generation processes in the silicon.

A schematic diagram of the bias supply circuit is given in Fig. 12.26b. This circuit monitors the output of the capacitance meter, compares the measured capacitance against the desired capacitance setting, and automa-

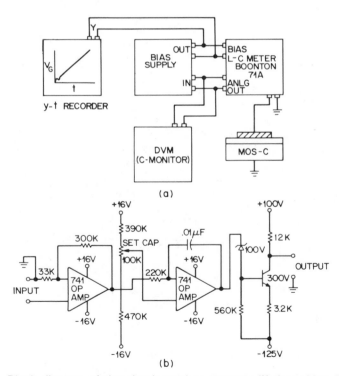

Fig. 12.26 Block diagram of the circuit used to measure lifetime. After Pierret and Small.[27] (b) Basic schematic of the bias supply circuit in (a). After Pierret and Small.[27] Copyright (1975), IEEE.

tically adjusts gate bias to maintain capacitance at a preset value. Gate bias is plotted as a function of time with an x-t recorder with the preset capacitance (potentiometer shown in Fig. 12.26b) as a variable parameter. Under the condition of constant capacitance, the generation rate g in the silicon surface depletion layer is given by

$$g = \frac{C_{ox}}{q} \frac{dV_G}{dt}. \tag{12.15}$$

Because g is constant for a given depletion layer width (constant capacitance), the V_G versus t dependence observed will be linear* except for a brief initial transient period.

Lifetime τ_o is related to g by the relation

$$\tau_o = \frac{n_i(w - w_o)A_g}{2g} \tag{12.16}$$

where n_i is the intrinsic carrier density, w is the depletion layer width in deep depletion, and w_o is the equilibrium depletion layer width in inversion. To obtain τ_o, a family of V_G versus t curves is measured with capacitance as parameter. Then g is calculated from the slope using (12.15) for each curve to get the corresponding value of w. The deep depletion capacitance C_D for each curve is calculated from $1/C_D = 1/C - 1/C_{ox}$ where C is the predetermined high frequency deep depletion capacitance. Then, $w = \epsilon_s A_g / C_D$. The single value for w_o is calculated from the measured equilibrium high frequency inversion capacitance in the same way.

REFERENCES

1 M. C. McGregor, J. F. Hersh, R. D. Catkosky, F. K. Harris, and F. R. Kotter, *IRE Transact. Instrum.*, 254–261 (December 1958); J. F. Hersh, *Gen. Radio Exp.*, **33**, 7 (1954).

2 E. H. Nicollian and A. Goetzberger, *Bell Syst. Tech. J.*, **46**, 1055 (1967).

3 F. M. Gardner, *Phase Lock Techniques*, Wiley, New York, 1966.

4 Princeton Applied Research, P. O. Box 2565, Princeton, N. J., Technical Note 114.

5 R. Castagne and A. Vapaille, *Surface Sci.*, **28**, 157 (1971).

6 M. Kuhn, *Solid-State Electron.*, **13**, 873 (1970).

7 D. R. Kerr, Conference on Properties on Use of MIS Structures, Grenoble, France, June 17–21, 1966.

8 G. F. Derbenwick, private communication.

9 M. Kuhn and E. H. Nicollian, *J. Electrochem. Soc.*, **118**, 370 (1971).

10 K. Ziegler and E. Klausmann, *Appl. Phys. Lett.*, **26**, 400 (1975).

11 C. N. Berglund, *IEEE Transact. Electron Devices*, **ED-13**, 701 (1966).

12 Princeton Applied Research, P. O. Box 2565, Princeton, N. J., Application Note 128A.

13 P. D. Tonner and J. G. Simmons, *Rev. Sci. Instrum.*, **51**, (10), 1378 (1980).

---

*Nonideal device behavior is signaled by the observation of a nonlinear V_G versus τ dependence.

14 K. H. Zaininger, *RCA Rev.*, **27**, 341 (1966).

15 A. Goetzberger and E. H. Nicollian, *J. Appl. Phys.*, **38**, 4582 (1967).

16 M. Morita, K. Tsubouchi, and N. Mikoshiba, *Appl. Phys. Lett.*, **33**, 745 (1978).

17 U. G. Graeme, G. E. Tobey, and L. P. Huelsman, Eds., *Operational Amplifiers*, McGraw-Hill, New York, 1971, Chapter 6.

18 R. L. Johnston, private communication.

19 W. L. Bond, "Micromanipulators," *Bell Lab. Rec.*, **34**, 90 (1956).

20 R. Hammer, *Rev. Sci. Instrum.*, **41**, 292 (1970).

21 S. Dushman, *Scientific Foundations of Vacuum Technique*, Wiley, New York, 1949.

22 W. L. Bond, *J. Am. Opt. Soc.*, **44**, 429 (1954).

23 R. J. Powell, *J. Appl. Phys.*, **40**, 5093 (1969).

24 E. H. Nicollian and C. N. Berglund, *J. Appl. Phys.*, **41**, 3052 (1970).

25 E. H. Nicollian, C. N. Berglund, P. F. Schmidt, and J. M. Andrews, *J. Appl. Phys.*, **42**, 5654 (1971).

26 *Handbook of Chemistry and Physics*, 37th ed., Chemical Rubber Publishing Company, Cleveland Ohio, 1956, pp. 2141–2144.

27 R. F. Pierret and D. W. Small, *IEEE Transact. Electron Devices*, **ED-22**, 1051 (1975).

13

Oxidation of Silicon—Oxidation Kinetics

13.1 INTRODUCTION

In this chapter we discuss the oxide itself and its growth mechanisms, structure, and oxidation kinetics. Oxide layers in silicon integrated circuit technology provide surface passivation for a silicon device (see Section 15.3.1), serve as a diffusion mask, serve to isolate one device from another, serve to insulate the gate electrode from the silicon in field effect devices, and serve to isolate multiple levels of device interconnection in an integrated circuit. Moreover, oxide layers that perform these functions must be produced at low cost. To fulfill the requirements of these various roles, several methods of oxide preparation have been devised. The most important ones are vapor phase reaction,[1] plasma anodization,[2,3] wet anodization,[4] and thermal oxidation in both dry and wet environments.[5]

For *passivation* of bipolar transistors and junction diodes, the oxide layer must ensure that the electrical characteristics of the device are dominated by bulk rather than surface properties. That is, the interface between silicon and its oxide must have a minimum, stable density of oxide fixed charge and interface traps, and the oxide must be sufficiently thick that potential gradients along the air-oxide interface have a minimum influence on the semiconductor.

As a *diffusion mask*, purity and continuity of the oxide layer are important. Both thermally grown oxides and oxides grown by vapor phase reaction achieve these goals.

Isolation of active devices in an integrated circuit requires a thick oxide layer with a low electrical conductivity that can be quickly grown. Anodic, thermal, and vapor phase reaction methods of oxide preparation all will suffice for this application.

For *gate oxides* in field effect devices such as the MOSFET and the CCD, the oxide-semiconductor interface again must have a minimum, stable density of oxide fixed charge and interface traps, but now stability under fields normal to the air-oxide interface is especially important. The oxide that meets these requirements best is thermally grown oxide. For these reasons, thermal growth is the key oxidation process in integrated circuit technology.

Finally, the oxide preparation methods used widely for electrical *isolation* of multiple levels of device interconnections are vapor phase reaction or RF sputtering. Of these, vapor phase reaction is the best because it can be applied over metal layers. For polysilicon gates and interconnections, thermal oxidation of the polysilicon can provide insulation for interconnection isolation.

There are two aspects to understanding the oxidation of silicon: (1) understanding the oxidation techniques well enough to grow oxides suitable for integrated circuit application and (2) understanding the interaction between device characteristics and the electrical and chemical properties of the oxide layer. Oxide fixed charge, interface traps, and ionic species in the oxide layer can each affect device characteristics. All are related intimately to the oxide growth process, although details of this relationship are not always clear.

Because thermal oxidation is the key process, with other oxidation methods serving mainly to increase the options available to the device fabricator, thermal oxidation of silicon is emphasized in this chapter.

Section 13.2 discusses the growth mechanism of both dry and wet thermal oxides. Insight into these mechanisms leads to the control of oxide growth, which is a practical concern. It is shown that an oxidizing species and not silicon moves across the oxide layer and that the oxidation reaction occurs at the silicon-oxide interface. However, just which oxidizing species moves through the oxide layer is unsettled at the present time. Section 13.3 discusses the structure of thermally grown oxide. The structure of the oxide, which determines its electrical and mechanical properties, is a noncrystalline form of SiO_2 (silica) characterized by short range order (microscopic dimensions) but not long range order (macroscopic dimensions). Section 13.4 discusses oxidation kinetics. The thickness of oxide grown at a given temperature and pressure for a given length of time can be predicted from oxidation kinetics. A phenomenological description of oxidation kinetics is given in this section because the chemistry that occurs during oxidation is not understood in sufficient detail to give a more fundamental description.

13.2 GROWTH MECHANISMS

13.2.1 The Oxidation Process

The basic oxidation process is the sharing of valence electrons between silicon and oxygen to form four silicon-oxygen bonds. Each bond is largely covalent with a small ionic component at room temperature. The ionic component becomes more important at elevated temperatures. A characteristic of the covalent bond is directionality. Its manifestation in the silica structure is discussed in Section 13.3.

During oxidation the top surface of the SiO_2 film will not be coplanar with the original silicon surface because a volume expansion occurs during oxidation. This expansion occurs because the density of SiO_2 (2.21 g/cm^3) is slightly less than the density of silicon (2.33 g/cm^3). Growth of an oxide of thickness x_o will consume a layer of silicon about 0.45 x_o thick, as can be calculated from the density and the molecular weight of silicon and SiO_2.

For oxidation in pure oxygen, which produces "dry" oxides, the stoichiometric chemical reaction producing the oxide film is

$$Si(solid) + O_2 \leftrightharpoons SiO_2(solid). \tag{1}$$

The stoichiometric chemical reaction for producing the oxide film in water vapor, "wet" oxides, is

$$Si(solid) + 2H_2O \leftrightharpoons SiO_2(solid) + 2H_2. \tag{2}$$

Reaction (1) describes the overall reaction between oxygen and silicon, and reaction (2) describes that between water vapor and silicon; however, there also may be elementary reactions in which intermediate species are produced during the oxidation process, particularly for the case of oxidation in water vapor.*

Reactions (1) and (2) occur when a silicon surface is exposed to an oxidizing ambient, usually at a pressure of 1 atm, and elevated temperatures, typically 900–1200°C. The rate of oxide growth is limited by the availability of the reactants, namely, oxidant molecules and silicon–silicon bonds. Initially, oxide growth rate is limited by the availability of silicon–silicon bonds, as there are more than enough oxidant molecules available at 1 atm pressure. Eventually, the supply of oxidant molecules becomes the

---

*Such elementary chemical reactions are important in understanding the oxidation process and are discussed later in this chapter.

limiting factor as an oxide layer grows, separating the oxidant molecules in the gaseous ambient from the silicon–silicon bonds. Now the oxidant must diffuse through the oxide layer to react at the Si–SiO$_2$ interface. For oxidation in dry oxygen, the oxide thickness at which diffusion of oxidant through the oxide becomes limiting is about 40 Å, whereas for oxidation in steam, the limiting thickness is about 1000 Å.

Silica films grown in pure oxygen and in steam have different chemical, mechanical, and electrical properties. Silica grown in steam is actually SiO$_2$ doped with water to some variable degree. In practice, the silica used in electronic device applications is SiO$_2$ doped with varying amounts of water, and any routine reference to the terms "oxide," "silica," or "SiO$_2$ layer" automatically includes this water content.

13.2.2　Implementation

In industrial practice today, silicon dioxide films are grown in a reactor such as that illustrated in Fig. 13.1. The reactor consists of a resistance heated furnace held at a temperature around 1000°C by a temperature controller, a cylindrical fused quartz tube in which the silicon wafers are placed, and a source of either pure dry oxygen or pure water vapor. In production furnaces, temperature can be held for short terms (days) to within ±0.1°C over the length of the flat zone (up to 2 ft long), and for long terms (months) to within ±0.5°C.

The loading end of the furnace tube protrudes into a vertical laminar flow (Whitfield) hood where a filtered flow of air is maintained. Flow is in the direction of the arrows shown in Fig. 13.1. The purpose of the hood is to reduce dust and particulate matter in the air surrounding the wafer to negligible levels. Such particulate matter is undesirable because it can

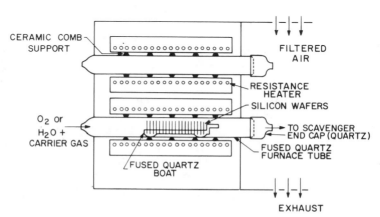

Fig. 13.1 Schematic cross section through a stack of two resistance heated oxidation furnaces. The silicon wafer loading area is shown in a Whitfield-type hood.

adhere, by electrostatic forces, to the silicon wafers prior to oxidation and contaminate either the oxide layer subsequently grown or the underlying silicon. One of the most common contaminants in airborne particulate matter is sodium, which can become incorporated in the silica as a positively charged mobile ion. Sodium ions in the oxide film can cause instability of device characteristics (see Section 15.3.3) and can affect how fast the oxide grows [see Section 13.4.5(b)].

Another source of sodium is the furnace lining. Sodium from this source can diffuse readily through the hot fused quartz tube and contaminate the oxide being grown on the wafers. In the early days of this technology, a silicon carbide or alumina liner, not shown in Fig. 13.1, was placed between the fused quartz tube and the furnace heater windings to act as a sodium barrier. Today, sodium contamination of the furnace lining is minimized by the furnace manufacturer. This is the type of furnace shown in Fig. 13.1. Despite such precautions, sodium remains one of the most prevalent and common electrically active contaminants in SiO_2.

Gases coming out of the furnace during oxidation are not exhausted into the Whitfield hood but are passed through a scavenger (not shown in Fig. 13.1) and then exhausted into the atmosphere. The scavenger is made of stainless steel or coated with Kynar to retard corrosion. When anhydrous HCl or Cl_2 is mixed with oxygen, which is sometimes done to produce sodium free oxides, the exhaust gases flow through a water spray scrubber, to remove chlorine from the gas stream by forming HCl.

In modern practice, contamination from the environment is avoided by making the pushrod an integral part of the wafer holder and using an automatic mechanism to push and pull the loaded assembly into and out of the oxidation furnace.

Another source of contamination is the oxidant gas. Controlling contamination requires the use of high purity oxygen. Distilled water, or deionized water passed through filters to remove particulate matter and organics are the sources of steam or water vapor for wet oxidation. Water vapor of extremely high purity also can be obtained pyrolytically by burning hydrogen gas, but filtered deionized water is the most widely used steam source in manufacture.

To prevent room air—which may contain impurities that adversely affect film quality—from entering the oxidation zone, the ambient gases in the furnace tube are held at a pressure just slightly above 1 atm. The oxidant gas flows at a rate of about 1 cm/sec past the wafers in the furnace at oxidation temperature.

To grow wet oxides, the water vapor can be steamed at slightly above 1 atm or can be mixed with a neutral carrying gas such as argon or nitrogen. Steam can be generated by boiling pure water in a fused quartz flask. Steam flows into the oxidation furnace through a hot pipe to prevent condensation. A mixture of argon or other inert gas and water can be produced by heating pure water in a fused quartz flask to temperatures below boiling to get water vapor pressures less than 1 atm and slowly bubbling the argon through the heated water. The argon becomes saturated with water vapor at a pressure determined by the temperature of the water in the flask. The saturated stream of argon and water then flows over the silicon wafers in the oxidation furnace to grow the oxide film. Oxide grows more rapidly in steam than in water vapor mixed with a nonoxidizing carrier gas because this mixing reduces the water vapor pressure. Therefore, use of a carrier gas provides better thickness control because growth is slower.

Oxides grown in pure dry oxygen have the very best electrical properties, but considerably more time is required for growing the same thickness of oxide at a given temperature in dry oxygen than in water vapor. For relatively thin oxides, such as the gate oxide in a MOSFET, typically ≤ 1000 Å thick, dry O_2 oxidation is commonly used. Such a film can be grown in a matter of hours. In many other applications, such as for bipolar transistors and for thick field or isolation oxide (>5000 Å) in MOS integrated circuits, oxidation in water vapor or steam is used and provides adequate passivation. Again, such a film can be grown in a matter of hours.

13.2.3 Interface at Which Oxidation Occurs

Because the silicon surface is highly reactive, a layer of oxide rapidly forms on exposure to an oxidant gas, and its rate of formation is reaction limited. However, as the oxide becomes thicker, its rate of formation becomes diffusion limited because the silicon and the oxidizing ambient are separated by the oxide layer. It was not known *a priori* whether the oxidation reaction proceeded at the Si–SiO$_2$ interface or at the SiO$_2$-gas interface. However, it has been demonstrated by experiment that the oxidizing species (oxygen) diffuses through the oxide layer and reacts with the silicon when it arrives at the Si–SiO$_2$ interface. Evidence for the inward migration of the oxidizing species and reaction at the Si–SiO$_2$ interface is given by infrared isotope shift experiments,[6,7] marker experiments,[8] radioactive tracer experiments,[5] and controlled etch studies.[9] We discuss the infrared isotope shift and marker experiments in some detail.

13.2.4 Infrared Isotope Shift Experiment

The infrared isotope shift experiment of Ligenza and Spitzer[6] is a method of detecting and following the progress of the oxidizing species as it moves through the silica layer. The experiment is based on measurements of the wavelength shifts of two Si—O infrared absorption bands when ^{18}O is substituted for ^{16}O (^{16}O is the abundant isotope of oxygen). The origin of this wavelength shift can be understood as follows. Infrared photons are strongly absorbed when they excite vibrations of the Si—O bond. To excite these vibrations, the infrared photons must be at a frequency corresponding to the vibration frequency of the bond. This vibration frequency, in turn, depends on the mass of the oxygen atom. Thus absorption of infrared energy at a particular frequency characteristic of an isotope is evidence for the presence of that isotope.

The mass dependence of the vibration frequency can be understood by viewing the Si—O molecule as a dumbbell with silicon and oxygen atoms as the end weights connected together by the bonding force, which acts as a spring. This dumbbell system will have a resonant frequency dependent on the masses of silicon and oxygen atoms and the bond strength between them. A photon that has a wavelength corresponding to this resonant frequency will be absorbed by imparting its energy to the dumbbell, causing it to vibrate. This loss of energy is detected as an absorption of the incident infrared beam. By substituting the heavier isotope of oxygen, ^{18}O, for the normal oxygen atom, ^{16}O, in the silica structure, the mass of the dumbbell is changed, but not the bond strength. Therefore, the resonant frequency of the dumbbell is changed so that ^{18}O can be distinguished from ^{16}O. In this way the movement of ^{18}O through the oxide layer can be monitored.

In the infrared isotope-shift method, the fundamental resonance vibrations of the silica lattice are measured. The transmission minimum of $Si\,^{16}O_2$ at 9.2 μm shifts to 9.6 μm for $Si^{18}O_2$ and a fraction of this shift proportional to the concentrations of either isotope occurs in a film containing both.[6] Therefore, measurement of the shift gives the extent of conversion. Only oxygen atoms chemically bonded to silicon atoms are of importance because the fundamental resonance vibrations of the silica are measured. The oxygen that does not participate in the formation of silica, that is, interstitial oxygen or water, does not have a pronounced effect on the transmission minimum. The method is clarified in Figure 13.2.

Figure 13.2 shows infrared transmission as a function of wavelength of three different SiO_2 films after removal of the silicon substrate by a chlorine gas etch. Curve 1 was measured on a 3300 Å thick SiO_2 film oxidized in pure ^{16}O at atmospheric pressure and 950°C. Curve 3 was measured on a sample produced by oxidation in $H_2^{16}O$ at a pressure of 120 atm and a temperature of 650°C. Comparison of curves 1 and 3 shows that the absorption spectrum is the same whether the oxide is grown in high pressure steam or at atmospheric pressure in pure oxygen. Curve 2 was measured on a sample oxidized in steam containing 90% $D_2^{18}O$. Comparison of curve 2 to curves 1 and 3 reveals that two of the transmission minima (absorption maxima) in the ^{18}O sample (curve 2) have undergone appreciable isotopic shifts. The minima at 9.25 and 21.9 μm for the samples grown in ^{16}O have shifted to 9.6 and 22.6 μm, respectively, for the sample grown in ^{18}O. (Note that silicon is transparent at these wavelengths.) Figure 13.3 is a plot of the ^{16}O concentration divided by the total $^{16}O + ^{18}O$ concentration as a function of wavelength of the transmission minima for each of the two absorption bands. Figure 13.3 shows that the isotope

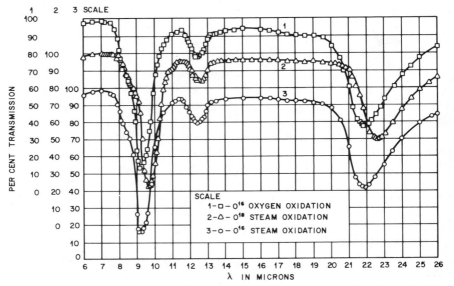

Fig. 13.2 Infrared transmission of three different SiO_2 films where the silicon has been removed. Film 1 was made by oxidation in ^{16}O, film 2 was oxidized in $D_2{}^{18}O$, and film 3 was oxidized in $H_2{}^{16}O$. After Ligenza and Spitzer.[6] Copyright (1960), Pergamon Press, Ltd. Reprinted with permission.

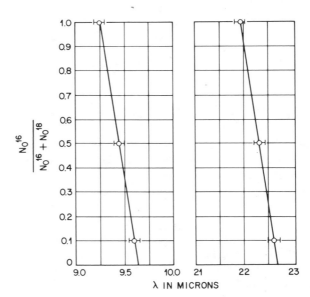

Fig. 13.3 Plot of ^{16}O concentration divided by the total $^{16}O + {}^{18}O$ concentration as a function of wavelength for the two infrared absorption bands. After Ligenza and Spitzer.[6] Copyright (1960), Pergamon Press, Ltd. Reprinted with permission.

652

wavelength shift depends linearly on the fractional isotope concentration.

To see whether an oxygen species or a silicon species moves through the film, a silicon sample was first oxidized to a thickness of 2000 Å in ^{16}O steam. Oxidation then was continued in ^{18}O steam until total oxide thickness was 3500 Å. The entire 3500 Å film then was found to be composed uniformly of $Si^{18}O_2$. The ^{16}O originally in the film had diffused to the oxide surface, evaporated, and been replaced by ^{18}O. The film then was etched to 2000 Å, but the infrared absorption bands still showed the film to be almost completely $Si^{18}O_2$. A similar specimen was prepared, except the order of the two oxidations was reversed. A 3000 Å oxidation in ^{18}O was followed by a 2500 Å ^{16}O oxidation. Now the entire 5500 Å film was found to be composed of $Si^{16}O_2$.

These experiments show that exposure of an SiO_2 film containing only one of the two oxygen isotopes, to steam containing the second isotope results not only in the growth of new oxide containing the second isotope, but also in the complete conversion of the already existing oxide film to an oxide consisting of the second isotope. This transformation occurs by way of an oxygen exchange mechanism[7] in which diffusion of one isotope proceeds by the exchange of position in the silica lattice with the other isotope. The net direction of motion is along the concentration gradient of each isotope.

Two conclusions can be drawn from these experiments of Ligenza and Spitzer: (1) an oxidizing species moves through the oxide and the reaction proceeds at the Si–SiO$_2$ interface, not at the SiO$_2$-gas interface (otherwise, silicon would be the moving species, the second isotope would be only at the SiO$_2$-gas interface, and the entire oxide could not be converted to the second isotope); and (2) the oxidizing species forms an Si—O bond because the absorption bands measured arise from the vibration of this bond.

13.2.5 Marker Experiment

Marker experiments by Jorgensen,[8] illustrated in Fig. 13.4, also show directly that the oxidizing species moves through the SiO_2 layer and reacts at the Si–SiO$_2$ interface. The marker was a narrow strip of platinum about

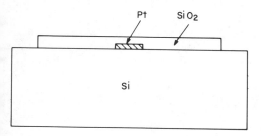

Fig. 13.4 Schematic cross section through the silicon specimen showing the position of the platinum and initial SiO₂ before oxidation. After Jorgensen.[8]

1 μm thick sputtered onto the surface of a mechanically polished 1000 Ω-cm silicon crystal. If silicon were the mobile species, the platinum marker would remain in contact with the silicon after growth of the oxide. If oxygen were the mobile species, oxide would grow underneath the platinum marker. To perform this experiment, a layer of SiO_2 was deposited over the marker and the complete assembly oxidized. The SiO_2 initially deposited over the platinum marker was necessary because oxygen easily permeates sputtered platinum at oxidation temperatures. Without the initial top layer of SiO_2, oxide formed during later oxidation always would be under the platinum regardless of which species diffused through the SiO_2. The position of the marker was determined by measuring the resistance between the silicon and platinum. If this resistance was unchanged by oxidation, the platinum remained at the Pt–Si interface and SiO_2 formed over the platinum. Moreover, at oxidation temperature, thermal conversion would reduce bulk silicon resistance. Therefore, contact resistance should decrease somewhat if silicon were the mobile species. If this resistance increased after oxidation, SiO_2 grew between the platinum and the silicon. Resistance before oxidation was 400 Ω, and after oxidation it increased to 60,000 Ω. Therefore, oxygen must be the diffusing species. The existence of an SiO_2 layer under the platinum also was observed by making a 10:1 section and viewing with an optical microscope.

13.2.6 Nature of the Diffusing Species

What is the diffusing species? Measurements of the pressure dependence of the silicon oxidation rate[10] and of the permeation of oxygen through fused silica[11] show that both the oxidation and the permeation rate are linearly proportional to the O_2 pressure. It follows from the law of mass action, namely, that the equilibrium constant depends linearly and not quadratically on oxygen pressure, that a species containing molecular oxygen is moving through the oxide film (see Section 13.4.1). This evidence does not indicate conclusively that the diffusing species is molecular oxygen, although this is the simplest species. Similarly, for oxidation in water vapor, the oxidation rate is linearly dependent on water vapor pressure,[10] suggesting that a species containing water is moving through the oxide film.* The oxygen isotope experiments of Ligenza and Spitzer,[6] described in Section 13.2.4, also support the picture of oxygen or water diffusing through the oxide.

In an attempt to describe the charge state of the oxidizing species, Jorgensen[8] performed an experiment in which an electric field was applied to a growing silica film by means of platinum electrodes attached to the

---

*At pressures of about 1 atm, solubility of water in silica[12-17] depends on the square root of water vapor pressure in the ambient gas. At pressures of several atmospheres, the solubility also depends linearly on pressure.[10]

opposite sides of a partially oxidized single crystal of silicon as shown in Fig. 13.5. Oxidation could proceed in this structure because oxygen could penetrate the platinum electrodes from the surrounding oxygen gas ambient at the oxidation temperature of 850°C. Electric fields of up to 10^4 V/cm were applied across the silica films and silica film thickness measured by interferometry (discussed in Section 14.3.3) as a function of oxidation time. It was found that the oxidation rate of the film on the right side in Fig. 13.5 was accelerated whereas the oxidation rate of the film on the left side in Fig. 13.5 was retarded. Jorgensen's interpretation of these results was that the diffusing species in the oxide was a negatively charged oxygen ion, and it was inferred that the diffusing species in the normal oxidation process also was a negatively charged oxygen ion.

(a) Charge State of the Diffusing Species

It was widely accepted in the early literature[10,18–24] that the Jorgensen experiment demonstrated that the diffusing species was a negatively charged oxygen ion. However, independent experimental evidence shows that no free oxygen ions remain throughout the oxide in detectable quantity after oxidation has ceased and the sample has been cooled to room temperature. This experimental evidence is:

1 No net negative charge in the oxide is detected by C-V measurements (measured at frequencies up to 1 MHz) (see Chapters 3, 4, and 12 for details of C-V measurements). In a C-V measurement negative charge due to oxygen ions in the oxide would cause a shift in the C-V curve along the voltage axis (see Section 10.2). The C-V method is sufficiently

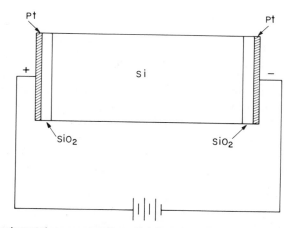

Fig. 13.5 Experimental apparatus for studying the effect of an electric field on the oxidation of silicon. After Jorgensen.[8]

sensitive to detect charge in as small a concentration as 1 ppb in the oxide, which is a sensitivity adequate to detect the net negative charge corresponding to the excess oxygen in the silica required for oxidation. This result rules out both O_2^- and O_2^{2-}.

2 No paramagnetic centers have been found in thermally grown silica films,[25,26] thus ruling out O_2^-, which is a paramagnetic center, but not O_2^{2-}, which is not a paramagnetic center. Therefore, either the oxygen ions become neutral on cooling or the oxygen diffuses as a neutral oxygen molecule, or if it diffuses as a negative ion, the oxygen ion current must be balanced by a hole current flowing in the opposite direction to maintain charge neutrality. Although the behavior of SiO_2 is not sufficiently understood to rule out the first possibility, it is more likely that the oxidant diffuses as a neutral species, either as a neutral molecule or as a negative ion balanced by a hole countercurrent.

There are several other possible explanations of the Jorgensen experiment.[8] One explanation, proposed by Raleigh,[27] is that by applying both an electric field across the growing silica film and by providing a path for electronic current flow through the battery as shown in Fig. 13.5, the electrolytic process of anodization occurs on the right side, where SiO_2 forms from its elements, and that the electrolytic process of electrolysis occurs on the left side, where SiO_2 is decomposed into its elements. The experimental assembly shown in Fig. 13.5 can be viewed as consisting of two electrolytic cells, O_2 (gas), Pt|SiO_2(solid), and |Si(solid), placed back to back. These cells are batterylike because a voltage will develop across each cell during the oxidation reaction. Because the oxidation reaction involves the exchange of electrons between silicon and oxygen, probably at the Si–SiO_2 interface, the following half cell reactions can occur

$$O_2(gas) + 4e \leftrightarrows 2O^{2-} \tag{3}$$

$$2O^{2-} + Si \rightleftarrows SiO_2 + 4e \tag{4}$$

and the total cell reaction is reaction (1). The direction of the arrows in reactions (1), (3), and (4) depends on the direction of current flow through the cell. The flow of a given current for a given length of time would, according to Faraday's law, form the same amount of SiO_2 on the right side as is decomposed on the left side.

The linear pressure dependence of the silicon oxidation rate and the oxygen permeation rate through vitreous silica suggest that molecular oxygen may be the diffusing species. No definitive experiments have been done to distinguish between the transport model of a diffusing neutral oxygen molecule exchanging electrons at the Si–SiO_2 interface and the model of a diffusing O_2^{2-} molecule balanced by a hole countercurrent. Very little is known about hole transport in silica, except that hole mobility is very low at room temperature. Thus it is difficult to construct a quantitative hole countercurrent model that could be tested experimentally.

In view of the preceding arguments, Jorgensen's experiment is inconclusive. The question of the charge state of the oxidizing species diffusing through the growing silica layer remains unsettled.

13.2.7 State of Water in Silica

To understand the oxidation process and the nature of oxide layers containing water, it is important to know how water is incorporated in silica. In effect, all oxide layers used in electron device applications contain water. The degree of hydration of an oxide layer is important because it can determine such practical matters as how well a photoresist film will stick to the oxide surface and whether device performance and stability will be satisfactory. The effect of water in the oxide on device performance and stability is discussed in Sections 11.5.1 and 15.4.

The initial water content in the oxide layer depends on oxidation temperature and water vapor pressure during oxidation. There always will be some trace of water vapor in an oxidation furnace. After the oxide film is grown and exposed to room ambient, water will diffuse into the oxide, typically several hundred angstroms deep, in a week or so. The water content of the oxide film will be determined eventually by the temperature and water vapor partial pressure of the room ambient.

Although most cases of practical interest involve exposure to low water vapor pressures (<1 atm), the experiments we discuss next were done at high water vapor pressures (>1 atm). However, the basic chemistry is the same in both cases, so the understanding gained from the high pressure experiments can be applied to low pressures.

The mechanical properties of silica are affected by the incorporation of water in the form of silanol groups. The detection of differences between "wet" and "dry" silica constitutes evidence for the formation of silanol groups in hydrated silica.

The existence of silicon hydroxide or silanol (SiOH) groups in hydrated silica has been inferred from infrared absorption measurements in the 2.7 μm band that detect the presence of the hydrogen-oxygen bond.[16,17,28,29] Spitzer and Ligenza[7] have shown that *both* water molecules and silanol groups exist simultaneously in hydrated oxides. Their argument is based on two different experiments: the first experiment establishes that water enters the silica as water molecules, and the second experiment shows that the interstitial water molecule reacts with a silicon-oxygen bridge to form a paired silanol group. The pair can then condense to reform an interstitial water molecule and a silicon-oxygen bridge.

(a) The First Experiment

In the first experiment, done in high pressure steam, the oxidation rate* is observed to depend linearly on water vapor pressure in the gas phase so that the oxidation rate is reaction limited rather than diffusion limited. The oxidation rate depends on the concentration of water in the silica phase. Therefore, a linear dependence of oxidation rate on pressure in the gas phase means that the concentration of water in the silica depends linearly on the water vapor pressure in the gas phase at high pressures. Then, by the law of mass action, a linear gas phase pressure dependence of water concentration in the silica means that water dissolves in the silica lattice as a water molecule.

Once in the silica lattice, the water molecule must occupy an interstitial site located near the silica surface. The water molecule then may diffuse deeper into the silica film by moving from its initial interstitial site to an adjacent interstitial position, and so on. Another likely possibility is the oxygen exchange reaction between a water molecule in an interstitial site and a neighboring silicon-oxygen bond in the silica lattice. The end result of this oxygen atom exchange process is the formation of an exchanged oxygen atom in the silicon-oxygen bridge of the silica lattice, originally part of the interstitial water molecule, and a re-formed water molecule. The oxygen atom, in the re-formed water molecule, was originally in the silicon-oxygen bridge of the lattice. The final model for the diffusion of water through silica will be shown to be *principally* interstitial diffusion of water molecules and, at a much slower pace, re-formation of water molecules as a result of the oxygen exchange process.

(b) The Second Experiment

The second experiment consists of oxidizing the silicon in $D_2^{16}O$ and then measuring the isotopic wavelength shift as a function of time, temperature, and pressure in a $D_2^{18}O$ ambient as the $D_2^{18}O$ diffuses into the original $Si^{16}O_2$ layer† (see Section 13.2.4).

The relationship that empirically fits the measured data is

$$s = s_\infty [1 - \exp(-\alpha(T)p_G t)] \tag{13.1}$$

where s is the measured wavelength shift, s_∞ is the maximum possible shift (0.4 μm), $\alpha(T)$ contains the temperature dependence, p_G is the water vapor pressure in the gas phase, and t is time. The wavelength shift is estimated from the change in position of the initial 9.25 μm transmission minimum. The wavelength of the transmission minimum for $Si^{16}O_2$ is 9.25 μm, as seen

---

*The oxidation rate is the rate of increase of oxide thickness of growing oxide and is determined by measuring oxide thickness as a function of oxidation time.

†The symbol D represents deuterium which has a nucleus consisting of a proton and a neutron. Substitution of hydrogen for deuterium does not change reactions (5) through (7) as it is only the electron in either deuterium or hydrogen that is important.

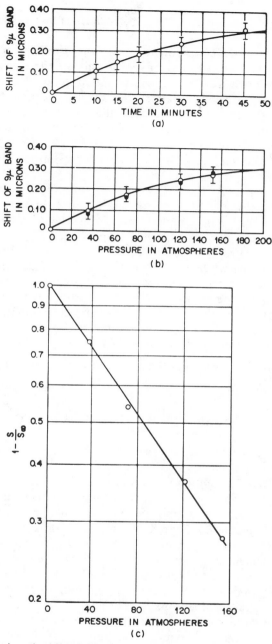

Fig. 13.6 (a) Wavelength shift of the 9.25 μm transmission minimum as a function of oxidation time in $D_2{}^{18}O$ at 650°C and 35 atm. Original thickness was 6000 Å of $Si^{16}O_2$ on front and back surface of the silicon wafer. (b) Wavelength shift of the 9.25 μm transmission minimum as a function of pressure at 650°C for 10 min. Front and back surfaces of the silicon were 6000 Å of $Si^{16}O_2$. Dark circles, 6000 Å; open circles, after thinning the oxide layer to 3000 Å. (c) Semilog plot of $1 - s/s_\infty$ versus p_G using data from curve b. After Spitzer and Ligenza.[7] Copyright (1961), Pergamon Press, Ltd. Reprinted with permission.

in Fig. 13.2. Figures 13.6a,b demonstrate the good fit of (13.1) to the experimental data. Figure 13.6a shows the shift of the transmission mimimum at constant pressure (35 atm) and temperature (650°C) as a function of exposure time to steam containing $D_2^{18}O$ for six samples. The dots represent the measured shifts, and the solid curve is a plot of (13.1) with $\alpha(T)p_G = 3.1 \times 10^{-2}$ min^{-1}. Figure 13.6b shows the shift of the transmission minimum at constant temperature (650°C) and time (10 min) as a function of pressure for four samples. The dark dots represent the measured shifts, and the solid curve is a plot of (13.1) with $\alpha(T)t = 9.5 \times 10^{-3}$ atm^{-1}. To show that the conversion of $Si^{16}O_2$ to $Si^{18}O_2$ occurred uniformly with depth in the silica film, the silica was etched from its original thickness of 6000 Å to 3000 Å on each of four samples and the shift in wavelength of the transmission minimum remeasured. The measured shifts for the 3000 Å thick silica films are represented by open circles in Fig. 13.6b. For each of the four samples, the wavelength shift for the two silica thicknesses agree within experimental error. Therefore, oxygen exchange occurs uniformly with depth. Because the pressure dependence is of crucial importance, Fig. 13.6c shows a semilog plot of $1 - s/s_\infty$ versus p_G. Experimental points taken from Fig. 13.6b lie on a straight line, so that the wavelength shift is an exponential function of the first power of pressure.

(c) Derivation of the Experimental Wavelength Shift

Next, we derive (13.1) from the ^{16}O to ^{18}O exchange reaction (5)

$$D_2^{18}O + Si—^{16}O—Si \rightarrow Si—^{18}O—Si + D_2^{16}O. \tag{5}$$

Starting with (5), we use D_2O rather than H_2O to be consistent with the experiments where D_2O was used.

The silica lattice in reaction (5) is represented by a single silicon-oxygen bridge (Si—O—Si) where the oxygen exchange reaction takes place. Reaction (5) is quite general and applies even if intermediate species form. The reason for this generality is that the reaction starts and ends with an interstitial water molecule regardless of how many intermediate species are formed. In particular, an interstitial water molecule will react with a silicon-oxygen bridge to form a paired silanol group. The slow step in this reaction is the condensation reaction of the pair to reform an interstitial water molecule and the silicon-oxygen bridge.

There are two intermediate reactions for the formation of a silanol pair. The first is a fast reaction between an interstitial water molecule and a silicon-oxygen bridge to form a silanol pair

$$D_2^{18}O + Si—^{16}O—Si \xrightarrow{\text{fast}} Si—^{16}OD + D^{18}O—Si. \tag{6}$$

The second is a slow condensation reaction within the silanol pair to form a silicon-oxygen bridge containing an ^{18}O atom and an interstitial water molecule containing an ^{16}O atom

$$\text{Si}\text{—}^{16}\text{OD} + \text{D}^{18}\text{O}\text{—}\text{Si} \xrightarrow{\text{slow}} \text{Si}\text{—}^{18}\text{O}\text{—}\text{Si} + \text{D}_2{}^{16}\text{O}. \qquad (7)$$

The two silicon ions in the Si—O—Si bridge are at a fixed distance from each other. If this bridge reacts with an interstitial water molecule to form a silanol pair, another oxygen ion and two hydrogen ions must be inserted between them. This reaction only can happen if the imposed strain is relieved by a shift in the positions of the ions in their immediate neighborhood to decrease the void volume of the silica. In effect, the silica would assume a more ordered quartzlike structure in the neighborhood of the silanol pairs. There is a difference of 20% observed in the void volume of quartz and silica, where the silica has a larger volume than the quartz.[7] According to this picture, the stability of silanol pairs, once formed, is due to a partial transition from the metastable silica structure to a thermodynamically stable quartz form. These arguments form the basis for assuming that the condensation of silanol pairs [reaction (7)] is the slow intermediate step in the reaction.[7]

To test this picture of water in silica, a mathematical relationship between isotopic wavelength shift, time, pressure, and temperature is derived on the basis of reaction (5). Then modifications of this mathematical relationship are made to take into account the formation and the condensation of silanol pairs described by reactions (6) and (7). The derived relationship will fit (13.1) obtained from experimental observations.

Because oxygen in the Si—O—Si bridges is immobile, the rate at which the concentration $C_{18}(t)$ of Si—^{18}O—Si bridges increases is

$$\frac{dC_{18}(t)}{dt} = \zeta C_{16}(t) C_{18}^*(t) - \zeta C_{18}(t) C_{16}^*(t) \qquad (13.2)$$

where ζ is the bimolecular rate constant for the oxygen exchange process,* $C_{16}(t)$ is the concentration of Si—^{16}O—Si bridges at time t, $C_{18}^*(t)$ is the concentration of $\text{D}_2{}^{18}\text{O}$ in the silica at time t, and $C_{16}^*(t)$ is the concentration of $\text{D}_2{}^{16}\text{O}$ in the silica at time t. A simplification of (13.2) is possible by considering values of $C_{16}^*(t)$ and $C_{18}^*(t)$ under experimental conditions. First, $C_{16}^*(t) = 0$ to a good approximation because this species diffuses out of the silica and becomes diluted in the gas phase. Second, $C_{18}^*(t) = C^*$, where C^* is the initial concentration in silica. The concentration of $\text{D}_2{}^{18}\text{O}$ in silica always equals its initial concentration because the concentration of $\text{D}_2{}^{18}\text{O}$ in the gas phase is in thermal equilibrium with the concentration of $\text{D}_2{}^{18}\text{O}$ in the silica phase, and the oxygen exchange occurs uniformly throughout the thickness of the silica

---

*In general, the value of ζ in the first term in (13.2) is different from that in the second because the masses of ^{16}O and ^{18}O are not the same. But this difference is too slight to be detected in these experiments, so they are considered to be equal.

as shown in Fig. 13.6b. Under these conditions, (13.2) becomes

$$\frac{dC_{18}(t)}{dt} = \zeta C^*[A_o - C_{18}(t)]$$ (13.3)

where A_o is the initial concentration of Si—^{16}O—Si and $[A_o - C_{18}(t)]$ is the concentration of Si—^{16}O—Si at time t. The solution of (13.3) is

$$C_{18}(t) = A_o[1 - \exp(-C^*\zeta t)].$$ (13.4)

Because the isotopic wavelength shift is proportional to $C_{18}(t)/A_o$, (13.4) becomes

$$s = s_\infty[1 - \exp(-C^*\zeta t)].$$ (13.5)

Equation (13.5) is identical to the empirical equation (13.1) with $C^*\zeta = \alpha(T)p_G$. To derive this pressure dependence, we assume Henry's law,* which states that in thermal equilibrium, the concentration of water in the gas phase is proportional to the concentration of water in the silica phase. We can express this proportionality mathematically

$$C^* = KC_g$$ (13.6)

where K is the distribution coefficient between water in the gas and silica phases and C_g is the water concentration in the gas phase. To calculate the dependence of C^* on p_G, note that $C_g = nL_o/V$, where n is the number of moles of water in the gas phase, V is the gas phase volume, and L_o is Avogadro's number. Treating water in the gas phase as an ideal gas, a good approximation for high temperature steam is $p_G V = nRT$, where R is the universal gas constant and T is temperature. Solving for n/V, we obtain $C_g = p_G L_o/RT$. Substituting this result into (13.6) yields

$$C^* = \frac{KL_o p_G}{RT}.$$ (13.7)

Thus $C^*\zeta = KL_o p_G \zeta/RT = \alpha(T)p_G$ so that $\alpha(T) = K\zeta L_o/RT$. Substituting (13.7) into (13.5) yields

$$s = s_\infty\left\{1 - \exp\left[-\frac{(KL_o \zeta p_G t)}{RT}\right]\right\}.$$ (13.8)

To evaluate $\alpha(T)$, consider first the temperature dependence of K, which is

$$K = \exp\frac{S}{R}\exp\frac{H}{RT}$$ (13.9)

where S and H are the entropy and the enthalpy (heat flow at constant pressure) of solution of water in silica, respectively.† Consider next the

---

*See Section 13.4.1 for the conditions under which Henry's law applies.
†Equation (13.9) arises from standard thermodynamic considerations that the first-order perturbation as well as all higher-order perturbations of free energy vanish.

temperature dependence of ζ, which is

$$\zeta = \zeta_o \exp\left(-\frac{E_A}{RT}\right) \tag{13.10}$$

where E_A, the activation energy of the oxygen exchange reaction, is assumed to be independent of temperature. Combining (13.9) and (13.10) yields

$$\alpha(T) = \frac{\zeta_o L_o}{RT} \exp\frac{S}{R} \exp\left(-\frac{H + E_A}{RT}\right) \tag{13.11}$$

and (13.8) becomes

$$s = s_\infty\left\{1 - \exp\left[-\frac{\zeta_o L_o}{RT} p_G t \exp\frac{S}{R} \exp\left(-\frac{H + E_A}{RT}\right)\right]\right\}. \tag{13.12}$$

Figure 13.7 is a plot of the wavelength shift as a function of temperature at constant pressure (35 atm) and time (20 min). Figure 13.8 is a plot of $\log[\alpha(T)Tp_G/L_o]$ versus $1/T$. Values of $\alpha(T)p_G/L_o$ used in Fig. 13.8 are calculated from (13.1) using the wavelength shifts from Fig. 13.7 for five different temperatures. The points lie quite close to a straight line. Interpretation of the linearity of the experimental curve in Fig. 13.8 in terms of (13.11) requires certain restrictions on the magnitude and temperature dependence of the activation energy of the oxygen exchange reaction E_A, the entropy S, and the enthalpy H of the solution of water in silica. Normally, both S and H would be functions of temperature. Thus, for (13.11) to predict a straight line in a plot such as that in Fig. 13.8, there is a requirement that either (1) $E_A \gg H + S$, where E_A is temperature independent over the temperature range measured, or (2) S varies as $1/T$, $E_A \gg H$, and where E_A is temperature independent over the temperature range measured. These are reasonable restrictions on the temperature dependences and magnitudes of S, H, and E_A. However, the magnitude of each quantity cannot be separately determined, and the two cases cannot

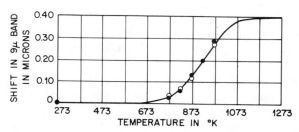

Fig. 13.7 Wavelength shift of the 9.25 μm transmission minimum as a function of temperature at a pressure of 35 atm and a time of 20 min. As in Fig. 13.6b, the black circles represent a silica thickness of 6000 Å, and the open circles give the shifts after the silica is thinned to 3000 Å. After Spitzer and Ligenza.[7] Copyright (1961), Pergamon Press, Ltd. Reprinted with permission.

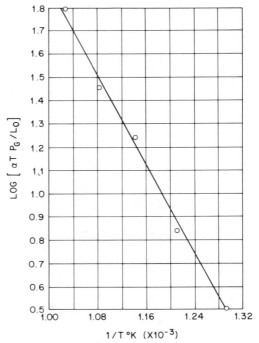

1/T °K (X10⁻³)

Fig. 13.8 Diagram illustrating $\log[\alpha(T)Tp_G/L_o]$ as a function of $1/T$. After Spitzer and Ligenza.[7] Copyright (1961), Pergamon Press, Ltd. Reprinted with permission.

be distinguished using the data presented by Spitzer and Ligenza.[7] The slope and intercept of the line in Fig. 13.8 were calculated by Spitzer and Ligenza[7] using the method of averages to obtain

$$\frac{\alpha(T)p_G}{L_o} = \frac{3.6 \times 10^6}{T} \exp\left(-\frac{21,400}{RT}\right). \tag{13.13}$$

The curve in Fig. 13.7 was calculated using (13.13) in (13.1) for $t = 20$ min. The agreement is again quite good.

To take into account the formation of silanol pairs, (13.12) must be modified. The oxygen exchange rate will be limited by the rate at which paired silanol groups decompose. This reaction will lead to an equation having the same form as (13.12) provided that the concentration of silanol pairs is small compared to A_o, the initial concentration of silicon-oxygen bridges. This statement must be true, in practice, because the total amount of water dissolved in the silica, in equilibrium at 35 atm and high temperatures, is quite small (a few percent). The modified form of (13.12) is

$$s = s_\infty \left\{ 1 - \exp\left[-\frac{\zeta_o L_o}{RT} p_G t \exp\left(\frac{S+S'}{R}\right) \exp\left(-\frac{H+H'+E_A}{RT}\right) \right] \right\} \tag{13.14}$$

where the quantities S' and H' are the entropy and enthalpy changes, respectively, associated with reaction (6). It is not possible to distinguish between (13.12) and (13.14) by infrared transmission measurements in the $9 \mu m$ absorption band. Therefore, the existence of silanol groups in hydrated silica must be inferred from infrared absorption measurements in the $2.7 \mu m$ band.

This picture of the state of water in silica applies at pressures down to 1 atm or less and at high temperatures, although the experimental verification was done at 35 atm and relatively low temperatures. The key evidence for the validity of this picture, at all temperatures and pressures of interest in device applications, is that Henry's law (see Section 13.4.1) is found to apply over the temperature range 500–1200°C and over the pressure range 0.1–150 atm. Because Henry's law applies, there is no dissociation of water and water must enter the silica as water molecules over these pressure and temperature ranges.

Oxygen also obeys Henry's law, so that, like water molecules, oxygen molecules must enter the silica by occupying interstitial sites and then move principally by interstitial diffusion and secondarily by an oxygen exchange reaction.

13.3 STRUCTURE

The basic structural unit of a thermally grown SiO_2 layer is a silicon ion surrounded tetrahedrally by four oxygen ions. This structural unit is illustrated in Fig. 13.9. The silicon-to-oxygen internuclear distance is 1.6 Å, and the oxygen-to-oxygen internuclear distance is 2.27 Å. These tetrahedra are joined together at their corners by oxygen bridges in a variety of ways to form the various phases or structures of SiO_2. Silica has several crystalline structures and an amorphous structure. The basic crystalline structures of SiO_2 are called *cristobalite, tridymite,* and *quartz.* As each of these structures has several modifications, SiO_2 can exist in a large variety of crystalline forms.[31-33]

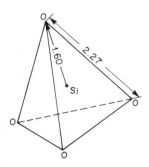

Fig. 13.9 Illustration of the basic structural unit of SiO_2.

The amorphous structure grows when silicon is thermally oxidized. Therefore, the amorphous structure is the most important one in integrated circuit applications and is our main concern here. The basic difference between the crystalline structures and an amorphous structure is that the crystalline structures have long range order, extending over many atoms, whereas an amorphous structure is not ordered on a macroscopic scale. Figure 13.10 illustrates this difference; Fig. 13.10a shows a two-dimensional schematic diagram of a crystalline structure made up of rings with six silicon atoms, and Fig. 13.10b shows a two-dimensional schematic diagram of an amorphous structure for comparison. In the amorphous structure there still is a tendency for the formation of characteristic rings with six silicon atoms.

Amorphous silica is in a metastable state that tends toward a crystalline or devitrified[31-33] stable state.* Even at room temperature, crystallization is occurring, but at an extremely slow and negligible rate. Crystallization is facilitated by the presence of impurities in the silica such as sodium that

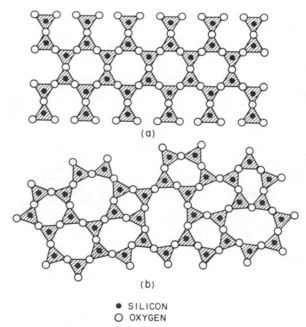

(a)

(b)

● SILICON
○ OXYGEN

Fig. 13.10 Two-dimensional schematic representation of (a) a quartz crystal lattice and (b) the amorphous structure of fused silica. After Stevels.[30] Reproduced by permission of the National Academy of Sciences, Washington, D.C.

---

*The amorphous structure is the vitreous or glassy phase of SiO_2. Devitrification is the process of changing from the glassy to the crystalline structure.

act as nucleation centers for the formation of crystallites. Thus silica films contaminated with sodium during growth at elevated temperatures often contain numerous small crystallites found by X-ray diffraction analysis. However, device-grade silica films are relatively free of impurities where crystallites could nucleate during oxidation.

The metastable state of the silica film in a device affects its stability only over very long times. This is fortunate because crystallization would produce severe mechanical strain and cracking of either the silica, the silicon, or both. Such strain or cracking would have a disastrous effect on the electrical properties of the devices on the chip. At typical device operating temperatures below 120°C, crystallization is an even slower process than the diffusion of the dopant impurities in the silicon. Consequently, the tendency of impurities to diffuse until they are uniformly distributed throughout the silicon chip would be a more likely cause of device failure than devitrification.

The amorphous structure of Fig. 13.10b is very open compared to the crystalline structure because only 43% of the space is occupied by SiO_2 molecules in the amorphous structure. This structural difference is reflected by the fact that the density of amorphous silica is about 2.20, compared with 2.65 for crystalline quartz. The relatively open structure of amorphous SiO_2 provides the means for a variety of electrically active impurities to enter and diffuse readily through the layer. Such diffusion is unacceptable, as discussed in Section 15.3. The amorphous structure of thermally grown SiO_2 has been determined by density and X-ray diffraction studies.[5,34-36]

13.4   OXIDATION KINETICS

In this section we consider the oxidation kinetics of silicon from 0.1–150 atm oxidant pressure and high temperatures (500–1200°C). The purpose is to gain an understanding of the oxidation process and to quantitatively predict the time, the temperature, and the pressure dependence of the thickness of oxide grown. Then the dependence of oxidation kinetics on impurities incorporated in the oxide is considered.

The first oxides were grown to improve the performance and stability of diodes and bipolar transistors.[37] This improvement was achieved with oxides grown in high pressure steam by Ligenza.[6,38] These high pressure steam oxides allowed the fabrication of the first MOSFET by Kahng and Atalla.[39] However, most integrated circuits are made by oxidation in 1 atm oxygen or steam, which is simpler to implement for high volume production.

The oxidation of silicon for electron device applications, at pressures of 1 atm and at high temperatures, was studied extensively by many workers. The earlier papers were by Yeh,[40] Karube et al.,[41] Deal,[42] and Edagawa et al.[34] Later, details of silicon oxidation kinetics were discussed for oxidation near atmospheric pressure in wet and dry oxygen by Deal and Grove,[10] in steam by Pliskin,[43] in ultradry oxygen by Burkhardt and Gregor,[44] in wet oxygen by Nakayama and Collins,[45] and in an RF heated furnace by Revesz and Evans.[46] Although many workers have made significant contributions, which are described here, our present understanding of the oxidation kinetics of silicon for electron device applications rests primarily on the basic work of Ligenza and Spitzer, Deal and Grove, and Pliskin.

In all this work, the observation of the oxidation process at high temperatures always began with a "real" silicon surface, usually one exposed to a hydrofluoric acid etch immediately prior to oxidation to remove the natural thin silica film (typically 20–50 Å) that may have been present. After this etch, a thin layer of fluorine atoms temporarily coats the silicon surface, thereby preventing oxidation in room air. Once in the oxidation furnace at high temperature, this layer of fluorine atoms is quickly driven off and oxidation begins.

13.4.1 Phenomenological Model

The oxidation kinetics of silicon can be illustrated by the simple phenomenological model developed by Deal and Grove shown in Fig. 13.11.[10] This model will be valid for oxide thicknesses above 300 Å, oxidation in dry oxygen, oxidant partial pressure of 1 atm or less, and

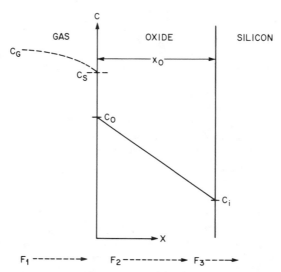

Fig. 13.11 Cross section of silicon with a growing SiO_2 layer illustrating the model for the thermal oxidation of silicon. The three fluxes, F_1 into the silica, F_2 through the silica, and F_3 at the Si–SiO_2 interface, are shown. After Grove.[48]

temperatures above 800°C. Oxides for most integrated circuit applications are grown over these thickness, pressure, and temperature ranges. Therefore, the simple model to be presented is of paramount importance. Modifications of this model are discussed in Section 13.4.2 for oxides below 300 Å thick, less than 800°C in temperature, and less than 1 atm in pressure.

Let the silicon be covered initially by an oxide layer of thickness x_o, as shown in Fig. 13.11. For oxidation to occur, the oxidizing species must go through three consecutive steps:

1 It moves from the oxygen-containing gas phase across the gas-oxide interface with flux F_1. (Flux is defined as the number of atoms or molecules crossing a unit area per unit time.)

2 It moves across the SiO_2 film already present toward the silicon with flux F_2.

3 It reacts with the silicon at the Si–SiO_2 interface with flux F_3.

In steady state, $F_1 = F_2 = F_3$. Each steady-state flow will be approximated next. A linear approximation of the gas phase flux F_1 is made by taking the flux of oxidant from the gas phase across the gas-oxide interface to be proportional to the difference between the equilibrium concentration of oxidant in the oxide C^*, the concentration in the oxide per cubic centimeter that would be in equilibrium with the partial pressure of oxidant in the gas p_G, and the actual concentration of oxidant at the outer surface of the oxide C_o

$$F_1 = h(C^* - C_o) \qquad (13.15)$$

where h is the gas phase transport coefficient in units of cm/sec. Equation (13.15) implies that F_1 is nonzero when the actual concentration of oxidant at the oxide-gas interface is below the equilibrium concentration in the oxide. Oxidation is a nonequilibrium process with the deviation of concentration from equilibrium the driving force.

The equilibrium concentration of oxidant in the oxide is related to the partial pressure of oxidant in the gas p_G by Henry's law[47]

$$C^* = kp_G \qquad (13.16)$$

where $k = KL_o/RT$ from (13.7). The oxidation process will depend strongly on whether Henry's law applies. Henry's law holds only in the absence of dissociation or association of the oxidant at the gas-oxide interface and when the chemical potential of the oxidant in the oxide is independent of pressure (dilute case). In fact, a linear pressure dependence of the oxidation rate constant is observed. The main exception to this observation is for oxide thicknesses less than about 300 Å grown in dry oxygen. Therefore, Henry's law [see (13.16)] applies for most oxides grown in device applications. For oxidation in both pure oxygen and water vapor, the linear pressure dependence of the oxidation rate, based on the law of mass action, supports the idea that no dissociation or association takes place at the gas-oxide interface, thus implying that the diffusing species is molecular oxygen.

The flux of oxidant across the oxide layer is assumed to follow Fick's law

$$F_2 = -D\frac{dC}{dx} \tag{13.17}$$

at any point x within the oxide layer, where D is the diffusion coefficient and dC/dx is the concentration gradient of the oxidizing species within the oxide. From the assumption of steady-state oxidation, F_2 must be the same at all points in the SiO_2 or $dF_2/dx = 0$. As a result of the condition for steady state, the concentration of oxidant within the growing SiO_2 is linear, as shown in Fig. 13.11. The flux F_2 is then

$$F_2 = \frac{D(C_o - C_i)}{x_o} \tag{13.18}$$

where C_i is the concentration of oxidant at the $Si-SiO_2$ interface.

Finally, the flux representing the oxidation reaction occuring at the $Si-SiO_2$ interface is assumed to be proportional to the concentration C_i of oxidant there

$$F_3 = k_s C_i \tag{13.19}$$

where k_s is the chemical surface-reaction rate constant in units of cm/sec for the oxidation of silicon. The concentration of oxidant at the $Si-SiO_2$ interface and at the gas-oxide interface, respectively, is obtained by using the condition $F_1 = F_2$ and $F_2 = F_3$ and solving these two simultaneous equations for C_i and C_o. Thus

$$C_i = \frac{C^*}{1 + \dfrac{k_s}{h} + \dfrac{k_s x_o}{D}} \tag{13.20}$$

and

$$C_o = \frac{\left(1 + \dfrac{k_s x_o}{D}\right)C^*}{1 + \dfrac{k_s}{h} + \dfrac{k_s x_o}{D}}. \tag{13.21}$$

There are two limiting cases in the oxidation of silicon. In the first case, the term $k_s x_o/D$ in (13.20) and (13.21) is very large (i.e., $D \ll k_s x_o$), so that from (13.20) and (13.21), $C_i \rightarrow 0$ and $C_o \rightarrow C^*$. This case is diffusion controlled because the diffusion flux, governed by D, becomes small compared to the silicon surface reaction flux governed by k_s. Here, the rate of oxidation is limited by the availability of oxidant at the $Si-SiO_2$ interface, which is controlled by the diffusion process. The diffusion controlled case usually occurs when the oxide is thick, typically for oxide thicknesses of 40–100 Å for dry O_2 and up to 1000 Å for steam. In this case the time needed to grow a given thickness of oxide increases *parabolically* with thickness.

In the other limiting case the term $k_s x_o / D$ is very small. From (13.20) and (13.21), $C_i = C_o$. This case is concentration-reaction controlled. Now more than enough oxidant is transported across the oxide layer, and the oxidation rate is controlled by the concentration of oxidant at the silicon surface and by the rapidity with which the oxidation reaction can proceed. This situation occurs when the oxide layer is thin and for oxidation in steam. In this case the time needed to grow a given thickness of oxide increases *linearly* with thickness. The two limiting cases that lead to parabolic and linear growth are illustrated in Fig. 13.12.

Generally in integrated circuit technology, the oxidation kinetics are linear-parabolic; that is, neither diffusion nor the reaction rate is dominant, but both are important. An exception is the growth of field oxide, which is diffusion controlled because of its great thickness ($x_0 > 5000$ Å).

To calculate the oxide growth rate, which is of practical interest for a device fabricator, we define the quantity N_1, the number of oxidant molecules incorporated into a unit volume of oxide. There are 2.2×10^{22} SiO_2 molecules/cm^3 in the oxide, and one O_2 molecule is incorporated into each SiO_2 molecule, as seen from reaction (1). For oxidation in water vapor, two H_2O molecules are incorporated into each SiO_2 molecule, as seen from reaction (2). Thus N_1 will be 4.4×10^{22} cm^{-3} for oxidation in water vapor.

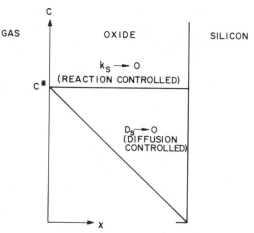

Fig. 13.12 Distribution of oxidant in the oxide layer for the reaction and diffusion limited cases. From Grove.[48]

Substituting (13.20) into (13.19), the flux of oxidant reaching the Si–SiO$_2$ interface is

$$N_1 \frac{dx_o}{dt} = F_3 = \frac{k_s C^*}{1 + k_s/h + k_s x_o/D}.$$ (13.22)

To solve this differential equation, a set of initial conditions is required. A general set is one in which the total oxide thickness x_o consists of two parts: (1) an initial layer of oxide of thickness x_i that might have been present on the silicon prior to the oxidation step and (2) the additional thickness grown. The initial condition for this second step is

$$x_o = x_i \quad at \quad t = 0.$$ (13.23)

Such a division of the oxide layer is important because it permits consideration of multiple oxidation steps and also because x_i can be regarded as the thickness of the layer grown before the approximations in this treatment become valid.

The solution of the differential equation, (13.22), subject to the boundary condition, (13.23), is

$$x_o^2 + A x_o = B(t + \tau)$$ (13.24)

where

$$A = 2D(k_s^{-1} + h^{-1}) \quad \text{(cm)}$$
$$B = 2DC^* N_1^{-1} \quad \text{(cm}^2\text{/sec)}$$

and

$$\tau \equiv \frac{x_i^2 + A x_i}{B}.$$

The quantity τ corresponds to a shift in the time coordinate that corrects for the presence of the initial oxide layer of thickness x_i. *Equation (13.24) is the general relationship from our model for the thermal oxidation of silicon.* The form of this mixed linear-parabolic relationship was first proposed by Evans.[49,50]

Solving the quadratic relationship (13.24) for x_o as a function of time t, we obtain

$$\frac{x_o}{A/2} = \left(1 + \frac{t + \tau}{A^2/4B}\right)^{1/2} - 1.$$ (13.25)

There are two limiting cases of (13.25). For long oxidation time (i.e., thick oxides) when $t \gg A^2/4B$ and $t \gg \tau$, (13.25) becomes

$$x_o^2 = Bt.$$ (13.26)

Equation (13.26) is called the *parabolic law* and B is called the *parabolic rate constant.* This limiting case is the diffusion-controlled case discussed previously. For short times (i.e., thin oxides) when $(t + \tau) \ll A^2/4B$, (13.25) becomes

$$x_o = \frac{B}{A}(t + \tau).$$ (13.27)

This relation is called the *linear law* and the quantity B/A is called the *linear rate constant*. The linear rate constant is related to the fundamental quantities by

$$\frac{B}{A} = \frac{k_s h C^*}{(k_s + h)N_1}.$$ (13.28)

This limiting case is the concentration-reaction-controlled case discussed previously.

13.4.2 Fit of the Model to Experiment

To see how well the general oxidation equation, (13.24), fits experiment, it is compared to the measurements made by Deal and Grove[10] of oxide thickness as a function of time at various oxidation temperatures.

The wet oxidizing atmosphere in these measurements was obtained by bubbling O_2 through a water bath held at 95°C, giving a partial pressure of water vapor of about 640 Torr. This oxidation amounts to oxidation in water rather than O_2 because the replacement of O_2 as the carrier gas with argon made no quantitative difference in the results. Temperature was controlled to ±1°C from 700 to 1200°C. Oxide thickness was measured by using multiple beam interferometric techniques (see Section 14.3).

For oxidation in water vapor, $x_i = 0$ at $t = 0$ at all temperatures. If (13.24) applies and $\tau = 0$, a plot of x_o versus t/x_o from this equation should yield a straight line with intercept $-A$ and slope B. Oxidation data for wet oxygen are plotted in Fig. 13.13. Figure 13.13 shows that x_o versus t/x_o yields straight lines with the absolute values of A increasing with decreasing temperature. At the same time, the slope B decreases with decreasing temperature. Values of A and B determined from Fig. 13.13 are listed in Table 13.1. The precision of these data was ±2% for B over the entire range of measurements. For larger values of A (low temperatures), the precision was ±2%. For higher temperatures, the precision of A was ± 12%.

In contrast to oxidation in wet oxygen, a plot of x_o versus t for oxidation in dry oxygen will not extrapolate to $x_o = 0$ at $t = 0$. To see what happens during the initial phase of oxidation in dry oxygen, measurements of x_o versus t at the low oxidation temperature of 700°C are shown in Fig. 13.14. For times of up to about 50 hr at 700°C, the initial phase of oxidation is faster than during the subsequent linear growth phase. The transition point between the initial and linear phases of oxidation is found by extrapolating the linear phase to $t = 0$. For the temperature range of 700–1200°C, the

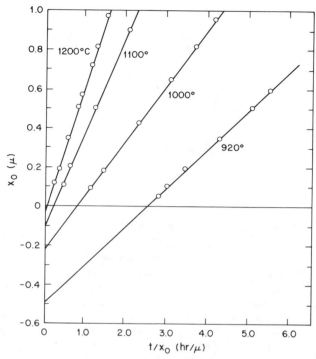

Fig. 13.13 Experimental curves of x_0 versus t/x_0 for oxidation in wet oxygen produced by slowly bubbling O_2 through a water bath held at 95°C. Slopes of the lines correspond to B in (13.24), and intercepts at $t/x_0 = 0$ correspond to $-A$ in (13.24). After Deal and Grove.[10]

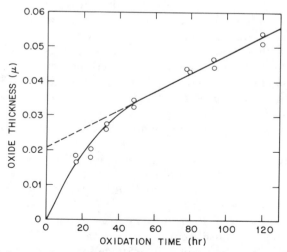

Fig. 13.14 Plot of x_0 versus t at 700°C in dry oxygen, showing the initial rapid growth phase followed by a linear growth phase. After Deal and Grove.[10]

Table 13.1 Rate Constants for Oxidation of Silicon in Wet Oxygen [$p_{H2O} = 640$ Torr; $x_o^2 + Ax_o = B(t + \tau)$][a]

Oxidation Temperature (°C)	A (μm)	B (μm²/hr)	B/A (μm/hr)	τ (hr)
1200	0.05	0.720	14.40	0
1100	0.11	0.510	4.64	0
1000	0.226	0.287	1.27	0
920	0.50	0.203	0.406	0

[a] After Deal and Grove.[10]

intercept is 230 ± 30 Å and is temperature independent. Thus, for oxidation in dry oxygen, the initial condition $x_i = 230$ Å must be used in (13.24). The corresponding values of τ can be estimated graphically by plotting x_o against t and extrapolating the linear phase through $x_o = 230$ Å to the time axis. The intercept of the extrapolated line at $t < 0$ will be equal to τ. This graphical procedure will work well at lower temperatures when there is a clear linear region. However, as temperature is increased and the curve becomes linear-parabolic or, at the highest temperatures, almost entirely parabolic, the linear part must be extracted from the data or from the definition of τ in (13.24). Although x_i is temperature independent, τ will vary with temperature [see (13.24)] because A and B are temperature dependent.

Figure 13.15 is a plot of x_o as a function of $(t + \tau)/x_o$ for oxidation in 1 atm of dry oxygen. Because oxidation at 800°C proceeds very slowly, values of $(t + \tau)/x_o$ are not included in Fig. 13.15. Therefore, only the line extrapolated from data taken at this temperature is shown in Fig. 13.15. The plots are straight lines again in Fig. 13.15 with intercept $-A$ and slope B. The resulting rate constants are given in Table 13.2, and the precision of the constants is the same as for wet oxidation.

According to (13.24), the parabolic rate constant B should be proportional to C^*, which, in turn, should be proportional to partial pressure of oxidant in the gas phase provided that Henry's law holds [see (13.16)]. However, A should be independent of partial pressure. From plots of x_o versus $(t + \tau)/x_o$ such as Fig. 13.14 but with pressure as parameter, A is found to be constant over a wide range of partial pressures. However, B is directly proportional to partial pressure at all experimental temperatures as predicted by (13.16) and (13.24). Both the linear pressure dependence of B and the pressure independence of A are evident in Fig. 13.16. The values of B and A, normalized to the 760 Torr (1 atm) value at the same temperature, are shown as functions of the partial pressure of the oxidizing species O_2 or H_2O, respectively, in the gas ambient.

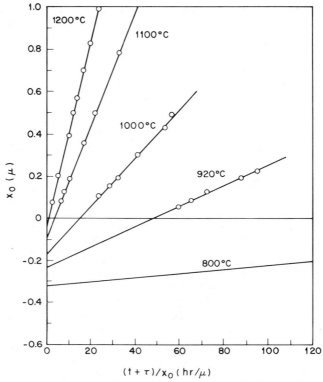

$(t + \tau)/x_0$ (hr/μ)

Fig. 13.15 Experimental curves of x_0 versus $(t + \tau)/x_0$ for oxidation in dry oxygen. Slopes of the lines correspond to B in (13.24), and the intercepts at $(t + \tau)/x_0$ correspond to $-A$ in (13.24). After Deal and Grove.[10]

Table 13.2 Rate Constants for Oxidation of Silicon in Dry Oxygen $[p_{O_2} = 760$ Torr; $x_0^2 + Ax_0 = B(t + \tau)]^a$

Oxidation Temperature (°C)	A (μm)	B (μm^2/hr)	B/A (μm/hr)	τ (hr)
1200	0.040	0.045	1.12	0.027
1100	0.090	0.27	0.30	0.076
1000	0.165	0.0117	0.071	0.34
920	0.235	0.0049	0.0208	1.40
800	1.340	0.0011	0.0030	9.0

^{a} After Deal and Grove.[10]

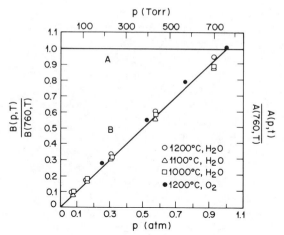

Fig. 13.16 Pressure dependence of A and B. Both A and B are normalized to their 760 Torr value at the same temperature. Data for wet and dry oxygen at 1100–1200°C are plotted from the data due to Flint.[53] After Deal and Grove.[10]

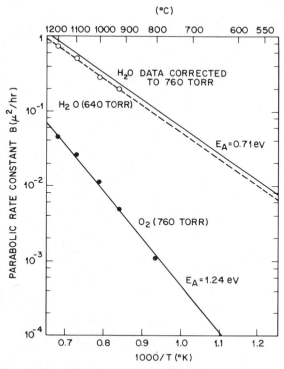

Fig. 13.17 Parabolic rate constant B as a function of $1/T$ for oxidation in both wet and dry oxygen. Activation energy for each oxidant is shown. After Deal and Grove.[10]

Because A is independent of pressure, B/A will have the same linear pressure dependence as B. Thus in the temperature range 1000–1200°C and the pressure range 0.1–1 atm, both the linear (B/A) and the parabolic (B) rate constants vary linearly with pressure for both H_2O and O_2. Henry's law, which predicts a linear pressure dependence of the oxidation rates, is valid over these temperature and pressure ranges.

In Fig. 13.17 log B is plotted against $1/T$ for both wet and dry oxidation. According to (13.24), the temperature dependence of B should be the same as that of D. The linear plot in Fig. 13.17 shows that B depends exponentially on temperature as does D. For dry O_2, the activation energy of B is 28.5 kcal/mole (1.24 eV) from Fig. 13.17, which is very close to the 27 kcal/mole (1.18 eV) for the diffusivity of O_2 through fused silica[11] shown in Fig. 13.18. The wet oxygen values in Fig. 13.17 are corrected from the experimentally used 640 Torr pressure to 760 Torr using the linear pressure dependence verified in Fig. 13.16. The activation energy for wet oxidation from Fig. 13.17 is 16.3 kcal/mole (0.7 eV), which is in good agreement with

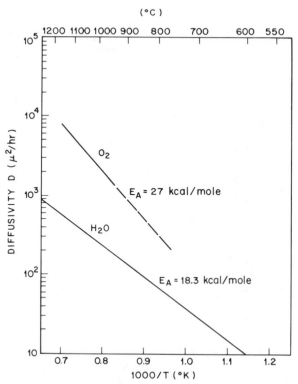

Fig. 13.18   Diffusivity of oxygen (from Norton[11]) and water (from Moulsen and Roberts[14]) in fused silica as a function of temperature. After Deal and Grove.[10]

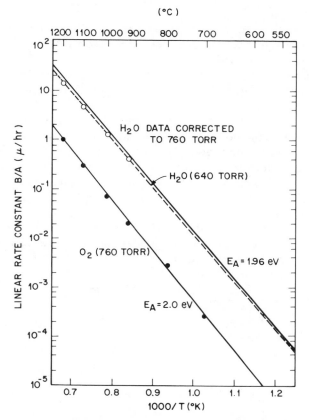

Fig. 13.19 Linear rate constant B/A as a function of $1/T$ for oxidation in wet and dry oxygen for a (111) silicon orientation. Activation energy for each oxidant is shown. After Deal and Grove.[10]

the value of 18.3 kcal/mole (0.8 eV) for the diffusivity of water in fused silica as a function of temperature shown in Fig. 13.18.

The temperature dependence of the linear rate constant B/A is shown in Fig. 13.19 to be exponential for both dry and wet oxidation. Figures 13.17 and 13.19 also show that oxidation rates measured at 1 atm follow the same exponential law over the entire temperature range 800–1200°C. Thus reduction of SiO_2 by silicon is negligible in these temperature and pressure ranges. In addition to oxidation reaction (1), SiO_2 can be reduced by silicon to form SiO, which diffuses away as a gas.[51,52] Such a reduction reaction would result in oxidation rates that increase more slowly with temperature than otherwise expected. However, reduction of SiO_2 by silicon is important only at temperatures near the melting point of silicon and at very low oxygen partial pressures.[52] In the temperature and pressure ranges generally used for oxidation, reduction of SiO_2 by silicon is negligible.

However, when the oxide is annealed in a nonoxidizing ambient, that is, low oxygen partial pressure at 1100°C or above, as is common in high temperature annealing, oxide thickness can be observed to decrease with time because of the reduction of SiO_2 by silicon.

From oxide thickness versus time, the rate constants obtained, as just explained from Figs. 13.13, 13.15, 13.17, and values of τ obtained as described in Section 13.4.2, we can illustrate the fit of (13.25) and its two limiting cases to a wide range of experimental data. This fit is done in Fig. 13.20. Figure 13.20 includes a large amount of experimental data taken by various experimenters under widely varying temperatures, partial pressures of oxidants, and both dry oxygen and water vapor as oxidants.[10,53–55] The solid lines in Fig. 13.20 represent the general relationship (13.25) and the dotted lines, the two limiting forms. From Fig. 13.20, the experimental results follow the predictions of our relatively simple model of oxidation kinetics over a wide range of experimental conditions. Thus (13.24) and (13.25) are valid descriptions of the oxidation kinetics and can be used to quantitatively predict oxide growth over most conditions met in practice.

The equilibrium concentration C^* of the oxidants O_2 and H_2O can be calculated by using (13.24) ($B = 2DC^*/N_1$). The parabolic rate constant B can be taken from Fig. 13.17 and diffusivities from Fig. 13.18. The resulting values of C^* at 1000°C are listed in Table 13.3.

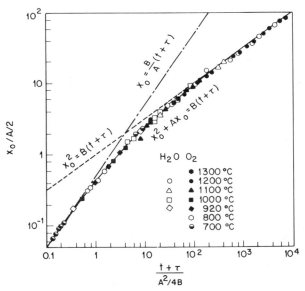

Fig. 13.20 General relationship for thermal oxidation of silicon. The solid line through the data points represents the general relationship (13.24); the dotted lines represent the two limiting forms. Experimental data were reduced using values of A and B determined from Figs. 13.16, 13.17, and 13.19. The values of τ correspond to $x_i = 0$ for wet oxygen and $x_i = 200$ Å for dry oxygen. After Deal and Grove.[10]

Table 13.3 Equilibrium Concentration of Oxidizing Species in SiO_2 at 1000°C and 1 atm[a]

Species	$C^*(cm^{-3})$
O_2	5.2×10^{16}
H_2O	3.0×10^{19}

[a] After Deal and Grove.[10]

At low pressures (0.1 Torr), the oxidation rate is governed by the gas flow rate.[44] However, the gas phase transport coefficient h is usually quite large at pressures of around 1 atm. Thus gas phase transport is not very important in controlling the oxidation rate, as is seen by considering the linear rate constant, which depends on both h and k_s. From (13.28), the linear rate constant is $B/A = k_s h C^* [(k_s + h) N_1]^{-1}$. If h is much larger than k_s, then $k_s h (k_s + h)^{-1} = k_s$. This result can be calculated from the known values of N_1 and C^* as previously described. The value of $k_s h (k_s + h)^{-1}$ is calculated to be 3.6×10^4 μm/hr for dry O_2 and 1.8×10^3 μm/hr for wet O_2 at 1000°C. The value of h, estimated on the basis of boundary layer considerations, is 10^8 μm/hr for typical gas flow conditions. Thus $k_s h (k_s + h)^{-1} = k_s$ and $B/A = k_s C^* N_1^{-1}$. The linear oxidation rate is then given by the relation

$$\nu = N_1 \frac{dx_o}{dt} = k_s C^* \qquad (13.29)$$

where ν is in molecules cm^{-2} sec^{-1}. The activation energies for B/A in Fig. 13.19 then reflect the temperature dependence of the interfacial reaction constant k_s. These values are in close agreement with the value of 42.2 kcal/mole (1.83 eV) required to break a Si—Si bond (as given by Pauling[56]).

13.4.3 Orientation Dependence of the Linear Oxidation Rate

The linear oxidation rate is observed to depend on the crystallographic orientation of the silicon surface[7,43] as expected because k_s in (13.29) depends on orientation. The values of B/A in Tables 13.1 and 13.2 were obtained for the (111) orientation. The parabolic rate constant B is independent of silicon surface orientation[43] as the parabolic oxidation rate is diffusion limited.

To see the experimentally determined dependence of the linear oxidation rate on orientation, it is convenient to express oxidation rate as a function of pressure. High pressure steam oxidation is better suited for measuring the linear oxidation rate than steam oxidation around atmospheric pressure because the steam oxidation rate at high pressure is linear for all temperatures and oxide thicknesses, whereas the oxidation rate is linear-parabolic at 1 atm steam. Therefore, 1 atm steam oxidation requires more care to ensure that measurements are made in the regime where the linear oxidation rate is dominant. The orientation dependence should be independent of pressure over a wide range because in the linear range, there always must be more water molecules available at the Si–SiO$_2$ interface than can react at any given time. Therefore, the results and the explanation should apply for high pressure as well as 1 atm steam oxidation. Substituting C^* from (13.16) into (13.29) yields

$$\nu = k_s k p_G \equiv \gamma p_G. \tag{13.30}$$

Table 13.4 shows $\bar{\gamma}$, the average value of γ, for three oxidations done at each of the temperatures listed using a different crystal in each case. Oxidation times were sufficiently long to yield silica film thicknesses of about 1500 Å at the lowest temperature to 10,000 Å at the highest temperature. Table 13.4 shows that the linear oxidation rate at all the temperatures listed follows the sequence $\nu_{110} > \nu_{311} > \nu_{111}$.

A further significant observation can be seen from Table 13.5, where $\bar{\gamma}_{311}/\bar{\gamma}_{110}$, $\bar{\gamma}_{111}/\bar{\gamma}_{110}$, and $\bar{\gamma}_{111}/\bar{\gamma}_{311}$ are listed as functions of temperature. The ratio $\bar{\gamma}_{111}/\bar{\gamma}_{311}$ remains approximately constant, with a mean value of 0.80 throughout the temperature range listed, whereas the ratios $\bar{\gamma}_{311}/\bar{\gamma}_{110}$ and $\bar{\gamma}_{111}/\bar{\gamma}_{110}$ tend to increase with increasing temperature. This observation suggests that the (311) and (111) planes oxidize with the same activation energy and that the activation energy for the (110) plane is smaller over the temperature range studied.

Ligenza[38] has proposed a plausible model to explain the observed silicon surface orientation dependence of the linear oxidation rate. This model is of importance not only because it attempts to explain the orientation dependence of the linear oxidation rate, but also because it describes the basic oxidation process itself. Ligenza[38] based his model on the results of high pressure (40–150 atm) and low temperature (500–800°C) steam oxidation.

Table 13.4   Dependence of Linear Oxidation Rate on Orientation[a]

Temperature (°C)	Pressure (atm)	$\bar{\gamma}_{110}$ (Å/min-atm)	$\bar{\gamma}_{311}$ (Å/min-atm)	$\bar{\gamma}_{111}$ (Å/min-atm)
500	150	0.0435	0.0277	0.0201
550	120	0.1041	0.0593	—
600	100	0.369	0.243	0.208
650	100	0.804	0.535	0.468
700	75	1.588	1.382	1.063
800	40	5.50	4.75	3.75

[a] After Ligenza.[38]

Table 13.5 Relative Values of Average Rate Constants[a]

Temperature (°C)	$\bar{\gamma}_{311}/\bar{\gamma}_{110}$	$\bar{\gamma}_{111}/\bar{\gamma}_{110}$	$\bar{\gamma}_{111}/\bar{\gamma}_{311}$
500	0.63	0.45	0.71
550	0.78	—	—
600	0.66	0.56	0.86
650	0.66	0.58	0.88
700	0.87	0.67	0.77
800	0.86	0.68	0.78

[a] After Ligenza.[38]

The simplest model for the oxidation reaction is a direct reaction between a water molecule in the silica and a silicon-silicon bond at the Si–SiO$_2$ interface, over which the H$_2$O is positioned, to form an Si—O—Si bridge and a hydrogen molecule. This reaction is one of the elementary chemical reactions involved in the oxidation of silicon by water. The reaction is

$$\begin{array}{cc} H & H \\ \diagdown & \diagup \\ & O \end{array}$$
$$H_2O(\text{silica}) + Si—Si \rightarrow Si—Si \rightarrow Si—O—Si + H_2(\text{silica}) \qquad (8)$$

where H$_2$O(silica) is the water dissolved in the silica,* Si—Si is the reaction site,

$$\begin{array}{cc} H & H \\ \diagdown & \diagup \\ & O \\ & Si—Si \end{array}$$

is the activated complex formed by the water molecule in position over an interfacial silicon–silicon bond, Si—O—Si is the bridge, and H$_2$(silica) is the hydrogen molecule dissolved in the silica. The H$_2$(silica) forms at the Si–SiO$_2$ interface at a rate equal to the oxidation rate and then rapidly diffuses away. The picture of the interface based on reaction (8) is one in which all the silicon atoms are partially bonded to silicon atoms in the substrate below and to oxygen atoms in the silica layer above.

*At high pressures, water molecules exist in the silica in significant numbers; at low pressures and high temperatures, water molecules exist in far smaller numbers.

The model for the orientation dependence of the linear oxidation rate is based on reaction (8). We derive a quantitative expression from (8) for the linear oxidation rate to relate the model to experiment. The concentration of water in the silica is C^*, and the concentration of *available* silicon–silicon bonds per square centimeter is N. The rate at which reaction (8) proceeds is proportional to C^*N. The oxidation rate is then

$$\nu = C^*Nk \exp\left(-\frac{E_A}{RT}\right) \tag{13.31}$$

where the factor of proportionality is $k \exp(-E_A/RT)$, E_A now is the activation energy for oxidation. The quantity kN in (13.31) can be interpreted as the linear oxidation rate constant B/A, where the orientation dependence is embodied in the factor N. From (13.7) and (13.9), we obtain

$$C^* = \frac{p_G L_o}{RT} \exp\frac{S}{R} \exp\left(-\frac{H}{RT}\right). \tag{13.32}$$

Combining (13.31) and (13.32), we obtain

$$\nu = \frac{k L_o \exp(S/R)}{RT} p_G N \exp\left(-\frac{E_A + H}{RT}\right). \tag{13.33}$$

The orientation dependence of ν arises from the orientation dependence of N and E_A. The quantity N is not simply the concentration of silicon–silicon bonds in the surface plane, which would be twice the surface atom concentration, but is a smaller quantity because not all the surface silicon bonds are available as reaction sites in any given instant of time. The reason is that the silicon-silicon bond is directional, so that its availability as a reaction site at any given instant depends on its angle relative to the surface plane, the positions of adjacent silicon atoms, and the dimensions of the water molecule. The water molecule is large enough that when reacting with some angled silicon-silicon bonds, it screens adjacent silicon-silicon bonds from other water molecules during the time it takes to complete the reaction. In other cases the water molecule will not be able to reach a surface silicon-silicon bond because the bond is blocked by adjacent silicon atoms. Access is blocked until sufficient oxidation of the more easily reached bonds in the vicinity exposes the bond in question for reaction. These geometric effects, which limit the availability of reaction sites, are called *steric hindrance.*

Steric hindrance causes the linear oxidation rate to be orientation dependent. Steric hindrance also causes a polished surface, which is not a true plane but has hills and valleys about 100 Å in size, to be smoothed into a plane by oxidation. Protruding silicon atoms in the hills oxidize more rapidly than do those in the valleys, as their bonds are more available, until a steric hindrance characteristic of a planar surface becomes controlling.

These statements are supported by experiment. The boundary between silicon and its oxide is found to be very sharp and very straight by cross-sectional transmission electron microscopy[57] (see Section 16.3). Structural defects at the silicon surface will not affect the oxidation rate unless a significant fraction of the surface atoms are involved, which is unlikely in device grade silicon.

Figure 13.21a shows how the concentration of available reaction sites can be determined on a (110) surface, and Fig. 13.21b shows how the concentration of available reaction sites can be determined on a (111) surface. The shaded circles in Fig. 13.21 represent silicon atoms in the interfacial plane bonded to oxygen atoms in the silica structure above them by bonds labeled "B" and bonded to silicon atoms in the silicon lattice beneath them represented by the unshaded circles by bonds labeled "C".

Bond A in Fig. 13.21a is a silicon-silicon bond parallel to the surface plane between two interface atoms. A water molecule represented by a dashed circle is shown in a position to react with an A bond in the (110) plane. Only four water molecules can be placed in a position to react with A bonds within the unit cell of the (110) plane represented by the straight dashed lines in Fig. 13.21a. Interfacial silicon atoms in the adjacent row (outside the unit cell) present no available bonds and prevent a water molecule from reaching an angled C bond until all the available A bonds in the immediate vicinity have reacted.

In Fig. 13.21b there are no parallel A bonds. A water molecule is shown in a position to react with a C bond. The two adjacent C bonds are screened by this water molecule from other water molecules that might come along. Thus at any given instant, only three water molecules can be positioned over three silicon-silicon bonds in the (111) plane of the unit cell, indicated in Fig. 13.21b by the straight dashed lines. The area of a unit cell in the (111) plane is $\frac{1}{2}\sqrt{3a^2}$, where a is the silicon lattice parameter equal to 5.431 Å. The total area of this unit cell is thus 2.54×10^{-15} cm^2. There are three reaction sites within this area, so there are 1.176×10^{15} available reaction sites/cm^2 in a (111) surface.

The density of available reaction sites for other orientations and silicon-silicon bond densities are calculated for comparison in this manner and listed in Table 13.6. Therefore, Table 13.6 gives the dependence on orientation of N in the preexponential constant of (13.33) for the linear oxidation rate.

To complete the picture, the orientation dependence of E_A in (13.33) also must be considered. The ease of formation of the activated complex between a water molecule and a silicon-silicon bond will depend on the relative position of the two.

Steric hindrance results in a higher activation energy. Consider the parallel bonds such as A in Fig. 13.21a, which have the most favorable positions on the surface for the formation of the activated complex. These bonds require the least energy to form. Silicon-silicon bonds that make an angle with the surface such as the C bonds in Fig. 13.21a have a less

Table 13.6 Calculated Properties of Four Silicon Crystal Planes (Lattice Parameter, $a = 5.431 \times 10^{-8}$ cm)[a]

Orientation	Area of Unit Cell (cm²)	Si Atoms in Area	Si Bonds in Area	Bond Density ($\times 10^{15}$ cm⁻²)	Available Bonds in Area	Available Bond Density N ($\times 10^{15}$ cm⁻²)	N Relative to (110)
(110)	$\sqrt{2}\,a^2$	4	8	1.918	4	0.959	1.000
(311)	$\frac{1}{8}\sqrt{11}\,a^2$	1.5	3	2.454	2	1.636	1.707
(111)	$\frac{1}{2}\sqrt{3}\,a^2$	2	4	1.568	3	1.176	1.227
(100)	a^2	2	4	1.355	2	0.677	0.707

[a] After Ligenza.[38]

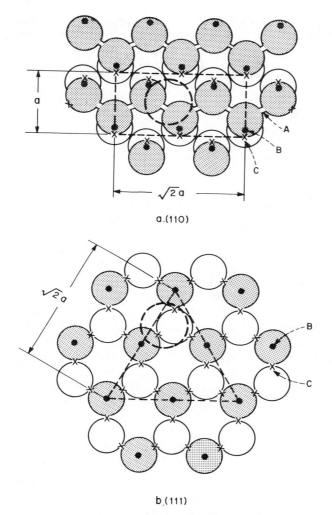

a.(110)

b.(111)

Fig. 13.21 (a) Silica-silicon interface in the (110) plane; (b) silica-silicon interface in the (111) plane. In both (a) and (b) the dimensions of the plane through the unit cell are indicated by straight dashed lines. The lattice parameter a is 5.431×10^{-8} cm. After Ligenza.[38] Copyright (1961), American Chemical Society.

favorable position for the formation of the activated complex because of other silicon atoms in the vicinity and require more energy to form.

From such arguments, a sequence of activation energies can be qualitatively determined from Table 13.6. Four of the eight bonds in the (110) plane and one of the three bonds in the (311) plane are parallel bonds. There are no parallel bonds in either the (111) or (100) planes, and the silicon-silicon bonds in the (100) plane make a larger angle with the surface

Table 13.7   Preexponential Constants and Apparent Activation Energies[a]

Orientation	$H + E_A$ (kcal/mole)	Preexponential Constant (molecule-°K-cm^{-2} -sec^{-1}-atm^{-1})	Preexponential Constant Relative to (110)
(110)	28.4 ± 0.5	$1.2 \pm 0.4 \times 10^{22}$	1.00
(311)	29.9 ± 0.7	$2.1 \pm 0.8 \times 10^{22}$	1.7
(111)	29.8 ± 0.7	$1.6 \pm 0.7 \times 10^{22}$	1.3

[a] After Ligenza.[38]

plane than do those in the (111) plane. From these arguments, the activation energies should follow the sequence $E_A(100) > E_A(111) > E_A(311) > E_A(110)$.

To check the prediction of this sequence and the sequence of N listed in Table 13.6 against experiment, the measured values of γ listed in Table 13.4 were analyzed by the method of least squares according to (13.33) to obtain a preexponential constant $[kL_o \exp(S/R)/R]N$ and the apparent activation energy $E'_A = H + E_A$. These quantities are listed in Table 13.7. The calculated sequence of the preexponential constant in the series (110):(311):(111) is 1.00:1.707:1.227 from Table 13.6, and the experimental sequence is 1.0:1.7:1.3 from Table 13.7. Quantitative agreement between the calculated and experimental sequence for N is quite good. The sequence of the apparent activation energies in Table 13.7 bears out the directly measured sequence listed in Table 13.5, where the activation energies for the (311) and (111) planes were identical and larger than those for the (100) plane. Moreover, the predicted sequence of the activation energies, which was $E_A(111) > E_A(311) > E_A(110)$, is close to the experimental sequence $E_A(111) = E_A(311) > E_A(110)$ from Tables 13.5 and 13.7.

The consistency of the experimental data with the calculations supports the assumption that the linear oxidation rate is the rate of reaction between water molecules and silicon-silicon bonds at the Si–SiO$_2$ interface. The calculation in Table 13.6 shows that N for the (100) plane is the smallest for the four planes listed; hence this plane would have the smallest preexponential constant. Furthermore, the (100) plane is predicted to have the largest activation energy. Therefore, the linear oxidation rate of the (100) plane would be the smallest, and the sequence would be $\nu_{110} > \nu_{311} > \nu_{111} > \nu_{100}$.

By measuring linear oxidation rate in one atmosphere steam, Pliskin[43] found that the (100) plane oxidized most slowly. However, the sequence he found was $\nu_{111} > \nu_{110} > \nu_{311} > \nu_{100}$. The measurements by Pliskin[43] were done at a significantly higher temperature (981°C) than by Ligenza,[38] which were done at 500–800°C. Pliskin[43] suggests that this discrepancy can be resolved by measuring linear oxidation rate at 1 atm of steam in the temperature range 500–800°C. This argument implies that the activation energies, which are temperature dependent, as seen from Table 13.5, will change sufficiently to make the low temperature sequence[38] $\nu_{110} > \nu_{311} > \nu_{111} > \nu_{100}$ become the high temperature sequence[43] $\nu_{111} > \nu_{110} > \nu_{311} > \nu_{100}$. Therefore, the activation energy sequence may change at higher temperatures. That is, the rate of the limiting surface reaction may depend on temperature (see Section 13.4.3).

The linear oxidation rate for oxidation in pure oxygen also depends on orientation because the oxygen molecule, like the water molecule, reacts directly with silicon-bonds at the $Si–SiO_2$ interface to form the oxide, and this reaction is similarly subject to steric hindrance. Irene[58] found that the linear oxidation rate in oxygen contaminated with only 1 ppm water vapor follows the sequence $\nu_{110} > \nu_{111} > \nu_{100}$, which agrees reasonably well with the sequence for water vapor found by both Ligenza[38] and Pliskin.[43]

Because the oxygen molecule is larger than the water molecule, the effect of steric hindrance on the activation energy for linear oxidation might be expected to be different in certain temperature ranges. However, the activation energy of the linear oxidation rate on a (111) surface obtained from Fig. 13.19 is about the same for both O_2 and H_2O in the temperature range 700–1200°C. Deal and Grove[10] measured the water vapor contamination during their oxidations in 1 atm oxygen to be 5 ppm or less. Similarly, Irene[58] found, at temperatures of 800–996°C, that the activation energy for the linear oxidation rate at 1 atm oxygen pressure contaminated with 1 ppm water vapor was about the same, although oxidation rates were larger for the higher level of water vapor contamination. In the results from both Deal and Grove[10] (Fig. 13.19) and Irene,[58] the activation energies found agree closely with the expected silicon-silicon bond energy. One conclusion from these results is that relative sizes of oxygen and water molecules do not make a significant difference in activation energy at temperatures above 700°C. Another conclusion is that even the small amounts of water vapor contamination present in these experiments is sufficient to mask any differences in steric hindrance between water and oxygen molecules. This conjecture is a likely one because even small quantities of water vapor contamination have been shown, as discussed in Section 13.4.5(a), to have a measurable effect on the linear oxidation rate. The experiments cited represent careful work; thus it will be quite difficult to measure differences, if any, in linear oxidation rates between O_2 and H_2O caused by steric hindrance to settle this point. However, from a practical point of view, the activation energies are the same, equal to the silicon-silicon bond energy.

13.4.4 Applications and Implications

There is an upper limit to how thick the oxide can be made. At thicknesses much greater than 1 μm, the oxide will tend to crack and be of poor quality. This poor quality results from the long time at elevated temperature required for oxidation. This long time allows the silica to crystallize or to devitrify. For oxide thicknesses greater than a few tenths of a micron, such as for a field oxide or for oxide isolation, steam oxidation commonly is used. However, oxides several microns thick can be grown without adverse effects in high pressure steam as described in Section 14.6.1.

Figures 13.17, 13.19, and 13.22 show that oxidation proceeds at a much faster rate in H_2O than in O_2 in both the linear and parabolic ranges. Figure 13.22 shows oxide thickness as a function of time with oxidation temperature as a parameter for oxidation in wet (95°C H_2O) oxygen (solid lines) and dry oxygen (dotted lines). These curves are typical curves useful for growing an oxide layer in device applications. The curves in Fig. 13.22 are described quantitatively by the oxidation kinetics obeying (13.24). Comparing the curves for oxidation in O_2 to those for oxidation in H_2O, a given thickness of oxide grows faster at a given temperature in steam than in dry O_2. This effect is also shown in Figs. 13.17 and 13.19. However, Fig. 13.18

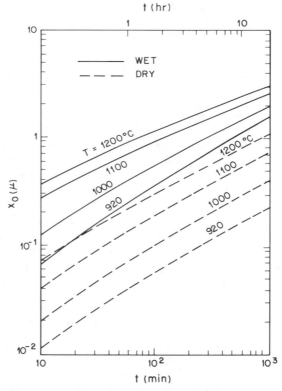

Fig. 13.22 Oxide thickness as a function of time with temperature as parameter. Solid lines represent oxygen; dotted lines represent steam. Silicon surface orientation is (111). From Wolfe,[59] based on the data due to Deal and Grove.[10]

shows that the diffusivity of H_2O in SiO_2 at any given temperature is less than the diffusivity of O_2 in SiO_2. This seeming paradox is resolved when it is recalled that the flux of oxidant and hence B is proportional to C^* [see (13.24)]. This concentration is three decades larger for H_2O than for O_2 from Table 13.3, while $D(H_2O) = 0.1D(O_2)$. It is the *greater equilibrium concentration* of water than of O_2 in the oxide, rather than the lower diffusivity, which is the major reason for the faster oxidation rate in the parabolic range. Another factor that contributes to faster oxidation in steam is that the linear rate of oxidation dominates to thicknesses of about 1000 Å in steam, whereas in oxygen alone the linear rate dominates only for the first 40–100 Å. Comparison of Figs. 13.17 and 13.19 shows that the linear oxidation rate for steam is considerably higher than the parabolic oxidation rate for oxygen at a given oxidation temperature. Because 1000 Å represents a large fraction of the oxide thickness usually grown for device application, a given oxide thickness can be grown faster in steam than in oxygen.

In the linear range Fig. 13.19 shows that the linear rate constant B/A is larger for oxidation in H_2O than in O_2 over the temperature range covered. No difference in activation energies for each oxidant have been experimentally detected, as discussed in Section 13.4.3. However, B/A [given by (13.28)] also depends on C^*, which, from Table 13.3, is much larger for H_2O than O_2. Thus the faster linear oxidation rate in H_2O also is due to the *greater equilibrium concentration* of H_2O in SiO_2.

For all practical purposes, the silica film can be considered to be stoichiometric throughout its entire thickness. The truth of this statement can be appreciated from Table 13.3, which shows that the equilibrium concentration of oxygen or water molecules in SiO_2 at 1000°C and 1 atm pressure is very small. The actual concentration of these molecules in the silica film must be even smaller because the concentration of oxidant at the gas-SiO_2 interface must be somewhat smaller than the equilibrium value for oxidation to proceed. Then the concentration of oxidant decreases from its value at the gas-SiO_2 interface through the silica film to the Si–SiO_2 interface as shown in Fig. 13.11. Therefore, the overall concentration of oxidant throughout the silica film is considerably less than the equilibrium concentrations in Table 13.3. Another fact evident from Table 13.3 is that neither oxygen nor water is very soluble in SiO_2, even at 1000°C and a pressure of 1 atm. This fact can be seen by comparing the equilibrium concentration of oxygen and water molecules listed in Table 13.3 with the concentration of SiO_2 molecules in the silica film, which is $2.2 \times 10^{22} \text{ cm}^{-3}$.

13.4.5   Influence of Impurities on Oxidation Rates

Both linear and parabolic oxidation rates are extremely sensitive to impurities. The impurities of greatest practical significance are (1) water vapor, because it is so prevalent and is such an important oxidant, (2) sodium, because it is a universal contaminant, (3) boron and phosphorus in high concentrations, because they are widely used as dopants in the silicon, and (4) chlorine, because it is sometimes used in the oxidation process to control sodium contamination. Oxidation rates in oxygen will depend on how effectively contamination by water vapor and sodium are kept under control, and oxidation rates in water vapor will depend on how effectively contamination by sodium is kept under control. Thus the accuracy of measuring oxidation rate constants depends on how well contamination by impurities is controlled. Moreover, oxidation rate constants, particularly for oxidation in wet oxygen or steam, where oxidation rates are most rapid, and at short oxidation times will vary in magnitude with how fast the silicon reaches oxidation temperature and how fast it is cooled below oxidation temperature. Because the degree to which these factors were taken into account by various workers was different, there is little agreement to be found in the literature on the values of oxidation rates and quantities such as oxidation rate constants and activation energies derived from the measured oxidation rates. In this light, the oxidation rate constants in Tables 13.1 and 13.2 can be regarded as typical of a well-controlled oxidation process used in practical integrated circuit fabrication. However, if a high degree of oxide thickness control is desired (<10%), the oxidation process should be calibrated by measuring thickness as a function of time at the temperature desired. The rest of this section will deal with the effects on oxidation rate of water vapor, sodium, boron, phosphorus, and Cl_2 or HCl.

(a) Water

Water has been discussed as an oxidant. Now, we treat it as an impurity in the oxygen gas stream in the "dry" oxidation process. It is difficult to measure accurately either the linear or parabolic oxidation rates for pure oxygen because it is nearly impossible, even with the utmost care, to reduce the amount of water vapor in the oxidation furnace to such low levels that oxidation rates are unaffected. The best that can be done is to measure the oxidation rates at the lowest level of water vapor contamination it is practical to achieve. Attempts have been made to control and minimize water contamination. Burkhardt and Gregor[44] and Revesz and Evans[46] measured oxidation rates in oxygen contaminated with less than 0.1 ppm of water vapor, whereas Irene[58] measured oxidation rates in oxygen contaminated with 1 ppm of water vapor. The extreme sensitivity of the oxidation rates to small amounts of water vapor contamination was quantitatively demonstrated by Irene,[58] who found that only 25 ppm of water vapor deliberately added to the oxygen stream resulted in measurable increases in both oxidation rates compared to a water vapor contamination of 1 ppm at temperatures of 800–996°C. It should not be surprising that such small quantities of water vapor contamination influence oxidation rates because both the linear and parabolic oxidation rates are significantly higher in water vapor than in oxygen for reasons described in Section 13.4.4.

To measure oxidation rates with 1 ppm or less of water vapor contamination, water vapor from three major sources must be minimized. The first source is water vapor contamination of the oxygen gas. The second source is hydrogen and hydrocarbon vapor, which also are common contaminants of oxygen gas. The problem with hydrogen and hydrocarbon vapors is that they oxidize in the high temperature oxygen ambient in the oxidation furnace to form water vapor and CO_2. Water vapor from these two sources can be minimized by passing the contaminated oxygen first through a preburner to oxidize the hydrogen and hydrocarbon vapors forming water and then through a cold trap or other drier to remove the water vapor before it enters the oxidation furnace. This procedure will be of no avail if the third source of water vapor is not minimized. This third source of water vapor contamination is the diffusion of water molecules from the room ambient through the fused quartz walls of the furnace tube to the oxidation zone. It is well known that water molecules readily diffuse through fused quartz at oxidation temperatures normally used (800–1300°C) in sufficient quantity to affect oxidation rates.[60] Burkhardt and Gregor[44] and Irene[58] minimized this source of water vapor by using a double-walled fused quartz furnace tube with inert gas such as argon or nitrogen flowing between the inner and outer tubes. This gas flow sweeps away most of the water molecules diffusing from the room ambient through the walls of the outer fused quartz tube. Revesz and Evans[46] went a step further, using a double-walled fused quartz tube with nitrogen cooling and RF heating of a pedestal holding the silicon sample, to confine the heating to the sample. None of these methods is necessary in manufacture, but these methods were important in attempts to establish that both oxidation rates at a given temperature and pressure are smallest in the driest attainable oxygen ambient.

(b) Sodium

Sodium is another impurity that, if present in the oxide at high concentrations, can increase both the linear and parabolic oxidation rates. The mechanism for these increased oxidation rates is that some of the silicon-oxygen bridges (Si—O—Si) must be broken to form Si—O—Na to allow sodium to enter the oxide. These broken silicon-oxygen bridges make it easier for oxygen molecules to enter the oxide also because less energy is required. As a result, both diffusion and the concentration of oxygen molecules in the oxide are enhanced. Revesz and Evans[46] found that doping the oxide with about 10^{20} sodium atoms/cm^3 resulted in an increase of both linear and parabolic oxidation rates by a factor of 2 or so over undoped controls in the temperature range 900–1200°C. Controlling water vapor concentration in the oxidation furnace to better than 0.1 ppm ensured that the effect of sodium on oxidation rates would be dominant and not masked by the strong influence of water vapor. As discussed in Section 15.3, such a large sodium concentration would result in intolerably poor device stability. Therefore, sodium levels in a well-controlled oxidation furnace are kept to low levels;* thus the effect of sodium contamination on oxidation rates is negligible.

(c) Boron and Phosphorus

Boron and phosphorus are two more impurities that, in very high concentrations, cause both the linear and parabolic oxidation rates to increase. The silicon must be doped to degeneracy with these dopant impurities before an increase in the oxidation rates can be observed. Therefore, the effect of boron and phosphorus on oxidation rates will be important over p^+n and n^+p planar junctions, where the oxide will be thicker than elsewhere on the wafer.

Ligenza[61] showed that the linear oxidation rate of silicon doped to degeneracy with phosphorus is larger than for silicon lightly doped with phosphorus at steam pressures of 25–500 atm and temperatures of 500–850°C. Deal and Sklar[62] and Pliskin[43] showed that both the linear and parabolic oxidation rates of silicon doped to degeneracy with either boron or phosphorus are larger than for silicon lightly doped with these impurities. Deal and Sklar[62] performed their experiments in 1 atm of dry and wet oxygen at temperatures of 920–1200°C. Figure 13.23a illustrates the dependence of the oxidation rate $U_{B,P}$ on impurity concentration in the silicon bulk for oxidation in dry oxygen at 1200°C. Figure 13.23b illustrates the dependence of the oxidation rate on the average impurity concentration in the silica film, with all other conditions the same as in Fig. 13.23a. A distinction is made between impurities in the silicon bulk and in the silica film because the solubility of boron is greater in silica than in silicon, so that the concentration of boron at the silicon surface drops below the level in the silicon bulk whereas the concentration of boron increases on the silica side of the Si–SiO$_2$ interface during oxidation. Phosphorus is more soluble in silicon than in silica, so its concentration increases at the silicon surface compared to the bulk. Very little phosphorus is incorporated in the silica during oxidation. The segregation of boron and phosphorus in the oxide and in the silicon is an important effect, discussed in Section 14.4.2.

Figures 13.24a,b show results for oxidation in wet oxygen at 920°C. Water vapor was introduced into the oxygen stream by slowly bubbling oxygen through a water bath held at 95°C before it entered the oxidation furnace. In Fig. 13.24a $U_{B,P}/U_{STD}$ is plotted against impurity concentration in the silicon bulk and in Fig. 13.24b, against impurity concentration at the silicon surface. If the results in Fig. 13.24b were plotted against average

---

*The term "low levels" here means about $10^{10}\,\mathrm{cm}^{-2}$ or less electrically active sodium and considerably more that is electrically inactive. The electrically active sodium is detected by MOS capacitor measurements described in Section 10.2.3.

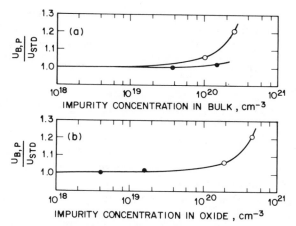

Fig. 13.23 (a) Dependence of normalized oxidation rate constant on impurity concentration in the bulk silicon and (b) on impurity concentration in the silica. Open circles represent boron; and black circles represent phosphorus. The value U_{STD} is the oxidation rate for an impurity concentration of 10^{16} cm^{-3}. Data are based on oxidation in dry oxygen at 1200°C. After Deal and Sklar.[62] Reprinted by permission of the publisher. The Electrochemical Society, Inc.

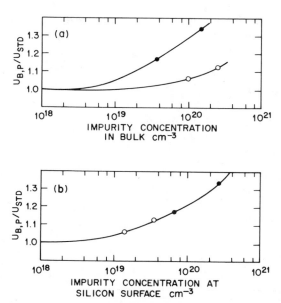

Fig. 13.24 (a) Dependence of normalized oxidation rate on impurity concentration in the bulk silicon and (b) on impurity concentration at the silicon surface. Open circles represent boron; black circles represent phosphorus. Data are based on oxidation in wet oxygen (95°C H$_2$O) at 920°C. After Deal and Sklar.[62] Reprinted by permission of the publisher, The Electrochemical Society, Inc.

695

impurity concentration in the oxide, the resulting curves would diverge even more than they do in this figure.

The oxidation rate of heavily doped silicon is chiefly dependent on the impurity concentration, either in the silica for diffusion limited oxidation or at the silicon surface for reaction limited oxidation. This statement is verified by the observation that wet oxygen data at 1200°C could be included in Fig. 13.23 for dry oxygen and dry oxygen data at 920°C in Fig. 13.24 for wet oxygen with reasonable agreement with the existing curves shown. That is, the curves in Figs. 13.23 and 13.24 depend not on water concentration in the oxide, but only on dopant impurity concentration.

The explanation of the increase in oxidation rate for increasing boron concentration is that boron is preferentially incorporated into the silica during oxide growth, thereby weakening the bond structure of the silica analogous to sodium; that is, silicon-oxygen bonds are broken. Therefore, oxygen or water molecules enter the silica more easily and diffuse through it more rapidly. The boron concentration in wet oxides is less than in drier oxides, so the increase in the parabolic oxidation rate is greater for dry oxides. This effect can be seen by comparing the boron curve in Fig. 13.23a to the boron curve in Fig. 13.24a at the highest concentration. This comparison is quite valid even though the boron curves in Fig. 13.23 and Fig. 13.24 are taken at different temperatures because the ratio U_B/U_{STD} is nearly temperature independent, as shown in Table 13.8.

The cause of the increased oxidation rate with increasing phosphorus concentration is different than for boron or sodium because very little phosphorus is incorporated into the silica layer during growth. Therefore,

Table 13.8 Rate Constants for Boron and Phosphorus

Oxidation Ambient	Oxidation Temperature (°C)	Standard Rate Constant $U_{STD}(\mu m/hr^{1/2})$	Boron Doped Silicon U_B/U_{STD}	Phosphorus Doped Silicon U_P/U_{STD}
Dry O_2	1200	0.206	1.21	1.02
Dry O_2	1100	0.153	1.29	1.03
Dry O_2	1000	0.093	1.31	1.03
Dry O_2	920	0.050	1.20	1.18
Wet O_2	1200	0.840	1.11	1.00
Wet O_2	1100	0.675	1.10	1.00
Wet O_2	1000	0.480	1.11	1.09
Wet O_2	920	0.325	1.13	1.30

[a] After Deal and Sklar.[62]
[b] Constants: U_B, rate constant for a boron concentration of 2.5×10^{20} cm^{-3}; U_{STD}, rate constant for a boron concentration of 10^{16} cm^{-3}; U_P, rate constant for a phosphorus concentration of 1.5×10^{20} cm^{-3}; U_{STD}, rate constant for a phosphorus concentration of 4×10^{15} cm^{-3}

the ease of entry and diffusion of oxidant are not significantly affected. There is no significant effect of phosphorus when oxidation is primarily diffusion limited, but when oxidation is mainly reaction limited or is linear-parabolic, there is an effect. Figure 13.25 shows that in dry oxygen at 1200°C oxidation is diffusion limited because the slope of the curve is one-half and there is virtually no dependence of oxidation rate on phosphorous concentration. Figure 13.25 also shows that at 920°C oxidation is mixed linear-parabolic because the slope is between one-half and unity and there is a marked dependence on phosphorus concentration. Figures 13.23 and 13.24 summarize the effect. At 1200°C there is hardly any effect on the parabolic oxidation rate due to phosphorus, but at 920°C there is marked effect at higher concentrations.

The concentration of boron atoms in the silica must become comparable (i.e., 1% or so) to the concentration of SiO_2 molecules before the diffusion rate and equilibrium concentration of oxygen in the silica become affected. Therefore, the parabolic oxidation rate is not affected except at high boron concentrations initially in the silicon. Similarly, the concentration of phosphorus atoms must become comparable to the number of silicon surface atoms before the linear oxidation reaction rate becomes affected. Thus, there is an effect whenever there is a high initial concentration of phosphorus in the silicon.

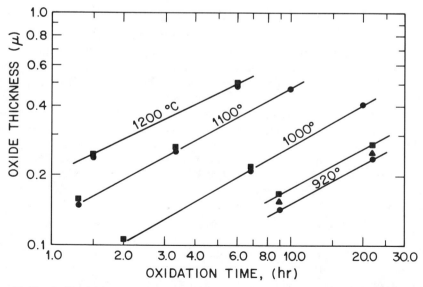

Fig. 13.25 Oxide thickness versus oxidation time for oxidation of phosphorus doped silicon in dry oxygen for three different phosphorus concentrations with temperature as parameter. Squares, $C_B = 1.5 \times 10^{20}$ cm^{-3}; triangles, $C_B = 3.7 \times 10^{19}$ cm^{-3}; circles, $C_B = 4.0 \times 10^{15}$ cm^{-3}. After Deal and Sklar.[62]

Ho et al.[84] measured an increase in the linear oxidation rate constant of more than an order of magnitude as the concentration of phosphorus in the silicon increased from 10^{19} cm^{-3} toward 10^{21} cm^{-3}. The parabolic rate constant increased only slightly for the same increase in phosphorus concentration. Phosphorus at the silicon surface causes an increase in the surface reaction rate, thereby increasing the linear oxidation rate. A mechanism for this increase was proposed by Ho and Plummer.[85]

The thermal oxidation rate of polysilicon is greater than single-crystal silicon because it is usually heavily doped with either phosphorus or boron and because it contains grain boundaries.* Polysilicon is deposited over thermally grown SiO$_2$ to act as a gate electrode in a MOSFET and to interconnect circuit elements in integrated circuits. It is, then, thermally oxidized for protection and to provide insulation for a second layer of metallization over it. The polysilicon must be heavily doped to make it a good enough conductor to function as a gate and interconnecting lead because polysilicon has a high resistivity (3×10^6 Ω-cm) as deposited. In n-channel technology the polysilicon is heavily doped with phosphorus during the source and drain diffusion, whereas in p-channel technology it is heavily doped with boron during the source and drain diffusion.

Thermal oxide grown on polysilicon has a higher electrical conductivity than thermal oxide grown on bulk silicon,[63–66] which may be an important factor in some applications.

(d) Chlorine

Chlorine is another impurity that has a large effect on oxidation rate. The presence of chlorine in the oxidation ambient increases the oxidation rate. Chlorine is deliberately added to the oxidation ambient to grow an oxide that has a minimum sodium ion concentration, a reduced interface trap density, and an increased dielectric breakdown strength. Because these factors are beneficial, resulting in increased device performance and stability, addition of chlorine to the oxidation ambient is a widespread practice. In this section we concentrate on the influence of chlorine on oxidation rate.

Chlorine is added to a dry oxygen carrier gas in small amounts, up to 6% by volume, either as chlorine gas, anhydrous HCl that oxidizes to form chlorine and water vapor, or is incorporated into an organic molecule, such as trichloroethylene, which oxidizes to release chlorine. Anhydrous HCl is most widely used to introduce chlorine into the oxidation ambient at present, although trichloroethylene also is coming into use primarily because it is easier to store. However, there can be problems of corrosion of the metallic parts of the oxidation apparatus when using HCl. Corrosion can become particularly rapid at higher HCl concentrations (>1%). Cor-

---

*The more rapid oxidation along grain boundaries is due to the larger number of *available* silicon-silicon bonds there.

rosion occurs because water vapor forms by the oxidation of HCl in the hot zone of the furnace and as the result of mixing HCl with atmospheric water vapor in the exhaust system. The latter would be a problem with Cl_2 as well.

Figure 13.26a, which is based on measurements by Hess and Deal,[67] shows a log plot of oxide thickness as a function of oxidation time for

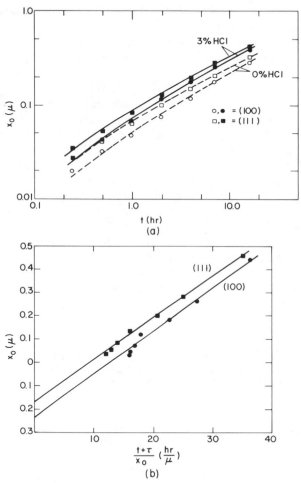

Fig. 13.26 (a) Oxide thickness x_0 versus oxidation time t for the oxidation of (100) and (111) oriented n-type silicon ($N_D = 10^{15}$ cm^{-3}) in 0 and 3% HCl–O_2 mixtures at 1000°C. After Hess and Deal.[67] (b) Lines represent a plot of (13.24) and circles are experimental points from Fig. 13.26a. Slopes of the lines correspond to the parabolic rate constant B and intercepts, to the constants $-A$ in (13.24). After Hess and Deal.[67] These figures were originally presented at the Fall 1976 Meeting of The Electrochemical Society, Inc. held in Las Vegas, Nevada.

(100)- and (111)-oriented silicon oxidized at 1000°C in both 0% and 3% $HCl-O_2$ mixtures. The figure shows that (1) oxide grows faster in the presence of chlorine and (2) oxide grows faster on a (111) surface than on a (100) surface. The data in Fig. 13.26a can be fitted to the linear-parabolic relation (13.24). As previously discussed, plots of x_o versus $(t + \tau)x_o$ should be straight lines with slopes B, the parabolic rate constant, and intercepts $-A$, from which the linear rate constant B/A can be determined. An example of such a plot for oxidation in 3% HCl/O_2 at 1000°C is shown in Fig. 13.26b. Values of B and B/A obtained by a least squares analysis in Fig. 13.26b are given in Table 13.9.

According to (13.24), B depends linearly on the diffusion coefficient of the oxidizing species diffusing through the oxide layer. Therefore, B will be independent of the orientation of the silicon surface, as was the case for oxidation in dry oxygen or steam. The slight differences in B with orientation seen in Table 13.9 are due to experimental error. The orientation dependence of the oxidation rate seen in Fig. 13.26a is attributable entirely to the orientation dependence of B/A, seen in Table 13.9. Table 13.9 also shows that both B and B/A increase with the addition of HCl to the oxidizing ambient.

With the addition of HCl to the oxidation ambient, water vapor will form by the gas phase equilibrium reaction: $2HCl + \frac{1}{2}O_2 \leftrightarrows H_2O + Cl_2$. However, the resulting water vapor in the oxidation ambient cannot account entirely for the increased oxidation rate observed. No water vapor is formed when Cl_2 is added to the oxidation ambient, yet the oxidation rate is observed to increase by a comparable amount.[67] This observation suggests that chlorine itself must play an important role in increasing the oxidation rate, but details of the chemistry of oxide growth in the presence of chlorine are not known.

Table 13.9 Parabolic (B) and Linear (B/A) Rate Constants for Thermal Oxidation of Silicon in 0 and 3% $HCl-O_2$ Mixtures at 1000°C[a]

Oxidation Ambient	Silicon Orientation	B ($\mu m^2/hr$)	B/A ($\mu m/hr$)
O_2			
	(100)	0.00913	0.047
	(111)	0.00956	0.080
$O_2 + 3\% HCl$			
	(100)	0.0187	0.078
	(111)	0.0177	0.11

[a] After Hess and Deal.[67]

At different oxidation temperatures and different concentrations of HCl or Cl_2 than shown in Fig. 13.26 and Table 13.9, large increases in B and B/A are observed.[68-70] The device fabricator must obtain, by measurement, the oxide thickness versus oxidation time grown in the oxidation apparatus for the oxidation temperature and chlorine concentration used.

(e) Other Impurities

The effects of other commonly used dopant impurities on oxidation rates can be predicted from their redistribution characteristics. Those impurities which are rejected by the oxide during growth, such as phosphorus, antimony, and arsenic will cause no increase in the parabolic range where the oxidation rate is primarily diffusion limited, but the oxidation rate should increase at high concentrations when the silicon surface reaction becomes the important rate limiting step such as in 1 atm steam below a thickness of 1000 Å. For those impurities that are incorporated in the oxide and diffuse slowly through the oxide, such as boron, the oxidation rate will increase at high concentrations in both the linear and parabolic ranges. Impurities such as gallium, aluminum, and indium, which deplete at the silicon surface and then rapidly diffuse out of the growing silica layer, will not affect the oxidation rate in either the linear or parabolic range even at initially high concentrations in the silicon.

13.4.6 Thin Oxides

The two reasons for discussing thin oxides are to alert the reader that (1) the kinetic theory developed previously in this section does not describe the growth of thin oxides, particularly when the oxide is grown in dry oxygen and (2) modern devices use oxides as thin as 300 Å or less. One example is the short channel MOSFET.[71-72] Even thinner oxides (25–100 Å thick) are important in a class of memory devices that use potential wells at the interface between SiO_2 and another insulator to store charge. The potential wells at the insulator-insulator interface are filled and emptied with electrons by tunneling between the wells and the silicon through the thin SiO_2 layer.[73-75] Another possible application of thin oxides is in semiconductor photovoltaic solar cells, where a thin oxide layer (on the order of 30 Å) between a metal electrode and the semiconductor has been found to increase the output voltage and efficiency of the cell.[76] The use of thin oxides in interface trap measurements are discussed in Section 5.10.

(a) Growth of Thin Oxides at Elevated Temperatures

One of the basic features of growing thin oxides is that their growth does not follow the kinetic theory previously developed in this section. This discrepancy is more pronounced for oxidation in dry oxygen than for oxidation in wet oxygen or in steam. The oxide growth always has an initial rapid phase followed at low temperatures by a linear phase and at higher temperatures by a linear-parabolic phase. Figure 13.14 illustrates the initial rapid growth phase followed by a linear growth phase at low temperature (700°C). The initial rapid growth phase does not follow our kinetic theory.

The next question is whether oxidation in the subsequent linear or linear-parabolic phase follows our kinetic theory. One crucial measurement is to measure oxidation rate as a function of pressure. This measurement was done by Van der Meulen,[77] who measured oxidation rates in dry oxygen for oxide thicknesses in the range 10–300 Å over a temperature range of 700–1000°C and a pressure range of 0.1–1.0 atm. In these ranges, Van der Meulen[77] found an initial rapid growth phase followed by a linear or linear-parabolic phase. The linear oxidation rate extracted from the measured data varied from a linear pressure dependence at the high end of the temperature range to nearly a square root pressure dependence at the low end. A possible reason for this observed pressure dependence is that Henry's law does not hold at lower temperatures. Henry's law would not be expected to hold when the oxide is very thin because the concentration of oxygen molecules in the growing SiO_2 film must be less than the equilibrium concentration at a given temperature until a sufficient number of SiO_2 molecules have formed. There is a serious problem, however, to be considered in this discussion. The problem is that the data due to Van der Meulen,[77] Goodman and Breece,[78] (who grew oxides up to 30 Å thick in dry oxygen at 600°C), and Deal and Grove[10] are not consistent with each other. By extrapolating, from the linear phase as in Fig. 13.14 for the linear-parabolic phase to $t = 0$, Deal and Grove[10] found $x_i = 230$ Å independent of temperature in the range 700–1200°C for oxidation in dry oxygen. Therefore, the initial rapid growth phase in the measurements by Deal and Grove[10] persisted up to oxide thicknesses of 300 Å or more. In contrast, Van der Meulen[77] and Goodman and Breece[78] interpret their data as going through both the initial and linear or linear-parabolic phases in 300 Å or less. The only clear conclusions to draw, then, are that for oxidation in dry oxygen, the kinetic theory developed earlier does not apply to the first 300 Å or so of growth, and there is no satisfactory explanation for this region.

For oxidation in wet oxygen or steam, Deal and Grove[10] and Pliskin[43] did not find an initial rapid growth phase over the temperature range 700–1200°C. Thus the kinetic theory is valid down to much smaller thicknesses for oxidation in steam or wet oxygen than is the case for oxidation in dry oxygen.

To grow thin oxides for device applications with a reasonable degree of control and in a reasonable length of time, the growth rate must be slower than the rates common for thick oxides, but faster than at room temperature. The optimum rate is established by growing the oxide at lower temperatures (500–800°C) and at oxidant pressures of 1 atm or less or at higher temperatures (1000°C) and very low oxidant pressures.

Goodman and Breece[78] were among the first to report growing thin oxides for device applications on freshly etched silicon surfaces. They concluded that it is possible to reproducibly grow thin oxides in either 1 atm oxygen or water vapor at 600°C and that the thin oxides produced would continue to grow under room ambient conditions at a much reduced rate.

Irene[79] has grown thin oxides (≤ 200 Å) at temperatures of 780–980°C in dry O_2 and H_2O–N_2 ambients and found that (1) oxidation is considerably more linear in dry O_2 than in H_2O–N_2, which exhibits parabolic growth, and (2) the oxides grown in H_2O–N_2 have a higher dielectric breakdown strength than oxides grown in dry O_2. Because oxides grown in dry O_2 were not grown at the same rate (they were grown faster), it is not clear whether the ambient species (wet or dry) or the oxidation rate is responsible for the differences observed (see the discussion in Section 15.5).

(b) Oxidation in Room Ambient

This section deals with the oxidation of silicon in room air (usually moist) at room temperature. Such films are always present on a silicon surface unless removed by etching in HF.

Archer[80] studied the growth of an oxide film in moist room air at room temperature following a rinse in hydrofluoric acid. The hydrofluoric acid rinse etches away any oxide that may have been initially present. By measuring oxide thickness using ellipsometry (see Section 14.3.2) as a function of time elapsed after the HF etch, he found that the data for times of up to 2×10^5 sec agreed well with the logarithmic relation $x_o = (a + b) \log(t + t_o)$, where a, b, and t_o are constants. Lukes[81] found that the qualitative dependence of the oxidation process in moist room air at room temperature was the same for a cleaved silicon surface as for an etched silicon surface. Figure 13.27 shows oxidation time versus film thickness for a cleaved silicon crystal with a (111)-oriented surface oxidized in air at room temperature. Figure 13.27 shows that the logarithmic law is followed up to thicknesses of about 35–40 Å. For thicknesses greater than this, the logarithmic law no longer holds. Instead, the experimental data appear to follow the law $x_o = a_1 + b_1(t - t_1)^{1/3}$, where a_1, b_1, and t_1 are constants. The fitting of the thin oxide growth data to a logarithmic law at small thicknesses and to a power law at larger thicknesses is purely empirical. Attempts have been made to describe the growth mechanism that gives rise to these laws. In proposing a growth mechanism, it is necessary not only that the mechanism lead to the observed growth law, but that it be unique. The lack of experimental data over a wide range of oxidation conditions makes it impossible to select one mechanism to derive the observed growth law. The crux of the problem is to determine the rate-limiting reaction at the silicon surface. At present, there is no unique underlying rate-limiting reaction known to result in a logarithmic followed by a cubic growth law.

Wet chemical treatments are important because silicon wafers are often exposed to such treatments in the manufacture of devices. An oxide film

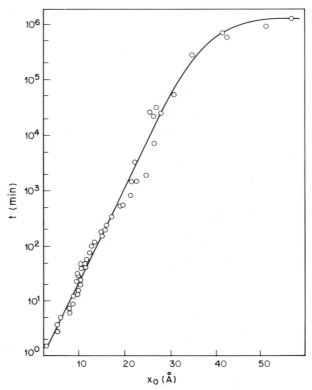

Fig. 13.27   Oxidation time versus oxide thickness for a cleaved (111) oriented single crystal of silicon oxidized in air at room temperature. After Lukes.[81]

can be grown on silicon by exposure to boiling deionized water containing dissolved oxygen.[82] Another way to grow a thin oxide on silicon is to expose the silicon surface to nitric acid at room temperature.[83] These wet chemical treatments all produce an oxide film about 10–20 Å thick. Oxide thicknesses of up to 100 Å also can be produced by exposing the silicon surface to boiling nitric acid, hot sulfuric acid, or a mixture of sulfuric acid and hydrogen peroxide at room temperature.

REFERENCES

1   E. L. Jordan, *J. Electrochem. Soc.*, **108**, 478 (1961).

2   J. R. Ligenza, *J. Appl. Phys.*, **36**, 2703 (1965).

3   J. R. Ligenza and M. Kuhn, *Solid-State Technol.*, **13**, 33 (1970).

4. P. F. Schmidt and W. Michel, *J. Electrochem. Soc.*, **104**, 230 (1957).

5. M. M. Atalla, in *Properties of Elemental and Compound Semiconductors*, Vol. 5, H. Gatos, Ed., Wiley-Interscience, New York, 1960, pp. 163–181.

6   J. R. Ligenza and W. G. Spitzer, *J. Phys. Chem. Solids*, **14**, 131 (1960).

7   W. G. Spitzer and J. R. Ligenza, *Phys. Chem. Solids*, **17**, 196 (1961).

8   P. J. Jorgensen, *J. Chem. Phys.*, **37**, 874 (1962).

9   W. A. Pliskin and R. P. Gnall, *J. Electrochem. Soc.*, **111**, 872 (1964).

10 B. E. Deal and A. S. Grove, *J. Appl. Phys.*, **36**, 3770 (1965).

11 F. J. Norton, *Nature*, **191**, 701 (1961); *Transactions of the Eighth Vacuum Symposium and Second International Congress*, Pergamon, New York, 1962, pp. 8–16.

12 A. J. Moulson and J. P. Roberts, *Transact. Br. Ceram. Soc.*, **59**, 388 (1960).

13 G. Stephenson and K. H. Jack, *Transact. Br. Ceram. Soc.*, **59**, 397 (1960).

14 A. J. Moulson and J. R. Roberts, *Transact. Faraday Soc.*, **57**, 1208 (1961).

15 G. Hetherington and K. H. Jack, *Phys. Chem. Glasses*, **3**, 129 (1962).

16 T. Bell, G. Hetherington, and K. H. Jack, *Phys. Chem. Glasses*, **3**, 141 (1962).

17 T. Drury and J. P. Roberts, *Phys. Chem. Glasses*, **4**, 79 (1963).

18 K. Motzfeldt, *Acta Chem. Scand.*, **18**, 1596 (1964).

19 P. Siffert and A. Coche, *IEEE Transact. Nucl. Sci.*, **11** (3), 244 (1964).

20 J. E. Thomas and D. R. Young, *IBM J. Res. Dev.*, **8**, 368 (1964).

21 M. Waldner, *J. Appl. Phys.*, **36**, 188 (1965).

22 A. G. Revesz, *IEEE Transact. Electron Devices*, **12**, 97 (1965).

23 M. Yamin, *IEEE Transact. Electron Devices*, **12**, 88 (1965).

24 P. F. Schmidt and J. E. Sandor, *Transact. AIME*, **233**, 517 (1965).

25 Y. Nishi, *Jap. J. Appl. Phys.*, **5**, 333 (1966).

26 R. Brücker, *J. Non-Cryst. Solids*, **5**, 177 (1971).

27 D. O. Raleigh, *J. Electrochem. Soc.*, **113**, 782 (1966).

28 A. E. Owen and R. W. Douglas, *J. Soc. Glass Technol.*, **43**, 159 (1959).

29 K. H. Beckman and N. J. Harrick, *J. Electrochem. Soc.*, **118**, 614 (1971).

30 J. M. Stevels, in *Non-Crystalline Solids*, V. D. Frechette, Ed., Wiley, New York, 1960, Chapter 17.

31 R. B. Sosman, *The Phases of Silica*, Rutgers University Press, New Brunswick, N.J., 1965.

32 H. Rawson, *Inorganic Glass-Forming Systems*, Academic, New York, 1967.

33 W. Eitel, *Silicate Science*, Vol. II, Academic, New York, 1965.

34 H. Edagawa, Y. Morita, S. Maekawa, and Y. Inuishi, *Jap. J. Appl. Phys.*, **2**, 765 (1963).

35 P. V. Pavlov and E. V. Shitova, *Sov. Phys. Crystallogr.*, **12**, 95 (1967).

36 N. Nagasima, *Jap. J. Appl. Phys.*, **9**, 879 (1970).

37 M. M. Atalla, E. Tannenbaum, and E. J. Scheibner, *Bell Syst. Tech. J.*, **38**, 749 (1959).

38 J. R. Ligenza, *J. Phys. Chem.*, **65**, 2011 (1961).

39 D. Kahng and M. M. Atalla, "Silicon-Silicon Dioxide Field Induced Surface Devices," paper presented at IRE-AIEE Solid-State Device Research Conference, Carnegie Institute of Technology, Pittsburgh, Pa., 1960; M. M. Atalla, U.S. Patent 3,206,670 (filed in 1960, issued in 1965); D. Kahng, U.S. Patent 3,120,230 (filed in 1960, issued in 1963).

40 T. H. Yeh, *J. Appl. Phys.*, **33**, 2849 (1962).

41 N. Karube, K. Yamamato, M. Kamidiyama, *Jap. J. Appl. Phys.*, **2**, 11 (1963).

42 B. E. Deal, *J. Electrochem. Soc.*, **110**, 527 (1963).

43 W. A. Pliskin, *IBM J. Res. Dev.*, **10**, 198 (1966).

44 P. J. Burkhardt and L. V. Gregor, *Transact. Metall. Soc. AIME*, **236**, 299 (1966).

45 T. Nakayama and F. C. Collins, *J. Electrochem. Soc.*, **113**, 706 (1966).

46 A. G. Revesz and R. J. Evans, *J. Phys. Chem. Solids*, **30**, 551 (1969).

47 L. D. Landau and E. M. Lifshitz, *Statistical Physics*, Pergamon, 1958, p. 280.

48 A. S. Grove, *Physics and Technology of Semiconductor Devices*, Wiley, New York, 1967.

49 U. R. Evans, *The Corrosion and Oxidation of Metals*, Arnold, London, 1960, pp. 819–859.

50 U. R. Evans, *Transact. Electrochem. Soc.*, **46**, 247 (1924).

51 E. A. Gulbranson, K. F. Andrew, and F. A. Brassart, *J. Electrochem. Soc.*, **113**, 834 (1966).

52 C. Wagner, *J. Appl. Phys.*, **29**, 1295 (1958).

53 P. S. Flint, Extended Abstracts of the May 1962 Electrochemical Society Meeting, Abstract 94, pp. 222–223.

54 H. C. Evitts, H. W. Cooper, and S. S. Flaschen, *J. Electrochem. Soc.*, **111**, 688 (1964).

55 C. R. Fuller and F. J. Strieter, Extended Abstracts of the May 1964 Electrochemical Society Meeting, Abstract 74, pp. 180–181.

56 L. Pauling, *The Nature of the Chemical Bond*, 3rd ed., Cornell University Press, Ithaca, N.Y., 1960, p. 85.

57 O. L. Krivanek, T. T. Sheng, and D. C. Tsui, *Appl. Phys. Lett.*, **32**, 437 (1978); O. L. Krivanek, D. C. Tsui, T. T. Sheng, and A. Kamgar, in *The Physics of SiO₂ and its Interfaces*, S. T. Pantelides, Ed., Pergamon, New York, 1978, pp. 356–361.

58 E. A. Irene, *J. Electrochem. Soc.*, **121**, 1613 (1974).

59 H. F. Wolf, *Silicon Semiconductor Data*, Pergamon, New York, 1969, Chapter 8.

60 F. L. Cristie, U.S. Govt. Report No. AD456798, 1964.

61 J. R. Ligenza, *J. Electrochem. Soc.*, **109**, 73 (1962).

62 B. E. Deal and M. Sklar, *J. Electrochem. Soc.*, **112**, 430 (1965).

63 H. Fizuka, T. Sato, F. Masuoka, K. Ohuchi, H. Hara, H. Tango, M. Ishikawa, and Y. Takeishi, *Suppl. J. Jap. Soc. Appl. Phys.*, **42**, 158 (1973).

64 S. A. Abbas and C. A. Barile, *13th Annual Proceedings, Reliability Physics Symposium*, Vol. 13, Las Vegas, Nevada, 1975, p. 1.

65 D. J. Di Maria and D. R. Kerr, *Appl. Phys. Lett.*, **27**, 505 (1975).

66 D. R. Kerr, Extended Abstracts of the October 1976 Electrochemical Society Meeting, Abstract 326, pp. 839–841.

67 D. W. Hess and B. E. Deal, Extended Abstracts of the October 1976 Electrochemical Society Meeting, Abstract 323, pp. 829–831.

68 R. J. Kriegler, Y. C. Cheng, and D. R. Colton, *J. Electrochem. Soc.*, **119**, 388 (1972).

69 K. Hirabayashi and J. Iwamura, *J. Electrochem. Soc.*, **120**, 1595 (1973).

70 Y. J. van der Meulen and J. G. Cahill, *J. Electron. Mater.*, **3**, 371 (1974).

71 R. H. Dennard, F. H. Gaensslen, H. N. Yu, V. L. Rideout, E. Bassous, and A. R. LeBlanc, *IEEE J. Solid-State Circ.*, **SC-9**, 256 (1974).

72 R. H. Dennard, F. H. Gaensslen, E. J. Walker, and P. W. Cook, *IEEE Transact. Electron Devices*, **ED-26**, 325 (1979).

73 H. C. Pao and M. O'Connell, *Appl. Phys. Lett.*, **12**, 260 (1968).

74 J. T. Wallmark and J. H. Scott, Jr., *RCA Rev.*, **30**, 335 (1969).

75 T. L. Chou, J. R. Szedon, and C. H. Lee, *Solid-State Electron.*, **10**, 897 (1967).

76 H. C. Card and E. S. Yang, *Appl. Phys. Lett.* **29**, 51 (1976).

77 Y. J. Van der Meulen, *J. Electrochem. Soc.*, **119**, 530 (1972).

78 A. M. Goodman and J. M. Breece, *J. Electrochem. Soc.*, **117**, 982 (1970).

79 E. A. Irene, *J. Electrochem. Soc.*, **125**, 1708 (1978).

80 R. J. Archer, *J. Electrochem. Soc.*, **104**, 619 (1957).

81 F. Lukes, *Surface Sci.*, **30**, 91 (1972).

82 W. Przyborski, J. Roed, J. Lippert, and L. Sarholt-Kristensen, *Rad. Effects*, **1**, 33 (1969).

83 R. E. Oakley and G. A. Godber, *Thin Solid Films*, **9**, 287 (1972).

84 C. P. Ho, J. D. Plummer, J. D. Meindl, and B. E. Deal, *J. Electrochem. Soc.*, **125**, 665 (1978).

85 C. P. Ho and J. D. Plummer, *J. Electrochem. Soc.*, **126**, 1516 (1979); ibid., **126**, 1523 (1979).

14

Oxidation of Silicon—Technology

14.1 INTRODUCTION

In this chapter we discuss how oxide films are prepared and how their thickness is measured, how oxidation influences bulk properties of the silicon, and how the oxide can be used as a mask against dopant impurities.

Section 14.2 describes oxidation technology. Although every detail of oxidation technology is not given, the overall principles should enable the reader to grow a thermal oxide suitable for integrated circuit and MOS capacitor applications.

Section 14.3 describes oxide thickness measuring techniques. The oxidation kinetics presented in Section 13.4 are based on measurements of oxide thickness as a function of time, temperature, and oxidant pressure. Because time, temperature, and pressure parameters are measured by standard methods, we concentrate on the measurement of oxide thickness.

Section 14.4 discusses the influence of thermal oxidation on silicon bulk electrical properties. Thermal oxidation has three effects on the electrical properties of silicon that are important in integrated circuit technology: (1) oxidation-induced stacking faults, which are structural defects at the silicon surface and are important because they can be electrically active, causing increased junction leakage and reduced minority carrier injection efficiency; (2) redistribution of ionized impurities at the silicon surface, which can alter the I-V characteristics of a MOSFET and the C-V characteristics of an MOS capacitor; and (3) strain in the oxide and in the silicon.

Section 14.5 discusses the effectiveness of a thermally grown oxide layer as a mask against dopant impurities. Masking is one of the key features of SiO_2 that makes it possible to diffuse dopant impurities into selected regions of the silicon substrate to make an integrated circuit.

Thermally grown SiO_2 has the best electrical characteristics for applications in the critical areas of devices used in integrated circuits. This was discovered from measurements of MOS capacitors made with these insulators. However, Sections 14.6 and 14.7 discuss other methods of oxide preparation, silicon nitride deposition, and oxygen doped silicon layers and their uses. Discussion of these oxides and insulators is outside the mainstream of this book. However, they are included and briefly discussed to give the reader perspective about oxides and insulators used in integrated circuit technology other than thermally grown SiO_2.

14.2 OXIDATION TECHNOLOGY*

A detailed description of how to grow an oxide film is beyond the scope of this book. However, the overall steps are outlined. The silicon crystal for electron device application is pulled into the form of a cylindrical single crystal ingot. Ingots that have diameters of up to 5 or even 6 inches can be grown on a commercial scale today. To prepare wafers from an ingot for oxidation, three steps are required: (1) saw, (2) chemically etch, and (3) polish. It was once customary to abrasively lap after sawing, but this thickness-adjusting step now is unnecessary because of improved thickness control in the sawing, etching, and polishing steps and because planar technology seldom requires precise substrate thicknesses.

14.2.1 Sawing and Etching

The crystallographic orientation of the wafer is determined during the sawing process. Orientation is established by having a birotational X-ray goniometer serve as a crystal holder on the sawing apparatus. A single

*This section follows the Bell Telephone Laboratories Manual entitled *Silicon Oxidation Guide*.

X-ray diffraction spectrometer and a Laue back reflectance camera are required for determining orientation prior to sawing. The saw is generally a circular disk with an edge charged with diamond dust. A diamond saw is used because silicon is a relatively hard material. The sawing operation mechanically damages the surface on each side of the wafer. Damaged surface layers have electrical properties significantly different from the bulk properties of single-crystal silicon. These damaged surface layers must be removed to eliminate adverse effects on device characteristics. Chemical etching of 25 μm of silicon from each side of the wafer removes the electrically active damage completely and neither leaves the wafers nonflat nor causes feather edges prone to chipping. The usual etchant consists of 4 parts by volume of nitric acid, 1 part hydrofluoric acid, and 3 parts acetic acid, American Chemical Society reagent grade. The nitric acid oxidizes the silicon surface, and the hydrofluoric acid dissolves the oxide formed.

14.2.2 Polishing

The polishing operation is the last and most critical step. It must leave one surface of the wafer flat, specular, and undamaged. Polishing is done by rubbing the wafer against a cloth saturated with an abrasive material such as Syton. Syton is powdered SiO_2, which is used because it is not foreign to the Si-SiO_2 system. After polishing, an oxide usually is grown to protect the silicon surface during storage.

14.2.3 Preoxidation Cleaning

After polishing residues have been cleaned off, the wafers are ready for the preoxidation cleaning step. The preoxidation cleaning step must be regarded as part of the oxidation process, as the wafers should be placed in the oxidation furnace immediately after preoxidation cleaning. The purpose of preoxidation cleaning is to remove both organic and inorganic contaminants that the wafers have picked up during handling and storage. If not removed, such contaminants can cause poor performance and stability of the devices subsequently made.

Many preoxidation cleaning procedures are possible. The cleaning procedure described here is practical and illustrates what cleaning procedures must accomplish. After degreasing in trichloroethylene and acetone, the wafers are exposed to a boiling mixture of distilled water, ammonium hydroxide, and hydrogen peroxide.[1] This solution removes organic contaminants that are attacked by the solvating action of the ammonium hydroxide and the oxidizing action of the hydrogen peroxide. In addition, some metal ions from groups I and II of the periodic table are complexed by the small concentrations of NH_3^+ found in equilibrium with NH_4^+. Next, a mixture of hydrochloric acid, hydrogen peroxide, and distilled water is used. Metal impurities are oxidized by the action of both the hydrogen peroxide and the chlorine in the solution to form stable chloride complexes. This solution removes metallic impurities from the silicon surface and prevents their displacement plating back onto the silicon.

There are two practical advantages of these two cleaning solutions: (1) they are composed of volatile components and thus leave no residues on the silicon; and (2) they present no environmental disposal problems. The hydrogen peroxide used must be 30%, *unstabilized, electronic grade*. Typical commercial peroxide stabilizers are sodium pyrophosphate and sodium stannite or ammine derivatives. Such stabilizers containing sodium should be avoided for high quality silica films for electron device applications. Hydrogen peroxide containing less than 200 ppb of sodium stabilizers is considered unstabilized. Particulate matter is removed by mechanical scrubbing and ultrasonic agitation. Oxide grown by the oxidizing agent during the cleaning process is removed by a dilute mixture of hydrofluoric acid and high resistivity de-ionized water as a final step. The adsorbed layer of HF molecules serves to repel particulates away from the silicon surface as the wafer is extracted from the solution and placed into the oxidation furnace. This layer of HF molecules renders the silicon surface hydrophobic so that no drying procedure, with the possibility of recontamination, is needed.

At no time during the cleaning process should the wafers be permitted to dry even partially because even the highest purity water or reagent has some submicron-size particles that become deposited on the wafter during drying, possibly remaining there throughout the remainder of the cleaning process. The oxidation apparatus also should be cleaned, as described in Section 15.3.2. Oxidation in an anhydrous $HCl-O_2$ mixture, another method for minimizing sodium ion contamination in the silica film, also is discussed in Section 15.3.2.

14.2.4   Oxidation Conditions Commonly Used in Manufacture

In practice, gate oxide (the oxide grown prior to deposition of the gate electrode in a MOSFET) is grown in dry oxygen at 1150°C to a thickness that is typically 0.1 μm or less. High temperature oxidation in dry oxygen produces an oxide film that has the desirable interfacial electrical properties described in Section 15.5. The characteristics of the MOSFET are very sensitive to the electrical properties of the $Si-SiO_2$ interface. Therefore, the best oxide possible is used for the gate oxide. It is practical to grow the gate oxide in dry oxygen even though the oxidation rate is slow because it is relatively thin (typically ≤ 0.1 μm).

The major functions of the field oxide, usually grown in steam to thicknesses greater than 0.5 μm, are to isolate metallization patterns, that carry current to various portions of the integrated circuit, from the silicon substrate and to act as a diffusion mask. Therefore, the electrical properties of the $Si-SiO_2$ interface are of secondary concern. To obtain the fastest growth rate and adequate interfacial properties, the field oxide usually is grown in atmospheric steam at 1050°C. This oxidation temperature is chosen to have the least effect on the channel stop* diffusions that have been done prior to growth of the field oxide.

In many cases, because bipolar transistor characteristics are not as sensitive to the electrical properties of the Si-SiO$_2$ interface as are MOS-FETs, a steam-grown oxide usually is adequate for bipolar manufacture.

14.3 SILICA FILM THICKNESS MEASUREMENT

Accurate measurement of the thickness of a silica film grown on a silicon substrate is important because (1) it is one of the fundamental measurements in the study of oxidation mechanisms and kinetics and (2) it is necessary to monitor thickness in manufacturing integrated circuits to keep performance of the integrated circuit within specifications.

There are a number of very accurate ways of measuring thickness of a silica film on a silicon substrate. We mention weight gain, ellipsometry, multiple-beam interferometry, optical interference, capacitance, and light section microscopy. Most of these methods have been used in the study of oxidation mechanisms and kinetics. However, they are not all equally suited for routine measurements, nor are they all nondestructive measurements, as is required for monitoring thickness during manufacture. Optical interference is widely used for thickness measurement during manufacture. Some other methods are briefly described, but emphasis is placed on optical interference methods.

14.3.1 Weight Gain Method

The weight gain method is based on the gain in weight of the silicon when it oxidizes; the added weight depends on the amount of oxygen incorporated in the film. This weight gain is measured with a vacuum microbalance, and thickness is calculated from the area of the oxidized silicon surface and the density of bulk silica. The assumption of bulk density for extremely thin films may not be valid, and this assumption places a lower limit on the accuracy with which thickness can be measured with this technique. The weight gain method is nondestructive but requires so much skill to use that it cannot be regarded as a routine method.

14.3.2 Ellipsometry

Archer[2] was the first to use ellipsometry to measure the thickness of silica films grown on a silicon surface. Measurement of film thickness by

*The purpose of a channel stop is to prevent inversion of the substrate outside the source-drain region in a MOSFET.

ellipsometry is based on the change in polarization of a beam of plane polarized light on reflection by a surface (oxidized silicon). The incident beam of visible light is plane polarized to have the plane of oscillation of its electric field vector inclined at 45° to the plane of incidence. The ellipticity of the reflected beam is characterized by the phase difference and amplitude ratio between the two components of electric field vector measured in and normal to the plane of incidence. The magnitude of this phase difference and amplitude ratio depends on the angle of incidence, optical constants of the reflecting surface, thickness, and index of refraction of the silica film. Thickness and index of refraction are extracted from a measurement of the phase difference and the amplitude ratio of the reflected beam compared to the incident beam. The ellipsometry method is nondestructive, but it is not routine as it requires considerable skill and extensive calculations to extract thickness values. Nonetheless, it has been widely used in industry.

14.3.3 Multiple-Beam Interferometry

Multiple-beam interferometry as developed by Tolansky[3] is another accurate thickness measuring method used in the study of oxidation kinetics. This method requires special sample preparation. A step must be etched in the silica film down to the silicon and a highly reflective material such as silver must be evaporated over the entire wafer to make reflectivity on both sides of the step the same. The evaporated silver film will accurately contour the underlying step in the silica. A smooth glass flat, coated with a highly reflecting film such as silver, is matched against the sample. Interference fringes result from multiple reflections of the light beam between the sample surface and the matching flat. Fringes on one side of the step will be displaced with respect to those on the other side and can be observed with a microscope. The difference between the number of fringes on one side of the step and the other side over a given wavelength interval will be proportional to the step height. The sample used in these measurements is no longer suitable for electronic application; hence multiple-beam interferometry is a destructive method. Mainly for this reason, multiple-beam interferometry is not used to monitor thickness in manufacturing.

14.3.4 Capacitance Method

Capacitance is widely used for measuring thickness. Chapter 12 gives a detailed description of capacitance measuring techniques. The capacitor is made by depositing a thin metal electrode ($\sim$2000–3000 Å thick), usually by evaporation, over the silica film. Capacitance can be measured either with a capacitance bridge or a lock-in amplifier tuned to the 90° phase component of the ac small-signal current flowing through the test capacitor. The silica film thickness is calculated from the measured capacitance, the area of the metal electrode, and the dielectric permittivity of the silica. To measure oxide thickness this way, three requirements must be met:

1   The dimensions of the capacitor plate (metal gate) must be large compared to the oxide thickness x_o to avoid a correction for the fringing field at the edge of the gate. This requirement is met easily by evaporating a metal electrode several hundred microns in diameter over the grown oxide layer, which is on the order of tenths of microns thick.

2   The area of the metal electrode must be known accurately. This area can be calculated from a measurement of the diameter of the metal electrode. The diameter can be measured accurately with a calibrated movable microscope stage and a cross-hair reticle in the eyepiece of the microscope.

3   The dielectric permittivity of the SiO_2 must be known. Then $x_o = C_{ox}A_g/\epsilon_{ox}$, where C_{ox} is the measured differential oxide layer capacitance (see Chapters 3 and 12), A_g is the gate area, and ϵ_{ox} is the dielectric permittivity* of SiO_2. The dielectric constant must be measured independently. The dielectric constant is insensitive to processing, impurity content, and oxide structure and for silica is 3.9.[4] Therefore, it need not be measured every time oxide thickness is measured by the capacitance method. Dielectric constant also is insensitive to frequency over a wide range. Thus oxide thickness can be measured by the capacitance method without significant error at frequencies from the audio range to several megahertz using the same value of ϵ_{ox}. For very thin oxides, there may be some error in using the dielectric constant valid for thick oxides.

The dielectric constant of SiO_2 also can be obtained from ellipsometer measurements for thick oxides. However, this value will be valid at visible light frequencies so that it will be in error if used in capacitance measurements made in the megahertz range or below.

The capacitance method is not as accurate or rapid as the optical methods. Because metal must be deposited over the silica, it is a destructive method. However, many integrated circuit chips incorporate a metallized test pattern that has a capacitor. The capacitor is used also to measure many other properties of the film. Alternatively, the oxide surface can be contacted with a drop of mercury in a mercury probe (see Section 12.8.3). This drop of mercury then acts as a gate electrode leaving the oxide underneath unaffected after the measurement.

---

*The dielectric permittivity is the product of the dielectric constant of SiO_2 and the permittivity of free space.

14.3.5   Optical Interference Method

The direct optical interference methods are simple, nondestructive, and of high accuracy suitable for routine measurement of the thickness of silica films grown on silicon substrates in the thickness range of several hundred to several thousand angstrom units. Almost all electron device applications will fall within this range. However, for thinner films, some of the other methods such as ellipsometry must be used and for thicker films a method such as light section microscopy would be faster and more convenient.

The optical interference method is based on interference between light reflected from the air-SiO_2 interface and the Si-SiO_2 interface of the silica film. The basic equation for optical interference phenomena for a film of thickness d is

$$d = (2n^*)^{-1}(g - P)\lambda \qquad (14.1)$$

where $n^* = (n_1^2 - \sin^2 \theta)^{1/2}$, n_1 is the refractive index of the silica film, θ is the angle of incidence, g is the order of the interference, P is the net phase shift, $P = \phi_f - \phi_s$, ϕ_f is the phase shift at the air-SiO_2 interface, ϕ_s is the phase shift at the Si-SiO_2 interface, and λ is the wavelength of the incident light. Equation (14.1) predicts minima in reflected intensity when $g = n + \frac{1}{2}$, where n is an integer, and maxima when $g = n$. Although (14.1) is not derived here, its derivation is based on the assumption that the refractive index and the attenuation coefficient of the silica film and silicon do not vary with wavelength and that effects of multiple reflections within the silica film can be neglected. In regions of the spectrum where the index of refraction and the attenuation coefficient of the silica and/or silicon vary with wavelength, the intensity maxima and minima will be slightly displaced from the wavelengths predicted by (14.1). For example, if the reflectivity of the silica film increases with decreasing wavelength, the intensity minima will be displaced to longer wavelengths and maxima to shorter wavelengths.

In optical interference based on (14.1), two methods can be used to determine silica film thickness: (1) observation of the angular dependence of the maxima and minima of the reflected light intensity at constant wavelength, when the angle of incidence is varied; and (2) observation of the wavelength dependence of the maxima and minima of the reflected light intensity at constant angle of incidence, when the wavelength is varied. Pliskin and Conrad[5] measured silica thickness by the first method, whereas Corl and Wimpfheimer[6] and Reizman and Van Gelder[7] measured silica film thickness by the second method. In practice, the constant wavelength, variable angle of incidence method seldom is used because commercially available instrumentation is lacking and a minimum silica film thickness of about 5000 Å is needed to obtain a pair of minima in an angular scan.*

The basic instrumentation required in the constant angle of incidence, variable wavelength method is a spectrophotometer. As wavelength is varied, the light intensity detected by a photodetector is recorded as a function of wavelength. Silica film thickness can be calculated from this intensity versus wavelength record using (14.1). To make this calculation easier and more accurate, a fringe chart can be used. Figure 14.1 is a fringe chart from Ref. 7 for SiO_2 on silicon in which silica film thickness is plotted as a function of wavelength with the order of the interference g as parameter. Figure 14.1 was prepared by measuring the wavelengths of the minima (the maxima also could be used, but the minima are usually sharper) in the reflection spectrum of silica films whose thickness was measured afterward by multiple-beam interferometry. Because film thicknesses were independently determined, the fringe chart includes all phase shifts, dispersion effects, and multiple reflection effects. No knowledge of index of refraction or net phase shift is needed. Index of refraction and phase shift, both as a function of wavelength, can be calculated from information

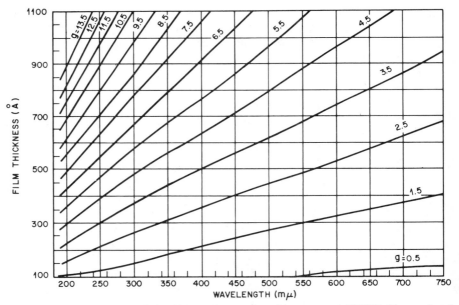

Fig. 14.1 Fringe chart for SiO_2. Minima for steam oxide grown at 1050°C. The angle of incidence θ is 26.5°. After Riezman and Van Gelder.[7] Copyright (1967), Pergamon Press, Ltd. Reprinted with permission.

---

*Thinner films can be accurately determined with this method by finding maxima and minima for more than one wavelength.[5]

implicit in the fringe chart using (14.1). Thus the use of the fringe chart for thickness determination involves no assumptions or approximations, as would be the case in the use of (14.1). The silica films used in constructing Fig. 14.1 were grown on 55 Ω-cm, p-type silicon in steam at 1050°C, but the fringe chart will be valid for oxides grown on silicon substrates of opposite conductivity type and different doping level and in both dry and wet oxygen because the optical constants of the Si–SiO$_2$ system are insensitive to the relatively minute differences between such samples.

To use the fringe chart for a film of unknown thickness, wavelength is scanned over the range given in the fringe chart, and intensity is recorded as a function of wavelength. Wavelengths at which all the minima occur are then marked as points on a straightedge that has the same wavelength calibration as the abscissa in the fringe chart. The straightedge is then placed on the fringe chart parallel to the wavelength axis and moved vertically until each point coincides with a line on the fringe chart. This vertical alignment determines the orders of the fringes, and the thickness is obtained directly from the ordinate.

The precision of this measurement is ± 25 Å in the 500–10,000 Å range of thickness, which is more than adequate for most electronic applications. This precision is limited by the ability to locate the wavelengths at which maxima and minima occur. Because the interference fringes are relatively broad, the location of the maxima and minima in the spectrophotometric trace is more dependent on the signal:noise ratio of the photodetector than the dispersive resolution of the optics in the instrument.

This method can be applied for other films such as silicon nitride and double layer films such as a layer of grown silica covered by a layer of silicon nitride. The reader is referred to Ref. 6 for details.

Commercial instrumentation is available, on the basis of this measurement method, which automatically gives silica film thickness without need for calculations or use of a fringe chart.

14.3.6 Color Chart

A very rapid but much less accurate method of determining silica film thickness is to use a color chart. A color chart is prepared by oxidizing silicon to various thicknesses that are then measured by optical interference or ellipsometry. Each wafer appears a different color when observed in white light as a result of destructive interference. The colors repeat for the different fringe orders except for minor changes in hue, particularly for oxide thicknesses around 8000 Å.[5] To obtain the correct thickness, the unknown is placed next to a slice on the color chart most nearly matching it in color. The color chart and the unknown are tilted to vary the angle of incidence. If the unknown passes through the same sequence of colors as its match on the chart, the thickness is that of the match; if not, the wrong order was chosen, and another match must be found.

To devise a simple, unambiguous system of color designations, the National Bureau of Standards has made available a set of color charts.*

14.4 INFLUENCE OF THERMAL OXIDATION ON SILICON BULK PROPERTIES

In integrated circuit technology, there are two important effects of thermal oxidation of silicon on bulk electrical properties of the silicon substrate: (1) oxidation induced defects, particularly stacking faults discussed in Section 14.4.1, and (2) ionized impurity redistribution, discussed in Section 14.4.2.

Strain is another effect of oxidation on silicon. Growth of an oxide film on silicon puts the silicon under strain at room temperature because of the mismatch in the coefficient of thermal expansion between silica and silicon. This strain does not directly affect the electrical properties of most devices used in integrated circuits, but if the silicon wafer is too thin, the mismatch strain can cause the wafer to warp or bow. Further strain is induced in the silicon by metallization on the SiO_2. Wafer warpage that results from these strains can pose difficulties in the photolithographic steps of the processing by causing distortions of the desired pattern. The warpage problem becomes worse as wafer diameter is increased, but warpage is avoided, and breakage during handling minimized by using wafers about 11–20 mils thick for a 2 inch diameter, about 15–20 mils thick for a 3 inch diameter, and about 21–25 mils thick for a 4 inch diameter. Larger thicknesses would work but would use more silicon than necessary, thus increasing costs. Mismatch strain induced in the silicon is not discussed further as it poses no serious problems.

14.4.1 Oxidation Induced Stacking Faults

A stacking fault is a structural defect in the silicon lattice consisting of an extra plane of silicon atoms bounded along the edges by dislocations.[8-21] The dislocations are the transition between the extra plane of silicon atoms and the normal silicon lattice structure. A more detailed discussion of structural defects in solids is given in Ref. 22. Stacking faults are important in integrated circuit technology because they can be electrically active, enhancing junction leakage currents and reducing minority carrier lifetime.

---

*Standard Material No. 2106 from the Office of Standard Reference Materials, National Bureau of Standards, Washington, D.C. 20234. This material is to be used in conjunction with NBS circular 553, "The ISCC-NBS Method of Designating Colors and a Directory of Color Names," *NBS Tech. News Bull.*, **49**, 70 (1965) available from the Superintendent of Documents, U.S. Government Printing Office, Washington, D.C. 20402.

The thermal oxidation process can induce stacking faults that extend from the silicon surface into the interior of the silicon, and these faults are called *oxidation induced stacking faults*. There also can be electrically active stacking faults in the silicon bulk. However, oxidation induced stacking faults usually constitute 95% or more of the total. We concentrate on oxidation induced stacking faults for this reason and because we are interested in growing oxides without inducing a large density of electrically active stacking faults. To achieve this aim, it helps to understand how stacking faults are created during thermal oxidation and how they become electrically active. Because of their importance in integrated circuit technology, there is an extensive literature on them. Starting with Thomas,[23] several workers have reported on the occurrence and properties of oxidation induced stacking faults.[8–21, 23–46]

The electrical activity of a stacking fault is markedly enhanced when it is decorated with impurities, particularly heavy-metal atoms, which decorate the stacking faults by clustering around them. Most, but not all, of the clustering is around the dislocation bounding the extra plane of silicon atoms. Such clustering is favored because the stacking fault is a distortion that results in an opening in the silicon lattice where impurity atoms easily can fit, thereby minimizing lattice strain. That is, impurity atoms located at a stacking fault distort the lattice much less than when they are located in a region free of structural defects. Thus impurity atoms diffusing about in the silicon lattice stay put when they arrive at a stacking fault. A cluster of impurity atoms builds up with time at elevated temperatures at the stacking fault. Some of the impurity atoms clustered about the stacking fault have energy levels near the center of the silicon bandgap, which makes them efficient generation and recombination centers. (See Section 4.3 for a discussion of generation and recombination processes.) If decorated stacking faults pass through the depletion region of a p-n junction, they will increase the leakage current of that junction by generation or decrease the injection efficiency of that junction by recombination. If the concentration of impurities decorating the stacking fault is sufficiently high, leakage current can be so large that the junction effectively is shorted.

Varker and Ravi[28] have shown conclusively that decorated oxidation induced stacking faults passing through a junction diode enhance its leakage current. This correlation combined an analysis of the I-V characteristics of a junction diode in an array of junction diodes with information derived from a scanning electron beam used in the electron beam induced current mode.[47] The presence of an electrically active stacking fault was revealed in the region of the junction.[28]

Oxidation induced stacking faults can extend from the silicon surface a distance of anywhere up to several micrometers into the silicon bulk. This length increases with increasing oxidation temperature, oxidation time, and oxygen pressure.[42, 43] The density of oxidation induced stacking faults can vary from none to $10^7 \, \mathrm{cm}^{-2}$ (on mechanically polished wafers) independent of oxidation temperature and time. Oxidation induced stacking fault density appears to depend on the density of nucleation sites available.

The undesirable effect on device characteristics of electrically active stacking faults can be minimized by either gettering heavy-metal impurities[48–51] or by reducing the number of stacking faults. To see how to minimize the number of oxidation induced stacking faults, we explain how they are formed.[35, 39, 44, 46] Although understanding is not complete, it has been agreed that (1) growth of stacking faults during thermal oxidation is related to the presence of excess unoxidized silicon at the $Si-SiO_2$ interfacial region and (2) stacking faults are heterogeneously nucleated by the coalescence of excess silicon atoms at nucleation centers. There are two complementary explanations of oxidation induced stacking fault formation: (1) excess silicon atoms, which are the source of stacking faults, are unoxidized silicon atoms in the very thin transition layer between the oxide and the silicon; or (2) there is a decreased vacancy concentration in the silicon near the $Si-SiO_2$ interface. Both explanations have experimental support and are complementary because excess silicon concentration means lower vacancy concentration at the $Si-SiO_2$ interface or vice versa. Sanders and Dobson[14] showed that vacancy concentration at the $Si-SiO_2$ interface is lower than in bulk silicon at 1100°C and that the ratio of vacancy concentration at the $Si-SiO_2$ interface to that in bulk silicon was 0.8.

The measurements described briefly in Section 15.4 suggest the existence of excess unoxidized silicon atoms in the transition layer. These unoxidized silicon atoms are free, having been severed from the silicon lattice during the oxidation reaction. There are few silicon interstitial atoms initially in the silicon lattice so that a concentration gradient will exist between free silicon atoms in the transition layer and the silicon lattice, driving them into the interstices of the silicon lattice. Because of their extremely high mobility, free silicon atoms in the transition layer quickly enter the silicon lattice interstices, causing the silicon lattice to become supersaturated with self-interstitials. Silicon lattice strain will be minimized if these interstitial silicon atoms coalesce into stacking faults. For coalescence to occur, nucleation sites are required. It is more or less accepted that suitable nucleation centers can be mechanical damage,[21, 23] ion implantation damage,[36] impurity inhomogeneities such as silicon-oxygen complexes,[20] carbon contamination,[18] HF contamination,[17, 34] and defect centers that appear as shallow pits[20] or hillocks[17] after chemical etching of the surface prior to oxidation. Such nucleation clusters also could be produced by heat treatment[11, 33] of the wafers prior to oxidation.*

*Such heat treatments are questionable. Raider et al.[52] have shown that heat treating silicon wafers in nitrogen (such as given in Ref. 32 to generate stacking faults) leads to a nitrogen-rich surface layer that acts as a barrier to further oxidation. Vacuum annealing in sealed quartz tubes (in which silicon powder was placed to prevent surface erosion) at 1000–1200°C[11] could lead to condensation of silicon on the wafer surface on subsequent cooling.

The observation that no stacking faults are formed above a critical temperature also is explained by the mechanism just described.[43] This explanation is based on the additional experimental observation that the critical oxidation temperature decreases as the oxidation rate is increased[53] by introducing increasing amounts of chlorine-containing species to the oxidizing ambient as described in Section 13.4.5(d).

Above the critical temperature the diffusion rate exceeds the oxidation rate. Thus if the vacancy mechanism of diffusion is applicable, if any excess silicon is produced during oxidation, it will diffuse away from the transition layer into the silicon faster than it is created, so that no stacking faults can form.[43]

Gettering will reduce the concentration of impurities, which can decorate stacking faults, but it will also reduce the number of nucleation centers. The number of oxidation induced stacking faults can be minimized by eliminating some of the nucleation centers by gettering prior to oxidation and by cleaning the silicon surface. However, the gettering techniques discussed in Refs. 48–51 do not completely eliminate the formation of stacking faults. Murarka et al.[45] have developed a simple and practical method claimed to eliminate oxidation induced stacking faults. The procedure is to clean the wafers by degreasing followed by a wet chemical cleaning consisting of a scrub and an $HCl-H_2O_2$ and an $NH_4OH-H_2O_2$ boil described in Section 14.2.3. The wafers must be immediately loaded into a clean fused quartz boat and the boat inserted into the hot zone of the oxidation furnace at oxidation temperature. The furnace ambient can be any inert gas except nitrogen or a mixture of 0.1% anhydrous HCl and an inert gas. Concentrations of HCl greater than 1% result in pitting of the silicon surface. In most cases placement for about 15 min in this ambient at oxidation temperature prior to oxidation will suffice. It is believed that this annealing procedure cleans the wafer by driving impurities either into the vapor phase where they are swept away by the gas flow or deep into the bulk silicon, thereby eliminating the nucleation sites necessary for formation of stacking faults. Also, an epitaxial layer, which has too few impurities to decorate stacking faults, can be used.

14.4.2 Dopant Impurity Redistribution

(a) Uniform Case

Dopant impurities near the silicon surface will be redistributed during thermal oxidation. Some of the dopant impurities will appear in the SiO_2, where they are seldom electrically active. Because dopant impurities are electrically active in the silicon, redistribution there is most important. Redistribution of dopant impurities in the silicon is important both from the standpoint of device characteristics and in the interpretation of measurements of interface trap properties and oxide fixed charge. Also, the oxidation rate is affected at high initial dopant impurity concentrations in the silicon. We describe the phenomenon in this section and its effect on measurements in Chapters 5, 7, 8, and 9.

Redistribution of a dopant impurity in the silicon depends on several factors.[54] When two solid phases are brought together, an impurity in one solid will redistribute between the two solids until equilibrium is reached. The ratio of the equilibrium concentration of the impurity in the silicon to the equilibrium concentration of the impurity in the SiO_2, called the *segregation coefficient*, is defined as

$$m = \frac{\text{equilibrium concentration of impurity in Si}}{\text{equilibrium concentration of impurity in } SiO_2}. \tag{14.2}$$

A second factor that influences impurity redistribution is that the impurity may be able to diffuse rapidly through the SiO_2 and escape. If the diffusivity of the impurity in SiO_2 is large, this factor will be important. A third factor in the redistribution process is that the oxide layer is growing, and thus the boundary between the silicon and the oxide is advancing into the silicon as a function of time. The relative rate of this advance compared to the diffusion rate of the impurity through the oxide is important in determining the extent of the redistribution. Note that redistribution in the silicon still will occur even if $m = 1$, and escape through the SiO_2 can be neglected. On the basis of the molecular weights and densities of silicon and SiO_2, the SiO_2 layer will be twice as thick as the silicon layer it replaced. Thus more impurity is required to bring up the concentration level in this larger volume, resulting in the depletion of the impurity from the silicon.

Four possible redistribution processes are illustrated in Fig. 14.2. These processes can be classed into two groups. In one group the oxide takes up the impurity, and in the other the oxide rejects the impurity. In each case, what happens depends on how rapidly the impurity can diffuse through the oxide. In case 1, $m < 1$, so that the oxide takes up the impurity. In this case the silicon surface is depleted of impurities. An example of case 1 is boron. Rapid diffusion of the impurity through the SiO_2 will increase the amount of depletion. An example is boron doped silicon heated in an H_2 ambient because hydrogen in the silica enhances the diffusivity of boron. In case 2, $m > 1$, so that the oxide rejects the impurity. If diffusion of the impurity through the SiO_2 is relatively slow, the impurity piles up near the silicon surface. An example of this case is phosphorus. When diffusion through the SiO_2 is rapid, so much impurity may escape from the solid to the gaseous ambient that the overall effect will be a depletion of the impurity. An example of this is gallium.

Table 14.1 lists the segregation coefficients m of widely used donor and acceptor impurities determined by redistribution[54-56] and oxide masking experiments.[57] For comparison, values of m predicted from solubility thermodynamic considerations[58] are also listed. The segregation coefficient

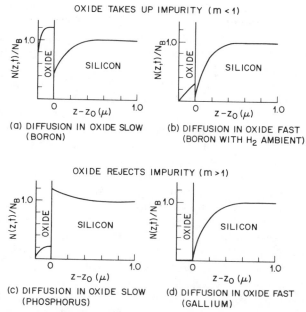

Fig. 14.2 Four different cases of impurity redistribution in silicon resulting from thermal oxidation. After Grove et al.[54]

Table 14.1 Segregation Coefficients of Impurities at the Si–SiO₂ interface[a]

	m		
Impurity	Thermodynamic Estimate	Redistribution Experiments	Oxide Masking Experiments[57]
Ga	$>10^3$		~20
B	$10^{-3}-10^3$	~0.3,[54, 55] 0.1[66]	~10^{-2}
In	$>10^3$		
P, Sb, As	$>10^3$	~10[54, 55]	

[a] After Grove et al.[54]

is temperature dependent. There are few measurements of this temperature dependence, but the temperature dependence of m for ion implanted boron has been measured and compared with measurements of other workers in Ref. 59.

To solve for the redistribution quantitatively, the diffusion equation must be solved with a moving boundary.[54] This solution is done for the case of an initially uniform impurity distribution in the silicon with two boundary conditions: (1) concentration of the impurity of the gas-SiO_2 interface is a constant (i.e., steady-state) N_g; and (2) far within the silicon, away from the Si–SiO_2 interface, the concentration of impurity approaches the bulk concentration N_B.

In addition, two matching conditions must be satisfied. First, the concentrations of dopant atoms on the two sides of the interface are assumed to be in the ratio m. That is, the system is idealized as two bulk phases, silica and silicon, with an abrupt interface. At locations very near the interface, exchange of dopant atoms between phases is assumed to be near equilibrium, so that use of m is justified. Second, under different oxidation conditions, the number of dopant atoms stored in the transition region varies. Therefore, we must allow the idealized planar boundary that replaces the true interface to store dopant atoms. The second matching condition relates the difference in dopant atom flux from the two phases to the number of dopant atoms stored at the interface. This condition is derived in Appendix XIV.

It is assumed that the thickness of growing SiO_2 increases as the square root of oxidation time, that is, $x_o = (Bt)^{1/2}$, where B is the parabolic rate constant. Finally, it is assumed that B does not change as the impurity concentration in the silicon changes during oxidation. Solutions of this boundary value problem are as follows.

1   In the silica film:

$$\frac{N_o(x, t) - N_g}{N_o - N_g} = \frac{\text{erf}[x/2(D_o t)^{1/2}]}{\text{erf}[(B/4D_o)^{1/2}]} \qquad 0 < x < x_o \qquad (14.3)$$

where $x_o(t)$ is the oxide thickness at time t, $N_o(x, t)$ is the impurity concentration at time t and distance x in the silica measured from the air-silica interface, N_o is the impurity concentration in the silica at the Si–SiO_2 interface, N_g is the impurity concentration at the air-silica interface, and D_o is the diffusion coefficient of the impurity in silica.

2   In the silicon:

$$\frac{N_s(z, t) - N_B}{N_s - N_B} = \frac{\text{erfc}[z/2(D_s t)^{1/2}]}{\text{erfc}[a(B/4D_s)^{1/2}]} \qquad z > z_o(t) \qquad (14.4)$$

where $z_o(t)$ is the thickness of silicon consumed by oxidation in time t, $N_s(z, t)$ is the impurity concentration in the silicon at time t and distance z measured from the initial silicon surface, N_s is the impurity concentration in the silicon at the Si–SiO_2 interface, N_B is the initial impurity concentration throughout the silicon, D_s is the diffusion coefficient of the impurity in silicon, and a is the ratio of the silicon thickness consumed during oxidation to silica thickness ($a = 0.45$).

The second matching condition gives, in addition, the ratio of impurity concentration at the silicon surface to the initial impurity concentration

$$\frac{N_s}{N_B} = \frac{1 + (N_g/N_B)b}{1 + (1/m - a)\pi^{1/2} \exp[a^2 B/4D_s] \operatorname{erfc}[a(B/4D_s)^{1/2}](B/4D_s)^{1/2} + b/m}$$

(14.5)

where

$$b = r \exp\left[(a^2 r^{2*} - 1)\frac{B}{4D_s r^2}\right] \operatorname{erfc}\left[a\left(\frac{B}{4D_s}\right)^{1/2}\right]\left\{\operatorname{erf}\left[\left(\frac{B}{4D_o}\right)^{1/2}\right]\right\}^{-1}$$

and

$$r = \left(\frac{D_o}{D_s}\right)^{1/2}.$$

Equations (14.3), (14.4), and (14.5) are derived in Appendix XIV following Grove et al.[54]

An interesting feature of (14.5) is that N_s/N_B at the moving Si–SiO$_2$ interface is independent of oxidation time. The reason is that both oxide thickness and diffusion of the impurity proceed with the square root of time. Therefore, the time variable disappears as a steady state is reached at the silicon surface described by (14.5). Equation (14.5) also shows that N_s is a function of m; the relative diffusivities of the impurity in silicon and SiO$_2$, D_s and D_o, respectively; and the ratio of the relative oxidation rate to the relative diffusivity B/D_s. Figure 14.3 shows values of N_s/N_B, calculated

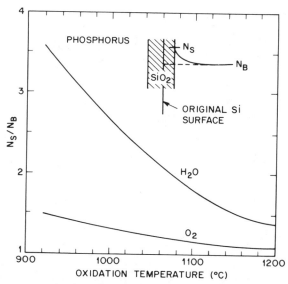

Fig. 14.3 Surface concentration of phosphorus in silicon after thermal oxidation. After Grove.[4]

from (14.5) using $m = 10$, as a function of oxidation temperature in both oxygen and steam for phosphorus. Figure 14.4 is similar to Fig. 14.3 except that it is for boron, and the value of $m = 0.3$ was used in (14.5). Figures 14.3 and 14.4 show that by oxidizing at a given temperature in H_2O rather than O_2, the faster oxidation rate in H_2O results in an increase in the extent of the redistribution. This result is evidenced by a higher surface concen-

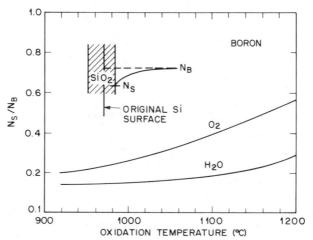

Fig. 14.4 Surface concentration of boron in silicon after thermal oxidation. After Grove.[4]

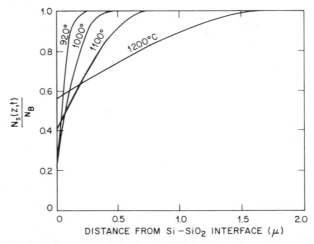

Fig. 14.5 Calculated boron concentration distribution in silicon after thermal oxidation at various temperatures in dry oxygen to a thickness of 0.2 μm for each temperature. After Deal et al.[55] Reprinted by permission of the publisher, The Electrochemical Society, Inc.

tration of phosphorus in Fig. 14.3 and a lower boron concentration in Fig. 14.4.

Figure 14.5 shows the boron profile in silicon oxidized at various temperatures to a thickness of 2000 Å. The boron profile is calculated for $m = 0.3$ using (14.4) to calculate $N_s(z, t)/N_B$. Note that as temperature is decreased, N_s/N_B decreases as shown in Fig. 14.4. However, because the diffusivity of boron in silicon decreases with decreasing temperature, the distance over which the profile deviates from the uniform profile case will be smaller.

The impurity deficit in the silicon after oxidation for a time t can be calculated from

$$Q_N(t) = -q \int_{z_0(t)}^{\infty} [N_s(z, t) - N_B] \, dz \qquad (14.6)$$

where $N_s(z, t)$ is given by (14.4).

Measurements have been made to verify the predicted impurity profiles using p-n junctions[43] and MOS capacitors.[55, 60] On each structure the profile was obtained from measured C-V curves. Details of how a doping profile can be extracted from such measurements are given in Ref. 60 and in Section 9.4. Figure 14.6 shows the boron profile in a silicon sample oxidized at 1100°C in oxygen to a thickness of 0.108 μm. This profile was measured by an MOS capacitor method[60] that permits accurate profile determination very close to the Si–SiO$_2$ interface. Figure 14.6 shows the good fit of the theory to an actual profile near the Si–SiO$_2$ interface. Because the measurement of the impurity profile was obtained with an x-y

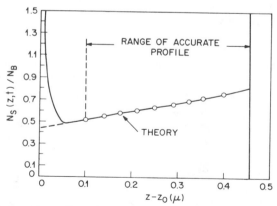

Fig. 14.6 Normalized impurity concentration $N_s(z, t)/N_B$ versus depth from the Si–SiO$_2$ interface measured on a sample doped with boron to a preoxidation concentration of 10^{17} cm^{-3} and thermally oxidized to a thickness of 0.108 μm in dry oxygen at 1100°C. Fundamental frequency was 5 MHz and the measurement was made at 300°K. The solid curve is the measured data, and the open circles were calculated. Adapted from Nicollian et al.[60] Copyright (1973), IEEE.

recorder, the solid curve in Fig. 14.6 represents the experimental data. The solid curve for $(z - z_o)$ less than 0.1 μm does not represent the actual profile. The curve in this region is subject to an exponentially increasing error because the actual profile cannot be resolved this close to the interface as discussed in Section 9.4. The open circles in Fig. 14.6 are the calculated points. The value of $N_s(z, t)/N_B$ is calculated from (14.4) using $N_B = 10^{17}$ cm^{-3}, and $N_s/N_B = 0.42$ is estimated by extrapolating the experimental curve to $z = z_o$ as indicated by the dashed line in Fig. 14.6. Values of B and D_s needed to evaluate $N_s(z, t)/N_B$ from (14.4) are $B = 3 \times 10^{-2}$ μm^2/hr, obtained from Fig. 13.17 for oxygen at 1100°C, and $D_s = 0.36$ μm^2/hr, obtained from Ref. 54 for the diffusion coefficient of boron in silicon. As seen from Fig. 14.6, (14.4) predicts the correct shape of the impurity distribution, which is linear over the depth range measured.

To check the consistency of the calculation, m was calculated from (14.5) using the experimental value $N_s/N_B = 0.42$. For boron, r in (14.5) is very small because the diffusivity of boron in silica can be considered negligible. As a result, $b = 0$ in (14.5), and the calculation with $b = 0$ yields the value $m = 0.2$. This value of m lies between the independently measured values of m of 0.1–0.3 listed in Table 14.1 for boron. Thus the calculation gives the right shape of the impurity distribution and a reasonable value of m, and it can be concluded that the theory accurately predicts the impurity distribution in the silicon. However, the theory has not been tested over a complete impurity profile because of the difficulty in measuring the profile from the silicon surface to the point where the impurity concentration equals N_B in any given sample. Moreover, almost all the profile measurements have been made after oxidation at high temperatures and virtually none after low oxidation temperatures, where the impurity profile would be much sharper as predicted in Fig. 14.5. However, the theory has been compared to measurements of profiles of impurities other than boron such as gallium, indium, phosphorus, arsenic, and antimony.[54,55] Although these profile measurements had a lower resolution[55] or did not measure the profile as close to the silicon surface[54] as the profile measurement in Fig. 14.6, they were consistent with the theory. For phosphorus, which piles up on the silicon side of the Si–SiO$_2$ interface, the concentration in the piled-up region can become very high. At very high concentrations, some of the phosphorus precipitates or forms complexes that are electrically neutral.[61] In such cases, the actual phosphorus profile cannot be measured by the methods used in obtaining Fig. 14.6, which measure only the ionized fraction. Although the theory still will hold, the diffusion coefficients of the electrically neutral phosphorus complexes will be different from those that can be used at lower concentrations.

(b) Nonuniform Case

The relatively simple problem of the redistribution of impurities from an initially uniform profile has been considered. This case is important in many devices and in many MOS capacitor admittance measurements. However, there are many cases of practical interest where the initial impurity distribution in the silicon is not uniform. In general, it is not possible to obtain a solution in closed form for the initially nonuniform distribution problem with a parabolic oxide growth rate. Therefore, the diffusion and boundary equations [(1) and (2) in Appendix XIV] must be numerically solved on a computer. Kato and Nishi[56] were the first to treat this problem both mathematically and experimentally. They obtained an approximate solution that fitted the measurements using the method of Green's functions. Their approximation was based on dividing the total oxidation time into several unequal time intervals and assuming that the oxidation velocity dx_o/dt is constant in each time interval making oxidation velocity independent of time. Another simplification is to use a coordinate system that moves with the Si–SiO$_2$ interface, in contrast to using a fixed coordinate system for solving the initially uniform impurity distribution case as done in Appendix XIV. Use of a moving coordinate system makes the boundary equation (1) only explicitly a function of time, whereas use of a fixed coordinate system makes the boundary equation depend both explicitly and implicitly on time. In the fixed coordinate system the implicit time dependence comes about because N_s and its first derivative are evaluated at $z = z_o(t)$. In contrast, N_s and its first derivative are evaluated at $z = 0$ in the coordinate system moving with the Si–SiO$_2$ interface. The implicit time dependence of the boundary equation in the fixed coordinate system introduces unnecessary difficulties in solving the nonuniform distribution problem, which is avoided by a transformation to a moving coordinate system. In contrast, for the uniform initial impurity distribution, the fixed coordinate system resulted in the simpler mathematics. Another advantage of the moving coordinate system is that the profile is calculated from $z = 0$ and measured from $z = 0$. Use of the fixed coordinate system requires the calculated profile to be shifted by z_o to correspond to the measured profile.

In the moving coordinate system, the diffusion equation will have an additional term related to the oxidation velocity

$$\frac{\partial N_s(z, t)}{\partial t} = D_s \frac{\partial^2 N_s(z, t)}{\partial z^2} + a \left(\frac{dx_o}{dt}\right) \frac{\partial N_s(z, t)}{\partial z}. \tag{14.7}$$

The boundary equation for the conservation of impurity atoms at the Si–SiO$_2$ interface derived in Appendix XV has the same form in the moving coordinate system as it did in the fixed coordinate system. Neglecting diffusion in the oxide, a realistic approximation for boron and phosphorus, the boundary equation in the moving coordinate system is

$$D_s \frac{\partial N_s(0, t)}{\partial z} = \left(1 - \frac{1}{ma}\right)\left(\frac{dx_o}{dt}\right) N_s(0, t). \tag{14.8}$$

Solutions to (14.7) with boundary conditions (14.8) and $N_I(z, 0) \to N_B$ as

$z \to \infty$, where $N_I(z, 0)$ is the initial impurity distribution, can be found as long as oxidation velocity is small compared to diffusion velocity. This condition will be true for boron and phosphorus and oxidation at high temperature in one atmosphere of dry oxygen. Because the Green's function method used by Kato and Nishi[56] and others[62,63] requires long computation time to evaluate for many problems of practical interest, it is not pursued further here. Rather, we describe, in Appendix XV, the calculation by Krambeck,[64] who solved (14.7) for an initial impurity distribution $N_I(z, 0)$ by the method of finite differences. The primary virtue of Krambeck's[64] calculation is that computer time, on a Honeywell 600 time sharing facility, for example, was less than 1 min delay for solving (14.7) for a simulated 2 hr oxidation. Moreover, this calculation agreed with analytically checkable solutions, such as an initially uniform impurity concentration, to within 1%, which is an accuracy acceptable for most device applications.

There may be specific problems where the Green's function method could result in shorter computation time than the method of finite differences. However, there has been no study to date comparing the two methods for a wide variety of problems. The method of finite differences is emphasized here mainly because it is simpler and has been found to be very practical.

Figure 14.7 shows the good agreement between Krambeck's calculation and the experimental data of Kato and Nishi,[56] who measured the redistributed impurity profile after oxidizing silicon samples that were prede-

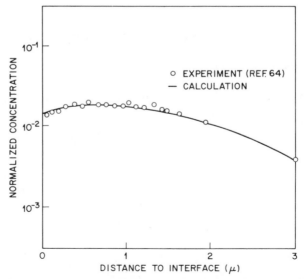

Fig. 14.7 Redistribution of a diffused boron layer after thermal oxidation. After Krambeck.[64] Reprinted by permission of the publisher, The Electrochemical Society, Inc.

posited and simultaneously diffused with boron. The predepositions were made in the temperature range 950°–1100°C for 20–30 min in a purified nitrogen atmosphere. The boron source was B_2O_3. The surface concentration N_{sp} was found by sheet resistivity and depth measurements to be 0.3×10^{20}–10^{20} cm^{-3}. Impurity concentration, prior to oxidation was $N_I(z, 0) = N_{sp}\,\mathrm{erfc}(z/2\sqrt{D_s t_p})$, where t_p was the predeposition-diffusion time. Prior to oxidation, the predeposited samples were dipped in HF to remove any boron glass. Drive-in was then done in 1 atm of dry oxygen in the temperature range 1120–1230°C. The solid curve in Fig. 14.7 was calculated by Krambeck[64] for $2\sqrt{D_s t_p} = 3.6 \times 10^{-2}\ \mu$m, $B = 3.7 \times 10^{-2}\ \mu$m^2/hr, $D_s = 0.7\ \mu$m^2/hr, an oxidation time of 0.86 hr, and $m = 0.125$. The open circles are the data of Kato and Nishi[56] obtained by successively etching away a thin layer of silicon and measuring resistivity of the silicon, using the four-point probe method, before each layer was etched off. Thus the application of the method of finite differences to the solution of the impurity redistribution problem by Krambeck[64] affords a useful method in the design of silicon devices.

The analysis given here has been done for constant B and D_s. For cases where the impurity concentration dependence of these quantities must be considered see Warner and Wilson[65] and their summary of the literature.

14.5 OXIDE MASKING AGAINST IMPURITIES

Frosch and Derick[66] found that SiO_2 is an effective barrier against many impurities commonly used in doping silicon. This discovery is important because it forms the basis of present-day integrated circuit technology. By etching holes in the oxide, using the remaining oxide as a mask against impurities, dopant impurities can be incorporated into a silicon substrate in selective areas to make the active p-n junction regions in an integrated circuit. To be effective as a mask, diffusivity of the dopant impurity in SiO_2 must be smaller than in silicon. The commonly used doping impurities for which diffusivity in SiO_2 is smaller than in silicon are phosphorus, boron, arsenic, and antimony, but not gallium. Thus SiO_2 will be an effective mask against phosphorus, boron, arsenic, and antimony. A brief description of the effectiveness of SiO_2 as a mask against phosphorus, boron, arsenic, antimony, and gallium with emphasis on phosphorus and boron is given.

Figure 14.8 shows the thickness x_m at which SiO_2 grown in dry oxygen fails to mask against P_2O_5 and B_2O_3 vapors as a function of time with diffusion temperature as parameter.[67,68] Phosphorus and boron are given as quantitative examples, but the SiO_2 masks against other impurities as well, some of which are discussed later. To continue with the case of phosphorus diffusion in SiO_2, which illustrates the masking effect, the diffusion process can be described as occurring in two steps. During the first step, SiO_2 is completely effective in masking against P_2O_5 in the gas phase, so no

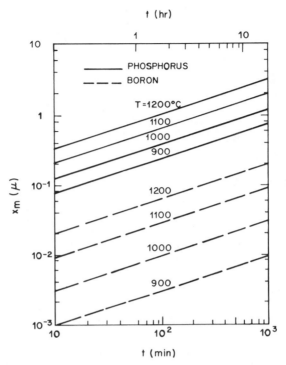

Fig. 14.8 Minimum thickness x_m of dry oxygen-grown SiO_2 required to mask against phosphorus and boron as a function of diffusion time with diffusion temperature as parameter. The phosphorus curves follow (14.12) and the boron curves, (14.13). From Wolfe,[113] based on the phosphorus data due to Sah et al.[67] and the boron data due to Horiuchi and Yamaguchi.[68] Copyright (1969), Pergamon Press, Ltd. Reprinted with permission.

phosphorus-rich layer forms in the silicon beneath the SiO_2 layer. In this first step, P_2O_5 reacts with SiO_2 to form a glass. As time goes on, the thickness of the glass increases until the entire SiO_2 layer is converted into a phosphosilicate glass. Now the second step begins. An oxide of phosphorus diffuses through the glass. When the phosphorus oxide reaches the glass-silicon interface, phosphorus atoms enter the silicon and diffuse away, and the phosphorus converts the silicon surface to n-type. A similar mechanism applies for boron.[68] The SiO_2 masks against boron until the entire SiO_2 layer is converted to a borosilicate glass. The two-step diffusion process just described is illustrated in Fig. 14.9. Figure 14.9a shows the beginning of the diffusion process in the SiO_2 with only the surface of the SiO_2 converted into a glass. Figure 14.9b shows the situation after most of the SiO_2 layer has been converted into a glass. Figure 14.9c shows the situation just after the entire SiO_2 layer has been converted into a glass. Finally, Fig. 14.9d shows the situation a considerable time after the SiO_2

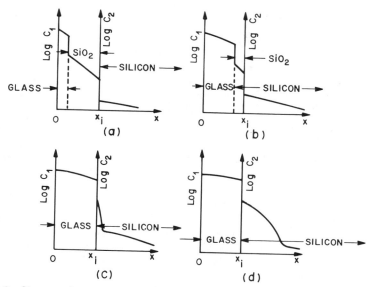

Fig. 14.9 Change of concentration profile of the diffusant with the growth of the glass: (a) beginning of the diffusion; (b) after an appreciable fraction of the SiO_2 has reacted to form the glass; (c) just after the entire SiO_2 layer has been converted to a glass; and (d) continuing diffusion for a considerable time after the entire SiO_2 layer has been converted to a glass. After Horiuchi and Yamaguchi.[68]

layer has been converted into a glass and the impurity concentration in the silicon has increased significantly. Growth of the glass is limited by diffusion of the diffusant into the SiO_2. To find the limiting oxide thickness x_m (given in Fig. 14.8) at which the oxide just fails to mask (Fig. 14.9c), Fick's law must be solved during a chemical reaction as done in Appendix XVI. The solution to this problem is

$$\frac{x_m}{2\sqrt{D_o t_m}} = L \qquad (14.9)$$

where x_m is the oxide thickness at which masking fails, t_m is the corresponding diffusion time, D_o is the diffusion coefficient of the diffusant, and L is a constant that depends on the boundary condition at the vapor-SiO_2 interface for the limiting case of complete reaction at the glass-SiO_2 interface. The parabolic law of film growth is predicted by (14.9). To render (14.9) into a form that can be compared to experimental data, we need the temperature dependence of D_o, which is

$$D_o = D_\infty \exp\left(-\frac{qE_A}{kT_m}\right) \qquad (14.10)$$

where E_A is the activation energy and T_m the diffusion temperature.

Substituting (14.10) into (14.9) yields the parabolic relation

$$\frac{x_m^2}{t_m} = 4L^2 D_\infty \exp\left(-\frac{qE_A}{kT_m}\right).$$ (14.11)

Fitting (14.11) to the experimental data for phosphorus yields the following parabolic law with an activation energy of 1.46 volts[67]

$$\frac{x_m^2}{t_m} = 1.7 \times 10^{-7} \exp(-1.46\beta) \quad (\text{cm}^2/\text{sec})$$ (14.12)

where $\beta = q/kT_m$. Equation (14.12) describes the curves for phosphorus given in Fig. 14.8. Similarily, for boron,[68] the parabolic relationship is

$$\frac{x_m^2}{t_m} = 4.9 \times 10^{-5} \exp(-2.80\beta)\,(\text{cm}^2/\text{sec}).$$ (14.13)

Equation (14.13) describes the curves for boron given in Fig. 14.8. The curves in Fig. 14.8 must be applied with caution because the diffusivity of boron and phosphorus are concentration dependent. However, Fig. 14.8 is conservative because it is based on the highest diffusivities for boron and phosphorus reported in the literature. The curves for boron were obtained under a high B_2O_3 vapor pressure, resulting in a high diffusivity of boron in the SiO_2,[68] and the curves for phosphorus were obtained under a high P_2O_5 vapor pressure, resulting in a high diffusivity of phosphorus in the SiO_2.[62] For phosphorus, the diffusion constants and activation energy reported in the literature vary widely. At 1100°C the diffusivities reported by Allen et al.[69] $(1.4 \times 10^{-15}\,\text{cm}^2/\text{sec})$ and by Thurston et al.[70] $(1.7 \times 10^{-15}\,\text{cm}^2/\text{sec})$ closely match but are about five times larger than the values given by Barry[71] $(2.9 \times 10^{-16}\,\text{cm}^2/\text{sec})$ and over 2 orders of magnitude smaller than the value given by Sah et al.[67] $(2.0 \times 10^{-13}\,\text{cm}^2/\text{sec})$. The reason for the high diffusivity found by Sah et al.[67] is the high vapor pressure of P_2O_5 used in their experiments.

For boron, the preexponential constant D_∞ in (14.10) is highly dependent on the boron concentration in the diffusion source, regardless of whether the boron source is a vapor or a glass. The constant D_∞ increases nearly three decades from low to high boron concentration, as is apparent from the data due to Brown and Kennicott[72] and Schwenker.[73] Horiuchi and Yamaguchi[68] reported an increase in boron diffusivity in SiO_2 at 1200°C from $3.2 \times 10^{-16}\,\text{cm}^2/\text{sec}$ to $6.2 \times 10^{-15}\,\text{cm}^2/\text{sec}$ caused by an increase in the B_2O_3 vapor pressure. Grove et al.[57] report that the diffusivity of boron in SiO_2 increases by three decades in a hydrogen ambient. Water vapor has no effect on the diffusivity of boron in SiO_2.[57] However, the diffusivity of boron in SiO_2 can be enhanced during oxidation in steam because hydrogen is generated as one of the final products of the oxidation reaction as shown in reaction (2) in Section 13.2.1.[74] Thus it is important to avoid a hydrogen ambient to obtain optimum masking by the SiO_2 film against boron. It is particularly important to oxidize in dry oxygen in silicon gate technology, if the polysilicon is boron doped, to prevent boron from the polysilicon from diffusing all the way through the gate oxide to the substrate.

Investigations of arsenic diffusion through SiO_2 from an arsenosilicate glass source by Wong and Ghezzo[75] gave results consistent with those reported by Hsueh.[76] The diffusivity of arsenic in SiO_2 was found to increase a small amount with concentration at 1100°C in argon from 2.6×10^{-16} cm²/sec to 3.5×10^{-15} cm²/sec compared to boron and phosphorus. The diffusion ambient affects the diffusivity of arsenic in SiO_2, which is drastically reduced in the presence of oxygen.[77]

The only data available on antimony diffusivity through SiO_2 at different temperatures were given by Thurston et al.[70] At diffusion temperatures of 1100–1200°C, the diffusivity of antimony increases from 9.9×10^{-17} cm²/sec to 1.5×10^{-14} cm²/sec.

The diffusivity of gallium through SiO_2 reported by Grove et al.[57] is 5.2×10^{-11} cm²/sec at 100°C. This value of diffusivity is two to six decades larger than the diffusivities of boron, phosphorus, arsenic, and antimony in SiO_2. In addition, the diffusivity of gallium in SiO_2 (5.2×10^{-11} cm²/sec)[57] is larger than in silicon (1.7×10^{-13} cm²/sec)[78] at 1100°C; hence SiO_2 is not effective as a mask against gallium.

14.6 OTHER METHODS OF OXIDE PREPARATION

14.6.1 High Pressure Steam Oxidation

High pressure (> 1 atm) steam oxidation has two advantages over steam oxidation at one atmosphere: (1) oxide thicknesses of several micrometers can be grown in a relatively short time with no cracking or devitrification; and (2) a slice in which diffusions have been undertaken to form p-n junctions can be oxidized without significant movement of the diffusants because of the low temperatures and short times required in high pressure steam oxidation. These advantages are not necessary in most practical applications, so high pressure steam oxidation has found only limited application. However, there has been interest in high pressure steam oxidation for growing the relatively thick field oxide. Growth of the field oxide is the longest single step in integrated circuit processing, and high pressure steam oxidation holds the promise of substantially reducing that time. The practice of high pressure steam oxidation is briefly discussed for those applications where its advantages are paramount.

There are two ways of implementing high pressure steam oxidation. The first is a steam generator feeding a closed tube that contains the silicon to be oxidized. Steam pressure is established in the closed tube by a pressure-sensitive valve that opens when steam pressure reaches a predetermined value. This arrangement works much like an ordinary steam boiler and is suitable for the lower pressure ranges up to about 10 or 20 atm. For higher

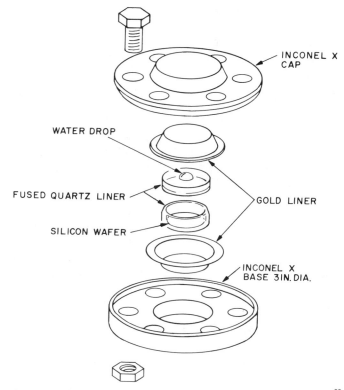

INCONEL X
CAP

WATER DROP

FUSED QUARTZ LINER

GOLD LINER

SILICON WAFER

INCONEL X
BASE 3IN.DIA.

Fig. 14.10   High pressure steam oxidation reactor. After Powell.[80]

pressures, the second method, a steam high pressure reactor, is used.[79] Figure 14.10 illustrates a cutaway of a typical steam reactor used for handling a single wafer for experimental purposes. This reactor will support a pressure of 5000 atm at room temperature and 100 atm at 890°C.[79] The reactor is made of Inconel X to give the system mechanical strength at elevated temperatures with six bolts to secure tight closure. Two gold liners, as shown, provide the seal and avoid contamination of the steam. Because high pressure steam at elevated temperatures is very corrosive, an inert metal such as gold must be used as the liner. If another more active metal were used, high pressure steam would dissolve impurities from the chamber walls, and these impurities would be incorporated into the silica layer being grown with adverse effects on the electrical properties of the underlying devices.

The ultimate steam pressure reached within the reactor is determined by the quantity of water placed into the reactor before sealing, oxidation temperature, and the volume of the reactor interior. This quantity of water can be calculated from *PVT* (pressure-volume-temperature) data for water above critical temperature and delivered to the reactor by means of a pipette with an accuracy of 0.5%. Typically, reactor volume is known to 0.2%, so pressure in the reactor is known to 1%. The drop in steam pressure in the reactor due to reaction with silicon is negligible when compared to the accuracy with which pressure is known for most of the film thicknesses of interest. A silica liner also is used, as shown in Fig. 14.10, to house the silicon slice and keep it from coming in contact with the liquid water drop placed into the reactor before sealing. Quality, uniformity, and reproducibility of the silica film are improved by keeping the liquid water drop from contacting the silicon slice, as the water drop can contain both dissolved and particulate impurities.[80] The reactor is then heated by placing it into a furnace held at oxidation temperature.

Growth of the silica film proceeds linearly with pressure and time as described previously. Experiments conducted at temperatures of 500–850°C and pressures of 25–400 atm for times of 0.5–8 hr and thicknesses of up to 7 μm gave linear kinetics.[79] Figure 14.11 illustrates oxide growth in high pressure steam at one temperature and pressure.

Film growth under conditions of very high pressure for long periods occurs at rates that are nonlinear. Furthermore, these rates are always smaller than those found at lower pressures. This behavior is due to the solubility of oxide in steam.[81] It is known that steam at pressures of 100 atm and above can dissolve the various forms of silicon dioxide. The overall trend for the solubility of the oxide for water is to increase with temperature and pressure. Above certain pressures the silica film does not grow, and an etching of silicon occurs, producing a surface approximating the appearance of a lapped surface. Table 14.2 shows the minimum pressure at a given temperature at which etching occurs.

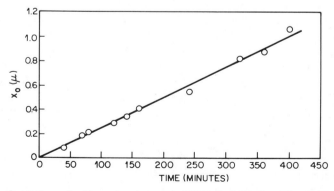

Fig. 14.11 Oxidation of silicon in steam at 650°C and 50 atm. The oxidation rate is 25.7 ± 0.8 Å/min. After Ligenza.[79] Reprinted by permission of the publisher, The Electrochemical Society, Inc.

Table 14.2 Minimum Pressure at
Which Etching of the Silicon by Steam
Occurs at Different Temperatures[a]

Temperature (°C)	Minimum Pressure $P_m(\pm 50 \text{ atm})$
500	500
650	400
750	200
850	150

[a] After Ligenza.[79]

High pressure steam reactors are commercially available, suitable for rapidly growing field oxides in production. Also, rapid oxidation in high pressure oxygen suitable for growing field oxides has been reported.[82-84]

14.6.2 Anodic and Plasma Grown Oxides

(a) *Anodization in a Liquid Electrolyte*

If a silicon wafer is made the anode of an electrolytic cell containing an electrolyte with an oxygen bearing component, which does not dissolve silicon or its oxide, and a current is passed between the anode and an inert cathode such as platinum, an oxide film will grow on the silicon surface. The oxide film grows as the current, which is predominantly electronic with a small ionic component, flows through the growing oxide layer under an applied electric field. The exact mechanism of the anodic oxidation of silicon is not well established at present.

Anodically grown SiO_2 films can be made comparable in quality to thermally grown SiO_2 provided that the anodic films are annealed. Structural differences between thermally and anodically grown and annealed films have been detected by X-ray diffraction methods[85] and infrared spectroscopy.[86] The anodic film appears to be truly amorphous, whereas the thermally grown oxide has short range order. Another difference is that impurities in the electrolyte and the material composing the solutes and solvents in the electrolyte can be incorporated into the growing anodic film. Nevertheless, the electrical properties of the Si–SiO_2 interface of anodic films can be made comparable to thermally grown films by postoxidation annealing treatments such as heating in H_2 at 500°C or heating in an inert ambient such as argon or helium at 1000-1200°C.[77] Because anodic films have not been studied as extensively as have thermally grown oxide films, it is not known whether anodic films after annealing will age in the same way as thermally grown films.

The major advantages of anodically grown oxide films are: (1) there is no impurity redistribution at the silicon surface as there is in thermal oxidation; and (2) implementation of the process and oxide thickness control are easy and low in cost.

In most cases it is not practical to grow silica films thicker than about 0.2 μm because the voltages become so high for greater thicknesses that the oxide breaks down and is of poor quality. A way of growing thicker oxides is by a combination of anodic and thermal oxidation. The anodic part consists of exposing silicon to an HF electrolyte at low voltages[87, 88] to form a layer of porous silicon containing pipes of SiO_2.[89] The depth of such a layer is limited by reaction time but can be made up to 10 μm deep. The thermal part consists of annealing in dry oxygen or steam, which rapidly converts the porous layer to a good insulator.

An application of this technique has been to provide an isolation region between bipolar transistors in an integrated circuit.[88] When the isolation region is initially p-type, the slice is anodized, thereby converting the diffused p^+ isolation region into porous silicon. The forming voltage for the p^+ region is much lower than for nearby n-type regions because of the high hole concentration in the p^+ layer. Therefore, the anodic reaction progresses selectively in the p^+ regions, leaving the n-regions unaffected. Thus no covering on the n-type regions is required. When the isolation region is n-type, a silicon nitride layer must be used to mask the p-type regions. Light of sufficient energy to create hole-electron pairs in the silicon must be used to raise hole density in the n-type isolation region sufficiently high for anodization to occur. After anodization the porous layer is thermally oxidized. Typically, a porous layer 5 μm thick will completely oxidize after 10 min at 1150°C.[90] To grow the same thickness of oxide thermally in bulk silicon would require 35 hr at 1150°C in steam. Thus oxidation of the porous layer is so much faster than for bulk silicon that only the porous regions are selectively oxidized. The isolation characteristics of the oxidized porous layer are not good for p-n junctions, for example, because leakage current is about 10 times higher.[90] However, the oxidized, porous layer isolation results in a smaller bipolar transistor than junction isolation. In some cases the higher leakage current of the oxidized porous layer may be tolerable, making it a useful technique.

There are two major disadvantages of anodization. First, postanodization annealing is required to obtain a device-grade film. This annealing is done at high temperatures (1000°C) in a hydrogen ambient. If optimum interfacial properties are desired, the overall process is a high temperature one, even though the oxide is grown anodically at room temperature. Therefore, anodic films are used for purposes other than passivation or gate oxide. Thermally grown oxides are almost exclusively used for passivation or where interfacial electrical properties are critical, such as the gate oxide of a MOSFET. Second, more impurities, which can have a detrimental effect on device performance and stability, will be incorporated in a carefully grown anodic oxide than in a carefully grown thermal oxide. The greater purity possible with thermal oxidation and, therefore, better equality for device applications in most cases outweighs the advantages of anodic oxides. However, anodic techniques are used in some practical applications, such as bipolar transistor isolation in an integrated circuit as just described; another is described next. In this latter application silicon nitride is converted to silicon dioxide to permit easy etching of holes to the underlying silicon.[91] The way this works is as follows. Holes are first cut in the silica layer by etching in buffered HF (see notes in Table 14.3) down to the silicon. Then a layer of silicon nitride is deposited, as described in Section 14.6.4, over the entire wafer surface to seal the oxide against alkali metal contamination. This nitride layer also covers the holes in the oxide previously cut. Nitride in these holes must be etched away to permit subsequent metallization to make electrical contact to the silicon while leaving the nitride layer intact over the oxide. One way of solving this problem is by anodization, which avoids a masking step. During anodization, any nitride in the contact holes, where electric field and hence ionic current density are highest, is converted into silicon dioxide. In those regions where silicon nitride covers the oxide, there will be little conversion because of the underlying oxide, which makes the electric field and the ionic current density there negligibly small. The converted nitride in the contact holes is then etched away in buffered HF, which does not etch the unconverted nitride in the surrounding areas.

The anodization process is briefly described next. Although various electrolytes can be used, such as boric acid, sulfuric acid, and HF, the best quality oxides are obtained by using either a solution of KNO_3 in tetrahydrofurfuryl alcohol or KNO_3 in glycol. Addition of a salt such as KNO_3 is necessary to increase the electrical conductivity of the electrolyte and to supply oxygen for the anodic reaction. Unfortunately, some potassium from such an electrolyte will become incorporated in the oxide film. Some of the potassium incorporated in the oxide film can be ionized to form mobile positive charge[92] and possibly cause device instability. However, the mobility of potassium in SiO_2 is one-tenth the mobility of sodium in SiO_2 for equal concentrations,[93] so incorporation of potassium in the anodized film may not be too serious in some cases.

There are two regimes in the current-voltage characteristics of the electrolytic cell during anodization. At low voltages the current that flows is small and predominantly electronic, caused by the discharge of ions in the electrolyte. No oxide grows until the voltage applied to the cell becomes large enough that a significant component of the leakage current is ionic. The ionic component rises rapidly with applied voltage. The ionic current efficiency is usually quite low—on the order of 1% of the total leakage current. The voltage can be increased up to a certain limit beyond which the oxide breaks down, and a good-quality film cannot be made. Typically, voltages applied for producing good-quality films are on the order of 250 V, with current densities on the order of 10 mA/cm^2.

Table 14.3 Properties of SiO$_2$ Prepared in the Various Ways Described in This Chapter[a]

Type of Oxide	Density (g/cm^3)	Index of Refraction n_1 ($\lambda = 5460$ Å)	Dielectric Constant (at 1 MHz Unless Otherwise Noted)	Resistivity (Ω-cm)	Dielectric Strength (10^6 V/cm)	Etch Rates[b] 1.8 Molar HF Etch (Å/min)	Buffered HF Etch (Å/min)	p_0-Etch (Å/min)
Thermal oxide								
Steam (97°C H$_2$O)								
Open tube	2.00–2.20	1.45–1.46	3.9	10^{15}–10^{17}	6.8–9.0	162	438	120
High pressure	2.32	1.48			6.8–9.0	156–162		
Dry oxygen	2.24–2.27	1.46–1.47	3.9	3×10^{15}–2×10^{16}	6.8–9.0	198	408	120
Anodic oxide								
Electrolytic	1.80	1.32–1.49			5.2–20	21,600	3000–4500	10,800–13,800
Plasma		1.46–1.47			10	174	540	144
Deposited oxide								
Silane			4.4–4.6 (10 kHz)		5		1200–6000	
Bulk silica glass	2.20	1.46	3.8	10^{15}–10^{16}	1–5			

[a] After Donovan.[99]

[b] Buffered HF etch: 10 cm^3 HF (48%), 100 cm^3 NH$_4$F solution (1 lb NH$_4$F/680 cm^3 de-ionized H$_2$O); p-etch: 15 cm^3 HF (48%), 10 cm^3 HNO$_3$ (70%), 300 cm^3 de-ionized H$_2$O.[112]

There are two simple experimental conditions under which anodization can be accomplished: (1) constant current and (2) constant voltage. During formation at constant current, each new layer of oxide of thickness dx requires an additional potential dV to be applied across the film to maintain the field in the oxide and, therefore, the current constant. If the potential drop in the previously existing part of the oxide film is not changed during further growth, the differential field strength dV/dx_o, is the actual electric field in the new layer just forming. Typically, dV/dx_o does not vary with increasing oxide thickness at constant current.

The relation between the rate of increase of potential and current density is derived next. The rate of increase of oxide thickness dx_o/dt is given by

$$\frac{dx_o}{dt} = \frac{JM}{ZF\rho} \tag{14.14}$$

where J is the ionic current density component, Z is the number of Faradays F required to form the molecular weight of oxide, M, and ρ is the density of the oxide film. If the differential field strength E is constant, the rate of increase of potential required to maintain constant current is

$$\frac{dV}{dt} = E\left(\frac{dx_o}{dt}\right) = \frac{JEM}{ZF\rho}. \tag{14.15}$$

In practice, a rough idea of the magnitude of dV/dt can be obtained by noting that E varies slowly with J, usually less than 10% for a decade change in J, so that dV/dt is about proportional to J.

At constant voltage, growth of the oxide film results in a continuous decrease in electric field, and thus current density decreases. Eventually, the rate of oxide growth becomes so small that the oxide thickness may be said to have reached a limiting value. This limiting value is not real because at infinite time the oxide would be infinitely thick. However, a limiting value exists in the sense that an inordinate amount of time would be needed to grow a thicker oxide.

(b) Plasma Anodization

In plasma anodization the liquid electrolyte is replaced by a gaseous plasma at a low pressure. It is thought that oxygen ions, produced in the gaseous plasma, drift through the growing oxide layer under an applied electric field and react with the silicon at the Si–SiO$_2$ interface to make new oxide. A property common to plasma anodization and anodization in a liquid electrolyte is that both oxidation rate and oxide thickness are electrically controlled. One important advantage of plasma anodization over anodization in a liquid electrolyte, for electron device applications, is the extreme purity of the oxide grown by plasma anodization resulting from freedom from solute, solvent, and other impurity incorporation into the oxide film. The plasma anodizing process is inherently self-purifying with regard to positively charged mobile impurity ions because the applied electric field repels such ions from the silicon surface. A high temperature (1000°C in H_2) annealing treatment is required to obtain interfacial electrical properties comparable to thermally grown oxide films. Thus the overall process is a high temperature one, although the plasma anodization part can be done at room temperature. Dielectric breakdown strength and dielectric constants also are the same as for thermally grown oxides.

Plasma grown oxides are not superior to thermally grown oxides, and both are high temperature processes, but the apparatus needed to plasma anodize is more expensive and requires more skill to use than that for thermal oxidation. Therefore, plasma anodization is not used in silicon integrated circuit manufacture. However, plasma anodization may be useful in certain research or development applications and is a candidate as a process for growing the field oxide in competition with high pressure steam oxidation.

There are three methods of plasma anodization: (1) the glow discharge method, which produces a plasma that is feeble but can be generated in large volumes; hence this method yields small oxidation rates on large areas; (2) the microwave method of producing a plasma,[94] where the microwave discharge produces an intense plasma of small volume that results in very high oxidation rates on small areas; and (3) a high current, low voltage, dc arc plasma oxidation method,[95] utilizing a thermionic cathode that spans the operation parameters of plasma density, volume, and gas pressure of the two other methods. Therefore, method 3 is more feasible than method 1 or 2.

The silicon surface can be sputter cleaned in the vacuum apparatus, necessary for plasma anodization, prior to anodization. In the high current, low voltage, plasma method, typical anodization with an oxygen pressure of 0.07 Torr and a constant current density of 34 mA/cm^2 produces 500 Å of SiO_2 in 10 min with the silicon at room temperature.

14.6.3 Deposited Silicon Dioxide

A number of different methods are used in the electronics industry to deposit insulating films. Some of these methods are: thermal vapor deposition; electrophoresis where charged particles of material are deposited; spin-on; spray-on; and drip-drain techniques that are used mainly for photoresist application and in optical devices; and silk screening, which is a method for applying thick films in a particular pattern. We discuss only chemical vapor deposition (CVD), as this method has such a wide variety of uses in integrated circuit technology.

Chemical vapor deposited oxides do not replace thermally grown oxides because the best electrical properties are obtained with thermally grown films. Rather, CVD oxides are used for masking, as a diffusion source, for electrical insulation between multilayer metallizations, or over a silicon gate, to obtain thick oxide layers and to act as a protective layer over the entire integrated circuit. A CVD film of SiO_2 covering the aluminum metallization in an integrated circuit can serve to inhibit electromigration of aluminum.* For these applications, CVD oxides have several advantages: (1) thicker films can be obtained at low temperatures and shorter times than can be obtained by thermal oxidation; (2) the silicon substrate is not consumed in the CVD process, and little impurity redistribution occurs in the silicon, as would occur in thermal oxidation; and (3) CVD films can be produced on materials other than silicon, such as the metal lead pattern on an integrated circuit.

In most integrated circuit applications there are variations in oxide thickness, such as between field and gate oxide, and steps between the metallization pattern and the oxide surface. To cover these steps with a smooth CVD oxide, the oxide is doped with phosphorus and heated to 1000 or 1100°C. At these temperatures the phosphosilicate glass will flow, forming a smooth topography over the various steps on the SiO_2 surface. However, in those regions of the integrated circuit where the electrical properties of the Si–SiO$_2$ interface are critical to device performance, such as under the gate of a MOSFET, a thermally grown film is used.

Silicon dioxide deposition methods can be divided into three main categories defined by their deposition temperature ranges. The choice of method depends on the temperature and ambient that can be tolerated and the properties of the films required. The use for each method is described along with the method itself. Descriptions of the wide variety of apparatus designs commercially used for the CVD of SiO_2 in integrated circuit processing is not given, except to illustrate some of the principles.

(a) Low Temperature Oxide

Low temperature silica can be deposited by the use of silane (SiH_4) and O_2 and N_2 as carrier gas.[96] The useful deposition temperature is 300–500°C. A relatively simple resistance heated deposition system is employed. The oxide film produced at these low temperatures can be used for masking or covering temperature-sensitive structures or layers on silicon, some metals, and ceramics. The basis for oxide deposition is the oxidation of silane

$$SiH_4(gas) + O_2(gas) \rightarrow SiO_2(solid) + H_2O(gas). \qquad (1)$$

Silane is a gas that ignites spontaneously in air to form a powdery deposit of colloidal silica that, as such, is useless in semiconductor technology. However, the reaction can be controlled by diluting the silane with an inert

---

*Electromigration is particularly severe in aluminum leads. It results from high current densities ($<10^6$ A/cm^2) in the lead. Such current densities can occur because of the narrow and thin dimensions of many leads used in integrated circuits. At such high current densities, the electron wind pushes aluminum atoms before it creating voids that ultimately result in an open circuit connection.

gas prior to reacting it with oxygen. Diluting the silane to 1% in nitrogen is satisfactory. Transparent and adherent films of silica can be deposited by the low temperature CVD method.

There are several disadvantages to low temperature CVD oxide films. First, the etch rates in buffered HF are high being about four to five times as fast as for thermally grown oxides. Such high etch rates can lead to severe undercutting of photoresist films or lifting if the photoresist films have poor adherence to the oxide film. The high etch rate (>4000 Å/min in buffered HF) can be reduced to as low as 1000 Å/min by annealing the oxide in oxygen at high temperatures. This anneal densifies the oxide, resulting in the lowered etch rate. However, the use of high temperatures for annealing destroys the low temperature feature of this method of oxide deposition. A second disadvantage of the low temperature CVD method is the tendency for silica dust produced during the reaction to build up on the reactor walls. These particles can break away and fall on the wafer during the deposition cycle and cause defects in the film. The low temperature CVD method also demands scrupulously cleaned surfaces, which is an ideal not often achieved in practice, to prevent "swirl" or watermark patterns visible in the deposited film.

Deposition rates of SiO_2 are dependent on the concentrations of silane, oxygen, and temperature. The deposition rate is nearly linearly dependent on silane concentration, as shown in Fig. 14.12a. The effect of oxygen on the deposition rate is shown in Fig. 14.12b. A maximum occurs in Fig. 14.12b at oxygen concentrations near 0.5%. The decrease in deposition rate for oxygen concentrations above 0.5% is caused by the formation of SiO_2 in the reaction chamber and its deposition on the chamber walls rather than on the silicon. The deposition rate is not a strong function of temperature around 350°C. The deposition rate changes from 435 Å/min at 300°C to 520 Å/min at 375°C.

Because the quality of low temperature CVD oxides is sensitive to surface cleanliness, pinhole density or the number of other film defects can vary over a wide range. Typical values of pinhole density are $50\ cm^{-2}$, measured with an electrophoretic tester.

(b) Intermediate-Temperature Oxide

Intermediate temperature (500–850°C) oxides may be deposited using pyrolysis (thermal decomposition) of tetraethylorthosilicate, $(C_2H_5O)_4Si$, in a hot wall tube furnace reactor.[97,98] The deposition temperature is normally around 765°C, and the carrier gas is nitrogen. Advantages of the intermediate temperature method are a short time cycle (e.g., a film 2000 Å thick can be deposited in 17 min), and large numbers of silicon wafers can be handled on a furnace paddle or rack.

The pinhole density of the oxide produced is largely determined by wafer surface cleanliness and by the condition of the furnace tube walls and how frequently they are cleaned. Intermediate temperature CVD

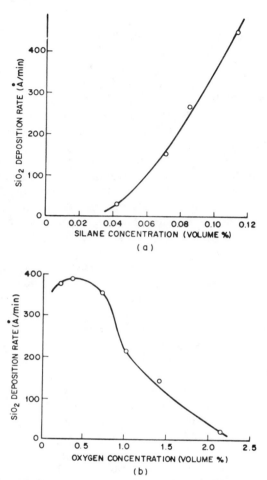

Fig. 14.12 (a) Effect of silane concentration on SiO₂ deposition rate. Oxygen concentration was 0.494% in nitrogen carrier gas. Total flow rate was 7.1 l/min, and temperature was 350°C. (b) Effect of oxygen concentration on SiO₂ deposition rate. Silane concentration was 0.0845% in nitrogen carrier gas. Total flow rate was 7.1 l/min, and temperature was 350°C. From the Bell Telephone Laboratories Silicon Oxidation Guide.

oxides have been used most often as a masking layer for the etching of silicon nitride films and for diffusion masking of germanium substrates.

(c) High Temperature Oxide

High temperature deposition of SiO_2 is usually done at temperatures of 850–1100°C using silane diluted to 1% in nitrogen, CO_2, and hydrogen carrier gas. Silicon dioxide is produced by the controlled oxidation of the silane according to reaction (1).

The high temperature deposition method is most useful for the top masking oxide for etching holes in silicon nitride films. Masking occurs because silicon dioxide is etched by buffered HF, which does not attack silicon nitride. The silicon nitride is rapidly etched by hot phosphoric acid, which also slowly attacks the silicon dioxide at a rate less than 10 Å/min. The oxide is used as a mask rather than photoresist because the photoresist polymer is attacked by hot phosphoric acid. The top oxide can be deposited at high temperature in sequence after deposition of the silicon nitride layer without opening the reactor. To simplify matters, the silicon dioxide can be deposited at the same temperature as the silicon nitride layer, which not only simplifies handling, but minimizes exposure of the silicon nitride surface to contaminants and thereby decreases the number of pinholes. Next, a pattern is defined on the top oxide by photoresist techniques and holes etched in the top oxide down to the silicon nitride with buffered HF. Now the top oxide acts as a mask for etching the silicon nitride with hot phosphoric acid.

High temperature CVD oxide films are rapidly etched in buffered HF, dissolving at a rate of 3000–4000 Å/min, or three to four times as rapidly as thermally grown SiO_2. In all the methods of producing CVD oxides, the total pressure in the reaction chamber was 1 atm. There are a number of advantages to depositing SiO_2 by the pyrolysis of tetraethylorthosilicate at low pressures:[100-102] (1) less particulates that can deposit on the wafers are generated; (2) up to 200 wafers can be processed at a time; (3) better thickness uniformity across each wafer is obtained; (4) a more specular

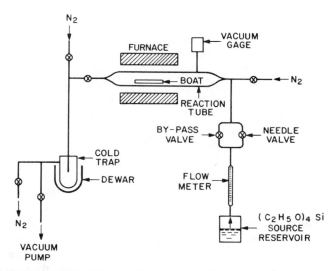

Fig. 14.13 Experimental system used for the deposition of pyrolitic oxides. After Oroshnik and Kraitchman.[100] Reprinted by permission of the publisher, The Electrochemical Society, Inc.

surface results; (5) less material is used; (6) and this method is less hazardous and lower in cost. For these reasons, low pressure CVD is finding more widespread use and commercial systems are coming onto the market.

Figure 14.13 illustrates an experimental system for low pressure CVD of SiO_2. Typically, the reaction chamber, maintained at 750°C, is pumped down to a pressure of 15–60 Torr,[87] and the needle valve is opened to admit the tetraethylorthosilicate vapor into the reaction chamber containing the wafers to be coated. On the completion of a run, the vapor flow is stopped and the reaction chamber pumped to remove the remaining vapor and gaseous decomposition products. The wafers can be removed after the reaction chamber is isolated from the cold trap and backfilled to 1 atm of nitrogen. Deposition rates of up to 300 Å/min could be obtained with the system shown in Fig. 14.13.

The details of how to prepare CVD silicon dioxide films and the descriptions of various commercially available reactors are beyond the scope of this book. Table 14.3 lists some useful properties of SiO_2 films prepared by all the methods described in this chapter.

14.6.4 Deposited Silicon Nitride

Although silicon nitride is not an oxide, it is so intimately connected with silicon dioxide in integrated circuit technology that it has been included in this section.

The two most important applications of silicon nitride in integrated circuit technology are (1) as a mask against alkali metal contamination and (2) as an oxidation mask.[103-106]

Used as an oxidation mask, it is possible to have oxide isolation rather than junction isolation resulting in reduced device size similar to the anodic oxide isolation described previously. The oxidation masking feature also allows self-alignment and simplified masking that translate into higher yields and low costs.

There are three methods used to deposit or to form silicon nitride films. The most commonly used is the chemical vapor deposition of the silicon nitride from a gaseous mixture of either SiH_4 or $SiCl_4$ diluted to 1–3% in nitrogen or hydrogen and ammonia.[106] The volatile silicon compound, silane or silicon tetrachloride, reacts with the ammonia at temperatures above 850°C to form silicon nitride. The source of nitrogen for the reaction is ammonia rather than N_2 because of the higher free energy of formation. The use of ammonia is advantageous kinetically because N_2 is more chemically inert as a result of the large bond energy in the molecule. A second method of depositing silicon nitride is by reactive sputtering of silicon in a low pressure nitrogen atmosphere.[107] A third method uses a radio frequency glow discharge to ionize the reactants to form silicon nitride films on either cold or heated substrates.[108, 109] Because of the difficulties in scaleup for large volume production and because of unsuitable film properties, the second and third methods are used only for specialized development work. Finally, radio frequency sputtering of silicon nitride may be used, but good target materials, which do not generate particles, are not generally available. Thus, the first CVD method remains the one most widely used in the manufacture of integrated circuits.

In the CVD method, typically, a 2000 Å thick film can be deposited in 7 min. The water and oxygen content of the reactor ambient should be less than 1 ppm to avoid forming SiO_2. A low etch rate, less than 15 Å/min in buffered HF, is indicative of a silicon nitride film effective as a diffusion barrier against alkali metals.

Details of the CVD process for forming silicon nitride films and a description of the reactors commerically available are beyond the scope of this book.

14.7 OXYGEN DOPED SILICON (SIPOS)*

High voltage integrated circuits, a newly emerging class of devices, and high voltage transistors pose a special set of problems involving the oxide. High voltage devices require lightly doped regions to obtain high avalanche breakdown voltages.[4] Therefore, high voltage devices are more sensitive than conventional low voltage devices to positive oxide fixed charge and mobile ionic contaminants, which can invert p-type and accumulate n-type surfaces, thus altering the avalanche breakdown voltage. A high density of charged interface traps also is undesirable. In an attempt to overcome these problems, in place of a thermally grown oxide film, undoped polysilicon films have been deposited on the device surface by chemical vapor deposition.[110] Such a polysilicon film is immune to ionic contamination and has no characteristic interfacial charge, although it has some interface traps. Unfortunately a large reverse bias leakage current is observed in a junction diode with the undoped polysilicon film.[110, 111] Most of this leakage current flows through the polysilicon film, which shunts the junction. The remainder of the leakage current is generated through traps at the polysilicon-silicon interface. Doping the polysilicon film with oxygen reduces the leakage current to acceptable levels.[110, 111]

Oxygen doped polysilicon films are deposited at 650°C using the gas mixture SiH_4-N_2O-N_2.[110, 111] The oxygen concentration in the deposited film can be controlled by changing the flow rate ratio of N_2O to SiH_4. A deposition rate of 500 Å/min was obtained for undoped films; this rate decreased with increasing amounts of N_2O. Polysilicon films containing 15–35 atomic % oxygen were found to be most suitable.[110, 111] Such films are

*Semi-insulating polycrystalline oxygen doped silicon.

chemically stable and have high resistivity and high dielectric strength. The resistivity increases from 3×10^6 Ω-cm for an undoped polysilicon film to 10^9 Ω-cm for a film doped with 15–35% oxygen.[110, 111]

REFERENCES

1   W. Kern and D. A. Puotinen, *RCA Rev.*, **31**, 187 (1970).

2   R. J. Archer, *J. Electrochem. Soc.*, **104**, 619 (1957).

3   S. Tolansky, *Surface Microtopography*, Interscience, New York, 1960.

4   A. S. Grove, *Physics and Technology of Semiconductor Devices*, Wiley, New York, 1967.

5   W. A. Pliskin and E. E. Conrad, *IBM J. Res. Dev.*, **8**, 43 (1964).

6   E. A. Corl and H. Wimpfheimer, *Solid-State Electron.*, **7**, 755 (1964).

7   F. Reizman and W. Van Gelder, *Solid-State Electron.*, **10**, 625 (1967).

8   G. R. Booker and W. J. Tunstall, *Phil. Mag.*, **13**, 71 (1966).

9   F. J. Wilhelm and M. L. Joshi, *J. Appl. Phys.*, **36**, 2572 (1965).

10   R. J. Jaccodine and C. M. Drum, *Appl. Phys. Lett.*, **1**, 29 (1966).

11   M. L. Joshi, *Acta Metall.*, **14**, 1157 (1966).

12   G. Dionne, *J. Appl. Phys.*, **39**, 2940 (1968).

13   J. E. Lawrence, *J. Appl. Phys.*, **40**, 360 (1969).

14   I. R. Sanders and P. S. Dobson, *Phil. Mag.*, **20**, 881 (1969).

15   P. Raichowdhury, *J. Electrochem. Soc.*, **118**, 1183 (1971).

16   K. V. Ravi, *J. Appl. Phys.*, **43**, 1785 (1972).

17   C. M. Hsieh and D. M. Maher, *J. Appl. Phys.*, **44**, 1302 (1973).

18   R. Ogden, R. R. Bradley, and B. E. Watts, *Phys. Status Solidi*, **26**, 135 (1974).

19   W. K. Tice and T. C. Huang, *Appl. Phys. Lett.*, **24**, 157 (1974).

20   K. V. Ravi and C. J. Varker, *J. Appl. Phys.*, **45**, 263 (1974).

21   R. Ogden, *Phys. Status Solidi*, **28**, K97 (1975).

22   J. Weertman and J. R. Weertman, *Elementary Dislocation Theory*, MacMillan, New York, 1964.

23   D. J. D. Thomas, *Phys. Status Solidi*, **3**, 2261 (1963).

24   R. H. Finch, H. J. Queisser, G. Thomas, and J. Washburn, *J. Appl. Phys.*, **34**, 406 (1963).

25   T. L. Chu and J. R. Gavaler, *J. Electrochem. Soc.*, **110**, 388 (1963).

26   H. J. Queisser and P. G. G. Van Loon, *J. Appl. Phys.*, **35**, 3066 (1964).

27   G. R. Booker and R. Stickler, *Phil. Mag.*, **12**, 1303 (1965).

28   C. J. Varker and K. V. Ravi, *J. Appl. Phys.*, **45**, 272 (1974).

29   A. W. Fisher and J. A. Amick, *J. Electrochem. Soc.*, **113**, 1054 (1966).

30   F. Barson, M. S. Hess, and M. M. Roy, *J. Electrochem. Soc.*, **116**, 304 (1969).

31   A. Mayer, *RCA Rev.*, **31**, 414 (1970).

32   Y. Sugita, T. Kato, and M. Tamura, *J. Appl. Phys.*, **42**, 5847 (1971).

33   S. Prussin, *J. Appl. Phys.*, **43**, 733 (1972).

34   C. M. Drum and W. Van Gelder, *J. Appl. Phys.*, **43**, 4465 (1972).

35   S. M. Hu, *J. Appl. Phys.*, **45**, 1567 (1974).

36   S. Prussin, *J. Appl. Phys.*, **45**, 1635 (1974).

37 C. M. Osburn and D. W. Ormond, *J. Electrochem. Soc.*, **121**, 1229 (1974).

38 S. M. Hu, *Appl. Phys. Lett.*, **27**, 165 (1975).

39 R. Conti, G. Corda, R. Mottenci, and C. Ghezzi, *J. Mater. Sci.*, **10**, 705 (1975).

40 H. Shiraki, *Jap. J. Appl. Phys.*, **14**, 747 (1975).

41 D. W. Small and R. F. Pierret, *Appl. Phys. Lett.*, **27**, 147 (1975).

42 S. P. Murarka and G. Quintana, *J. Appl. Phys.*, **48**, 46 (1977).

43 S. P. Murarka, *J. Appl. Phys.*, **48**, 5020 (1977).

44 S. P. Murarka, *Phys. Rev. B*, **16**, 2849 (1977).

45 S. P. Murarka, H. J. Levinstein, R. R. Marcus, and R. S. Wagner, *J. Appl. Phys.*, **58**, 4001 (1977).

46 P. E. Freeland, K. A. Jackson, C. W. Lowe, and J. R. Patel, Abstract AG15, *Bull. Am. Phys. Soc.*, **21**, 229 (1976); paper presented at March 1976 Meeting of American Physical Society at Atlanta (1976).

47 P. F. Kane and G. B. Larrabee, Eds., *Characterization of Solid Surfaces*, Plenum, New York, 1974.

48 S. P. Murarka, *J. Electrochem. Soc.*, **123**, 765 (1975). A comprehensive list of references is given in this paper.

49 G. A. Rozgonyi, P. M. Petroff, and M. H. Read, *J. Electrochem. Soc.*, **122**, 1725 (1975).

50 P. M. Petroff, G. A. Rozgonyi, and T. T. Sheng, *J. Electrochem. Soc.*, **123**, 565 (1976).

51 T. E. Seidel, R. L. Meek, and A. G. Cullis, *J. Appl. Phys.*, **46**, 600 (1975).

52 S. I. Raider, R. A. Gdula, and J. R. Petrak, *Appl. Phys. Lett.*, **27**, 150 (1975).

53 S. P. Murarka, private communication.

54 A. S. Grove, O. Leistiko, and C. T. Sah, *J. Appl. Phys.*, **35**, 2695 (1964).

55 B. E. Deal, A. S. Grove, E. H. Snow, and C. T. Sah, *J. Electrochem. Soc.*, **112**, 308 (1965).

56 T. Kato and Y. Nishi, *Jap. J. Appl. Phys.*, **3**, 377 (1964).

57 A. S. Grove, O. Leistiko, and C. T. Sah, *J. Phys. Chem. Solids*, **25**, 985 (1964).

58 C. D. Thurmond, in *Properties of Elemental and Compound Semiconductors*, H. G. Gatos, Ed., Interscience, New York, 1960, p. 121.

59 S. P. Murarka, *Phys. Rev. B*, **12**, 2502 (1975).

60 E. H. Nicollian, M. H. Hanes, and J. R. Brews, *IEEE Transact. Electron Devices*, **ED-20**, 380 (1973).

61 E. Tannenbaum, *Solid-State Electron.*, **2**, 123 (1961).

62 S. Margalit, A. Neugroschel, and A. Bar-Lev, *IEEE Transact. Electron Devices*, **ED-19**, 861 (1972).

63 G. Masetti, S. Solmi, and G. Soncini, *Electron. Lett.*, **9**, 226 (1973).

64 R. H. Krambeck, *J. Electrochem. Soc.*, **127**, 588 (1974).

65 D. D. Warner and C. L. Wilson, *Bell Syst. Tech. J.*, **59**, 1 (1980).

66 C. J. Frosch and L. Derick, *J. Electrochem. Soc.*, **104**, 547 (1957).

67 C. T. Sah, H. Sello, and D. A. Tremere, *J. Phys. Chem. Solids*, **11**, 288 (1959).

68 S. Horiuchi and J. Yamaguchi, *Jap. J. Appl. Phys.*, **1**, 314 (1962).

69 R. B. Allen, H. Bernstein, and A. D. Kurtz, *J. Appl. Phys.*, **31**, 334 (1960).

70 M. O. Thurston, J. C. C. Tsai and K. D. Kang, "Diffusion of Impurities into Silicon Through an Oxide Layer," Ohio State University Research Foundation, AD-261201, Contract DA-36-039-SC-83874, final report, Columbus, Ohio, March 1961.

71 M. L. Barry, *J. Electrochem. Soc.*, **117**, 1405 (1970).

72 D. M. Brown and P. R. Kennicott, *J. Electrochem. Soc.*, **118**, 293 (1971).

73 R. O. Schwenker, *J. Electrochem. Soc.*, **118**, 313 (1971).

74 J. C. Clemens, private communication.

75 J. Wong and M. Ghezzo, *J. Electrochem. Soc.*, **119**, 1414 (1972).

76 Y. W. Hsueh, *Electrochem. Technol.*, **6**, 361 (1968).

77 M. Ghezzo and D. M. Brown, *J. Electrochem. Soc.*, **120**, 110 (1973).

78 A. D. Kurtz and C. L. Gravel, *J. Appl. Phys.*, **29**, 1456 (1958).

79 J. R. Ligenza, *J. Electrochem. Soc.*, **109**, 73 (1962).

80 R. J. Powell, private communication.

81 G. R. B. Elliott, University of California, Radiation Laboratories UCRL-1831 (1952); E. L. Brady, *J. Phys. Chem.*, **57**, 706 (1953); A. S. Berezhnoi, "Silicon and its Binary Systems," Consultants Bureau, New York, 1960.

82 R. J. Zeto, C. G. Thornton, E. Hryckowian, and C. D. Bosco, *J. Electrochem. Soc.*, **122**, 1409 (1975).

83 S. Marshall, R. J. Zeto, and C. G. Thornton, *J. Electrochem. Soc.*, **122**, 1411 (1975).

84 R. J. Zeto, Extended Abstracts of the May 1979 Electrochemical Society Meeting, Abstract 96, pp. 254–255.

85 P. F. Schmidt and M. J. Rand, *Solid-State Commun.*, **4**, 169 (1966).

86 W. S. Pliskin and H. S. Lehman, *J. Electrochem. Soc.*, **112**, 1018 (1965).

87 D. R. Turner, *J. Electrochem. Soc.*, **105**, 402 (1958).

88 A. Uhlir, *Bell Syst. Tech. J.*, **35**, 333 (1958).

89 H. Yamanaka, M. Sakamoto, and K. Hamonon, *Jap. J. Appl. Phys.*, **14**, 303 (1975).

90 S. Nakajimo, Y. Watanabe, T. Yokoyama, and K. Kato, *Proceedings of the Sixth Conference on Solid State Devices*, Tokyo, 1974, B-2-(3), p. 303.

91 P. F. Schmidt and D. R. Wonsidler, *J. Electrochem. Soc.*, **114**, 603 (1967).

92 J. S. Logan and D. R. Kerr, "Migration Rates of Alkali Ions in SiO_2 Film," IEEE Device Research Conference, Princeton, N. J., June 1965; A. G. Revesz, *J. Electrochem. Soc.*, **114**, 629 (1967).

93 M. H. Woods and R. Williams, *J. Appl. Phys.*, **44**, 5506 (1973).

94 J. R. Ligenza, *J. Appl. Phys.*, **36**, 2703 (1965).

95 J. R. Ligenza and M. Kuhn, *Solid-State Technol.*, **13**, 33 (1970).

96 N. Goldsmith and W. Kern, *RCA Rev.*, **28**, 153 (1967).

97 E. L. Jordan, *J. Electrochem. Soc.*, **108**, 478 (1961).

98 J. Klerer, *J. Electrochem. Soc.*, **112**, 503 (1965).

99 R. P. Donovan, *RTI Report ASD-TDR-63-316*, Vol. vii, June 1965.

100 J. Oroshnik and J. Kraitchman, *J. Electrochem. Soc.*, **115**, 649 (1968).

101 E. Tanikawa, O. Takayama, K. Maeda, *Fourth International Conference on Chemical Vapor Deposition*, Electrochemical Society Meeting, Boston, Mass., 1973, p. 261.

102 E. Tanikawa, T. Okabe, and K. Maeda, *Jap. J. Elec. Soc.*, **41**, 491 (1973).

103 J. A. Appels, E. Kooi, M. M. Patten, J. J. Schatorje, and W. H. C. G. Verkuijen, *Phillips Res. Rep.*, **25**, 118 (1970).

104 J. A. Appels and M. M. Patten, *Phillips Res. Rep.*, **26**, 157 (1971).

105 D. Peltzer and B. Herndon, *Electronics*, **44**, 53–55 (1971).

106 T. L. Chu, C. H. Lee, and G. A. Gruber, *J. Electrochem. Soc.*, **114**, 717 (1967).

107 S. M. Hu and L. V. Gregor, *J. Electrochem. Soc.*, **114**, 826 (1967).

108 H. F. Sterling and R. C. G. Swann, *Solid-State Electron.*, **8**, 653 (1965).

109 R. C. G. Swann, R. R. Mehta, and T. P. Cauge, *J. Electrochem. Soc.*, **114**, 713 (1967).

110 T. Aoki, T. Matsushita, H. Yamoto, H. Hayashi, M. Okayama, and Y. Kawana, Extended Abstracts of the May 1975 Electrochemical Society Meeting, Abstract 148, pp. 352–353.

111 T. Matsushita, T. Aoki, T. Ohtsu, H. Yamoto, H. Hayashi, M. Okayama, and Y. Kawana, *IEEE Transact. Electron Devices*, **ED-23**, 826 (1976).

112 W. A. Pliskin and R. P. Gnall, *J. Electrochem. Soc.*, 111, 872 (1964).

113 H. F. Wolfe, *Silicon Semiconductor Data*, Pergamon, New York, 1969, p. 601.

15

Control of Oxide Charges

15.1 INTRODUCTION

Integrated circuit technology has progressed to the point' where interface traps, oxide fixed charge, oxide traps, and mobile sodium ions no longer pose major problems. In this chapter we show how these quantities affect device performance and stability and how their effects have been minimized by applying the understanding described in Chapters 2–14. The devices discussed in this chapter are the junction diode, the bipolar transistor, the MOSFET, and the gate controlled diode. An understanding of these devices is assumed and can be obtained from the books by Sze[1] and Grove,[2] for example.

To optimize device performance and stability, one must be able to (1) control the oxide and interfacial properties during fabrication and (2) keep these properties stable during device operation. Both of these objectives are addressed in this chapter.

The use of accelerated aging experiments is described in Section 15.2. Section 15.3 deals with the control of sodium ion drift. Section 15.4 discusses effects of oxide fixed charge and interface traps on device characteristics and the empirical methods used to control oxide fixed charge and interface trap level densities. Section 15.5 discusses the oxidation ambients that give the best results for gate and field oxides. Section 15.6 deals with the negative bias-temperature instability, and Section 15.7 with ionizing radiation.

In principle, the term "oxide fixed charge" refers to localized charged centers that cannot change their charge state by exchange of mobile carriers with the silicon. The term "interface trap charge" refers to charge localized on interface trap sites, sites that can change charge state by exchange of carriers with the silicon.

In practice, oxide fixed charge and interface trap charge are not distinguished according to this definition. Instead, an operational definition is used that employs the depletion and weak inversion portion of C-V curves. Thus oxide fixed charge, which is invariant under bias changes, rigidly translates the C-V curve in gate bias. Operationally, any shift of the depletion-to-weak inversion portion of the C-V curve is attributed to oxide fixed charge. Interface trap charge distorts the C-V curve in the chosen bias ranges. Operationally, interface trap charge is inferred from this distortion.*

Reconciliation of the definition in principle with the operational definition is complicated in two ways. First, some of the interface traps that can exchange mobile carriers with the silicon may have energy levels in or near the allowed energy bands of the silicon. Changes in charge state of these levels will not occur in depletion or weak inversion, where the Fermi level is remote from the band edges. Hence the operational definition includes these interface traps with oxide fixed charge, whereas the definition in principle does not. Second, spatial nonuniformity of oxide fixed charge can distort the C-V curves. Such distortion may be included as interface trap charge in the operational definition unless a test for charge nonuniformity is made using both high and low frequency C-V curves. For a Poisson distribution of localized oxide fixed charge, this second complication is not important in depletion and weak inversion but causes difficulty in interpreting C-V measurements near the band edges.

In this chapter we have adopted the operational definition. Consequently, some ambiguity enters the interpretation of the annealing behavior of oxide fixed charge, which may be composed of various chemical centers.

Because much of the chemistry involving the creation and annihilation of interface trap and oxide fixed charge centers is unknown, the methods of controlling the densities of these centers are largely empirical. Attempts at understanding the origin of these centers and their chemistry are mostly speculative. These speculations are discussed in Section 16.5.

---

*Alternatively, the density of interface trap levels near the center of the bandgap can be measured using the conductance method.

15.2 ACCELERATED AGING

Accelerated aging is used to cause device degradation in times short enough to allow reasonable observation times. In accelerated aging a particular physical or chemical phenomenon is made to occur faster than it otherwise would. Acceleration is accomplished by applying a higher than normal stress to the sample, such as heating, or by applying an electric field, or both.

Accelerating a particular phenomenon facilitates study and in most cases is necessary even to make a study possible. But the usefulness of accelerated aging depends on predicting the rate of device degradation under normal operating stress from the degradation under elevated stress. If the stress dependence of the phenomenon is known, this prediction can be made. Once it is known whether a given degradation mechanism will be a serious problem at operating stress, means can be found to circumvent, minimize, or eliminate it.

The MOS capacitor is used to study degradation mechanisms related to the MOS system because elimination of a mechanism in the MOS capacitor eliminates this mechanism in transistors and diodes as well.

The application of accelerated aging described here differs from that used in failure analysis of devices coming off a production line.[3] In production failure analysis, large numbers of devices may be subjected to accelerated aging and statistics used to determine the number that pass or fail the test. Degradation occurs because some process in the fabrication sequence has gone out of control, and the objective is to identify this process. Only a few samples are used in the accelerated aging described in this chapter, and the objective is to study the failure mechanism itself. Understanding then leads to control of the failure mechanism and to means for its identification and correction when a process in the fabrication sequence goes out of control.

15.3 IONIC DRIFT

15.3.1 Passivation

The properties of the semiconductor surface are frequently much different from those of the bulk, and the surface properties can influence device performance. A p-n junction illustrates one facet of this influence. The junction depletion layer must reach the surface, and there its width largely is determined by surface charges. Surface defect centers contribute to leakage current. In some cases a ten-thousandth (0.0001) of a monolayer of electrically active contamination on a semiconductor surface can dominate the electrical properites. Surface-dominated characteristics normally are less stable than are bulk-dominated characteristics.

Semiconductor surface properties are affected markedly by small amounts of contamination picked up from the environment during processing or device operation. In a humid ambient a film of water can build up on the semiconductor surface. Because water has a high dielectric constant (80), surface contaminants become ionized. When bias is applied to the device, an ionic current flow occurs along the semiconductor surface. As a result, a potential builds up along the surface that can affect the electrical properties of the device. Because of sensitivity to contaminants and to humidity, it was found that semiconductor device characteristics could change in an uncontrolled manner.

The first attempt to remedy this problem was to seal the device hermetically in a metal can. This remedy proved costly and bulky. It has proved more practical to passivate the semiconductor surface. Passivation minimizes undesirable surface effects and is achieved by forming a thin insulating layer on the semiconductor surface.

The most successfully passivated semiconductor is silicon with a thermally grown SiO_2 layer. The naturally occurring layer of oxide on silicon is only about 40 Å thick; therefore, potentials built up along it profoundly affect the silicon underneath. However, if an oxide several thousand angstroms thick is grown, ions and potentials on the oxide surface are too far away from the silicon surface to have much effect. An oxide thickness of 5000 Å will effectively passivate.

Atalla et al.[4] were first to effectively passivate diodes and bipolar transistors with a thermally grown SiO_2 film. Planar technology, introduced by Hoerni,[5] was the practical embodiment of oxide passivation in silicon diodes and bipolar transistors. In planar technology a diode is made by diffusing impurities into the silicon through a hole cut into an oxide several tenths of a micrometer thick. The surface portion of the metallurgical junction forms underneath the oxide because the impurity diffuses laterally as well as normally into the surface. That is, the depletion layer edge intersects the thermally grown Si–SiO_2 interface rather than the naturally oxidized silicon surface.

Planar technology superseded the older mesa technology* because of its ease of fabrication and the beneficial effects of passivation. In mesa technology the junction is formed by diffusion all over the wafer, and then an array of mesas of height considerably greater than the junction depth are etched in the silicon to form the individual diodes.

Figure 15.1 illustrates how oxide passivation improves the performance of bipolar transistors in a comparison of planar and mesa transistors. The planar transistor has a 5000 Å thermally grown SiO_2 film, whereas the mesa transistor has a natural 40 Å oxide film. The upper curves illustrate the reduction in collector leakage current, which should be as low as possible for best performance. Figure 15.1(a) shows (1) lower collector junction

*Oxide passivation was not a part of this technology.

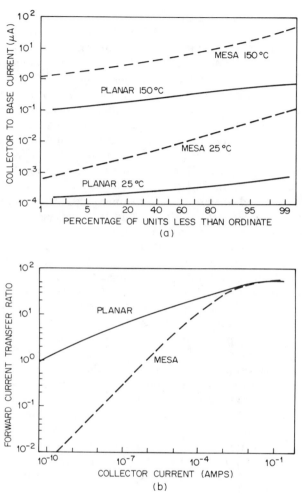

Fig. 15.1 Effect of oxide passivation on bipolar transistor characteristics. (a) Collector-to-base current versus percentage of units less than ordinate for planar (passivated) and mesa (unpassivated) bipolar transistors measured with a base-collector bias of 30 V at 25 and 150°C. After Hoerni.[5] Copyright (1960), IEEE. (b) Forward current transfer ratio versus collector current for planar and mesa transistors. After Hoerni.[5] Copyright (1960), IEEE.

leakage currents for all passivated (planar) units tested at both 25 and 150°C and (2) greater uniformity of leakage current from unit to unit in the passivated transistors compared to the mesa transistors. Greater uniformity results in greater yields in manufacture. Figure 15.1b shows the increased current gain of the passivated transistor, which should be as high as possible. At low collector currents particularly, the passivated transistors exhibit much higher gain. Both performance and yield are improved by passivation.

Collector leakage current and current gain are improved by passivation because the density of traps at the silicon surface is markedly reduced when an SiO_2 film is thermally grown. These interface traps at the silicon surface contribute to diode leakage current by increasing electron-hole generation, and they reduce current gain by increasing electron-hole recombination, both described in Section 4.3. Interface trap level density under the thin natural oxide is very high and sensitive to the ambient, but the passivating film has a lower interface trap level density, not as ambient sensitive. Thus the improved device characteristics should be more stable than those of unpassivated devices.

About the same time as the invention of planar technology, the MOS-FET was demonstrated by Kahng and Atalla.[6] The groundwork now was laid for a class of devices depending on thermally grown SiO_2 that would form the heart of the coming integrated circuit technology. Attempts were made to mass produce these devices. However, the characteristics of MOSFETs, oxide passivated diodes, and bipolar transistors all were found to be notoriously unstable. The stability expected from passivation could not be exploited and a reliable MOSFET cound not be made, thus seriously restricting the widespread use of these devices. Kerr et al.[7] and Yamin[8] showed that these instabilities were caused by ionic drift in the SiO_2 under the action of an applied field. Then, in a classic paper, Snow et al.[9] identified the culprit as sodium ions by deliberately enhancing and by eliminating sodium contamination in the SiO_2 layer of an MOS capacitor.

15.3.2 Sodium

The most important mobile ions in SiO_2 are ions of positively charged alkali metals. Of all the alkali metals, sodium is the most important because of its rapid transport in SiO_2 and its abundance in the environment. In fact, the human body is one of the major sources of sodium contamination. Other alkali metals from reagents or gate metal evaporation sources, for example, can be introduced into the SiO_2 during manufacture. For equal concentrations, Li^+ is the most mobile alkali metal ion in SiO_2, followed by Na^+, then K^+, and on down the column of alkali metals from the smallest to the largest. Lithium is not an important contaminant, despite its rapid transport in SiO_2, because of its scarcity in the environment. The electrical effect of K^+ is an order of magnitude less than that for Na^+ for equal concentrations because of its slower transport in SiO_2. Consistent with this picture, Cs^+ is completely immobile in SiO_2.[10]

It has been suggested that protons introduced into the SiO_2 by water or alcohol used in the wet processing during device fabrication could cause ionic instabilities.[11,12] Although protons may be mobile in SiO_2, it is not known whether they discharge at the electrodes. If protons become neutral at the electrodes, their electrical effects would be transient. Only if they remained charged at the electrodes would their electrical effects rival those of sodium ions. Regardless of whether protons are introduced, the introduction of sodium into SiO_2 from alcohol or other organic solvents has been demonstrated repeatedly.[13,14] Although the interaction between protons and sodium in SiO_2 may have a strong influence on the migration of both species, it is now well established that the exposure of SiO_2 films to boiling water and alcohol alone does not result in ionic instability.

Sodium ion densities in the low 10^{10} cm^{-2} range measured at 200°C can be achieved by the processing described later in this section. Such a low sodium ion density would cause a threshold voltage shift of a few tenths of a volt in a MOSFET that has a gate oxide thickness of 1000 Å and correspondingly less threshold voltage shift for thinner gate oxides. Such a small instability is unimportant for present integrated circuit applications but may be important in VLSI. For many bipolar transistor applications, sodium ion densities in the low 10^{11} cm^{-2} range would be tolerable. For example, 2.6×10^{11} ions/cm^2 will not invert a p-type silicon surface for silicon doping densities above about 4×10^{16} cm^{-3} and $x_o > 1000$ Å according to Fig. 10.28. Thus the base region of an n-p-n transistor, typically doped in the 10^{17} cm^{-3} range, would not become inverted. In a p-n-p transistor such a sodium ion density might invert the collector region if it were doped lighter than 4×10^{16} cm^{-3}.

15.3.3 Influence of Sodium on Device Characteristics

(a) MOSFET

The main influence of sodium ions on MOSFET characteristics is to alter threshold voltage. The effect is more pronounced in n-channel devices than in p-channel because the gate bias for n-channel is positive, driving sodium ions from the gate towards the silicon. p-Channel MOSFETs are less sensitive to sodium ions because gate bias is negative, keeping the sodium ions near the metal-oxide interface, where they induce a very small image charge in the silicon. Because p-channel devices are less sensitive to sodium ions than are n-channel devices, p-channel devices were first to be manufactured. Later, as described in Section 15.3.4, practical means were developed to control sodium contamination, and n-channel devices replaced p-channel devices in integrated circuits because of their superior performance.*

The basic mechanism of sodium ion drift in the MOSFET is the same as in the MOS capacitor. That is, the number of sodium ions that drift is limited by ion emission from the electrode at which the ions are initially located.

*Improved performance of n-channel over p-channel occurs mainly because electron mobility exceeds hole mobility.

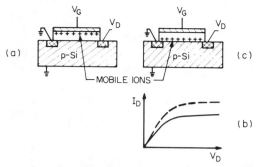

Fig. 15.2 (a) Cross section of an *n*-channel MOSFET with sodium ions at the metal. (b) The solid line is the drain current-drain-voltage characteristic when sodium ions are at the metal-SiO₂ interface. The dashed curve is when sodium ions are at the Si–SiO₂ interface ($V'_G = V_G - Q_m/C_{ox}$). (c) Cross section of the MOSFET with sodium ions at the Si–SiO₂ interface.

Figure 15.2 illustrates the sodium ion instability in an *n*-channel MOS-FET. Figure 15.2 is a cross section of the MOSFET showing sodium ions initially at the metal-oxide interface. A corresponding drain current-drain-voltage characteristic is shown as the solid curve in Fig. 15.2b. Figure 15.2c shows sodium ions drifted to the Si–SiO₂ interface, and the dashed curve in Fig. 15.2b is the resulting characteristic at the same gate bias as the solid curve.

The positive sodium ions at the Si–SiO₂ interface induce a negative image charge in the silicon, thereby reducing the positive gate bias required to produce a channel of a given conductivity or a given channel current at a given drain voltage. Figure 15.2 shows that sodium ion drift can cause the characteristics of a MOSFET to change enough to cause a circuit to malfunction.

(b) Planar Bipolar Transistor

In a planar bipolar transistor no metal gate imposes a field across the oxide. Rather, sodium ions drift in the junction fringing field. The most important effect of sodium is to invert the base region of an *n-p-n* transistor. When the entire base region becomes inverted, the inversion layer provides a current shunt from emitter to collector, thus degrading current gain. A *p-n-p* transistor is much less sensitive to sodium because only the collector region can become inverted increasing collector leakage current because of increased depletion layer volume. In both types of transistor, collector junction avalanche breakdown voltage is reduced by sodium ion migration. This effect is described in Section (c) below.

The sensitivity of planar bipolar transistors to sodium ion drift depends on their application. For most applications, even the *n-p-n* transistor is far less sensitive to sodium than an *n*-channel MOSFET because the base region typically is doped to about $10^{18}\,\mathrm{cm}^{-3}$. To strongly invert such a highly doped base region, about 3.5×10^{12} ions/cm^2 would be required. The substrate of an n-channel MOSFET is typically doped to a concentration in the $10^{15}\,\mathrm{cm}^{-3}$ range, and sodium ions alter threshold voltage seriously, even at densities too low to cause inversion.

To illustrate the effects of sodium ion drift on bipolar transistor characteristics, consider an n-p-n phototransistor. The sensitivity to sodium of this device rivals that of the n-channel MOSFET because of the combined requirements of a very low dark current and a large base area.

Figure 15.3 illustrates the mechanism of sodium ion drift across the base of an n-p-n phototransistor accelerated by temperature and field. We assume that the initial sodium ion concentration, which can be as high as 2–$3 \times 10^{12}\,\mathrm{cm}^{-2}$, does not invert the base typically doped to $10^{18}\,\mathrm{cm}^{-3}$. When a high positive bias is applied between collector and emitter with the base floating, a fringing field is established in the oxide at the collector-base and emitter-base junctions, as shown in Fig. 15.3a. At elevated temperature the

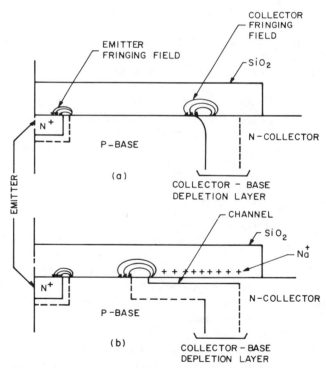

Fig. 15.3 (a) Schematic cross section of an n-p-n phototransistor showing emitter-base and collector-base fringing fields in the SiO$_2$ at the start of aging. (b) Schematic cross section of the phototransistor showing the fringing field in the SiO$_2$ at the leading edge of the channel growing out from the collector-base junction.

fringing field of the reverse biased collector-base junction rapidly sweeps sodium ions toward the base.

Sodium ions are swept from a volume of oxide that has a height comparable to the depletion layer width at the silicon surface, a width equal to the depletion layer width, and a length equal to the collector-base junction perimeter. The collector-junction fringing field concentrates the sodium ions from this volume into a much smaller volume along the base surface, resulting in a larger sodium ion density there. If this sodium ion density is sufficient to invert the base surface, growth of a channel from the collector-base junction is initiated. The depletion layer at the surface moves to the channel-base junction as shown in Fig. 15.3b.

As aging is continued, the fringing field in the oxide at the channel-base junction sweeps sodium ions ahead of it, concentrating them as before, so the inversion layer extends. If additional sodium ions can diffuse into the oxide from an external source during aging, the additional sodium ions are concentrated by the fringing field until the entire base is inverted.

Because current flows down the channel as a result of collector-to-emitter bias, a voltage drop exists between the collector-base junction and the growing edge of the inversion layer. There is a small fringing field in the oxide due to this voltage that sweeps sodium ions previously concentrated along the channel toward the growing edge of the channel. If no additional sodium enters the oxide, the strength of the channel decreases, and eventually the channel is eliminated. Finally, a condition is established in which there is a pile of sodium ions built up somewhere on the base surface with no connecting channel to the collector-base junction. The channel stops progressing toward the emitter, and the process is self-quenching.

Figure 15.4 is a plot of dark current as a function of collector-to-emitter bias with aging time at 120°C as parameter. The self-quenching effect is evidenced by the dark current, which initially rises with aging time and then decreases to its original value. Dark current increases as the channel grows out from the collector-base junction, because either (1) the collector-base depletion layer volume increases, thus increasing generation current, and/or (2) the channel reaches a defect at the silicon surface, a defect that has a high trap concentration and thus a high generation current. Such a defect acts as a miniature source of carriers. The characteristics then become similar to a MOSFET.

The self-quenching effect illustrated in Fig. 15.4 applies for moderate sodium contamination without additional external sources. For high sodium concentrations and external sources of contamination, the entire base surface from collector to emitter will become inverted. The device then will exhibit MOSFET characteristics with channel currents as high as milliamperes.

Initially, when a bias is applied between collector and emitter, there is a small fringing field in the oxide at the forward biased emitter-base junction

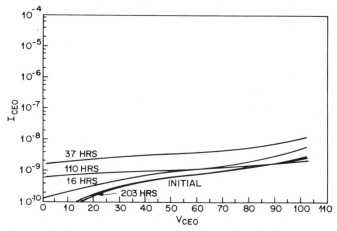

Fig. 15.4 Dark current I_{CEO} as a function of collector-emitter bias V_{CEO}, with aging time at 120°C as parameter.

that sweeps sodium ions from the base surface to the emitter surface. This may cause the high current gain of the transistor to increase slightly in the early stages of aging. Current gain will decrease drastically if the entire base becomes inverted.

(c) Planar Junction Diode

Sodium ion drift in a planar p-n junction diode changes both avalanche breakdown voltage and reverse bias leakage current. When the diode is aged at elevated temperatures with reverse bias applied, the junction fringing field in the oxide causes sodium ions to drift, as explained in Section (b) above. For the n^+p diode, sodium ions will be concentrated over the p-region after drift, thus reducing the electric field at the silicon surface. The avalanche breakdown voltage increases as a result. Leakage current also will increase because the junction depletion layer volume increases. For the p^+n diode, the sodium ions are concentrated over the p^+ region after drift, thereby increasing the surface field. The avalanche breakdown voltage decreases as a result. Junction leakage current decreases somewhat because of the reduced depletion layer volume.

15.3.4 Elimination of the Sodium Problem

(a) Sources of Sodium Contamination

To prepare uncontaminated oxides, it is necessary to know which steps in the processing introduce sodium into the oxide. The critical sources of sodium during processing are (1) gate or contact metalization,[15] (2) oxidation and high temperature annealing (3) high temperature processes such as diffusion and photoresist bake, (4) chemical reagents used in cleaning,[15] and (5) general handling of samples. The sodium content of silicon wafers is very low,[16] so the silicon itself is not a source of sodium.

(b) Cleanliness During Processing

In early MOS technology evaporation of aluminum introduced sodium because the tungsten evaporation filament usually contained sodium.* When the aluminum melted during evaporation, it formed a molten alloy with tungsten, causing sodium to evaporate along with aluminum. This problem was remedied by using sodium free tungsten or tantalum filaments and high purity aluminum.

There also were two sources of sodium in the oxidation furnace: (1) the furnace insulation and (2) the fused quartz furnace tube in which the silicon was oxidized. Sodium in the furnace insulation could easily diffuse through the fused quartz oxidation tube at oxidation temperatures to contaminate the growing oxide. The first solution to this problem was to use a liner, impervious to sodium, between the furnace and the fused quartz tube. High density alumina or silicon carbide[17] were found to be suitable liners. Today, sodium free furnaces are commercially available, so liners are not needed. The fused quartz furnace tube was a source of sodium particularly in the presence of steam, which continuously leached sodium from the fused quartz tube.[18] Because of its high dielectric constant (80) water is an extremely good solvent and easily picks up sodium from anything that it contacts. This is why "wet" grown oxides often contain more sodium than do "dry" grown ones. Higher levels of sodium in oxides grown in O_2–H_2O mixtures also may be due to sodium carried into the furnace tube by the mixture made in a bubbler. The solution to these problems is to use high purity fused quartz for the furnace tube and the bubbler flask or to use burned hydrogen rather than a bubbler.

Cleaning the furnace tube with an HCl–O_2 mixture at elevated temperature reduces sodium to negligible levels.[19] Figure 15.5 illustrates the effectiveness of HCl cleaning and shows gate current peaks measured at 250°C by the triangular voltage sweep method described in Section 10.2.3(c). In Fig. 15.5a the oxide was grown in a fused quartz furnace tube that had been given an HF rinse followed by a purge at 1150°C in dry oxygen for 24 hr. The area under the gate current peak corresponds to the available sodium ion density at 250°C. Even an available sodium ion density of 5×10^{10} cm^{-2} at 250°C, as measured in Fig. 15.5a, is small enough for many applications.

The furnace tube then was cleaned by flowing a 6% HCl–O_2 (mole ratio) mixture through it at 1150°C for 2 hr. Figure 15.5a applies to an oxide grown in this tube after HCl cleaning. This oxide has a sodium ion density at 250°C of 1.3×10^9 cm^{-2}. This sodium ion density is at the limit of

---

*This sodium was incorporated in the tungsten to make it more ductile for drawing into wire.

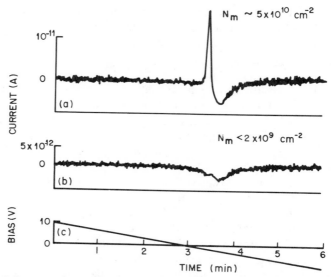

Fig. 15.5 (a) Gate current peak measured at 250°C on an oxide grown after cleaning the fused quartz furnace tube in HF and purging with O_2. After Kriegler.[20] Copyright (1974), IEEE. (b) Gate current peak measured at 250°C on an oxide grown after HCl cleaning of the furnace tube. After Kriegler.[20] Copyright (1974), IEEE. (c) Ramp gate bias as a function of time. After Kriegler.[20] Copyright (1974), IEEE.

detectability at 250°C, and for practical purposes sodium has been eliminated from this oxide. This discussion applies to any tube in which high temperature processing is done.

The mechanism of HCl cleaning is thought to be the oxidation of HCl in the HCl–O_2 mixture to form water vapor and chlorine gas, followed by the reaction between chlorine and sodium in the furnace tube to form NaCl. Sodium chloride is volatile at high temperatures and is swept out in the gas stream.

If the sodium in a furnace tube rises to an unacceptable level, HCl cleaning is an effective way of returning it to negligible proportions while keeping the furnace tube at oxidation temperature.*

There is a further benefit of cleaning sodium from the furnace tube. A sodium free, fused quartz tube will have a much longer life than will a contaminated one. The reason is that fused quartz is in an amorphous metastable state, but its stable form is crystalline (the crystobalite phase). At elevated temperatures the tendency of fused quartz to crystallize or

---

*Furnace tubes are kept continuously at oxidation temperature to maximize the number of wafers that can be oxidized per day. Besides, the tube may crack if it is cooled down from oxidation temperature.

devitrify is enhanced. Sodium acts as a nucleation center for the formation of crystallites, thereby accelerating devitrification. Devitrification is undesirable because it causes a volume change that results in flaking and cracking. The flaking and cracking turns the tube from clear to milky white, indicating that the furnace tube has to be replaced.

Sodium also can be introduced into the oxide during photoresist steps. The photoresist bake, usually done at 150°C, is a temperature sufficiently high for sodium to diffuse into the oxide. Electrical effects that arise from such contamination have been found to depend to a considerable extent on the anion connected with the sodium.[16, 21] These effects also are dependent on device geometry. This aspect of the problem has not been explored sufficiently to warrant further discussion.

Almost all chemical reagents contain some sodium, except distilled water. The last rinse the sample receives should be distilled water.[15] Finally, because the human body is a major source of sodium, care must be taken not to touch or breathe on samples or to handle samples with tools that have been touched.

Although sodium from the critical processing steps may have been eliminated initially, sodium invariably will build up in time to intolerable levels. This is true particularly in silicon gate technology, where the growth of gate oxide is followed by a number of high temperature processes before fabrication of the device is completed. Therefore, sodium contamination after high temperature processing steps such as oxidation usually is monitored to keep the processing under control and to pinpoint quickly where trouble lies. Monitoring is done on a wafer that sees the given processing and then is withdrawn. Metal gates are evaporated to form an MOS capacitor array, and the sodium level is measured, usually by the C-V method. Monitoring of sodium levels in all chemicals used also may be necessary [see Section 10.2.3(d)].

(c) Gettering

The approach just described avoids sodium contamination during the various stages of processing. Avoidance requires very tight control over the critical steps where sodium can be introduced, thus increasing costs. Consequently, gettering has developed as a low cost method of eliminating sodium contamination during processing and during device life. Two effective gettering methods are widely used. One getter is a layer of phosphosilicate glass grown or deposited on top of the thermally grown SiO_2 layer.[22] The other is oxide growth in a dilute mixture of HCl and dry oxygen.[21, 23]

A phosphosilicate glass layer can be grown by exposing the thermally grown SiO_2 layer to P_2O_5 vapors at 1000°C. Neutron activation and radioactive tracer studies [see Section 11.4.1(a)] have shown that the phosphosilicate glass acts as a getter by chemically binding sodium in an electrically neutral form.[24]

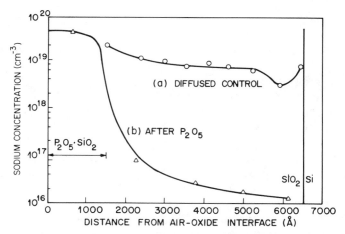

Fig. 15.6. (a) Sodium profile in the oxide of a control sample contaminated with sodium. After Yon et al.[25] (b) Sodium profile in the oxide of a sodium contaminated sample with phosphosilicate glass getter. After Yon et al.[25] Copyright (1966), IEEE.

Figure 15.6 is a sodium profile in SiO_2 measured by the radio tracer method showing the gettering effect of phosphosilicate glass.[25] This experiment was done by diffusing sodium into two different oxides at 800°C for 1 hr. Curve *a* is the profile measured after sodium diffusion into the control sample. A layer of phosphosilicate glass was deposited at 950°C over the oxide of the other sample. Curve *b* is the sodium profile after deposition of the glass. The sodium ion density at 140°C has been reduced from 10^{12} cm^{-2} to 10^{11} cm^{-2} by the phosphosilicate glass.[25]

A problem with phosphosilicate glass interposed between gate metal and gate oxide is that it can polarize[26] under high electric fields and high temperatures, thus introducing its own electrical instability. The instability due to polarization becomes negligible[26] if the thickness of the glass is kept below 1 μm and if relatively low phosphorus concentrations are used.

Another problem is that phosphosilicate glass can react with moisture to form phosphoric acid, which can corrode aluminum metallization. Again, this effect can be minimized by reducing the phosphorus content in the glass.

In silicon gate technology[27] phosphosilicate glass is not interposed between silicon gate and oxide. Rather, the glass is formed over both silicon gate and field oxide to a thickness of a micrometer or so and then made to flow at high temperature (~ 1100°C) to eliminate sharp corners. Aluminum interconnections are then deposited over the glass, which has holes cut in it at the appropriate places to allow contact to either the underlying gate or the diffused regions. The phosphosilicate glass now acts as a getter for sodium coming from the environment during device life.

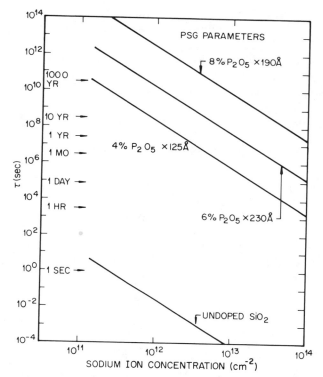

Fig. 15.7 Time for a flatband voltage shift of -0.1 V at 80° C as a function of sodium concentration with phosphorus content of the glass as parameter. Oxide thickness was 1000 Å. After Balk and Eldridge.[28] Copyright (1969), IEEE.

Figure 15.7 shows the resistance of very thin phosphosilicate glass layers to sodium penetration.[28] In this figure, the time required for a flatband voltage shift of -0.1 V after drifting sodium ions at 80°C under an oxide field of 2×10^6 V/cm into a 1000 Å thick oxide is plotted as a function of sodium surface concentration with phosphorus mole concentration in the glass as parameter. An example from Fig. 15.7 is the 125 Å thick layer of 4% glass, which will provide protection against 10^{12} ions/cm^2 for 10 years at 80°C. Thus a very thin, dilute composition phosphosilicate glass is a very effective sodium getter during device life.

Growing the oxide in a HCl–O$_2$ mixture* is one way of producing stable oxides. In this method sodium is gettered in two ways: (1) the HCl getters sodium in the ambient of the furnace tube in a way similar to cleaning the furnace tube in HCl just described; and (2) chlorine is incorporated into the oxide near the Si–SiO$_2$ interface, neutralizing sodium

*Chlorine gas or other chlorine containing species can be used as described in Section 13.4.5(d).

ions when they *arrive* there. This conclusion is based on a comparative analysis of the ionic gate current versus voltage characteristics of MOS capacitors. These capacitors were prepared with gettered and ungettered oxides deliberately contaminated with sodium. In the gettered oxides sodium ions do not drift back toward the gate under negative bias-temperature aging, after positive bias-temperature aging has drifted the sodium ions to the Si–SiO$_2$ interface. That is, the sodium is either immobilized, neutralized, or both. However, there is virtually no flatband voltage shift. Therefore, the sodium must be neutralized and may be immobilized as well. Etch-off experiments show that, in fact, the sodium also is immobilized by the chlorine.

Growing the oxide in an HCl–O$_2$ mixture has the advantages of getting sodium not only during oxidation, but also during subsequent high temperature processing and during low temperature operation of the completed integrated circuit.

The gettering effect on an oxide grown in a chlorine containing ambient[29] occurs even at low temperatures, as illustrated in Fig. 15.8. The curves in Fig. 15.8 were obtained with the triangular voltage sweep method at 250°C. Also listed with each trace in this figure is the corresponding flatband

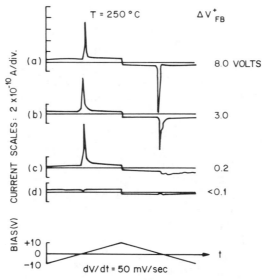

Fig. 15.8 Gate current as a function of gate bias measured at 250°C: (a) contaminated ungettered oxide; (b) oxide grown in a 3% HCl–O$_2$ mixture and contaminated with sodium to the same level as (a); (c) oxide grown in a 6% HCl–O$_2$ mixture and contaminated with sodium to the same level as (a); (d) oxide grown in a 6% HCl-O$_2$ mixture with no deliberate sodium contamination. In all these traces the small jog in gate current when the ramp changes from positive to negative slope is caused by the changing silicon surface capacitance. All the oxides were grown on phosphorus doped, 10 Ω-cm, (100) silicon. Sweep rate was 50 mV/sec. After Kriegler.[29]

voltage shift obtained with the C-V method at 250°C. In trace a the sample was oxidized at 1150°C for 30 min in dry O_2, contaminated with sodium by the evaporation of NaOH onto the oxide surface, metallized, and then measured. After positive bias-temperature aging, the flatband voltage shift was 8 V, which, for an oxide thickness in this sample of 1000 Å, corresponds to a mobile ion density of $1.55 \times 10^{12} \, cm^{-2}$ at 250°C. The left peak is for ion drift from the metal to the silicon and the right peak, for drift from the silicon to the metal. Traces b and c in Fig. 15.8 were measured on a sample grown in 3% and 6% HCl-O_2 mixtures, respectively, and then contaminated with the same amount of sodium as the oxide in trace a.

In trace a the same number of sodium ions moved back and forth from one interface to the other. In trace b fewer ions returned to the metal-SiO_2 interface from the Si–SiO_2 interface than went from the metal-SiO_2 interface to the Si–SiO_2 interface. Trace c shows that virtually all the sodium ions drifted to the Si–SiO_2 interface became neutralized and remained there. There is a negligible flatband voltage shift observed in the sample of trace c. Comparison of the gate current peaks and the flatband voltage shift of trace a to traces b and c indicates that (1) the flatband voltage decreases with increasing HCl concentration, (2) the area under the gate current peak for drift from the metal to the silicon does not depend significantly on HCl concentration, and (3) the area under the gate current peak decreases with increasing HCl concentration for drift from the silicon to the metal. Therefore, the presence of chlorine in the oxide does not affect sodium ion drift from the metal to the silicon, but once at the Si–SiO_2 interface, the sodium ions lose their charge and cannot drift back. The sodium ions most likely lose their charge by becoming chemically bound to chlorine related sites located near the Si–SiO_2 interface.[30] The details of the chemistry of this neutralization effect are unknown. Trace c in Fig. 15.8 shows that sodium is chemically bound to chlorine related sites strongly enough that diffusion of sodium at 250°C back toward the metal-SiO_2 interface is unlikely even though a large concentration gradient exists.

The gate currents observed in traces b and c for ion drift from the metal to the silicon are not related to the motion of a chlorine associated species rather than sodium ions because no gate current is observed in trace d.

Figure 15.9a shows the sodium profile in the oxide after drifting it to the Si–SiO_2 interface. The sodium profile was obtained with the radioactive tracer method by measuring the ^{22}Na activity remaining in the oxide after successively removing small sections of oxide. Figure 15.9b shows that chlorine in an oxide grown in a 5% HCl-O_2 mixture at 1100°C ends up mainly at the Si–SiO_2 interface. Although measured on different samples by different workers, comparison of Figs. 15.9a and 15.9b does show that the drifted sodium ions and the chlorine are both located near the Si–SiO_2 interface. This gives further support to the idea of chemical binding of sodium at chlorine related sites.

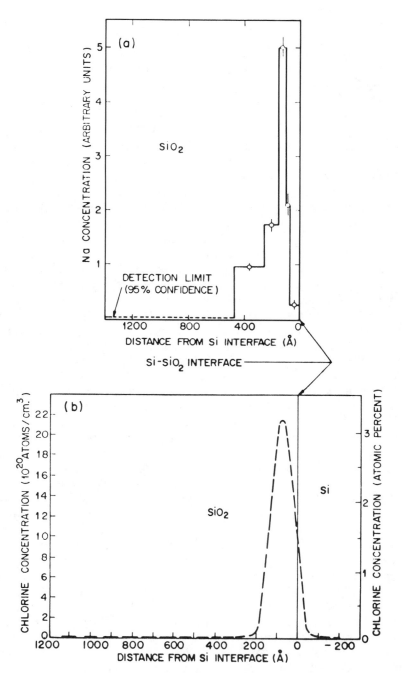

Fig. 15.9 (a) Sodium distribution in an oxide grown in a 6% HCl–O₂ mixture to 1400 Å. The oxide was then contaminated by evaporation of radioactive ²²NaCl. Sodium ions were drifted to the Si–SiO₂ interface under positive gate bias at 250° C. After Kriegler.[29] (b) Chlorine profile in the oxide measured by secondary ion mass spectrometry. After Deal et al.[31] Reprinted by permission of the publisher, The Electrochemical Society, Inc.

Effective gettering by chlorine in HCl grown oxides requires that the HCl be mixed with *dry* oxygen. Growing the oxide in an HCl–H_2O mixture does not produce the gettering effect. No chlorine is found at the Si–SiO_2 interface in oxides grown in HCl–H_2O mixtures.

Also, sodium gettering is not achieved with HCl alone.[30] That is, the HCl must be incorporated during a dry oxidation to produce the gettering effect. Therefore, the oxygen provides Cl_2 according to the reaction $2HCl + \frac{1}{2}O_2 \rightleftarrows H_2O + Cl_2$ and helps to incorporate chlorine in a form that is active as a getter.

One important characteristic of HCl gettering is that gettering rapidly diminishes with decreasing oxidation temperature, even if oxidation time is increased to allow growth of oxides of identical thickness.[30] Oxides grown at 1150°C will be effectively gettered, and increasing the temperature to 1200°C provides little improvement. However, at 1050°C the gettering effect virtually disappears. The diminishing gettering effect means either the unavailability of chlorine, or a failure to form a chemical complex which does the gettering.

There is an optimal HCl concentration in the HCl–O_2 mixture. This optimum is about 6%, but this is not a critical value. For values much below this concentration, the gettering will be less effective as seen from Fig. 15.8. For concentrations much above this value, the silicon surface will

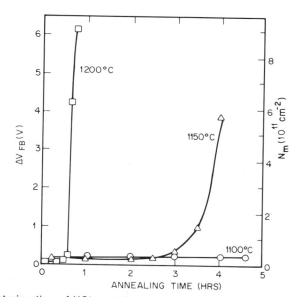

Fig. 15.10 Deterioration of HCl gettering by helium annealing. The oxides were grown to 1400 Å in 6% HCl–O_2 at 1150°C, and ΔV_{FB} was determined from a high frequency *C-V* curve. Sodium contamination level measured on a standard (ungettered) oxide was 1.1×10^{12} ions/cm^2 for all three samples. After Kriegler.[30]

become pitted and gas bubbles will be incorporated under the oxide.[32] Both of these effects are to be avoided.

Following oxidation in an HCl–O$_2$ mixture, some processing usually must be done at high temperatures in an ambient without any chlorine-bearing species. To provide gettering action during this processing, the chlorine must remain incorporated in the oxide and not diffuse out. A deterioration of the gettering effect at elevated temperatures was measured by annealing in helium as a function of time at three elevated temperatures and is shown in Fig. 15.10.[30]

These results indicate that the chlorine complex can break down and the chlorine diffuse out of the oxide during heating to temperatures above 1150°C in an inert gas ambient. However, there will be no significant deterioration of gettering action during the processing following oxidation because this processing is all done at temperatures below 1150°C for times less than a few hours (see Fig. 15.10). Figure 15.10 also shows that there is no deterioration of gettering action due to the outward diffusion of chlorine during device operation at temperatures below 100°C.

Finally, oxides grown in dry oxygen and subsequently contaminated with sodium can be cleaned by exposure to HCl at elevated temperatures. Although this process is not a standard part of integrated circuit processing, it can be used to salvage wafers that have gone out of control at an early stage of the processing.

(d) Sodium Barrier Layers

Another approach to prevent sodium contamination of the oxide during device life is to protect it with a film impervious to sodium. One such film is the phosphosilicate glass already discussed in Section (c) above. Other films that do not getter sodium but are impervious to sodium are relatively thick layers of silicon nitride,[33] aluminum oxide,[34] polysilicon, or metal deposited over the oxide. Silicon nitride and aluminum oxide have a denser structure than SiO$_2$—so dense, in fact, that sodium does not penetrate more than 50–100 Å into these insulators. Dalton and Drobek[35] showed by electron diffraction and radioactive tracer profiling that the smaller the crystallite size in a sodium nitride film, the more effective the film became as a sodium barrier. Figure 15.11 illustrates this effect.

Silicon nitride or aluminum oxide films are not interposed between the gate and the oxide in a MOSFET but are deposited over the gate to simply cover the entire chip and act as an encapsulant.

(e) Summary

These methods give the processor a variety of effective ways to deal with the sodium problem. Most widely used in silicon gate technology are strict cleanliness in processing, process monitoring to maintain control, and the combination of periodic cleaning of furnace tubes with HCl, oxidation

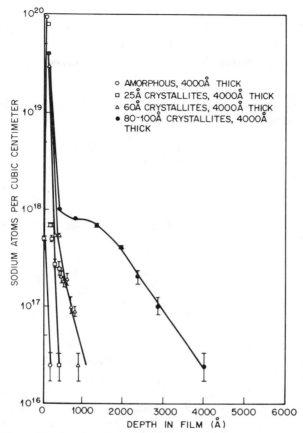

Fig. 15.11 Sodium diffusion in silicon nitride with crystallite size in the nitride as parameter. After Dalton and Drobek.[35] Reprinted by permission of the publisher, The Electrochemical Society, Inc.

of gate oxides in $HCl-O_2$ mixtures to getter sodium during processing, and a phosphosilicate glass layer over the entire wafer to getter sodium during device life. Chlorine in the oxide also will getter sodium during device life.

15.4 INFLUENCE OF OXIDE FIXED CHARGE AND INTERFACE TRAPS ON DEVICE CHARACTERISTICS

Oxide fixed charge and interface traps characteristic of thermally grown SiO_2 can adversely affect device performance if their densities are not controlled. The influence of these electrically active centers on device characteristics is compounded by the fact that their densities can change with time during device life, thus posing a stability problem. Unless the oxide is relatively free of sodium, sodium drift will mask changes in oxide fixed charge density Q_f and interface trap level density D_{it}. Therefore, changes in these densities are next in importance after sodium drift.

15.4.1 MOSFET

The characteristics affected by increases in oxide fixed charge and interface traps are:

1 Threshold voltage is affected. Because oxide fixed charge is always positive, threshold voltage is increased in p-channel devices and decreased in n-channel devices. Charged interface traps will tend to oppose the effect of oxide fixed charge on both p-channel and n-channel devices.

2 Drain junction breakdown voltage is decreased for a p-channel device and increased for an n-channel device. Changes in drain junction avalanche breakdown voltage may be undesirable because the range of drain bias over which the device can be operated may become reduced. For p-channel, the drain junction will be p^+n. Positive oxide fixed charge in the oxide under the gate causes, in the absence of gate bias, an accumulation of electrons at the surface of the n-type substrate, forming a p^+n^+ junction there that has a lower avalanche breakdown voltage than the p^+n junction in the bulk. For n-channel, the drain junction will be n^+p. In the absence of gate bias, positive oxide fixed charge in the oxide under the gate depletes the surface of the p-type substrate of holes. The electric field near the corner of the junction at the silicon surface will be reduced below even that obtained in a planar junction with no surface fields. Avalanche breakdown voltage there will be higher than it would be without any oxide fixed charge. In both p- and n-channel devices, charged interface traps will oppose the effect of oxide fixed charge on avalanche breakdown voltage.

3 Gain or transconductance is reduced. Although considerable variation in gain can be tolerated in an amplifier circuit with negative feedback, there will be limits over which gain can vary. In switching circuits, lowered gain is undesirable because it slows down switching times. Reduction in gain occurs in two ways. First, a change in gate bias produces a change in interface trap level occupancy *and* a change in channel charge density. Therefore, a larger change in gate bias is required to produce a given change in channel conductance when interface trap level density is large than when it is small. Second, gain or transconductance depends on the mobility of free carriers in the channel. Free carrier mobility is reduced by the presence of charged interface traps and oxide fixed charge. This reduction, which is most pronounced at low temperatures and low fields, is caused by scattering of channel carriers by oxide fixed charge and charged interface traps.

4 High drain junction leakage currents are undesirable because they decrease the sensitivity of a MOSFET to a weak signal and more power is consumed by the device than necessary for optimum performance. Drain junction leakage current is increased by an increase in D_{it}. The interface traps act as generation centers. The charge values Q_f and Q_{it} determine how near the Fermi level at the silicon surface is to midgap, where interface traps generate leakage current most efficiently.* Leakage current flows when reverse bias is applied to the junction, thereby increasing the electric field across the depletion layer above its equilibrium value. The applied field sweeps holes to the p-region and electrons to the n-region of the junction, making the hole and electron concentrations in the depletion layer much lower than the equilibrium concentrations. Because there are few carriers available, hole and electron capture by interface traps will be negligible. The dominant process will be emission. To maintain steady-state current, each interface trap alternately emits a hole and an electron; thus interface traps are most efficient when the Fermi level at the silicon surface is near midgap. Emission of a hole corresponds to the transition of an electron from the silicon valence band to an interface trap level near the Fermi level, and the emission of an electron corresponds to a transition of an electron from an interface trap level near the Fermi level to the silicon conduction band. Thus an electron makes a transition from the valence band to the conduction band though an interface trap level by emission.

5 Flicker noise is dominated by interface traps for D_{it} above about $10^{11}\,\mathrm{cm^{-2}\,eV^{-1}}$.[36] Interface trap dominated flicker noise[37] occurs because interface trap level occupancy fluctuates randomly. The resulting fluctuation in interface trap charge causes a corresponding fluctuation in channel charge density and thus channel conductance. At midgap D_{it} less than $10^{11}\,\mathrm{cm^{-2}\,eV^{-1}}$, flicker noise is determined by the volume of the channel. Increasing channel volume decreases flicker noise power.[36]

15.4.2 Bipolar Transistor

The influence of interfacial charges and traps on bipolar transistor characteristics are:

1 Common emitter current gain is reduced at low collector currents for large Q_f and D_{it}. High current gain requires that most of the carriers injected from the emitter into the base get to the collector. The injected carriers constitute an excess concentration over the equilibrium concentration in the base. There will be an excess of free carriers in the

*If the Fermi level at the silicon surface is not near midgap, leakage current will be generated by bulk traps.

emitter-base depletion layer as well. In this case capture is more important than emission. An interface trap will capture an injected carrier, and when it captures a carrier of the opposite type, recombination will occur. Suppose electrons are the injected carrier, and an injected electron is captured by an interface trap. Capture corresponds to a transition of an electron from the silicon conduction band to an interface trap level near the Fermi level. Next, a hole is captured by the interface trap, which corresponds to a transition of an electron from an interface trap level near the Fermi level to the silicon valence band. In effect, the electron has made a transition from the conduction band to the valence band through an interface trap level. As a result of recombination in both the emitter-base junction depletion layer and the base region, some injected electrons do not reach the collector. Consequently, current gain is reduced over what it would be without recombination. Recombination current depends exponentially on the forward bias applied to the emitter-base junction. If Q_f and Q_{it} are high, band bending will be different for a given forward bias from what it would be without these charges. Thus these charge densities also will affect current gain. When collector current is much larger than the recombination current, gain is not as strongly affected.

2 Collector-junction leakage current will increase with increasing D_{it} in the same way as described for the drain junction of a MOSFET.

3 Oxide fixed charge and charged interface traps will influence the collector junction avalanche breakdown voltage in the same way as described for the drain junction of a MOSFET. However, because oxide fixed charge is always positive, collector avalanche breakdown voltage will be decreased for an n-p-n transistor and increased for a p-n-p transistor for large Q_f.

4 Flicker noise in the bipolar transistor is caused by, interface traps randomly capturing and emitting injected carriers flowing across the emitter-base depletion layer and the base region. Flicker noise power will increase with increasing D_{it}.

15.4.3 Control of Interface Trap Level and Oxide Fixed Charge Densities

There are four conditions under which changes in Q_f and D_{it} can occur:

1 Ambient-dependent annealing will cause changes in both Q_f and D_{it}. The value of D_{it} can be changed by annealing in different ambients without bias. The value of Q_f is the charge density measured after annealing out interface traps. Usually, this anneal is a low temperature postmetallization anneal. Therefore, the low temperature annealing properties of Q_f are unknown. However, after annealing out the inter-

face traps, Q_f can be varied significantly by annealing at high temperatures ($>600°C$) in a nonoxidizing ambient. The value of D_{it} can change with low temperature annealing (200–600°C) without bias by amounts up to one order of magnitude. The key agent responsible for the low temperature annealing is the transfer of hydrogen, usually provided by water vapor, into or out of the oxide layer. The chemistry by which interface traps are activated and annihilated by hydrogen is unknown. Stability of D_{it} during device operation can be achieved by sealing the oxide with a layer that is impervious to water and hydrogen, such as a silicon nitride layer, a silicon gate, or a metal layer.

2   Hot carrier injection will cause oxide trapping, resulting in a change in oxide charge. This change is positive for hole injection and negative for electron injection. An increase in D_{it} also occurs, as described in Sections 11.2 and 11.3. There are several ways to avoid electron injection:

(a)   Devices can be operated at voltages that are too low for hot carrier injection.[38] For example, the energy barrier between the silicon conduction band and the SiO_2 is 3.2 eV (ignoring image force lowering or tunneling). An electron must overcome this barrier to enter the SiO_2. An electron moving between the source and the drain of an n-channel MOSFET will gain energy from the potential it sees. There will be no hot carrier injection if the electron sees less than 3.2 V.

(b)   A buried channel can be used to keep hot carriers away from the Si–SiO_2 interface.

(c)   Junctions can be made so that avalanche breakdown does not occur near the silicon surface.

(d)   The geometry of the structure can be designed to avoid high fields.

It also is possible to reduce trapping once injection occurs. The concentration of water in the oxide can be reduced. This reduction can be accomplished in two ways. The water can be gettered with phosphosilicate glass. According to Iwamatsu and Tarui,[39] negative oxide charge buildup caused by electron injection can be reduced by implanting 3×10^{13} phosphorus ions/cm^2 into the oxide surface followed by oxidation in dry oxygen at 1050°C for 10 min to form the glass. Also, these workers found a reduction of positive oxide charge buildup caused by hole injection on implanting 3×10^{12} phosphorus ions/cm^2 all the way to the Si–SiO_2 interface, followed by the same oxidation step as before. There may be an increase in D_{it} in this solution to the electron trapping problem because of the gettering of water by the phosphosilicate glass. Reduced negative oxide charge buildup caused by electron injection also is observed with silicon gate as shown by comparing curves a in b in Fig. 15.12. This is because the silicon gate of the MOS capacitor measured to obtain curve b in Fig. 15.12 was

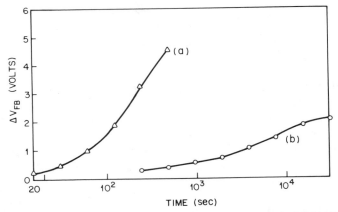

Fig. 15.12 Plot of ΔV_{FB} versus time during which a gate current of 9.9×10^{-5} A/cm^2 of avalanche injected electrons flowed ($x_o = 1000$ Å; and $N_A = 8 \times 10^{16}$ cm^{-3}.): (a) aluminum gate; (b) polysilicon gate.

deposited at a total pressure of 5 μm at 650°C, which probably drove some of the water out of the oxide. Also, the thinner the oxide, the lower the observed flatband voltage shift.[40] Technology has not advanced to the point where a well-understood, optimal method can be recommended for the minimization or elimination of oxide charge buildup caused by the injection of electrons or holes into the oxide.

3   Ionizing radiation increases positive oxide charge and interface trap level densities. Such exposure is common in manufacturing integrated circuits when aluminum is deposited using electron gun sources, electron and X-ray lithography, sputter and plasma deposition of materials onto the oxide surface, ion implantation, and cutting of material with the use of beams. The effects of ionizing radiation depend on dose, device geometry (e.g., the gate oxide region may be shielded by a silicon gate), and subsequent processing. Positive oxide charge and interface trap level densities created by ionizing radiation can be reduced to their preexposure levels by annealing at several hundred degrees Celsius for periods of up to 30 min in an inert ambient.

4   Bias-temperature aging accelerates a negative-bias instability.[41,42–46] the so-called slow trapping instability. As a result of this accelerated aging, an irreversible change occurs in Q_f and D_{it}. For negative gate bias, both positive Q_f and D_{it} increase. For positive gate bias, both positive Q_f and D_{it} increase, but to a much lesser extent.[44] Therefore, negative-bias instability is more important for p-channel than for n-channel devices. The chemistry involved in this effect is not well understood, but Sinha et al.[46] have shown that this instability can be significantly reduced by a high temperature (800–900°C) hydrogen anneal prior to gate deposition.

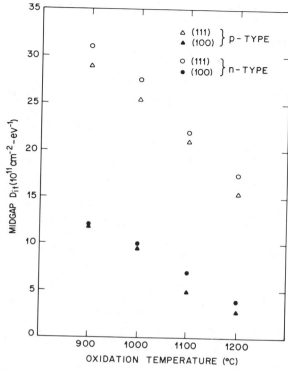

Fig. 15.13 Interface trap level density near midgap as a function of oxidation temperature. Oxides were grown at 700–1000 Å in dry oxygen. Acceptor density was about 10^{15} cm^{-3}. After Razouk and Deal.[47]

The control of D_{it} and Q_f obtained after thermal oxidation are described in Sections (a) and (b) below. Radiation hardening is described in Section 15.7 and the negative bias instability, in Section 15.6.

(a) Control of Interface Trap Level Density

In an as-grown oxide, interface trap level density depends on oxidation temperature,[47] oxidation ambient (wet or dry), oxygen pressure, and silicon substrate orientation. It is independent of impurity type and doping density. Figure 15.13 shows D_{it} near midgap obtained by using the high-low frequency capacitance method described in Section 8.2.6 as a function of oxidation temperature for (111) and (100) silicon substrate orientations. Oxidation was in dry oxygen, followed by a fast cool in oxygen to avoid further oxidation at lower temperatures. Figure 15.13 shows that midgap D_{it} decreases with increasing oxidation temperature and that midgap D_{it} is less on a (100) surface than on a (111) surface. Midgap D_{it} in Fig. 15.13 is unacceptably high, even at the highest oxidation temperature on the (100) sample. The value of D_{it} is minimized by using a high oxidation temperature and a (100) orientation, but it must be reduced further by an anneal.

There are two annealing methods effective for reducing D_{it} after thermal oxidation: (1) the low temperature postmetallization anneal[48, 49] and (2) the high temperature postoxidation anneal.[41] Both of these empirical methods are widely used.

Because D_{it} varies little in magnitude across the silicon bandgap, it is convenient to measure it at only one energy. Midgap is chosen as the bandgap energy for measuring D_{it}, and it is assumed that the variation of midgap D_{it} with annealing treatments is representative of the variation of D_{it} at all other energies in the bandgap.

(b) Postmetallization Anneal

In the low temperature postmetallization anneal, the oxide is grown in either steam or oxygen. Pure aluminum is deposited over the oxide. Aluminum is essential as explained shortly. The sample then is annealed at 350–500°C in a hydrogen or nonoxidizing ambient for up to 30 min. The aluminum then is etched away to form gate electrodes. Annealing can also be done after gate electrodes have been defined, such as during bonding of the device to the header or package.

Before annealing, a steam-grown SiO_2 layer has a typical D_{it} in the low $10^{11} \, cm^{-2} \, eV^{-1}$ range near midgap, and a dry oxygen-grown SiO_2 layer has a typical D_{it} in the low $10^{12} \, cm^{-2} \, eV^{-1}$ range near midgap. After either a low or a high temperature anneal, D_{it} will be in the low $10^{10} \, cm^{-2} \, eV^{-1}$ range near midgap. Figure 15.14 shows D_{it}, measured by the conductance method,

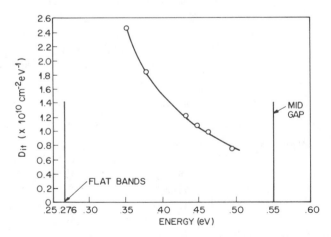

Fig. 15.14 Interface trap level density as a function of bandgap energy measured with respect to the valence band edge at the Si–SiO₂ interface after a low temperature postmetallization anneal. Oxide thickness was 837 Å, aluminum gate area (circular) was $7.1 \times 10^{-3} \, cm^2$, and acceptor density was $10^{15} \, cm^{-3}$. After Nicollian et al.[50] Copyright (1969), Pergamon Press, Ltd. Reprinted with permission.

as a function of silicon band gap energy after a low temperature post-metallization anneal.[50] Comparison of Fig. 15.14 to Fig. 7.5 shows the dramatic reduction of D_{it} near midgap produced by this anneal. For most device applications, such low values of D_{it} will have a negligible adverse effect on device performance.

Figure 15.15 shows midgap D_{it} as a function of oxidation temperature after low temperature post-aluminum metallization anneals in hydrogen.[47] An unannealed sample is included for comparison. In Fig. 15.15, the lowest value of midgap D_{it} is obtained on a (100) substrate oxidized at 1200°C followed by a 450°C anneal in a 25% hydrogen-nitrogen mixture for 10 min. Midgap D_{it} in the $10^{10}\,cm^{-2}\,eV^{-1}$ range is acceptable for most device applications.

Figure 15.16 shows the influence of annealing temperature and percent hydrogen in the ambient on midgap D_{it} in low temperature post-metallization annealing. Percent H_2 in the ambient increases with the

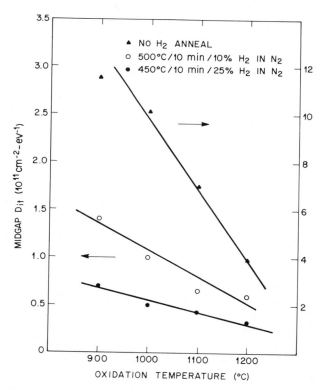

Fig. 15.15 Midgap interface trap level density as a function of oxidation temperature with low temperature postaluminum metallization anneal. Samples were p-type, oriented in the (111) direction. After Razouk and Deal.[47]

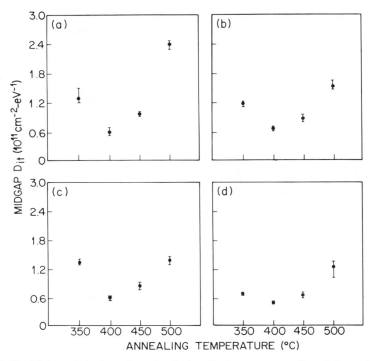

Fig. 15.16 Midgap interface trap level density versus annealing temperature with percent hydrogen in the ambient gas as parameter. Midgap D_{it} was measured using the high-low frequency capacitance method in Section 8.2.5, and annealing time was 10 min. Gates were aluminum, n-type silicon orientation was (111), and the oxide was grown to a thickness of 2000 Å in dry oxygen. Concentrations: (a) 0% H_2 in N_2; (b) 10% H_2 in N_2; (c) 25% H_2 in N_2; (d) 50% H_2 in N_2. After Razouk and Deal.[48]

sequence $a \rightarrow d$. The optimum temperature for minimum midgap D_{it} is about 400°C.* Hydrogen in the gas mixture results in a somewhat lower midgap D_{it} at all temperatures as seen by comparing Fig. 15.16a for zero percent H_2 with Figs. 15.16b–d. However, midgap D_{it} is very weakly dependent on the hydrogen percentage in the gas mixture, as seen in Figs. 15.16b–d.

Deal et al.[51] proposed a mechanism to explain low temperatures postmetallization annealing. Water, present even in oxides grown in dry oxygen, chemically reacts with aluminum to form aluminum oxide and atomic hydrogen. Some of the atomic hydrogen then diffuses to the Si–SiO$_2$ interface and chemically reacts with the interface traps, making them electrically inactive. Although this neutralization may occur for

---

*Annealing temperature for minimum midgap D_{it} may vary with initial oxidation conditions and cooling rate from oxidation temperature.

interface trap levels distributed in energy throughout the silicon bandgap, it is verified by measurement only over the center half of the gap for reasons explained in Section 8.3.5.

Several experiments support this model: (1) a silicon nitride layer placed between the aluminum gate and the SiO_2 is found to block the migration of hydrogen,[51] and thus no reduction in midgap D_{it} is found after annealing;[51] (2) atomic hydrogen was found to be more effective in annihilating interface traps than molecular hydrogen for samples with no gate; and (3) by using several metals, it has been found that the more active metals such as aluminum and magnesium result in low midgap D_{it} whereas the less active metals such as gold and platinum result in very little reduction of midgap D_{it} after low temperature postmetallization annealing.[51]

Using the atomic hydrogen model of interface trap annealing, Fig. 15.16 can be explained. First, the minima in the curves is explained by proposing a slower reaction between atomic hydrogen and interface traps at lower temperatures and a thermal decomposition of the annealed centers at higher temperatures.[48] Thus the decrease in midgap D_{it} as temperatures increase toward the minimum is due to a more nearly completed reaction at higher temperature for fixed annealing time. The increase in midgap D_{it} with temperatures beyond the minimum is due to increased dissociation of the centers.[48] That is, the number of annealed interface traps present when the reaction has reached completion is less at higher temperatures.

According to this model, if the annealing time is extended to allow annealing to go to completion, D_{it} will become time independent at any temperature. This steady-state value of D_{it} should increase monotonically with temperature because higher temperatures favor dissociation. In particular, midgap D_{it} might increase at higher temperatures for annealing times longer than 10 min if dissociation has not been allowed to reach completion in this time. These dependencies of midgap D_{it} on annealing time at various temperatures have yet to be reported.

The weak dependence on the percentage of molecular hydrogen in the ambient is likely due to the small amount of atomic hydrogen in equilibrium with molecular hydrogen at each temperature value. This atomic hydrogen entering the oxide annihilates interface traps as effectively as the atomic hydrogen generated by the reaction of water in the oxide and aluminum but is present in such low amounts that it has little effect.

The experimentally observed behavior of midgap D_{it} with temperature and time is consistent with an underlying chemical reaction between hydrogen and the interface traps.

(c) Postoxidation Anneal

The second method for reducing D_{it} is high temperature postoxidation annealing. In this process an oxide first is grown in dry oxygen at 1150°C or less and then annealed at the same temperature for 30 min in either hydrogen or inert gas. The annealing usually is done in the oxidation furnace immediately after oxidation has been completed. After this annealing, polysilicon or any suitable metal can be deposited on the oxide. The mechanism of interface trap annihilation in a high temperature post oxidation anneal is not understood, but hydrogen from the ambient or small traces of water vapor[52] in an inert gas ambient may have some effect.

(d) Control of Oxide Fixed Charge Density

After D_{it} has been reduced by one of the annealing methods described in Section (a) above, the C-V curve still does not correspond to an ideal C-V curve. Rather, it is shifted parallel to the ideal C-V curve along the voltage axis more than can be accounted for by the work function difference.

Additional shift usually is attributed to oxide fixed charge Q_f. However, an unknown part of this shift is caused by interface trap levels near the band edges. Band edge interface trap levels shift the C-V curve in depletion and weak inversion and distort the C-V curve in accumulation and inversion. However, the distortion is slight and difficult to measure, leaving the shift as the major effect. Therefore, shifts from both band edge interface trap levels and from oxide fixed charge are lumped together.

Even with oxide fixed charge and band edge interface trap charge lumped together in the experimentally measured Q_f, determination of the annealing behavior of Q_f is not easy. For example, the low temperature annealing behavior of the lumped Q_f must be separated from that of D_{it}. To do this, one must know not only how D_{it} anneals, but also how Q_{it} anneals. As (5.22) shows, Q_{it} depends not only on D_{it}, the sum of donor and acceptor interface trap level densities, but on the donor and acceptor interface trap level densities individually. Therefore, the donor and acceptor interface trap level densities both must be determined during annealing if Q_{it} is to be found. Only then can the behavior of Q_f be inferred. Although such a program is possible in principle, using the width of the conductance peak to find the donor and acceptor interface trap densities [see Section 7.3.6(c)], no such program has been carried out, and the low temperature annealing behavior of Q_f is unknown.

Therefore, attention is focused on the annealing behavior of the lumped Q_f in samples where D_{it} is negligible throughout the annealing experiment. That is, Q_f can be studied only in samples subject to either a post-metallization anneal or a high temperature inert gas or hydrogen anneal. Moreover, because no donor or acceptor analysis has been made, conclusions about Q_f are reliable only under the requirement $Q_{it} \ll Q_f$.

Within these limitations it has been found that oxide fixed charge density depends on oxidation temperature, oxidation ambient (wet or dry), oxygen pressure in the oxidizing ambient, and silicon substrate orientation. The value of Q_f can be increased by negative bias-temperature aging and decreased by short term (30 min) annealing in nitrogen or argon.

In integrated circuit technology Q_f is not controlled by any of these methods. Rather, threshold voltage, the most important MOSFET parameter affected by Q_f, is adjusted by doping the substrate in the gate region more heavily with an ion implant to make n-channel enhancement mode MOSFETs, for example. This procedure permits threshold voltage to be precisely and uniformly tailored to a desired value for every MOSFET on the integrated circuit chip. In a typical process for making n-channel MOSFETs the gate oxide is grown to 500 Å at 1000°C in dry oxygen, resulting in threshold voltages of about 0.2–0.3 V ($N_A = 2 \times 10^{15}$ cm^{-3} and $Q_f/q \approx 1.3 \times 10^{11}$ cm^2). After an implant dose of 5×10^{11} cm^{-2} of 50 keV boron ions, all the MOSFETs operate in the enhancement mode with a threshold voltage of 1 V.

Such precise tailoring could not be accomplished by minimizing Q_f, either by optimizing oxidation conditions, or by annealing. However, it is important to discuss the oxidation and annealing conditions that can be used to minimize Q_f for three reasons: (1) these methods are useful for junction diodes and bipolar transistors; (2) understanding how to minimize Q_f by these methods is a necessary step in understanding the fundamental nature of Q_f; and (3) these methods allow greater versatility in the design of experiments.

Figure 15.17a shows a family of C-V curves measured on several samples relatively free of sodium ions oxidized at different temperatures in dry oxygen.[41] Aluminum gates were deposited, and the samples were given a low temperature postmetallization anneal in either nitrogen or argon to reduce D_{it}. For these curves, $D_{it} \leq 5 \times 10^{10}$ cm^{-2} eV^{-1}, making Q_{it}/C_{ox} much less than the observed shift in the C-V curves. Therefore, the observed approximately parallel shift may be taken as due to Q_f.

Figure 15.17a shows that the lower the oxidation temperature, the greater the shift in the C-V curve along the voltage axis in the negative voltage direction. This direction of the voltage shift means that oxide fixed charge is positive. Figure 15.17b summarizes data taken on oxides grown in dry oxygen from curves such as those shown in Fig. 15.17a. Oxide fixed charge density was calculated from the relation $Q_f = C_{ox}(\Delta V + W_{ms})$, where C_{ox} is oxide layer capacitance per unit area, work function difference W_{ms} between aluminum gate and silicon is -0.25 V for $N_D = 10^{16}$ cm^{-3} and -0.95 V for $N_A = 10^{16}$ cm^{-3}, and ΔV is the observed voltage shift. The lowest Q_f is obtained at the highest oxidation temperature on both p-type and n-type samples.

Similar results are obtained for oxidation in wet oxygen,[41] except that Q_f decreases much more slowly with increasing oxidation temperature than for oxidation in dry oxygen as shown in Fig. 15.18. Figure 15.18 also shows that minimum Q_f at 1200°C is about 4×10^{11} cm^{-2} for oxidation in wet oxygen, whereas Fig. 15.17b shows that minimum Q_f at 1200°C is 10^{11} cm^{-2} for oxidation in dry oxygen. Oxidation in dry oxygen is to be preferred for minimum Q_f.

Another parameter that determines Q_f is the rate at which the sample is cooled down in oxygen from oxidation temperature.[41] For a given oxidation

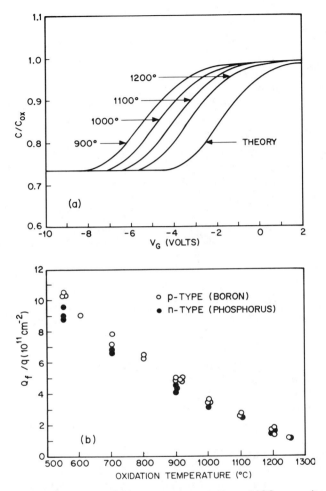

Fig. 15.17 (a) High frequency C-V curves measured on MOS capacitors that have oxides grown at different temperatures in dry oxygen. Above 920°C the oxides were grown at the temperature indicated. Below 920°C the oxides were grown at 1200°C and then immediately subjected to further growth at the lower temperature until the final value of Q_f was reached. This time varied from less than 10 min at 900°C to nearly 1 hr at 550°C. All samples were rapidly quenched from oxidation temperature to freeze in the value of Q_f characteristic of that temperature. Oxide thickness was 2000 Å, donor density was 1.4×10^{16} cm^{-3}, and silicon orientation was (111). The C-V curve marked "theory" was calculated with $Q_f = Q_{it} = W_{ms} = 0$. After Deal et al.[41] Reprinted by permission of the publisher, The Electrochemical Society, Inc. (b) Oxide fixed charge density as a function of oxidation temperature for both n-type and p-type substrates. After Deal et al.[41] Reprinted by permission of the publisher, The Electrochemical Society, Inc.

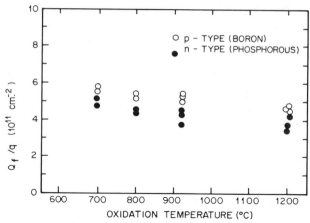

Fig. 15.18 Oxide fixed charge density as a function of oxidation temperature for both *n*-type and *p*-type, (111)-oriented silicon oxidized in wet oxygen (95°C H_2O). After Deal et al.[41] Reprinted by permission of the publisher, The Electrochemical Society, Inc.

temperature, the lowest Q_f will be obtained for the fastest cooling rate. Slower cooling permits oxidation to continue at low temperatures, resulting in a higher Q_f as seen from Fig. 15.17*b*. Cooling rapidly does not permit the sample to oxidize much at low temperatures, so Q_f remains low. However, the risk of wafer warping, which increases with wafer diameter, may make it impractical to use fast cooling. Converting from oxygen to an inert gas (e.g., argon) at oxidation temperature and then cooling down slowly is one solution to this problem.

The relation of Q_f to oxidation conditions and to cooling or annealing in dry nitrogen or argon is summarized in the "Deal triangle" shown in Fig. 15.19. The hypotenuse represents a Q_f versus oxidation temperature curve such as that in Fig. 15.17*b*. High temperature anneals in inert gas are represented by the vertical bars in the diagram. The temperature of such an anneal is shown in the abscissa of Fig. 15.19. The base of the triangle represents an inert gas cool down.

Before *C-V* measurements were made for Fig. 15.19, the oxidized wafers were annealed at a given temperature in an inert ambient followed by a 450°C low temperature postaluminum metallization anneal to reduce D_{it} to negligible proportions in those samples oxidized at temperatures below 1000°C. Although the only difference in annealing seen by the wafers was the inert ambient postoxidation anneal, there are no experimental data to decide whether Q_f is affected by the low temperature postmetallization anneal common to all the samples. Neither is it known whether the low temperature anneal differently affects samples with different high temperature histories. Predictions of Q_f annealing behavior from Fig. 15.19 are secure only for samples subject to the same low temperature postmetallization anneal used to obtain the figure.

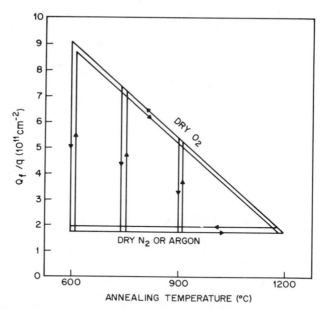

Fig. 15.19 Oxide fixed charge density as a function of annealing temperature for annealing times of one hour or less showing oxidation-annealing paths. Based on (111) silicon samples. After Deal et al.[41] Reprinted by permission of the publisher, The Electrochemical Society, Inc.

For samples given a final low temperature postmetallization anneal, Fig. 15.19 shows that a low Q_f can be obtained by either a high temperature oxidation (1200°C) followed by cooling in an inert ambient or a low temperature oxidation ($< 1000°C$) followed first by an anneal at the same temperature (downward vertical arrows) in either dry nitrogen or argon and then cooling down in the inert gas ambient.[41]

The annealing behavior in dry nitrogen shown in Fig. 15.19 is observed only for relatively short annealing times. This time is 15 min or less at 1100°C and several hours or more at 950°C.[53] That is, Q_f first decreases according to Fig. 15.19 and then increases as a function of annealing time.* Thus Fig. 15.19 is valid only for relatively short annealing times.

The value of Q_f depends on both oxidation temperature and oxygen pressure during growth. The dependence of Q_f on oxidation temperature at fixed oxygen pressure (1 atm) was shown in Fig. 15.17b. The dependence of Q_f on oxygen pressure at fixed oxidation temperature (1100°C) is shown in

---

*This increase in Q_f may be related to the experimental observation by Raider et al.[54] that at high temperatures (984–1200°C) nitrogen diffuses through the SiO_2 and chemically reacts with silicon at the Si–SiO_2 interface. This reaction is inhibited by oxidants[54] such as water vapor[54,55] because oxidation is so rapid compared with nitridation that few silicon atoms are available to react with the nitrogen.

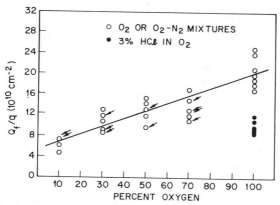

Fig. 15.20 Oxide fixed charge density versus the percent oxygen in a dry oxygen-nitrogen oxidation ambient. An oxide 1000 Å thick was grown at 1100°C on phosphorus doped, (100), 2–5 Ωm-cm silicon wafers. Oxidation times were 410, 140, 90, 64, and 44 min for 10, 30, 50, 70, and 100% oxygen ambients, respectively. Pull out time from the hot zone in the oxidizing ambient was 30 sec. Gates and back contact were aluminum. The wafers were annealed after metallization at 450°C for 0.5 hr in hydrogen to reduce D_{it}. After Muraka.[56]

Fig. 15.20. Figure 15.20 shows that Q_f increases with increasing partial pressure of oxygen in an ambient[56] of dry nitrogen mixed with oxygen. Figure 15.20 represents MOS capacitor measurements on several wafers. The same uncertainties regarding the effect of the low temperature post-metallization anneal on Q_f given in the discussion of Fig. 15.19 apply. Measurements on some wafers resulted in the same Q_f, and such points are indicated on the graph by arrows pointing to the reproducible value of Q_f. The number of arrows gives the number of additional measurements, yielding the same value of Q_f as the circled point. There is no satisfactory explanation of this behavior.

(e) Orientation Dependence of Oxide Fixed Charge and Interface Traps

Figure 15.21 shows the orientation dependence of Q_f and D_{it}. These curves were obtained by steam oxidation of a silicon sample made in the shape of a hemisphere. In a hemisphere various crystallographic orientations are exposed.[57] Figure 15.21 shows that both Q_f and D_{it} produced by thermal oxidation have the same silicon surface orientation dependence. The lowest values of Q_f and D_{it} are obtained for (100) orientation, and the highest are obtained for (111) orientation. For this reason, the (100) orientation is the most widely used orientation in integrated circuit technology. This orientation dependence is consistent with the findings using the photo I-V method described in Section 11.4.3(a) that oxide fixed charges and interface traps are located close to the Si–SiO₂ interface.

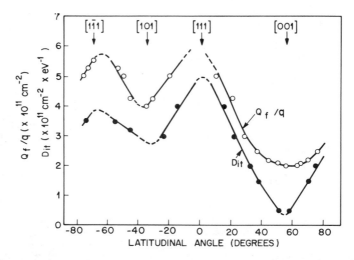

Fig. 15.21 Oxide fixed charge density and interface trap level density as a function of silicon surface orientation. Oxide thickness was 1500 Å, and acceptor density 2.5 × 10^{15} cm^{-3}. After Arnold et al.[57]

(f) Doping Density and Dopant Dependence

For given oxidation conditions, Fig. 15.22 shows no significant difference between D_{it} near midgap on n-type and p-type silicon substrates.[47] Therefore, it can be concluded that midgap D_{it} is independent of silicon conductivity type. Moreover, it is found that D_{it} is independent of acceptor density from $N_A = 4 \times 10^{17}$ cm^{-3} to 2.5×10^{14} cm^{-3}, which was the range measured. Hence dopant impurities such as boron incorporated in the oxide during oxidation or dopants such as phosphorus that pile up in the silicon during oxidation are not related to the interface traps seen in thermally oxidized silicon.

15.5 OXIDATION AMBIENT

The oxidation of silicon in dry oxygen, wet oxygen, and steam was described in Chapters 13 and 14. For the critical regions of a MOSFET such as the gate oxide or the passivation of a high gain, low current bipolar transistor, oxidation in dry oxygen followed by an anneal to reduce D_{it} produces an oxide with the best *electrical* properties. Because gate oxides are usually thin (≤ 1000 Å), the greater oxidation time in dry oxygen compared to steam or wet oxygen is not a disadvantage. However, for field oxides that are relatively thick (> 5000 Å), oxidation in dry oxygen is prohibitively slow. Therefore, steam oxidation is used predominantly for field oxides where speed of growth rather than optimum interfacial and

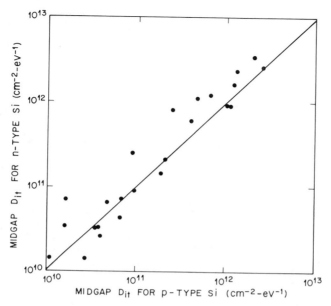

Fig. 15.22 Interface trap level density near midgap measured by the high-low frequency capacitance method described in Section 8.2.6 on (111) samples. Midgap D_{it} was varied by oxidizing the wafers at different temperatures in dry oxygen. After Razouk and Deal.[47]

oxide properties is paramount. To further speed growth of the field oxide, the most time-consuming step in the entire integrated circuit process sequence, high pressure steam or oxygen (>1 atm) is coming into wide-spread use.

Oxidation in dry oxygen is inherently a purer process than oxidation in water vapor because water has such a high dielectric constant (80). Thus water is a good solvent, leaching out impurities from whatever surface it contacts. However, oxidation in dry oxygen results in a high interface trap level density distributed over the silicon bandgap, as shown in Fig. 15.23a.[58] After a low temperature anneal, D_{it} is reduced in magnitude and the interface trap energy level distribution has the characteristic U shape shown in Fig. 15.23b.[58] An interface trap energy level distribution similar to Fig. 15.23b also would have been produced by a high temperature anneal.

As shown in Fig. 7.5, the U-shaped distribution also is characteristic of oxides grown in steam without any annealing. However, D_{it} in Fig. 7.5 is greater than that in Fig. 15.23b. A high temperature postoxidation anneal or a low temperature postmetallization anneal would reduce D_{it} to values similar to those in Fig. 15.23b. The final interface is similar in both wet and dry oxides. Therefore, dry oxygen is preferred to water vapor as the oxidation ambient for gate oxides only because dry oxygen is an inherently cleaner ambient than water vapor or steam.

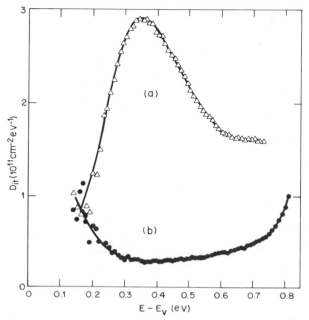

Fig. 15.23 Interface trap level density as a function of silicon bandgap energy with respect to the valence band edge measured using the high-low frequency capacitance method described in Section 8.2.6. The substrate was p-type, (100) silicon with $x_o =$ 1020 Å. (a) After oxidation in dry oxygen; (b) after annealing at 450°C with aluminum gate electrodes. After Johnson et al.[58] Copyright (1978), Pergamon Press, Ltd. Reprinted with permission.

15.6 NEGATIVE BIAS-TEMPERATURE INSTABILITY

The negative bias-temperature instability, sometimes called "slow trapping," is revealed by accelerated aging with negative gate bias at elevated temperatures. This threat to the stability of a p-channel MOSFET is one of a class of electrochemical processes that can occur in thermally grown SiO_2. Accelerating the negative bias-temperature instability results in an increase of positive oxide fixed charge and interface trap level densities for both n-type and p-type silicon substrates. The effect is greater when initial $Q_f + Q_{it}$ is large. The mechanism for this effect has not been established.

The negative bias-temperature instability can be aggravated by ionizing radiation.[44,59] For example, an oxide damaged by ionizing X-rays during the electron-gun evaporation of aluminum and then annealed to completely eradicate the electrical effects of radiation damage will exhibit a greater change in $Q_f + Q_{it}$ and midgap D_{it} after a given negative bias-temperature aging test than will the same oxide that had never been subjected to ionizing radiation.[44]

One of the marked features of negative bias-temperature aging is the development of a broad, readily identifiable peak in D_{it} near midgap,[44] as revealed by the high-low frequency capacitance method. The energy at which D_{it} peaks (0.47–0.75 eV) depends on the gate metal used (even though the oxide thickness was 1000 Å) and whether the oxide was grown in steam or dry oxygen.[44]

Deal et al.[41] were first to report the negative bias-temperature instability. Figure 15.24 shows the negative bias-temperature instability measured with an MOS capacitor. Curve A in Fig. 15.24a is a theoretical high frequency C-V curve. Curve B is the initial high frequency C-V curve measured prior to aging a "sodium free" oxide. Curve C is measured after 2 min at 400°C under an applied oxide field of -2×10^6 V/cm. Curve C has shifted along the voltage axis to more negative voltages, and its shape is distorted compared to curves A and B. The shift of the C-V curve along the voltage axis indicates an increase of positive oxide charge density. The stretchout of the C-V curve along the voltage axis or distortion in its shape can be caused by an increase in D_{it}. Both result from negative-bias temperature aging.

Distortion of the shape of a C-V curve is not a definite indication that D_{it} has increased because distortion in shape can also be produced by lateral nonuniformities of interface charge, as explained in Section 6.2. The two mechanisms have been distinguished here by measuring surface recombination velocity with a gate controlled diode and D_{it} by the conductance method (described in Section 5.5). Both methods show an increase in D_{it} after negative bias-temperature aging. Thus the distortion in the shape of curve C compared to curve B in Fig. 15.24a is due at least in part to the formation of interface trap levels during aging. Figure 15.24b shows a summary of data from curves such as those in Fig. 15.24a. The shift of the midpoint of the C-V curve arbitrarily is used to calculate $Q_f/q + Q_{it}/q$. Figure 15.24b shows that $Q_f/q + Q_{it}/q$ increases linearly with the electric field applied during aging and that the slope and intercept at zero field depend on the initial values of $Q_f/q + Q_{it}/q$.

Changes in $Q_f + Q_{it}$ and D_{it} saturate with time. Saturation occurs in less than 2 min at 400°C, so $Q_f + Q_{it}$ in Fig. 15.24b is a saturation value at each aging field. At lower temperatures, saturation occurs more slowly (≤ 15 min), but sufficient buildup of $Q_f + Q_{it}$ and D_{it} can occur at low temperatures to cause device failure.*

Sinha et al.[46] showed that annealing in hydrogen at 800–900°C *prior* to metallization significantly reduces the negative bias-temperature instability. Such an anneal also reduces the negative bias-temperature instability in oxides after exposure to ionizing radiation. Figure 15.25 illustrates the benefits of this anneal. Figure 15.25a shows a set of high-low frequency

---

*Failure is said to occur when the characteristics deviate from the specifications set for the device.

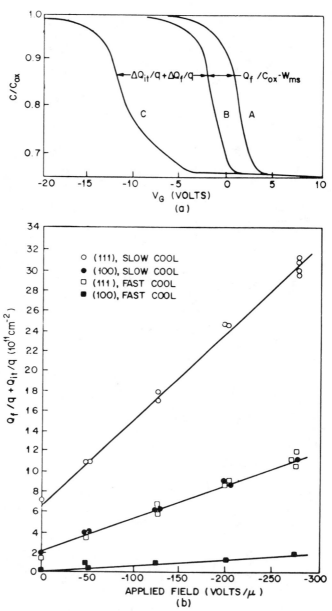

Fig. 15.24 (a) Plot of C/C_{ox} versus gate bias measured before and after negative bias-temperature aging. Oxide was grown to 2000 Å in dry oxygen, and acceptor density was 1.4×10^{16} cm^{-3}. After Deal et al.[41] Reprinted by permission of the publisher, The Electrochemical Society, Inc. (b) Oxide fixed charge density Q_f/q and interface trap charge density Q_{it}/q versus electric field applied during negative bias-temperature aging. Aging time was 2 min at 400°C. Curves with different values of Q_f/q and Q_{it}/q were obtained by using samples of different orientation, (111) and (100), and using various cooling rates from oxidation temperature. After Deal et al.[41] Reprinted by permission of the publishers, The Electrochemical Society, Inc.

C-V curves that did not have any high temperature anneal. A shift and a distortion in shape are observed as expected. Figure 15.25b shows a set of high-low frequency C-V curves measured on a sample that had a 30 min anneal in hydrogen at 900°C prior to aluminum metallization. Comparison of the curves in Fig. 15.25a to those in Fig. 15.25b shows the significant

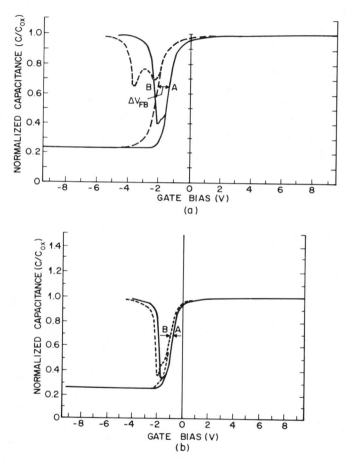

Fig. 15.25 (a) High and low frequency C-V curves measured on an MOS capacitor that had received no high temperature H_2 anneals. Oxide was grown to 1000 Å in an HCl cleaned fused quartz furnace tube at 1100°C in dry oxygen. Silicon had a donor density of $\sim 10^{15}$ cm^{-3} and a (111) surface orientation. The gate was aluminum. Curves A (solid lines) initial, curves B (dotted lines) after aging at 250°C for 15 min with $V_G = -23.5$ V. After Sinha and Smith.[45] Reprinted by permission of the publisher, The Electrochemical Society, Inc. (b) High and low frequency C-V curves measured on an MOS capacitor that had received a 30 min anneal in hydrogen at 900°C prior to aluminum gate metallization. Sample parameters were the same as those in (a). Curves A (solid lines) initial, and curves B (dotted lines) after aging at 250°C for 15 min with $V_G = -23.5$ V. After Sinha et al.[46] Copyright (1978), Pergamon Press, Ltd. Reprinted with permission.

improvement brought about by the high temperature hydrogen anneal. Although these results were obtained on samples with an aluminum gate, similar results would be obtained with a silicon gate.

It was also found that although a 30 min anneal at 450°C in hydrogen *after* aluminum metallization reduced D_{it} near midgap, it slightly enhanced the negative bias-temperature instability.[44]

Finally, in samples that had a boron doped polysilicon gate, $Q_f + Q_{it}$ exhibited an abnormal voltage asymmetry after the high temperature premetallization hydrogen anneal. That is, rather than a large increase in $Q_f + Q_{it}$ and D_{it} after negative bias-temperature aging and a small increase after positive bias temperature aging, the increases of these densities were slightly greater for positive than negative bias-temperature aging. Because the polysilicon gate samples were given a high temperature H_2 anneal after deposition of the polysilicon, and because boron migration through SiO_2 is enhanced by annealing in a hydrogen ambient, the asymmetry may be caused by boron from the gate reaching the Si–SiO_2 interface.

15.7 IONIZING RADIATION

Ionizing radiation creates positive oxide charge and interface traps as described in Section 11.6. Experiment shows that permanent charge trapping occurs in the oxide for photon energies above 8.8 eV. This trapping involves simple ionization processes, rather than displacement events, which require secondary electrons with energies of the order of 150 keV. The trapped charge density is very sensitive to the oxide growth process as well as other processing to which the oxide may be subjected after growth. Charge buildup saturates at large total ionizing radiation doses, and the saturation value of the charge is process dependent. The radiation induced charge is stable at room temperature, but it can be removed by heating to near 300°C or by low energy (< 9 eV) photon irradiation or by avalanche injection of electrons. These processes inject electrons into the oxide from the metal or the silicon. Recombination of these electrons with trapped holes neutralizes the oxide charge.

Section 15.4 explained the changes in device characteristics due to increasing Q_f and D_{it} during processing. In this section these effects are illustrated by experiments in which positive oxide charge and D_{it} are increased by ionizing radiation. Then radiation hardened oxides* are discussed.

---

*A radiation hardened oxide is one whose electrical properties are insensitive to ionizing radiation.

15.7.1 MOSFET

Figure 15.26 shows the drain current I_D versus gate bias V_G characteristics of two MOSFETs before and after exposure to ionizing radiation. Figure 15.26a shows a p-channel MOSFET and Fig. 15.26b, is an n-channel MOSFET. In both Figs. 15.26a and 15.26b, curve 1 corresponds to a dose of 7.5×10^4 rads, curve 2 to 1.5×10^5 rads, curve 3 to 3×10^5 rads, and curve 4 to 6×10^5 rads. Each device was irradiated with $+1$ V applied to the gate and zero source and drain bias.

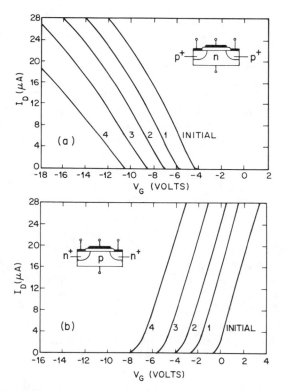

Fig. 15.26 (a) Forward transfer characteristics of a p-channel MOSFET measured before and after four successive doses of X-ray photons. The doses were: (1) 7.5×10^4 rads, (2) 1.5×10^5 rads, (3) 3×10^5 rads, and (4) 6×10^5 rads. The device was irradiated with $+1$ V applied to the gate and zero source and drain bias. Gate oxide thickness was 1200 Å, and donor density was 5×10^{15} cm^{-3}. These characteristics were measured with a drain bias of 25 mV and zero source bias. After Snow et al.[60] Copyright (1967), IEEE. (b) Forward transfer characteristics of an n-channel MOSFET measured before and after four successive doses of X-ray photons. The radiation doses and biasing conditions during irradiation were the same as those in (a). Gate oxide thickness was 1000 Å, and acceptor density was 10^{16} cm^{-3}. These characteristics were measured under the same biasing conditions as those in (a). After Snow et al.[60] Copyright (1967), IEEE.

For both devices in Fig. 15.26, the threshold voltage V_T, the gate bias required to induce an inversion layer between source and drain, is seen to shift along the voltage axis to higher negative gate biases with increasing dose. This shift is a direct consequence of the radiation induced buildup of positive charge in the oxide. Figure 15.26 also shows that the slope of the I_D versus V_G curves decreases with increasing radiation dose. This decrease in slope is analogous to the distortion of the C-V characteristics discussed previously and is due to an increase in D_{it}. Because the transconductance g_m of a MOSFET is defined by[1,2]

$$g_m = \left(\frac{\partial I_D}{\partial V_G}\right)_{V_D = constant} \tag{15.1}$$

the decrease in slope represents a decrease in transconductance.

15.7.2   Planar Junction Diode[59]

In a planar p-n junction diode the most important radiation induced changes resulting from the buildup of positive charge in the oxide and an increase in D_{it} are an increase in reverse current and changes in avalanche breakdown voltage. Figure 15.27 shows reverse current versus bias measured initially and after radiation on a p^+n gate controlled diode. The initial reverse current was in the picoampere range. Following radiation exposure, the junction had a soft reverse characteristic and reverse current, measured at a reverse bias of 20 V, had increased by 5 orders of magnitude. In addition, zero bias junction capacitance increased by a factor of 5. The mechanism causing increased reverse current involves inversion of the surface of the p^+ region by radiation induced positive oxide charge. Carriers supplied to this inversion layer result in a channel current when the junction is reverse biased. This current results from either tunnel or avalanche breakdown of the field induced junction between the inversion layer and the underlying silicon, as shown in the inset of Fig. 15.27 labeled "after irradiation." For the device in Fig. 15.27, breakdown occurs by tunneling at about 1 V. The radiation induced increase in zero bias junction capacitance is due to the capacitance of the field induced junction, whose depletion layer width is of the order of 100 Å, in parallel with the capacitance of the metallurgical junction. If the surface dopant concentration of the p^+ region is greater than about 5×10^{18} cm^{-3}, tunnel current also flows when the junction is forward biased, giving rise to an excess forward current. This excess forward current plays a role in radiation induced current gain degradation of bipolar transistors.

Channel currents similar to the above also occur in n^+p junctions that are surrounded by a p^+ annular channel-stop region. Omission of the p^+ region does not solve the problem. Without it the inversion layer, induced over the p-region by positive oxide charge, is effectively short-circuited to the substrate at the ends of the wafer containing the diode, where there is a

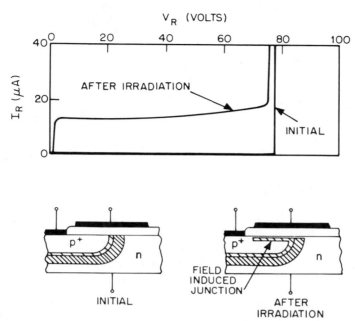

Fig. 15.27 Reverse current versus reverse bias characteristics of a p^+n gate controlled diode before and after X-irradiation. The radiation dose was 9×10^5 rads with $+45$ V applied to the gate and zero bias to the junction. The surface dopant concentration of the p^+ region was 2.5×10^{18} cm^{-3}, and oxide thickness was 6000 Å. After Snow et al.[60] Copyright (1967), IEEE.

region of high damage produced when the wafer was cut. Therefore, a channel current again results.

A decrease of avalanche breakdown voltage also can be seen in Fig. 15.27. This effect is illustrated in more detail in Figs. 15.28a,b for n^+p and p^+n gate controlled diodes, respectively. Figure 15.28 shows avalanche breakdown voltage as a function of gate bias before and after exposure to ionizing radiation. Figure 15.28 shows that gate biases that tend to invert the substrate result in increased avalanche breakdown voltage compared to the $V_G = 0$ case, whereas gate biases which tend to invert the diffused region result in decreased avalanche breakdown voltage. After radiation the characteristics are shifted to the left along the V_G axis in Fig. 15.28 as a result of positive oxide charge buildup. Thus the avalanche breakdown voltage measured at $V_G = 0$ decreases for p^+n diodes and increases for n^+p diodes, as indicated by the arrows in Figs. 15.28a, b. In addition, the slopes of the characteristics in Fig. 15.28 decrease with radiation dose resulting from an increase in D_{it}.

The change in avalanche breakdown voltage for a given radiation dose depends on the slope of the initial avalanche breakdown voltage versus

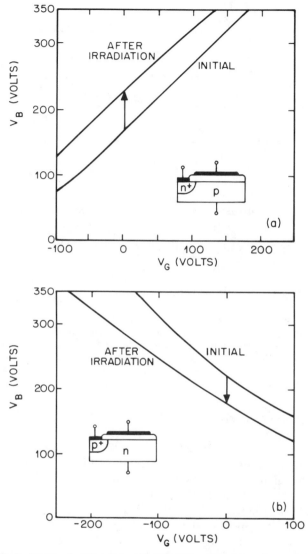

Fig. 15.28 (a) Avalanche breakdown voltage versus gate bias characteristics of a p^+n gate controlled diode before and after X-irradiation. The radiation dose was 3×10^5 rads with $+45$ V applied to the gate and zero junction bias. Donor density was 5×10^{14} cm^{-3}, and oxide thickness was 8000 Å. After Snow et al.[60] Copyright (1967), IEEE. (b) Avalanche breakdown voltage versus gate bias characteristics of an n^+p gate controlled diode before and after X-irradiation. The radiation dose was 1.5×10^5 rads with $+10$ V applied to the gate and zero junction bias. Acceptor density was 5×10^{14} cm^{-3}, and oxide thickness was 1200 Å. After Snow et al.[60] Copyright (1967), IEEE.

gate bias characteristic of the device. The higher the initial slope, the larger the change. This slope depends on substrate doping and oxide thickness. It is unity at low doping concentrations and thin oxides and decreases with increasing doping concentration and oxide thickness. Thus, for a given radiation exposure, the change in avalanche breakdown voltage will be larger in a high voltage diode than in a low voltage diode.

15.7.3 Planar Bipolar Transistor[60]

In the planar bipolar transistor an increase in D_{it} and positive oxide charge density both degrade current gain. The effects of these two quantities on current gain can be separated by employing a gate over the emitter-base junction.

Figures 15.29a,b show common emitter current gain h_{FE} versus dose for an n-p-n and a p-n-p planar bipolar transistor, respectively. Large low current h_{FE} decreases are evident, even at moderate doses for the n-p-n transistor. The dependence of h_{FE} on dose and differences between p-n-p and n-p-n transistors can appear very complicated unless the effect of changes in silicon band bending due to radiation induced positive oxide charge are separated from the effect of changes in surface recombination velocity due to the creation of interface traps. The only method of accomplishing such a separation is by measuring the h_{FE} versus V_G characteristics of gate controlled transistors. This is not pursued further here.

Table 15.1 gives a summary of the effects of radiation induced positive oxide charge and interface traps on the parameters of several widely used semiconductor devices. The most serious of these effects are shown in capital letters. Besides the planar p-n junction diode, planar bipolar transistor, and the MOSFET considered in the previous discussion, the JFET is included for comparison.

Table 15.1 shows that positive oxide charge density and increased D_{it} can give rise to reverse current degradation in all device types considered. In addition, positive oxide charge results in threshold voltage shifts in MOS-FETs and, together with radiation induced interface traps, a decrease of h_{FE} at low current levels in bipolar transistors. The JFET, in contrast, does not have a degradation mode other than reverse current degradation. This is a common degradation mode of all the other devices listed in Table 15.1, but the only important degradation effect in the JFET. Thus in ionizing radiation environments the most radiation tolerant device is the JFET.

15.7.4 Hardening Techniques

Our next task is to describe ways of hardening MOSFETs and planar bipolar transistors against ionizing radiation. The discussion in this section applies primarily to the use of devices and integrated circuits in a radiation

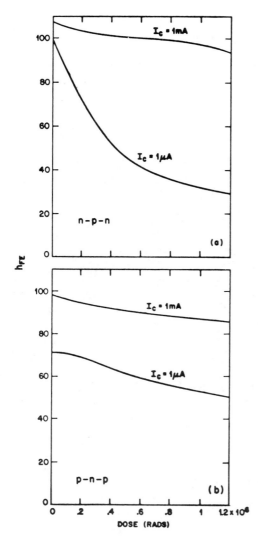

Fig. 15.29 (a) Common emitter current gain h_{FE} measured at $V_G = 0$ and collector currents of 1 mA and 1 μA as a function of dose for an n-p-n gate controlled bipolar transistor. The surface dopant concentration of the base was $1.5 \times 10^{18} \text{ cm}^{-3}$, and oxide thickness was 6000 Å. After Snow et al.[60] (b) Common emitter current gain h_{FE} measured with $V_G = 0$ and collector currents of 1 mA and 1 μA, as a function of dose for a p-n-p gate controlled transistor that has the same base concentration as that in (a) and an oxide thickness of 8000 Å. After Snow et al.[60] Copyright (1967), IEEE.

environment, and does not apply to reducing radiation damage during processing.

(a) Impurities

Impurities such as sodium increase the sensitivity of an oxide to radiation because radiation ionizes some of the neutral sodium.* Water-related traps, described in Section 11.5.1, trap electrons created by the ionizing radiation. This trapping counters the buildup of positive oxide charge by

---

*The sodium incorporated in SiO_2 seldom is all ionized, as explained in Section 10.2.3.

Table 15.1 Summary of the Effects of the Radiation-Induced Oxide Charge and Interface Traps on Device Parameters.[a]

Device Phenomenon	p-n Junction diode	Bipolar transistor	MOS transistor	Junction field-effect transistor
Oxide Charge	REVERSE CURRENT DEGRADATION	REVERSE CURRENT DEGRADATION	REVERSE CURRENT DEGRADATION	REVERSE CURRENT DEGRADATION
	breakdown voltage change	breakdown voltage change	breakdown voltage change	breakdown voltage change
		h_{FE} DECREASE AT LOW-CURRENT LEVELS	THRESHOLD VOLTAGE CHANGE	
Interface traps	reverse current increase	reverse current increase	reverse current increase	reverse current increase
		h_{FE} DECREASE AT LOW-CURRENT LEVELS	g_m decrease	

[a] After Snow et al.[60]

hole capture. However, impurities such as sodium and water-related traps play a secondary role at most. The primary effect is the buildup of positive oxide charge and interface trap level densities by hole capture, as described in Section 11.6. Other impurities can be incorporated to act as electron traps or recombination centers, to reduce positive oxide charge buildup. Incorporation of these impurities is described later in this section.

Hughes et al.[61] have performed an extensive set of experiments to investigate the radiation sensitivity of oxides as a function of sodium contamination in the oxide. The oxides used by Hughes et al.[61] were grown in dry oxygen at 1000°C and then implanted with sodium ions. Sodium concentration in the oxide was then determined with the ion microanalyzer. In this method a beam of monoenergetic ions is used to sputter secondary ions and other molecular species from the surface of a specimen. The ionized particles then are collected and analyzed by mass spectrometric methods.

The radiation induced flatband voltage shift determined from MOS capacitor measurements is shown as a function of sodium concentration in Fig. 15.30 for a fixed radiation dose of 10^6 rads and a gate bias of $+10$ V applied during radiation exposure. In these experiments as well as those in Ref. 62, sodium atoms, ionized by radiation, drift in the applied electric field to the Si–SiO$_2$ interface.

Although Fig. 15.30 implies that there is a correlation between ΔV_{FB} and sodium concentration, the positive charge buildup corresponding to ΔV_{FB} is not due to all the sodium ions. For example, the largest ΔV_{FB} in Fig. 15.30 is 35 V for a sodium concentration of 10^{15} cm^{-2}. This sodium level corresponds to a volume density of 3×10^{19} cm^{-3}. The estimated charge

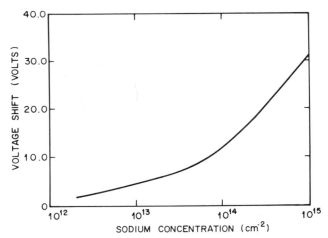

Fig. 15.30 Voltage shift as a function of implanted sodium ions for a dose of 10^6 rads at a gate voltage of $+10$ V. After Hughes et al.[61] Copyright (1972), IEEE.

density necessary to produce the observed 35 V shift is about $4 \times 10^{18}\,cm^{-3}$. Therefore, a maximum of about 10% of the sodium in the oxide contributes to the observed charge. Of course, not all the neutral sodium in the sample is ionized by radiation, and the ion implantation may have introduced damage in the oxide, thus preventing an accurate evaluation of the charge buildup due to sodium.

The maximum sodium concentration in Fig. 15.30 is much higher than the largest amounts usually found in SiO_2. Typically, the maximum sodium concentrations found are of the order of about $7 \times 10^{12}\,cm^{-2}$ because of the limited solid solubility of sodium in SiO_2. Only by ion implantation can higher concentrations be obtained in SiO_2.

The most significant implication in Fig. 15.29 is that one mechanism for charge buildup is ionization of neutral sodium in the SiO_2 and migration of Na^+ to the $Si–SiO_2$ interface.

(b) Processing

There are two approaches to radiation hardening: (1) to find processing techniques that minimize the presence and formation of intrinsic traps and impurity levels in the SiO_2; and (2) to introduce traps that capture electrons into the SiO_2 deliberately, thus offsetting the positive trapped hole charge, or to introduce traps that act as recombination centers, to cause recombination to take place rather than trapping. We start our discussion with the first method, reducing intrinsic and extrinsic trapping centers* by processing. Because the exact chemical nature of the intrinsic traps is not known, the processing required to harden SiO_2 must be determined empirically. To reduce extrinsic trapping, it is necessary simply to use processing that minimizes impurity contamination such as sodium.

Aubuchon[63] performed a comprehensive study of the effects of processing variations on the radiation sensitivity of MOSFETs and MOS capacitors. The results were:

1 Positive charge buildup is less for dry oxides grown at 1000 than at 1150°C. In fact, the lower the oxidation temperature, the more radiation resistant the oxide. However, lowering oxidation temperature increases Q_f. Thus threshold voltage of a MOSFET must be controlled by an ion implant.

2 Nitrogen annealing of the dry oxide at 1000°C to reduce Q_f increases radiation sensitivity. Only annealing done below 800°C does not adversely affect hardness.

3 (100) Orientation is harder than (111) orientation.

4 Wet oxides are more radiation sensitive than dry oxides, probably because of their greater sodium content.

---

*Intrinsic trapping centers are characteristic of thermally grown SiO_2, whereas extrinsic trapping centers are associated with known impurities such as sodium.

By varying these processing conditions. Aubuchon[63] was able to improve hardness levels. Other experiments show that oxides grown in HCl are harder than those grown in dry O_2. Such optimization of processing has been observed by other workers as well. These results indicate that attention to fabrication can lead to MOS devices less sensitive to ionizing radiation.

(c) Impurity Doping

The second method of hardening SiO_2 is by impurity doping. Several investigators, including Hughes et al.,[61] Lindmayer,[64] Donovan et al.,[65,66] Peel and Kinoshita,[67] and others have shown that the radiation sensitivity of the oxide can be altered by doping the oxide with various impurities by either diffusion or ion implantation. Lindmayer[64] and Hughes et al.[61] have shown that the type of gate metal used influences radiation hardness. Experiments using aluminum, molybdenum, and chromium for the gate showed that the best results are obtained using chromium. Related experiments showed that chromium diffuses into the oxide and influences the hardness level by either trapping electrons or aiding recombination of electrons and holes created by ionizing radiation.

Aluminum implanted in the oxide, using an optimized dry oxidation process, was found to increase device hardness for positive gate bias. For implants of 10^{14} cm^{-2} at energies of 10 and 20 keV and a radiation dose of 10^6 rads, ΔV_{FB} was about 2 V, compared with a shift of 4 V without aluminum, for $V_G = +10$ V applied during radiation. For a 30 keV implantation energy, ΔV_{FB} was much larger than unimplanted samples for either polarity of gate bias.

The results indicate that aluminum implantation must be restricted to n-channel devices that are not subject to negative gate bias and that an optimum implantation energy must be used to produce hardened devices. The increase in hardness has been tentatively attributed to the reduction in the number of nonbridging oxygens by the aluminum. This assumption is based on UV absorption experiments performed on doped fused silica and on silicate glasses, which show that an absorption band at 8.5 eV is caused by nonbridging oxygens in the structure. These nonbridging oxygen ions are increased by adding sodium. Aluminum alters the bonding perhaps by bridging the nonbridging oxygen ions to reduce optical absorption at 8.5 eV.

Donovan et al.[66] investigated the radiation tolerance of oxide layers implanted with nitrogen. The results show that ΔV_{FB}, after a radiation dose of 6×10^{14} electrons/cm^2 at 20 keV with $V_G = +10$ V applied, could be reduced from 90 to 50 V.

(d) Other Insulators

Insulators other than SiO_2, containing large concentrations of recombination centers, have been used to fabricate radiation hardened MOS devices. These insulators include Al_2O_3, silicon nitride, and composite insulators such as oxynitrides. Although the radiation induced voltage shift is generally less for these insulators, they exhibit other problems such as current leakage, threshold voltage instabilities, charge injection from the electrodes, and low junction avalanche breakdown voltages. In addition, the level of radiation hardness can vary from lot to lot in manufacture because of the lack of control of the properties of these insulators. The use of insulators other than SiO_2 and doped SiO_2 has been limited to specific manufacturers and research laboratories and has not found acceptance throughout the industry for the factors just cited.

(e) Gettering

Phosphorus pentoxide has been used as a getter applied between the gate oxide and the gate metal. In some instances an increase in radiation hardness has been observed. However, instability caused by the polarizability of phosphosilicate glass reduces the attractiveness of this material for device hardening.

(f) Circuit Hardening

Circuit hardening techniques also can be used to extend the region of acceptable MOS circuit operation in a radiation environment. These techniques include proper component selection and circuit design techniques. In choosing the basic device components, p-channel MOSFETs are more desirable than n-channel because ΔV_{FB} is inherently much less than for corresponding n-channel devices. This point can be seen from Fig. 11.32 by noting that p-channel devices operate with $-V_G$ whereas n-channel devices operate with $+V_G$. Device selection can be based on sample radiation tests or radiation followed by annealing to select the hardest devices when specifications are severe.

One of the major problems in operating a circuit in a radiation environment is the electrical transients induced in the circuit by radiation. Special circuit design techniques can be used in some instances to minimize the effects of these transients. For example, circuits can be designed so that a low bias level or a clocked gate voltage is applied during radiation exposure, as ΔV_{FB} is linearly proportional to V_G applied during radiation. Circuits can be designed to operate properly with large changes in signal level or threshold voltage. Thus ΔV_{FB} produced by radiation exposure can be compensated for by an equal and opposite change in another part of the circuit or system. This method of circuit design, in addition to requiring innovation, may require that radiation induced changes be about equal for all devices in the circuit—a requirement difficult to achieve in practice. In addition to ΔV_{FB}, the effective time delay and noise immunity changes for MOSFETs and planar bipolar transistors must be considered in integrated circuit design. The best solution would be to harden the devices as much as possible.

(g) Shielding

Another solution to the radiation problem is to use shielding. However, for penetrating radiation, shielding may have to be thick and heavy, which is a distinct disadvantage for applications in a radiation environment such as space.

(h) Comparison of Radiation Sensitivity

To compare the radiation sensitivity of various oxides, we define a sensitivity factor S_R as

$$S_R \equiv \frac{|\Delta V_{FB}|}{|V_G|}. \tag{15.2}$$

Early devices using dry oxides had sensitivity factors of 10–15 for $+V_G$ and S_R of about unity for $-V_G$ during radiation exposure, as shown in Table 15.2. The sensitivity factor for the latest, hardened devices is less than 0.3 for $+V_G$. This factor of 50 increase in device hardness has been achieved by improvements in oxide processing, including a reduction of impurity content.

Table 15.2 Radiation Sensitivity as a Function of Processing[a]

S_R	Conditions	
15.0	Unhardened oxide; n - channel.	oxide grown at
1.0	Unhardened oxide; p - channel.	1100°C in dry O_2
0.4	Hardened oxide; n - channel.	
0.2	Hardened oxide with aluminum implant; n - channel.	
0.0	Hardened oxide; p - channel.	

[a] After Gwyn.[68]

Of primary importance is intrinsic hole trapping. The effect of impurities on radiation sensitivity is of secondary importance. The hardness level for MOS devices has been improved mainly by optimizing oxide growth and postoxidation annealing treatments.

Although hardened devices have been produced in the laboratory, the ability to produce routinely reliable MOS devices of specified hardness has not been demonstrated. The methods and the procedures required for routinely producing hardened devices, such as those fabricated by Aubuchon,[63] must be precisely defined to enable their use at other laboratories and commercial facilities.

REFERENCES

1   S. M. Sze, *Physics of Semiconductor Devices*, Wiley, New York, 1969.

2   A. S. Grove, *Physics and Technology of Semiconductor Devices*, Wiley, New York, 1967.

3   G. A. Dodson and B. T. Howard, *Proceedings of the Seventh National Symposium on Reliability and Quality Control in Electronics*, 1961, pp. 262–272.

4   M. M. Atalla, E. Tannenbaum, and E. J. Scheibner, *Bell Syst. Tech. J.*, **38**, 749 (1957).

5   J. A. Hoerni, "Planar Silicon Transistors and Diodes," paper presented at the Electron Devices Meeting, Washington, D.C., October 1960.

6   D. Kahng and M. M. Atalla, "Silicon-Silicon Dioxide Field-Induced Surface Devices," paper presented at IRE-AIEE Solid-State Device Research Conference, Carnegie Institute of Technology, Pittsburgh, Pa., 1960.

7   D. R. Kerr, J. S. Logan, P. J. Burkhardt, and W. A. Pliskin, *IBM J. Res. Dev.*, **8**, 376 (1964).

8   M. Yamin, *IEEE Transact. Electron Devices*, **ED-12**, 88 (1965).

9   E. H. Snow, A. S. Grove, B. E. Deal, and C. T. Sah, *J. Appl. Phys.*, **36**, 1664 (1965).

10 M. H. Woods and R. Williams, *J. Appl. Phys.*, **44**, 5506 (1973).

11 S. R. Hofstein, *IEEE Transact. Electron Devices*, **ED-13**, 222 (1966).

12 S. R. Hofstein, *IEEE Transact. Electron Devices*, **ED-14**, 749 (1967).

13 B. Yurash and B. E. Deal, *J. Electrochem. Soc.*, **115**, 1191 (1968).

14 S. I. Raider and R. Flitsch, *J. Electrochem. Soc.*, **118**, 1011 (1971).

15 B. E. Deal, U.S. Patent 3,426,422 (filed in 1965, issued in 1969).

16 T. M. Buck, F. G. Allen, J. V. Dalton, and J. D. Struthers, *J. Electrochem. Soc.*, **114**, 862 (1967).

17 R. Schmidt, *J. Electrochem. Soc.*, **115**, 193c (1968).

18 T. E. Burgess and H. M. Donega, *J. Electrochem. Soc.*, **116**, 1313 (1969).

19 R. J. Kriegler, Y. G. Cheng, and D. R. Colton, *J. Electrochem. Soc.*, **119**, 388 (1972).

20 R. J. Kriegler, *12th Annual Proceedings Reliability Physics*, 1974, pp. 250–258. IEEE Catalog No. 74CH0839-1 PHY.

21 A. B. Kuper, *Surface Sci.*, **13**, 172 (1969).

22 D. R. Kerr, J. S. Logan, P. J. Burkhardt and W. A. Pliskin, *IBM J. Res. Dev.*, **8**, 376 (1964).

23 P. H. Robinson and F. P. Heiman, *J. Electrochem. Soc.*, **118**, 141 (1971).

24 W. A. Pliskin, D. R. Kerr, and J. A. Perri, *Thin Glass Films, Physics of Thin Films*, Vol. 4, Academic, New York, 1967, pp. 257–324.

25 E. Yon, W. H. Ko, and A. B. Kuper, *IEEE Transact. Electron Devices*, **ED-13**, 276 (1966).

26 E. H. Snow and B. E. Deal, *J. Electrochem. Soc.*, **113**, 263 (1966).

27 J. C. Sarace, R. E. Kerwin, D. L., Klein, and R. Edwards, *Solid-State Electron.*, **11**, 653 (1968).

28 P. Balk and J. M. Eldridge, *Proc. IEEE*, **57**, 1558 (1969).

29 R. J. Kriegler, *Appl. Phys. Lett.*, **20**, 449 (1972).

30 R. J. Kriegler, *Thin Solid Films*, **13**, 11 (1972).

31 B. E. Deal, A. Hurrle, and M. J. Schulz, *J. Electrochem. Soc.*, **125**, 2024 (1978).

32 J. Stach and R. E. Tressler, "Microstructures and Properties of Silicon Oxides Grown in $HCl–O_2$ Ambients," paper given at 1978 IEEE Interface Specialists Conference, Miami Beach, Fla.

33 T. E. Burgess, J. C. Baum, F. M. Fowkes, R. Holmstrom, and G. A. Shim, *J. Electrochem. Soc.*, **116**, 1005 (1969).

34 S. K. Tung and R. E. Caffrey, in *Thin Film Dielectrics*, F. Vratny, Ed., Electrochemical Society, New York, 1969, pp. 286–296.

35 J. V. Dalton and J. Drobek. *J. Electrochem. Soc.*, **115**, 865 (1968).

36 R. Voss, paper presented at the 1978 Gordon Research Conference on Metal-Insulator-Semiconductor Systems, Kimball Union Academy, Meriden, N. H.

37 A. L. McWhorter, *Semiconductor Surface Physics*, University of Pennsylvania Press, Philadelphia, Pa., 1957, pp. 207–228.

38 T. H. Ning, P. W. Cook, R. H. Dennard, C. M. Osburn, S. E. Schuster, and H. N. Yu, IEEE International Electron Devices Meeting, Technical Digest, 1978, pp. 472–475.

39 S. Iwamatsu and Y. Tarui, Extended Abstracts of the May 1979 Electrochemical Society Meeting, Abstract 88, pp. 228–229.

40 D. R. Young, E. A. Irene, D. J. Di Maria, and R. F. DeKeersmaeker, *J. Appl. Phys.*, **50**, 6366 (1979).

41 B. E. Deal, M. Sklar, A. S. Grove, and E. H. Snow, *J. Electrochem. Soc.*, **114**, 266 (1967).

42 A. Goetzberger and H. E. Nigh, *Proc. IEEE*, **54**, 1454 (1966).

43 R. Hofstein, *Solid-State Electron.*, **10**, 657 (1967).

44 A. Goetzberger, A. D. Lopez, and R. J. Strain, *J. Electrochem. Soc.*, **120**, 90 (1973).

45 A. K. Sinha and T. E. Smith, *J. Electrochem. Soc.*, **125**, 743 (1978).

46 A. K. Sinha, H. J. Levinstein, L. P. Adda, E. N. Fuls, and E. I. Povilonis *Solid-State Electron.*, **21**, 531 (1978).

47 R. R. Razouk and B. E. Deal, Extended Abstracts of the May 1979 Electrochemical Society Meeting, Abstract 135, pp. 363–365; *J. Electrochem. Soc.*, **126**, 1573 (1979).

48 R. R. Razouk and B. E. Deal, private communication.

49 P. Balk, Extended Abstracts, Electronics Division; *Electrochem. Soc.*, **14**, (1), 237 (1965).

50 E. H. Nicollian, A. Goetzberger, and A. D. Lopez., *Solid-State Electron.*, **12**, 937 (1969).

51 B. E. Deal, E. L. MacKenna, and P. L. Castro, *J. Electrochem. Soc.*, **116**, 997 (1969).

52 F. Montillo and P. Balk, *J. Electrochem. Soc.*, **118**, 1463 (1971).

53 S. D. Brotherton, D. R. Tomb, and J. W. Clancy, *Internat. J. Electron.*, **31**, 629 (1971); B. E. Deal, *J. Electrochem. Soc.*, **121**, 198C (1974).

54 S. I. Raider, R. A. Gdula, and J. R. Petrak, *Appl. Phys. Lett.*, **27**, 150 (1975).

55 E. A. Irene, *J. Electrochem. Soc.*, **125**, 1708 (1978).

56 S. P. Murarka, *Appl. Phys. Lett.*, **34**, 587 (1979).

57 E. Arnold, J. Ladell, and G. Abowitz, *Appl. Phys. Lett.*, **13**, 413 (1968).

58 N. M. Johnson, D. J. Bartelink, and M. Schulz, in *The Physics of SiO₂ and its Interfaces*, S. T. Pantelides, Ed., Pergamon, New York, 1978, Chapter 8, pp. 421–427.

59 K. G. Aubuchon, E. Harari, and P. Chang, Annual Report on Contract No. N00014-72-C-0424, Department of the Navy, Office of Naval Research, Arlington, Va., October 1974.

60 E. H. Snow, A. S. Grove, and D. J. Fitzgerald, *Proc. IEEE*, **55**, 1168 (1967).

61 H. L. Hughes, R. D. Baxter, and B. Phillips, *IEEE Transact. Nucl. Sci.*, **NS-19**, 256 (1972).

62 D. V. McCaughan and V. T. Murphy, *IEEE Transact. Nucl. Sci.*, **NS-19**, 249 (1972).

63 K. G. Aubuchon, *IEEE Transact. Nucl. Sci.*, **NS-18**, 117 (1971); Final Technical Report, Contract No. N00014-71-C-0079, March 1972, Hughes Research Laboratories.

64 J. Lindmayer, *IEEE Transact. Nucl. Sci.*, **NS-18**, 91 (1971).

65 R. P. Donovan and M. Simons, *J. Appl. Phys.*, **43**, 2897 (1972).

66 R. P. Donovan, M. Simons, and L. V. Monteith, research report prepared by Research Triangle Institute under Contract No. NAS 1–8156 for Langley Research Center, Rept. No. N70–26555, May 1970.

67 J. L. Peel and G. Kinoshita, *IEEE Transact. Nucl. Sci.*, **NS-19**, 271 (1972).

68 C. W. Gwyn, Sandia Report SLA-73-0013, "Ionizing Radiation Effects in the Insulator Region of MOS Devices," January 1973.

16

Models of the Interface

16.1 INTRODUCTION

Electrical measurements of the MOS system are of practical importance in understanding device behavior, in the control of integrated circuit fabrication, in developing integrated circuit processing technology, and in aging and reliability studies of integrated circuits. These topics are among the main themes of this book. However, the electrical measurements do not provide a complete understanding. For example, the microscopic origins of interface traps and oxide charges cannot be identified from electrical measurements alone. Nor can electrical measurements alone identify the chemical reactions that may occur during processing or annealing of interface traps and oxide fixed charges. Supplementary information is needed to construct a chemical and structural picture of the interfacial region. With such a picture, control of interface trap level and oxide charge densities could be firmly based, rather than empirically based as at present.

Electrical measurements have been more widely used than chemical measurements in characterizing the MOS system for two reasons: (1) the electrical measurements can be directly related to device characteristics;

814

and (2) the defect densities giving rise to interface traps and oxide fixed charges are so small that the electrical methods are the only ones with sufficient sensitivity. To illustrate how small the quantities to be measured can be, interface trap level densities over the center of the silicon bandgap in the 10^{10} cm^{-2} eV^{-1} range can be routinely produced. As there are roughly 10^{15} silicon surface atoms/cm^2, such a low interface trap level density corresponds to only one interface trap level within the bandgap for every 10^5 silicon surface atoms, which is below the sensitivity of most analytical chemical methods.

New analytical chemical measurement instruments of greater sensitivity have been developed recently. These analytical tools are being used to supplement the electrical measurements to measure the chemical composition and the structure of the boundary between silicon and its thermally grown oxide. Ellipsometry, Rutherford backscattering, low energy ion backscattering, Auger electron spectroscopy, X-ray photoelectron spectroscopy, electron spin resonance, and optical reflectance are some of the analytical methods being used to explore the chemical composition. Transmission electron microscopy can be used to study the structure of the Si–SiO$_2$ interface. X-ray photoelectron spectroscopy also can be used to study structure. However, this method measures bonding, which is related to structure, but not necessarily in a unique way.

In most of these analytical methods the region being studied is altered by the measurement, there are many artifacts, and in many cases sensitivity to the atomic structure and the underlying chemistry is insufficient. All these factors make interpretation difficult. At present, use of these methods requires considerable experience and judgment. Results are tentative. Nonetheless, we attempt to summarize the conclusions that have emerged from these analytical studies.

In Section 16.2 the information available from electrical measurements is reviewed. Morphology of the Si–SiO$_2$ interface is discussed in Section 16.3 and composition, in Section 16.4. Finally, a speculative picture of the Si–SiO$_2$ interface is presented in Section 16.5.

16.2 SPATIAL LOCATION OF INTERFACE TRAPS AND OXIDE FIXED CHARGE

Before describing morphology and composition of the boundary between silicon and its oxide, we summarize the experimental evidence from electrical measurements. The earliest experiment[1] designed to locate electrically active centers in the oxide was a measurement of interface trap and oxide fixed charge densities using the high frequency C-V method. As the oxide was reduced in thickness by etching off 50–100 Å steps, no change in density was observed. However, for oxide thicknesses below 200 Å, C-V curves could not be measured because of short circuits between the gate metal and the silicon or because of oxide breakdown at low voltages. Therefore, this experiment located the electrically active centers within 200 Å of the Si–SiO$_2$ interface.

Using the photo I-V method,[2,3] which can probe much closer to the Si–SiO$_2$ interface than an etch-off experiment, and a sample estimated from C-V measurements to have an oxide fixed charge density of about 5×10^{11} cm^{-2}, 90% of these electrically active centers were found to be within 34 Å of the interface.

Additional evidence that interface traps are located very close to the interface was provided by Preier,[4] who calculated the interface trap contribution to the admittance of the MOS capacitor according to the assumption that interface traps were distributed into the oxide and were charged and discharged by tunneling. It was found that interface traps had to be within 10 Å* of the Si–SiO$_2$ interface to obtain a good fit to the admittance measurements by Nicollian and Goetzberger.[5]

Photothermal probing of the Si–SiO$_2$ interface[6,7] showed that interface traps were located very close to the Si–SiO$_2$ interface in quantitative agreement with the calculation in Ref. 4. These photothermal measurements also suggest that oxide fixed charges have energy levels outside the silicon bandgap.

If oxide fixed charges have energy levels outside the bandgap, their occupancy obviously cannot be changed with gate bias. That is, under gate bias variations, the occupancy of energy levels changes because the position of the Fermi level varies. However, the Fermi level can be varied only over the bandgap. Consequently, interface trap levels within the bandgap can change occupancy, but oxide fixed charge levels outside the bandgap cannot.

It may be possible for atomically similar centers to have energy levels that extend over a very wide range, not limited to the bandgap. Therefore, oxide fixed charges and interface traps could be atomically similar, differing only in the position of their energy levels. However, photothermal experiments do not rule out oxide fixed charges located in the oxide deeper than interface traps† nor centers that are atomically dissimilar.

Thus various experiments locate both interface traps and oxide fixed charges very close to the Si–SiO$_2$ interface. Questions remain about their atomic structure and their relation to the morphology and nonstoichiometry of the Si–SiO$_2$ boundary.

---

*This number is based on the reasonable assumptions that the attenuation constant for the wave function of an electron penetrating into the oxide is independent of energy and has a value equal to about 5×10^7 cm^{-1}.

† Such centers could conceivably be regarded as oxide fixed charge if they were located so deep in the oxide that their time constants were extremely long. Long time constants would mean that occupancy could not be changed with gate bias.

16.3 MORPHOLOGY

How rough is the Si–SiO$_2$ boundary? Krivanek et al.[8,9] reported cross-sectional transmission electron studies of the Si–SiO$_2$ interface on (100), (111), and (011) orientations. The oxides were thermally oxidized in dry air at 1100°C to thicknesses of 1000–1500 Å and then hydrogen annealed at 380°C. Figure 16.1a is a large scale view of the Si–SiO$_2$ interface for a (100) orientation. Figure 16.1b is a lattice image of the framed part of a. Resolution in the silicon is 3 Å. The (111) planes in the silicon can be seen. Resolution in the silica is not known.

Figure 16.1b shows that a transition from single-crystal silicon to an amorphous material occurs. However, the abruptness of the transition cannot be inferred from the sharpness of the boundary between the region where the silicon lattice is clear, and the region where it is no longer visible. The sharpness of this boundary indicates how rapidly the resolution of the measurement degrades as the material departs from an ideal crystal. That is, this boundary indicates a transition in the resolution of the measurement. However, we do not know how this resolution depends on the spatial and chemical disorder at the interface. Consequently, it is not possible to determine from Fig. 16.1 what portion of the unresolved region is amorphous SiO$_2$, what portion is disordered silicon, and what portion is a

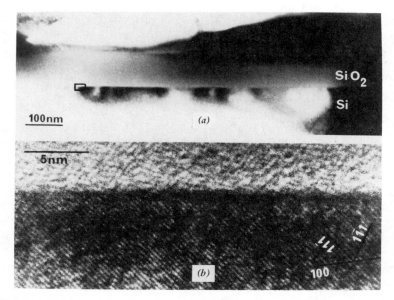

Fig. 16.1 (a) Large scale view of the Si–SiO$_2$ interface. (b) Lattice image of the framed part of (a) showing the principal crystallographic planes in the silicon. Contrast was obtained by defocusing the electron beam and tilting the sample with respect to the beam. After Krivanek et al.[8]

transitional layer between these two solids. It is not possible even to locate the Si–SiO$_2$ interface in this diagram.

Other features common to all the samples of Krivanek et al.[9] were long range undulations of the boundary of resolution about 4–8 Å high with a wavelength of 200–500 Å, and one-plane-high atomic steps separated by 20–40 Å. These results are consistent with a fairly long range disturbance of the composition or, more likely, the crystallinity of the silicon near the Si–SiO$_2$ interface.

The results due to Blanc et al.[10] using electron microscopy with a resolution in the silicon of about 10 Å agree with those of Krivanek et al.[8,9] However, Chen and Sugano[11] reported a transmission electron microscope study of the Si–SiO$_2$ interface at variance with Refs. 8–10. Chen and Sugano found particles at the interface that they identified as crystalline silicon clusters, which varied from 15–50 Å to 40–50 Å in size as oxidation temperature was increased from 1100 to 1200°K. Typical cluster density was 6×10^{11} cm^{-2}. Sugano[12] also found that the density of these crystallites depended on the final oxidation rate. No clusters could be detected for final oxidation rates in dry oxygen below 10 Å/min, but the cluster density was found to increase rapidly for final oxidation rates larger than this value. Irene[13] found similar silicon clusters whose density and size increased with oxidation temperature or when the films were exposed to high temperature inert gas annealing. Oxidation in HCl–O$_2$ ambients also appeared to increase the silicon cluster density.[13] The presence of these clusters correlates with an increase in the SiO$_2$ dielectric breakdown defect density.[14] Thus the disagreement between Refs. 8–10 and 11 may be caused by different oxidation conditions.

16.4 CHEMICAL COMPOSITION

An interesting argument for an abrupt Si–SiO$_2$ interface can be based on the silicon-oxygen phase diagram.[15,16] The argument is that compositional variations, such as SiO, are useful for adjusting to interfacial forces. If a variety of suboxides are stable, solid phases, then any or all of these phases can exist at the interface. However, in the silicon-oxygen system, only silicon and SiO$_2$ solid phases coexist at equilibrium under the range of temperatures and pressures of interest. Therefore, a rather large interfacial force is needed to stabilize an otherwise volatile phase such as SiO. Such a force could be exerted by the interface over only a short distance, if at all.

This argument is not quantitative and does not take into account the fact that the amorphous silica film is metastable, that is, that it has a higher free energy than the crystalline phase, β-cristobalite. Nevertheless, it is suggestive and distinguishes the Si–SiO$_2$ interface from interfaces between other materials where a variety of stable stoichiometries are readily available.

Most of the thermally grown oxide layer is stoichiometric* SiO_2 throughout its thickness. However, one may question whether a very narrow nonstoichiometric region exists at the boundary where oxidation took place. If such a region exists, what is its structure? Pantelides and Long[17] constructed a continuous random network model for the interface taking into account bond-stretching forces, bond-bending forces, and van der Waals forces. They found that at $T = 0°K$, geometric and energetic constraints can be satisfied by an abrupt boundary with *no* nonstoichiometric region. It is likely, however, that the excess oxygen associated with the oxidation reaction will remain after cool down, along with some incompletely oxidized silicon near the boundary.

The earliest analytical results suggesting an abrupt nonstoichiometric boundary region were those due to Grunthaner and Maserjian[18] obtained with the use of X-ray photoelectron spectroscopy. Support for a nonstoichiometric boundary region was found by Aspnes and Theeten[19,20] from ellipsometric data over photon energies of 1.5–5.8 eV. The data were fitted with a variety of models. A poor fit was obtained if an abrupt, stoichiometric interface was assumed. Similarly, a poor fit resulted on assuming a mechanical mixture of silicon and silica. However, a good fit was possible using a variety of silicon-oxygen combinations. The best fit was obtained for the chemically mixed $SiO_{0.4}$ composition. Such a mixture of chemically different Si—O combinations shifts the energy position of optical absorption. A mechanical mixture of Si and SiO_2 changes only the amplitudes of Si and SiO_2 absorption.

The discussion of Aspnes and Theeten also illustrates some of the difficulties involved in determining the thickness of the interfacial region. For their experiment, the interface is the region with an optical response *different* from that of a juxtaposition of stoichiometric SiO_2 and crystalline silicon. Thus a disturbed layer of silicon near the interface is included as part of their interface. Also, any portion of the SiO_2 layer that does not behave like bulk SiO_2 is part of their interface.† Finally, fine-grained lateral variations in composition across the interface cannot be distinguished from in-depth variations normal to the interface, particularly for a very thin interfacial region.

In general, each experiment designed to measure the interfacial region incorporates its own definition of what is included in this region and its own problems in distinguishing lateral variations across the interface from variations normal to the interface.

---

*The compound SiO_x is stoichiometric only when $x = 2$.
†Deviations in stoichiometry of the SiO_2 between $SiO_{1.5}$ and SiO_2 cause little change in the optical response. With the present experimental accuracy, such deviations in stoichiometry could extend as far as 40 Å into the SiO_2.[19,20]

16.4.1 Width of the Interface

Various measurement methods have been used to study the chemical composition and the thickness of the region over which the chemical composition varies near the boundary. A transitional region of non-stoichiometric oxide always has been found, but its width δ tends to vary roughly in proportion to the depth resolution of the method used. From Rutherford backscattering,[21-23] $\delta = 600$ Å and later 4–7 Å; from low energy ion backscattering,[24] $\delta = 20$ Å; from Auger electron spectroscopy,[25,26-28] $\delta = 6$–35 Å; and from X-ray photoelectron spectroscopy,[29,30-32] $\delta = 5$–20 Å. Another method used was optical reflectance.[33-34] All these methods are under development, and considerable skill and judgment are required for their interpretation. The interested reader can learn more about these methods in Refs. 35 and 36 and read a review of the results of probing the Si–SiO$_2$ interface with these methods in Ref. 37.

The spread in the width estimated for the nonstoichiometric region probed by these analytical methods arises from three basic problems of interpretation: (1) the morphology of the boundary—that is, there may be undulations of the boundary comparable to those discussed in Section 16.3, that are a significant fraction of the values of δ and hence should affect the results (in methods where etching or ion milling is used to thin the oxide to allow construction of a depth profile of chemical composition, some undulation of the surface is generated by the thinning process); (2) the resolution of these methods, which is in some cases a significant fraction of δ; and (3) the difficulty in the deconvolution of δ from the raw data. With few exceptions, there are no critical assays of errors inherent in deconvolution. These errors are particularly severe in the case of structural or chemical interpretation of data.

To probe the interfacial region with these methods, it is necessary to cut through the oxide layer to the vicinity of the interface. The two commonly used methods for cutting through the oxide are ion milling and wet chemical etching. In ion milling, the oxide is cut away by bombarding it with an energetic ion beam. Ion milling completely destroys all chemical and structural information to depths of several tens of angstroms, even after removal of only a monolayer of oxide at the lowest ion energies practical. However, the elemental ratio of the original composition can be estimated to within a depth resolution of a few tens of angstroms if careful allowance is made for differential sputter rates and knock-in mixing effects.[27] Likewise, in Auger analysis, the electron beams used cause decomposition of the SiO$_2$ and electron stimulated desorption. In X-ray photoelectron spectroscopy, the use of nonmonochromatic X-ray beams requires high beam intensities, resulting in ionization damage and a secondary electron flux, which, as in Auger analysis, can damage the oxide and cause uncontrollable charging. Wet chemical etching is more gentle than ion milling, but the etching must be uniform.

Regardless of the significance of the "width" of the interfacial region, it is clear that a profile in depth of the interface is uncertain when obtained using techniques that cannot determine the width of this region. For construction of a microscopic model of the interface, a further difficulty is that lateral variations cannot be resolved but are averaged over a large area. Thus an average interfacial region is measured, which may not resemble the actual interfacial region in any particular location.

In another type of experiment, Williams and Goodman[38] measured the contact angle of a water drop on a thermally grown SiO_2 layer as a function of oxide thickness. For oxide thicknesses less than about 30 Å, they found the contact angle of the water drop varied monotonically from that characteristic of SiO_2 to that characteristic of silicon.* One explanation of these results is that the oxide becomes increasingly silicon rich as the interface is approached. As a consequence, the surface properties of the oxide gradually give way to those of silicon. This interpretation suggests a layer of nonstoichiometric oxide about 30 Å thick, in qualitative agreement with the analytical methods just discussed. However, there may be other explanations of the water drop experiment.

In summary, measurements of the chemical composition of the interfacial layer, of morphology of the interfacial layer, and of the location of the electrically active centers all are consistent with a transition layer at the interface no wider than about 30 Å.

16.5 SPECULATIVE MODEL OF THE Si–SiO₂ INTERFACE

Modeling of the entire Si–SiO₂ interfacial region is essential for a better understanding of this interface. Such understanding would be of technological importance if it led to better control of the properties of the Si–SiO₂ interface, control based on science rather than on empirical art.

In view of the theoretical and experimental difficulties facing a model for the Si–SiO₂ interface, a description of all the primitive existing models and calculations is not given. Rather, a few models that seem reasonable to the authors are reviewed very briefly.

16.5.1 Interfacial Region

An appealing picture of the Si–SiO₂ interface proposed by Grunthaner and Maserjian[32] and Grunthaner et al.[39] is one of an interfacial region comprised of single-crystal silicon followed by (1) a monolayer in which there is some SiO_2, Si_2O_2, and Si_2O (i.e., incompletely oxidized silicon†), (2) a strained region of SiO_2 roughly 10–40 Å deep, and (3) the remaining stoichiometric, strain free, amorphous SiO_2.

---

*Silica is hydrophillic and thus the contact angle is small, whereas silicon is hydrophobic, making the contact angle large.

†The relative amounts of these suboxides depend on processing.

During oxidation or other high temperature treatments, structural defects form and accommodate impurities such as heavy metals, alkali metals, hydrogen, chlorine, and carbon, to name only a few. Defects and impurities tend to accumulate in the strained region.

16.5.2 Interface Traps

Interface traps can be produced in several different ways: (1) thermal oxidation in dry oxygen or steam, (2) plasma oxidation, (3) avalanche injection of electrons or holes into the SiO_2, (4) the diffusion of metals such as chromium to the Si–SiO_2 interface, and (5) the exposure of the MOS system to ionizing radiation. More than one model for interface trap formation may be needed.

A U-shaped interface trap level distribution is characteristic of steam oxidation and of device-quality gate oxides grown in dry oxygen and then annealed by one of the methods described in Section 15.4.3. The U-shaped distribution is shown in Figs. 7.5 and 15.23. There have been three interface trap models proposed to explain this characteristic U-shaped distribution qualitatively: the coulombic, bond, and defect models.

(a) Coulombic Model

The first proposal was the coulombic model due to Goetzberger et al.[40] In their model, charges in the oxide induce potential wells in the silicon and quantum levels within these wells are identified with interface trap levels. For the most part, such interface trap levels are located in energy near the silicon band edges. Midgap levels are expected only if many charges are clustered together, producing a deep potential well. Because such clusters are unlikely, the density of midgap levels will be smaller than the density of shallow levels, giving a U-shaped distribution. However, the predicted interface trap level density probably falls very rapidly with distance from the band edges, resulting in too low a midgap D_{it}.

(b) Bond Models

A second proposal is that the interface trap level distribution is produced by a distribution of bond angles or by stretched bonds at the silicon surface. Local strain or local nonstoichiometry at the Si–SiO_2 interface could cause these bond distortions.

A specialized form of bond disorder model was proposed by Laughlin.[41] This model involves disorder only on the silicon side of the interface and requires some source of strain, such as the SiO_2, to cause this disorder. In this model,[41] distortions of the Si—Si bonds produce an antibonding level in the upper half of the silicon bandgap and a bonding level in the lower half of the silicon bandgap. Thus the interface trap levels at the two band edges are related. Shallow levels are produced at both band edges by slight deviations from normal whereas levels nearer midgap are produced by larger deviations. Because small deviations are more likely than large deviations, a U-shaped distribution of energy levels could result.

A model of bond angle distortions was presented by Laughlin et al.[42] In this model, both the silicon and the SiO_2 are represented as Bethe lattices. Distortion of the Si—O—Si bond angle by as much as 30° is expected to create energy levels near the silicon conduction band edge, while distortion of Si—Si bond angles in the silicon tends to produce energy levels near the silicon valence band edge. However, these conclusions were based upon arguments by analogy, and were not quantitative.

Sakurai and Sugano[43] have presented a bond stretching model for D_{it}. Using an idealized tight-binding model for the silicon and a Bethe lattice for the SiO_2, they find that stretched Si—O bonds at the interfacial plane cause trap levels in the lower half of the bandgap and that oxygen vacancies or stretched Si—Si bonds cause levels in the upper half of the gap. The silicon dangling bond produces levels near midgap. These defects are illustrated schematically in Fig. 16.2.

Figure 16.2a shows the silicon lattice schematically as a rectangular grid and the SiO_2 as trees of Si—O_4 tetrahedra attached to the silicon lattice. Figures 16.2b–d show a stretched Si—O bond, a stretched Si—Si bond or oxygen vacancy, and a silicon dangling bond. A stretched bond such as that in Fig. 16.2c cannot be distinguished from a defect model—the third proposal for interface traps.

No reliable calculation of capture cross sections nor interface trap level density as a function of position in the silicon bandgap are available on the

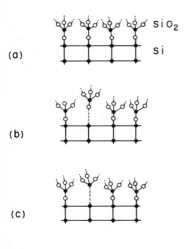

(a)

(b)

(c)

(d)

Fig. 16.2 Several possible interfacial defects in the model due to Sakurai and Sugano.[43] (a) Silicon lattice and perfect SiO_2 overlayer; (b) stretched Si—O bond; (c) stretched Si—Si bond or oxygen vacancy; (d) silicon dangling bond (schematic).

basis of any of these models. The calculation of D_{it} by Sakurai and Sugano[43] drops too rapidly with distance from the band edges. However, this prediction is based on a very simple estimate of the effect of disorder on their model, an estimate that neglects clustering of strained bonds and interactions between them. A more complete estimate might improve the calculated D_{it}. Experimental verification of any proposed bond distribution is difficult.

(c) Defect Models

A third proposal is that defects within or near the interfacial region cause interface trap levels. Such defects range from stacking faults and micropores to various atomic or molecular fragments left as a residue of imperfect oxidation. A problem with modeling interface traps is that the atomic structure and the chemical composition of the interfacial layer are not known. However, the results of several workers indicate the presence of nonstoichiometry.[19,20,27,28,31,32] No relation has been experimentally established between nonstoichiometry at the interface and the electrically active centers.

Several specific defects may be responsible for interface traps produced by thermal oxidation. Four different types of defect that *could* exist at or near the Si–SiO$_2$ interface that *might* produce interface traps are (1) excess silicon[28] (trivalent silicon[1,44,45]), (2) excess oxygen[16,46] (nonbridging oxygen), (3) impurities,[47] and (4) states in oxide charge induced potential wells.[40]

As a consequence of oxidation, there may be *trivalent silicon atoms* or silicon in excess of the stoichiometric amount in the SiO$_2$ at the interface, probably related to incomplete oxidation of the silicon or to generation of vacancies in the silicon during oxidation.[48] In this type of defect the silicon atom shares three of its four valence electrons with neighboring silicon atoms. Its fourth unsatisfied valence bond acts as a hole trap becoming positively charged after capturing a hole and remaining neutral when empty.

In the normal SiO$_2$ lattice, oxygen atoms share electrons with adjacent silicon atoms, forming a "bridge" between the two silicon atoms. When one of the Si—O bonds is broken, the formerly bridging oxygen atom is now called a *nonbridging oxygen*. It shares one of its two valence electrons with the remaining silicon atom, and its unsatisfied valence bond acts as an electron trap becoming negatively charged by electron capture and remaining neutral when empty.

Nonbridging oxygen, oxygen in excess of the stoichiometric amount in the oxide, may be present near the interface for three reasons: (1) excess oxygen associated with the oxidation reaction must be present and may form nonbridging oxygen defects; (2) the strain in the region near the interface might be relieved by formation of this defect; and (3) there are water-related electron traps near the Si–SiO$_2$ interface that might be related to nonbridging oxygen defects.

Both trivalent silicon and nonbridging oxygen centers are known to react with hydrogen. In this reaction the unsatisfied valence bond is satisfied for both types of defect; hence such a reaction is likely to render these defects electrically inactive. For nonbridging oxygen centers, reaction with water seems to render them neutral, as described in Section 11.5.1. Thus both types of center are consistent with the observation that interface traps are annihilated by annealing in hydrogen or water vapor.

There is some tentative experimental evidence that trivalent silicon exists in SiO_2 and may act as interface traps. Nishi[49] and Poindexter et al.[50] interpret their electron spin resonance signals as due to trivalent silicon. In their measurements the observed spin resonance signal is found to increase or decrease in the same manner as midgap interface trap level density under oxidation or annealing variations. Thus the spin resonance signal appears to be related to midgap interface trap level density. They conclude that the most likely source of the spin resonance signal observed in their samples is trivalent silicon. However, conclusive evidence for the species responsible for the observed spin resonance signal awaits the detection of hyperfine structure which is a difficult measurement. The trivalent silicon defect is consistent with a bond disorder model.

The third type of defect is an *impurity atom* at the interface. Because strain at the interface creates a potential minimum for impurities, the interface can accommodate many of the impurities incorporated in the relatively open SiO_2 lattice. There is also a strained region at the metal-SiO_2 interface. Strained regions at either interface are likely to act as sinks for impurities incorporated in the oxide.

The fourth type of defect is an *oxide charge* near the interface that induces an attractive coulombic potential well in the silicon.[40] The interface trap model based on this defect has been discussed earlier in this section. There is evidence from low temperature measurements that sodium ions drifted to the Si–SiO₂ interface induce an interface trap band near the conduction band edge.[51]

Using conductance measurements on thin oxides, Kar and Dahlke[47] obtained evidence that metal impurities can cause interface traps. These workers diffused the metal evaporated onto the thin oxide surface all the way to the Si–SiO₂ interface by heating the sample.

Avalanche injection experiments, described in Section 11.5.1, display interface trap buildup. These interface traps appear to be donor type. The exact mechanism of their formation is unknown.

Another defect, associated with the cleavage of strained Si—O bonds, is the E' center.[52,53] This center consists of a nonbridging oxygen and a trivalent silicon atom formed when an Si—O bond is broken. Such a defect is likely to result from radiation damage.

At present, all these models remain qualitative. Moreover, there is no independent experimental support for any model. Therefore, these models remain speculative.

16.5.3   Oxide Fixed Charge

Trivalent silicon could act as a positive oxide fixed charge if it were located some distance from the interface. However, Poindexter et al.[50] find that their electron spin resonance signal varies with midgap interface trap level density, varied by varying the pull rate out of the oxidation furnace [see Section 15.4.3(a)], but not with oxide fixed charge or band edge interface trap level density, varied by high temperature annealing [see Section 15.4.3(d)]. Thus Poindexter's measurements are correlated only with interface trap levels near the middle of the bandgap.

Nonbridging oxygen sites could act as negatively charged centers when the water-related traps with which they are likely to be associated capture an energetic electron. However, because electrons are seldom present in the conduction band of the SiO_2, the formation of negatively charged centers may be improbable. More probable may be the loss of an electron from a nonbridging oxygen center near the $Si-SiO_2$ interface to the silicon, making it a positively charged center and hence a candidate for oxide fixed charge.

16.5.4   Annealing Experiments and Interface Trap Models

To check proposed models for interface traps, comparison can be made with annealing behavior. However, annealing experiments can be ambiguous because it is not always clear what species in the ambient is responsible for the annealing effect. In the case of high temperature annealing of interface trap level density, it has not been established whether it is hydrogen or trace quantities of water vapor that annihilate interface traps. By analogy with low temperature annealing, where atomic hydrogen appears critical, it is thought that molecular hydrogen is critical to high temperature hydrogen annealing. But trace quantities of water vapor appear critical in high temperature annealing in an otherwise inert ambient.[54] Therefore, there is a question as to whether the annealing agent in a hydrogen ambient is trace water vapor or hydrogen. The distinction is important for any proposed model because even a trace of water vapor will oxidize, whereas hydrogen reduces.

It is not known whether interface traps are related to oxide fixed charge. Whereas the annealing behavior of interface trap levels in the central portion of the bandgap is easily monitored, by using either capacitance or conductance methods, monitoring the annealing of oxide fixed charge is more complicated. In fact, one cannot distinguish between annealing of oxide fixed charge density and annealing of interface trap level density near the band edges, as is now discussed.

Flatband voltage V_{FB} is affected by both interface trap charge and oxide fixed charge. If both donor and acceptor type interface traps are present, a measured reduction in interface trap level density $D_{it} = D_{it}^a + D_{it}^d$ (where a represents acceptor and d, donor) may be accompanied by a shift in flatband voltage of either sign. That is, elimination of donor interface traps reduces positive charge, whereas elimination of acceptor interface traps reduces negative charge. Therefore, interface trap charge density $Q_{it} = Q_{it}^d - Q_{it}^a$ may shift either way, depending on the charge balance of donor- and acceptor-type interface traps. Such a shift can mask any change in oxide fixed charge density.

By measuring the variance of band-bending fluctuations σ_s^2, the sum $Q_f + Q_{it}^d + Q_{it}^a$ can be determined. As discussed in Section 7.3.6(d), by combining measurements of D_{it} and σ_s^2, one can determine over a limited region of the bandgap whether donor or acceptor interface traps dominate and set a bound on how many interface traps of the subdominant type are present. Therefore, the V_{FB} shift expected from interface trap annealing can be estimated for interface trap levels in part of the bandgap. Unfortunately, it is not possible to provide this information throughout the bandgap. Therefore, it is not possible to distinguish the annealing behavior of Q_f from that of band edge interface trap levels.

Some assumptions must be made to interpret annealing experiments. For example, one may suppose that interface trap levels near the band edge are of the same type and have the same annealing behavior as those observable in the central portion of the bandgap. However, such assumptions are speculative at the moment. The commonly stated belief that oxide fixed charge is not affected by low temperature annealing is based not on experiment, but on speculation. Model predictions of how Q_f anneals cannot be checked until a model for interface traps is established.

REFERENCES

1 B. E. Deal, M. Sklar, A. S. Grove, and E. H. Snow, *J. Electrochem. Soc.*, **114**, 266 (1967).

2 C. M. Berglund and R. J. Powell, *J. Appl. Phys.*, **42**, 573 (1971).

3 J. R. Brews, *J. Appl. Phys.*, **44**, 379 (1973).

4 H. Preier, *Appl. Phys. Lett.*, **10**, 361 (1967).

5 E. H. Nicollian and A. Goetzberger, *Bell Syst. Tech. J.*, **46**, 1055 (1967).

6 R. F. Pierret, *Solid-State Electron.*, **19**, 577 (1976).

7 R. F. Pierret and B. B. Roesner, *Solid-State Electron.*, **19**, 593 (1976).

8 O. L. Krivanek, T. T. Sheng, and D. C. Tsui, *Appl. Phys. Lett.*, **32**, 437 (1978).

9 O. L. Krivanek, D. C. Tsui, T. T. Sheng, and A. Kamgar, in *The Physics of SiO₂ and Its Interfaces*, S. T. Pantelides, Ed., Pergamon, New York, 1978, Chapter 7, pp. 356–361.

10 J. Blanc, C. J. Buiocchi, M. S. Abrahams, and W. E. Ham, *Appl. Phys. Lett.*, **30**, 120 (1977).

11 J. J. Chen and T. Sugano, in *The Physics of SiO₂ and Its Interfaces*, S. T. Pantelides, Ed., Pergamon, New York, 1978, Chapter 7, pp. 351–355.

12 T. Sugano, private communication.

13   E. A. Irene, in *The Physics of SiO₂ and Its Interfaces*, S. T. Pantelides, Ed., Pergamon, New York, 1978, Chapter 4, pp. 205–209.

14   C. M. Osburn and D. W. Ormond, *J. Electrochem. Soc.*, **119**, 597 (1972).

15   A. S. Berezhnoi, *Silicon and Its Binary Systems*, Consultants Bureau, New York, 1960, p. 104.

16   R. Schmidt, *J. Electrochem. Soc.*, **116**, 249C (1969); Abstract, 344 RNP.

17   S. T. Pantelides and M. Long, in *The Physics of SiO₂ and Its Interfaces*, S. T. Pantelides, Ed., Pergamon, 1978, Chapter 7, pp. 339–343.

18   F. J. Grunthaner and J. Maserjian, *IEEE Transact. Nucl. Sci.*, **NS-24**, 2108 (1977).

19   D. E. Aspnes and J. B. Theeten, *Phys. Rev. Lett.*, **43**, 1046 (1979).

20   D. E. Aspnes and J. B. Theeten, *J. Electrochem. Soc.*, **127**, 1359 (1980).

21   T. W. Sigmon, W. K. Chu, E. Lugujjo, and J. W. Meyer, *Appl. Phys. Lett.*, **24**, 105 (1974).

22   A. Benninghoven, *Surface Sci.*, **35**, 427 (1973).

23   L. C. Feldman, I. Stensgaard, P. J. Silverman, and T. E. Jackman, in *The Physics of SiO₂ and Its Interfaces*, S. T. Pantelides, Ed., Pergamon, New York, 1978, Chapter 7, pp. 344–350; N. W. Chang, L. C. Feldman, P. J. Silverman, and I. Stensgaard, *Appl. Phys. Lett.*, **35**, 859 (1979).

24   W. L. Harrington, R. E. Honig, A. M. Goodman, and R. Williams, *Appl. Phys. Lett.*, **27**, 644 (1975).

25   J. S. Johannessen, W. E. Spicer, and Y. E. Strausser, *J. Appl. Phys.*, **47**, 3028 (1976).

26   C. R. Helms, C. M. Garner, J. Miller, I. Lindau, S. A. Schwarz, and W. E. Spicer, *Proceedings of the Seventh International Congress and Third International Conference on Solid Surfaces*, No. 2241, Vienna, 1977.

27   C. R. Helms, N. M. Johnson, S. A. Schwarz, and W. E. Spicer, in *The Physics of SiO₂ and Its Interfaces*, S. T. Pantelides, Ed., Pergamon, New York, 1978, Chapter 7, pp. 366–372.

28   C. C. Chang, in *Semiconductor Characterization Techniques*, P. A. Barnes and G. A. Rozgonyi, Electronics Division Proceedings 1978 (The Electrochemical Society, Princeton, N. J.), No. 78-3, p. 106.

29   R. A. Clarke, R. L. Tapping, M. A. Hooper, and L. Young, *J. Electrochem. Soc.*, **122**, 1347 (1975).

30   T. Hattori and T. Nishina, in *The Physics of SiO₂ and Its Interfaces*, S. T. Pantelides, Ed., Pergamon, New York, 1978, Chapter 7, pp. 379–383.

31   S. I. Raider and R. Flitsch, ibid., Chapter 7, pp. 384–388.

32   F. S. Grunthaner and J. Maserjian, ibid., Chapter 7, pp. 389–395.

33   H. R. Phillipp, *J. Phys. Chem. Solids*, **32**, 1935 (1971).

34   H. R. Phillipp, *J. Appl. Phys.*, **43**, 2835 (1972).

35   P. F. Kane and G. B. Larrabee, Eds., *Characterization of Solid Surfaces*, Plenum, New York, 1974.

36   Proceedings of ARPA/NBS Workshop IV, Surface Analysis for Silicon Devices, NBS, Gaithersburg, Maryland, April 23–24, 1975 (issued March 1976).

37   J. Blanc, in *Semiconductor Characterization Techniques*, P. A. Barnes and G. A. Rozgonyi, Eds., Electronics Division Proceedings (The Electrochemical Society, Princeton, N.J.), No. 78-3, 1978, p. 139.

38   R. Williams and A. M. Goodman, *Appl. Phys. Lett.*, **25**, 531 (1974).

39   F. J. Grunthaner, P. J. Grunthaner, R. P. Vasquez, B. F. Lewis, J. Maserjian, and A. Madhukar, *J. Vacuum Sci. Technol.*, **16**, 1443 (1979); *Phys. Rev. Lett.*, **43**, 1683 (1979).

40   A. Goetzberger, V. Heine, and E. H. Nicollian, *Appl. Phys. Lett.*, **12**, 95 (1968).

41   R. B. Laughlin, MIS Gordon Conference, Tilton, N.H., 1980.

42 R. B. Laughlin, J. D. Joannopoulos, and D. J. Chali, in *The Physics of SiO₂ and Its Interfaces*, S. T. Pantelides, Ed., Pergamon, New York, 1978, Chapter 6, pp. 321–327.

43 T. Sakurai and T. Sugano, Technical Report of the Institute Electronics and Communication Engineers in Japan, **ED-64**, 35 (1980); *J. Appl. Phys.*, **52**, 2889 (1981).

44 E. Kooi, *The Surface Properties of Oxidized Silicon*, MacMillan, New York, 1967.

45 C. M. Svensson, in *The Physics of SiO₂ and Its Interfaces*, S. T. Pantelides, Ed., Pergamon, New York, 1978, Chapter 6, pp. 328–332.

46 S. I. Raider and A. Berman, *J. Electrochem. Soc.*, **125**, 629 (1978).

47 S. Kar and W. E. Dahlke, *Solid-State Electron.*, **15**, 221 (1972); ibid., 95 (1968).

48 J. R. Ligenza, private communication.

49 Y. Nishi, *Jap. J. Appl. Phys.*, **10**, 51 (1971).

50 E. H. Poindexter, E. R. Ahlstrom, and P. J. Caplan, in *The Physics of SiO₂ and Its Interfaces*, S. T. Pantelides, Ed., Pergamon, New York, 1978, Chapter 4, pp. 227.

51 A. Hartstein and A. B. Fowler, *Phys. Rev. Lett.*, **34**, 1435 (1975); A. Hartstein and A. B. Fowler, *Surface Sci.*, **73**, 19 (1978); A. Hartstein, A. B. Fowler, and M. Albert, *Institute of Physics Conference Series*, No. 43, 1979, Chapter 26.

52 D. L. Griscom, E. J. Freibele, and G. H. Sigel, Jr., *Solid-State Commun.*, **15**, 479 (1974).

53 F. J. Feigl, W. B. Fowler, and K. L. Yip, *Solid-State Commun.*, **14**, 225 (1974).

54 F. Montillo and P. Balk, *J. Electrochem. Soc.*, **118**, 1463 (1971).

APPENDIX I

Derivation of Trap Admittance

(a) Electron Capture and Emission Admittance

In this appendix the electron capture and emission admittance is calculated from (4.21)

$$i_n(x, t) = qn_T c_n [1 - f(t)]n(x, t) - qn_T e_n f(t) \ (\text{A/cm}^3).$$ (1)

Expressing $f(t)$ as the sum of a dc and an ac part, we obtain

$$f(t) = f_o + \delta f$$ (2)

where f_o is the Fermi function established by the gate bias and δf is the change caused by the ac gate voltage. Similarly, breaking the electron density into an ac and dc part results in

$$n(x, t) = n_o(x) + \delta n(x)$$ (3)

where $n_o(x)$ is the electron density established at point x by the gate bias and $\delta n(x)$ is the change in electron density at point x caused by the ac gate voltage. Substituting (2) and (3) into (1) and making the small-signal approximation by neglecting second-order terms, we obtain

$$i_n(x, t) = qn_T c_n[(1 - f_o)n_o(x) + (1 - f_o)\delta n(x) - n_o(x)\delta f] - qe_n n_T (f_o + \delta f).$$ (4)

Because there is no dc current flow, the system is in thermal equilibrium with regard to the applied gate bias. Thus the dc components of the capture and emission rates must be equal, that is, $r_{ao}(x) = r_{bo}(x)$. This equilibrium condition can be used to eliminate e_n from (4). Equating the dc components of the capture and emission rates, we obtain

$$qn_T c_n(1 - f_o)n_o(x) = e_n n_T f_o.$$ (5)

Solving (5) for e_n and substituting into (4) yields

$$i_n(x, t) = qn_T c_n \left[(1 - f_o)\delta n(x) - n_o(x)\frac{\delta f}{f_o} \right].$$ (6)

The net current density also can be expressed as

$$i_n(x, t) = qn_T \frac{df(t)}{dt}. \tag{7}$$

Equating (7) to (6) yields

$$\frac{df(t)}{dt} = c_n(1 - f_o)\delta n(x) - c_n n_o(x) \frac{\delta f}{f_o}. \tag{8}$$

The small-signal variation of the Fermi function is $\delta f = f_m \exp(j\omega t)$. Combining this relation with (2) results in

$$\frac{df(t)}{dt} = j\omega \, \delta f. \tag{9}$$

Substituting (9) into (8) and solving for δf yields

$$\delta f = \frac{f_o(1 - f_o)\delta n_o(x)}{n_o(x)[1 + j\omega f_o/c_n n_o(x)]}. \tag{10}$$

Substituting (10) into (6) results in

$$i_n(x, t) = \frac{j\omega q n_T f_o(1 - f_o)\delta n(x)}{[1 + j\omega f_o/c_n n_o(x)]n_o(x)}. \tag{11}$$

The ratio $\delta n(x)/n_o(x)$ is equal to the ac potential at x. This equality can be seen as follows. From (2.37), $n(x) = N_D \exp[v(x)]$, so that

$$\frac{\delta n(x)}{n_o(x)} = \frac{N_D \exp[v(x)]\delta v(x)}{N_D \exp[-v(x)]} = \delta v(x) \tag{12}$$

where N_D is the donor concentration and $v(x)$ is the band bending at the silicon surface in units of kT/q. Expressing $\delta v(x)$ in volts, we have

$$\delta v(x) = \frac{q}{kT} \, \delta\psi(x). \tag{13}$$

Equation (12) becomes

$$i_n(x, t) = j\omega \frac{q^2 n_T f_o(1 - f_o)\delta\psi(x)}{kT[1 + j\omega f_o/c_n n_o(x)]}. \tag{14}$$

Equation (14) can be written as

$$i_n(x, t) = Y_n(x, t)\delta\psi(x) \tag{15}$$

where

$$Y_n(x, t) = j\omega \frac{q^2 n_T f_o(1 - f_o)}{kT[1 + j\omega f_o/c_n n_o(x)]}. \tag{16}$$

(b) Hole Capture and Emission Admittance

In this appendix the hole capture and emission admittance is calculated from (4.26)

$$i_p(x, t) = qn_T\{c_p p(x)f(t) - e_p[1 - f(t)]\}. \tag{17}$$

Separating the hole density into a dc and an ac part results in

$$p(x, t) = p_o(x) + \delta p(x). \tag{18}$$

Substituting $f(t) = f_o + \delta f$ and (18) into (17) and making the small-signal approximation by neglecting second-order terms, we obtain

$$i_p(x, t) = qn_T\{c_p[p_o(x)(f_o + \delta f) + f_o\, \delta p(x)] - e_p(1 - f_o - \delta f)\}. \tag{19}$$

From the condition of thermal equilibrium with respect to gate bias, the dc voltage component of the capture rates are equal, that is, $r_{co}(x) = r_{do}(x)$, so that

$$c_p p_o(x) n_T f_o = e_p n_T(1 - f_o). \tag{20}$$

When (20) is used to eliminate e_p, (19) becomes

$$i_p(x, t) + qn_T c_p\left\{[p_o(x)(f_o + \delta f) + f_o\delta p(x)] - p_o(x)f_o\left[1 - \frac{\delta f}{1 - f_o}\right]\right\}. \tag{21}$$

Expressing the current density in terms of the Fermi function as before, and taking the time dependence of the Fermi function into account we obtain

$$\delta f = \frac{f_o(1 - f_o)\delta p(x)}{p_o(x)\left[\dfrac{j\omega(1 - f_o)}{c_p p_o(x)} - 1\right]}. \tag{22}$$

Substituting (22) into (21) yields

$$i_p(x, t) = \frac{j\omega qn_T f_o(1 - f_o)\delta p(x)}{p_o(x)\left[\dfrac{j\omega(1 - f_o)}{c_p p_o(x)} - 1\right]}. \tag{23}$$

As before, $\delta p(x)/p_o(x) = \delta v(x) = (q/kT)\delta\psi(x)$, and (21) becomes

$$i_p(x, t) = j\omega = \frac{q^2 n_T f_o(1 - f_o)\delta\psi(x)}{kT[j\omega(1 - f_o)/c_p p_o(x) - 1]} \tag{24}$$

and the admittance is

$$Y_p(x, t) = j\omega\,\frac{q^2 n_T f_o(1 - f_o)}{kT[j\omega(1 - f_o)/c_p p_o(x) - 1]}. \tag{25}$$

APPENDIX II

Evaluation of Several Integrals Involved in Calculating the Admittance

(a) Calculation of the Integral Involved in the Generation-Recombination Conductance

The integral is H_G from (4.81)

$$H_G = \int_0^1 df \left[j\omega f(1-f) + \frac{f}{\tau_p} + \frac{(1-f)}{\tau_n} \right]^{-1} \tag{1}$$

where

$$\tau_n \equiv (c_n n)^{-1} \tag{2}$$

$$\tau_p \equiv (c_p p)^{-1}. \tag{3}$$

To evaluate this integral, it is convenient to change variables of integration. Let $x = f + \frac{1}{2}$; then

$$H_G = \int_{-1/2}^{1/2} dx \left[-j\omega x^2 + x \left(\frac{1}{\tau_p} - \frac{1}{\tau_n} \right) + \frac{j\omega}{4} + \frac{1}{2} \left(\frac{1}{\tau_p} + \frac{1}{\tau_n} \right) \right]^{-1}. \tag{4}$$

The denominator is rewritten as

$$-j\omega \left[x^2 - x \left(\frac{1}{j\omega \tau_p} - \frac{1}{j\omega \tau_n} \right) - \frac{1}{2} \left(\frac{1}{2} + \frac{1}{j\omega \tau_p} + \frac{1}{j\omega \tau_n} \right) \right]$$

$$= -j\omega (x - r_+)(x - r_-) \tag{5}$$

where

$$r_+ - r_- = \left[1 + 2 \left(\frac{1}{j\omega \tau_p} + \frac{1}{j\omega \tau_n} \right) + \left(\frac{1}{j\omega \tau_p} - \frac{1}{j\omega \tau_n} \right)^2 \right]^{1/2} \tag{6}$$

and

$$r_+ r_- = -\frac{1}{4} \left[1 + 2 \left(\frac{1}{j\omega \tau_p} + \frac{1}{j\omega \tau_n} \right) \right]. \tag{7}$$

Substituting (5) in (4), we find

$$H_G = -\frac{1}{j\omega(r_+ - r_-)} \ln\left[\frac{1 - 2(r_+ - r_-) - 4r_+ r_-}{1 + 2(r_+ - r_-) - 4r_+ r_-}\right]. \tag{8}$$

In particular, consider several special cases.

Case 1   Near midgap, where recombination-generation is most efficient, we obtain

$$\tau_p^{-1} = \tau_n^{-1} = \tau^{-1} \tag{9}$$

and

$$r_+ - r_- = \left(1 + \frac{4}{j\omega\tau}\right)^{1/2} \tag{10}$$

$$r_+ r_- = -\frac{1}{4}\left(1 + \frac{4}{j\omega\tau}\right) \tag{11}$$

$$H_G = -\frac{2\tau}{j\omega\tau(1 + 4/j\omega\tau)^{1/2}} \ln\left[\frac{(1 + 4/j\omega\tau)^{1/2} - 1}{(1 + 4/j\omega\tau)^{1/2} + 1}\right]. \tag{12}$$

In the limit $\omega\tau \ll 4$, we find

$$\lim_{\omega\tau \to 0} (H_G) = \tau. \tag{13}$$

Case 2   In depletion, $\tau_n^{-1} \gg \tau_p^{-1}$ for n-type material, and

$$r_+ - r_- \cong 1 + \frac{1}{j\omega\tau_n} + \left(\frac{\tau_n}{\tau_p}\right)\frac{1}{j\omega\tau_n}\frac{1 - \dfrac{1}{j\omega\tau_n}}{1 + \dfrac{1}{j\omega\tau_n}}. \tag{14}$$

Then

$$H_G \cong -\tau_n \frac{1 + j\omega\tau_n}{(1 + j\omega\tau_n)^2 - (1 - j\omega\tau_n)(\tau_n/\tau_p)} \ln\left[\frac{(\tau_n/\tau_p)}{(1 + j\omega\tau_n)^2 + (j\omega\tau_n)(\tau_n/\tau_p)}\right]$$

$$\cong -\frac{\tau_n}{1 + j\omega\tau_n} \ln\left[\frac{\tau_n/\tau_p}{(1 + j\omega\tau_n)^2}\right]. \tag{15}$$

Similarly, for weak inversion where $\tau_p^{-1} \gg \tau_n^{-1}$, (15) applies with τ_n and τ_p interchanged.

Case 3   Limit of low frequencies, $\omega\tau_n$, $\omega\tau_p \ll 1$

$$r_+ - r_- \cong (j\omega\tau_p)^{-1} - (j\omega\tau_n)^{-1} \tag{16}$$

$$H_G \cong -\frac{\tau_p \tau_n}{(\tau_n - \tau_p)} \ln\left(\frac{\tau_p}{\tau_n}\right). \tag{17}$$

which, for $\tau_p = \tau_n = \tau$, agrees with (13) and, for $\tau_n^{-1} \gg \tau_p^{-1}$, agrees with the low frequency limit of (15). To evaluate H_{C_n} and H_{C_p} in this limit, see (31) and (32).

Case 4 Limit of high frequencies, $\omega\tau_n$, $\omega\tau_p \gg 1$. From (6) and (7), in this case we find

$$r_+ - r_- \approx \left[1 + 2\left(\frac{1}{j\omega\tau_p} + \frac{1}{j\omega\tau_n}\right)\right]^{1/2} \tag{18}$$

and

$$r_+ r_- \approx -\frac{1}{4}\left[1 + 2\left(\frac{1}{j\omega\tau_p} + \frac{1}{j\omega\tau_n}\right)\right]. \tag{19}$$

Consequently, H_G is given by (12) with the substitution

$$\frac{2}{\tau} = \frac{1}{\tau_p} + \frac{1}{\tau_n}. \tag{20}$$

When the high frequency limit of (12) is taken, H_G becomes

$$H_G \approx (j\omega)^{-1}2[\ln(\omega\tau) + j\tan^{-1}(\omega\tau)]$$

$$\approx \frac{\pi}{\omega} - j\frac{2}{\omega}\ln(\omega\tau). \tag{21}$$

(b) Calculation of the Integrals Involved in the Capacitances C_{TX_p}, C_{TX_n} from (4.69)

These integrals are

$$H_{C_p} = \int_0^1 df\, fD^{-1}(f) \tag{22}$$

and

$$H_{C_n} = \int_0^1 df(1-f)D^{-1}(f) \tag{23}$$

with the denominator $D(f)$ given by

$$D(f) \equiv j\omega f(1-f) + \frac{f}{\tau_p} + \frac{(1-f)}{\tau_n}. \tag{24}$$

We note that the derivative of the denominator D is

$$\frac{dD}{df} = -2j\omega f + \left(j\omega + \frac{1}{\tau_p} - \frac{1}{\tau_n}\right). \tag{25}$$

Therefore, we rewrite H_{C_p} as

$$H_{C_p} = -\frac{1}{2j\omega}\int_0^1 df\frac{\left[-2j\omega f + j\omega + \frac{1}{\tau_p} - \frac{1}{\tau_n}\right]}{D}$$

$$+\frac{1}{2}\int_0^1 df\frac{\left[1 + \frac{1}{j\omega\tau_p} - \frac{1}{j\omega\tau_n}\right]}{D}.$$

That is

$$H_{C_p} = -\frac{1}{2j\omega}\ln\left(\frac{\tau_n}{\tau_p}\right) + \frac{1}{2}\left[1 + \frac{1}{j\omega\tau_p} - \frac{1}{j\omega\tau_n}\right]H_G. \tag{26}$$

Similarly, we rewrite H_{C_n} as

$$H_{C_n} = H_G - H_{C_p}$$

$$= -\frac{1}{2j\omega}\ln\left(\frac{\tau_p}{\tau_n}\right) + \frac{1}{2}\left[1 + \frac{1}{j\omega\tau_n} - \frac{1}{j\omega\tau_p}\right]H_G. \tag{27}$$

When the various limiting forms of H_G, (12), (15), and (17) are used, the corresponding special forms of H_{C_p}, H_{C_n} are obtained.

Case 1   Near midgap, $\tau_n = \tau_p = \tau$; thus

$$H_{C_p} = H_{C_n} = \frac{1}{2}H_G \tag{28}$$

where H_G is given by (8).

Case 2   In depletion, $\tau_n^{-1} \gg \tau_p^{-1}$ for n-type material; that is

$$H_{C_p} \cong -\frac{1}{2j\omega}\ln\left(\frac{\tau_n}{\tau_p}\right) + \frac{1}{2}\left(1 - \frac{1}{j\omega\tau_n}\right)H_G$$

$$\cong -\frac{\tau_n}{1+j\omega\tau_n}\ln\left(\frac{\tau_n}{\tau_p}\right)\left[1 + \frac{\tau_n/\tau_p}{(1+j\omega\tau_n)^2}\right] - \frac{(\tau_n/\tau_p)(1-j\omega\tau_n)}{(1+j\omega\tau_n)^3}\tau_n$$

$$+ \left[-\frac{1}{2j\omega}\left(\frac{1-j\omega\tau_n}{1+j\omega\tau_n}\right) + \frac{2(\tau_n/\tau_p)}{(1+j\omega\tau_n)^3}\tau_n\right]\{\ln[1+(\omega\tau_n)^2] + 2j\tan^{-1}\omega\tau_n\}$$

$$\cong -\frac{\tau_n}{1+j\omega\tau_n}\ln\left(\frac{\tau_n}{\tau_p}\right) - \frac{1}{2j\omega}\left(\frac{1-j\omega\tau_n}{1+j\omega\tau_n}\right)\{\ln[1+(\omega\tau_n)^2] + 2j\tan^{-1}(\omega\tau_n)\}. \tag{29}$$

$$H_{C_n} \cong -\frac{1}{2j\omega}\ln\left(\frac{\tau_p}{\tau_n}\right) + \frac{1}{2}\left(1 + \frac{1}{j\omega\tau_n}\right)H_G$$

$$\cong \frac{1}{2j\omega}\{\ln[1+(\omega\tau_n)^2] + 2j\tan^{-1}(\omega\tau_n)\}$$

$$+ \tau_n\left(\frac{\tau_n}{\tau_p}\right)\frac{1}{(1+j\omega\tau_n)^2}\left[1 - 2\ln(1+j\omega\tau_n)^2 + \ln\left(\frac{\tau_n}{\tau_p}\right)\right]$$

$$\cong \frac{1}{2j\omega}\{\ln[1+(\omega\tau_n)^2] + 2j\tan^{-1}(\omega\tau_n)\}. \tag{30}$$

In weak inversion H_{C_n} and H_{C_p} are interchanged and τ_n and τ_p are interchanged as well.

The first form of (29) and (30) includes terms of order $(\tau_n/\tau_p) \ll 1$; in the second form, these terms have been dropped.

Case 3   We now discuss the low frequency limit, for which (17) is inadequate for finding H_{C_n} and H_{C_p} because the next term in $\omega\tau_p$ and $\omega\tau_n$ is needed to produce a nonzero result. It is simpler to return to (22) and (23) in the limit

$\omega \to 0$ to find

$$H_{C_p} \cong \int_0^1 df\, f \left[\frac{f}{\tau_p} + \frac{(1-f)}{\tau_n} \right]^{-1}$$

$$= [\tau_p^{-1} - \tau_n^{-1}]^{-1} \int_0^1 df - \frac{\tau_n^{-1}}{[\tau_p^{-1} - \tau_n^{-1}]} \int_0^1 df [\tau_n^{-1} + (\tau_p^{-1} - \tau_n^{-1})f]^{-1}$$

$$= (\tau_p^{-1} - \tau_n^{-1})^{-1} \left[1 - \frac{\tau_p}{(\tau_n - \tau_p)} \ln\left(\frac{\tau_n}{\tau_p}\right) \right]. \tag{31}$$

Similarly, in the limit of low frequencies

$$H_{C_n} = H_G - H_{C_p}$$

$$= -(\tau_p^{-1} - \tau_n^{-1})^{-1} \left[1 - \left(\frac{\tau_p}{\tau_n - \tau_p} + 1 \right) \ln\left(\frac{\tau_n}{\tau_p}\right) \right].$$

$$= (\tau_n^{-1} - \tau_p^{-1})^{-1} \left[1 - \frac{\tau_n}{\tau_p - \tau_n} \ln\left(\frac{\tau_p}{\tau_n}\right) \right] \tag{32}$$

which is H_{C_p} with τ_n and τ_p interchanged.

APPENDIX III

Equivalent Circuit in Figure 4.7

To find the admittance of the circuit in Fig. 4.7, we redraw this circuit in Fig. III-1, showing a representative branch due to traps at position x connected by dashes to the main circuit. The Δx factor has been suppressed to simplify notation, and the oxide capacitance must be connected in series to the semiconductor portion of the circuit, which is all that is shown in Fig. III-1.

The ac semiconductor band-bending variation δv_s is applied across the capacitive divider at the top of the figure, drawing the ac semiconductor current I_s. Two internal loops are indicated, one carrying the total current* ΣI_{out} that passes through the inversion layer capacitance C_I and splits into components I_{out}, which vary from one bulk trap branch to the next. The other loop carries current ΣI_{in}, which also divides into different individual components I_{in} in each bulk trap branch.

Summing voltages in the capacitive divider, we find

$$\delta v_s = \frac{I_s}{j\omega C_D} - \frac{\Sigma I_{out}}{j\omega C_{x_c}} - \frac{\Sigma I_{in}}{j\omega C_{w-x_c}}. \tag{1}$$

Across the inversion layer capacitance, we have

$$\delta v_I = \frac{\Sigma I_{out}}{j\omega C_I}. \tag{2}$$

Using the definition

$$Y_s \equiv \frac{I_s}{\delta v_s} \tag{3}$$

we obtain from (1)

$$Y_s = j\omega C_D + \left(\frac{C_D}{C_{x_c}}\right)\frac{\Sigma I_{out}}{\delta v_s} + \left(\frac{C_D}{C_{w-x_c}}\right)\frac{\Sigma I_{in}}{\delta v_s} \tag{4}$$

where we have used the relation

$$C_D^{-1} = C_{w-x_c}^{-1} + C_{x_c}^{-1}.$$

*The summation indicates a sum over all bulk trap branches; that is, over all depths x.

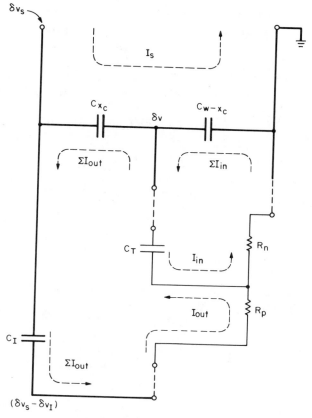

Fig. III-1 Semiconductor portion of equivalent circuit in Fig. 4.7. Of all the branches contributed to the circuit by bulk traps at various depths, only a typical branch is shown. This typical branch, contributed by bulk traps at a depth x, is connected by dashed lines to indicate that many such branches are involved. The summations on the loop currents in the main circuit indicate all these branches have been summed over.

To obtain Y_s, we must find ΣI_{in} and ΣI_{out}. To do this, we obtain I_{in} and I_{out} in a typical branch. We sum voltages in the loop for I_{in} and I_{out}

$$\delta v = I_{in}[R_n + (j\omega C_T)^{-1}] - I_{out}(j\omega C_T)^{-1} \tag{5}$$

$$\delta v_s - \delta v_I - \delta v = -I_{in}(j\omega C_T)^{-1} + I_{out}[R_p + (j\omega C_T)^{-1}]. \tag{6}$$

Solving (5) and (6) for I_{in} and I_{out}, we obtain

$$I_{in} = [\delta v(j\omega C_T G_n + G_n G_p) + (\delta v_s - \delta v_I - \delta v)G_n G_p]D^{-1} \tag{7}$$

$$I_{out} = [\delta v\, G_n G_p + (\delta v_s - \delta v_I - \delta v)(j\omega C_T G_p + G_n G_p)]D^{-1} \tag{8}$$

where D is defined by

$$D \equiv G_n + G_p + j\omega C_T. \tag{9}$$

If we now sum over all the trap branches (integrate over x), (7) and (8) become

$$\Sigma I_{in} = \delta v [j\omega C_{TXn} + G_{gr}] + (\delta v_s - \delta v_I - \delta v)G_{gr} \tag{10}$$

$$\Sigma I_{out} = \delta v \, G_{gr} + (j\omega C_{TXp} + G_{gr})(\delta v_s - \delta v_I - \delta v) \tag{11}$$

where C_{TXp}, C_{TXn}, and G_{gr} are introduced according to (4.61), and (4.80) and δv has been evaluated at crossover.

Returning to (4), the semiconductor admittance Y_s becomes

$$Y_s = j\omega C_D + \left(\frac{C_D}{C_{x_c}}\right)\left[\left(\frac{\delta v}{\delta v_s}\right)G_{gr} + \left(1 - \frac{\delta v_I}{\delta v_s} - \frac{\delta v}{\delta v_s}\right)(j\omega C_{TXp} + G_{gr})\right]$$

$$+ \left(\frac{C_D}{C_{w-x_c}}\right)\left[\left(\frac{\delta v}{\delta v_s}\right)j\omega C_{TXn} + \left(1 - \frac{\delta v_I}{\delta v_s}\right)G_{gr}\right]$$

$$= j\omega C_D + j\omega C_I \left(\frac{\delta v_I}{\delta v_s}\right)$$

$$+ \left(\frac{C_D}{C_{w-x_c}}\right)\left[\left(\frac{\delta v}{\delta v_s}\right)j\omega(C_{TXn} + C_{TXp}) - \left(1 - \frac{\delta v_I}{\delta v_s}\right)j\omega C_{TXp}\right]. \tag{12}$$

To obtain the second form of (12), C_D/C_{x_c} was replaced by $(1 - C_D/C_{w-x_c})$, and the resulting term with unit coefficient [which is $\Sigma I_{out}/\delta v_s$, according to (11)] was replaced using (2). Finally, note that (2) and (11) imply

$$\frac{\delta v_I}{\delta v_s} = \frac{\left(1 - \dfrac{\delta v}{\delta v_s}\right)(j\omega C_{TXp} + G_{gr}) + \left(\dfrac{\delta v}{\delta v_s}\right)G_{gr}}{j\omega C_I + j\omega C_{TXp} + G_{gr}}$$

or

$$1 - \frac{\delta v_I}{\delta v_s} = \frac{j\omega C_I + \dfrac{\delta v}{\delta v_s} j\omega C_{TXp}}{j\omega C_I + j\omega C_{TXp} + G_{gr}}$$

$$= \frac{\delta u_{Fp}}{\delta v_s} \tag{13}$$

according to (4.79). If (13) is used to replace $\delta v_I/\delta v_s$ by $(1 - \delta u_{Fp}/\delta v_s)$ in (12), the resulting Y_s is that of (4.87), derived algebraically from an approximate version of the small-signal equations.

APPENDIX IV

Diffusion and Depletion Layer Resistances

(a) Diffusion Resistance

In this section we calculate the resistance for diffusion controlled minority carrier response,[1,2] for example, of n-type silicon where the minority carriers are holes. The resistance for the diffusion of holes across the quasi-neutral region of the silicon substrate in the MOS capacitor is given by

$$R_d = -\frac{\delta\phi_{Fp}(w)}{I_p(w)} \tag{1}$$

where the Si–SiO$_2$ interface is taken at $x = 0$, the depletion layer edge at $x = w$, $\delta\phi_{Fp}(w)$ is the ac quasi-Fermi level variation at the depletion layer edge due to the ac gate voltage, and $I_p(w)$ is the current per unit area flowing by diffusion into the depletion layer from the quasi-neutral region. The minus sign in (1) ensures that R_d is positive because the current I_p flows into the depletion region and hence in the negative x-direction.

To evaluate (1), it is necessary to find the dependence of p on x. To do this, the continuity equation for holes must be solved. The continuity equation for holes is

$$\frac{\partial p}{\partial t} = -\frac{(p-p_B)}{\tau_p} + D_p \frac{\partial^2 p}{\partial x^2} \tag{2}$$

where the first term on the right describes generation and recombination of holes in the quasi-neutral region of the silicon and the second term on the right describes the diffusion of minority carriers across the quasi-neutral region of the silicon, τ_p is the lifetime (in seconds), D_p is the hole diffusion coefficient (in cm^2/sec), and p_B is the equilibrium hole density (in cm^{-3}). The time dependence of p is given by $p(x, t) = p_B + \delta p(x) \exp(j\omega t)$, which, when substituted into (2), gives

$$j\omega \, \delta p = -\frac{\delta p}{\tau_p} + D_p \frac{\partial^2 \delta p}{\partial x^2} \, . \tag{3}$$

Solving (3) with the boundary conditions $\delta p = -\delta p_w$ at $x = w$ and $\delta p = 0$ at $x = \infty$, we obtain

$$\delta p = -\delta p_w \exp\left[-(x-w)\left(\frac{1+j\omega\tau_p}{D_p\tau_p}\right)^{1/2}\right]. \tag{4}$$

Consider the case where $\omega\tau_p \ll 1$, which is reasonable for device-grade material that has a lifetime of 1–10 μsec and for the frequency range of up to 1 MHz or so. With this approximation, (4) becomes

$$\delta p = -\delta p_w \exp\left[-\frac{x-w}{(D_p\tau_p)^{1/2}}\right]. \tag{5}$$

Using (5), the ac current at $x = 0$ entering the depletion layer by diffusion is

$$I_p = -qD_p \left.\frac{\partial \delta p}{\partial x}\right|_{x=w}$$

$$= -\left(\frac{qD_p}{L_p}\right)\delta p_w \tag{6}$$

where $L_p = (D_p\tau_p)^{1/2}$ is the diffusion length. However, the Einstein relation states

$$D_p = \mu_p \frac{kT}{q}. \tag{7}$$

Therefore

$$I_p = -\left(\frac{kT}{q}\right)\frac{q\mu_p\delta p_w}{L_p}. \tag{8}$$

Now, for this quasi-equilibrium, small-signal case, we have

$$\delta p_w = p_B\beta \, \delta\phi_{Fp} \tag{9}$$

where $\beta = (kT/q)^{-1}$. Substituting (8) and (9) into (1) yields

$$R_d = \frac{L_p}{q\mu_p p_B}. \tag{10}$$

In explanation of (9), note that (3) assumes no drift and, therefore, no field in the quasi-neutral region. Hence the variation in p is entirely due to quasi-Fermi level variation. Now $p_B N_D = n_i^2$, so that $p_B = n_i^2/N_D$. Substituting for p_B, (10) provides

$$R_d = \frac{L_p N_D}{q\mu_p n_i^2}. \tag{11}$$

The diffusion conductance corresponding to (11) is

$$G_d = \frac{q\mu_p n_i^2}{L_p N_D}. \tag{12}$$

(b) Calculation of Depletion Layer Resistance[2]

In this section we calculate the resistance of the depletion layer to the flow of holes and show that this resistance can be neglected.

The resistance of the depletion layer to the flow of holes is given by

$$R_p = \int_0^w \frac{dx}{\sigma_p(x)} \quad (\Omega/cm^2) \tag{13}$$

where $\sigma_p(x) = q\mu_p p(x)$ is the hole conductivity [in $(mhos\text{-}cm)^{-1}$] at x in the depletion layer and $p(x)$ is the hole density (in cm^{-3}) at x. We write, using $p(x)n(x) = n_i^2$, that $\sigma_p(x) = q\mu_p n_i^2/n(x)$, where $n(x)$ is the electron density (in cm^{-3}) at x. Then (13) becomes

$$R_p = (q\mu_p n_i^2)^{-1} \int_0^w n(x)dx. \tag{14}$$

Using the relation $n(x) = N_D \exp[v(x)]$ from (2.37), we rewrite (14) as

$$R_p = (q\mu_p n_i^2)^{-1} N_D \int_0^w \exp[v(x)]dx. \tag{15}$$

To integrate (15), it is necessary to know the dependence of band bending on x. For this dependence, we use the depletion approximation $\psi_s = qN_D w^2 (2\epsilon_s)^{-1}$, from which

$$v(x) = \frac{1}{2}\left(\frac{w-x}{\lambda_n}\right)^2 \equiv y^2. \tag{16}$$

If we change variables from x to y, (15) becomes

$$R_p = (2)^{1/2} \frac{\lambda_n}{q\mu_p n_i^2} N_D \int_0^{w/\lambda_n} \exp[y^2]dy. \tag{17}$$

Typically, $w/\lambda_n \gg 1$, so that the upper limit of the integral in (17) can be considered to be infinity. Thus the value of the integral in (17) is $(\pi/2)^{1/2}$. Therefore, (17) becomes simply

$$R_p = (\pi)^{1/2} \frac{\lambda_n}{q\mu_p n_i^2} N_D \tag{18}$$

or

$$R_p = \frac{(\pi)^{1/2}\lambda_n}{q\mu_p p_B}. \tag{19}$$

Forming the ratio R_d/R_p using (10) from Section (a) above for R_d gives

$$\frac{R_d}{R_p} = \frac{L_p}{\pi^{1/2}\lambda_n}. \tag{20}$$

The values of the quantities in (20) for typical silicon MOS structures are $L_p = 10^{-2}$ cm for $\tau_p = 10 \, \mu$ sec and $\lambda_n = 4.0 \times 10^{-6}$ cm for $N_D = 10^{16}$ cm^{-3}. With these values, $R_d/R_p \cong 1.4 \times 10^3$, so that R_p is negligible with respect to R_d.

REFERENCES

1   C. G. B. Garrett, *Phys. Rev.*, **107**, 478 (1957).
2   K. Lehovec and A. Slobodskoy, *Solid-State Electron.*, **7**, 59 (1964).

APPENDIX V

Calculation of Δ

In this appendix we calculate (4.150) for Δ in the high frequency capacitance. The value of Δ is obtained from the constraint of constant inversion layer charge, following the analysis by Brews.[1] For p-type, the constant inversion layer charge condition, (4.148), can be rewritten, transforming variables from x to v, using $dx = (dx/dv)dv = \lambda_n \, dv/F(v, u_B)$ as

$$\int_0^{v_s} dv \, \frac{\exp(v + u_{Fn}) - \exp(u_{Fn})}{F(v, u_B, u_{Fn})} = \int_0^{v_{so}} dv_o \, \frac{[\exp(v_o + u_B) - \exp(u_B)]}{F(v_o, u_B)}. \tag{1}$$

Introducing the small-signal quantities from (4.141) and (4.142) into (1), we obtain

$$\frac{\exp(v_{so} + u_B) - \exp(u_B)}{F(v_{so}, u_B)} = -\int_0^{v_{so}} dv \, [\exp(v) - 1] \frac{\partial}{\partial u_{Fn}} \left[\frac{\exp(u_{Fn})}{F(v, u_{Fn}, u_B)} \right] \frac{\partial u_{Fn}}{\partial v_s} \tag{2}$$

where the derivative with respect to u_{Fn} is evaluated at $u_{Fn} = u_B$. By carrying out the differentiation, we transform (2) to

$$\frac{\partial v_s}{\partial u_{Fn}} = \frac{F(v_{so}, u_B)}{\exp(v_{so}) - 1} \int_0^{v_{so}} dv_s \, \frac{\exp(v_s) - 1}{F(v_s, u_B)} \left[1 - \left(\frac{n_i}{N_A} \right)^2 \frac{\exp(v_s) - v_s - 1}{F^2(v_s, u_B)} \right]. \tag{3}$$

For better accuracy in strong inversion, when $\partial v_s/\partial u_{Fn} \approx 1$, (3) can be rewritten as

$$\frac{\partial v_s}{\partial u_{Fn}} = 1 + \Delta. \tag{4}$$

To do this, we remove the square brackets in (3) and obtain two integrals. Taking the first of these integrals, we rewrite it as

$$\int_0^{v_{so}} dv_s \, \frac{\exp(v_s) - 1}{F(v_s, u_B)} = \int_0^{v_{so}} dv_s \, \exp(v_s) \frac{1 - \exp(-v_s)}{F(v_s, u_B)}. \tag{5}$$

This integral is done by parts to obtain

$$\int_0^{v_{so}} dv_s \frac{\exp(v_s) - 1}{F(v_s, u_B)} = \exp(v_s) \frac{1 - \exp(-v_s)}{F(v_s, u_B)} \Big|_0^{v_{so}} - \int_0^{v_{so}} dv_s \left[\frac{1}{F(v_s, u_B)} - \frac{\exp(v_s) - 1}{F^3(v_s, u_B)} \right.$$

$$\times \left. \left\{ 1 - \exp(-v_s) + \left(\frac{n_i}{N_A}\right)^2 [\exp(v_s) - 1)] \right\} \right]$$

$$= \frac{\exp(v_{so}) - 1}{F(v_{so}, u_B)} - \frac{1}{[1 + (n_i/N_A)^2]^{1/2}} - \int_0^{v_{so}} F^{-3}(v_s, u_B) \Big\{ F^2(v_s, u_B)$$

$$- \exp(v_s) - \exp(-v_s) + 2 - \left(\frac{n_i}{N_A}\right)^2 [\exp(v_s) - 1]^2 \Big\}. \qquad (6)$$

Combining the integral in (6) with the second integral in (3), which has $(n_i/N_A)^2$ as a factor, we find that (3) is given by (4), where

$$\Delta = - \frac{F(v_{so}, u_B)}{\exp(v_{so}) - 1} \left[\beta - \int_0^{v_{so}} dv_s \frac{\exp(v_s) - \exp(-v_s) - 2v_s + (n_i/N_A)^2 h(v_s)}{F^3(v_s, u_B)} \right]$$

$$(7)$$

with $h(v_s) = \exp(v_s)(v_s - 2) + v_s + 2$ and $\beta = [1 + (n_i/N_A)^2]^{-1/2}$. To obtain (4.150) in the text, the term $h(v_s)$ has been dropped and β has been taken to be unity. Both neglected terms are of the order of $(n_i/N_A)^2$ at all values of bias from flatbands to strong inversion.

Note that (5) and hence (4.150) are integrated from flatbands to v_{so}. For $0 < v_s < u_B$, u_{Fn} is not uniform,[2] so that (5), which is based on a uniform u_{Fn}, is not valid. Therefore, strictly speaking, the limits of integration should be from u_B to v_{so}. However, electron density is so small in the region $(0 < v_s < u_B)$ that electrons will have a negligible influence on capacitance. Therefore, it makes little difference to the final result whether the integration is carried out from 0 to v_{so} or from u_B to v_{so}.

REFERENCES

1   J. R. Brews, *J. Appl. Phys.*, **45**, 1276 (1974).
2   G. Baccarani and M. Severi, *IEEE Transact. Electron Devices*, **ED-21**, 122 (1974).

APPENDIX VI

Derivation of Covariance of Charge for Poisson Distribution

In this appendix (6.63) is derived, namely

$$\langle [Q(\boldsymbol{R}) - \langle Q \rangle][Q(\boldsymbol{R}') - \langle Q \rangle] \rangle = q \langle Q \rangle \delta(\boldsymbol{R} - \boldsymbol{R}') \tag{1}$$

where $Q(\boldsymbol{R})$ is the charge density per unit area at position $\boldsymbol{R}$ on the Si–SiO$_2$ interface, and where angular brackets $\langle \cdots \rangle$ indicate an area average that, for a Poisson distribution, can be replaced by a statistical average. To derive (1), we note

$$\langle [Q(\boldsymbol{R}) - \langle Q \rangle][Q(\boldsymbol{R}') - \langle Q \rangle] \rangle = \langle Q(\boldsymbol{R})Q(\boldsymbol{R}') \rangle - \langle Q \rangle^2 \tag{2}$$

which follows simply by multiplying the square brackets and averaging each of the resulting terms.

To perform the averages in (2), first suppose that $Q(\boldsymbol{R})$ is the charge at position $\boldsymbol{R}$ due to an array of N point charges

$$Q(\boldsymbol{R}) = q \sum_{n=1}^{N} \delta(\boldsymbol{R} - \boldsymbol{R}_n). \tag{3}$$

The probability that there are N charges present in a Poisson distribution over an area A is P_N, where

$$P_N = \left[\frac{A \langle Q \rangle}{q} \right]^N \exp \left[-\frac{A \langle Q \rangle}{q} \right] / N!. \tag{4}$$

The probability that one of these charges lies in the elementary area $d^2 R_n$ about position $\boldsymbol{R}_n$ is $p(R_n)d^2 R_n = d^2 R_n / A$. Therefore, the probability that there are N charges in elementary areas located at the N positions $\{\boldsymbol{R}_n\}$ is $P_N p(R_1) \cdots p(R_n) \cdots p(R_N) d^2 R_1 \cdots d^2 R_n \cdots d^2 R_N$. To evaluate the average $\langle Q(\boldsymbol{R})Q(\boldsymbol{R}') \rangle$ in (2), we multiply $Q(\boldsymbol{R})Q(\boldsymbol{R}')$ by the above probability, using (3), and sum over all possible N and integrate over all areas.

The result is

$$\langle Q(\mathbf{R})Q(\mathbf{R}')\rangle = q^2 \sum_{N=0}^{\infty} P_N \int d^2R_1 \, p(R_1) \cdots \int d^2R_N \, p(R_N)$$

$$\times \sum_{n=1}^{N} \delta(\mathbf{R} - \mathbf{R}_n) \sum_{m=1}^{N} \delta(\mathbf{R}' - \mathbf{R}_m). \tag{5}$$

In the product of sums of (5), we note two types of term: those for which $\mathbf{R}_n = \mathbf{R}_m$ and those for which $\mathbf{R}_n \neq \mathbf{R}_m$. If these terms are separated, (5) becomes

$$\langle Q(\mathbf{R})Q(\mathbf{R}')\rangle = q^2 \sum_{N=0}^{\infty} P_N \int d^2R_1 \, p(R_1) \cdots \int d^2R_n \, p(R_N)$$

$$\times \left[\sum_{n=1}^{N} \delta(\mathbf{R} - \mathbf{R}_n)\delta(\mathbf{R}' - \mathbf{R}_n) + \sum_{n,m(\neq n)} \delta(\mathbf{R} - \mathbf{R}_n)\delta(\mathbf{R}' - \mathbf{R}_m) \right]. \tag{6}$$

Using the delta-functions, one integration can be performed for each term in the first sum. The remaining integrations are then simply of the form

$$\int d^2R_n \, p(R_n) = 1. \tag{7}$$

Each of the N terms contributes the same result A^{-1}. Similarly, for the second sum, two integrations can be performed using the delta-functions. Then the remaining integrals are done using (7). In this case, there are $N(N-1)$ terms, all of which contribute the same result A^{-2}. Consequently, (6) becomes

$$\langle Q(\mathbf{R})Q(\mathbf{R}')\rangle = q^2 \sum_{N=1}^{\infty} P_N \frac{N}{A} \delta(\mathbf{R} - \mathbf{R}') + q^2 \sum_{N-2}^{\infty} P_N \frac{N(N-1)}{A^2}. \tag{8}$$

In (8) the first summation begins at $N = 1$ because no contribution to $Q(\mathbf{R})$ occurs if there are zero charges. Similarly, the second summation begins at $N = 2$ because we cannot have charges at two distinct positions $\mathbf{R}_n \neq \mathbf{R}_m$ unless at least two charges are present.

Now, using (4), we have

$$\sum_{N=1}^{\infty} P_n \left(\frac{N}{A}\right) = \frac{\langle Q \rangle}{q} \sum_{N=1}^{\infty} \frac{[A\langle Q \rangle / q]^{N-1}}{(N-1)!} \exp\left[-\frac{A\langle Q \rangle}{q}\right] \tag{9}$$

where we note, setting $N' = N - 1$, that the sum is simply $\exp[A\langle Q \rangle / q]$. Also

$$\sum_{N=2}^{\infty} P_N \frac{N(N-1)}{A^2} = \left[\frac{\langle Q \rangle}{q}\right]^2 \sum_{N=2}^{\infty} \frac{[A\langle Q \rangle / q]^{N-2}}{(N-2)!} \exp\left[-\frac{A\langle Q \rangle}{q}\right] = \left[\frac{\langle Q \rangle}{q}\right]^2 \tag{10}$$

where again, setting $N' = N - 2$, the sum is simply $\exp[A\langle Q \rangle / q]$. Therefore, (8) becomes

$$\langle Q(\boldsymbol{R})Q(\boldsymbol{R}')\rangle = q\langle Q\rangle\delta(\boldsymbol{R} - \boldsymbol{R}') + \langle Q\rangle^2 \tag{11}$$

and (2) becomes

$$\langle [Q(\boldsymbol{R}) - \langle Q\rangle][Q(\boldsymbol{R}') - \langle Q\rangle]\rangle = q\langle Q\rangle\delta(\boldsymbol{R} - \boldsymbol{R}') \tag{12}$$

which is the desired result.

APPENDIX VII

Derivation of Green's Function

In this appendix the formula for the band bending due to one-point charge is calculated for the point charge at position (r', x'). The band bending is evaluated at a point (r, x). The plane $x = 0$ is the silica-silicon interface, $-x_o < x < 0$ is in the silica, $x > 0$ is in the silicon, and $r = (y, z)$ is the vector position of points on any plane $x = $ constant.

The equation defining the band bending of the point charge $G(r - r'; x, x')$, is

$$\nabla \cdot [\epsilon(x)\nabla G] = -q\delta(x - x')\delta(r - r') \tag{1}$$

where $\epsilon(x) = \epsilon_{ox}$ (for $x < 0$) and $= \epsilon_s$ (for $x > 0$).

The boundary conditions on G are that G vanishes both at the gate $(x = -x_o)$ and at the depletion layer edge $(x = x_D)$. The depletion layer width x_D is approximated as a constant, and no modulation of the depletion layer width by the point charge is included. This constant depletion layer width approximation should be valid in depletion or in inversion where the depletion layer edge is remote from the Si–SiO$_2$ interface.

Equation (1) is solved by a Fourier transform technique. We define $G_k(x, x')$ by

$$G(r - r'; x, x') = \int \frac{d^2k}{(2\pi)^2} G_k(x, x') \exp[jk \cdot (r - r')]. \tag{2}$$

Substituting (2) in (1), we find an equation for $G_k(x, x')$

$$\frac{d}{dx}\left[\epsilon(x)\frac{dG_k}{dx}\right] - \epsilon(x)k^2 G_k = -q\delta(x - x') \tag{3}$$

where use has been made of a result for the two-dimensional delta function

$$\delta(r - r') = \int \frac{d^2k}{(2\pi)^2} \exp[jk \cdot (r - r')]. \tag{4}$$

Integrating (3) over an infinitesimal interval containing $x = x'$, we obtain a discontinuity condition on G_k

$$\epsilon(x')\left[\frac{dG_k}{dx}\bigg|_{x=x'(+)} - \frac{dG_k}{dx}\bigg|_{x=x'(-)}\right] = -q \tag{5}$$

where the notation $x'(+)$ and $x'(-)$ refers to evaluation of the derivative for an x value infinitesimally larger (smaller) than $x = x'$.

Consider the case $x' > 0$ so that $\epsilon(x') = \epsilon_s$. The case $x' < 0$ is analogous. Then for $x \neq x'$, (3) provides

$$\frac{d^2G_k}{dx^2} - k^2G_k = 0. \tag{6}$$

Solutions to this equation are arbitrary linear combinations of $\sinh(kx)$ and $\cosh(kx)$. The correct combination must satisfy the discontinuity condition (5) and the boundary conditions. We try the following combinations

$$\begin{aligned}
G_k &= A_k \sinh[k(x + x_o)] && \text{(in the oxide)} \\
&= B_k \sinh(kx) + C_k \cosh(kx); && (0 \leq x \leq x') \\
&= D_k \sinh k(x_D - x); && (x' \leq x \leq x_D).
\end{aligned} \tag{7}$$

The choice (7) automatically satisfies $G_k = 0$ at gate and depletion layer edge. To ensure that G_k is a continuous function of position x, the coefficients A_k to D_k must satisfy

$$A_k \sinh(kx_o) = C_k \tag{8a}$$

$$B_k \sinh(kx') + C_k \cosh(kx') = D_k \sinh[k(x_D - x')]. \tag{8b}$$

Continuity of the displacement $\epsilon(x)(dG_k/dx)$ at the Si–SiO$_2$ interface $x = 0$ provides

$$\epsilon_{ox}A_k \cosh(kx_o) = \epsilon_s B_k. \tag{8c}$$

The discontinuity condition, (5), provides

$$k\epsilon_s[B_k \cosh(kx') + C_k \sinh(kx') + D_k \cosh k(x_D - x')] = q. \tag{8d}$$

Equations (8a–d) provide four linear equations in the four unknowns and thus determine A_k, B_k, C_k, and D_k. To solve these equations substitute (8a) and (8c) in (8b) to obtain for A_k

$$A_k\left[\frac{\epsilon_{ox}}{\epsilon_s} \sinh(kx')\cosh(kx_o) + \cosh(kx')\sinh(kx_o)\right] = D_k \sinh[k(x_D - x')]. \tag{9}$$

Substitute (8a) and (8c) in (8d) to find another relation between A_k and D_k

$$A_k\left[\frac{\epsilon_{ox}}{\epsilon_s} \cosh(kx_o)\cosh(kx') + \sinh(kx_o)\sinh(kx')\right] + D_k \cosh k(x_D - x') = \frac{q}{k\epsilon_s}. \tag{10}$$

Combining (9) and (10), we find

$$A_k \left[\frac{\epsilon_{ox}}{\epsilon_s} \cosh(kx_o)\sinh(kx_D) + \cosh(kx_D)\sinh(kx_o) \right] = -\frac{q}{k\epsilon_s} \sinh[k(x_D - x')]$$

(11)

that is

$$A_k = \frac{q}{k} \frac{\sinh[k(x_D - x')]}{\Delta'_k}$$

(12)

where

$$\Delta'_k = \epsilon_{ox} \cosh(kx_o)\sinh(kx_D) + \epsilon_s \cosh(kx_D)\sinh(kx_o).$$

(13)

Similarly, using (12) in (9), we obtain

$$D_k = \frac{q}{k} \frac{(\epsilon_{ox}/\epsilon_s)\sinh(kx')\cosh(kx_o) + \cosh(kx')\sinh(kx_o)}{\Delta'_k}.$$

(14)

From (8a) and (8c), C_k and B_k follow immediately. Returning to (7), we have

$$G_k(x, x') = \left(\frac{q}{k} \right) \sinh[k(x_D - x')] \frac{\sinh[k(x + x_o)]}{\Delta'_k} \qquad \text{(in the oxide)}$$

$$= \frac{q}{k\epsilon_s} [\epsilon_{ox}\cosh(kx_o)\sinh(kx)$$

$$+ \epsilon_s \sinh(kx_o)\cosh(kx)] \frac{\sinh[k(x_D - x')]}{\Delta'_k}; \qquad (0 \le x \le x')$$

$$= \frac{q}{k\epsilon_s} [\epsilon_{ox} \cosh(kx_o)\sinh(kx')$$

$$+ \epsilon_s \sinh(kx_o)\cosh(kx')] \frac{\sinh[k(x_D - x)]}{\Delta'_k}; \qquad (x' \le x \le x_D).$$

(15)

The solution for $G_k(x; x')$ in the regions $(0 \le x \le x')$ and $(x' \le x \le x_D)$ are identical except for the interchange of x and x'. Equation (15) determines G_k and hence $G(\mathbf{r} - \mathbf{r}'; x, x')$ from (2).

A generalization of G_k to include the effects of inversion layer screening and interface traps can be accomplished by including these two sources of charge as a charge sheet located at the silica-silicon interface.

Then (1) becomes

$$\nabla \cdot [\epsilon_s(x)\nabla G] - \delta(x)[C_{it} + C_I]G = -q\delta(x - x')\delta(\mathbf{r} - \mathbf{r}').$$

(16)

The term in $(C_{it} + C_I)$ expresses the change in interface charge due to the change in surface band bending by the amount G. A more careful introduction of C_I that shows some limitations of (16) was made in Ref. 3 of Chapter 6. Using (16), we replace (8c) by

$$k[\epsilon_{ox}A_k \cosh(kx_o) - \epsilon_s B_k] = -(C_{it} - C_I)A_k \sinh(kx_o).$$

(17)

Rewriting (17) to resemble (8c) more closely yields

$$\epsilon_{ox}A_k\left[\cosh(kx_o)+(C_{it}+C_I)\frac{\sinh(kx_o)}{k\epsilon_{ox}}\right]=\epsilon_s B_k. \tag{18}$$

Therefore, the previous result for G_k is modified by replacing Δ'_k of (13) with Δ_k, where

$$\Delta_k=\epsilon_{ox}\cosh(kx_o)\sinh kx_D$$

$$+(C_{it}+C_I)\sinh(kx_o)\frac{\sinh(kx_D)}{k}+\epsilon_s\cosh(kx_D)\sinh(kx_o). \tag{19}$$

When G_k is taken from (15) with Δ'_k replaced by Δ_k of (19), the result is the form of Green's function used in Section 6.3.1(b).

APPENDIX VIII

Alternative Derivation of Parallel-Array Formula for Capacitance

To show that (6.89) is consistent with (6.85), we find $\sigma^2(x)$ and $\langle v(x) \rangle$ for this case and relate them to σ_s^2 and $\langle v_s \rangle$ (their values at $x = 0$). We introduce

$$P(v_s) = (2\pi)^{-1/2} \sigma_s^{-1} \exp\left\{-\frac{[v_s - \langle v_s \rangle]^2}{2\sigma_s^2}\right\} \tag{1}$$

and show that

$$\frac{d\langle \beta Q_s \rangle}{dV_G} = \int\limits_{-\infty}^{\infty} dv_s\, P(v_s) \frac{C_{ox} C_s(v_s)}{C_{ox} + C_s(v_s)} \tag{2}$$

which is the same as (6.85) provided that $\langle v_s \rangle$ and σ_s^2 in (1) are chosen to approximate the Poisson distribution P_n as described in Section 6.3.1 using (6.34) and (6.44).

In the quasi-uniform model the band bending $v(x)$ is related to v_s approximately as

$$v(x) = v_s \left(1 - \frac{x}{w}\right)^2. \tag{3}$$

Consequently, a small fluctuation in $v(x)$ is related to the corresponding fluctuation in v_s by

$$dv = dv_s \left(1 - \frac{x}{w}\right) \tag{4}$$

where we have used the results for the one-dimensional case

$$dw = \frac{\lambda_n^2}{w} dv_s \tag{5}$$

and

$$v_s = \frac{w}{2\lambda_n^2}. \tag{6}$$

Consequently, using (3), we rewrite (4) as

$$\langle v(x) \rangle = \langle v_s \rangle \left(1 - \frac{x}{w} \right)^2 \tag{7}$$

$$\sigma(x) = \sigma_s \left(1 - \frac{x}{w} \right). \tag{8}$$

Substituting (7) and (8) in (6.88) and placing the result in (6.87), we find

$$
\begin{aligned}
\langle \beta Q_s \rangle &= \frac{\epsilon_s}{\lambda_n^2} \int_{-\infty}^{\infty} dv_s P(v_s) \int_0^{\infty} dx \{ \exp[v(x)] - 1 \} \\
&= \frac{\epsilon_s}{\lambda_n^2} \int_{-\infty}^{\infty} dv_s \, P(v_s) \int_0^{v_s} dv \, \frac{\exp(v) - 1}{(dv/dx)} \\
&= \int_{-\infty}^{\infty} dv_s \, P(v_s) \beta Q_s(v_s). \tag{9}
\end{aligned}
$$

Differentiating with respect to gate bias, we obtain (2), as required.

APPENDIX IX

Admittance of Interface Dipoles

In this appendix we derive (7.1) for the admittance of the potential well model in Fig. 7.3.

Let the wells closest to the gate electrode be called the *oxide wells* with occupancy N_o and the wells closest to the silicon, the *silicon wells*, with occupancy N_s. Imagine the probability of transfer per ion from the oxide wells to the semiconductor wells to be w_{os}. Then the rate of loss from the oxide wells is

$$\frac{dN_o}{dt} = -N_o w_{os} + N_s w_{so} = -\frac{dN_s}{dt} \tag{1}$$

and

$$\frac{d}{dt}(N_s - N_o) = -(N_s - N_o)(w_{os} + w_{so}) + N_T(w_{os} - w_{so}). \tag{2}$$

Let the total number of ions involved be N_T

$$N_o + N_s = N_T. \tag{3}$$

Then, at equilibrium, denoted by zero superscripts, (2) provides

$$\frac{N_s^o - N_o^o}{N_T} = \frac{w_{os}^o - w_{so}^o}{w_{os}^o + w_{so}^o}. \tag{4}$$

We model w_{os} and w_{so} as follows. Assume that Boltzmann statistics apply. If the potential of the oxide wells is ψ_o and of the silicon wells is ψ_s, then at equilibrium, we obtain

$$\frac{N_o^o}{N_T} = \frac{\exp(-\beta\psi_o)}{\exp(-\beta\psi_o) + \exp(-\beta\psi_s)} \tag{5}$$

$$\frac{N_s^o - N_o^o}{N_T} = \tanh\frac{\beta(\psi_o - \psi_s)}{2}. \tag{6}$$

Now apply a small-signal ac field variation at the dipoles $F_s \exp(j\omega t)$. Then the potential ψ_o oscillates as $\psi_o(t) = \psi_o^o + (F_s d/2)\exp j\omega t$ and ψ_s as $\psi_s(t) = \psi_s^o - (F_s d/2)\exp j\omega t$, where d is the dipole length. We suppose that raising the potential of the oxide wells tends to reduce their occupancy

$$w_{os} = w_{os}^o \left[1 + \beta \frac{F_s d}{2} \exp j\omega t \right] \tag{7}$$

$$w_{so} = w_{so}^o \left[1 - \beta \frac{F_s d}{2} \exp j\omega t \right]. \tag{8}$$

Then (7.2) becomes

$$j\omega(\delta N_s - \delta N_o) = N_T(\beta F_s d)(w_{so}^o + w_{os}^o)\exp(j\omega t) + (\delta N_o - \delta N_s)(w_{os}^o + w_{so}^o) \tag{9}$$

that is

$$\delta N_s - \delta N_o = \frac{\beta N_T(F_s d)(w_{so}^o + w_{os}^o)}{[j\omega + w_{os}^o + w_{so}^o]} \exp(j\omega t). \tag{10}$$

The net ac current is then $qZ[\delta N_s - \delta N_o]$, where Z is the charge of the ion, and the admittance becomes

$$Y = j\omega \frac{qZ(\delta N_s - \delta N_o)}{\delta v_s - \delta v_o} = j\omega\beta qZN_T \frac{w_{os}^o + w_{so}^o}{j\omega + w_{os}^o + w_{so}^o}. \tag{11}$$

Let

$$w_{so}^o = w_o \exp\frac{\beta(\psi_s - \psi_o)}{2}$$

and

$$w_{os}^o = w_o \exp\frac{\beta(\psi_o - \psi_s)}{2}. \tag{12}$$

Then equations (6)–(8) are satisfied, and (11) becomes

$$Y = \beta qN_T \frac{j\omega}{1 + j\omega\tau} \tag{13}$$

where the time constant τ is

$$\tau^{-1} \equiv 2w_o \cosh\frac{\beta(\psi_o - \psi_s)}{2}. \tag{14}$$

APPENDIX X

Derivation of $N(\omega, w)$

In Chapter 9 the apparent doping profile was expressed in (9.38) as

$$N_\omega(w) = -2 \left[\frac{1 - C_{LF}/C_{ox}}{1 - C_m(\omega)/C_{ox}} \right] \left[q\epsilon_s \frac{d}{dV_G} C_m^{-2}(\omega) \right]^{-1}. \qquad (1)$$

We show here that (1) is equivalent to

$$N_\omega(w) = \left\{ C_P(\omega) q\epsilon_s \frac{d}{d\psi_s} C_P^{-1}(\omega) \right\}^{-1}. \qquad (2)$$

Here

$$C_m^{-1}(\omega) = C_{ox}^{-1} + C_P^{-1}(\omega) \qquad (3)$$

and

$$C_P(\omega) \equiv C_D + C_{it}(\omega). \qquad (4)$$

To do this, we note first that differentiation of Gauss's law provides

$$C_{ox}(dV_G - d\psi_s) = -(C_D + C_{it})d\psi_s \qquad (5)$$

where slow variations in gate bias dV_G and band bending $d\psi_s$ are considered. The interface trap capacitance C_{it} in (5) is the low, or zero frequency, capacitance. Rearranging (5), we find

$$
\begin{aligned}
d\psi_s &= C_{ox}(C_{ox} + C_D + C_{it})^{-1} \, dV_G \\
&= \left[1 - \frac{C_P(0)}{C_{ox} + C_P(0)} \right] dV_G \\
&= \left(1 - \frac{C_{LF}}{C_{ox}} \right) dV_G \, .
\end{aligned} \qquad (6)
$$

Consequently, (1) can be rewritten as

$$N_\omega(w) = -2 \left\{ \left[1 - \frac{C_m(\omega)}{C_{ox}} \right] q\epsilon_s \frac{d}{d\psi_s} C_m^{-2}(\omega) \right\}^{-1}. \qquad (7)$$

Also, the chain rule of differentiation provides

$$\frac{d}{d\psi_s} C_m^{-2} = 2C_m^{-1} \frac{d}{d\psi_s} C_m^{-1}$$

$$= 2C_m^{-1} \frac{d}{d\psi_s} C_P^{-1}(\omega) \tag{8}$$

where the last step uses (3) and the bias independence of C_{ox}. Substituting (8) in (7), we find

$$N_\omega(w) = -\left\{ C_P^{-1}(\omega) q\epsilon_s \frac{d}{d\psi_s} C_P^{-1}(\omega) \right\}^{-1} \tag{9}$$

which is the desired result, (9.44).

APPENDIX XI

Derivation of Reflectance and Transmittance of a Thin Oxide MOS Capacitor

In this appendix R and T from (10.25) in Section 10.3.2 are derived following Powell.[1] Introducing the refractive indices of the gold electrode, N_1 of the SiO_2, N_2 and of the silicon, N_3 where all indices are complex to allow for absorption, we define the Fresnel coefficients

$$r_1 = \frac{1 - N_1}{1 + N_1} \tag{1}$$

$$r_2 = \frac{N_1 - N_2}{N_1 + N_2} \tag{2}$$

$$r_3 = \frac{N_2 - N_3}{N_2 + N_3}. \tag{3}$$

The ratio of incident to reflected wave amplitudes r is[2]

$$r = \frac{r_1 + r_2 \exp(2\delta_1) + r_3 \exp 2(\delta_1 + \delta_2) + r_1 r_2 r_3 \exp(2\delta_2)}{1 + r_1 r_2 \exp(2\delta_1) + r_1 r_3 \exp 2(\delta_1 + \delta_2) + r_2 r_3 \exp 2\delta_2}$$

$$= \frac{E_{\text{reflected}}}{E_{\text{incident}}}. \tag{4}$$

Also, the ratio of incident to transmitted (into the substrate) wave amplitudes t is

$$t = \frac{(1 + r_1)(1 + r_2)(1 + r_3)\exp(\delta_1 + \delta_2)}{1 + r_1 r_2 \exp(2\delta_1) + r_1 r_3 \exp 2(\delta_1 + \delta_2) + r_2 r_3 \exp(2\delta_2)}$$

$$= \frac{E_{\text{transmitted}}}{E_{\text{incident}}}. \tag{5}$$

In (4) and (5) the phase factors δ_1 and δ_2 are related to the thickness of the gold electrode d_1 and of the silica d_2 by

$$\delta_1 = -j \frac{2\pi}{\lambda} N_1 d_1 \tag{6}$$

$$\delta_2 = -j \frac{2\pi}{\lambda} N_2 d_2. \tag{7}$$

The power ratios are then

$$R = rr^* \tag{8}$$

and

$$T = n_3 tt^* \tag{9}$$

where n_3 is the real part of N_3 and asterisks denote complex conjugation. Finally, the absorptance A is given by

$$A = 1 - R - T. \tag{10}$$

Powell[1] evaluated (8)–(10) numerically for the case of no loss in the SiO_2, as a function of photon energy. The silicon optical constants were taken from a Kramers–Kronig analysis of data in the literature.[3] A formula due to Malitson[4] for fused SiO_2 was used for the thermally grown SiO_2. The optical constants of gold were taken from several sources.[5,6]

REFERENCES

1 R. J. Powell, *J. Appl. Phys.*, **40**, 5093 (1969).
2 O. S. Heavens, *Optical Properties of Thin Solid Films*, Academic, New York, 1955.
3 H. R. Phillip and H. Ehrenreich, *Phys. Rev.*, **129**, 1550 (1963).
4 I. H. Malitson, *J. Opt. Soc. Am.*, **55**, 1205 (1965).
5 L. G. Shultz, *J. Opt. Soc. Am.*, **44**, 357, 362 (1954).
6 D. Beaglehole, *Proc. Phys. Soc. (Lond.)*, **85**, 1007 (1965).

APPENDIX XII

Derivation of (10.45)

This appendix derives (10.45) following the work of Powell.[1] This equation relates the second derivative of the photocurrent to photon energy and is used to show that as photon energy is increased, curves of photocurrent versus gate bias change from concave downward to concave upward, the switch occurring for $\hbar\omega = E_b$, where $\hbar\omega$ is the photon energy and E_b is the barrier to injection.

To begin, we assume that the quantum yield of electrons per photon Y is a function only of the energy above the barrier; that is

$$Y = Y(E - E_b). \tag{1}$$

The barrier depends on gate bias as

$$E_b = E_{bo} - KV^{1/2} \tag{2}$$

as follows when image-force lowering is included in determining the barrier height and position.

Now the quantum yield is given approximately by[1]

$$Y = C \int_0^{\hbar\omega - E_b} dE\, N_i(E)(\hbar\omega - E_b - E) \tag{3}$$

where $N_i(E)$ is the number of electrons at energy E and C is a proportionality factor obtained by approximating the probability of injection of an electron as $C(\hbar\omega - E_b - E)$.

Using (3) and differentiating with respect to E_b yields

$$Y' = -C \int_0^{\hbar\omega - E_b} dE\, N_i(E) \tag{4}$$

and

$$Y'' = CN_i(\hbar\omega - E_b). \tag{5}$$

The quantity Y' can be expressed in terms of the average value $\bar{N}_i$ of

$N_i(E)$ in the energy range $0 < E < \hbar\omega - E_b$

$$Y' = -C(\hbar\omega - E_b)\bar{N}_i. \tag{6}$$

From (3), it can be seen that Y is a function of $\hbar\omega - E_b$. Consequently, differentiating Y with respect to voltage, using (2), we find

$$\frac{\partial Y}{\partial V} = \left(\frac{\partial Y}{\partial E_b}\right)\left(\frac{\partial E_b}{\partial V}\right)$$

$$\frac{\partial^2 Y}{\partial V^2} = \frac{\partial^2 Y}{\partial E_b^2}\left(\frac{\partial E_b}{\partial V}\right)^2 + \frac{\partial Y}{\partial E_b}\frac{\partial^2 E_b}{\partial V^2}$$

$$= Y''\frac{1}{4}K^2V^{-1} + Y'\frac{1}{4}KV^{-3/2}. \tag{7}$$

Rearranging terms yields

$$\left(\frac{4V^{3/2}}{K}\right)\frac{\partial^2 Y}{\partial V^2} = KV^{1/2}Y'' + Y'. \tag{8}$$

Using (5) and (6), we rewrite (8) as

$$\left(\frac{4V^{3/2}}{K}\right)\frac{\partial^2 Y}{\partial V^2} = KCV^{1/2}N_i(\hbar\omega - E_b) - (\hbar\omega - E_b)C\bar{N}_i$$

$$= C\bar{N}_i\left[\frac{K}{\xi}V^{1/2} - (\hbar\omega - E_{bo}) - KV^{1/2}\right] \tag{9}$$

where ξ is defined by

$$\xi \equiv \frac{\bar{N}_i}{N_i(\hbar\omega - E_b)}. \tag{10}$$

The parameter $\xi \leq 1$, and for injection from Si into SiO_2 from the valence band, $\xi = 1$.

The photocurrent I is proportional to the quantum yield Y

$$I = \left|\frac{A(\hbar\omega)}{C}\right|Y. \tag{11}$$

Therefore, taking the second derivative of (11) with respect to voltage, (9) provides

$$\frac{4V^{3/2}}{[K\bar{N}_i\xi A(\hbar\omega)]}\frac{d^2 I}{dV^2} = K\left[\frac{1}{\xi} - 1\right]V^{1/2} - (\hbar\omega - E_{bo}) \tag{12}$$

which is (10.45).

REFERENCE

1   R. J. Powell, *J. Appl. Phys.*, **41**, 2424 (1970).

APPENDIX XIII

Formula for Photocurrent in the Presence of Scattering in the Image-Force Potential Well

In this appendix equation (11.20) for the photocurrent I_o is derived. The approach of Powell and Berglund[1] is used. It is assumed that any injected electron that is scattered is lost. It is also assumed that no electron is scattered more than once during its passage from the point of injection to the barrier maximum, located a distance x_{mo} from the injecting interface.

Using a mean free path l the probability, $p\,dx$, of a scattering event in distance dx is

$$p\,dx = \frac{dx}{l}. \tag{1}$$

Therefore, the number of electrons lost in distance dx, $-dn$, is

$$-dn = n(x)\frac{dx}{l}. \tag{2}$$

Therefore, solving (2), we have

$$n(x) = n(0)\exp\left(-\frac{x}{l}\right) \tag{3}$$

where $n(0)$ is the number injected at $x = 0$. The number of electrons reaching the barrier maximum, which then escape to form the photocurrent, is now

$$n(x_{mo}) = n(0)\exp\left[-\frac{x_{mo}}{l}\right]. \tag{4}$$

It remains to relate $n(0)$ to the photon flux and frequency. Following Powell,[2] the photocurrent at the injecting contact I_{inj} is given by

$$I_{inj} = \frac{PY}{\hbar\omega} \tag{5}$$

where P is the absorbed light power in watts, I_{inj} is the current in amperes, $\hbar\omega$ is the photon energy in electron volts, and Y is the quantum yield in electrons per photon.

The quantum yield Y is a function of the escape probability of the electrons and the energy distribution of electrons at the injecting interface. The escape probability for an electron of energy E is approximated reasonably well by $C(E - E_b)$, where C is a constant and $(E - E_b)$ is the energy of the electron relative to the barrier energy. Then the yield Y is

$$Y \equiv C \int_{E_b}^{\infty} n(E - \hbar\omega)(E - E_b) \, dE \qquad (6)$$

where $n(E - \hbar\omega)$ is the number of electrons with energy E when photons of energy $\hbar\omega$ are incident, supposing that these photons simply translate the electron distribution uniformly to higher energy.

For example, we might suppose that the initial electron distribution is a step function, extending from lower energies up to some maximum energy, which could be the top of the valence band in silicon; then

$$\begin{aligned} n(E - \hbar\omega) &= n \quad (E < E_v + \hbar\omega) \\ &= 0 \quad (E > E_v + \hbar\omega) \end{aligned} \qquad (7)$$

where E_v is the energy of valence band maximum, and Y from (6) becomes

$$\begin{aligned} Y &\cong Cn \int_{E_b}^{E_v + \hbar\omega} (E - E_b) \, dE \\ &= Cn \frac{(E_v + \hbar\omega - E_b)^2}{2}. \end{aligned} \qquad (8)$$

Similarly, we might suppose (as Powell[2] determined experimentally for injection from the valence band of silicon to silica) that

$$\begin{aligned} n(E - \hbar\omega) &= \gamma[E_v - E + \hbar\omega] \quad E < E_v + \hbar\omega \\ &= 0 \quad E > E_v + \hbar\omega \end{aligned}$$

where γ is a proportionality factor. Then (6) becomes

$$\begin{aligned} Y &= C\gamma \int_{E_b}^{E_v + \hbar\omega} (E_v - E + \hbar\omega)(E - E_b) \, dE \\ &= C\gamma \int_{0}^{E_v + \hbar\omega - E_b} dx \, x(\hbar\omega + E_v - E_b - x) \end{aligned} \qquad (9)$$

(where we have substituted $x = E - E_b$)

$$\begin{aligned} &= C\gamma \left[(\hbar\omega + E_v - E_b) \frac{x^2}{2} - \frac{x^3}{3} \right]_0^{E_v + \hbar\omega - E_b} \\ &= C\gamma \frac{(\hbar\omega + E_v - E_b)^3}{6}. \end{aligned} \qquad (10)$$

In general, if $n \propto (E_v - E + \hbar\omega)^r$, then $Y \propto (E_v - E_b + \hbar\omega)^{r+2}$. Consequently, using (5), we obtain

$$I_{inj} \propto \frac{P}{\hbar\omega} (E_v - E_b + \hbar\omega)^{r+2} \qquad (11)$$

and, using (4) with the proportionality constant in (11) rewritten as some unspecified function of $\hbar\omega$, $A(\hbar\omega)$, we have for the photocurrent at the barrier

$$I_o = A(\hbar\omega)(E_v - E_b + \hbar\omega)^{r+2} \exp\left(-\frac{x_{mo}}{l}\right) \qquad (12)$$

whenever $n(0) \propto (E_v - E)^r$. This agrees with (11.20) if the power $p = r + 2$ and $E_{bo} + q\,\Delta\phi = E_b - E_v$.

REFERENCES

1   R. J. Powell and C. N. Berglund, *J. Appl. Phys.*, **42**, 4390 (1971).
2   R. J. Powell, *J. Appl. Phys.*, **41**, 2424 (1970).

APPENDIX XIV

Impurity Redistribution—Uniform Case

In this appendix we solve the diffusion problem, following Grove et al.,[1] of the redistribution of donor or acceptor impurities between the silicon and the silica during thermal oxidation at high temperature. The initial impurity distribution in the silicon is assumed to be uniform. These results are used in Chapter 14. After oxidation, the impurity distribution in the silica is given by the solution of the diffusion equation

$$\frac{\partial N_o(x, t)}{\partial t} = D_o \frac{\partial^2 N_o(x, t)}{\partial x^2}; \qquad 0 < x < x_o \tag{1}$$

where $N_o(x, t)$ is the impurity concentration in the silica at time t and distance x from the gas-SiO_2 interface, D_o is the diffusion coefficient of the impurity in the silica, and x_o is the oxide thickness. The impurity concentration in the silicon, after oxidation, will be given by the solution of the diffusion equation

$$\frac{\partial N_s(z, t)}{\partial t} = D_s \frac{\partial^2 N_s(z, t)}{\partial z^2}; \qquad z > z_o(t) \tag{2}$$

where $N_s(z, t)$ is the impurity concentration in the silicon at time t and distance z from the initial silicon surface and $z_o(t)$ locates the Si–SiO_2 interface at time t relative to the initial silicon surface. To take into account the outward movement of the silica layer from the silicon surface, different distance coordinates are used. The silica layer moves outward because the silica occupies a larger volume than the silicon it replaces. By using two separate coordinate systems, whose origins are fixed with respect to the moving Si–SiO_2 interface, the diffusion equation can be expressed in its simplest form, where the physical processes are not obscured by mathematical complexity. The alternative coordinate system would become the one whose origin moved within the Si–SiO_2 interface. In this case the diffusion equation would have an extra term in it to take into account the velocity of the moving interface. In the fixed coordinate system the coordinate labeled x is used to denote distance, within the silica film, measured from the gas-SiO_2 interface; and the coordinate labeled z is used to denote distance, within the silicon, measured from the initial silicon surface. These fixed coordinates are illustrated in Fig. XIV-1.

The solution of (1) and (2) for a moving boundary is obtained for the case of an initially uniform distribution of impurity in the silicon

$$N_s(z, 0) = N_B \tag{3}$$

and the two boundary conditions: (1) the concentration of impurity at the gas silica interface N_g is constant, where

$$N_o(0, t) = N_g \tag{4}$$

and (2) far within the silicon, away from the Si–SiO$_2$ interface, the impurity concentration reaches the bulk concentration; thus

$$N_s(z, t) \to N_B \quad \text{as } z \to \infty. \tag{5}$$

In addition, two matching conditions at the Si–SiO$_2$ interface must be satisfied. The first condition is that the concentration of impurities at the two sides of the Si–SiO$_2$ interface must be in the ratio m

$$\frac{N_s(z_o, t)}{N_o(x_o, t)} = m. \tag{6}$$

The second matching condition determines the quantity of the impurity stored at the moving Si–SiO$_2$ interface. To derive this matching condition, we start with an expression for the total amount of impurity within the solid Si–SiO$_2$ system at a given time, which is

$$Q(t) = \int_0^{x_o(t)} N_o(x, t)\,dx + \int_{z_o(t)}^{\infty} N_s(z, t)\,dz. \tag{7}$$

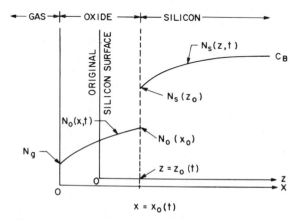

Fig. XIV-1 The diffusion of an impurity into oxide covered silicon. A coordinate system is shown that is fixed with respect to the moving Si–SiO₂ boundary. The symbols are defined in the text. After Grove et al.[1]

Next we consider the flow of impurity atoms through the system during oxidation. Because all the impurity is initially in the silicon, the concentration gradient creates a flux of impurity atoms moving out of the silicon through the growing silica layer into the gas ambient. The rate of escape of impurity atoms from the silica surface depends on the flux of impurity atoms arriving at the gas-SiO$_2$ interface, which is $D_o(\partial N_o/\partial x)_{x=0}$ and is equal to the rate of decrease of $Q(t)$. Therefore, we can write

$$\frac{dQ(t)}{dt} = -D_o \left(\frac{\partial N_o}{\partial x}\right)_{x=0}. \tag{8}$$

To obtain $dQ(t)/dt$, we differentiate (7), obtaining

$$\frac{dQ}{dt} = \int_0^{x_o} \left(\frac{\partial N_o}{\partial t}\right) dx + N_o(x_o, t)\frac{dx_o}{dt} + \int_{z_o}^{\infty} \left(\frac{\partial N_s}{\partial t}\right) dz - N_s(z_o, t)\frac{dz_o}{dt}. \tag{9}$$

The integrals in (9) can be evaluated by substituting the diffusion equations (1) and (2) into (9) to transform the time partial derivatives of the concentration into distance partial derivatives, which then permit the integration, with respect to the distance coordinates, to be carried out. The first integral in (9), using (1), becomes

$$\int_0^{x_o} \frac{\partial N_o}{\partial t} dx = D_o \int_0^{x_o} \frac{\partial^2 N_o}{\partial x^2} dx = D_o \left(\frac{\partial N_o}{\partial x}\right)_{x=x_o} - D_o \left(\frac{\partial N_o}{\partial x}\right)_{x=0}. \tag{10}$$

The second integral in (9), using (2), becomes

$$\int_{z_o}^{\infty} \frac{\partial N_s}{\partial t} dz = D_s \int_{z_o}^{\infty} \frac{\partial^2 N_s}{\partial z^2} dz = -D_s \left(\frac{\partial N_s}{\partial z}\right)_{z=z_o}. \tag{11}$$

The term $D_s(\partial N_s/\partial z)_{z=\infty} = 0$ because of the second boundary condition, (5), which states that $N_s(z, t) \to N_B$ as $z \to \infty$. Combining (8) through (11) yields

$$D_o \left(\frac{\partial N_o}{\partial x}\right)_{x=x_o} + N_o(x, t)\left(\frac{dx_o}{dt}\right) - D_s \left(\frac{\partial N_s}{\partial z}\right)_{z=z_o} - N_s(z_o, t)\left(\frac{\partial z_o}{\partial t}\right) = 0. \tag{12}$$

Using the matching condition at the Si–SiO$_2$ interface, (6), to rearrange (12) yields

$$N_s(z_o, t)\left[\frac{dz_o}{dt} - \frac{1}{m}\left(\frac{dx_o}{dt}\right)\right] = D_o \left(\frac{\partial N_o}{\partial x}\right)_{x=x_o} - D_s \left(\frac{\partial N_s}{\partial z}\right)_{z=z_o}. \tag{13}$$

However, $z_o = ax_o$, where a relates the thickness of silicon oxidized z_o to the thickness of the silica layer formed x_o. The value of a calculated from the densities and molecular weights of silicon and silica is 0.45. Substituting

$dz_o/dt = a\, dx_o/dt$ into (13) yields

$$N_s(z_o, t)\frac{dx_o}{dt}\left(a - \frac{1}{m}\right) = D_o\left(\frac{\partial N_o}{\partial x}\right)_{x=x_o} - D_s\left(\frac{\partial N_s}{\partial z}\right)_{z=z_o}. \qquad (14)$$

Equation (14), called the *boundary equation*, states the quantity of impurity stored at the Si–SiO$_2$ interface. According to (14), the flux of dopant atoms flowing from the silica to the Si–SiO$_2$ interface exceeds the flux from the Si–SiO$_2$ interface into the oxide. The excess dopant atoms are stored in an infinitesimally thin sheet at the Si–SiO$_2$ interface. Of course, such a sheet is an idealization of the transition region in which oxidation occurs, which really must be of nonzero thickness.

Finally, it is assumed that the thickness of the growing silica layer increases according to the parabolic law $x_o(t) = (Bt)^{1/2}$. This rate is also the one at which the Si–SiO$_2$ interface moves. It is assumed that as the impurity concentration changes during oxidation, B does not change. It is reasonable to use the parabolic law for the oxide growth rate because in many practical cases where relatively thick oxides are grown at high temperatures and at one atmosphere of dry oxygen, the oxide film grows according to the parabolic law almost immediately. This analysis does not apply to the very early phases of oxidation, but these phases can be ignored without introducing significant error for very high temperatures and 1 atm of dry oxygen. Therefore, we can solve our problem from $t = 0$ to $t = \infty$. The solutions of this boundary value problem are[2] (1) in the silica film

$$\frac{N_o(x, t) - N_g}{N_o(x_o, t) - N_g} = \frac{\mathrm{erf}\left[\dfrac{x}{2(D_o t)^{1/2}}\right]}{\mathrm{erf}\left[\left(\dfrac{B}{4D_o}\right)^{1/2}\right]}; \qquad 0 < x < x_o \qquad (15)$$

and (2) in the silicon

$$\frac{N_s(z, t) - N_B}{N_s(z_o, t) - N_B} = \frac{\mathrm{erfc}\left[\dfrac{z}{2(D_s t)^{1/2}}\right]}{\mathrm{erfc}\left[a\left(\dfrac{B}{4D_s}\right)^{1/2}\right]}; \qquad z > z_o. \qquad (16)$$

In addition, the conservation of impurity atoms at the Si–SiO$_2$ interface, (14), gives the interfacial impurity concentration, which on the silicon side is

$$\frac{N_s(z_o, t)}{N_B} = \frac{1 + (N_g/N_B)b}{1 + (1/m - a)\pi^{1/2}\exp(a^2 B/4D_s)\,\mathrm{erfc}[a(B/4D_s)^{1/2}](B/4D_s)^{1/2} + b/m} \qquad (17)$$

where

$$b = r\exp\left[(a^2 r^2 - 1)\frac{B}{4D_s r^2}\right]\mathrm{erfc}\left[a\left(\frac{B}{4D_s}\right)^{1/2}\right]\mathrm{erf}\left[\left(\frac{B}{4D_o}\right)^{1/2}\right]$$

and

$$r = \left(\frac{D_o}{D_s}\right)^{1/2}.$$

Noting that in (17) the silicon surface concentration is independent of time, we can write $N_s(z_o, t) = N_s$. Similarly, because $N_o(x_o, t)$ is related to $N_s(z_o, t)$ by (6), $N_o(x_o, t) = N_o$.

REFERENCES

1   A. S. Grove, O. Leistiko, and C. T. Sah, *J. Appl. Phys.*, **35**, 2695 (1964).

2   Mathematically, this problem is similar to a moving boundary change-of-state heat conduction problem studied by Neumann; see H. W. Carslow and J. C. Jaeger, *Conduction of Heat in Solids*, 2nd ed., Oxford University Press, 1960, Chapter XI.

APPENDIX XV

Impurity Redistribution–Nonuniform Case

In this appendix we solve the diffusion problem for the redistribution of phosphorus and boron in the silicon during thermal oxidation in 1 atm of dry oxygen when the initial impurity distribution in the silicon $N_I(z, 0)$ is not uniform. The solution presented is the method of finite differences used by Krambeck,[1] and the symbols used are the same as in Appendix XIV and in the text.

The first step in the method of finite differences is to divide the silicon into a number of spatial segments and time intervals. It is necessary to confine the total number of segments to a finite length. Furthermore, only the region within a micron or two of the Si–SiO$_2$ interface is of interest in device applications. Therefore, it is desirable to concentrate the spatial segments near the interface to increase both resolution and precision. Both of these objectives can be achieved by transforming the distance z, which extends over the interval $0 \le z \le \infty$, to a new variable u, which is defined in the interval $0 \le u \le 1$. The particular transformation to choose is hyperbolic because this transformation gives the best compromise between a high density of segments at the interface and a reasonable density deeper in the silicon. This transformation is

$$z = \frac{1}{A} \frac{u}{1-u} \qquad (1)$$

where A is a scaling constant. Transforming the diffusion equation, (14.7), using (1) yields

$$\frac{\partial N_s}{\partial t} = D_s A^2 (1 - u)^4 \frac{\partial^2 N_s}{\partial u^2} - 2D_s A^2 (1 - u)^3 \frac{\partial N_s}{\partial u} + a \left(\frac{dx_o}{dt}\right) A(1 - u)^2 \frac{\partial N_s}{\partial u}. \qquad (2)$$

Similarly, transforming the boundary equation, (14.8), using (1) yields

$$AD_s \frac{\partial N_s(0, t)}{\partial u} = \left(1 - \frac{1}{ma}\right)\left(\frac{dx_o}{dt}\right) N_s(0, t). \qquad (3)$$

To solve (2), the interval, $0 \le u \le 1$, is divided into L equal segments and the value of N_s at each node designated; $N_{s0}, N_{s1}, \ldots, N_{sL}$. Time is divided into appropriate increments also, but these are not necessarily equal. In using the method of finite differences, the time derivative is taken to be constant over a given time interval. Because the derivative actually changes with time, it is necessary to decide what value should be adopted for the calculation. The only way to obtain a stable solution is to choose the value at the end of the increment, and this is the choice that was made. When (2) is rewritten in terms of finite differences in this way, it becomes

$$\frac{(N_{sj,n+1} - N_{sj,n})}{\Delta t} = D_s A^2 (1-u)^4 \frac{N_{sj+1,n+1} + N_{sj-1,n+1} - 2N_{sj,n+1}}{(\Delta u)^2}$$

$$- D_s A^2 (1-u)^3 \frac{N_{sj+1,n+1} - N_{sj-1,n+1}}{\Delta u}$$

$$+ a\left(\frac{dx_o}{dt}\right) A(1-u)^2 \frac{N_{sj+1,n+1} - N_{sj-1,n+1}}{2\Delta u} \qquad (4)$$

where the first subscript (j) to N_s indicates a position on the transformed spacial segment and the second subscript (n) indicates a time. From (2), the right side of (4) equals the value of the time derivative at the end of the time interval, and (4) sets the average time derivative equal to this value. There are ($L-1$) of these equations, corresponding to $j=1$ through $j=(L-1)$, but there are a total of $L+1$ unknowns. The additional equation needed expresses the time invariance of the solution at $z = \infty$, namely

$$N_{sL,n+1} = N_{sL,n} \qquad (5)$$

where $N_{sL} = N_B$ is the impurity concentration at $z = \infty$, which cannot change with time. Finally, the boundary equation (14.8), becomes

$$AD_s \frac{N_{sL,n+1} - N_{s0,n+1}}{\Delta u} = \left(1 - \frac{1}{ma}\right)\left(\frac{dx_o}{dt}\right) N_{s0,n+1}. \qquad (6)$$

This set of ($L+1$) equations can be solved simultaneously for the values of N_s at the time ($n+1$). The resulting values of N_s can be used to obtain the value of N_s at the next time interval, again using (4)–(6). To obtain results with at least 1% accuracy, the following parameters can be used: $L = 500$, $t = 10^{-4} n^2$ (hr), and A (cm^{-1}) $= (0.02/D_s)^{1/2}$ (5×10^4), where D_s is in μm^2/hr.

The selection of a nonlinear mapping of time does not affect the solution of (4) at all, and it allows the solution to proceed expeditiously even though the diffusion process slows as time goes on.

The specific method of solving (4) is as follows. The first step is to write (4) with all the unknowns on the left

$$N_{sj-1,n+1} A_j + N_{sj,n+1} B_j + N_{sj+1,n+1} C_j = E_j \qquad (7)$$

where

$$A_j = \frac{-D_s A^2 (1-u)^4}{(\Delta u)^2} - \frac{D_s A^2 (1-u)^3}{\Delta u} + a\left(\frac{dx_o}{dt}\right) \frac{A(1-u)^2}{2\Delta u} \qquad (8)$$

$$B_j = \frac{2D_s A^2 (1-u)^4}{(\Delta u)^2} + \frac{1}{\Delta t} \tag{9}$$

$$C_j = \frac{-D_s A(1-u)^4}{(\Delta u)^2} + \frac{D_s A^2 (1-u)^3}{\Delta u} - a\left(\frac{dx_o}{dt}\right)\frac{A(1-u)^2}{2\Delta u} \tag{10}$$

$$E_j = \frac{N_{sj,n}}{\Delta t}. \tag{11}$$

Thus E_j is determined by the impurity distribution at the beginning of the time increment. The other two equations necessary are (5) and

$$\frac{N_{s0,n+1}}{N_{s1,n+1}} = \frac{AD_s/\Delta u}{\left(1 - \frac{1}{ma}\right)(dx_o/dt) + (AD_s/\Delta u)}. \tag{12}$$

The tridiagonal nature of the matrix associated with these equations makes a solution easy to obtain. The technique as shown by Strain and Schryer,[2] for example, begins with the calculation of two intermediate sets of numbers

$$G_j = \frac{A_j}{B_j + C_j G_j - 1} \tag{13}$$

and

$$H_j = \frac{E_j - C_j H_j - 1}{B_j + C_j G_j - 1} \tag{14}$$

where G_0 is the ratio $N_{s0,n+1}/N_{s1,n+1}$, given by (12), and $H_0 = 0$. Values of G and H are calculated for all values of j from 1 to $(L-1)$. Next, these values of G and H are used to find the new values of concentration

$$N_{sj,n+1} = G_j N_{sj+1,n+1} + H_j. \tag{15}$$

This calculation gives values for concentration at the end of the interval. These values can be used to obtain new values for E_j in (11) and the process repeated to obtain values of N_s at the end of the next time interval.

REFERENCES

1   R. H. Krambeck, J. Electrochem. Soc., **121**, 500 (1974).
2   R. J. Strain and N. L. Schryer, Bell Syst. Tech. J., **50**, 1721 (1971).

APPENDIX XVI

Conversion of SiO₂ to Glass

Following Sah et al.,[1] we derive the parabolic law [(14.9) in the text] for the conversion of the SiO_2 layer to a glass in the presence of either phosphorus or boron. Let $x_i(t)$ denote the moving boundary between the growing glass and the SiO_2 measured from the original gas-SiO_2 interface. Figure XVI-1 shows a schematic of the concentrations of diffusants in the glass phase and unreacted SiO_2. The values D_1 and C_1 are the diffusion coefficient and the concentration of the phosphorus or boron oxide in the glass, respectively, whereas D_2 and C_2 are the diffusion coefficient and the concentration of the dopant oxides in the unreacted SiO_2, respectively. Fick's law in the glass phase is

$$D_1 \frac{\partial^2 C_1}{\partial x^2} = \frac{\partial C_1}{\partial t}; \qquad 0 < x < x_i(t) \tag{1}$$

whereas in the unreacted SiO_2, Fick's law is

$$D_2 \frac{\partial^2 C_2}{\partial x^2} = \frac{\partial C_2}{\partial t} + \frac{\partial S_2}{\partial t}; \qquad x_i(t) < x < \infty. \tag{2}$$

In (2) the reaction of the phosphorus or boron oxide with SiO_2 is assumed to be in local equilibrium. Therefore, the concentration of the reacted substance S_2 is directly proportional to the concentration C_2 of the substance free to diffuse. Thus $S_2 = RC_2$, and (2) becomes

$$D_2 \frac{\partial^2 C_2}{\partial x^2} = (1 + R) \frac{\partial C_2}{\partial t}; \qquad x_i(t) < x < \infty. \tag{3}$$

The following boundary and initial conditions must be satisfied. The impurity has a different solubility in the glass and the SiO_2. Therefore, the first boundary condition at the moving glass-SiO_2 interface is

$$C_2 = mC_1 \ x = x_i(t), \quad t > 0 \tag{4}$$

where m is the segregation coefficient of the impurity. The second boundary condition takes into account the continuity of concentration gradient

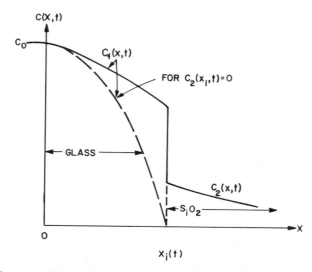

Fig. XVI-1 Concentration profile of glass forming diffusant in the glass phase and in the unreacted SiO₂. After Sah et al.[1] Copyright (1959), Pergamon Press, Ltd. Reprinted with permission.

and boundary velocity; thus

$$D_1 \frac{\partial C_1}{\partial x} + C_1 \frac{dx_i}{dt} = (1+R)^{-1} D_2 \frac{\partial C_2}{\partial x} + C_2 \frac{dx_i}{dt}; \qquad x = x_i(t), \quad t > 0. \quad (5)$$

The rate of increase of the phosphorus or boron glass thickness is assumed to be proportional to the net rate of the number of diffusing atoms crossing the interface located at $x_i(t)$. Assuming that the amount of diffusant crossing the interface is completely reacted, we have

$$N \frac{dx_i}{dt} = -D_1 \frac{\partial C_1}{\partial x} - C_1 \frac{dx_i}{dt}; \qquad x = x_i(t), \quad t > 0 \quad (6)$$

where N is the concentration of the reacted phosphorus or phosphorus oxide, or boron or boron oxide in the glass.

The remaining initial and boundary conditions are

$$C_1 = C_o \qquad \text{at} \quad x = 0, \quad t > 0 \quad (7)$$

$$C_2 = 0 \qquad \text{at} \quad x = \infty, \quad \text{all } t \quad (8)$$

$$C_1 = C_2 = 0 \quad \text{at} \quad x > 0, \quad t < 0. \quad (9)$$

The solution of (1) and (2), satisfying the boundary and initial conditions (3)–(9), is

$$\frac{x_i}{\sqrt{D_1 t}} = L = \text{constant} \quad (10)$$

where $x_i = x_m$ is the limiting oxide thickness. The quantity L is obtained from

$$\frac{m[\sqrt{\pi}L \exp(L^2) \operatorname{erf}(L)]N - C_o}{\sqrt{\pi}L \exp(L^2) \operatorname{erf}(L) - 1} = \frac{[\sqrt{\pi}rL \exp(r^2L^2) \operatorname{erfc}(rL)]N}{\sqrt{\pi}rL \exp(r^2L^2) \operatorname{erfc}(rL) + 1}. \quad (11)$$

The diffusant crossing the glass-SiO₂ interface is assumed to be completely reacted so that $C_2(x_i) = 0$, and

$$C_1(x, t) = C_o\left[1 - \frac{\operatorname{erf}(x/2\sqrt{D_1 t})}{\operatorname{erf}(L)}\right] \quad (12)$$

and

$$\frac{C_o}{N} = \sqrt{\pi}L \exp(L^2) \operatorname{erf}(L). \quad (13)$$

The solution of the problem, after the SiO₂ has been completely converted to a glass, is also treated by Sah et al.[1] but is not reproduced here.

REFERENCE

1   C. T. Sah, H. Sello, and D. A. Tremere, *J. Phys. Chem. Solids*, **11**, 288 (1959).

APPENDIX XVII

List of Abbreviations

CCD—charge coupled device

C-V—capacitance versus voltage

CVD—chemical vapor deposition

emf—electromotive force

G-V—conductance versus voltage

ITIC—isothermal transient ionic current

I-V—current versus voltage

JFET—junction field effect transistor

MOS*—metal oxide semiconductor

MOSFET*—metal oxide semiconductor field effect transistor

ppm—parts per million

SIMS—secondary ion mass spectroscopy

SIPOS—semi-insulating polycrystalline oxygen doped silicon

SRH—Shockley-Read-Hall (theory)

TSIC—thermally stimulated ionic current

TVS—triangular voltage sweep

VLSI—very large scale integration

VUV—vacuum ultraviolet

*These abbreviations include silicon gate as well as metal gate.

Subject Index

Accelerated aging, 756
  see also Bias-temperature aging
Acceptor:
  defined, 36
  degeneracy, 38
  interface trap, 308, 309-311
  reactions involving, 308
Accumulation:
  and avalanche injection, 498
  and band-bending diagram, 51, 52, 82,
    498
  capacitance in, 84
  and carrier degeneracy, 90
  and C-V curve, 82
  test for, 374
Accumulation layer, defined, 41
Activation energy:
  for holes in oxide, 565
  and orientation, table, 688
  for oxidation, 684
  for oxide charge annealing, 542-543
  for parabolic rate constant, 677
  for potassium in oxide, 443
  and sodium emission, 434-435
  for sodium in oxide, 442-443
  and steric hindrance, 685
  of transition frequency, 141
  for water diffusion, 542-543, 678
Admittance:
  measured, *see* Measured admittance
  silicon:
    bulk traps, *see* Silicon admittance, bulk
      traps
    interface traps, *see* Silicon admittance,
      interface traps
  see also Capacitance; Conductance
Admittance bridge:
  accuracy, 584
  frequency calibration, 585

and gate bias, 587-588
  sensitivity, 584
  and signal amplitude, choice of, 585-587
Aging, 756
  see also Bias-temperature aging
Aluminum:
  barrier height, 473
  as gate electrode, 630-631
  as oxide trap, 532, 548
  and postmetallization anneal, 785
  and radiation hardness, 808
  and workfunction difference, 466
Annealing:
  by electron injection, 573
  at high temperatures, 785-786, 795-798
  of hole traps, 573
  of irradiation damage, 562-563, 572-573,
    575-576
  in low temperature hydrogen, 784
  postmetallization, 633, 782-785
  thermal, 563, 575-576
  using ultraviolet light, 562-563
  of water related traps, 542-544
Arsenic:
  diffusivity, 735
  and oxidation rate, 701
  oxide masking, 735
  as oxide trap, 532, 549
  segregation coefficient, 723
Automatic instrumentation, 588
Avalanche:
  and breakdown, 776, 778
  and C-V curve, 376-377
  and doping density, 378
  at edge of gate, 378-380
  and hot carrier injection, *see* Avalanche
    injection
  and irradiation, 800-802
  uniform, 378-380

Symbol Index

This index lists the page numbers not only where symbols are defined, but also where particularly useful formulas or evaluations are given (italicized page numbers). Some symbols have multiple meaning - the reader should pick the page most nearly preceding the section being read.

The index is ordered so Roman symbols precede Greek symbols and both precede numbers. A lower case letter precedes its capital, and subscripts precede arguments precede superscripts. For example, $D_{it}^{a}(\zeta_s)$ is ordered as though it were Ditζsa. Various bars, brackets etc. precede letters. Thus our alphabet begins: a, $\bar{a}$, $<a>$, A, $\bar{A}$, $<A>$, b, etc. . . . .